ROUTLEDGE LIBRARY EDITIONS: SCIENCE AND TECHNOLOGY IN THE NINETEENTH CENTURY

Volume 7

DISCOVERIES AND INVENTIONS OF THE NINETEENTH CENTURY

T0262956

DISCOVERIES AND INVENTIONS OF THE NINETEENTH CENTURY

ROBERT ROUTLEDGE

Routledge
Taylor & Francis Group

LONDON AND NEW YORK

First published in 1901 by George Routledge and Sons Ltd

This edition first published in 2019
by Routledge
2 Park Square, Milton Park, Abingdon, Oxon OX14 4RN

and by Routledge
52 Vanderbilt Avenue, New York, NY 10017

Routledge is an imprint of the Taylor & Francis Group, an informa business

© 1901 Robert Routledge

British Library Cataloguing in Publication Data
A catalogue record for this book is available from the British Library

ISBN: 978-1-138-39006-5 (Set)
ISBN: 978-0-429-02175-6 (Set) (ebk)
ISBN: 978-1-138-39275-5 (Volume 7) (hbk)
ISBN: 978-1-138-39278-6 (Volume 7) (pbk)
ISBN: 978-0-429-40205-0 (Volume 7) (ebk)

Publisher's Note
The publisher has gone to great lengths to ensure the quality of this reprint but points out that some imperfections in the original copies may be apparent.

Disclaimer
The publisher has made every effort to trace copyright holders and would welcome correspondence from those they have been unable to trace.

PLATE I.

THE GREAT WHEEL IN ACTION.

DISCOVERIES AND INVENTIONS

OF THE

NINETEENTH CENTURY

BY

ROBERT ROUTLEDGE, B.Sc.,

FOURTEENTH EDITION

CONTAINING FOUR HUNDRED AND FIFTY-SIX ILLUSTRATIONS

LONDON
GEORGE ROUTLEDGE AND SONS, LIMITED
BROADWAY, LUDGATE HILL
1901

PREFACE.

IN the following pages an attempt has been made to present a popular account of remarkable discoveries and inventions which distinguish the XIXth century. They distinguish it not merely in comparison with any previous century, but in comparison with all the centuries that have preceded, in regard to far-reaching intellectual acquisitions, and to material achievements, which together have profoundly affected our ways of thinking and our habits of life. In the latter, the enormously increased facilities of locomotion and international communication due to railways and steam navigation have wrought the greatest changes. These inventions depending primarily upon that of the steam engine, this first claims our notice, although properly assignable to a period preceding our era by a few years. Again, much of our material advancement is connected with improvements in the manufacture of iron and its applications in the form of steel, which have been especially the work of the last half of the century. So great has been the progress in this department, that for the present edition it has been found necessary to re-write altogether the article devoted to it. Our social conditions have also been greatly modified by the celerity of verbal intercourse afforded by the telegraph and the telephone, and these inventions have received appropriate notice in this work. In every branch of science also we have reason to be proud of the discoveries our era can claim, for they vastly excel in number and are not inferior in range to those of all the ages taken together. From so large a field, selection was of course necessary ; and the instances selected have been those which appeared to some extent typical, or those which seemed to have the most direct bearing on the general advance of our time. The topics comprise chiefly those great applications of mechanical engineering and arts, and of physical and chemical science, in which every intelligent person feels concerned ; while some articles are devoted to certain purely scientifi discoveries that have excited general interest.

The author has aimed at giving a concise but clear description of the several subjects; and that without assuming on the part of the reader any knowledge not usually possessed by young persons of either sex who have received an ordinary education. The design has been to treat the subjects as familiarly as might be consistent with a desire to impart real information ; while the popular character of the book has not been considered a reason for regarding accuracy as unnecessary. On the contrary, pains have been taken to consult the best authorities ; and it is only because the sources of information to which the author is under obligation are so many, that he cannot acknowledge them in detail.

The present edition has been revised throughout, and such changes have been made as were required to bring the matter into accordance with the progress that has taken place since this book was first published in 1876. But details given in the former editions have at the same time been retained where they served to indicate the successive stages of improvement. It would, for example, be impossible in a section on steam navigation, to omit some notice of the *Great Eastern,* and therefore the drawings and the account of the construction of that remarkable ship that appeared in the first edition, have been left with but slight alterations in the present volume, although the vessel has since been broken up. On the other hand, two sections are devoted to projects which the XIXth century has not seen realised ; but the XXth century will in all probability shortly witness the completion of one or other of the great canal schemes ; and if the first submarine tunnel is destined not to be one connecting England with the Continent, it will be one uniting Great Britain with her sister isle.

1899.

FOR permission to make use of illustrations in this volume the author's and publishers' thanks are due to the several proprietors of *The Graphic* (for Plates I., XI., and XII.)—of *The Engineer* (for sketch design of the Great Wheel, map and views of the Tower Bridge)—of *The Scientific American* (map of North Sea Canal); also to Mr. Walter B. Basset (for Plate V.)—to "The Cassier Magazine Company" (for Edison's Kinetographic Theatre and the Hotchkiss Gun)—to "The Century Company" (for portrait of M. Tesla, from a photograph by Sarony) —to "The Incandescent Gas Light Company" (for cuts of burners, etc.)—to *The Engineering Magazine*, and *The Engineering News*, both of New York—to the Remington Company—to Mr. W. W. Greener, of Birmingham (for cuts of rifles, etc., from his comprehensive book on "The Gun")—to *The Photogram*, Limited—to the Proprietors of *Nature* —to the Linotype Company—and to Captains Hadcock and Lloyd (for illustrations of modern artillery from their great work on the subject).

CONTENTS.

LIST OF ILLUSTRATIONS.

xi

LIST OF PLATES.

———

Wind, Steam, and Speed (after TURNER).

INTRODUCTION.

ONLY by knowledge of Nature's laws can man subjugate her powers and appropriate her materials for his own purposes. The whole history of arts and inventions is a continued comment on this text; and since the knowledge can be obtained only by observation of Nature, it follows that Science, which is the exact and orderly summing-up of the results of such observation, must powerfully contribute to the well-being and progress of mankind.

Some of the services which have been rendered by science in promoting human welfare are thus enumerated by an eloquent writer: " It has lengthened life; it has mitigated pain; it has extinguished diseases; it has increased the fertility of the soil; it has given new securities to the mariner; it has furnished new arms to the warrior; it has spanned great rivers and estuaries with bridges of form unknown to our fathers; it has guided the thunderbolt innocuously from heaven to earth; it has lighted up the night with the splendour of the day; it has extended the range of the human vision; it has multiplied the power of the human muscles; it has accelerated motion; it has annihilated distance; it has facilitated intercourse, correspondence, all friendly offices, all dispatch of business; it has enabled man to descend to the depths of the sea, to soar into the air, to penetrate securely into the noxious recesses of the earth, to traverse the land in cars which whirl along without horses, to cross the ocean in ships which run ten knots an hour against the wind. These are but a part of its fruits, and of its first-fruits; for it is a philosophy which never rests, which has never attained, which is never perfect. Its law is progress. A point which yesterday was invisible is its goal to-day, and will be its starting-point to-morrow."—(MACAULAY).

Thus every new invention, every triumph of engineering skill, is the embodiment of some scientific idea; and experience has proved that discoveries in science, however remote from the interests of every-day life they may at first appear, ultimately confer unforeseen and incalculable benefits on mankind. There is also a reciprocal action between science and its application to the useful purposes of life; for while no advance is ever made in any branch of science which does not sooner or later give rise to a corresponding improvement in practical art, so on the other hand every advance made in practical art furnishes the best illustration of scientific principles.

The enormous material advantages which this age possesses, the cheapness of production that has placed comforts, elegancies, and refinements unknown to our fathers within the reach of the humblest, are traceable in a high degree to the arrangement called the " division of labour," by which it is found more advantageous for each man to devote himself to one kind of work only; to the steam engine and its numerous applications; to increased knowledge of the properties of metals, and of the methods of extracting them from their ores; to the use of powerful and accurate tools; and to the modern plan of manufacturing articles by processes of copying, instead of fashioning everything anew by manual labour. Little more than a century ago everything was slowly and imperfectly made by the tedious toil of the workman's hand; but now marvellously perfect results of ingenious manufacture are in every-day use, scattered far and wide, so that their very commonness almost prevents us from viewing them with the attention and admiration they deserve. Machinery, actuated by the forces of nature, now performs with ease and certainty work that was formerly the drudgery of thousands. Every natural agent has been pressed into man's service: the winds, the waters, fire, gravity, electricity, light itself.

But so much have these things become in the present day matters of course, that it is difficult for one who has not witnessed the revolution produced by such applications of science to realize their full importance. Let the young reader who wishes to understand why the present epoch is worthy of admiration as a stage in the progress of mankind, address himself to some intelligent person old enough to remember the century in its teens; let him inquire what wonderful changes in the aspect of things have been comprised within the experience of a single lifetime, and let him ask what has brought about these changes. He will be told of the railway, and the steam-ship, and the telegraph, and the great guns, and the mighty ships of war—

> " The armaments which thunderstrike the walls
> Of rock-built cities, bidding nations quake,
> And monarchs tremble in their capitals."

He will be told of a machine more potent in shaping the destinies of our race than warlike engines—the steam printing-press. He may hear of a chemistry which effects endless and marvellous transformations; which from dirt and dross extracts fragrant essences and dyes of resplendent hue. He may hear something of a wonderful instrument which can make a faint beam of light, reaching us after a journey of a thousand years, unfold its tale and reveal the secrets of the stars. Of these and of other inventions and discoveries which distinguish the present age it is the purpose of this work to give some account.

STEAM ENGINES.

T O track the steps which led up to the invention of the Steam Engine, and fully describe the improvements by which the genius of the illustrious Watt perfected it at least in principle, are not subjects falling within the province of this work, which deals only with the discoveries and inventions of the present century. But as it does enter into our province to describe some of the more recent developments of Watt's invention, it may be desirable to give the reader an idea of his engine, of which all the more recent applications of steam are modifications, with improvements of detail rather than of principle.

Watt took up the engine in the condition in which it was left by Newcomen; and what that was may be seen in Fig. 2, which represents Newcomen's atmospheric engine—the first practically useful engine in which a piston moving in a cylinder was employed. In the cut, the lower part of the cylinder, c, is removed, or supposed to be broken off, in order that the piston, h, and the openings of the pipes, d, e, f, connected with the cylinder, may be exhibited. The steam was admitted beneath the piston by the attendant turning the cock k, and as the elastic force of the steam was only equal to the pressure of the atmosphere, it was not employed to raise the piston, but merely filled the cylinder, the ascent of the piston being caused by the weight attached to the other side of the beam, which at the same time sent down the pump-rod, m; and when this was at its lowest position, the piston was nearly at the top of the cylinder, which was open. The attendant then cut off the communication with the boiler by closing the cock, k, at the same time opening another cock which allowed a jet of cold water from the cistern, *g*, to flow through the opening, *d*, into the cylinder. The steam which filled the cylinder was, by contact with the cold fluid, instantly condensed into water; and as the liquefied steam would take up little more than a two-thousandth part of the space it occupied in the gaseous state, it followed that a vacuum was produced within the cylinder; and the weight of the atmosphere acting on the top of the piston, having no longer the elastic force of the steam to counteract it, forced the piston down, and thus raised the pump-bucket attached to

FIG. 2.—*Newcomen's Steam Engine.*

the rod, *m*. The water which entered the cylinder from the cistern, together with that produced by the condensation of the steam, flowed out of the cylinder by the opening, *f*, the pipe from which was conducted downwards, and terminated under water, the surface of which was at least 34 ft. below the level of the cylinder; for the atmospheric pressure would cause the cylinder to be filled with water had the height been less. The improvements which Watt, reasoning from scientific principles, was enabled to effect on the rude engine of Newcomen, are well expressed by himself in the specification of his patent of 1769. It will be observed that the machine was formerly called the "fire engine."

"My method of lessening the consumption of steam, and consequently fuel, in fire engines, consists of the following principles :—*First.* That vessel in which the powers of steam are to be employed to work the engine (which is called the cylinder in common fire engines, and which I

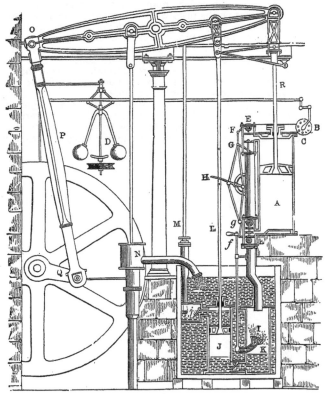

FIG. 3.—*Watt's Double-action Steam Engine.*

call the steam-vessel), must, during the whole time the engine is at work, be kept as hot as the steam that enters it ; first, by enclosing it in a case of wood, or any other materials that transmit heat slowly ; secondly, by surrounding it with steam or other heated bodies ; and thirdly, by suffering neither water nor any other substance colder than the steam to enter or touch it during that time.—*Secondly.* In engines that are to be worked either wholly or partially by condensation of steam, the steam is to be condensed in vessels distinct from the steam-vessels or cylinders, although occasionally communicating with them,—these vessels I call condensers ; and whilst the engines are working, these condensers ought to be kept at least as cold as the air in the neighbourhood of the engines by the application of water or other cold bodies.—*Thirdly.* Whatever air or other elastic vapour is not condensed by the cold of the condenser, and may impede the working of the engine, is to be drawn out of the steam-vessels or condensers by means of pumps, wrought by the engines themselves or otherwise.—*Fourthly.* I intend in many cases to employ the expansive force of steam to press on the pistons, or whatever may be used instead of them, in the same manner in which the pressure of the atmosphere is now employed in common fire engines. In cases where cold water cannot be had in plenty, the engines may be wrought by this force of steam only, by discharging the steam into the air after it has done its office.—*Lastly.* Instead of using water to render the pistons and other parts of the engines air and steam-tight, I employ oils, wax, resinous bodies, fat of animals, quicksilver, and other metals in their fluid state."

From the engraving we give of Watt's double-action steam engine, Fig. 3, and the following description, the reader will realize the high degree of perfection to which the steam engine was brought by Watt. The steam is conveyed to the cylinder through a pipe, B, the supply being regulated by the

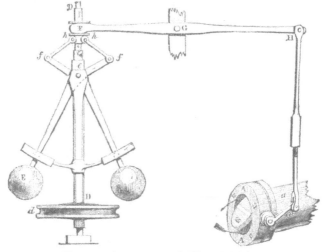

FIG. 4.—*Governor and Throttle-Valve.*

throttle-valve, acted on by rods connected with the governor, D, which has a rotary motion. This apparatus is designed to regulate the admission of steam in such a manner that the speed of the engine shall be nearly uniform; and the mode in which this is accomplished may be seen in Fig. 4, where D D is a vertical axis carrying the pulley, *d*, which receives a rotary motion from the driving-shaft of the engine, by a band not shown in the figures. Near the top of the axis, at *e*, two bent rods work on a pin, crossing each other in the same manner as the blades of a pair of scissors. The two heavy balls are attached to the lower arms of these levers, which move in slits through the curved guides intended to keep them always in the same vertical plane as the axis, D D. The upper arms are jointed at *ff* to rods hinged at *h h* to a ring not attached to the axis, but allowing it to revolve freely within it. To this ring at F is fastened one end of the lever connected with the throttle-valve in a manner sufficiently obvious from the cut. The position represented is that assumed by the apparatus when the engine is in motion, the disc-valve, *z*, being partly open. If from any cause the velocity of the engine increases, the balls diverge from increased centrifugal force, and the effect is to draw down the ring at F, and, through the system of levers, to turn the disc in the direction of the arrows, and diminish the supply of steam. If, on the other hand, the speed of the engine is checked, the balls fall towards the axis, and the valve is opened wider, admitting steam more freely, and so restoring its former speed to the engine. On one side of the cylinder are two hollow boxes, E E, Fig. 3, communicating with the cylinder by an opening near the middle of the box. Each of these steam-chests is divided into three compartments by conical valves attached to rods connected with the lever, H. These valves are so arranged that when the upper part of the cylinder is in communication with the boiler, the lower part is open to the condenser, I, and *vice versâ*. The top of the cylinder is covered, and the piston-rod passes through an air and steam-tight hole in it; freedom of motion, with the necessary close fitting, being attained by making the piston-rod pass through a *stuffing-box*, where it is closely surrounded with greased tow. The piston is also packed, so that, while it can slide freely up and down in the cylinder, it divides the latter into two steam-tight chambers. In an engine of this kind, the elastic force of the steam acts alternately on the upper and lower surfaces of the piston; and the condenser, by removing the steam which has performed its office, leaves a nearly empty space before the piston, in which it advances with little or no resistance. On the rod which works the air-pump, two pins are placed, so as to move the lever, H, up and down through a certain space, when one pin is near its highest and the other near its lowest position, and thus the valves are opened and closed when the piston reaches the termination of its stroke. In the condenser, I, a stream of cold water is constantly playing, the flow being regulated by the handle, *f*. The steam, in condensing, heats the cold water, adding to its bulk, and at the same time the air, which is always contained in water, is disengaged, owing to the heat and the reduced pressure. Hence it is necessary to pump out both the air and the water by the pump, J, which is worked by the beam of the engine. In his engines Watt adopted the heavy fly-wheel, which tends to equalize the movement, and render insensible the effects of those variations in the driving power and in the resistance which always occur. In the action of the engine itself there are two positions of the piston, namely, where it is changing its direction, in which there is no force whatever communicated to the piston-rod by the steam. These positions are known as the "dead points," and in a rotatory engine occur twice in each revolution. The resistance also is liable to great variations. Suppose, for example, that the engine is employed to move the shears by which thick plates of iron are cut. When a plate has been cut, the resistance is removed, and the speed of the engine increases; but this increase, instead of taking place by a sudden start, takes place gradually, the power of the engine being in the meantime absorbed in imparting increased velocity to the fly-wheel. When another plate is put between the shears, the power which the fly-wheel has gathered up is given out in the slight diminution of its speed occasioned by the increased resistance. But for the fly-wheel, such changes of velocity would take place with great suddenness, and the shocks and strains thereby caused would soon injure the machine. This expedient, in conjunction with that admirable contrivance, the "governor," renders it possible to set the same engine at one moment to forge an anchor, and at the next to shape a needle. One of the most ingenious of Watt's improvements is what is termed the "parallel motion," consisting of a system of jointed rods connecting the head of the piston-rod, R, with the end of the oscillating beam. As, during the motion of the engine, the former moves in a straight line, while the latter describes a circle, it would be impossible to connect them directly. Watt accomplished this by hinging rods together in form of a parallelogram, in such a manner that, while three of the angles describe circles, the fourth moves in nearly a straight line. Watt was himself surprised at the regularity of the action. "When I saw it work for the first time, I felt truly all the pleasure of novelty, *as if I was examining the invention of another man.*"

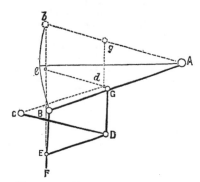

A B is half the beam, A being the main centre; B E, the main links, connecting the piston-rod, F, with the end of the beam; G D, the air-pump links, from the centre of which the air-pump-rod is suspended; C D moves about the

fixed centre, C, while D E is movable about the centre D, itself moving in an arc, of which C is the centre. The dotted lines show the position of the links and bars when the beam is at its highest position.

FIG. 4a.— *Watt's Parallel Motion.*

Many improvements in the details and fittings of almost every part of the steam engine have been effected since Watt's time. For example, the opening and closing of the passages for the steam to enter and leave the cylinder is commonly effected by means of the slide-valve (Fig. 5). The steam first enters a box, in which are three holes placed one above the other in the face of the box opposite to the pipe by which the steam enters. The uppermost hole is in communication with the upper part of the cylinder, and the lowest with the lower part. The middle opening leads to the condenser, or to the pipe by which the steam escapes into the air. A piece of metal, which may be compared to a box without a lid, slides over the three holes with its open side towards them, and its size is such that it can put the middle opening in communication with either the uppermost or the lowest opening, at the same time giving free passage for the steam into the cylinder by leaving the third opening uncovered. In A, Fig. 5, the valve is admitting steam below the piston, which is moving upwards, the steam which had before propelled it downwards now having free exit. When the piston has arrived at the top of the cylinder, the slide is pushed down by the rod connecting it with the eccentric into the position represented at B, and then the opposite movement takes place. The slide-valve is not moved, like the old pot-lid valves, against the pressure of the steam, and has other advantages, amongst which may be named the readiness with which a slight modification renders it available for using the steam "*expansively.*" This expansive working was one of Watt's inventions, but has been more largely applied in recent times. In this plan. when the piston has performed a part of its stroke, the steam is shut off. and the piston is then urged on by the expansive force of the steam enclosed in the cylinder. Of course as the steam expands its pressure decreases; but as the same quantity of steam performs a much larger amount of work when used expansively, this plan of cutting off the steam is attended with great economy. It is usually effected by the modification of the slide-valve, shown at C, Fig. 5, where the faces of the slides are made of much greater width than the openings. This excess of width is called the "*lap*," and by properly adjusting it, the opening into the cylinder may be kept closed during the interval required, so that the steam is not allowed to enter the cylinder after a certain length of the stroke has been performed. The slide-valve is moved by an arrangement termed the eccentric. A circular

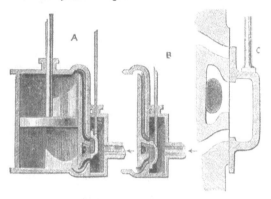

FIG. 5.—*Slide Valve.*

disc of metal is carried on the shaft of the engine, and revolves with it. The axis of the shaft does not, however, run through the centre of the disc, but towards one side. The disc is surrounded by a ring, to which it is not attached, but is capable of turning round within it. The ring forms part of a triangular frame to which is attached one arm of a lever that communicates the motion to the rod bearing the slide. Expansive working is often employed in conjunction with *superheated steam*, that is, steam heated out of contact with water, after it has been formed, so as to raise its temperature beyond that merely necessary to maintain it in the state of steam, and to confer upon it the properties of a perfect gas. Experience has proved that an increased efficiency is thus obtained.

The actual power of a steam engine is ascertained by an instrument called the Indicator, which registers the amount of pressure exerted by the steam on the face of the piston in every part of its motion. The indicator consists simply of a very small cylinder, in which works a piston, very accurately made, so as to move up and down with very little friction. The piston is attached to a strong spiral spring. so that when the steam is admitted into the cylinder of the indicator the spring is compressed, and its elasticity resists the pressure of the steam, which tends to force the piston up. When the pressure of steam below the piston of the indicator is equal to that of the atmosphere, the spring is neither compressed nor extended; but when ʻhe steam-pressure falls below that of the atmosphere, as it does while the steam is being condensed, then the atmospheric pressure forces down the piston of the indicator until it is balanced by the tension of the now stretched spring. The extension or compression of the spring thus measures the difference between the pressure of the atmosphere and that of the steam in the cylinder of the engine, with which the cylinder of the indicator freely communicates.

From the piston-rod of the indicator a pencil projects horizontally, and its point presses against a sheet of paper wound on a drum, which moves about a vertical axis. This drum is made to move backwards and forwards through a part of a revolution, so that its motion may exactly correspond with that of the piston in the cylinder of the steam engine. Thus, if the piston of the indicator were to remain stationary, a level line would be traced on the paper by the movement of the drum; and if the latter did not move, but the steam were admitted to the indicator, the pencil would mark an upright straight line on the paper. The actual result is that a figure bounded by curved lines is traced on the paper, and the curve accurately represents the pressure of the steam at every point of the piston's motion. The position of the point of the pencil which corresponds with each pound of pressure per square inch is found by trial by the maker of the instrument, who attaches a scale to show what pressures of steam are indicated.

If the pressure per square inch is known, it is plain that by multiplying that pressure by the number of square inches in the area of the piston of the engine, the total pressure on the piston can be found. The pressure does not rise instantly when the steam is first admitted, nor does it fall quite abruptly when the steam is cut off and communication opened with the condenser. When the steam is worked expansively, the pressure falls gradually from the time the steam is shut off. Now, the amount of work done by any force is reckoned by the pressure it exerts multiplied into the space through which that pressure is exerted. Therefore the work done by the steam is known by multiplying the pressure in pounds on the whole surface of the piston into the length in feet of the piston's motion through which that pressure is exerted. The trace of the pencil on the paper—*i.e.*, the *indicator diagram*—shows the pressures, and also the length of the piston's path through which each pressure is exerted, and therefore it is not difficult to calculate the actual work which is done by the steam at every stroke of the engine. If this be multiplied by the number of strokes per minute, and the product divided by 33,000, we obtain what is termed the *indicated horse-power* of the engine. The work done per minute is divided by 33,000, because that number is taken to represent the work that a horse can do in a minute: that is, the average work done in one minute by a horse would be equal to the raising of the weight of 1,000 lbs. thirty-three feet high, or the raising of thirty-three pounds 1,000 feet high. The number, 33,000, as expressing the work that could be done by a horse in one minute, was fixed on by Watt, but more recent experiments have shown that he over-estimated the power of horses, and that we should have to reduce this number by about one-third if we desire to express the actual average working power of a horse. But the power of engines having come to be expressed by stating the horse-power on Watt's standard, engineers have kept to the original number, which is, however, to be considered as a merely artificial unit or term of comparison between one engine and another; for the power of a horse to perform work will vary with the mode in which its strength is exerted. The source of the power which does the work in the steam engine is the combustion of the coal in the furnace under the boiler. The amount of work a steam engine will do depends not only on the quantity of steam which is generated in a given time, but also upon the pressure, and therefore the temperature at which the steam is formed.

The water constantly evaporating in the boiler of a steam engine is usually renewed by forcing water into the boiler against the pressure of the steam by means of a small pump worked by the engine. In the engraving of Watt's engine this pump is shown at M. But recently the feed-pump has been to a great extent superseded by a singular apparatus invented by M. Giffard, and known as Giffard's Injector. In this a jet of steam from the boiler itself supplies the means of propelling a stream of water directly into the boiler. Fig. 6 is a section of this interesting apparatus through its centre, and it clearly shows the manner in which

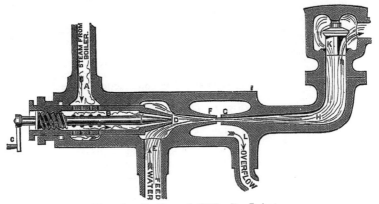

FIG. 6.— *Section of Giffard's Injector.*

the current of steam is made to operate on the jet of water. The steam from the boiler passes through the pipe A and into the tube B through the holes. The nozzle of this tube is of a conical shape, and its centre is occupied by a rod pointed to fit into the conical nozzle, and provided with a screw at the other end, so that the opening can be regulated by turning the handle, C. At D the jet of steam comes in contact with the water which feeds the boiler, the arrangement being such that the steam is driven into the centre of the stream of water which enters by the pipe E, and is propelled by the steam jet through another cone, F, issuing with such force from the orifice of the latter that it is carried forward through the small opening at G into the chamber H. Here the water presses on the valve K, which it raises against the pressure of the steam and enters the boiler. The water issuing from the cone, F, actually traverses an open space which is exposed to the air, and where the fluid may be seen rushing into the boiler as a clear jet, except a few beads of steam which may be carried forward in the centre, the rest of the steam having been condensed by the cold water. The steam, of course, rushes from the cone, B D, with enormous velocity, which is partly communicated to the water. The pipe, L, is for the water which overflows in starting the apparatus, until the pressure in H becomes great enough to open the valve. The supplies of water and of steam have to be adjusted according to the conditions of pressure in the boiler, and according to the temperature of the feed-water. It is found that wher the feed-water is at a temperature above 120° Fahrenheit, the injector wil/ not work: the condensation of the steam is therefore necessary to the result. For, as the steam is continually condensed by the cold water, it rushes from D with the same velocity as into a vacuum, and the water is urged on by a momentum due to this velocity. We must observe, moreover, that the net result of the operation is a lessening of the pressure in the boiler; for the entrance of the feed-water produces a fall of temperature in the boiler, and the bulk of steam expended is fourteen times the bulk of the water injected: thus, although the apparatus before actual trial would not appear likely to produce the required result, the effect is no more paradoxical than in the case of the feed-pump. The injector has been greatly improved by Mr. Gresham, who has contrived to make some of the adjustments self-acting, and his form of the apparatus is now largely used in this country. The injector is applicable to stationary, locomotive, or marine engines.

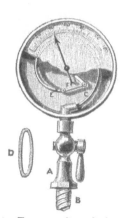

FIG. 7.—*Bourdon's Pressure Gauge.*

Steam boilers are now always provided with one of *Bourdon's* gauges, for indicating the pressure of the steam. The construction of the instrument will easily be understood by an examination of Fig. 7. The gauge is screwed into some part of the boiler, where it can always be seen by the person in charge. The stop-cock A communicates with the curved metallic tube C, which is the essential part of the contrivance. This tube is of the flattened form shown at D, having its greatest breadth perpendicular to the plane in which the tube is curved, and it is closed at the end E, where it is attached to the rod F, so that any movement of E causes the axle carrying the index-finger, F, to turn, and the index then moves along the graduated arc. The connection is sometimes made by wheelwork, instead of by the simple plan shown in the figure. The front plate is represented as partly broken away, in order to show the internal arrangement, which, of course is not visible in the real instrument, where only the index-finger and graduated scale are seen, protected by a glass plate.

When a curved tube of the shape here described is subjected to a greater pressure on the inside than on the outside, it tends to become straighter, and the end E moves outward; but when the pressure is removed, the tube resumes its former shape. The graduations on the scale are made by marking the position of the index when known pressures are applied. The amounts of pressure, when the gauges are being graduated, are known by the compression produced in air contained in another apparatus. Gauges constructed on Bourdon's principle are applied to other purposes, and can be made strong enough to measure very great pressures. such as several

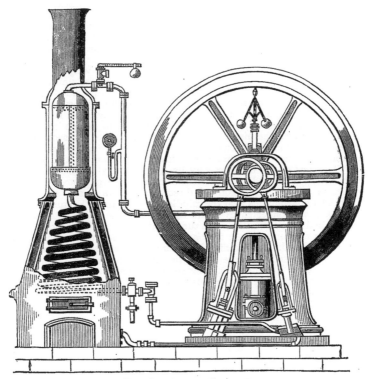

FIG. 8.—*Steam Generator.*

thousand pounds on the square inch; they may also be made so delicate as to measure variations of pressure below that of the atmosphere. The simplicity and small size of these gauges, and the readiness with which they can be attached, render them most convenient instruments wherever the pressure of a gas or liquid is required to be known.

A point to which great attention has been directed of late years is the construction of a boiler which shall secure the greatest possible economy in fuel. Of the total heat which the fuel placed in the furnace is capable of supplying by its combustion, part may be wasted by an incomplete burning of the fuel, producing cinders or smoke or unburnt gases, another part is always lost by radiation and conduction, and a third portion is carried off by the hot gases that escape from the boiler-flues. Many contrivances have been adopted to diminish as much as possible this waste of heat, and so obtain the greatest possible proportion of available steam power from a given weight of fuel. Boilers wholly or partially formed of tubes have recently been much in favour. An arrangement for quickly

generating and superheating steam is shown in Fig. 8, in connection with
a high-pressure engine.

Steam engines are constructed in a great variety of forms, adapted to
the purposes for which they are intended. Distinctions are made according as the engine is fitted with a condenser or not. When steam of a low
pressure is employed, the engine always has a condenser, and as in this
way a larger quantity of work is obtainable for a given weight of fuel, all
marine engines—and all stationary engines, where there is an abundant
supply of water and the size is not objectionable—are provided with condensers. High-pressure steam may be used with condensing engines, but
is generally employed in non-condensing engines only, as in locomotives
and agricultural engines, the steam being allowed to escape into the air
when it has driven the piston to the end of the stroke. In such engines
the beam is commonly dispensed with, the head of the piston-rod moving
between guides and driving the crank directly by means of a connecting-
rod. The axis of the cylinder may be either vertical, horizontal, or inclined. A plan often adopted in marine engines, by which space is saved,
consists in jointing the piston-rod directly to the crank, and suspending
the cylinder on trunnions near the middle of its length. The trunnions are
hollow, and are connected by steam-tight joints, one with the steam-pipe
from the boiler, and the other with the eduction-pipe. Such engines have
fewer parts than any others; they are lighter for the same strength, and
are easily repaired. The trunnion joints are easily packed, so that no leakage takes place, and yet there is so little friction that a man can with one
hand move a very large cylinder, whereas in another form of marine engine,
known as the side-lever engine, constructed with oscillating beams, the
friction is often very great.

THE LOCOMOTIVE.

THE first locomotive came into practical use in 1804. Twenty years
before, Watt had patented—but had not constructed—a locomotive
engine, the application of steam to drive carriages having first been suggested by Robinson in 1759. The first locomotives were very imperfect,
and could draw loads only by means of toothed driving-wheels, which
engaged teeth in rack-work rails. The teeth were very liable to break off,
and the rails to be torn up by the pull of the engine. In 1813, the important discovery was made that such aids are unnecessary, for it was found
that the "bite" of a smooth wheel upon a smooth rail was sufficient for all
ordinary purposes of traction. But for this discovery, the locomotive might
never have emerged from the humble duty of slowly dragging coal-laden
waggons along the tramways of obscure collieries. The progress of the
locomotive in the path of improvement was, however, slow, until about
1825, when George Stephenson applied the blast-pipe, and a few years
later adopted the tubular boiler. These are the capital improvements
which, at the famous trial of locomotives, on the 6th of October, 1829,
enabled Stephenson's "Rocket" to win the prize offered by the directors
of the Liverpool and Manchester Railway. The "Rocket" weighed 4½
tons, and at the trial drew a load of tenders and carriages weighing 12¾
tons. Its average speed was 14 miles an hour, and its greatest, 29 miles

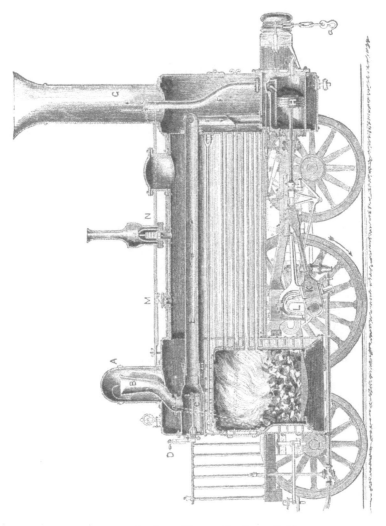

FIG. 9.—*Section of Locomotive* (A.D. 1837).

an hour. This engine, the parent of the powerful locomotives of the present day, may now be seen in the Patent Museum at South Kensington.
Since 1829, numberless variations and improvements have been made in

the details of the locomotive. In weight, dimensions, tractive power and speed, the later locomotives vastly surpass the earlier types.

Fig. 9 represents the section of a locomotive constructed *c.* 1837. The boiler is cylindrical; and at one end is placed the fire-box, partly enclosed in the cylindrical boiler, and surrounded on all sides by the water, except where the furnace door is placed, and at the bottom, where the fuel is heaped up on bars which permit the cinders to drop out. At the other end of the boiler, a space beneath the chimney called the smoke-box is connected with the fire-box by a great number of brass pipes, open at both ends, firmly fixed in the end plates of the boiler. These tubes are from $1\frac{1}{4}$ in. to 2 in. in diameter, and are very numerous—usually about one hundred and eighty, but sometimes nearly double that number. They therefore present a large heating surface to the water, which stands at a level high enough to cover them all and the top of the fire-box. The boiler of the locomotive is not exposed to the air, which would, if allowed to come in contact with it, carry off a large amount of heat. The outer surface is therefore protected from this cooling effect by covering it with a substance which does not permit the heat to readily pass through it. Nothing is found to answer better than felt; and the boiler is accordingly covered with a thick layer of this substance, over which is placed a layer of strips of wood $\frac{3}{4}$ in. thick, and the whole is surrounded with thin sheet iron. It is this sheet iron alone that is visible on the outside. The level of the water in the boiler is indicated by a gauge, which is merely a very strong glass tube; and the water carried in the tender is forced in as required, by a pump (not shown in the Fig.). The steam leaves the boiler from the upper part of the *steam-dome*, A, where it enters the pipe, B; the object being to prevent water from passing over with the steam into the pipe. The steam passes through the *regulator*, C, which can be closed or opened to any extent required by the handle, D, and rushes along the pipe, E, which is wholly within the boiler, but divides into two branches when it reaches the smoke-box, in order to conduct the steam to the cylinders. Of these there are two, one on each side, each having a slide-valve, by means of which the steam is admitted before and behind the pistons alternately, and escapes through the blast-pipe, F, up the chimney, G, increasing the draught of the fire by drawing the flame through the longitudinal tubes in proportion to the rush of steam; and thus the rate of consumption of fuel adjusts itself to the work the engine is performing, even when the loads and speeds are very different. Though the plane of section passing through the centre of boiler would not cut the cylinders, one of them is shown in section. H is the piston; K the connecting-rod jointed to the crank, L, the latter being formed by forging the axle with four rectangular angles, thus, ⌐⌐; and the crank bendings for the two cylinders are placed in planes at right angles to each other, so that when one is at the "dead point," the other is in a position to receive the full power of the piston. There are two safety valves, one at M, the other at N; the latter being shut up so that it cannot be tampered with.

Locomotives are fitted with an ingenious apparatus for reversing the engines, which was first adopted by the younger Stephenson, and is known as the "link motion." The same arrangement is employed in other engines in which the direction of rotation has to be changed; and it serves another important purpose, namely, to provide a means by which steam may be employed expansively at pleasure. The link motion is represented in Fig. 10, where A, B, are two eccentrics oppositely placed on the driving-shaft, and their rods joined to the ends of the curved bar or link, C D. A slit extends nearly the whole length of this bar, and in it works the stud E, forming part of the lever, F, G, movable about the fixed joint, G, and having its extremity, F, jointed to the rod H, that moves the slide-valve. The weight

FIG. 10.—*Stephenson's Link Motion.*

of the link and the eccentric rods is counterpoised with a weight, K, attached to the lever, I K, which turns on the fixed centre, L. This lever forms one piece with another lever, L M, with which it may be turned by pulling the handle of O P, connected with it through the system of jointed rods. When the link is lowered, as shown in the figure, the slide-valve rod will follow the movement of the eccentric, B, while the backward and forward movement of the other eccentric will only be communicated to the end of C, and will scarcely affect the position of the stud E at all. By drawing the link up to its highest position, the motion due to eccentric A only will be communicated to the slide-valve rod, which will therefore be drawn back at the part of the revolution where before it was pushed forward, and *vice versâ;* hence the engine will be reversed. When the link is so placed that the stud is exactly in the centre, the slide-valve will re-receive no motion, and remain in its middle position, consequently the engine is stopped. By keeping the link nearer or farther from its central position, the throw of the slide-valve will be shorter or longer, and the steam will be shut off from entering the cylinder when a smaller or larger portion of the stroke has been performed.

Although Fig. 9 represents with sufficient clearness all the essential parts of a locomotive, it should be observed that as actually constructed for use on the different lines of railway the machine is greatly modified in the arrangement and proportions of its parts. A greater number of adjuncts and subsidiary appliances are also provided for the more effective and convenient working of the engine, and for giving control over the movement of the train, and these, in fact, conduce much, to the greater economy and safety with which trains are now run. As the circumstances and conditions under which railways are worked vary much in different parts of the world, the locomotive has to be designed to meet the requirements of each case, and its general appearance, details and dimensions are accordingly much diversified. From among the many types of recent locomotives we select for illustration and a short description the form of express passenger engine that has lately been designed by Mr. T. W. Worsdell, the engineer of the North Eastern Railway, and this will give the opportunity of noticing some of the newest improvements, which are embodied in this engine. See Plate II.

The plan of causing the steam to work expansively has already been mentioned on pages 8 and 9, as used by cutting off the steam when part of the stroke of the piston has been made. Another mode by which the expansive principle has long been made use of in stationary and marine engines is to allow the steam from the boiler to enter first a smaller cylinder and from that, at the end of the stroke, to pass into a larger one in which, as it expands, it exercises a diminished pressure. This arrangement has been called the compound or double-cylinder engine, and was known to possess certain advantages where high pressure steam was made use of. Indeed, in marine engines the principle of "triple expansion" is now quite commonly adopted—that is, the steam passes successively into three cylinders of successively greater diameter. Mr. Webb, the locomotive engineer of the London and North Western Railway, appears to have been the first to make the "compounding" system a practical success as applied to the locomotive. In Mr. Webb's arrangement there are three cylinders, two smaller ones for the high-pressure steam from the boiler, and between these a single large low-pressure cylinder which receives the steam that has done its work from both the smaller cylinders. In Mr. Worsdell's engine the original and simpler locomotive construction of two cylinders has been adhered to, and thus the general plan of the engine is unchanged except in the larger size of the low-pressure cylinder. In the present engine the stroke is 24 in.; the high-pressure cylinder has its internal diameter 20 in. and the low-pressure cylinder a diameter of 28 in. The boiler-shell is made of steel, the fire-box is of copper, and there are 203 brass tubes, $1\frac{3}{4}$ in. diameter and 10 ft. 11 in. long, connecting the fire-box with the smoke-box. The frame, and indeed most parts of the engine, are also made of steel. The driving-wheels, which here are a single pair, have a diameter of 7 ft. $7\frac{1}{2}$ in. The total "wheel-base" is nearly 21 ft., and it will be observed that the forepart of the engine is supported on a four-wheeled *bogie*. The *bogie* is capable of a certain amount of horizontal motion by turning round a swivel, but this movement is controlled by springs, so that, notwithstanding the length of the frame, the engine is enabled to take curves with great facility, while its motion is perfectly steady even at the highest speeds. The working pressure of the steam in the boiler is 170 lbs. on the square inch. The steam which leaves the high-pressure cylinder is conveyed to

PLATE II.

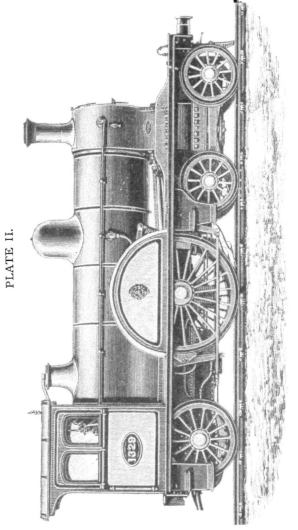

NORTH EASTERN RAILWAY LOCOMOTIVE.

the low-pressure cylinder by a pipe that is led round the inside of the smoke-box, and thus enters the larger cylinder after taking up heat that would otherwise be wasted, so that its elastic force is fully maintained. This circumstance, no doubt, contributes to the very marked economy of fuel that has been effected by the compound engines. How great the economy is found in the working will be seen by the following results, which are taken from the actual records. The same train was taken over the same rails in ordinary quick passenger traffic for several journeys which, as performed in the same time by the compound engine and by another otherwise similar non-compound engine, required for the compound, 25,254 lbs. of coal ; for the non-compound, 32,104 lbs. ; or, the consumption of coal by the former was 28 lbs. per mile ; by the latter, 36 lbs. per mile. This represents a saving of about 21 per cent. of the

FIG. 10*a*.—*G.N.R. Express Passenger Locomotive.*

fuel. As the steam enters the high-pressure cylinder first, it would not be possible to start the engine if it had stopped at one of the "dead-points" on that side, without a special arrangement for admitting the steam directly to the other cylinder in such cases. This, of course, is required only for the first stroke, and Mr. Worsdell and M. Von Borries have contrived for this purpose an ingenious valve, brought into operation when required by a touch from the engineer, and then immediately adjusting itself automatically, so as to restore the steam connections to their normal condition.

Another type of the high-speed passenger engines used for express trains on several of the great English railways is well represented by one of the Great Northern Company's locomotives, as depicted in Fig. 10*a*. In this there are a single pair of driving wheels of very large diameter, namely, 8 ft. 2 in., so that each complete movement of the pistons will carry the engine forwards a length of nearly 26 ft. There are outside cylinders, and therefore the driving axle is straight, and the leading wheels are in two pairs, mounted on a *bogie* which is capable of a certain amount of independent horizontal rotation.

The Stephenson's link motion, described on page 17, has lately been often supplanted by another arrangement known as Joy's valve gear, which leaves the crank axle unencumbered with eccentrics, and, as taking up less space, is generally now preferred for locomotives and also for marine engines. Its principle is very simple, and will be readily understood from the diagram in Fig. 10*b*, where *c* is the spindle of the slide-valves as in Fig. 5, but capable, we shall now suppose, of a horizontal movement only. Jointed to it at D is a rod D E attached to a block at E, which can move only within a slot in the strong bar E F in a circular segment, the centre of which is at D. The bar we suppose for the moment to be immovable, and disposed symmetrically to C D. Now let an alternate up and down motion along the circular segment be given to block E, and the effect will be to leave the centre, D, unchanged in position, and, therefore, in that case the valve will not be moved at all. Now this reciprocating movement is given to the block E by a system of levers

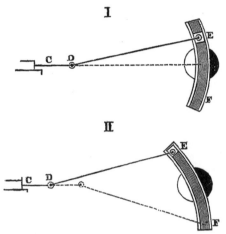

FIG. 10*b*.—*Joy's Valve Gear.*

(not here shown), jointed to the connecting-rod (K, Fig. 9) in such a manner that the rod D E is compelled to follow the movement of the connecting-rod, but the end E must always travel in the circular segment. We have hitherto supposed this segmental piece to be fixed, but the engineer has the power of so turning it as to tilt either the upper or lower part towards D. If, for instance, the guiding segment is fixed as at II, the block in rising will push in the valve-spindle, and in descending draw it out, as the length of the rod D E is invariable. But if the guides be turned over so as to bring F nearer D than E, the same movement of the block will give the reverse motions to the valve-spindle.

From the great rapidity with which the machinery of the locomotive moves, the different parts require to be carefully balanced in order to prevent dangerous oscillations. For example, the centrifugal force of the massive cranks, etc., is balanced by inserting between the spokes of the driving wheels certain counterpoises, the weights and positions of which are finally adjusted by trial. The engine is suspended by chains and set in motion, and a pencil attached to one corner of the frame marks on a horizontal card the form of the oscillation, usually by an oval figure. When the diameter of this figure is reduced to about $\frac{1}{16}$ inch, the adjustment is considered complete.

The power of a locomotive, of course, depends on the pressure of the steam and the size of the cylinder, &c. ; but a very much lower limit than is imposed by these conditions is set to the power of the engine to draw loads by the adhesion between the driving wheels and the rails. By the term "adhesion," which is commonly used in this case, nothing more is really meant than the friction between surfaces of iron. When the resistance of the load drawn is greater than this friction, the wheels turn round and slip on the rails without advancing. The adhesion depends upon the pressure between the surfaces, and upon their condition. It is greater in proportion as the weight supported by the driving-wheels is greater, and when the rails are clean and dry it is equal to from 15 to 20 per cent. of that part of the weight of the engine which rests on the driving-wheels ; but when the rails are moist, or, as it is called, "greasy," the tractive power may be only 5 per cent. of the weight ; about one-tenth may be taken as an average. Suppose that 30 tons of the weight of a locomotive are supported by the driving-wheels, that locomotive could not be employed to drag a train of which the resistance would cause a greater pull upon the coupling-links of the tender than they would be subject to if they were used to suspend a weight of 3 tons. The number of pairs of wheels in a locomotive varies from two to five ; most commonly there are three pairs ; and one, two, or all, are driven by the engine, the wheels being coupled accordingly ; very often two pairs are coupled.

The pressure at which the steam is used in the locomotive is sometimes very considerable. A pressure equal to 180 lbs. on each square inch of the surface of the boiler is quite usual. The greater economy obtained by the employment of high-pressure steam acting expansively in the cylinder, points to the probability of much higher pressures being adopted. There is practically no limit but the power of the materials to resist enormous strains, and there is no reason, in the nature of things, why steam of even 500 lbs. per square inch should not be employed, if it were found otherwise desirable. It need hardly be said that locomotives are invariably constructed of the very best materials, and with workmanship of the most perfect kind. The boilers are always tested, by hydraulic pressure, to several times the amount of the highest pressure the steam is required to have, and great care is bestowed upon the construction of the safety-valves, so that the steam may blow off when the due amount of pressure is exceeded. The explosion of a locomotive is, considering the number of engines in constant use, a very rare occurrence, and is probably in all cases owing to the sudden generation of a large quantity of steam, and not to an excessive pressure produced gradually. Among the causes capable of producing explosive generation of steam may be mentioned the deposition of a hard crust of stony matter, derived from the water ; this crust allows the boiler to be over-heated, and if water should then find its way into contact with the heated metal, a large quantity of steam will be abruptly generated. Or should the water in the boiler become too low, parts of the boiler may become so heated that on the admission of fresh water it would be suddenly converted into steam. When an explosion does take place, the enormous force of the agent we are dealing with when we bottle up steam in an iron vessel, is shown by the effects produced. Fig. 11 is from a photograph taken from an exploded locomotive, where we may see how the thick plates of iron have been torn like paper, and the tubes, rods, and levers of the engine twisted in inextricable confusion.

Locomotive engines for propelling carriages on common roads were

FIG. 11.—*Locomotive after Explosion.*

invented many years ago, by Gurney, Anderson, Scott Russell, Hancock, and others.   One designed by Hancock is represented in Fig. 12.   Such engines do not appear to have found much favour, though the idea has

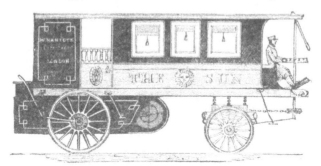

FIG. 12.—*Hancock's Steam Omnibus.*

been successfully realized in the traction engines lately introduced.   Probably the application of steam power to the propulsion of vehicles along common roads fell into neglect on account of the superior advantages of railways, but the common road locomotive is at present receiving some attention.   In the tramways which are now laid along the main roads in most large cities we see one-half of the problem solved.   It is not so much mechanical difficulties that stand in the way of this economical system of locomotion, as the prejudices and interests which have always to be over-come before the world can profit by new inventions. The engines can be made noiseless, emitting no visible steam or smoke, and they are under more perfect control than horses. But vestries and parochial authorities offer such objections as that horses would be frightened in the streets, if the engine made a noise ; and if it did not, people would be liable to be run over, and the horses be as much startled as in the other case. But horses would soon become accustomed to the sight of a carriage moving without equine aid, however startling the matter might appear to them at first ; and the objection urged against the noiseless engines might be alleged against wooden pavements, india-rubber tires, and many other improvements. It is highly probable that in the course of a few years the general adoption of steam-propelled vehicles will displace horses, at least upon tramways. The slowness with which inventions of undeniable utility and of proved advantage come into general use may be illustrated by the fact of some great English towns and centres of engineering industry not having made a single tramway until, in all the populous cities of the United States, and in almost every European capital, tramways had been in successful operation for many years. [1890.]

Some time has elapsed since the foregoing paragraph was written for an earlier edition of this work, and during that period there has been an advance in both practice and opinion ; so that now it has become highly probable that before the century ends a great change may be witnessed in our modes of locomotion, even on ordinary roads. Already every town of importance throughout the United Kingdom has been provided with excellent tramways, along which, in not a few instances, horseless vehicles roll smoothly, to the great convenience of the general public, while not one of the difficulties and dangers to general street traffic has been ex-perienced that were so confidently predicted by those who were unable to perceive that an innovation might be an improvement. The now uni-versally-popular bicycle has been continually receiving improvements, of which there appears to be no end, and as the machine and all the con-trivances connected with it are so familiar to everyone, there is no need here to do more than to refer to them, because they have led the way to great improvements in ordinary carriages.

The steam-propelled vehicle for common roads has just been mentioned as an invention belonging to the first half of the century, and the reasons it did not find favour have been alluded to. There exists in the United Kingdom a law concerning horseless carriages travelling on highways, which was passed to apply to traction engines, and enacts that other than horse vehicles are not to go along a road at a greater speed than four miles an hour, and only two miles an hour through a town, and moreover they are to be preceded by a man bearing a red flag, etc. But a bill has been introduced (1895) into the legislature to amend this law, and permit the British people to use on their common roads such light self-propelled carriages as are becoming popular in France, as may be seen from the following account :—

On Tuesday, 11th June, 1895, a race was started from Versailles to Bordeaux and back, a distance of 727 miles or more for the double journey. The first prize was the substantial sum of 40,000 francs (£1,600), to which was attached the condition of the carriage seating four persons, and other prizes were also to be awarded to various kinds of automatic vehicles. No fewer than sixty-six vehicles were entered for competition, and these were variously supplied with motive power from steam, electricity, or petroleum spirit. The starting place was Versailles at 12.9 p.m., and at 10.32 on Wednesday morning MM. Panhard & Levassor's petroleum carriage arrived at Bordeaux, whence, after a stop of only four minutes, the return journey was begun, but shortly afterward an accident caused a delay of one hour, but the carriage made the whole distance at the average of 14·9 miles per hour. In this and three other carriages belonging to the same firm, the propeller was the Daimler motor. Though this carriage was the first to accomplish the trip it received only the second prize, the condition of seating *four* persons not having been complied with. The first prize fell to a four-seated vehicle by Les Fils de Pengeot Frères, a firm who carried off besides the third and fourth prizes. These carriages were also driven by so-called petroleum motors. These motors are really gas engines on the principle to be presently mentioned, but the gas is produced by the vapourisation of a volatile constituent of petroleum (benzoline). The Daimler motor is a compact combination of two cylinders connected with a chamber containing the explosive mixture of gas and air. The pistons perform their in and out strokes simultaneously, but their working strokes alternately.

PORTABLE ENGINES.

THE application of steam power to agricultural operations has led to the construction of engines specially adapted by their simplicity and portability for the end in view. The movable agricultural engines have, like the locomotives, a fire-box nearly surrounded by the water, and horizontal tubes, and are set on wheels, so that they may be drawn by horses from place to place. There is usually one cylinder placed horizontally on the top of the boiler ; and the piston-rod, working in guides, is, as in the old locomotive, attached by a connecting-rod to the crank of a shaft, which carries a fly-wheel, eccentrics, and pulleys for belts to communicate the motion to the machines. Engines of this kind are also much used by contractors, for hoisting stones, mixing mortar, &c. These engines are made with endless diversities of details, though in most such simplicity of arrangement is secured, that a labourer of ordinary intelligence may, after a little instruction, be trusted with the charge of the engine; while their economy of fuel, efficiency, and cheapness are not exceeded in any other class of steam engine.

Besides the steam engines already described or alluded to, there are many interesting forms of the direct application of steam power. There are, for example, the steam roller and the steam fire-engine. The former is a kind of heavy locomotive, moving on ponderous rollers, which support the greater part of the weight of the engine. When this machine is made to pass slowly over roads newly laid with broken stones, a few repetitions of the process suffice to crush down the stones and consolidate the materials, so as at once to form a smooth road. Steam power is applied to the fire engine, not to propel it through the streets, but to work the pumps which force up the water. The boilers of these engines are so arranged that in a few minutes a pressure of steam can be obtained sufficient to throw an effective jet of water. The cut at the end of this chapter represents a very efficient engine of this kind, which will throw a jet 200 feet high, delivering 1,100 gallons of water per minute. It has two steam cylinders and two pumps, each making a stroke of two feet. These are placed horizontally, the pumps and the air reservoir occupying the front part of the engine, while the vertical boiler is placed behind. The steam cylinders, which are partly hid in the cut by the iron frame of the engine, are not attached to the boiler, which by this arrangement is saved from injurious strains produced by the action of the moving parts of the mechanism. There are seats for eight firemen, underneath which is a space where the hose is carried. A first-class steam fire-engine of this kind, completely fitted, costs upwards of £1,300.

A cheap and very convenient prime mover has lately come into use, which has certain advantages over even the steam engine. Where a moderate or a very small power is required, especially where it is used only at intervals, the *gas engine* is found to be more convenient. It is small and compact, no boiler or furnace is required, and it can be started at any moment. As now made, it works smoothly and without noise. The piston is impelled, not by the expansive force of steam, but by that of heated air, the heat being generated by the explosion of a mixture of common coal gas and air within the cylinder itself. Thus a series of small explosions has the same effect as the admissions of steam through a valve. A due quantity of gas and air is introduced into the cylinder, and is ignited by the momentary opening of a communication with a lighted gas jet outside. But the machine is provided with a regulator or governor, which so acts on the valve mechanism that this communication is made at each stroke only when the speed of rotation falls below a certain assigned limit, and thus the number of the explosions is less than the number of strokes, unless its work absorbs the machine's whole energy, which, according to the size of the engine, may be from that of a child up to 30-horse power.

THE STEAM HAMMER.

BEFORE the invention of the steam hammer, large forge hammers had been in use actuated by steam, but in an indirect manner, the hammer having been lifted by cams and other expedients, which rendered the apparatus cumbersome, costly, and very wasteful of power, on account of the indirect way in which the original source of the force, namely, the pressure of the steam, had to reach its point of application by giving the blow to the hammer. Not only did the necessary mechanism for communicating the force in this roundabout manner interfere with the space necessary for the proper handling of the article to be forged, but the range of the fall of the hammer being only about 18 in., caused a very rapid decrease in the energy of the blow when only a very moderate-sized piece of iron was introduced. For example, a piece of iron 9 in. in diameter reduced the fall of the mass forming the hammer to one-half, and the work it could accomplish was diminished in like proportion. Besides, as the hammer was attached to a lever working on a centre, the striking face of the hammer was parallel to the anvil only at one particular point of its fall ; and again, as the hammer was always necessarily raised to the same height at each stroke, there was absolutely no means of controlling the force of the blow.   When we reflect on the fact that the rectilinear motion of the piston in the cylinder of the engine had first to be converted into a rotary one, by beams, connecting-rod, crank, &c., and then this rotary movement transformed into a lifting one by the intervention of wheels, shafts, cams, &c., while all that is required in the hammer is a straight up-and-down movement, the wonder is that such an indirect and cumbersome application of power should have for so many years been contentedly used.   But in November, 1839, Mr. Nasmyth, an eminent engineer of Manchester, received a letter from a correspondent, informing him of the difficulty he had found in carrying out an order received for the forging of a shaft for the paddle-wheels of a steamer, which shaft was required to be 3 ft. in diameter.   There was in all England no forge hammer capable of executing such a piece of work.   This caused Mr. Nasmyth to reflect on the construction of forge hammers, and in *a few minutes* he had formed the conception of the steam hammer. He immediately sketched the design, and soon afterward the steam hammer was a *fait accompli*, for Mr. Nasmyth had one at once executed and erected at his works, where he invited all concerned to come and witness its performances.   Will it be believed that four years elapsed before this admirable application of steam power found employment outside the walls of Mr. Nasmyth's workshops ?   After a time he succeeded in making those best able to profit by such an invention aware of the new power—for such it has practically proved itself, having done more to revolutionize the manufacture of iron than any other inventions that can be named, except, perhaps, those of Cort and Bessemer.   The usual prejudice attending the introduction of any new machine, however obvious its advantages are afterward admitted to be, at length cleared away, and the steam hammer is from henceforth an absolute necessity in every engineering workshop, and scarcely less so for some of the early stages of the process of manufacturing crude wrought iron.   Whether blows of enormous energy or gentle taps are required, or strokes of every gradation and in any order, the steam hammer is ready to supply them.

A steam hammer of the smaller kind is represented in Fig. 13.   The general mode of action will easily be understood.   The steam is admitted below the piston, which is thus raised to any required height within the limits of the stroke.   When the communication with the boiler is shut off and the steam below the piston is allowed to escape, the piston, with the mass of iron forming the hammer attached to the piston-rod, falls by its own weight.   This weight, in the large steam hammers, amounts to several tons ; and the force of the blow will depend jointly upon the weight of the hammer, and upon the height from which it is allowed to fall.   The steam is admitted and allowed to escape by valves, moved by a lever under the control of a workman.   By allowing the hammer to be raised to a greater or less height, and by regulating the escape of the steam from beneath the piston, the operator has it in his power to vary the force of the blow.   Men who are accustomed to work the valves can do this with great nicety. They sometimes exhibit their perfect control over the machine by cracking a nut on the anvil of a huge hammer ; or a watch having been placed— face upwards—upon the anvil, and a moistened wafer laid on the glass, a practised operator will bring down the ponderous mass with such exactitude and delicacy that it will pick up the wafer, and the watch-glass will not even be cracked.   The steam hammer has recently been improved in several ways, and its power has been more than doubled, by causing the steam, during the descent, to enter above the piston and add its pressure to the force of gravity. Probably one of the most powerful steam hammers ever constructed is that recently erected at the Royal Gun Factory at Woolwich, for the purpose of forging great guns for the British Navy. It has been made by Nasmyth & Co., and is in shape similar to their other steam hammers.   Its height is upwards of 50 ft., and it is surrounded with furnaces and powerful cranes, carrying the huge iron tongs that are to grasp the glowing masses.   The hammer descends not merely with its own weight

FIG. 13.—*Nasmyth's Steam Hammer.*

of 30 tons ; steam is injected behind the falling piston, which is thus driven down with vastly enhanced rapidity and impulse. Of the lower portion of this stupendous forge, nothing is visible but a flat table of iron—the anvil —level with the floor of the foundry.   But more wonderful, perhaps, than anything seen aboveground, is the extraordinarily solid foundation beneath. Huge tablets of foot-thick castings alternate with concrete and enormous baulks of timber, and, lower down, beds of concrete, and piles driven deep into the solid earth, form a support for the uppermost plate, upon which the giant delivers his terrible stroke.   Less than this would render it unsafe to work the hammer to its full power.   As the monster works—

soberly and obediently though he does it—the solid soil trembles, and
everything movable shivers, far and near, as, with a scream of the steam,
our 'hammer of Thor' came thundering down, mashing the hot iron into
shape as easily as if it were crimson dough, squirting jets of scarlet and
yellow yeast. The head of the hammer, which of course works vertically,
is detachable, so that if the monster breaks his steel fist upon coil or anvil,
another can be quickly supplied. These huge heads alone are as big as a
sugar-hogshead, and come down upon the hot iron with an energy of more
than a thousand foot-tons. By the courteous permission of Major E.
Maitland, Superintendent of the Royal Gun Factories, we are enabled to
present our readers with the view of the monster hammer which forms the
Plate III.

Mr. Condie, in his form of steam hammer, utilizes the mass of the cylinder
itself to serve as the hammer. The piston-rod is hollow, and forms a pipe,
through which the steam is admitted and discharged, and the piston is stationary, the cylinder moving instead—between vertical guides. A hammer
face is attached to the bottom of the cylinder by a kind of dovetail socket,
so that if the striking surface becomes injured in any way, another can
easily be substituted. The massive framework which supports the moving
parts of Condie's hammer has its supports placed very far apart, so as to
leave ample space for the handling of large forgings.

FIG. 14.—*Merryweather's Steam Fire-Engine.*

PLATE III.

THE GREAT STEAM HAMMER, ROYAL GUN FACTORY, WOOLWICH.

FIG. 15.—*A Foundry.*

IRON.

"IRON and coal," it has been well said, "are kings of the earth"; and this is true to such an extent that there is scarcely an invention claiming the reader's attention in this book but what involves the indispensable use of these materials. Again, in their production on the large scale it will be seen that there is a mutual dependence, and that this is made possible only by means of the invention we have begun with; for without the steam engine the deep coal mines could not have the water pumped out of them,—it was indeed for this very purpose that the steam engine was originally contrived,—nor could the coal be efficiently raised without steam power. Before the steam engine came into use iron could not be produced or worked to anything like the extent attained even in the middle of the nineteenth century, for only by steam power could the blast be made effective and the rolling mill do its work. On the other hand, the steam engine required iron for its own construction, and this at once caused a notable increase in the demand for the metal. Once more, the engine itself supplies no force; for without the fuel which raises steam from the water in the boiler it is motionless and powerless, and that fuel is practically *coal*. In consequence of thus providing power, and also of supplying a requisite for the production of iron, coal has acquired supreme industrial importance, so that all our great trades and places of

densest population are situated in or near coal-fields. But what we
have further to say about coal may be conveniently deferred to a
subsequent article, while we proceed to treat of iron, and of the contrivances in which it plays an essential part.

Iron has also been called "the mainspring of civilization," and the
significance of the phrase is obvious enough when we consider the
enormous number and infinite variety of the things that are made of it :
the sword and the ploughshare ; all our weapons of war and all our
implements of peace ; the slender needle and the girders that span wide
rivers ; the delicate hair-spring of the tiny watch and the most tenacious
of cables ; the common utensils of domestic life and the huge battle-
ships of our fleets ; the smoothest roads, the loftiest towers, the most
spacious pleasure palaces. Such extensive applications of iron for
purposes so diverse have been rendered possible only by the greater
facility and cheapness of production, together with the better knowledge
of the properties of the substance and increased skill in its treatment, that
have particularly distinguished our century. Apart again from the
constructive uses of iron, it enters essentially into another class of inventions of which the age is justly proud, namely, those which utilize
electricity in the production of light, mechanical power, and chemical
action ; for it is on a quality possessed by iron, and by *iron alone*, that
the generation of current by the electric dynamo ultimately depends.
This peculiar property of iron, which was first announced by Arago in
1820, and has since proved so fertile in practical applications, is that a
bar of the metal can, under suitable conditions, be instantly converted
into the most powerful of magnets, and as quickly demagnetized. What
these conditions are will be explained when we come to treat of
electricity.

Besides the unique property of iron just referred to, and its superlative utility in arts and industries, there are other circumstances that
give a peculiar interest to this metal. It is the chief constituent of many
minerals, and traces or small quantities are found in most of the materials
that make up the crust of the earth ; it is present also in the organic
kingdoms, being especially notable in the blood of vertebrate (*back-boned*)
animals, of which it is an essential component. Notwithstanding its
wide diffusion, iron is not found *native*, that is, as *metal*, but has to be
extracted from its *ores*, which are usually dull stony-looking substances,
as unlike the metal as can be conceived. In this respect it differs from
gold, which is not met in any other than the metallic state, in the form of
nuggets, minute crystals or branching filaments, and from metals such as
silver, copper, and a few others which also are occasionally found native.
It is true that rarely small quantities of metallic iron have been met with
in the form of minute grains disseminated in volcanic rocks ; but in
contrast with the practical absence of metallic iron from terrestrial
accessible materials is the fact that masses of iron, sometimes of nearly
pure metal, occasionally descend upon the earth from inter-planetary
space. These are *aerolites*, of which there are several varieties, some
consisting only of crystalline minerals without any metallic iron, others of
a mixture of minerals and metals, but the most common are of iron,
always alloyed with a small quantity of nickel, and usually containing
also traces more or less of a few other metals and known chemical
elements. The iron in some specimens has been found to amount to 93
per cent. of the whole. These *aerolites*, or *meteorites*, as they are also

called, are of irregular shape and vary greatly in size, which however is sometimes very considerable : one found in South America was calculated to weigh 14 tons, another discovered in Mexico, 20 tons. There is in the British Museum a good specimen of an iron meteorite, which is represented in Fig. 16, where it will be observed that a portion has been cut off to form a plane surface, which when polished and etched by an acid, reveals a crystalline structure quite peculiar and distinctive, so that such meteorites can be recognized with certainty, even if they did not

FIG. 16.—*Aerolite in the British Museum.*

possess surface characters which are easily observed and identified when once a specimen has been examined. The fall of meteorites to the surface of the earth is comparatively rare, but it has been witnessed by even scientific observers ; as when Gassendi, the French astronomer, saw in Provence the fall of a meteorite weighing 59 lbs. In the *Transactions of the Royal Society* for 1802 may be found a detailed account of an instance in England of the fall which took place in Yorkshire, on the 13th December 1795, of a stone 56 lbs. in weight. Aerolites become ignited or incandescent by reason of the great velocity with which they pass through the atmosphere, whereby the air in front of them is condensed and heated, the heat often being sufficient to liquefy or even vaporize the solid matter. The so-called shooting stars are with good reason believed to be nothing but such incandescent aerolites, and the aerolites themselves are regarded as small asteroids, or scattered planetary dust, portions of which occasionally coming within the sphere of the earth's attraction are drawn to its surface. Meteoric iron is too rare to be of any value as a source of iron, but certain specimens have been found in which the metal was malleable and of excellent quality. From such meteorites the natives of India and other places have, it is said, sometimes forged weapons of wonderful temper and keenness, and we may well imagine that when such weapons have been made from iron that had actually been observed to fall from the sky, they would be regarded as endowed with magical powers, so that we may perhaps ascribe to such circumstances the origin of some of the legends about enchanted swords, etc. It is significant also that in some Egyptian inscriptions of the very highest antiquity, the word indicating iron has for its literal meaning *stone of the sky*.

But as nature has hardly provided man with the *metal* iron, he has been obliged to find the art of extracting it from substances which are utterly unlike the metal itself. In this case, as in many others, the art has been discovered and practised ages before any scientific knowledge of the nature of the processes employed had been acquired. The idea prevails that there are such difficulties in extracting this metal; that elaborate and complex appliances, not unlike those in use in modern times, were requisite for the purpose; and therefore that the use of iron is compatible only with a somewhat late period in man's history, and implies a comparatively advanced stage of civilization. Now there undoubtedly are facts which tend to confirm this view; for instance, the Spaniards who first colonized North America found the natives perfectly familiar with the use of copper, but without any acquaintance with iron, although the region abounded with the finest ferruginous minerals; and, again, the archæologists who have examined the relics of ancient civilizations and of pre-historic peoples about the shores of the Mediterranean, find in the earliest of these relics weapons and implements of rudely chipped stones, followed later by the use of better-shaped and polished stones; hence the periods represented by these, they have respectively designated by the terms *palæolithic* and *neolithic*—the old and the new stone ages. At some later time the stone of these implements was gradually replaced by *bronze*, which is a mixture of copper and tin, while as yet iron does not occur in any form among the remains. In the latest layers, however, articles of iron are found, and it is inferred that this metal came into use only after bronze had been known for an indefinite period; hence these later pre-historic periods have come to be respectively called *the bronze age* and *the iron age*. No doubt this succession really occurred in the localities where the observations were made, but it would not be justifiable to assume that the same was the case in every part of the world, for much would depend on such circumstances as the presence or absence of the essential minerals. We may also set against the supposed difficulty of obtaining iron from the ores, the still greater complexity of the methods required for the production of copper and of tin. Besides this is the fact that the ores of tin are found but in very few places in the world, and of these only the Cornwall mines, so well known in ancient times, would be likely to furnish a supply to the places where pre-historic bronzes are found ; this implies that navigation and commerce must have already made considerable progress. On the other hand, iron has been produced and worked for untold ages by the negro races all over Central Africa, and the method of treating the ore has no doubt been that which is there still practised by certain scarcely civilized tribes, and it is as simple as any metallurgical operation can possibly be, requiring merely a hole dug in a clay bank, wherein the fuel and minerals are piled up, and the mere wind supplies sufficient blast to urge the fire to the needful temperature, or air is blown in from rude bellows made of a pair of skins alternately raised and compressed. These very primitive furnaces have in some places developed into permanent clay structures, seven or eight feet in height. The natives of Central Africa have therefore long known the method of extracting iron, as well as of forging and casting it.

The nature and value of what has been done during the century in the treatment of iron would not be intelligible without some description of the ordinary processes of extracting the metal from the ores ; and a scientific understanding of these implies some acquaintance with chemistry. Not because metallurgy has been developed from chemistry, for the fact is rather the reverse; indeed, as we have seen, the art of extracting iron from its ores was practised ages before chemistry as a science was dreamt of. Although we may assume that many of our readers have sufficient knowledge of chemistry to attach distinct ideas to such few chemical terms as we shall have occasion to use, yet it may be of advantage to others to have some preliminary notes of the character of the chemical actions, and of some properties of the substances that will have to be referred to. It is certainly the case that people in general, and even people very well informed in other subjects, have but the vaguest notions of the nature of chemical actions, and of the meaning of the terms belonging to that science. For example, one of our most popular and justly esteemed writers, treating of the very subject of iron extraction, calls the ore a *matrix*, thereby implying that the iron as metal is disseminated in detached fragments throughout the mass, which is a conception inconsistent with the facts. The reader will be in a more advantageous position for understanding the relation of the ores of iron to the metal, if he will follow in imagination, or still better in reality, a few observations and experiments like the following—of which, however, he is recommended not to attempt the chemical part unless he is himself practically familiar with the performance of chemical operations, or can obtain the personal assistance of someone who is. Taking, say, a few common iron nails, let him note some obvious properties they possess : they have weight—are hard and tough so that they cannot be crushed in a mortar—are opaque to light—if a smooth surface be produced on any part, it will show that peculiar shiny appearance which is called metallic lustre, in this case without any decided colour—they are not dissolved by water as sugar or salt is—and are attracted by a magnet. If several of the nails be heated to bright redness they may be hammered on an anvil into one mass, and this may be flattened out into a thin plate, or it may be shaped into a slender rod and then drawn out into wire ; or otherwise the nails may be converted into the small fragments called iron filings. In these several forms the nails, as nails, will have ceased to exist; but the material of which they were formed will remain

unchanged, and each and every part of it however large or small will
continue to exhibit all the properties noted above as belonging to the
substance of the nails, which in the cases supposed has undergone
merely *physical* change of shape. Treating our nails in yet another way,
we may proceed to subject them to a *chemical* change, by an experiment
very simple in itself, but involving certain precautions, by neglect of
which the tyro in chemical operations would incur some personal risks ;
these might however be obviated by using only very small quantities of
the materials (a mere pinch of iron filings and a few drops of sulphuric
acid), when the results would still be sufficiently observable. A few of
the iron nails having been placed in a flask of thin glass, we pour upon
them a mixture of oil of vitriol (sulphuric acid) and water, which has previously been prepared by gradually adding 1 measure of the acid to 5
measures of water. The action that takes place is greatly accelerated by
heat, and indeed the contents should be heated to boiling by standing
the flask on a layer of fine sand spread on an iron plate and gently heated
from below. The nails will soon disappear, being completely dissolved
by the acid liquid, and the turbid solution should be filtered through
filtering paper as rapidly as possible and while still hot. This turbid and
dirty looking condition is due to foreign matters in the nails, for these
never consist of *pure* iron. The filtered liquid is set aside to cool in a
closed vessel, in which after a time will be found a deposit of crystals of
a pale bluish-green colour. The liquor above these having been poured
off, the crystals are to be rinsed with a very small quantity of *cold* water,
and then dried between folds of blotting-paper, after which they are
ready for examination. The quantity of the diluted acid put into the
flask should have a certain proportion to the weight of the nails ; about 5
fluid ounces to 1 ounce of iron will be found convenient, for if less is used
the nails will not be entirely dissolved, and an excess will tend to keep the
crystals in solution instead of depositing them when cold. The nails—
as such—will now have passed out of existence : can we say that the iron
that formed them exists in the crystals ? Certainly not as the *metal* iron,
for every property of the metal will have disappeared. The crystals are
brittle, can be crushed in a mortar—they are translucent—they show no
metallic lustre, but only glassy surfaces—they are readily dissolved by
water—they are not attracted by a magnet. The most powerful lens will
fail to show the least particle of iron in them ; they have in their properties no assignable relation to the metal of the nails, but are matter of
quite another sort ; and be it noted that this entire *otherness* is the
special and characteristic sign of *chemical change.* So complete is the
transformation in the case we have been considering that it would never
have been said that iron was contained in these crystals, but rather that
the metal had for ever passed out of existence, but for one circumstance ;
and that is, that by subjecting the crystals to certain processes of *chemical
analysis* we can again obtain from them the iron in metallic state. Nay
more, we should find the weight of metal so obtained to be exactly equal
to that of the pure iron dissolved from the original nails, supposing of
course that we operated upon the whole of the crystalline matter so produced. The inference therefore is that although every property of the
iron appeared to be absent from the crystals, the iron entering in them
retained there its original weight, and the correct statement of the change
would be, that in the crystals the iron had lost all its original properties
SAVE ONE, namely, its weight, or gravitating force, if we choose to call it

so, a property belonging to it in common with every material substance. Chemical analysis can also separate from the crystals their other constituents and weigh them apart—so much water and so much sulphuric acid—and when to these weights that of the iron is added, the sum exactly makes up the weight of the crystals.

A still simpler experiment, which may be performed by anyone with the greatest ease, may serve as a further illustration of the profound nature of the change in the properties of bodies brought about by chemical combination, and it will also serve as the occasion of directing attention to a remarkable circumstance that invariably characterizes such changes, and one that should always be present in our minds when we are considering them. A yard of flat *magnesium* wire can be bought for a few pence, and after its metallic character has been observed in the silvery lustre disclosed by scraping the dull white surface, a few inches is to be held vertically by a pair of tongs, or by inserting one extremity in a cleft at the end of a stick, then the lower part is brought into contact with a candle or gas flame. The metal will instantly burn with a dazzlingly brilliant light, and some white smoke (really fine white solid particles) will float into the air ; but if a plate be held under the burning metal, some of the smoke will settle upon it, together with white fragments that have preserved some shape of the metallic ribbon, but which a touch will reduce into a fine white powder, identical with the well-known domestic medicine called "calcined magnesia"—a substance totally different from the *metal magnesium.* The reader will scarcely require to be told that in this burning the metal is entering into combination with the oxygen of the air—by which that invisible gas somehow becomes fixed in these solid white particles, so entirely unlike itself. But this experiment might be so arranged that the quantities of magnesium and oxygen entering into the magnesia could be weighed. For this purpose special appliances would be required in order to ensure complete combustion of the metal, for in the experiment as just described some small particles are liable to be shielded from the oxygen by a covering of magnesia, and the arrangement would have to be such that the *whole* of the white powder could be gathered up and weighed. In the absence of such appliances, and of a delicate balance, together with the skill requisite for their use, the reader must for the time be contented to take our word for what would be the result. In every experiment the magnesia would be found heavier than the metal burned in the proportion of 5 to 3 ; in other words, magnesia always contains (so the phrase runs) 3 parts of magnesium combined with 2 of oxygen : never more nor less. A definite proportion between the weights of the constituent substances characterizes every chemical combination, and when this is once determined in a single sample of the compound, it is determined for every portion of the same, wherever found or however produced. But each compound has its own particular proportion, that is, the quantitative relations are different for each. For example, the two constituents of water, hydrogen and oxygen, are combined in the ratio of 1 to 8, etc. ; and oxygen combines with metals in a ratio different in each case. Then occasionally the same ratio of constituents occurs in compounds of different composition. The elementary student is apt to suppose that this is *because* of the *law* which he finds stated, probably in almost the first page of his text-book: "Every compound contains its elements in definite and invariable proportions" ; and even well-educated people entertain

the idea of the fact being "governed by" or "obeying" the law just
quoted,—a misconception arising from the other use of the word "law,"
as signifying an enactment. The real case however is the converse;
namely, that a multitude of facts like that above stated have governed
the law, and caused it to be what it is—the general statement of many
observed facts.

We have assumed that the reader's chemical knowledge had already
made him aware that in every case of ordinary combustion the oxygen
of the atmosphere is in the act of entering into combination with the
burning body : as with the magnesium, so with a coal fire, a gas flame,
or a burning candle ; only in these last cases the products of the combustion pass away invisibly. The candle by burning disappears from
sight, but its matter is not lost, and as in the case of magnesium, the compounds it forms weigh more than the unburnt candle. The experiment
is commonly shown in courses of elementary lectures on chemistry, of so
burning a candle that the invisible products are retained in the apparatus,
instead of being dissipated in the atmosphere, and the increase of weight
of the burnt candle over the original one is demonstrated by the balance.
Important as is the part played by oxygen in all chemical actions on the
earth, the composition of the atmosphere was not understood until the
end of the eighteenth century, and it was well on into the nineteenth
before the quantities of its constituents were accurately determined.
Now everyone knows that air is mainly made up of a *mixture* of the
two gases *oxygen* and *nitrogen.* A *mixture* of two or more things is very
different from a chemical *combination* of them ; for in the former each
ingredient retains its own properties. (See *Air* in Index.) Nitrogen
being an inert gas that takes no part in combustion, or in the ordinary
chemical actions of the air, acts therein simply as a diluent of the oxygen.
It is necessary in relation to our present subject to bear this in mind, as
well as the relative quantities of the two gases in air. For our immediate
purpose we may neglect the minor constituents of air—such as watery
vapour, carbonic acid, etc., of which the total weight does not exceed
one hundredth part of the whole—and consider air as a mixture of 23 parts
by weight of oxygen with 77 of nitrogen, or calculated in volumes, 21
measures of oxygen with 79 of nitrogen. Compounds of oxygen with
nearly every one of the other seventy or more chemical elements are
known, and these compounds, which are called *oxides,* are arranged by
chemists under five or six classes, forming as they do *basic radicles, acid
radicles, saline oxides,* etc. With some of these compounds belonging to
different classes, we must make acquaintance after noticing the elementary
substance with which the oxygen is united.

We begin with *carbon,* which forms the chief constituent of all our combustibles. Some specimens of graphite, plumbago, or "black-lead" consist
of almost pure carbon (98 per cent.), and some varieties of wood charcoal
exceptionally contain 96 per cent. ; but in ordinary charcoal the percentage is much less. Coal, the most familiar of our solid fuels, varies greatly
in composition, carbon being the predominating constituent, in amount
from 57 to 93 per cent. Coke, another fuel much used in metallurgical
operations, is made by heating coal without access of air, when a large
quantity of gaseous substances is expelled. Coke burns with an intense
and steady heat without emitting any visible smoke, but it does not ignite
as readily as coal. Carbon forms two different compounds with oxygen :
both are invisible gases, but they differ in the proportions of the con-

stituents, and present different properties. When carbon (coal, coke, or charcoal) is completely burnt, that is, with an abundant supply of air, the product is *carbonic acid* gas, in which 3 parts of carbon are combined with 8 of oxygen : when, on the other hand, the carbon is burnt with a sufficiently restricted access of air, the result is *carbonic oxide* gas, in which 3 parts of carbon are united with only 4 of oxygen. The reader will here observe that the former contains just twice as much oxygen as the latter for the same quantity of carbon. This fact and numberless others like it are expressed or summed up by another *law* of chemical combination which states that when two elements combine in several different proportions these are invariably such that the ratios in the several compounds will be found to have exact and simple numerical relations ; that is, such as may, when reduced to their lowest terms, be expressed by the simple integers 1, 2, 3, etc., as 1 : 2, 3 : 2 ; ... 8 : 9, etc. It comes to the same thing if we compare together the weights A and A' which are united in each compound with any one identical weight of B, giving of course the ratio A : B ÷ A' : B. For instance, in the case just given, of carbon and oxygen, 3 : 4 ÷ 3 : 8 = 2 : 1. This, which is simply stating the facts, is called the *law of multiple proportions*. On a later page will be found another illustration (see Index, *Nitrogen and Oxygen Compounds*), and its expression in terms of the *atomic theory*, which goes behind the facts (so to speak), but is extremely useful by comprehending many other groups of facts in chemistry and in other sciences. Carbonic acid gas is of course incombustible, but carbonic oxide gas burns by uniting with the additional proportion of oxygen and becoming carbonic acid. On the other hand, carbonic acid gas passing over red-hot coals takes up from them the additional proportion of carbon, and is, we may say, *unburnt* into carbonic oxide. When we see a pale blue flame flickering over the bright embers in a fire grate, it is carbonic oxide burning back again by taking more oxygen from the air above the coals. Carbonic oxide combines directly with two or three of the metals, as, for instance, it forms a volatile compound with nickel, at a certain temperature, and this is decomposed again at a higher temperature. The like takes place with iron, although in very small quantities, but the observation throws some light on the processes of reduction. Carbonic oxide is neither acid nor basic, but carbonic acid is an acid oxide, and as such unites with oxides of the basic class to form another range of compounds. Thus, for example, the oxide of the metal *calcium* is quicklime, which is strongly basic, and this directly combines with carbonic acid, forming a *neutral* substance called in systematic chemistry *calcium carbonate*, or more commonly but less correctly, carbonate of lime, familiar to everyone in the compact state as limestone, and marble, and in a more or less pulverent condition as chalk. When any of these is heated to redness, carbonic acid is expelled and quicklime remains. Like most oxides, quicklime forms a compound with water, the combination being attended with the extrication of much heat, the compact quicklime swelling and crumbling into *slaked lime*. The chemist's term for a compound of a *basic* oxide with water is *hydrate*, while that of an *acid* oxide with water is for him properly an *acid*, or in order to particularly distinguish this class, an *oxy-acid*. It was however the older practice to give the name of acid to the oxide alone, and this naming having found its way into popular language is much more familiar to the non-scientific reader. The systematic names of the two compounds of carbon and oxygen are *carbon monoxide* and *carbon dioxide*, but we shall use here the more familiar terms carbonic oxide and carbonic acid.

We have now to call attention to a substance which contributes by far the largest part to the solid crust of our globe. It is called *silica*, from *silic-*, the Latin word for flint (without case suffix) : it is seen in flint, and very pure in rock crystal, quartz, agate, and calcedony. It forms the essential part of every kind of sand and sandstone, and is the principal ingredient of clay, granite, slate, basalt, and many other minerals. Silica is the oxide of a quasi-metal called *silicon*, which can be obtained from silica with difficulty, and only by roundabout processes, presenting itself in different conditions according to the process used. Silica is an acid oxide, and it readily unites with most of the basic oxides when heated with them, forming a class of compounds of different properties which are much modified in admixtures containing two or more. Very few of these *silicates* are soluble in water, most of them are not : they are all fusible at various temperatures, except silicate of alumina, of which fireclay is chiefly constituted. Alumina, it should be stated, is the oxide of the metal aluminium. The silicates of lime and of magnesia fuse only with great difficulty ; but the silicates of iron and of manganese are easily fused, and silicate of lead still more so. Glass is a mixture of silicates, often of lime, soda, and alumina ; sometimes of lead and potash mainly ; porcelain and pottery consist chiefly of silicate of alumina with varying proportions of silicates of iron, of lime, etc.

It now remains only to mention two non-metallic elements that are nearly always present in crude iron, but which the metallurgist strives to eliminate, as they are in general very injurious to the quality of the material even when their amount is very small. The first is *sulphur*, well known as *brimstone*, also as *flowers of sulphur*, a yellow coloured solid, which burns in the air. The product of the combustion is an invisible gas of a readily recognized pungent odour : this is an acid-forming oxide containing equal weights of sulphur and oxygen. There is another oxide in which the weight of oxygen is one and a half times that of the sulphur, and this is the radicle of the very active *sulphuric acid* or *oil of vitriol*. Sulphur, like oxygen, unites with most of the other elements, forming compounds called *sulphides*. Of these the iron compound called *pyrites* is the best known, and its occurrence in coal prevents the use of that material as fuel in contact with iron or other metals. *Phosphorus* is an element that occurs naturally only in combination ; in its separated state it is a very inflammable solid. It combines directly with other substances and is taken up by some fused metals in large quantities. In many cases a very small proportion of it existing in a metal greatly modifies the properties of that metal. Phosphorus forms several oxides, and these are radicles of powerful acids, among which is *phosphoric acid* that combines with basic oxides to form *phosphates*.

We have now, in the few last paragraphs, set before the reader the minimum of chemical knowledge that will enable him to follow the *rationale* of such processes of the modern treatment of iron and its ores as we can here give an outline of. Although there are numberless minerals from which some iron can be extracted, the name of *iron ore* is confined to such as contain a sufficient amount to make the extraction commercially profitable, and this requires that the mineral should be capable of yielding at least one-fifth of its weight. The ores are very abundant in many parts of the world, and they consist mainly of *oxides* and their *hydrates*, or of *carbonate*, or of carbonate mixed with clay and silicates, sometimes also with coaly matters in addition. The carbonate iron ores are often mixed with oxides. Each class of ore is liable to be contaminated with phosphates and with sulphur. The richest ore is the *magnetic iron ore*, which is found in enormous masses in Sweden, Russia, and North America. It is an oxide containing 72·41 per cent. of iron. *Red hæmatite* and *specular ore* are varieties of another oxide with 70 per cent. of iron: the former is a very pure ore when compact. It is found in Lancashire, Cumberland, and South Wales, and much has been imported from Spain, while America has abundant supplies near Lake Superior. Specular iron ore forms brilliant steel-like crystals which show the red colour of hæmatite only when scratched or powdered. Elba was famous for this ore, which occurs also in Russia and Sweden, and large deposits are met with in both North and South America. Brown hæmatite is a hydrate of the former, containing 60 per cent. of iron ; it abounds in France and Spain, where some kinds are associated with a noteworthy quantity of *phosphate* of iron. *Spathic* or *sparry iron* ore is, when pure, a collection of nearly colourless transparent crystals, consisting of carbonate of iron ; it contains about 48 per cent. of iron, and also some of the metal *manganese*, which last circumstance makes it, as we shall see, particularly suitable for producing certain kinds of steel—indeed it is sometimes called *steel ore*. Large beds of it occur in Styria and Carinthia. *Clay iron stone*, or *clayband*, has been extensively mined in Britain. It is found abundantly in Staffordshire, Yorkshire, Derbyshire, and South Wales. It consists of carbonate of iron intimately mixed with clay. The quantity of iron in some samples falls as low as 17 per cent., but it rises with variations to as much as 50 per cent. Much of its importance arises from the fact of its occurring in beds alternating with layers of coal, limestone, and clay, so that the same pit is sometimes able to supply firebricks for building the furnace, fuel for the smelting, and limestone for the *flux*,—a combination of advantages that for long enabled iron to be produced in England cheaper than elsewhere. The like is true of the *blackband ore*, which, in addition to the same ferruginous composition as the last, contains also so much combustible or bituminous material that it can be *calcined* (roasted) without additional fuel. The deposits of *blackband* in Lanarkshire and Ayrshire, which were discovered only in 1801, have given great industrial importance to the district. Yet another British ore must be noticed, namely, the *Cleveland ironstone* of the North Riding of Yorkshire. This is a carbonate of a grey or bluish colour caused by the presence of a little iron silicate. It contains also a considerable amount of phosphorus.

How simple is the operation of obtaining iron from the ore has already been stated—that it is necessary only to surround lumps of ore by fuel in a fire urged by a natural or artificial blast, and then to hammer the mass extracted from the furnace so as to weld together the scattered particles of the metal, and at the same time squeeze out the associated slag and cinders, in order to obtain a coherent malleable piece, which can be reheated in a smith's fire, and forged into any required form. It is no wonder therefore that iron was so produced by the ancient Britons ; at any rate Cæsar found them well provided with iron implements and weapons. No doubt the Romans brought their more advanced skill to the working of the metal ; but in the matter of treating the original ore, the methods they pursued on an extensive scale in Britain were of the rude kind already described. Indeed in localities where the Romans were known to have carried on their operations, the remains of their workings are almost always found on high ground, so that it may be inferred that they relied upon the winds to fan their fires; and their operations were incomplete and wasteful. .The most extensive of them appear to have been in Sussex and Monmouthshire, in which last county there are places where the ground is in large areas covered by their cinders and refuse, and in this about 30 or 40 per cent. of iron occurs, so that for some centuries this material was found capable of being profitably reworked as a source of the metal. Iron continued to be produced in England during the middle ages with charcoal for fuel, but its export was forbidden, and whatever steel was required had to be imported from abroad. Afterwards German artisans were brought over for making steel, and soon afterwards the importation of shears, knives, locks, and other articles was prohibited. The native production of iron continued, and this consumed the forests so rapidly for the supply of charcoal, that various Acts were passed to restrain the ironmakers, in order to preserve the timber. In spite of these, the arts of smelting and working iron advanced apace: bellows were used for the blast, and then the works were brought down into the valleys, where water power could be employed to work them. The scarcity of charcoal fuel caused many attempts to supply its place with pit coal, but these met with small success, partly on account of the coal containing so much sulphur, and partly from the difficulty of obtaining with it a sufficiently high temperature, especially as the blowing apparatus was as yet very imperfect. At length, in the first half of the seventeenth century, the problem was solved by Dud Dudley, whose process was kept secret, but is believed to have consisted in supplying coal at the top of a higher furnace, in such a manner that the coal was converted into *coke* by the heat of the escaping gases before it reached the reducing zone of the furnace. This innovation was violently opposed by the charcoal smelters, who persecuted the inventor in every way, until their resistance was successful. But before the middle of the next century coke was regularly used in iron smelting, the process having been made successful by Darby at Coalbrookdale, and then many new applications of cast iron came into vogue. Coke being a substance burning less freely than charcoal, bellows were found inadequate to give the necessary blast, and were displaced by *blowing cylinders*, actuated at first by water wheels, but this uncertain and comparatively feeble source of power was soon superseded by the steam engine, the "fire engine," for which, as we have seen, Watt obtained his patent in 1769. The furnaces were not then all engaged in producing the fusible metal now called *cast* or *pig iron* as are the huge blast furnaces we see at the present time. Indeed it was much to the disgust of the old iron smelter that occasionally his product turned out to be of the fusible kind, unworkable by the hammer, which therefore he regarded as worthless. At what date *cast iron* was first used is uncertain; but probably it was not long before the fourteenth century. The furnaces in use up to that time were small square walled-in structures only 3 or 4 feet high, and their effect would not greatly exceed that of a smith's forge : but as improved blowing apparatus gave more power, they soon became enlarged into oval or round brick towers from 10 to 15 feet high, and they, like the small furnaces, could be made to yield either smith iron or steel by modifying the charge and the manner of applying the blast ; while furnaces of dimensions exceeding a certain limit could no longer be trusted

Fig. 17.—*Blast Furnace (Obsolete Type).*

to turn out malleable metal, but they produced instead the cruder substance we call white pig iron, and this requires much subsequent treatment before it is converted into malleable or "merchant iron." Nevertheless the demand for cast iron as such, and more particularly the
adoption of improved methods of deriving malleable iron from it, caused
further increase in the size and numbers of blast furnaces, until in the
early part of our century 30 feet was not an unusual height, the highest
one in England in 1830 attaining 40 feet. The total make of pig iron in
England was in that year nearly 700,000 tons, perhaps about fifty times
as much as it was a century before, and thirty years later (1860) it had
risen to nearly 4,000,000 tons. These figures show the extraordinary
expansion of the British iron manufacture in the earlier part of the century ;
and the still more extensive applications of iron during the next twenty
years had the effect of almost doubling the produce in 1880, and of

increasing also threefold the amount of foreign metal imported, raising it
to 2,500,000 tons. The reader will now, it is hoped, be prepared to follow
with some interest a brief account of the principal inventions which have
brought about results of such importance.

Deferring for the moment any description of the latest blast furnaces,
we invite his attention to Fig. 17, which represents the furnace used in
the first half of our century, but which now is of an obsolete type, Fig. 18
being the section and plan of the same. The lower part of Fig. 17 shows

FIG. 18.—*Section and Plan of Blast Furnace (Obsolete Type).*

where the molten metal has been allowed to run out of the furnace into channels made in dry sand ; first a main stream, then branches to right and left, each of these with smaller offsets on each side of it. These smaller channels are the moulds for the *pigs*, so called because of the fancied resemblance of their position with regard to the branch that supplied them, to the litter of a sow. They are easily broken off from the larger mass, and then form pieces about 3 ft. long with a ⛝-shaped section, 4 in. wide, the weight being from 60 to 80 lbs. This is iron of the crudest kind, and though it is often referred to as "cast iron," it is, as a matter of fact, not used in this state for any castings, except those of the very roughest and largest kind : a certain amount of purification is requisite in most cases. This is given by fusing the metal—along with some form of oxide and often other matters—in a *cupola furnace*, which is like a small blast furnace, being from 8 ft. to 20 ft. high and uses coke for fuel with a cold blast.

So far from being simply iron, pig contains a large and variable proportion of other matters amounting often to 10 or 12 per cent. ; and these confer upon it its fusibility. The principal one is carbon, which is found in the metal partly in the state of chemical combination with it, and partly in the form of small crystals similar to those of graphite or plumbago, disseminated through the mass. When there is a comparatively small proportion of the carbon combined with the iron, the substance is grey, and it can be filed or drilled or turned in a lathe. In white cast iron the combined carbon predominates, or is sometimes accompanied by scarcely any graphitic carbon ; it is brittle and so very hard that a file makes no impression. It fuses at a lower temperature than the other varieties. A third kind is the *mottled* cast iron, which shows a large coarse grain when broken, and distinct points of separate graphite particles ; it is tougher than the others, and therefore when cannon were made of cast iron this variety was preferred. The following table giving the percentage composition of four samples of crude cast iron will show their diversities.

	White.	White.	Mottled.	Grey.
Iron..	88·81	89·304	93·29	90·376
Combined carbon.....................	4·94	2·457	2·78	1·021
Graphite, or uncombined carbon.	......	0·871	1·99	2·641
Silicon....................................	0·75	1·124	0·71	3·061
Sulphur	trace	2·516	trace	1·139
Phosphorus	0·12	0·913	1·23	0·928
Manganese	5·38	2·815	trace	0·834

The reader will observe that the last item in the table above is a substance that he has not yet made the acquaintance of, namely, *manganese*. This is a metal which in many of its chemical relations much resembles iron, and ferruginous ores usually contain a greater or less proportion of it. Manganese is of great importance in the manufacture of steel, as we shall presently see ; but as a separate metal it has no application, and is obtainable in the metallic state with much difficulty. One of its oxides has however very extensive applications in the chemical arts, and others form acid radicles, which in combination with potash or soda give rise to useful products. The well-known "Condy's fluid" is a solution of one of these.

We have seen how malleable iron or steely iron may be directly obtained from the ores, but it has been found that on the large scale it is necessary and more economical to operate on the pig iron produced by the blast furnaces in such a manner as to remove the greater part of the foreign substances.

The first step in the conversion of the pig iron usually taken has been, and to a certain extent even is still, to remelt the metal in what is termed a *finery furnace*, a kind of forge in which a charcoal fire is urged by a cold blast, and so regulated that an excess of oxygen is supplied, or rather more than would suffice to convert all the carbon of the fuel into carbonic acid ; although this is perhaps not absolutely necessary, as carbonic acid would itself supply oxygen by suffering reduction to carbonic oxide. At any rate the melted metal is exposed to an oxidizing atmosphere and constantly stirred. Many different arrangements of the furnace and details of the process have been used. For instance, where the finest quality of malleable iron was not aimed at, coke has been the fuel employed, and many shapes of furnaces, etc., have been contrived, and various additions of ores, oxides, etc., made to the charge, according to local practice and the nature of the crude iron. One marked effect of the operation is the final removal of nearly all the *silicon*, which is burnt or oxidized into *silica*, and this at once unites with oxide of iron, which is also formed, to produce a readily fusible slag of silicate of iron, and in the production of this silicate any sand attached to the pig will also take part. Much of the carbon, amounting sometimes to more than half, is also eliminated as carbonic oxide, and of what is left but little remains in the graphitic state. The action on the phosphorus is usually less marked, but there is always a notable reduction of the quantity. The sulphur is also lessened in some degree, although when coke is used, the fuel has the disadvantage of itself containing sulphur, phosphates, and other deleterious matters. Sometimes a little lime is added to the charge to take up the sulphur from the coke. The operation lasts some hours, the fused metal being frequently stirred with an iron rod, until it assumes a pasty granular condition, when the workman gradually collects it upon the end of the rod into a ball of about three-quarters of a cwt. in weight. These balls, or *blooms* as they are called, are removed from the furnace while still intensely hot, and at once submitted to powerful pressure by means of some suitable mechanical arrangement, the effect being to squeeze out the liquid slag and force the particles of metal together by which the whole becomes partially welded into a more compact mass. Then this mass is, while still hot, either hammered with gradually increased force of the strokes, or in the more modern practice, passed between iron rollers (these we shall presently describe), by which it is shaped into a bar. The bars are afterwards cut into lengths, reheated without contact of fuel, again hammered or re-rolled ; and this process is several times repeated when the best product is required. During the first treatment of the blooms, and also in the subsequent hammering or rolling, the oxygen of the atmosphere acts on the surface of the glowing metal, so as to cover it with thin scales of oxide, and these, carried into the interior of the mass, will give up their oxygen to any residual silicon, carbon, etc., producing a little more slag, carbonic oxide, phosphate of iron, etc., which by the pressure of the hammers or rolls are ultimately forced out of the metal. It will be observed that in producing the pig iron the chemical action is the separation of oxygen from the metal, while conversely an oxidizing action is set up in the finery and subsequent treatment, in order to burn off the foreign ingredients. But this cannot be done without at the same time re-oxidizing some of the iron itself, of which therefore there is always a considerable loss, by its formation into slag (silicate), cinder, foundry scale (oxide), etc. The quantity of iron lost depends of course on many conditions, such as the care exercised in the operations, but it occurs in all the processes that have been devised for the conversion in question, even in the most modern : its amount may be taken to range between 10 and 20 per cent. The reader is requested to bear in mind the nature of the chemical actions that have just been described, for in even the most recently invented processes the principle is the same in nature and effect. So completely can the foreign

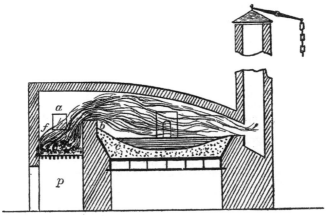

FIG. 19.—*Section of a Reverberatory Furnace.*

elements be eliminated by this, or some analogous process, such as we shall presently mention, that the finest Swedish bar iron contains more than 99½ per cent. of the metal, and in some cases only a very little carbon and a mere trace of phosphorus remain, amounting together to less than 1 part in 2000. Such metal is made from very pure ore, containing no sulphur and scarcely any phosphorus, while charcoal is the fuel used in all the operations. As already mentioned, the objection to the use of coke is the sulphur, phosphates, and siliceous matters it contains. Toward the close of the eighteenth century an invention came into use which obviated the disadvantages of the cheaper fuel for converting crude iron. This was the puddling furnace, brought into use after much experimenting by Henry Cort in 1784. In it the pig iron is fused in a *reverberatory furnace*, the form of which will be understood from Fig. 19, which is a diagram showing such a furnace in section, where f is the fire, a an aperture at which the fuel is introduced, p the ash pit, b is a low wall of refractory material called the "bridge," over which the flame passes, and is by the low arched roof reflected or *reverberated* downwards upon the charge, *c*, which is laid on a *hearth*, or iron floor, having spaces below it where air circulates in order to prevent it becoming too hot. In Cort's original arrangement the bed of the hearth was formed of sand, which gave rise to much inconvenience by producing a quantity of the very fusible silicate of iron, that speedily attacked the masonry of the furnace, and therefore a very important improvement was devised some years later by S. B. Rogers, who made the bed of his furnace of a layer of oxide of iron, spread on a cast iron plate 1½ inches thick. In later times it has become usual to cover the iron hearth with certain other refractory mixtures varied according to circumstances, of oxide, ore, cinder, lime, etc. There is one of these mixtures significantly designated "*bull-dog*" by the workmen. We may mention here that it has, in more recent times, when very high temperatures are obtainable, been found unnecessary to cause even the flame to come into contact with the substances on the hearth, inasmuch as the heat radiated from the flame and the intensely heated roof of the furnace suffices, so that in consequence of this the roofs are now constructed nearly flat. In the puddling furnace the melted metal is constantly stirred, and no little skill is required to regulate the fire by the damper on the chimney, and to admit the proper amount of air to mix with the flame. The pig iron softens and melts gradually, until at length it becomes perfectly liquid, at which stage it swells up and appears to boil owing to the escape of carbonic oxide in numerous jets, which burn with the characteristic pale blue flame. The puddler then briskly stirs the mass to cause more complete oxidation of the carbon, silicon, etc., by bringing the superficially formed oxide of iron into the interior. As the iron loses its carbon, it assumes much the texture of porridge, consisting of pasty lumps of malleable iron implexed with the liquid slag (silicate of iron, etc.) which drips from the spongy balls as the puddler collects them at the end of his stirring rod, as in the finery operation. The next thing is to run the mass immediately between powerful rolls (puddling rolls) by which the slag is squeezed out, as before, and finally through the *finishing rolls* that shape it into bars or plates.

When a comparatively impure pig iron is used or when a better quality of malleable metal is desired, the crude iron is submitted to a preliminary treatment before puddling. This treatment, by a technical distinction, called *refinery*, is practically identical with the *finery* process already described, except that instead of being collected into blooms, the fluid metal is run out to form a layer 2 or 3 inches thick, and this, before becoming quite solid, is suddenly cooled by having water thrown over it, the result being a white, hard, brittle mass, which broken into pieces is ready for the puddling furnace.

The operation that has been described is known as *hand puddling*, in contradistinction to later methods in which it has been sought to substitute some form of machine that will produce the same result automatically, such as revolving furnaces, etc. It has been found difficult to maintain these in good working order, and in England at least mechanical puddling has never found much favour, but in the great iron works of Creusot, in France, large revolving furnaces were in use about 1880, which could turn out 20 tons of converted iron in 24 hours, whereas the old hand puddling furnaces could in the same period produce only 2½ or 3 tons, with two sets of men, the puddler and one assistant. Of these mechanical furnaces it is unnecessary to give any account, especially as the puddling process itself has nearly gone out of use, having been superseded by more economical methods.

The use of rolls for treating the product of the puddling furnace, and for making it into bars, was also an invention of Henry Cort's, for which he obtained a patent in 1783. This was in many respects an immense improvement on the older system of hammering; it is still practised, and by it shapes can be given to the metal scarcely possible on the older system, while the tenacity of the metal is increased by the uniformity given to the grain. The difference of chemical composition between cast and wrought iron the reader has already been made acquainted with, and there is quite as great a difference in their textures. The former, when broken across, shows a distinctly crystalline structure, which we may compare to that of loaf-sugar, while the latter exhibits grain, not unlike that of a piece of wood. This fibrous structure depends upon the mechanical treatment of the iron, and in rolled bars the fibres always arrange themselves parallel to the length of the bar. Fig. 20 shows this fibrous structure in a piece of iron where a portion has been wrenched off. Like wood, wrought iron has much greater tenacity along the fibres than across them; that is, a much less force is required to tear the fibres asunder than to break them transversely. Consequently, to obtain the greatest advantage from the strength of wrought iron, the metal must be so applied that the chief force may act upon it in the direction of the fibres. Near the beginning of our article on IRON BRIDGES (*q.v.*) the reader will find some illustrations of the very different resisting powers of cast and wrought iron.

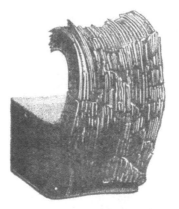

FIG. 20.—*Fibrous Fracture of Wrought Iron.*

Nothing in the way of inventions can be compared to those of Cort's as to the effect they have had in promoting the iron industry, until we reach a period some years after the middle of our century; but we must not neglect to recognize the scarcely inferior importance of Rogers' improvement. Singularly enough, neither of these men reaped any benefit from his inventions. Cort died in the last year of the eighteenth century, quite a poor man, having been supported only by a niggardly pension of some £160 from the Government, and leaving his family in indigent circumstances. Yet a most eminent authority on iron questions (Sir W. Fairbairn) estimated—some time about the middle of our era—that the two inventions of Cort's alone, the rolling-mill and the reverberatory puddling furnace, had by that time added to the wealth of Great Britain by an amount equivalent to six hundred million pounds sterling. For many ironmasters had profited by these inventions, amassing very great fortunes, in some instances also acquiring titles of honour. Clearly to Cort and Rogers may be applied the *sic vos non vobis* saying.

We shall now turn to the improvements that have been effected in the blast furnace, and of these none perhaps has been more marked than that made by Neilson, when in 1828 he substituted heated air for the ordinary cold air that had before always supplied the blast. It will be remembered that the heat is due to the combination of only the oxygen of the air with the carbon of the coke, but the greater part of the air—the four-fifths of nitrogen—take no part in the action, beyond abstracting a large proportion of the heat ; but when the air is heated to a high temperature before entering the furnace, the cooling effect of the nitrogen is greatly obviated, and consequently a much higher temperature is obtained at the place of combustion, and the requisite intensity of heat is at once produced, which is most effective in completing the fusion and separation from each other of the slags and iron, and also in accomplishing the reduction of the oxide. But Neilson found that the net result of burning some fuel to heat the air before entering the furnace was a great economy of the total fuel required for smelting the ore. He had to encounter many difficulties in carrying his invention into practice ; the iron ovens first used for heating the air were rapidly oxidized ; and when thick cast iron pipes were substituted, these were liable to leak at the joints on account of the expansions and contractions caused by changes of temperature. Then the new invention had as usual to contend with established prejudices and misconceptions ; but it soon came into use in Scotland, where it effected a great saving ; inasmuch as it was found possible to use with the hot blast raw coal of a certain kind, plentiful in Scotland, because the heat retained by the ascending gases sufficed to convert the coal at the top of the charge into coke.

It will be remembered that the active agent in the reduction of the ore is the carbonic oxide gas formed by the incomplete combustion of the carbon of the fuel ; or what comes to the same thing, the absorption by carbonic acid first produced of another proportion of carbon. The carbon oxide robs the iron oxide of its oxygen to become itself changed into carbonic acid. In reality however the action is more complex than this in its chemical relations ; for instance, metallic iron will under certain circumstances act conversely on carbonic acid, and rob it of half its oxygen. The net result of the reactions between carbon, iron, iron oxide, and these gases depends mainly upon the temperature and pressure and upon the relative quantities of each substance present. In the gases escaping from the blast furnace there is always a large quantity (nearly one-third) of carbonic oxide. At the blast furnaces in work during the first half of our century the combustible gases were allowed to burn to waste as they issued from the top of the furnace, in the manner shown in

Fig. 17, and at night the flames used to form a weird and striking feature in the prospect of an iron-smelting region.

Instead of allowing the escaping gases to burn to waste, it became the practice about,1860,and so continues, to draw them off and burn them under steam boilers or use their flames for heating the blast. An effective method of withdrawing the gases is shown in Fig. 21, which is a section through the upper part of a smelting furnace, with the "cup and cone" arrangement. The mouth of the furnace is covered by a shallow iron cone *a,* open at the bottom, into which fits another cone *b,* attached to a chain *c,* sustained by an arm of the lever *d,* which is firmly held in position by the chain *e,* and is also provided with a counterpoise *f.* When the mouth of the furnace is thus closed, the gases find an exit by the opening *g,* seen behind the cones, and leading into a downward passage, through which they are drawn by the draught of a tall chimney to the place where they are burnt. The charge for the furnace is filled into the hopper *a,* and at the proper time the chain, *e,* is slackened when the weight of the material resting on the suspended cone overcomes that of the counterpoise, and the charge slides down over the surface of the cone *b,* which is immediately drawn up again by the counterpoise, so that the opening into the air is at once closed.

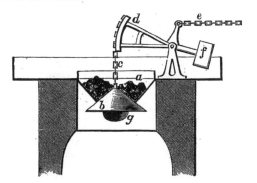

FIG. 21.—*Cup and Cone.*

The march of improvement in the blast furnace has been characterized particularly in Britain and the United States by a great increase of dimensions, which is found to promote economy in fuel, etc. In the former country the furnace of the latter part of our century is commonly from 70 to 80 feet high, and some have even been built with a height of more than 100 feet, while in the States the tendency to build very high furnaces is still more marked. A single large furnace may turn out as much as 1,500 tons of pig iron in a week, and some in America, it is said, actually produce as much as 2,500 tons. The more usual output of a blast furnace is however much less than these amounts ; but if we say only one-half, or even one-third of these quantities, a state of things is indicated very different from what obtained about 1837, when the best Welsh furnaces produced only 200 tons a week. If we go back to the beginning of the century, the difference is much more marked, for the

4

blast furnaces of that period could turn out only about 30 tons in a week.

The proportions of fuel, ore, and limestone charged into the furnace vary greatly according to the composition of the ore, the quality of iron aimed at, and the practice of each manufacturer. It is usual previously to calcine the carbonate ores and others also, in order to expel the carbonic acid and the moisture, of which last all contain a considerable amount : and sometimes the limestone is mixed with the ore to undergo this preliminary process. The charge being conveyed from the roasting kilns to the blast furnace while still hot effects an obvious economy of fuel in the latter. In the case of hæmatite ore the quantities of materials in one charge may be something like 54 cwt. of ore, 9 cwt. of limestone, and 33 cwt. of coke. It is quite common to use mixtures of different kinds of ore, so as to modify the quality of the product according to particular requirements. The use of the limestone is to take up silica, and the slag is found to consist mainly of silicates of lime and alumina. The amount flowing from a blast furnace of course varies much according to the conditions, and is larger than would commonly be supposed ; for the production of one ton of pig iron involves the production of from $\frac{1}{2}$ to $1\frac{1}{2}$ tons of slag.

Fig. 22 represents in section the later type of blast furnace, which of course is circular in plan. Its height may be taken as 80 feet, and the diameter at the widest part of the interior as $22\frac{1}{2}$ feet, narrowed to 20 feet near the top. The lowest portion, c, is called the *crucible*, the bottom of which is the *hearth*, both formed of the most refractory materials obtainable. The conical widening, B, above the crucible is the *boshes*, and at the top is seen the "cup and cone" apparatus already described, A, surmounted by the short cylindrical iron mouth, through apertures in which the charges are tipped from the gallery, D, these having been raised there in small trucks by hydraulic or other elevators. The escaping gases leave the furnace by the exit, E, which leads into the "down-come," G, and they are conducted from it to the "regenerative stoves" and dealt with as presently to be described. Our section represents the masonry of the furnace as sustained by pillars, P, at the outside of the lower part ; these pillars support a strong ring of iron plates upon which the wall rests. This arrangement has the advantage of allowing the workmen the greatest freedom of access to parts about the crucible, which require much attention. Here, at the lowest part, is an aperture from which the liquid iron is allowed to run out every five or six hours, it being plugged in the meantime by clay and sand. The slag being much lighter than the iron, floats above it, and runs off at a higher level over the *tympstone*. Opening into the hearth are several orifices to admit the hot blast from the nozzles of the *tuyères*, which of course do not project into the furnace itself ; but they are so near to the region of intensest heat that they would be rapidly destroyed unless they were surrounded by a casing through which a current of water is constantly running. The *tuyères*, of which there may be 3 or 5, are supplied from the pipe seen at K. The earlier plans of heating the air did not permit of a very high temperature being given to the hot blast, about 600° F. being the limit ; but the "regenerative" stoves can supply a blast of more than 1,600° F., or not far below the melting point of silver. Another great increase has been in the pressure of the blast ; 2 or 3 lbs. per square inch sufficed in the earlier practice ; but the lofty

modern furnaces have to be supplied with the blast at a pressure of
10 lbs. per square inch, and over.   Even when comparatively low
pressures were the rule, a large ironworks required much blowing power.
The works formerly at Dowlais, in South Wales, for instance, had an
engine of 650 horse-power for the blowing engine, in which a piston of
12 feet diameter moved in a cylinder 12 feet in length.   The quantity of
air that passes into a blast furnace amounts to thousands of tons per
week, its weight being much greater than that of all the ore, coke, and
limestone put together.

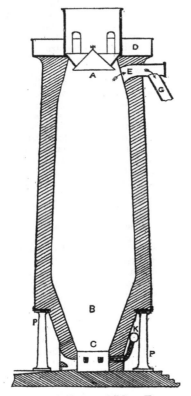

FIG. 22.—*Section of Blast Furnace.*

It need scarcely be said that great care and expense are bestowed on
the construction of these furnaces.   Only the best and most refractory
materials, such as firebricks, are used for the lining, and the exterior is a
casing of solid masonry, strengthened with iron bands.   When a new
furnace is finished it takes a month or six weeks to put it into operation ;
but when this is done it will remain in action night and day continuously

for a long period—perhaps for eight or ten years—before the necessity for repairs requires a "blow out." And the blow out and restarting, without the cost of repairs, entail an outlay of several hundred pounds.

The gases leaving the throat of the furnace consist mainly of nitrogen and a little carbonic acid, together with about one-third of their volume of the combustible gases, carbonic oxide, and some hydrogen ; but these last do not leave the furnace in an ignited state, because the oxygen there has already been consumed. They are conducted by the "down-come" pipe, G, Fig. 22, to a point at which, by means of a valve, they can be directed to one or other of two circular towers entirely filled with firebricks, arranged chequer-wise, so as to form innumerable passages between them. The furnace gases are admitted at the bottom of the Cowper tower, or "regenerative stove," into a flue to which a regulated quantity of air has access, and there they are fired : the flame ascending the flue to the upper part of the tower, thence descends, communicating its heat to the firebricks, which soon acquire a very high temperature, especially where the flame first enters, and the burnt gases leave the tower for a tall chimney, leaving most of their heat in the firebricks. When this action has continued for a sufficient time, the connection of the regenerator with the throat of the furnace is cut off, and the escaping gases are directed into the other regenerator, and at the same time the blast from the blowing engine is made to ascend among the firebricks of the first, where gaining increasing temperature as it ascends—the stove being hottest at the top—the air leaves the tower to be conducted to the *tuyères* at such high temperature as already mentioned. While the one regenerator is thus heating the blast, the other is in its turn accumulating heat from the flames of the escaping gases ; and thus they are worked alternately, the action being constantly reversed after suitable intervals.

When iron is combined with a much smaller proportion of carbon than in cast iron, and contains little or no graphitic or uncombined carbon, we have the very useful compound known as steel. In the earlier half of the century it was customary to distinguish steel from malleable iron on the one hand, and cast iron on the other. If the compound contained from 0·5 to 1·5 per cent. of carbon, it was called steel by some authorities, while others extended these limits a little on either side. Later it was found that the presence of elements other than carbon can confer steely properties on iron, and indeed it is possible to have a metal containing no carbon, but possessing the characteristic properties of steel. Sir Joseph Whitworth proposed to classify a piece of metal according to its tensile strength, without any regard to either its chemical composition or its mode of manufacture : if it could not bear more than 30 tons per square inch it should be considered iron, but if it had a higher tensile strength, it should then be regarded as steel. To estimate the engineering value a figure depending upon the elongation or stretching of the specimen before breaking was to be added to the number of tons of the breaking load. This stretching power of steel is in some cases of as much importance as the tensile strength : the ordnance maker, for instance, considers a steel with a breaking strength of 53 tons under an elongation of 5 per cent. as *for his purposes* to be rejected : while a specimen showing a breaking strain of only 30 tons along with an elongation of 35 per cent., on 2 inches of length, he will regard as good. The tensile strength of steel depends in part on its composition, in part on the mode of manufacture, and in part on the subsequent treatment. The *average* tensile strength of a wrought iron bar per square inch of section is about 25 tons (30 is the maximum) ; while the like average for steel is 43 tons, and some kinds of cast steel will bear nearly 60 tons. Steel bars of a certain temper subjected by Sir Joseph Whitworth to a process of hardening in oil showed a tensile strength of even 90 tons per square inch. These figures will suffice to show the great utility of steel in structures and machines. But steel has besides a characteristic property which makes it extremely valuable in a great variety of applications, namely, its capability of being *tempered.* If a piece of steel is heated to dull redness and suddenly cooled by plunging it into cold water, it becomes so extremely hard that it cannot be acted on by a file ; nay, its hardness may be made to rival that of the diamond, which is the hardest substance known. Now by a second operation this hardness can be reduced to any required degree : this is done by re-heating the metal to a certain moderate degree between 430° F. and 630° F. and again cooling it by immersion in some cooling medium. In this "letting down" process, it is the highest temperature that produces the greatest softening, and the properties of the tempered steel will depend upon the precise degree to which the metal has been re-heated. For example, if the product be required for making into sword blades, or watch-springs, and to possess much elasticity, the proper temperature is between 550° F. and 570° F. ; but if the steel is to be suitable for saws the temperature must range within a few degrees of 600° F., according to the fineness of the tool intended ; a lower temperature would give a metal too hard for them to be sharpened with a file. On the other hand, sharp cutting instruments and tools for working metals are obtained hard by tempering at lower degrees than springs. In practice the index of the temperature is taken from the colour of the film of oxide that gradually forms on a polished surface of the metal as the heat is raised, and begins by a very pale yellow (at 430° F.), passing through deeper shades into brown, then through purple into deep blue (at 570° F.), etc. The reader will now see why watch and clock springs have their deep blue colour, and he can observe for himself the whole series of colours by very gradually heating a piece of polished steel over a small flame.

If we compare the chemical composition of wrought iron and of cast iron with that of steel as regards the content of carbon, we see at once that steel holds an intermediate position, so that if in the puddling furnace we could arrest the decarbonization at a certain point we should obtain steel ; or if, on the other hand, we could put back into chemical combination with the decarbonized wrought iron a due percentage of carbon we should in that way also obtain steel. And it will be observed that the oldest primitive furnaces could not have failed sometimes to have produced steel as the net or final result of such actions. In fact, steel always has been and still is produced on one or other of these two principles, applied in divers ways, but severally and distinctly directed to that end. Of the many more or less modified processes of steel-making that have been in use, we need here but briefly mention a few which were *the* processes of the first sixty years of our century, and are to a considerable extent still in operation, although eclipsed in importance by two other processes that, since the date referred to, have been supplying the metal in enormously increased quantities, and which will have to be particularly described.

The most usual of the older processes of steel-making, still carried on at Sheffield and elsewhere, is known as the *cementation process* : it consists in heating bars of the best wrought iron in contact with charcoal, at a high temperature, for three or four weeks. At Sheffield the iron bars and charcoal are packed in alternate layers into troughs 14 ft. long by 3½ ft. deep and wide, constructed of slabs of siliceous sandstone 6 in. thick. The last layer of charcoal at the top is covered to a certain depth with a layer of refractory matter, and the flames from a furnace beneath are made to envelop the stone troughs or *pots*, as they are technically called, for a period of a week or more according to the thickness of the bars operated upon. These are generally 3 in. broad and from five-to six-eighths of an inch thick. When it is found by withdrawing a test bar for examination that the operation is complete, the fire is gradually diminished and the whole allowed to cool slowly, which requires about a fortnight. Instead of only charcoal, a mixture of powdered charcoal or soot with a little salt has been used by some makers—which mixture, technically called *cement powder*, has given its name to the process. In some works 16 tons or more of iron are treated in one operation. The bars are found unchanged in form, but increased in weight by perhaps 27 lbs. per ton, for carbon has combined with the iron, being apparently transferred in the iron from one particle to another. The surface of the bars becomes rough and uneven from a multitude of blebs or blisters, and hence they are called *blister bars*, and the steel of which they now consist is named *blister steel*. In this conversion we may suppose that the iron at its outer surface first enters into combination with carbon taken from the carbonic oxide gas, which would be produced by combustion of the charcoal with the limited quantity of air in its interstices, and the oxygen thus set free would immediately seize again on the surrounding charcoal, and by repeated changes of this kind in which the oxygen acts as a carrier of carbon to the iron, in which it is transferred inwards from particle to particle. The cause of the blisters has been much discussed : probably the cause is the formation and escape of a volatile compound of carbon and sulphur at the surface of the soft metal ; for it is known that nearly the whole of the little sulphur in the wrought iron disappears in the cementation process. Blister steel is never homogeneous, for near the surface it always contains more carbon than within ; the bars are therefore broken up into short lengths which are carefully assorted, bound together with wire, heated, welded together under a hammer or by rolling, and finally formed into a bar, which is stamped with the outline of a pair of shears, and is then known as *shear steel*, because this product was generally found the most suitable for making the shears used in dressing cloth.

Another method of dealing with the blister steel is to charge crucibles or pots having covers with 50 or 100 lbs. weight of the broken-up bars, and subject the crucibles to a strong heat in a reverberatory furnace, when the metal melts, and at the proper moment the contents of a great number of pots are almost simultaneously poured into a mould to form an *ingot*. The result is a very uniform steel of the finest texture, known and highly esteemed as *cast steel* or *crucible steel*. This steel is much more fusible than iron, but less so than cast iron.

The production of steel by arresting at a certain stage the decarbonizing of cast iron in the puddling furnace requires much experience on the part of the workman, who has to learn when the desired point has been reached by certain indications, such as the appearance of the flame, or by the examination of a small sample of the fluid metal withdrawn and rapidly cooled. Various additions to the charge in definite proportions are generally made, such as scales of iron oxide, or a quantity of an oxide ore (hæmatite, etc.) or other materials, the most essential for a good product consisting of a little manganese in some form. The result is *puddled steel*; and this, like blister steel, can be converted into cast steel by fusion in crucibles, running into ingot moulds, and subsequent treatment by hammering, pressing, rolling, etc. In 1864 puddled steel was described as an article of great commercial importance, but this it soon lost by the introduction of simpler, cheaper, and more reliable processes. The methods and improvements proposed for the production of steel have been exceedingly numerous, as is shown by the records of the English Patent Office alone, which contain up to the end of 1856 specifications of ninety-two patents for different steel-manufacturing processes, while from 1857 to 1865, the epoch-marking period of steel making, seventy-four more patents were obtained for this purpose. It would be quite beyond our limits to make special reference to these, and to the numerous patents which have since been granted, but there is one of great importance in steel-making which must be mentioned, and that is the patent for the employment in the cementation process of carbide of manganese, taken out by J. M. Heath in 1839. This made England almost independent of the former large importations of Swedish and Russian iron, and it caused an immediate reduction of £40 in the price per ton of good steel, effecting a saving which up to 1855 is calculated at not less than £2,000,000. Heath was one of those who fail to benefit by their inventions, for his was boldly appropriated by another person who took advantage of a verbal flaw in the specification, and Heath did not obtain any redress from the law courts until, after ten years' litigation, a majority of Exchequer judges reversed all the previous decisions against him (1853). In the meantime the man had died, but as the patent was about to expire his widow was on petition granted an extension of it for seven years. The nature of the influence of manganese on steel-making has not been fully explained, and there is some diversity of opinion on the subject, as it is said—on the one hand, merely to remove or counteract the injurious effects of sulphur or phosphorus ; on the other, to impart to the steel greater ductility, strength, and power of welding, tempering, etc.

The manufacture of *crucible* or *cast steel* has been carried on at Essen in Prussia by the firm of A. Krupp & Co., on a scale surpassing anything attempted elsewhere,—theirs being the largest steel-works in the world, and remarkable for the variety and excellence of its products. It began in so small a way that it is said only a single workman was employed. To the Great Exhibition of 1851, at London, Krupp's firm sent a block of crucible cast steel weighing 2¼ tons, a larger mass of the metal than had ever been shown before, and looked upon with no little astonishment, for at that time steel was a precious commodity, the price of refined steel ranging from £45 to £60 per ton. At the next London Exhibition, in 1862, the Essen Works showed a block of cast steel 20 tons in weight, and at the Vienna Exhibition of 1873, one of 52 tons. This casting, which was first made of a cylindrical shape, was forged into an octagonal form under an immense steam-hammer, larger than the Woolwich hammer described on a previous page, for the weight of the moving part is no less than 50 tons. This huge mass of cast steel was of the finest quality ; the forging into the prismatic form was to show its malleability, for it was intended for the body of a gun to have a bore of 14 inches. Since the period referred to, ingots of more than 100 tons have been cast. That shown at Vienna was the product of some 1,800 crucibles, each containing 65 lbs. of melted steel, which had to be poured into the mould in a regular and continuous stream, so that the metal might solidify into a perfectly uniform mass. Such work can be done only by trained men, who act in regular ranks with military precision, and in pairs emptying their crucibles into channels previously assigned, then filing off to the other end of the rank to receive another crucible, while the pair of men who were behind are pouring out theirs, and so on in succession. The crucibles are emptied into a number of channels formed of iron lined with fire-clay, and leading down into the mould. Many precautions have to be taken to ensure the regular progress of the operations, and all the time required to fill the huge moulds may be counted by minutes.

The headpiece to our chapter on Fire-Arms gives but a very inadequate idea of the magnitude of the Essen Works about 1870. A better notion will be obtained from a few figures which we select from a list giving some of the contents of the Essen Works in 1876. There were 1,109 furnaces of various kinds, of which 250 were for smelting ; 77 steam hammers, 294 steam engines, 18 rolling mills, 365 turning lathes, and 700 other machine tools ; 24 miles of ordinary gauge railway for traffic within the works ; together with 10 miles of narrow gauge railway ; 38 miles of telegraph lines, with 45 Morse apparatus, etc. (J. S. Jeans' *Steel: its History, etc.*, 1880). These figures belong, be it observed, to the state of things in 1876 ; but we learn from a later authority that in 1894 these works employed 15,000 men, and we must suppose that the plant has been proportionately increased since the earlier period, when 10,000 men were employed.

In the year 1854 a regular system of records began to be kept of the amounts of coal and ores raised in Great Britain, and also of the quantities of the various metals produced. These show that in 1894 very nearly three times as much coal was raised as in 1854, and that in the same period the quantity of British pig iron smelted annually had increased fourfold ; these increases look small when compared with the expansion of the steel production in Britain within the same period of forty years, for this had enlarged *thirtyfold*. This extraordinary development is attributable to the introduction of two processes by either of which various steels of excellent quality, and adapted to a great range of applications, can be produced cheaply and with certainty. These processes are respectively known as the Bessemer and the Open Hearth, and the reader should observe that with the main principles involved in these he has already been made acquainted.

Henry Bessemer, who first saw the light in England in 1813, may be said to have been born an inventor, for his father was one before him— a Frenchman employed in the royal mint at Paris, afterwards appointed by the Revolutionary authorities to superintend a public bakery ; on an accusation of giving short weight, thrown into prison, from which, and probably from the guillotine, he escaped, and found employment in the English mint. Subsequently he devised some notable improvements in the art of producing letterpress type, and for many years carried on a prosperous business as a typefounder. The son developed inventive faculties at a very early age : in lathe engraving, dies, dating stamps, etc. His name became familiar to everyone by his production of the metallic powder long known as " Bessemer's Gold Paint." It became known to Bessemer that the raw material of this substance, which was then sold at £5, 10s. per lb., really cost only about one shilling per lb., and he set himself to discover its composition and mode of manufacture. He succeeded in this so well that he could produce the article at the insignificant cost of four shillings a pound, and his first order for a supply of it was at the rate of £4 per lb., and the business was continued, realising profits of something like 1,000 per cent. at first. For this article no patent was taken out, but Bessemer himself, assisted by two trustworthy workmen, carried on the manufacture in secret, and he some time afterwards rewarded the fidelity of his men by handing over the business to them as a free gift. Then he took out patents for improvements in the manufacture of oils, varnishes, sugar, plate glass, etc. Several of his machines for these purposes were shown at the London Exhibition of 1851. Bessemer is said to have obtained altogether some 150 patents, including those granted for inventions connected with our subject. He may be regarded as the type of the very fortunate inventor, since on the patents of the one process we are going to describe he ultimately obtained royalties to the value of more than £1,057,000, and this irrespective of profits derived from commercially working it himself.

At the time of the Crimean War, Bessemer had some experiments made at Vincennes with cylindrical projectiles he had devised for firing from smooth-bore guns, yet so as to impart to the projectile at the same time rotation about its axis. The experiments were successful, but it was pointed out that the guns of cast iron then in use would not bear heavy projectiles, and he was induced, at the suggestion of the Emperor Napoleon III., to undertake some researches with the view of finding metal more suitable for artillery. Bessemer, having then little knowledge of the metallurgy of iron, applied himself on his return to England to the study of the best books on the subject, visited the principal iron-working districts, and began a series of experiments at a small experimental installation he set up in London. There, after repeated failures, he did at length succeed in producing a metal much tougher than the cast iron then used, and a small model gun was submitted to the Emperor, who encouraged Bessemer to persevere with his experiments ; which he did, though the expense was a great tax on his capital, continued as the experiments were for two years and a half. But by this time he had acquired a knowledge of many important facts, and these gradually led him to the experimental realization of the idea he had conceived, but only after many trials in which several thousand pounds were expended. At length the agenda of the British Association for the Cheltenham meeting of 1856 announced that a paper would be read by H. Bessemer, entitled "The Manufacture of Iron and Steel without Fuel." It will be easily understood that a title in such terms would give rise to much derisive incredulity ; and we may imagine the iron-masters on that occasion crowding into Section G, while asking each other in the spirit of certain philosophers of old, "What will this babbler say?" Some of what he did say may here be quoted, as at once explanatory and historically memorable.

"I set out with the assumption that crude iron contains about 5 per cent. of carbon ; that carbon cannot exist at a white heat in the presence of oxygen without uniting therewith and producing combustion; that such combustion would proceed with a rapidity dependent on the amount of surface of carbon exposed; and lastly, that the temperature which the metal would acquire would be also dependent on the rapidity with which the oxygen and carbon were made to combine; and consequently, that it was only necessary to bring the oxygen and carbon together in such a manner that a vast surface should be exposed to their mutual action, in order to produce a temperature hitherto unattainable in our largest furnaces.

FIG. 23.—*Experiments at Baxter House.*

"With a view of testing practically this theory, I constructed a cylindrical vessel of 3 ft. in diameter and 5 ft. in height, somewhat like an ordinary cupola furnace (see Fig. 23). The interior is lined with fire-bricks, and at about 2 in. from the bottom of it I inserted five *tuyère* pipes, the nozzles of which are formed of well-burned fire-clay, the orifice of each *tuyère* being about three-eighths of an inch in diameter; they are so put into the brick lining (from the outer side) as to admit of their removal and renewal in a few minutes, when they are worn out. At one side of the vessel, about half-way up from the bottom, there is a hole made for running-in the crude metal, and on the opposite side there is a tap-hole, stopped with loam, by means of which the iron is run out at the end of the process. In practice this converting vessel may be made of any convenient size, but I prefer that it should not hold less than one nor more than five tons of fluid iron at each charge; the vessel should be placed so near to the discharge hole of the blast furnace as to allow the iron to flow along a gutter into it. A small blast cylinder is required capable of compressing air to about 8 lbs. or 10 lbs. to the square inch. A communication having been made between it and the *tuyères* before named, the converting vessel will be in a condition to commence work ; it will however on the occasion of its first being used after re-lining with firebricks be necessary to make a fire in the interior with a few baskets of coke, so as to dry the brickwork and heat up the vessel for the first operation, after which the fire is to be all carefully raked out at the tapping-hole, which is again to be made good with loam : the vessel will then be in readiness to commence work, and may be so continued without any use of fuel until the brick lining, in the course of time, becomes worn away, and a new lining is required. I have before mentioned that the *tuyères* are situated nearly close to the bottom of the vessel, the fluid metal will therefore rise some 18 in. or 2 ft. above them; it is therefore necessary, in order to prevent the metal from entering the *tuyère* holes, to turn on the blast before allowing the fluid crude iron to run into the vessel from the blast furnace. This having been done, and the metal run in, a rapid boiling up of the metal will be heard going on within the vessel, the metal being tossed violently about and dashed from side to side, shaking the vessel by the force with which it moves ; from the throat of the converting vessel flame will immediately issue, accompanied by a few bright sparks such as are always seen rising from the metal when running into the pig-beds. This state of things will continue for about fifteen minutes, during which time the oxygen in the atmospheric air combines with the carbon contained in the iron, producing carbonic oxide, or carbonic acid gas, and at the same time evolving a powerful heat. Now, as this heat is generated in the interior of, and is diffused in innumerable fiery bubbles through, the whole fluid mass, the metal absorbs the greater part of it, and its temperature becomes immensely increased, and by the expiration of the fifteen minutes before named that part of the carbon which appears mechanically mixed and diffused throughout the crude iron has been entirely consumed : the temperature however is so high that the chemically combined carbon now begins to separate from the metal, as is at once indicated by an immense increase in the volume of flame rushing out of the throat of the vessel. The metal in the vessel now rises several inches above its natural level, and a light frothy slag makes its appearance and is thrown out in large foam-like masses. This violent eruption of cinder generally lasts about five or six minutes, when all further appearance of it ceases, a steady and powerful flame replacing the shower of sparks and cinder which always accompanies the boil. The rapid union of carbon and oxygen which thus takes place adds still further to the temperature of the metal, while the diminished quantity of carbon present allows a part of the oxygen to combine with the iron, which undergoes combustion and is converted into an oxide. At the excessive temperature that the metal has now acquired, the oxide as soon as formed undergoes fusion, and forms a powerful solvent of those earthy bases that are associated with the iron ; the violent ebullition which is going on mixes most intimately the scoria and metal, every part of which is thus brought in contact with the fluid oxide, which will thus wash and cleanse the metal most thoroughly from the silicon and other earthy bases which are combined with the crude iron, while the sulphur and other volatile matters which cling so tenaciously to iron at ordinary temperatures are driven off, the sulphur combining with the oxygen and forming sulphurous acid gas.

" The loss in weight of crude iron during its conversion into an ingot of malleable iron was found, on a mean of four experiments, to be 12½ per cent., to which will have to be added the loss of metal in the finishing rolls. This will make the entire loss probably not less than 18 per cent. instead of about 28 per cent., which is the loss on the present system. A large portion of this metal is however recoverable by heating with carbonaceous gases the rich oxides thrown out of the furnace during the boil. These slags are found to contain innumerable small grains of metallic iron, which are mechanically held in suspension in the slags and may be easily recovered.

" I have before mentioned that after the boil has taken place a steady and powerful flame succeeds, which continues without any change for about ten or twelve minutes, when it rapidly falls off. As soon as this diminution of flame is apparent the workman will know that the process is completed, and that the crude iron has been converted into pure malleable iron, which he will form into ingots of any suitable size and shape by simply opening the tap-hole of the converting vessel and allowing the fluid malleable iron to flow into the iron ingot moulds placed there to receive it. The masses of iron thus formed will be free from any admixture of cinder, oxide, or other extraneous matters, and will be far more pure and in a forwarder state of manufacture than a pile formed of ordinary puddle bars. And thus it will be seen that by a single process, requiring no manipulation or particular skill, and with only one workman, from three to five tons of crude iron pass into the condition of several piles of malleable iron in from thirty to thirty-five minutes, with the expenditure of about a third part the blast now used in a finery furnace, with an equal charge of iron, and with the consumption of no other fuel than is contained in the crude iron.

" To those who are best acquainted with the nature of fluid iron, it may be a matter of surprise that a blast of cold air forced into melted crude iron is capable of raising its temperature to such a degree as to retain it in a perfect state of fluidity after it has lost all its carbon and is in the condition of malleable iron, which, in the highest heat of our forges, only becomes softened into a pasty mass. But such is the excessive temperature that I am enabled to arrive at with a properly shaped converting vessel and a judicious distribution of the blast, that I am enabled not only to retain the fluidity of the metal, but to create so much surplus heat as to remelt all the crop-ends, ingot-runners, and other scrap that is made throughout the process, and thus bring them, without labour or fuel, into ingots of a quality equal to the rest of the charge of new metal. . . .

" To persons conversant with the manufacture of iron, it will be at once apparent that the ingots of the malleable metal which I have described will have no hard or steely parts, such as are found in puddled iron, requiring a great amount of rolling to blend them with the general mass, nor will such ingots require an excess of rolling to expel cinder from the interior of the mass, since none can exist in the ingot, which is pure and perfectly homogeneous throughout, and hence requires only as much rolling as is necessary for the development of fibre ; it therefore follows that, instead of forming a merchant bar, or rail, by the union of a number of separate pieces welded together, it will be far more simple and less expensive to make several bars or rails from a single ingot. Doubtless this would have been done long ago had not the whole process been limited by the size of the ball which the puddler could make.

"The facility which the new process affords of making large masses will enable the manufacturer to produce bars that, in the old mode of working, it was impossible to obtain; while at the same time it admits of the use of more powerful machinery, whereby a great deal of labour will be saved and the process be greatly expedited. . . . I wish to call the attention of the meeting to some of the peculiarities which distinguish cast steel from all other forms of iron, viz., the perfectly homogeneous character of the metal, the entire absence of sand-cracks or flaws, and its greater cohesive force and elasticity, as compared with the blister steel from which it is made,—qualities which it derives solely from its fusion and formation into ingots, all of which properties malleable iron acquires in like manner by its fusion and formation into ingots in the new process; nor must it be forgotten that no amount of rolling will give the blister steel, although formed of rolled bars, the same homogeneous character that cast steel acquires by a mere extension of the ingot to some ten or twelve times its original length. . . .

"I beg to call your attention to an important fact connected with the new process which affords peculiar facilities for the manufacture of cast steel. At that stage of the process immediately following the boil the whole of the crude iron has passed into the condition of cast steel of ordinary quality. By the continuation of the process the steel so pro- duced gradually loses its small remaining portion of carbon, and passes successively from hard to soft steel, and from soft steel to steely iron, and eventually to very soft iron; hence, at a certain period of the process, any quality of metal may be obtained. There is one in particular which by way of distinction I call semi-steel, being in hardness about midway between ordinary cast steel and soft malleable iron. This metal possesses the advantage of much greater tensile strength than soft iron; it is also more elastic, and does not readily take a permanent set, while it is much harder and is not worn or indented so easily as soft iron; at the same time it is not so brittle or hard to work as ordinary cast steel. These qualities render it eminently well adapted to purposes where lightness and strength are specially required, or where there is much wear, as in the case of railway bars, which from their softness and lamellar texture soon become destroyed. The cost of semi-steel will be a fraction less than iron, because the loss of metal that takes place by oxidation in the converting vessel is about $2\frac{1}{2}$ per cent. less than it is with iron; but as it is a little more difficult to roll, its cost per ton may fairly be considered to be the same as iron; but as its tensile strength is some 30 or 40 per cent. greater than bar iron, it follows that for most purposes a much less weight of metal may be used than that so taken. The semi-steel will form a much cheaper metal than any we are at present acquainted with. These facts have not been elicited from mere laboratory experiments, but have been the result of working on a scale nearly twice as great as is pursued in our largest iron works, the experimental apparatus doing 7 cwt. in thirty minutes, while the ordinary puddling furnace makes only $4\frac{1}{2}$ cwt. in two hours, which is made into six separate balls, while the ingots or blooms are smooth, even prisms, 10 in. square by 30 in. in length, weighing about equal to ten ordinary puddle balls."

The startling novelty of the methods and results described in this paper had the effect of paralyzing discussion at the time. But soon the voice of detraction was heard ; many iron-masters ridiculed the idea of producing iron and steel without fuel, and indeed it may have been observed, the title of the paper notwithstanding, that first the silicon and carbon, and then the iron itself, really supplied the fuel. And we must remember that malleable iron in a molten state was then deemed an impossibility, for the hottest furnaces then known could not effect the fusion, however prolonged their action might be, yet Bessemer was to obtain five tons in this condition in the short space of half an hour with no other aid than cold air. Then it was said that Bessemer's process of forcing air into melted cast iron had no claim of novelty, for it had been tried before and found valueless. Some iron-masters on trying experiments on a small scale and with imperfect appliances met with failures, and discredited the process at once ; but five large establishments paid for licences sums amounting to £26,500 within three weeks of the reading of the paper. At the works of the Dowlais Iron Co., in South Wales, who were the first licensees, the first converter was set up under Bessemer's personal superintendence, and at the first operation five tons of iron were produced direct from the blast furnace pig. This apparently satisfactory result proved quite otherwise when this iron came to be practically tested ; for it was found quite useless ! It was both *"cold-short"* and *"red-short,"* to use the technical terms,—the former of which means that although the sample may be welded, it is when cold brittle and rotten ; the latter means that at a low red heat it breaks and crumbles under the hammer. Further trials were made, new experiments instituted, but the success that attended Bessemer's early experiments could not be repeated, and as yet no one knew the reason why. Now it so happened that in the preliminary experiments an exceptionally pure pig iron had been made use of containing little or no phosphorus or sulphur, substances very deleterious in iron, and still more so in steel. With the capital obtained by the sale of his licences Bessemer quietly set to work to investigate the cause of his non-success, making daily experiments with a ton or two of metal at a time. These experiments extended over a period of two and a half years, and upon them Bessemer and his partner spent about £16,000, besides the £4,000 the preliminary researches had cost. But all difficulties were at length overcome, and the process was now found capable of turning out pure iron and steel when the pure pig iron of Sweden was used in the converter. In the meantime the licensees had made no attempts practically to carry out the process, which began to be denounced as visionary : it was "a mare's nest"; it was "a meteor that had passed through the metallurgical world, but had gone out with all its sparks." When Bessemer again brought the subject before the public, he found that no one believed in it ; everyone said, " Oh, this is the thing that made such a blaze two or three years ago, and which was a failure." Neither iron-makers nor steel-makers would now take it up. Bessemer and his partner thereupon joined with three other gentlemen to establish at Sheffield a steel-works of their own, where the invention should be carried into full practice. In due time works were erected, and they commenced to sell steel, receiving at first very paltry orders, for such quantities as 28 lbs. or 56 lbs. ; but the orders soon became larger, and afterwards very much larger, for they were underselling the Sheffield manufacturers by £20 a ton, and their steel was undistinguishable from the higher priced article. Bessemer had now bought his licences back again, and in the course of his second set of experiments had patented each improvement as it occurred to him, finally bringing the mechanical apparatus to the degree of efficiency requisite for practical working, without which his primary idea would have been valueless. Before directing the reader's attention to the form the apparatus had assumed, we may transcribe what Mr. Jeans, in the work above referred to, has told about the commercial success of the Bessemer steel-making firm :—

"On the expiration of the fourteen years' term of partnership of this firm the works, which had been greatly increased from time to time out of revenues, were sold by private contract for exactly twenty-four times the amount of the whole subscribed capital, notwithstanding that the firm had divided in profits during the partnership a sum equal to fifty-seven times the gross capital, so that by the mere commercial working of the process, apart from the patent, each of the five partners retired after fourteen years from the Sheffield works with eighty-one times the amount of his subscribed capital, or an average of nearly cent. per cent., every two months,—a result probably unprecedented in the annals of commerce."

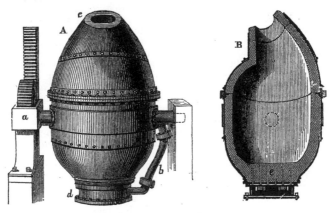

FIG. 24.—*Bessemer Converter.*

A, Front view, showing the mouth, *c* ; B, Section.

The form of the Bessemer apparatus as it finally left the inventor's hands may now be considered : but in certain details and arrangements some modifications, dictated by the experience and requirements of individual establishments, have been made, leaving the principles of the apparatus unchanged. Thus instead of making the converting vessel turn on trunnions, it is sometimes constructed fixed, the fluid metal after conversion being let out at a tap-hole : the number and size of the *tuyères* are varied ; and so with the disposition of the air chamber or *tuyère* box, the pressure of the blast, the capacity of the converter itself, etc. In capacity converters vary between $2\frac{1}{2}$ tons and 10 tons ; one of medium size is shown in elevation and section in Fig. 24, and may be described as an egg-shaped vessel about 15 ft. high and 6 ft. diameter inside. It is strongly made of wrought iron in two parts bolted together, and is lined inside with some thick infusible coating, of which more is to be said presently. The converter swings on trunnions, one of which is hollow, and admits the blast by the pipe *b* to the base of the vessel, whence it passes through the passages shown at *e*. The thickness of the lining at *e* may perhaps be 20 in., and passages for the air are perforated in fire-clay *tuyères*, of which there may be seven, each with seven perforations of half an inch diameter. To the other trunnion is attached a toothed wheel which engages the teeth of a rack receiving motion from hydraulic pressure. The iron for the operation is melted in a furnace having its hearth above the level of the converter; and to receive its charge the latter is turned so that the molten cast iron may be poured in from a trough until its surface is nearly on a level with the lowest of the *tuyères*. The blast having been turned on, the hydraulic power is set to work and the converter is slowly brought to an upright position. The pressure of the current of air prevents any of the fluid metal from entering the blow-holes. The blast of cold air is continued until all the silicon and carbon have been removed by oxidation. If the production is to be steel, there is then added to the contents of the converter, placed in position to receive it, a certain weight of melted cast iron of a special constitution, and the blow is resumed for a few minutes; or in more recent practice this special metal is added to the fluid metal run out of the converter into a spacious ladle in known quantity. On this addition an intense action takes place, attended by an extremely brilliant flame and a throwing out of cinder or slag. The metal thus added to the decarbonized iron is a carbonized alloy of iron and manganese obtained from an ore naturally containing the latter metal, and scarcely any phosphorus or sulphur. The charcoal pig from this ore is called *spiegeleisen* (German=mirror-iron) from its brilliant reflecting facets; it contains from 12 to 20 per cent. of manganese, with about 5 per cent. of carbon, and a considerable proportion of silicon. An exact chemical analysis of the particular spiegeleisen having been previously made, it is known what proportion of it is to be added to the decarbonized iron in order to convert this into a steel with any required content of carbon. The manganese probably acts by combining with oxide of iron diffused through the mass, and together with the silicon forming the very easily separated slag which is ejected.

The whole series of operations connected with the Bessemer process may be easily followed by the help of Fig. 25, which is taken from a beautiful model in the Museum of Practical Geology. This model, which was presented to the museum by Mr. Bessemer himself, represents every part of the machinery and appliances of the true relative sizes. C is the trough, lined with infusible clay, by which the liquid pig iron is conveyed to the converters, A. The hydraulic apparatus by which the vessels are turned over is here below the pavement, but the rack which turns the pinion on the axis of the converter is shown at B. The vessel into which the molten steel is poured from the converter is marked E, and this vessel is swung round on a crane, D, so as to bring it exactly over the moulds, placed in a circle ready to receive the liquid steel, which on cooling is turned out in the form of solid ingots. The valves which control the blast, and those which regulate the movements of the converter through the hydraulic apparatus, are worked by the handle seen at H. The

FIG. 25.—*Model of Bessemer Steel Apparatus.*

crane, or revolving table, D, is also under perfect control, so that the crude pig iron is converted into steel, and the moulds are filled with a rapidity and ease that are positively marvellous to a spectator.

The development of the Bessemer process soon had the effect of so reducing the price of steel that this material came into use for almost every purpose for which iron had previously been employed, such as railway bars, girders, etc., for bridges, boiler plates, etc., for all which "steely iron" containing only 0·12 to 0·40 per cent. carbon proved admirably adapted. The practical success of the Bessemer process had not long been demonstrated commercially by the inventor and his partners at Sheffield before other firms began the manufacture : so that in 1878 there were in Great Britain alone twenty-seven establishments making Bessemer steel and using 111 converters. It may give an idea of the magnitude the Bessemer steel manufacture had attained even at that time if we quote the cost of erecting a complete plant for two 5-ton converters : it was £44,400, as given in a detailed estimate. In all these cases pig iron from ores free from phosphorus and sulphur had to be used, for as we have seen the converter failed to eliminate these vitiating elements. Imported pig ores had in general to be used, or pig from the limited supply of British hæmatite ores in West Cumberland. The Barrow Hæmatite Steel Company engaged in the production of Bessemer steel on a very large scale, having by 1878 erected no fewer than sixteen converters of the capacity of 6 tons each. In the meanwhile many efforts were made to discover some method of eliminating phosphorus, so that the ordinary qualities of British pig iron, and iron derived in any part of the world from the coarse phosphorized ores, might be available for the converter. Many of the methods then devised proved correct in principle and feasible in practice ; but as, for sundry reasons, none of them came extensively into use, we need not here allude to them further.

The solution of the problem was announced in 1879. Some years before, G. J. Snelus had come to the conclusion that with a siliceous lining it would be impossible to eliminate phosphorus in the Bessemer converter, and that some refractory substance of a basic character must be sought for in order that the slag produced should be in a condition to absorb the phosphoric acid as fast as it is produced. He patented in 1872 the use of magnesian limestone as a material for the lining ; as that substance when intensely heated became very hard and stony, being in that condition quite unaffected by water. Two young chemists, Messrs. Thomas and Gilchrist, apparently without being aware of Mr. Snelus's conclusions, had also convinced themselves that the chief deficiency in the Bessemer process was due to the excess of silica in the slag, and in 1874 they began to try the effect of basic linings, and also of basic additions, such as lime, etc., to the charge in the converter, so that the lining itself should not be worn out by entering into the slag. Their results proved that phosphorus could be eliminated when the slag contained excess of a strong base. An example of an operation at Bolckow, Vaughan, & Company's Eston works with the highly phosphorized Cleveland pig iron may be quoted. The basic-lined converter received first 9 cwt. of lime, then 6 tons of metal. When the blast at 25 lbs. pressure was turned on, the silicon began at once to burn ; for three minutes the carbon was not affected, but for fourteen minutes longer it regularly diminished, the silicon keeping pace with it. After the blow had been continued for thirteen minutes from the commencement, the converter was turned down to allow of the further introduction of $19\frac{1}{2}$ cwt. of a mixture of two parts of lime with one of oxide of iron. So long as 1·5 per cent. of carbon remained in the metal the phosphorus was untouched, and at the end of the blow, *i.e.* when the flame dropped, only one-third of it had been eliminated ; it still formed 1 per cent. of the metal. The blast continued for another two minutes brought it down to $\frac{1}{4}$ per cent., and in one more minute only a trace was left. Most of the sulphur was got rid of at the same time. From Cleveland pig, thus de-phosphorized in the Bessemer converter, large quantities of steel rails were rolled for the North Eastern Railway Company, and were found entirely satisfactory, being as good as those made from the Cumberland hæmatite steel. This de-phosphorized process has been brought into operation wherever phosphoric ores are dealt with, and it has been applied with equal success in the "open hearth" furnaces, of which we have now to speak.

All discoveries and all inventions may be traced back to preceding discoveries and inventions in an endless series, and it is only by its precursors that each in its turn has been made possible. If we take one of the greatest marvels brought into existence at nearly the close of our epoch, namely, "wireless telegraphy," we may follow up links of a chain connecting it with the recorded observations of an ancient Greek (Thales) who flourished seven centuries before our era, and even these may not have been original discoveries of his. And it will have been gathered from what has already been said that steel must have been produced, however unwittingly, at the earliest period at which man began to reduce iron from its ores. So the very latest, and for many purposes the most extensively practised, process of modern steel-making, brought indeed to working perfection mainly by the perseverance and scientific insight of two individuals, is the result of the observation and the accumulated experience of former generations. The observations and experience here alluded to are chiefly those that follow two lines : one concerning the properties of the metal itself, the other relating to the means of commanding very high temperatures on a great scale. On this occasion we are able almost to lay a finger on some proximate links of the chain. Réaumur, the French naturalist, made steel in the early part of the eighteenth century by melting cast iron in a crucible, and in this liquid metal he dissolved wrought iron, the product being, as the reader will now easily understand, the intermediate substance, steel ; and this was obtained of course at a temperature which was incapable of fusing wrought iron by itself. He published in 1722 a treatise on "The Art of converting Iron into Steel, and of softening Cast Iron." For this, and certain other metallurgical discoveries, Réaumur received a life-pension equivalent to about £500 per annum,—a treatment very different from that dealt out by the British to Henry Cort. The action in Réaumur's crucible is precisely that used on the large scale in Siemens' open hearth. But this last became possible only when Siemens had worked out his "regenerative stove" or heat accumulator, the development of an idea suggested by a Dundee clergyman in 1817.

A general notion of the Siemens' regenerative stove will have been already gained from the account given before of its application to the modern type of blast furnace. Of the inventor himself, C. William Siemens, it may be observed that he was one of a family of brothers, all remarkable for their scientific attainments, and in many of his researches and processes he was aided by his brothers Frederick and Otto. In our article on "Electric Power and Lighting" there will be found some notice of a few of Siemens' inventions pertaining to those subjects. A still more admirable invention of his is the electric pyrometer, an instrument of the utmost utility for measuring, with an accuracy previously unapproachable, the high temperature of furnaces, etc. Indeed there are few departments of science, pure or applied, which have not been enriched by the researches and contrivances of this distinguished man, whose merits were acknowledged by the bestowal upon him of the highest scientific and academical honours, and also of a title, for he became Sir William Siemens.

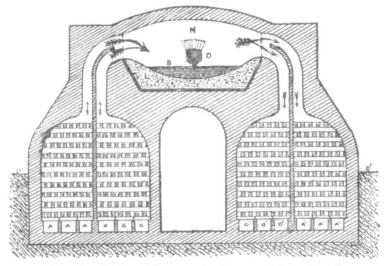

FIG. 26.—*Section of Regenerative Stoves and Open Hearth*

Siemens was much engaged from 1846 in conjunction with his brother Frederick in experimental attempts, continued over a period of ten years, at the construction of the regenerative gas furnace. At length, in 1861, he proposed the application of his furnace to an "open hearth," and during the next few years some partial attempts to carry out his process were made, and he himself had established experimental works at Birmingham in order to mature his processes, while Messrs. Martin of Sireuil, in France, having obtained licences under Siemens' patents, gave their attention to a modification of his process, by which they succeeded in producing excellent steel. Siemens having in 1868 proved the practicability of his plans by converting at his Birmingham works some old phosphorized iron rails into serviceable steel, a company was formed, and in 1869 the Landore Siemens' Steel Works were established at Landore in Glamorganshire, and a few years after, these had sixteen Siemens open hearth melting furnaces at work, giving a total output of 1,200 tons of steel per week. The number of furnaces was subsequently increased. Extensive works specially designed for carrying out the Siemens and the Siemens-Martin process were shortly afterwards erected at other places, as at Newtown, near Glasgow, Panteg in Wales, etc. In Great Britain the open hearth process gradually gained upon the Bessemer, until in 1893, when the total output of both kinds amounted to nearly 3,000,000 tons, this was almost equally divided between them, and since that period the steel made by the former has greatly surpassed in amount that made by the latter.

How the regenerative stove, or heat accumulator, works, and how it is applied in the open hearth process, the reader may learn by aid of the diagram Fig. 26, in which however no representation of the disposition of the parts in any actual furnace is given, nor any details of construction beyond what is necessary to make the principle clear. On the right and on the left of the diagram will be seen a pair of similar chambers which are shown as partly below the level of the ground s s', such being a usual disposition. The outer walls of these chambers are thick and the interior is entirely lined with the most refractory fire-bricks, of which also is formed the partition in between each pair of compartments, as well as the passages from the top of each opening on the furnace H. Each chamber or compartment is filled with rows of fire-bricks, laid chequerwise so as to leave a multitude of channels between. At the bottom of the chamber on the left let us suppose atmospheric air to be admitted by the channels A, A, A, and a combustible gas which we may take to be a mixture of carbonic oxide with some hydrogen is admitted in the same way to the second compartment on the left through the passages G, G, G. Supposing the apparatus quite cold in the first instance, the gas would ascend into the furnace H as shown by the arrows, because it might be drawn by an up-draught in a chimney connected with the six chambers shown at the bottom of the right, and it would also tend to rise up into the space H by its lighter specific gravity, and there it could be set on fire, when a volume of flame would pass across to the right, a plentiful supply of air rushing in through the air chamber from A, A, A, and the products of the combustion, mainly hot carbonic acid gas and hot nitrogen gas, in passing through the right-hand chambers, would make the bricks in both compartments very hot after a time, for the current would divide itself between the two passages, as indicated by the divided arrow. We have not shown the valves by which the workman is able, by merely pulling a lever, to shut off the air supply from A, A, A, and of gas from G, G, G, and put these channels into direct communication with the up-draught chimney, at the same time supplying gas at G', G', G', and air at A', A', A'. These rise up among the now heated bricks each in its own compartment, but mix where they enter the furnace H, now hot enough to set them on fire, and the gaseous products of combustion, hotter now than before, descend among the fire-bricks of the left-hand compartments, heating them in turn. After another period, say half an hour, the valves are again reversed, and again gas and air both heated burn in the space H, and their products supply still more heat to the right-hand compartments. And so the action may be continued with a great temperature each time produced by the combustion of the combining bodies at increasingly higher temperatures. Thus, if cold gas and air by combination give rise to 500° of heat, when the same combine, at say the initial temperature of

400°, the result would be a temperature of 900°; if burnt at this latter degree, then 900°+500° would be reached, and so on. It would seem as if there were no limit to the temperatures obtainable in this way. But the nature of the materials of which the furnace is constructed imposes a limit, for even the most refractory matters yield at length, and the working would come to an end by the fusing of the brick-work. The diagram is a section through the length of the hearth (for it is usually oblong in plan), and the low arch above H being exposed to the fiercest heat, is formed of the most refractory "silica bricks," that is, bricks made of coarsely ground silica held together with a little lime ; yet this extremely resisting material is acted upon, and the arch has to be renewed every few months or sometimes weeks. The hearth itself is supported by massive iron plates, shown in the diagram by the thick lines, above which is laid a deep bed L, of quartz sand or ganister, or where required a *basic* lining, beaten hard down, and forming a kind of basin with sides sloping down in all directions to a point immediately below the centre of the fire-brick door D, where is the aperture for tapping, stopped by a mixture of sand and clay until the metal is ready for drawing off, when it runs outside into an iron spout lined with sand and is received into the ingot moulds. B in the figure represents the "bath," as it is called, of molten metal, which, in the larger furnaces, where 20 tons of metal is operated on at once, may occupy an area of 150 square ft.

It need hardly be mentioned that there has to be a certain adjustment between the volumes of air and of gas that pass into the regenerative stoves, in order that the best effect may be obtained. Besides the limit of temperature occasioned by the nature of the materials, there is a chemical reason why the regenerative stoves cannot increase the temperature indefinitely. It is noticed that when the temperature of the furnace has become very high indeed, the flame over the hearth assumes a peculiar appearance, being interrupted by dark spaces. These are attributable to what is called in chemistry "dissociation," — in this case the dissociation of carbonic acid gas, which by the heat alone separates into carbonic oxide and oxygen gases. In the same way these gases refuse to combine if brought together heated beyond a certain temperature. This phenomenon of dissociation is a general one, for it is found that for any pair of substances there is a characteristic range of temperature above or below which they refuse to combine. The gas used in these stoves is either unpurified coal gas, or that produced by passing steam over red-hot coal or coke.

We have spoken of the Siemens and the Siemens-Martin open hearth processes. In the latter a charge of pig iron, say 1½ tons, is first melted on the hearth, then about 2 tons of wrought iron is added in successive portions, and in like manner nearly as much scrap steel (*i.e.* turnings, etc.), the final addition being half a ton of spiegeleisen containing 12 per cent. of manganese. A furnace of corresponding dimensions will allow of three charges every twenty-four hours. In the Siemens process it is not wrought iron or steel scrap that is mainly used to decarbonize the pig, but a pure oxide ore. This is thrown into the bath of molten metal in quantities of a few cwts. at a time, when a violent ebullition occurs. When samples of the metal and of the slag are found to be satisfactory, spiegeleisen or ferro-manganese is added, and the charge is cast. This process takes a rather longer time than the former, but gives steel of

FIG. 26a.—*Rolling Mill.*

more uniform character. ⸴ In both processes, phosphorus is oxidized at
the high temperature attained and passes into the slag, which last floats
of course on the molten metal and is from time to time tapped off as the
action proceeds.

Fig. 26a shows a rolling mill with what is called a "two-high" train for
finishing bars by passing them between the grooves cut in the rolls to
give the required section. The rolls in the illustration turn in one direction only, and therefore the bars after emerging from the larger grooves
have to be drawn back over the machine and set into a smaller pair from
the same side. This inconvenience is avoided in the "three-high train,"
on which three rolls revolve, and the bars can be passed through them
from one side to the other alternately. The celerity with which a glowing
steel ingot is without re-heating converted into a straight steel rail 60 or
100 feet long, by passing a few times backwards and forwards between
the rolls, is very striking. These rolls are made of solid steel, and in
some cases have a diameter of 26 inches or more.

IRON IN ARCHITECTURE.

EVERYONE knows how much iron is used in those great engineering structures that mark the present age, and of which a few examples will be described in succeeding articles. One other feature of the nineteenth century is the use of iron in architecture. Some have, indeed, protested against the use of iron for this purpose, and would even deny the name of architecture to any structure obviously or chiefly formed of that material. Stone and wood, they say, are the only proper materials, because each part must be wrought by hand, and cannot be cast or moulded ; and further, iron being liable to rust, suggests decay and want of permanence, and these are characters incompatible with noble building. All this can rest only on a relative degree of truth—as, for instance, machinery is used to dress and shape both wood and stone, and the permanence of even the latter is as much dependent on conditions as that of iron. Iron used in architecture is hideous when applied in shapes appropriate only to stone ; but when it is disposed in the way suggested by its own properties, and receives ornament suitable to its own nature, the result is harmonious and graceful, and the structure may display beauties that could be attained by no other materials. Be that as it may, the great and lofty covered spaces that are required for our railway stations and for other purposes could have been obtained only by the free use of iron, and everyone can recall to mind instances of such structure not devoid of elegance, in spite of the absence—the proper absence—of the Classic " orders " or Gothic " styles." The first notable instance of the application of iron on a large scale was the erection of the " Crystal Palace," in Hyde Park, for the great Exhibition of 1851. It was taken down and re-erected at Sydenham, and there it has become so well known to everyone that any description of it is quite unnecessary in this place.

As another conspicuous example of what may be done with iron, the Eiffel Tower at Paris may be briefly described.

The idea of erecting a tower 1,000 feet high was not of itself new. It had been entertained in England as early as 1833, in America in 1874, and in Paris itself in 1881. It has been reserved for M. Gustave Eiffel, a native of Dijon, who commenced to practise as an engineer in 1855, to realize this ambitious project. He has long been occupied in the construction of great railway bridges and viaducts, and in these he has adopted a system peculiar to himself of braced wrought-iron piers without masonry or cast-iron columns. He also was the first French engineer to erect bridges of great span without scaffolding. In the Garabit viaduct he planned an arch of 541 feet, crossing the Truyère at a height of nearly 400 feet above it. One result of M. Eiffel's studies in connection with these lofty piers was his proposal to erect the tower for the Paris Exhibition of 1889. This proposal met with great opposition on the part of many influential people in Paris—authors, painters, architects, and others protesting with great energy against the modern Tower of Babel, which was, as they said, to disfigure and profane the noble stone buildings of Paris by the monstrosities of a machine maker, etc. etc. The Eiffel Tower is now constructed, and no one has heard that it has dishonoured the monuments of Paris, for it has been instead a triumph of French skill, the glory of its designer, and the wonder of the Exhibition.

Fig. 26*b.*—*The Eiffel Tower in course of construction.*

The tower rests on four independent foundations, each at the angle of a square of about 330 feet in the side, and it may be noted that the two foundations near the Seine had to be differently treated from the other two, where a bed of gravel 18 feet thick was found at 23 feet below the surface, and where a bed of concrete, 7 feet thick, gave a good foundation. The foundations next the river had to be sunk 50 feet below the surface to obtain perfectly good foundations. Underlying the whole is a deep stratum of clay; but this is separated from the foundations by a layer of gravel of sufficient thickness. Above this are beds of concrete, covering an area of 60 square metres, and on the concrete rests a pile of masonry. Each of the four piles is bound together by two great iron bars, 25 feet long and 4 inches diameter, uniting the masonry by means of iron cramps, and anchoring the support of the structure, although its stability is already secured by its mere weight. The tower is of curved pyramidal form, so designed that it shall be capable of resisting wind pressure, without requiring the four corner structures to be connected by diagonal bracing. The four curved supports are, in fact, connected with each other only by girders at the platforms on the several stages, until at a considerable length they are sufficiently near to each other to admit the use of the ordinary diagonals. The work was begun at the end of January, 1887, and M. Eiffel notes how the imagination of the workmen was impressed by the notion of the vast height of the intended structure. Not steel, but iron is the material used throughout, and the weight of it is about 7,300 tons, without reckoning what is used in the foundations, and in the machinery connected with the lifts, etc. It has long ago been found that stone would be an unsuitable material for a structure of this kind, and it is obvious that only iron could possibly have been used to build a tower of so vast a height and within so short a space of time, for it was completed in April, 1889. A comparison of heights with the loftiest stone edifices may not be without interest. The highest building in Paris is the dome of the Invalides, 344 feet; Strasburg Cathedral rises to 466 feet; the Great Pyramid to 479 feet; the apex of the spire in the recently completed Cathedral at Cologne to 522 feet. These are overtopped by the lofty stone obelisk the Americans have erected at Washington, which attains a height of more than 550 feet. Such spires and towers have been erected only at the cost of immense labour. But iron, which can be so readily joined by riveting, lends itself invitingly to the skill of the constructor, more particularly by reason of the wonderful tensile strength it possesses. It is scarcely possible to convey any adequate idea of the great complicated network of bracings by which in the Eiffel Tower each standard of the columns is united to the rest to form one rigid pile. The horizontal girders unite the four piers in forming the supports of the first storey some 170 feet above the base. The arches which spring from the ground and rise nearly to the level of these girders are not so much intended to add to the strength of the structure as to increase its architectural effect. The first storey stands about 180 feet above the ground, and is provided with arcades, from which fine views of Paris may be obtained. Here there are spacious restaurants of four different nationalities. And in the centre of the second storey (380 feet high) is a station where passengers change from the inclined lifts to enter other elevators that ascend vertically to the higher stages of the tower. On the third storey, 900 feet above the ground, there is a saloon more than 50 feet square, completely shut in by glass, whence a vast panorama may be contemplated. Above this again are laboratories and scientific observatories, and, crowning all, is the lighthouse, provided with a system of optical apparatus for projecting the rays from a powerful electric light. This

FIG. 26c.—*The Eiffel Tower.*

light has been seen from the Cathedral at Orléans, a distance of about 70 miles.

The buildings of the Paris Exhibition of 1889 are themselves splendid examples, not only of engineering skill, but of good taste and elegant design in iron structures and their decorations. The vast *Salle des Machines* (machinery hall) exceeds in dimensions anything of the kind in existence, for it is nearly a quarter of a mile long, and its roof covers at one span its width of 380 feet, rising to a height of 150 feet in the centre. This great hall is to remain permanently, as well as the other principal galleries with their graceful domes.

The Eiffel Tower having proved one of the most striking features of the great Paris Exhibition, and of itself a novelty sufficient to attract visitors to the spot, and having, long before the Exhibition closed, completely defrayed the expense of its construction, with a handsome profit besides, its success has naturally provoked similar enterprises,—as, for instance, at Blackpool, a seaside resort in Lancashire, there has been erected an open-work metal tower, resembling the Paris structure, but of far less altitude.

Tall Buildings in American Cities.

In several of the great cities of the United States, the last few years have witnessed a novel and characteristic development of the use of iron in architecture. In many structures on the older continent, this material has been frankly and effectively employed, forming the obvious framework of the erection, even when the leading motive was quite other than a display of engineering skill. The Crystal Palace at Sydenham and other erections have been referred to, in which iron has taken its place as the main component of structures designed more or less to fulfil æsthetic requirements: the guiding principle in "tall office buildings" in the cities of the Western continent is, on the contrary, avowedly utilitarian. Iron has, of course, long been used in the form of pillars, beams, etc., in ordinary buildings, and it is only the extraordinary extension of this employment of it, after the lift or elevator had been perfected, and the ground-space in great commercial centres was daily becoming more valuable, that has led to the erection of structures of the "sky-scraper" class in American cities. For a given plot at a stated rent, a building of many stories, let throughout as offices, will obviously bring to its owner a greater return than one of few stories. The elevators now make a tenth story practically as accessible as a third storey, and the tall building readily fills with tenants. No claim for artistic beauty has been advanced for these structures, which aim simply at being places of business, and if provision be made for sufficient floor-space and daylight, and for artificial lighting, heating, and ventilation, together with the ordinary conveniences of modern life, and ready elevator service, nothing more is required by the utilitarian spirit, that seeks only facilities for money-getting. These tall buildings are usually erected on plots disproportionately small, and the architectural effect is apt to be bizarre and incongruous, especially when the structure shoots up skyward in some comparatively narrow street amid more modest surroundings. They are really engineering structures, but invested with features belonging to edifices of quite another order of construction. If they are necessities of the place and period, and are "come to stay," it cannot be doubted but that decoration of an appro-

PLATE IV.

THE AMERICAN TRACT SOCIETY BUILDING.

FIG. 26*d.*—*St. Paul Building, N.Y.*

priate and harmonious character will, in course of time, be evolved along
with them, when the conventionality that clings to architecture shall be
broken through, and a new style appear, as consistent, and therefore as
beautiful, in relation to the "tall office building," as were those of the
Greek temple and the Gothic minster in their free and natural adaptation.

Here, apparently, is the opportunity for the advent of a new and characteristic style. There is great ingenuity displayed in the arrangement
and internal finish of these buildings. But besides the somewhat novel
application of iron, the most notable circumstances regarding them are the
tendency to make them of greater and greater height, and the wonderfully
short time in which, upon occasion, they can be run up. Chicago has
recently been noted for its tall edifices, among which may be named *The
Reliance Building*, erected upon a site only 55 feet in breadth, but rising
in fourteen stories to the height of 200 feet, and presenting the appearance
of a tower. There are no cast iron pillars, but the whole metal framework is of rolled steel, the columns consisting of eight angle-sections,
bolted together in two-story lengths, adjoining columns breaking joint
at each floor, and braced together with plate girders, 24 inches deep,
bolted to the face of the columns, with which they form a rigid connection.
Externally, the edifice shows nothing but white enamelled terra-cotta and
plate glass. This building was originally a strongly-built structure of
five stories, the lower one being occupied as a bank. The foundations
and the first story were taken out, and prepared for the lofty edifice, the
superstructure being the while supported on screws. Then the three
upper stories were taken down, and the building was continued from the
second story, which was filled with tenants while the building was in
course of erection above.

Still more lofty edifices have been going skyward in other places.
Already in New York there are a great number of lofty piles due to the
introduction of the lifts or elevators, by which an office on the tenth floor
is made as convenient as one on the second. These buildings usually
receive the name of the owners of the structure, who occupy, perhaps,
only one floor. To mention only a few. There is the American Tract
Society building, with its twenty-three stories, 285 feet high, which is
one of the latest and handsomest of these tall piles in the city. See
Plate IV. Still loftier is the St. Paul building, fronting the New York Post-
Office at the junction of Park Row and Broadway. This structure is
splayed at the angle between Ann Street and Broadway, where its width
is 39½ feet, while its *loftiest* part has frontages of about 30 feet along
each of these thoroughfares. The height is no less than 313 feet above
the pavement, and the number of stories is twenty-five. This building is
faced with light yellow limestone, and although it was commenced only
in the summer of 1895, it was expected to be ready for occupation by the
autumn of 1896. Even this great height is overtopped by the Manhattan
Life Insurance Company's building, rising 330 feet, and remarkable as
perhaps beyond previous record of quickness in building a gigantic
structure. Obviously, the foundations of such a building must be most
seriously considered, prepared and tested, before the great bulk of the
building is begun, and in the *New York Engineering Magazine* one of
the architects has given a full account, with complete illustrations, of all
the works, from the rock foundation to the completed edifice. A description of the foundation work, though most interesting for the professional
engineer, would probably have little attraction for the general reader ;

but its importance may be inferred from the fact of its having taken nearly six months for its completion, while the huge superstructure required only eight months. The eighteenth tier of beams was reached in "three months from the time the foundations were ready on which to set the first piece of steel, composing the bolsters that support the cantilever system. . . . The substructure, which starts in bed-rock and continues to the cellar-floor, consists of fifteen piers, varying in size from 9 feet in

FIG. 26e.—*Manhattan Insurance Co.'s Building, in course of erection.*

diameter, to 21 feet 6 inches by 25 feet square. . . . The number of bricks used in the piers amounted to 1,500,000. From this it may be seen that a good-sized building was sunk out of sight before any part of the superstructure could be begun." An open court within the main structure, special framing for the arrangements of the company's offices on the sixth floor, the great height and weight of the tower, and the requisite provision for wind-bracing, delayed in some degree a regular advance of the stories;

but within three months no less than 5,800 tons were placed in position. There were girders weighing 40 tons, many columns of 10 and 12 tons, and cantilevers of 80 tons weight and 67 feet long. Strange to say, that in a building of this magnitude, where such masses had to be raised 300 feet into the air, there was not a single accident involving loss of life. When four stories of the steel framework had been put up, the bricklayers were set to work, and they followed the frame-setters throughout. After

FIG. 26*f.*—*Manhattan Insurance Co.'s Building, nearly completed.*

the masons came the pipe-layers, with their ten miles of pipes, followed by electricians, fixing their thirty-five miles of communicating wires. Thirty thousand cubic feet of stone was cut and set on the Broadway front in eighty days. Then craftsmen of the different trades followed each other, or worked in harmony together, story after story upwards : the engineers for boilers, heating, and elevators, the plumbers, the decorators, the carpenters and cabinet-makers, the plasterers, the marble and tile workers, the gasmen, etc. In fine, every story was completely finished and ready for occupation in eight months after the start from the foundations.

The shortness of the time in which these lofty buildings were run up is not less remarkable than the completeness of their fittings, which comprise everything requisite for communication within the premises and in connection with the outer world. The elevators or lifts are the perfection of mechanism in their way, and act with wonderful smoothness and regularity; of these are usually two at least, as well as an ample staircase. Notwithstanding all these appliances, some disastrous and fatal conflagrations have occurred at buildings erected on the "tall" principle; and as "business premises" of even 380 feet high are projected, the authorities have been considering the desirability of restricting the heights. It has been proposed that offices should not exceed in height 200 feet; hotels, 150 feet; and private houses, 75 feet.

BIG WHEELS.

THE Paris example of an engineering feat upon an unprecedented scale having proved sufficiently captivating for the general public to ensure for itself a great commercial success, even amid the attractions of an International Exhibition, was not lost upon the enterprising people of the States when the "World's Fair" at Chicago was in preparation in 1893. It was then that Mr. G. W. G. Ferris, the head of a firm of bridge constructors at Pittsburg, conceived the idea of applying his engineering skill to the erection of a huge wheel, revolving in a vertical plane, with cars for persons to sit in, constituting, in fact, an enormous "merry-go-round," as the machine once so common at country fairs was called. The novelty of the Chicago erection was, therefore, not the general idea, but the magnitude of the scale, which, for that reason, involved the application of the highest engineering skill, and the solution of hitherto unattempted practical problems. Several thousand pounds were, in fact, expended on merely preliminary plans and designs. The great wheel at Chicago was 250 feet in diameter, and to its periphery were hung thirty-six carriages, each seating forty persons. At each revolution, therefore, 1,440 people would be raised in the air to the height of 250 feet, and from that elevation afforded a splendid prospect, besides an experience of the peculiar sensation like that of being in a balloon, when the spectator has no perception of his own motion, but the objects beneath appear to have the contrary movement, that is to say, they seem to be sinking when he is rising, and *vice versâ*. The axle of the Chicago wheel was a solid cylinder, 32 inches in diameter and 45 feet long; on this were two hubs, 16 feet in diameter, to which were attached spoke rods, 2½ inches in diameter, passing in pairs to an inner crown, which was concentric with the outer rim, but 40 feet within it. The inner and outer crowns were connected together, and the former joined to the crown of the twin wheel by an elaborate system of trusses and ties, which, however, left an open space between the rims of 20 feet from the outside. These last were formed of curved riveted hollow beams, in section 25½ inches by 19 inches, and between them, slung upon iron axles through the roofs, were suspended, at equal intervals, the thirty-six carriages, each 27 feet long; and weighing 13 tons without its passengers, who added 3 tons more to the weight. The wheel with its passengers was calculated to weigh about

6

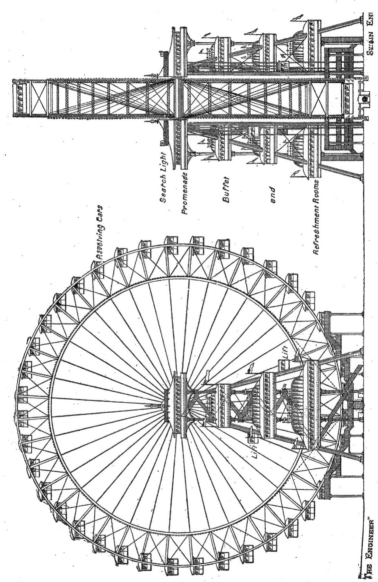

Search Light

Promenade

Buffet

and

Refreshment Rooms

Strain End

Revolving Cars

Lift

Lift

The Engineer

Fig. 263.—*Original Design for the Great Wheel.*

1,200 tons, and it rested on two pyramidal skeleton towers of iron-work 140 feet high, having bases 50 feet by 60 feet. The wheel was moved by power applied at the lowest point, the peripheries of both the rims having great cogs 6 inches deep and 18 inches apart, which engaged a pair of large cog-wheels, carried on a shaft 12 inches in diameter.

This curious structure was not begun until March, 1893, yet it was set in motion three months afterwards, having cost about £62,500. The Company had to hand over to the Exhibition one half of the receipts after the big wheel had paid for its construction, but even then they realised a handsome profit, and at the close of the World's Fair, they sold the machine for four-thirds of its cost, in order that it might be re-erected at Coney Island.

No sooner had the great Ferris wheel at Chicago proved a financial success than an American gentleman, Lieutenant Graydon, secured a patent for a like machine in the United Kingdom ; and as it has now become almost a matter of course that some iron or steel structure, surpassing everything before attempted, should form a part of each great exhibition, a Company was at once formed in London, under the title of "The Gigantic Wheel and Recreation Towers Co., Limited," to construct and work at the Earl's Court Oriental Exhibition of 1895, a great wheel, similar in general form to that of Chicago. But the design of the London wheel had some new features, as will be seen from the sketches, Fig. 26c (from *The Engineer* of 20th April, 1894), and, moreover, having been planned of larger dimensions than its American prototype, presented additional engineering problems of no small complexity. After due deliberation the scheme of the work was entrusted to Mr. Walter B. Bassett, a talented young engineer, connected with the firm of Messrs. Maudslay, Sons, & Field, and already experienced in designing iron structures. Under this gentleman, with the assistance of Mr. J. J. Webster in carrying out some of the details, the work has been so successfully accomplished that the "Great Wheel" of 1895 may be cited as one of the crowning mechanical triumphs of the nineteenth century. The original design has not been followed so far as regards the lower platforms for refreshment rooms, &c. Plate V., for which we are indebted to Mr. Basset, is a photographic representation of the actual structure.

The wheel at Earl's Court exceeds the Ferris wheel in diameter by 50 feet, being 300 feet across. It is supported on two towers, 175 feet high, each formed by four columns 4 feet square, built of steel plates with internal diaphragms, and surmounted by balconies that may be ascended in elevators raised by a weight of water, which, after having been discharged into a reservoir under the ground level, is again pumped up to the top of the towers. Between the balconies on each tower there is also a communication *through the axle* of the wheel, which, instead of being solid as at Chicago, is a tube of 7 feet diameter, and 35 feet long, made in sections, riveted together, of steel 1 inch thick, and weighing no less than 58 tons. The raising and fixing in its high place of such a mass of metal required specially ingenious devices, which have been greatly appreciated by professional engineers. But for these devices, the erection of scaffolding in the ordinary way of proceeding would have entailed an outlay simply enormous. The axle is stiffened by projecting rings, and, between pairs of these, the spoke rods are attached by pins 3 inches in diameter. The axle was the production of Messrs. Maudslay, Field & Co. ; all the rest of the metal work was made at the Arrol Works at

Glasgow, and the carriages were constructed by Brown, Marshall & Co., of Birmingham. The Earl's Court wheel is turned by a mechanism different from that of the Chicago wheel, for whereas the latter was provided with cogs, the former has two chains, each 1,000 feet long and 8 tons weight, surrounding the periphery of the wheel on either side. The chains go over drums in the engine-shed, from which they pass underground to guide-pulleys, and as they unwind from the Great Wheel, they again go over guide-pulleys to lead them back to the drums. These chains are firmly held throughout in the jaws of V-shaped grooves, and there are arrangements for taking up the slack. The drums are actuated by wheel gearing, connected with two horizontal Robey steam engines, each of 50 horse-power, one on either side, capable of being worked singly or together. It is, however, found sufficient to use the engine of one side only, and even then to work it at but 16 horse-power, and the operation can be controlled by one man, who has also the command of a brake. Both starting and stopping are accomplished with the greatest smoothness and absence of strain or jar. There are forty carriages, each 25 feet long, 9 feet wide, and 10 feet high. Each will accommodate forty passengers, and these enter at the ends from eight platforms at different heights from the ground, so arranged as to be on the level of the eight lowest carriages while the wheel is stationary. The passengers who have had their ride leave at the other end of the carriages by eight similar platforms on the other side of the wheel. After the change of passengers in one set of eight carriages, the wheel is turned through exactly one-fifth of a revolution, which has the effect of bringing the next eight carriages to the level of the platforms, and it is again brought to a standstill whilst the change of passengers is taking place ; and so on, until the whole freight of say 1,600 persons has been changed during the five stoppages in one revolution, for which about thirty-five minutes are required, and the process of emptying and filling eight carriages at once is repeated. There are first and second class carriages, the charge for the former being two shillings, and for the latter one shilling ; so that, reckoning 800 passengers of each class, one turn would bring to the treasury the handsome sum of £120.

The sensations experienced in a journey on the Great Wheel are, as already mentioned, comparable to those enjoyed by the aërial voyagers in a balloon, where all perception of proper motion is lost, and it is the world beneath that seems to recede and float away, presenting the while a strangely changing panorama. Many people who have never made a balloon ascent yet know the calm delight of floating in a boat without effort down some placid stream, unconscious of any motion beyond that vaguely inferred from the silent apparent gliding by of the banks. Very similar are, in part, the feelings of the passenger who is almost imperceptibly carried up into the air in a carriage of the Great Wheel, but the vertical direction of the movement, and the gradual expansion of the horizon as the vertex is approached, lend an unwonted novelty to the situation. From the Earl's Court Wheel the view is both interesting and extensive, for on a clear day the prospect stretches as far as the Royal Castle of Windsor.

The "Gigantic Wheel" at Earl's Court was inaugurated on the 11th July, 1895, in the presence of an assemblage of 5,000 people, including many distinguished personages, who were all treated to a ride. Plate I. shows a portion of the wheel and carriages as in motion.

PLATE V.

GENERAL VIEW OF THE GREAT WHEEL AT EARL'S COURT.

FIG. 27.—*Sir Joseph Whitworth.*

TOOLS.

O F the immense variety of tools and mechanical contrivances employed in modern times, by far the greatest number are designed to impart to certain materials some definite shape. The brickmaker's mould, the joiner's plane, the stonemason's chisel, the potter's wheel, are examples of simple tools. More elaborate are the coining press, the machine for planing iron, the drilling machine, the turning lathe, the rolling mill, the Jacquard loom. But all such tools and machines have one principle in common—a principle which casual observers may easily overlook, but one which is of the highest importance, as its application constitutes the very essence of the modern process of *manufacture* as distinguished from the slow and laborious mode of making things by hand. The principle will be easily understood by a single example. Let it be required to draw straight lines across a sheet of paper. Few persons can take a pen or pencil, and do this with even an approach to accuracy, and at best they can do it but slowly and imperfectly. But with the aid of a ruler any number of straight lines may be drawn rapidly and surely. The former case is an instance of *making* by hand, the latter represents *manufacturing*, the ruler being the tool or machine. Let it be observed that the ruler has in itself the kind of form required—that is to say, straightness—and that in using it we copy or transfer this straightness to the mark made on the paper. This is a

simple example of the *copying principle,* which is so widely applied in
machines for manufacturing; for, in all of these, materials are shaped or
moulded by various contrivances, so as to reproduce certain definite forms,
which are in some way contained within the machine itself. This will be
distinctly seen in the tools which are about to be described.

Probably no one mechanical contrivance is so much and so variously
applied as the *Screw.* The common screw-nail, which is so often used by
carpenters for fastening pieces of metal on wood, or one piece of wood to
another, is a specimen of the screw with which everybody is familiar. The
projection which winds spirally round the nail is termed the *thread* of the
screw, and the distance that the thread advances parallel to the axis in
one turn is called the *pitch.* It is obvious that for each turn the screw
makes it is advanced into the wood a depth equal to the pitch, and that
there is formed in the wood a hollow screw with corresponding grooves

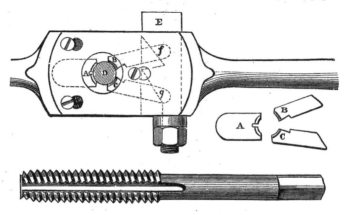

FIG. 28.—*Whitworth's Screw Dies and Tap.*

and projections. Screws are formed on the ends of the bolts, by which
various parts are fastened together, and the hollow screws which turn on
the ends of the bolts are termed *nuts.* The screws on bolts and nuts, and
other parts of machines, were formerly made with so many different pitches
that, when a machine constructed by one maker had to be repaired by
another, great inconvenience was found, on account of the want of uniformity in the shape and pitch of the threads. A uniform system was many
years ago proposed by Sir Joseph Whitworth, and adopted by the majority
of mechanical engineers, who agreed to use only a certain defined series of
pitches. The same engineer also contrived a hand tool for cutting screws
with greater accuracy than had formerly been attained in that process.
A mechanic often finds it necessary to form a screw-thread on a bolt, and
also to produce in metal a hollow screw. The reader may have observed
gasfitters and other workmen performing the first operation by an instrument having the same general appearance as Fig. 28. This contains
hard steel *dies,* which are made to press on the bolt or pipe, so that when
the *guide-stock* is turned by the handles, the required grooves are cut out.

The arrangement of these dies in Sir Joseph Whitworth's instrument is shown in Fig. 28, which represents the central part of the guide-stock: A, B, C are the steel dies retained in their places, when the instrument is in use, by a plate which can be removed when it is necessary to replace one set of dies by another, according to the pitch of thread required. The figure also shows the set of dies, A, B, C, removed from the guide-stock. D is the work, pressed up against the fixed die, A, by B and C, the pressure being applied to these last as required by turning the nut, thus drawing up the key, E, so that the inclined planes, *f*, *g*, press against similar surfaces forming the ends of the dies. For producing the hollow screws, *taps* are provided, which are merely well-formed screws, made of hard steel and having the threads cut into detached pieces by several longitudinal grooves, as represented in the lower part of Fig. 28.

The method of forming screws by dies and taps is, however, applicable only to those of small dimensions, and even for these it is not employed

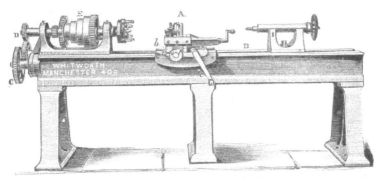

FIG. 29.—*Screw-cutting Lathe.*

where great accuracy is required. Perfect screws can only be cut with a lathe, such as that represented in Fig. 29. In this we must first call the reader's attention to the portion of the apparatus marked A, which receives the name of the *slide-rest*. The invention of this contrivance by Maudsley had the effect of almost revolutionizing mechanical art, for by its aid it became possible to *produce true surfaces in the lathe*. Before the slide-rest was introduced, the instrument which cut the wood or metal was held in the workman's hand, and whatever might be his skill and strength, the steadiness and precision thus obtainable were far inferior to those which could be reached by the grip of an iron hand, guided by unswerving bars. The slide-rest was contrived by Maudsley in the first instance for cutting screws, but its principle has been applied for other purposes. This principle consists in attaching the cutting tool to a slide which is incapable of any motion, except in the one direction required. Thus the slide, A, represented in Fig. 29, moves along the *bed* of the lathe, B, carrying the cutter with perfect steadiness in a straight line parallel to the axis of the lathe. There are also two other slides for adjusting the position of the cutter; the handle, *a*, turns a screw, which imparts a transverse motion to the piece, *b*. and the tool receives another longitudinal movement from the handle, *c.* The pieces are so arranged that these movements take place in straight lines in precisely the required direction, and without permitting the tool to be unsteady, or capable of any rocking motion. In Whitworth's lathe, between the two sides of the bed, and therefore not visible in the figure, is a shaft placed perfectly parallel to the axis of the lathe. One end of this shaft is seen carrying the wheel, C, which is connected with a train of wheels, D, and is thus made to revolve at a speed which can be made to bear any required proportion to that of the mandril, E, of the lathe, by properly arranging the numbers of the teeth in the wheels; and the machine is provided with several sets of wheels, which can be substituted for each other. The greater part of the length of this shaft is formed with great care into an exceedingly accurate screw, which works in a nut forming part of the slide-rest. The effect, therefore, of the rotation of the screw is to cause the slide-rest to travel along the bed of the lathe, advancing with each revolution of the screw through a space equal to its pitch distance. There is an arrangement for releasing the nut from the guiding-screw, by moving a lever, and then by turning the winch the slide-rest is moved along by a wheel engaging the teeth of a rack at the back of the lathe. Now, if the train of wheels, C D, be so arranged that the screw makes one revolution for each turn of the mandril, it follows that the cutting tool will move longitudinally a distance equal to the pitch of the guiding-screw while the bar placed in the lathe makes one turn. Thus the point of the cutter will form on the bar a screw having the same pitch as the guiding-screw of the lathe.

Here we have a striking illustration of the copying principle, for the lathe thus produces an exact copy of the screw which it contains. The screw-thread is traced out on the cylindrical bar, which is operated upon by the combination of the circular motion of the mandril with the longitudinal movement of the slide-rest. By modifying the relative amounts of these movements, screw-threads of any desired pitch can be made, and it is for this purpose that the *change wheels* are provided. If the thread of the guiding-screw makes two turns in one inch, one revolution of the wheel C will advance the cutter half an inch along the length of the bar. If the numbers of teeth in the wheels be such that the wheel D makes ten revolutions while C is making one, then in the length of half an inch the thread of the screw produced by the cutter will go round the core ten times, or, in technical language, the screw will be of $\frac{1}{20}$ inch pitch.

Since a screw turning in a nut advances only its pitch distance at each revolution, a finely-cut screw furnishes an instrument well adapted to impart a slow motion, or to measure minute spaces. Suppose a screw is cut so as to have fifty threads in an inch, then each turn will advance it $\frac{1}{50}$ in.; half a turn $\frac{1}{100}$ in.; a quarter of a turn, $\frac{1}{200}$, and so on. It is quite easy to attach a graduated circle to the head of the screw, so that, by a fixed pointer at the circumference, any required fraction of a revolution may be read off. Thus if the circle had two hundred equal parts, we could, by turning the screw so that one division passed the index, cause the screw to advance through $\frac{1}{200}$ of $\frac{1}{50}$ inch, or $\frac{1}{10000}$ part of an inch. This is the method adopted for moving the cross-wires of the instruments for measuring very small spaces under the microscope. Sir Joseph Whitworth, who has done so many great things in mechanical art, was the first mechanician to perceive the importance of extreme accuracy of workmanship, and he invented many beautiful instruments and processes by which this accuracy might be attained. Fig. 30 represents one of his measuring machines, intended for practical use in the workshop, to test the dimensions of pieces of metal where great precision is required. The base of the machine is constructed of a rigid cast iron bed bearing a fixed headstock, A, and a movable one, B, the latter sliding along the bed, C, with a slow movement, when the handle, D, is turned. This slow motion is produced by a screw on the axis, *a*, working in the lower part of the headstock, just as the slide-rest is moved along the bed of the lathe. The movable headstock, when it has been moved into the position required, is firmly clamped by a thumb-screw. The face of the bed is graduated into inches and their subdivisions. Here it should be explained that the machine is not intended to be used for ascertaining the absolute dimensions of objects, but for showing by what

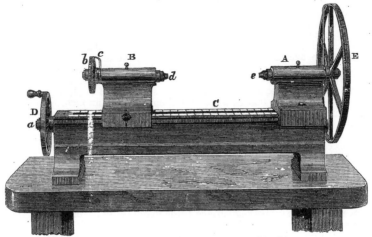

FIG. 30.—*Whitworth's Measuring Machine.*

fraction of an inch the size of the work measured differs from a certain standard piece. Each headstock carries a screw of $\frac{1}{20}$ inch pitch, made with the greatest possible care and accuracy. To the head of the screw in the movable headstock is attached the wheel, *b*, having its circumference divided into 250 equal parts, and a fixed index, *c*, from which its graduations may be counted. An exactly similar arrangement is presented in connection with the screw turning in the fixed headstock, but the wheel is much larger, and its circumference is divided in 500 equal parts. It follows, therefore, that if the large wheel be turned so that one division passes the index, the bar moves in a straight line $\frac{1}{500}$ of the $\frac{1}{20}$ of an inch, that is, $\frac{1}{10000}$ of an inch. The ends of the bars, *d* and *e*, are formed with perfectly plane and parallel surfaces, and an ingenious method is adopted of securing equality of pressure when comparisons are made. A plate of steel, with perfectly parallel faces, called a *gravity-piece*, or *feeler*, is placed between the flat end of the bar and the standard-piece, and the pressure when the screw-reading is taken must be just sufficient to prevent this piece of steel from slipping down, and that is the case when the steel remains suspended and can nevertheless be easily made to slide about by a touch of the finger. Thus any piece which, with the same screw-readings, sustains the gravity-piece in the same manner as the standard, will be of exactly the same length; or the number of divisions through which the large wheel must be turned to enable it to do so tells the difference of the dimensions in ten-thousandth parts of an inch. By this instrument, therefore, gauges, patterns, &c., can be verified with the greatest precision, and pieces can be reproduced perfectly agreeing in their dimensions with a standard piece. Thus, for example, the diameters of shafting can be brought with the greatest precision to the exact size required to best fit their bearings.

In another measuring machine on the same principle the delicacy of the measurement has been carried still farther, by substituting for the large divided wheel one having 200 teeth, which engage an endless screw or worm. This will easily be understood by reference to Fig. 31, where a similar arrangement is applied to another purpose. Imagine that a wheel like P, Fig. 31, but with 200 teeth, has taken the place of E in Fig. 30, and that the wheel, T, on the axis of the endless screw is shaped like E, Fig. 30. One turn of the axis carrying the endless screw, therefore, turns the wheel through $\frac{1}{200}$ of a revolution, and as this axis bears a graduated head, having 250 divisions, the screw having 20 threads to the inch, is, when one division passes the index, advanced through a space equal to $\frac{1}{250} \times \frac{1}{200} \times \frac{1}{20}$, or $\frac{1}{1000000}$ of an inch; that is, the one-millionth part of an inch. This is an interval so small that ten times its length would hardly be appreciated with the highest powers of the microscope, and the machine is far too delicate for any practical requirements of the present day. It will indicate the expansion caused by heat in an iron bar which has merely been touched with the finger for an instant, and even the difference of length produced by the heat radiated from the person using it. A movement of $\frac{1}{1000000}$ of an inch is shown by the gravity-piece remaining suspended instead of falling, and the piece falls again when the tangent-screw is turned back through $\frac{1}{250}$ of a revolution, a difference of reading representing a possible movement of the measuring surface through only $\frac{3}{1000000}$ of an inch. This proves the marvellous perfection of the workmanship, for it shows that the amount of play in the bearings of the screws does not exceed one-millionth of an inch.

A good example of a machine-tool is the *Drilling Machine,* which is used for drilling holes in metal. Such a machine is represented in Fig. 31, where A is the strong framing, which is cast in a single piece, in order to render it as rigid as possible. The power is applied by means of a strap round the speed pulley, B, by which a regulated speed is communicated to the bevel wheel, C, which drives D, and thus causes the rotation of the hollow shaft, E. In the lower part of the latter is the spindle which carries the drilling tool, F, and upon this spindle is a longitudinal groove, into which fits a projection on the inside of E. The spindle is thus forced to rotate, and is at the same time capable of moving up and down. The top of the spindle is attached to the lower end of the rack, G, by a joint which allows the spindle to rotate freely without being followed in its rotation by the rack, although the latter communicates all its vertical movements to the spindle, as if the two formed one piece. The teeth of the rack are engaged by a pinion, which carries on its axis the wheel H, turned by an endless screw on the shaft, I, which derives its motion by means of another wheel and endless screw from the shaft, K. The latter is driven by a strap passing over the *speed pulleys*, L and M, and thus the speed of the shaft K can be modified as required by passing the strap from one pair of pulleys to another. The result is that the rack is depressed by a slow movement, which advances the drill in the work, or, as it is technically termed, gives the *feed* to the drill. By a simple piece of mechanism at N the connection of the shafts K and I can be broken, and the handle O made to communicate a more rapid movement to I, so as to raise up the drill in a position to

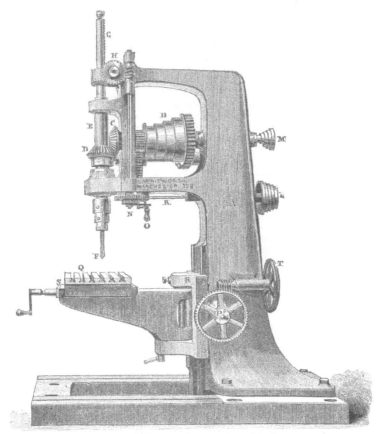

FIG. 31.—*Whitworth's Drilling Machine.*

begin its work again, or to bring it quickly down to the work, and then the arrangement for the self-acting feed is again brought into play. By turning the wheel, P, the table, Q, on which the work is fastened, is capable of being raised or lowered, by means of a rack within the piece R, acted on by a pinion carried on the axle, P. The table also admits of a horizontal motion by the slide S, and may besides be swung round when required.

The visitor to an engineer's workshop cannot fail to be struck with the operation of the powerful *Lathes* and *Planing Machines,* by which long thicl flakes or shavings of iron are removed from pieces of metal with the same apparent ease as if the machine were paring cheese. The figure on the opposite page represents one of the larger forms of the planing machine as constructed by Sir J. Whitworth. The piece of work to be planed is firmly bolted down to the table, A, which moves upon the V-shaped surfaces running its whole length, and accurately fitting into corresponding grooves in a massive cast iron bed. The bevel wheel, of which a portion is seen a B, is keyed on a screw, which extends longitudinally from end to end o the bed. This screw works in nuts forming part of the table, and as i turns in sockets at the ends of the bed, it does not itself move forward, bu imparts a progressive movement to the table, and therefore to the piece o metal to be planed. As this table must move backwards and forwards there must be some contrivance for reversing the direction of the screw's rotation, and this is accomplished in a beautifully simple manner by ar arrangement which a little consideration will enable any one to understand It will be observed that there are three drum-pulleys at C. Let the reade: confine, for the present, his attention to the nearest one, and picture to him self that the shaft to which it is attached is placed in the same horizonta plane as the axis of the screw and at right angles to it, passing in front o bevel wheel B. A small bevel wheel turning with this shaft, and engaging the teeth of the wheel B, may, it is plain, communicate motion to the screw Now let the reader consider what will be the effect on the *direction* of the rotation of B of applying the bevelled pinion to the nearer or to the farthei part of its circumference, supposing the direction of the rotation of thi: pinion to be always the same. He will perceive that the direction in one case will be the reverse of that in the other. The shaft to which the neares pulley is attached carries a pinion engaging the wheel at its farther edge and therefore the rotation of this pulley in the same direction as the hand: of a watch causes the wheel B to rotate so that its upper part moves toward: the spectator. The farthest pulley, *a,* turns with a hollow shaft, through which the shaft of the nearest pulley simply passes, without any connectior between them, and this hollow shaft carries a pinion, which engages the teeth of B at the nearer edge, and, in consequence, the rotation of the farther pulley, *a,* in the direction of the hands of a watch, would cause the upper part of B to be moving from the spectator. The middle pulley, *b* runs loosely on the shaft, and the driving-strap passes through the guide *c,* and it is only necessary to move this, so as to shift the strap from one drum to another, in order to reverse the direction of the screw and the motion. This shifting of the strap is done by a movement derived from the table itself, on which are two adjustable stops, D and E, acting on ar arrangement at the base of the upright frame when they are brought up to it by the movement of the table, so as not only to shift the strap, but also to impart a certain amount of rotation to upright shaft, F, in each directior alternately. The piece which carries the tools, G and H, is placed horizon-tally, and can be moved vertically by turning the axis, I, thus causing ar equal rotation of two upright screws of equal pitch, which are contained within the uprights and work in nuts, forming part of the tool-box. The pieces carrying the tools are moved horizontally by the screws which are seen to pass along the tool-box, and these screws receive a certain regu-lated amount of motion at each reversal of the movement of the table from the mechanism shown at K. The band-pulley, L, receives a certain amount of rotation from the same shaft, and the catgut band passing round the tops

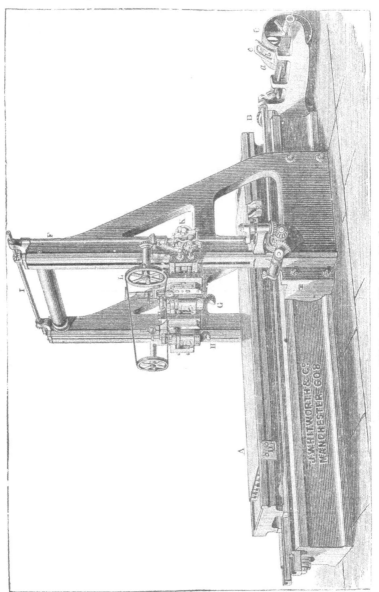

FIG. 32.—*Whitworth's Planing Machine*

of the cylinders which carry the cutters is drawn in alternate directions at the end of each stroke, the effect being to turn the cutters half round, so as always to present their cutting edges to the work. There are also contrivances for maintaining the requisite steadiness in the tools and for adjusting the depth of the cut. The cutting edge of the tools is usually of a V-shape, with the angle slightly rounded, and the result of the process is not the production of a plane, but a grooved surface. But by diminishing the amount of horizontal *feed* given to the cutters, the grooves may be made finer and finer, until at length they disappear, and the surface is practically a plane. Planing machines are sometimes of a very large size. Sir J. Whitworth has one the table of which is 50 ft. in length, and the machine is capable of making a straight cut 40 ft. long in any article not exceeding 10 ft. 6 in. high or 10 ft. wide.

The copying principle is evident in this machine; for the plane surface results from the combination of the straightness of the bed with the straightness of the tranverse slide along which the tools are moved. It should, moreover, be observed that it is precisely this machine which would be employed for preparing the straight sliding surfaces required in the construction of planing and other machines, and thus one of these engines becomes the

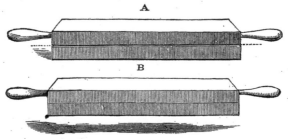

FIG. 33.—*Pair of Whitworth's Planes, or Surface Plates.*

parent, as it were, of many others having the same family likeness, and so on *ad infinitum.* Thus, having once obtained perfectly true surfaces, we can easily reproduce similar surfaces. But the reader may wish to know how such forms have been obtained in the first instance; how, for example, could a perfectly plane surface be fashioned without any standard for comparison? This was first perfectly done by Sir J. Whitworth, forty-five years ago. Three pieces of iron have each a face wrought into comparatively plane surfaces; they are compared together, and the parts which are prominent are reduced first by filing, but afterwards, as the process approaches completion, by scraping, until the three perfectly coincide. The parts where the plates come in contact with each other are ascertained by smearing one of them with a little oil coloured with red ochre: when another is pressed against it, the surfaces of contact are shown by the transference of the red colour. Three plates are required, for it is possible for the prominences of No. 1 exactly to fit into the hollows of No. 2, but in that case *both* could not possibly exactly coincide with the surface of No. 3; for if one of them did (say No. 1), then No. 3 must be exactly similar to No. 2, and consequently when No. 2 was applied to No. 3, hollow would be opposed to hollow and prominence to prominence. A little reflection will show that only when the three surfaces are truly plane will they exactly and entirely coincide with each other. The planes, when thus carefully prepared, approach to the perfection of the ideal mathematical form, and they are used in the workshop for testing the correctness of surfaces, by observing the uniformity or otherwise of the impression they give to the surface when brought into contact with it, after being covered by a very thin layer of oil coloured by finely-ground red ochre.

Fig. 33 represents a small pair of Whitworth's planes. When one of these is placed horizontally upon the other, it does not appear to actually come in contact with it, for the surfaces are so true that the air does not easily escape, but a thin film supports the upper plate, which glides upon it with remarkable readiness (A). When, however, one plate is made to slide over the other, so as to exclude the air, they may both be lifted by raising the upper one (B). This effect has, by several philosophers, been attributed to the mere pressure of the atmosphere; but recent experiments of Professor Tyndall's show that the plates adhere even in a vacuum. The adhesion appears therefore to be due to some force acting between the substances of the plates, and perhaps identical in kind with that which binds together the particles of the iron itself.

FIG. 34.—*Interior of Engineer's Workshop.*

FIG. 35.— *The Blanchard Lathe.*

THE BLANCHARD LATHE.

THIS machine affords a striking example of the application of the copying principle which is the fundamental feature of modern manufacturing processes. It would hardly be supposed possible, until the method
had been explained, that articles in shape so unlike geometrical forms as
gun-stocks, shoemakers' lasts, &c., could be turned in a lathe. The mode
in which this is accomplished is, however, very simple in idea, though in
carrying that idea into practice much ingenious contrivance was required.
The illustration, Fig. 35, represents a Blanchard's lathe, very elegantly constructed by Messrs. Greenwood and Batley, of Leeds. The first obvious
difference between an ordinary lathe and Blanchard's invention is that in
the former the work revolves rapidly and the cutting-tool is stationary, or
only slowly shifts its position in order to act on fresh portions of the work,
while in the latter the work is slowly rotated and the cutting-tools are made to
revolve with very great velocity. Again, it will be observed that the headstock of the Blanchard lathe, instead of one, bears *two* mandrels, having
their axes parallel to each other. One of these carries the pattern, c, which
in the figure has the exact shape of a gun-stock that is to be cut in the piece
of wood mounted on the nearer spindle. One essential condition in the
arrangement of the apparatus is that the pattern and the work having been
fixed in similar and parallel positions, shall always continue so at every point
of their revolutions. This is easily accomplished by placing exactly similar
toothed wheels on the two axles, and causing these to be turned by one and
the same smaller toothed wheel or pinion. The two axles must thus always
turn round in the same direction and with exactly the same speed, so that the
work which is attached to one, and the pattern which is fixed on the other,
will always be in the same phase of their revolutions. If, for example, the
part of the wood which is to form the upper part of the gun-stock is at the

bottom, the corresponding part of the pattern will also be at the bottom, as in the figure, and both will turn round together, so that every part of each will be at every instant in a precisely similar position. The wood to be operated upon is, it must be understood, roughly shaped before it is put into the lathe. The toothed wheels and the pinion which drives them are in the figure hid from view by the casing, *h*, which covers them. The pinion receives the power from a strap passing over *f*. The cutters are shown at *e*; they are placed radially, like the spokes of a wheel, and have all their cutting edges at precisely one certain distance from the axis on which they revolve, so that they all travel through the same circle. These cutting-edges, it may be observed, are very narrow, almost pointed. The shaft carrying the cutters is driven at a very high speed, by means of a strap passing over *k* and *i*. The number of revolutions made by the cutters in one second is usually more than thirty. The great peculiarity of the lathe consists in the manner in which the position of the cutters is made to vary. The axle which carries them rotates in a kind of frame, which can move backwards and forwards, so that the cutters may be readily put at any desired distance from the axis of the work. Their position is, however, always dependent on the pattern, for, fixed in a similar frame, *b*, which is connected with the former, is a small disc wheel, *a*, having precisely the same radius as that of the circle traced out by the cutters, and this disc is made by a strong spring to press against the pattern. The cutters, being fixed in the same rocking-frame which carries this guiding-wheel, must partake of all its backward and forward motions, and as the cutting-wheel and the guide-wheel are so arranged as to have always the same relative positions to the axes of the two headstocks, it follows that the edges of the cutters will trace out identically the same form as the circumference of the guiding-disc. The latter is, of course, not driven round, but simply turns slowly with the pattern by friction, for it is pressed firmly against the pattern by a spring or weight acting on the frame, in order that the cutters may be steadily maintained in their true, but ever-varying, position. The rocking-frame receives a slow longitudinal motion by means of the screw, *n*, so that the cutters are carried along the work, and the guide along the pattern.

The whole arrangement is self-acting, so that when once the pattern and the rough block of wood have been fixed in their positions, the machine completes the work, and produces an exact repetition of the shape of the pattern. It is plain that any kind of forms can be easily cut by this lathe, the only condition being that the surface of the pattern must not present any re-entering portions which the edge of the guide-wheel cannot follow. The machine is largely used for the purposes named above, and also for the manufacture of the spokes of carriage-wheels. The limits of this article will not permit of a description of the beautiful adjustments given to the mechanism in the example before us, particularly in the arrangement for driving the cutters in a framework combining lateral and longitudinal motions; but the intelligent reader may gather some hints of these by a careful inspection of the figure. The machine is sometimes made with the frame carrying the guide-wheel and cutters, not rocking but sliding in a direction transverse to the axes of the head-stocks. It is extremely interesting to see the Blanchard lathe at work, and observe how perfectly and rapidly the curves and form of the patterns seem to grow, as it were, out of the rudely-shaped piece of wood, which, of course, contains a large excess of material, or, in the picturesque and expressive phrase of the workmen, *always gives the machine something to eat.*

7

FIG. 36.—*Vertical Saw.*

SAWING MACHINES.

WITH the exception of the last, all the machines hitherto described in the present article are distinguished by this—they are tools which are used to produce other machines of every kind. Without such implements it would be impossible to fashion the machines which are made to serve so many different ends. Another peculiarity of these tools has also been referred to, namely, that they are especially serviceable, and indeed essential, for the reproduction of others of the same class. Thus, the accurate leading-screw of the lathe is the means used to cut other accurate screws, which shall in their turn become the leading-screws of other lathes, and a lathe which forms a truly circular figure is a necessary implement for the construction of another lathe which shall also produce truly circular figures. In these tools, therefore, we find the copying principle, to which allusion has been already made, as the great feature of all machines; but in order to bring this principle still more clearly before the reader, we have described in the Blanchard Lathe a machine of a somewhat different class, because it embodied a very striking illustration of the principle in question. We are far from having described all the implements of the mechanical engineer, or even all the more interesting ones; for example, we have given no account of the powerful lathes in which great masses of iron are turned, or of the analogous machines, which, with so much accuracy, shape the internal surfaces of the cylinders of steam engines, of cannons, &c. The history of the steam engine tells us of the difficulties which Watt had to contend with in the construction of his cylinders, for no machine at that time existed capable of boring them with an approach to the precision which is now obtained.

The kind of general interest which attaches to the tools we have already described is not wanting in yet another class of machine-tools, namely, those employed in converting timber into the forms required to adapt it for the uses to which it is so extensively applied. And for popular illustration, this class of tools presents the special advantage of being readily understood as regards their purpose and mode of action, while their sim-

FIG. 37.—*Circular Saw.*

plicity in these respects does not prevent them from showing the advantages of machine over hand labour. Everybody is familiar with the up-and-down movement of a common saw, and in the machine for sawing balks of timber into planks, represented in Fig. 36, this reciprocating motion is retained, but there are a number of saws fixed parallel to each other in a strong frame, at a distance corresponding to the thickness of the planks. The saws are not placed with their cutting edges quite upright, but these are a little more forward at the top, so that as they descend they cut into the wood, but move upwards without cutting, for the teeth then recede from the line of the previous cut, while in the meantime the balk is pushed forward ready for the next descent of the saw-frame. This pushing forward, or *feeding*, of the timber is accomplished by means of ratchet-wheels, which are made to revolve through a certain space after each descent of the saw-frame, and, by turning certain pinions, move forward the carriage on which the piece of timber is firmly fixed, so that when the blades of the saws are beginning the next descent they are already in contact with the edge of the former cut. To prevent the blades from moving with injurious friction in the saw-cuts, these last are made of somewhat greater width than the thickness of the blades, by the simple plan of bending the teeth a little on one side and on the other alternately. The rapidity with which the machine works, depends of course on the kind of wood operated upon, but it is not unusual for such a machine to make more than a hundred cuts in the minute. The figure shows the machine as deriving its motion by means of a strap passing over a drum, from shafting driven by a steam engine. This is the usual plan, but sometimes the steam power is applied directly, by fixing the piston-rod of a steam cylinder to the top of the saw-frame, and equalizing the motion by a fly-wheel on a shaft, turned by a crank and connecting-rod.

A very effective machine for cutting pieces of wood of moderate dimensions is the *Circular Saw*, represented in Fig. 37. Here there is a steel disc, having its rim formed into teeth ; and the disc is made to revolve with very great speed, in some cases making as many as five hundred turns in a minute, or more than eight in a second. On the bench is an adjustable straight guide, or fence, and when this has been fixed, the workman has only to press the piece of wood against it, and push the wood at the same time towards the saw, which cuts it at a very rapid rate. Sometimes the circular saw is provided with apparatus by which the machine itself pushes the wood forwards, and the only attention required from the workman is the fixing of the wood upon the bench, and the setting of the machine in gear with the driving-shaft. Similar saws are used for squaring the ends of the iron rails for railways, two circular saws being fixed upon one axle at a distance apart equal to the length of the rails. The axle is driven at the rate of about 900 turns per minute, and the iron rail is brought up parallel to the axle, being mounted on a carriage, and still red hot, when the two ends are cut at the same time by the circular saws, the lower parts of which dip into troughs of water to keep them cool.

FIG. 38.—*Pit-Saw.*

FIG. 39.—*Box Tunnel.*

RAILWAYS.

TOWARDS the end of last century, tramways formed by laying down narrow plates of iron, were in use at mines and collieries in several parts of England. These plates had usually a projection or flange on the inner edge, thus— ∟, in order to keep the waggons on the track, for the wheels themselves had no flange, but were of the kind used on ordinary roads. These flat tramways were found liable to become covered up with dirt and gravel, so that the benefit which ought to have been obtained from their smoothness was in a great measure lost. *Edge rails* were, therefore, substituted, and the wheels were kept on the rails by having a *flange* cast on the inner edge of the rim. The rails were then always made of cast iron, for, although they were very liable to break, the great cost of making them of wrought iron prevented that material from being used until 1820, when the method of forming rails of malleable iron by rolling came into use. The first time a tramway was used for the conveyance of passengers was in 1825, when the Stockton and Darlington Railway was opened—a length of thirty-seven miles. It appears that the carriages were at first drawn by horses, although locomotives were used on this and other colliery lines for dragging, at a slow rate, trains of mineral waggons. At that time engineers were exercising their ingenuity in overcoming a difficulty which never existed by devising plans for giving tractive power to the locomotive through the instrumentality of rack-work rails. It never occurred to them to first try whether the adhesion of the smooth wheel to the smooth rail was not sufficient for the purpose. During the first quarter of the present century the greater part of the goods and much passenger traffic was monopolized by the canals. It is quoted, as a proof of the careless manner in which this service was performed, that the transport of bales of cotton from Liverpool to Manchester sometimes occupied twice the length of time required in their voyage across the Atlantic. When an Act of Parliament authorizing the construction of a railway between Liverpool and Manchester

FIG. 40.—*Coal-pit, Salop.*

was applied for, the canal companies succeeded in retarding, by their influence, the passing of that Act for two years. It was passed, however, in 1828, and the construction of the line was proceeded with. This line was at first intended only for the conveyance of goods, especially of cotton and cotton manufactures, and the waggons were to be drawn by horses. When the line was nearly finished the idea of employing horses was, at the instigation of Mr. George Stephenson, abandoned in favour of steam power. The directors were divided in opinion as to whether the carriage should be dragged by ropes wound on large drums by stationary engines, or whether locomotives should be employed. Finally, the latter plan was adopted, and it was also suggested that passengers might be carried. The directors offered a prize for the best locomotive, and the result has been already mentioned. In the light of our experience since that time, it is curious to read of the doubts then entertained by skilful engineers about the success of the locomotive. In a serious treatise on the subject, one eminent authority hoped "that he might not be confounded with those hot-brained enthusiasts who maintained the possibility of carriages being driven by a steam engine on a railway at such a speed as twelve miles an hour." When the "Rocket" had accomplished the unprecendented velocity of twenty-nine miles an hour, and the railway was opened for passengers as well as goods, the thirty stage coaches daily plying between Liverpool

FIG. 41.—*Sankey Viaduct.*

and Manchester found their occupation gone, and all ceased to run except one, which had to depend on the roadside towns only, while the daily number of passengers between the two cities rose at once from 500 to 1,600. In that delightful book, Smiles's " Life of George Stephenson," may be found most interesting details of the difficulties attending the introduction of railways, especially with regard to the construction of this first important line. Mr. Smiles relates how the promoters of the scheme struggled against "vested interests ;" how the canal proprietors, confident at first of a secure and continuous enjoyment of their monopoly, ridiculed the proposed railway, and continued their exorbitant charges and tardy conveyance, pocketing in profits the prime cost of their canal about every three years; how, roused into active opposition, they did all in their power to thwart the new scheme; how the Lord Derby and the Lord Sefton of that day, and other landowners, offered every resistance to the surveyors; how the Duke of Bridgewater's farmers would not allow them to enter their fields, and the Duke's gamekeepers had orders to shoot them; how even a clergyman threatened them with personal violence, and they had to do their work by stealth, while the reverend gentleman was conducting the services in his church; how newspaper and other writers declared that the locomotives would kill the birds, prevent cows from grazing and hens from laying, burn houses, and cause the extinction of the race of horses. All the civil engineers scouted the idea of a locomotive railway, and Stephenson was held up to derision as an ignoramus and a maniac by the "most eminent lawyers," and the most advanced and "respectable" professional C.E.s of the time. An article appeared in the " Quarterly Review," very favourable to the construction of railways, but remarking in reference to a proposed line between London and Woolwich : " What can be more palpably absurd and ridiculous than the prospect held out of locomotives travelling *twice as fast as stage coaches !* We should as soon expect the people of Woolwich to suffer themselves to be fired off upon one of Congreve's *ricochet* rockets as trust themselves to the mercy of a machine going at such a rate. We will back old Father Thames against the Woolwich Railway for any sum. We trust that Parliament will, in all railways it may sanction, limit the speed to *eight or nine miles an hour*, which we entirely agree with Mr. Sylvester is as great as can be ventured on with safety." This passage, which reads so strangely now, may be seen in the "Quarterly Review" for March, 1825. But still more curious appear now the reports of the debates in Parliament, and of the evidence taken before the Parliamentary Committee, in which we find the opinions and fears of the best informed men of that period, and trace the frantic efforts of the holders of the "vested interests" to retain them, however obstructive of the public good.

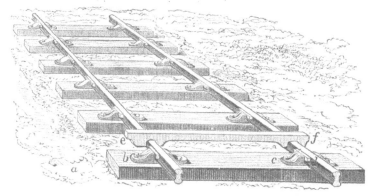

FIG. 42.—*Rails and Cramp-gauge.*

When it has been decided to construct a railway between two places, the laying-out of the line is a subject requiring great consideration and the highest engineering skill—for the matter is, on account of the great cost, much more important than the setting-out of a common road. The idea of a perfect railroad is that of a straight and level line from one terminus to another ; but there are many circumstances which prevent such an idea from being ever carried into practice. First, it is desirable that the line should pass through important towns situated near the route ; and then the cost of making the roadway straight and level, in spite of natural obstacles, would be often so great, that to avoid it detours and inclines must be submitted to—the inconvenience and the increased length of road being balanced by the saving in the cost of construction. It is the business of the engineer who lays out the line to take all these circumstances into consideration, after he has made a careful survey of the country through which the line is to pass. The cost of making railways varies, of course, very much according to the number and extent of the tunnels, cuttings, embankments, or other works required. The average cost of each mile of railway in Great Britain may be stated as about £35,000. The road itself when the rails are laid down is called *the permanent way*, perhaps originally in distinction to the temporary tramways laid down by the contractors during the progress of the works. The permanent way is formed first of

RAILWAYS. 105

ballast, which is a layer of gravel, stone, or other carefully chosen material, about 2 ft. deep, spread over the roadway. Above the ballast and partly em-bedded in it are placed the *sleepers,* which is the name given to the pieces of timber on which the rails rest. These timbers are usually placed trans-versely—that is, across the direction of the rails, in the manner shown in Fig. 42. This figure also represents the form of rails most commonly adopted, and exhibits the mode in which they are fastened down to the sleepers by means of the iron *chairs, b c,* the rail being firmly held in its place by an oak wedge, *d.* These wedges are driven in while the rails are maintained at precisely the required distance apart by the implement, *e f,* called a

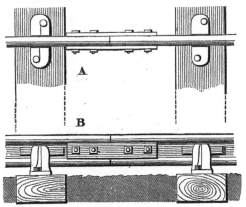

FIG. 43.—*Fish-plate.*

cramp gauge, the chairs having previously been securely attached to the sleepers by bolts or nails. The double T form of rail has several important advantages, such as its capability of being reversed when the upper surface is worn out, and the readiness with which the ends of the rails can be joined by means of *fish-plates.* These are shown in Fig. 43, where in A we are supposed to be looking down on the rails, and in B to be looking at them sideways. In Fig. 44 we have the rail and fish-plates in section. The holes in the rails through which the bolts pass are not round but oval, so that a certain amount of play is permitted to the ends of the rails.

It may easily be seen on looking at a line of rails that they are not laid with the ends quite touching each other, or, at least, they are not usually in contact. The reason of this is that space must be allowed for the ex-pansion which takes place when a rise in the temperature occurs. If the rails are laid down when at the greatest temperature they are likely to be subject to, they may then be placed in actual contact; but in cold weather a space will be left by their contraction. For this reason it is usual when rails are laid to allow a certain interval; thus rails 20 ft. long laid when the temperature is 70°, are placed with their ends $\frac{1}{20}$th of an inch apart, at 30° $\frac{1}{10}$th of an inch apart, and so on. The neglect of this precaution has some-times led to damage and accidents. A certain railway was opened in June, and after an excursion train had in the morning passed over it, the midday heat so expanded the iron, that the rails became in some places elevated 2 ft. above the level, and the sleepers were torn up; so that, in order to admit of the return of the train, the rails had to be hastily relaid in a kind of zigzag.   In June, 1856, a train was thrown off the metals of the North-Eastern Railway, in consequence of the rails rising up through expansion.

The distance between the rails in Great Britain is 4 ft. 8½ in., that width having been adopted by George Stephenson in the construction of the earlier lines. Brunel, the engineer of the Great Western, adopted, however, in the construction of that railway, a gauge of 7 ft., with a view of obtaining greater speed and power in the engines, steadiness in the carriages,

FIG. 44.—*Section of Rails and Fish-plates.*

and increased size of carriages for bulky goods.   The proposal to adopt this gauge gave rise to a memorable dispute among engineers, often called "The Battle of the Gauges."   It was stated that any advantages of the broad gauge were more than compensated by its disadvantages.   The want of uniformity in the gauges was soon felt to be an inconvenience to the public, and a Parliamentary Committee was appointed to consider the subject.   They reported that either gauge supplied all public requirements, but that the broad gauge involved a great additional outlay in its construction without any compensating advantages of economy in working ; and, as at that time 2,000 miles of railway had been constructed on the narrow gauge, whereas only 270 miles were in existence on the broad gauge, they recommended that future railways should be made the prevailing width of 56½ in.   The Great Western line had engines, bridges, tunnels, viaducts, &c., on a larger scale than any other railway in Britain.   The difference of gauge was after a time felt to involve so much inconvenience that lines which adopted the 7-ft. gauge have since relaid the tracks at the more common width.   At the present day we find the Great Western Railway completely reconstructed on the narrow gauge system, in order that trains may run without interruption in connection with other lines.

The wheels of railway carriages and engines differ from those of ordinary carriages in being fastened in pairs upon the axles, with which they revolve (see Fig. 45). The tire of the wheel is conical, the slope being about 1 in 20; that is, in a wheel 5 in. broad the radius of the outer edge is $\frac{1}{4}$ in. less than that of the inner; and the rails are placed sloping a little inwards. The effect of this conical figure is to counteract any tendency to roll off the rails; for if a pair of wheels were shifted a little to one side, the parts of the tires rolling upon the rails being then of unequal circumference, would cause the wheels to roll towards the other side. The conical shape produces

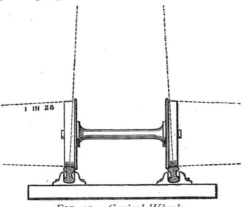

FIG. 45.—*Conical Wheels.*

this kind of adjustment so well that the flanges do not in general touch the rails. They act, however, as safeguards in passing over curves and junctions. In curves the outer line of rails is laid higher than the inner, so that in passing over them the train leans slightly inwards, in order to counteract what is called the centrifugal force, to which any body moving in a curve is subject. This so-called force is merely the result of that tendency which every moving body has to continue its motion in a straight line. A very good example of the effect of this may be seen when a circus horse is going rapidly round the ring. The inclination inwards is still more perceptible when a rider is standing on the horse's back, as shown in Fig. 46. The earth's attraction of gravity is pulling the performer straight down, and the centrifugal force would of itself throw her outwards horizontally. The resultant or combined effect of both acts is seen in the exact direction in which she is leaning, and it presses her feet on the horse's back, the animal itself being under similar conditions. It is obvious that

FIG. 46.—*Centrifugal Force.*

the amount of centrifugal force, and therefore of inward slope, will increase with the speed and sharpness of the curve, and on the railways the rails are placed so that the slope counteracts the centrifugal force when the train travels at about the rate of twenty miles per hour.

A very important part of the mechanism of a railway is the mode of passing trains from one line of rails to another. Engines and single carriages are sometimes transferred by means of *turn-tables*, but the more general plan is by *switches*, which are commonly constructed as shown in Fig. 47. There are two rails, A and B, tapering to a point and fixed at the other end, so that they have sufficient freedom to turn horizontally. A train passing in the direction shown by the arrow would continue on the main line, if the points are placed as represented; but if they be moved

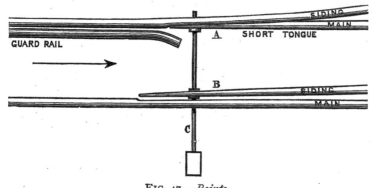

FIG. 47.—*Points.*

so that the *long tongue* is brought into contact with the rail of the main line, then the train would run on to the side rails. These *points* are worked by means of a lever attached to the rod, C, the lever being either placed near the rails, or in a *signal-box*, where a man is stationed, whose sole duty it is to attend to the points and to the signals. The interior of a signal-box near an important junction or station is shown in Fig. 48, and we see here the numerous levers for working the points and the signals, each of these having a connection, by rods or wires, with the corresponding point or signal-post. The electric telegraph is now an important agent in railway signalling, and in a signal-box we may see the bells and instruments which inform the pointsman whether a certain section of the line is "blocked" or "clear." The signals now generally used on British railways are made by the semaphore, which is simply a post from which an arm can be made to project. When the driver of the train sees the arm projecting from the left-hand side of the post, it is an intimation to him that he must stop his train; when the arm is dropped half-way, so as to project 45° from the post, it is meant that he must proceed cautiously; when the arm is down the line is clear. These signals, of course, are not capable of being seen at night, when their place is supplied with lamps, provided with coloured glasses—red and green—and also with an uncoloured glass. The lamp may have the different glasses on three different sides, and be turned round so as to present the required colour; or it may be made to do so without turning, if provided with a frame having red and green glasses, which can be moved

FIG. 48.—*Signal-box on the North London Railway*

like spectacles in front of it. The meanings of the various coloured lights and the corresponding semaphore signals are these:

| . . . White . . .	*All right* . . .	Go on.
↑ . . . Green . . .	*Caution*	Proceed slowly.
⌐ . . . Red	*Danger*	Stop.

A very clear account of the mode of working railway signals on what is now called the *block system*, together with a graphic description of a signal-box, was given in a paper which appeared some years ago in "The Popular Science Review," from the pen of Mr. Charles V. Walker, F.R.S., the telegraph engineer to the South-Eastern Railway Company, who was the first to organize an efficient system of electric signalling for railways. We may remark that the signalling instruments on the South-Eastern line, and indeed on all the lines at the present day, address themselves both to the ear and to the eye, for they consist of—first, bells, on which one, two, or more blows are struck, each series of blows having its own particular meaning; and, second, of a kind of miniature signal-post, with arms capable of being moved by electric currents into positions similar to those of the arm of an actual signal-post, so that the position of the arms is made always to indicate the state of the line. One arm of the little signal-post—the left —is red, and it has reference to *receding* trains; the other—viz., the right —arm is white, and relates to *approaching* trains. Mr. Walker thus describes the signalling

"The ordinary position of the arms of the electro-magnetic telegraph semaphores will be down; that is to say, when the line is clear of all trains, and business begins, say in early morning, all the arms will be down, indicating that no train is moving. When the first train is ready to start, say from Charing Cross, the signalman will give the proper bell-signal to Belvidere—two, three, or four blows, according as the train is for Greenwich, for North Kent, or Mid-Kent, or for the main line; and the Belvidere man will acknowledge this by one blow on the bell in reply, and without raising the Charing Cross red or left arm. This is the signal that the train may go on; and when the train has passed, so that the Charing Cross man can see the tail lights, he gives the out signal a second time, which the Belvidere man acknowledges, at the same time raising the red arm at Charing Cross, behind the train, and so protecting it until it has passed him at Belvidere, when he signals to that effect to Charing Cross, at the same time putting down the red arm there, as an indication that the line is again clear. While these operations are going on for down trains, others precisely similar, but in the reverse direction, are going on for up trains. . . . One and the same pressure on the key sends a bell signal and raises or depresses the semaphore arm as the case may require, a single telegraph wire only being required for the combined system, as for the more simple bell system." In one of the signal-boxes on the South-Eastern line, Mr. Walker states, on a certain day 650 trains or engines were signalled and all particulars accurately entered in a book, the entries requiring the writing down of nearly 8,000 figures: an illustration of the amount of work quietly carried on in a signal-box for the advantage and security of the travelling public.

Mr. Walker also gives us a peep into the inside of one of the signal-boxes, thus: "The interior of a large signal-box exhibits a very animated scene, in which there are but two actors, a man and a boy, both as busy as bees, but with no hurry or bustle. The ruling genius of the place is the strong, active, intelligent signalman, standing at one end of the apartment, the monarch for the time being of all he surveys. Immediately before him in one long line, extending from side to side, is a goodly array of levers, bright and clean from constant use and careful tending, each one labelled for its respective duty. Before him to the right and left are the various electro-magnetic semaphores, each one in full view and adjusted in position to the pair of roads to which it is appropriated, and all furnished with porcelain labels. Directly in front of him is a screen, along which are arranged the various semaphore keys; and on brackets, discreetly distributed, are the bells and gongs, the twin companions each of its own semaphore. Before the screen are the writing-desk and books, and here stands the youngster, the ministering spirit, all on the alert to take or to send electric signals and to record them, his time and attention being devoted alternately to his semaphore keys and to his books, being immediately under the eye and control of the signalman. This is no place for visitors, and the scenes enacted here have little chance of meeting the public gaze; indeed, the officers whose duties take them hither occasionally are only too glad to look on, and say as little as may be, and not interrupt the active pair, between whom there is evidently a good understanding in the discharge of duties upon the accurate performance of which so much depends. Looking on, the man will be seen in command of his rank and file: signals come, are heard and seen by both man and boy; levers are drawn and withdrawn, one, two, three, or more; the arms and the lamps on the gigantic masts outside, of which there are three, well laden, are displayed as required; distant signals are moved, points are shifted and roads made ready; telegraph signals are acknowledged; and on looking out—for the box is glazed throughout—trains are seen moving in accordance with the signals made; and on the signal-posts at the boxes, right and left—for here they are within easy reach of each other— arms are seen up and down in sympathy with those on the spot, and with the telegraph signals that have been interchanged. There is no cessation to this work, and there is no confusion in it; one head and hand directs the whole, so that there are no conflicting interests and no misunderstandings; all is done in perfect tranquillity, and the great secret is that one thing is done at a time. All this, which is so simple and so full of meaning to the expert, is to the uninitiated intricate and vague; and though he cannot at first even follow the description of the several processes, so rapidly are they begun and ended, yet, as the cloud becomes thin, and his ideas become clearer, he cannot fail to be gratified, and to be filled with admiration at the great results that are brought about by means so simple."

FIG. 49.—*Post Office Railway Van.*

Most of the carriages used on railways are so familiar to everyone that it is unnecessary to give any description of them. We give a figure of one which, though of early type, has special features of interest, being the well-designed Travelling Post Office, Fig. 49. In such vans as that here represented letters are sorted during the journey, and for this purpose the interior is provided with a counter and with pigeon-holes from end to end. When the train stops bags may, of course, be removed from or received into the van in the ordinary manner; but by a simple mechanism bags may be delivered at a station and others taken up while the train continues its journey at full speed. A bar can be made to project from the side of the carriage, and on this the bag is hung by hooks, which are so contrived that they release the bag when a rod, projecting from the receiving apparatus, strikes a certain catch on the van. The bag then drops into a netting, which is spread for its reception; and in order to receive the bags taken up, a similar netting is stretched on an iron frame attached to the van.

This frame is made to fold up against the side of the carriage when not in use.   When the train is approaching the station where the bag is to be taken up, this frame is let down, and a projecting portion detaches the bags, so that they drop into the net, from which they are removed into the interior of the vehicle.   These travelling post offices are lighted with gas, and are padded at the ends, so that the clerks may not be liable to injury by concussions of the carriages.

England has had to borrow from the United States not a few hints for such adaptations and appliances as tend to promote the comfort and convenience of travellers by rail, especially on what we insularly call long journeys.   Some of these vehicles on the American railways are luxurious hotels upon wheels ; they contain accommodation for forty persons, having a kitchen, hot and cold water, wine, china and linen closets, and more than a hundred different articles of food, besides an ample supply of tablecloths, table napkins, towels, sheets, pillow-cases, &c.   Then there are other Pullman inventions, such as the "palace" and the "sleeping" cars, in which the traveller who is performing a long journey makes himself at home for days, or perhaps for a week, as, for instance, while he is being carried across the American continent from ocean to ocean at the easy rate of twenty miles an hour on the Pacific and other connecting lines.   Mr. C. Nordhoff, an American writer, giving an account of his journey to the Western States, writes thus : " Having unpacked your books and unstrapped your wraps in your Pullman or Central Pacific palace car, you may pursue all the sedentary avocations and amusements of a parlour at home ; and as your housekeeping is done—and admirably done —for you by alert and experienced servants ; as you may lie down at full length, or sit up, sleep, or wake at your choice ; as your dinner is sure to be abundant, very tolerably cooked, and not hurried ; as you are pretty certain to make acquaintances in the car ; and as the country through which you pass is strange and abounds in curious and interesting sights, and the air is fresh and exhilarating—you soon fall into the ways of the voyage ; and if you are a tired business man or a wearied housekeeper, your careless ease will be such a rest as certainly most busy and overworked Americans know how to enjoy.   You write comfortably at a table in a little room called a 'drawing-room,' entirely closed off, if you wish it, from the remainder of the car, which room contains two large and comfortable arm-chairs and a sofa, two broad clean plate-glass windows on each side (which may be doubled if the weather is cold), hooks in abundance for shawls, hats, &c., and mirrors at every corner.   Books and photographs lie on the table.   Your wife sits at the window sewing and looking out on long ranges of snow-clad mountains or on boundless ocean-like plains.   Children play on the floor or watch at the windows for the comical prairie dogs sitting near their holes, and turning laughable somersaults as the car sweeps by.   The porter calls you at any hour you appoint in the morning ; he gives half an hour's notice of breakfast, dinner, or supper; and while you are at breakfast, your beds are made up and your room or your section aired.   About eight o'clock in the evening—for, as at sea, you keep good hours—the porter, in a clean grey uniform, comes in to make up the beds. The two easy-chairs are turned into a berth ; the sofa undergoes a similar transformation ; the table, having its legs pulled together, disappears in a corner, and two shelves being let down furnish two other berths. The freshest and whitest of linen and brightly-coloured blankets complete the outfit ; and you undress and go to bed as you would at home.

An important general truth may find a familiar illustration in the subject now under notice. The truth in question may be expressed by saying that, in all human affairs, as well as in the operations of nature, the state of things at any one time is the result, by a sort of growth, of a preceding state of things. And in this way it is certainly true of inventions, that they never make their appearance suddenly in a complete and finished state—like Minerva, who is fabled to have sprung from the brain of Jupiter fully grown and completely armed ; but rather their history resembles the slow and progressive process by which ordinary mortals attain to their full stature. We have already seen that railways had their origin in the tramways of collieries ; and, in like manner, the railway carriage grew out of the colliery truck and the stage coach ; for when railway carriages to convey passengers were first made, it did not occur to their designers that anything better could be done than to place coach bodies on the frame of the truck ; and accordingly the early railway carriages were formed by mounting the body of a stage coach, or two or three such bodies side by side, on the timber framework which was supported by the flanged wheels. The cut, Fig. 56, is from a painting in the possession of the Connecticut Historical Society, and it represents the first railway train in America on its trial trip (1831), in which sixteen persons took part, who were then thought not a little courageous. Here we see that the carriages were regular stage coaches, and the same was the case in England. But it is very significant that, to this day, the stage coach bodies are traceable in many of the carriages now running on English lines, especially in the first-class carriages, where, in the curved lines of the mouldings which are supposed to ornament the outside, one may easily recognize the forms of the curved bodies of the stage coaches, although there is nothing whatever, in the real framing of the timbers of the railway carriage, which has the most distant relation to these curves. Then again, almost universally on English lines, the old stage coach door-handles are still retained on the first-class carriages, in the awkward flat oval plates of brass which fold down with a hinge. Many other points might be named which would show the persistence of the stage coach type on the English railways. The cut, Fig. 56, proves that the Americans set out with the same style of carriages ; but North America, as compared with the Old World, is *par excellence* the country of rapid developments, and there carriages, or cars, as our Transatlantic cousins call them, have for a long time been made with numerous improvements, and in forms more in harmony with the railway system, than the conservatism of English ideas, still cleaving to the stage coach type, permitted to be attempted in this country.

Railway travellers in the United States had long enjoyed the benefit of comforts and convenience in the appointments of their carriages long before any change had been effected in the general arrangements of the vehicles provided by the railway companies in England. It is now indeed a considerable number of years since this state of things has been altered in the older country ; as all the great lines, following the example of the Midland Company, who first adopted the Pullman cars, have constructed luxurious vehicles in which every elegance and comfort are placed within the reach of the English traveller, and these improvements are highly appreciated by all who have long journeys to make by day or night.

The elegance and comfort of the arrangements are almost too obvious to require description. We see the luxuriously padded chairs, which, by turning on swivels, permit the traveller to adjust his position according to his individual wishes, so that he can, with ease, place his seat either to gaze

8

.directly on the passing landscape, or turn his face towards his fellow-travellers opposite or on either side. · The chairs are also provided with an arrangement for placing the backs at any required inclination, and the light and refined character of the decorations of the carriage should not .escape the reader's notice.   Pullman Cars of another kind, providing sleeping accommodation for night journeys, are also in use on the Midland line, and they are fitted up with the same thoughtful regard to comfort as the Parlour Car.

The great engineering feats which have been accomplished in the construction of railways are numerous enough to fill volumes. We give, therefore, only a short notice of one or two recently constructed lines which have features of special interest, concluding with a brief account of such remarkable constructions as the railway by which the traveller may now go up the Rigi, and the railways which ascend Vesuvius and Mt. Pilatus.

THE METROPOLITAN RAILWAYS.

WHEN the traffic in the streets of London became so great that the ordinary thoroughfares were unable to meet public requirements, the bold project was conceived of making a railway under the streets. The construction of a line of railway beneath the streets of a populous city, amidst a labyrinth of gas-pipes, water-mains, sewers, &c., is obviously an undertaking presenting features so remarkable that the London Underground Railway cannot here be passed over without a short notice. Its construction occupied about three years, and it was opened for traffic in 1863. The line commencing at Paddington, and passing beneath Edgware Road at right angles, reaches Marylebone Road, under the centre of which it proceeds, and passing beneath the houses at one end of Park Crescent, Portland Place, it follows the centre of Euston Road to King's Cross, where connection with the Great Northern and Midland system is effected. Here the line bends sharply southwards, and proceeds to Farringdon Street Station, the original terminus. A subsequent extension takes an easterly direction and reaches Aldgate Station, the nominal terminus. The crown of the arch which covers the line is in some places only a few inches beneath the level of the streets; in other places it is several feet below the surface, and, in fact, beneath the foundations of the houses and other buildings. The steepest gradient on the line is 1 in 100, and the sharpest curve has a radius of 200 yards. The line is nearly all curved, there not being in all its length three-quarters of a mile of straight rails. The difficulties besetting an undertaking of this kind would be tedious to describe, but may readily be imagined. The line traverses every kind of soil—clay, gravel, sand, rubbish, all loosened by previous excavations for drains, pipes, foundations, &c.; and the arrangements of these drains, water and gas-pipes, had to be reconciled with the progress of the railway works, without their uses being interfered with even for a time. Of the stations the majority have roofs of the ordinary kind, open to the sky; but two of them, namely, Baker Street and Gower Street, are completely underground stations, and their roofs are formed by the arches of brickwork immediately below the streets. The arrangements at these stations show great boldness and

FIG. 50.—*Gower Street Station, Metropolitan Railway.*

inventiveness of design. The booking offices for the up line are on one side of the road, and those for the down line on the other. Fig. 50 represents the interior of the Gower Street Station, and the other is very similar. In each the platforms are 325 ft. long and 10 ft. broad, and the stations are lighted by lateral openings through the springing of the arch which forms the roof. This arch is a portion of a circle of 32 ft. radius, with a span of 45 ft. and a rise of 9 ft. at the crown. The lateral openings are arched at the top and bottom, but the sides are flat. The width of each is 4 ft. 9 in., and the height outside 6 ft., increasing to 10 ft. at the ends opening on the platform. The openings are entirely lined with white glazed tiles, and the outward ends open into an area, the back of which is inclined at an angle of 45°, and the whole also lined with white glazed tiles, and covered with glass, except where some iron gratings are provided for ventilation. The tiles reflect the daylight so powerfully that but little gas is required for the illumination of the station in the day-time. The arched roofs of these stations are supported by piers of brickwork, 10 ft. apart, 5 ft. 6 in. deep, and 3 ft. 9 in. wide. In the spaces between the piers vertical arches, like parts of the brick lining of a well, are wedged in, to resist the thrust of the earth, and a straight wall is built inside of this between the piers, to form the platform wall of the station. The tops of the piers are connected by arches, and are thus made to bear the weight of the arched roof, which has 2 ft. 3 in. thickness of brickwork at the crown, and a much greater thickness towards the haunches.

The benefit derived by the public from the completion of the Metropolitan Railway was greatly increased by the subsequent construction of another railway—" The Metropolitan District," which, joining the Metropolitan at Paddington, makes a circuit about the west-end of Hyde Park, and passing close to the Victoria Terminus of the London, Chatham, and Dover and the Brighton and South Coast Railways, reaches Westminster Bridge, and then follows the Thames Embankment to Blackfriars Bridge, where it leaves the bank of the river for the Mansion House, Mark Lane and Aldgate stations. This line, taken in conjunction with the Metropolitan, forms the "*inner circle*" of the railway communication in London. The circuit was for a long time incomplete at the east by the want of connection between the Mansion House Station and that of Moorgate Street, although these stations are but little more than half a mile apart. A line connecting these two points has lately been constructed at great cost, and the public now possess a complete circle of communication. The number of trains each day entering and leaving some of the stations on the Metropolitan system is very great. Moorgate Street Station—a terminus into which several companies run—may have about 800 trains arriving or departing in the course of a day.

THE PACIFIC RAILWAY.

THE remarkable development of railways which has taken place in the United States has its most striking illustration in the great system of lines by which the whole continent can be traversed from shore to shore. The distance by rail from New York to San Francisco is 3,215 miles, and the journey occupies about a week, the trains travelling night and day. The traveller proceeding from the Eastern States to the far west has the choice of many routes, but these all converge to Omaha. From this point the Pacific Railroad will convey him towards the land of the setting sun. The map, Fig. 51, shows the course of this railway, which is the longest in the world. It traverses broader plains and crosses higher mountains than any other. Engineering skill of the most admirable kind has been displayed in the laying-out and in the construction of the line, with its innumerable cuttings, bridges, tunnels, and snow-sheds.

The road from Omaha to Ogden, near the Great Salt Lake—a distance of 1,032 miles—is owned by the Union Pacific Company, and the Central Pacific joins the former at Ogden and completes the communication to San Francisco, a further length of 889 miles—the whole distance from Omaha to San Francisco being 1,911 miles. The Union Pacific was commenced in November, 1865, and completed in May, 1869. There are at Omaha extensive workshops provided with all the appliances for constructing and repairing locomotives and carriages, and these works cover 30 acres of ground, and give employment to several thousand men. The population of Omaha rose during the making of the railway from under 3,000 in 1864 to more than 16,000 in 1870, and it is now a flourishing town. A little distance from Omaha the line approaches the Platte River, and the valley of this river and one of its tributaries is ascended to Cheyenne, 516 miles from Omaha, the line being nowhere very far from the river's course. Cheyenne is 5,075 ft. higher above the sea than Omaha, the elevation of which is 966 ft. The Platte River is a broad but very shallow stream, with a channel continually shifting, owing to the vast quantity of sand which its muddy waters carry down. This portion of the line passing through a district where leagues upon leagues of fertile land await the hand of the tiller, has opened up vast tracts of land—hedgeless, gateless green fields, free to all, and capable of receiving and supporting millions of human inhabitants.

Cheyenne, a town of 3,000 inhabitants, is entirely the creation of the railways, for southward from Cheyenne a railway passes to Denver, a distance of 106 miles, through rich farming and grazing districts. Seven miles beyond Cheyenne the line begins to ascend the Black Hills by steep gradients, and at Granite Canyon, for example, the rise in five miles is 574 ft., or about 121 ft. per mile. Many lime-kilns have been erected in this neighbourhood, where limestone is very abundant. A little beyond this point the road is in many places protected by snow-sheds, fences of timber, and rude stonework. At Sherman, 549 miles from Omaha, the line attains the summit of its track over the Black Hills, and the highest point on any railway in the world, being 8,242 ft. above the level of the sea. Wild and desolate scenery characterizes the district round Sherman, and the hills, in places covered with a dense growth of wood, will furnish an immense supply of timber for years to come. The timber-sheds erected over the line, and the fences beside it are not so much on account of the depth of snow that falls, but to prevent it from blocking the line by being drifted into the cuts by the high wind. A few miles beyond Dale Creek at Sherman is the largest bridge on the line. It is a trestle bridge, 650 ft. long and 126 ft. high, and has a very light appearance—indeed, to an English eye unaccustomed to these *impromptu* timber structures, it looks unpleasantly light. From Sherman the line descends to Laramie, which is 7,123 ft. above the sea level and 24 miles from Sherman, and here the railway has a workshop, for good coal is found within a few miles. A fine tract of grazing land, 60 miles long and 20 miles broad, stretches around this station, and it is said that nowhere in the whole North American continent can cattle be reared and fattened more cheaply. The line, now descending the Black Hills, crosses for many miles a long stretch of rolling prairie, covered in great part with sage-bush, and forming a tableland lying be-

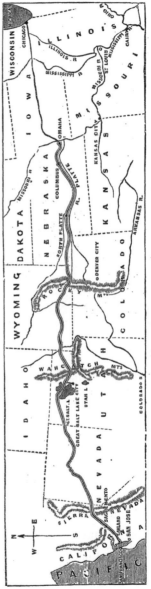

FIG. 51.—Map of the Route of the Pacific Railway.

FIG. 52.—*Trestle Bridge.*

tween the western base of the Black Hills and the eastern base of the
snowy range of the Rocky Mountains, which latter reach an elevation of
from 10,000 to 17,000 ft. above the sea level and are perpetually covered
with snow. Such tablelands are termed in America "parks." Before the
line reaches the summit of the pass by which it crosses the range of the
snowy mountains, it traverses some rough country among the spurs of the
hills—through deep cuts and under snow-sheds, across ravines and rivers,
and through tunnels. At Percy, 669 miles, is a station named after Colonel
Percy, who was killed here by the Indians when surveying for the line. He
was surprised by a party of the red men, and retreated to a cabin, where
he withstood the attack of his assailants for three days, killing several of
them; but at length they set fire to the cabin, and the unfortunate Colonel
rushing out, fell a victim to their ferocity. Near Creston, 737 miles from
Omaha, the highest point of the chief range is reached, though at an
elevation lower by 1,212 ft. than the summit of the pass where the line
crosses the Black Hills, which are the advanced guard of the Rocky
Mountains. Here is the water-shed of the continent, for all streams rising
to the east of this flow ultimately into the Atlantic,—while these, having
their sources in the west, fall into the Pacific. Before reaching Ogden the
line passes through some grand gorges, which open a way for the iron

horse through the very hearts of the mountains, as if Nature had foreseen railways and providently formed gigantic cuttings—such as the Echo and Weber Canyons, which enable the line to traverse the Wahsatch Mountains.

Echo Canyon is a ravine 7 miles long, about half a mile broad, flanked by precipitous cliffs, from 300 to 800 ft. high, and presenting a succession of wild and grand scenery. In Weber Canyon the river foams and rushes along between the mountains, which rise in massive grandeur on either side, plunging and eddying among the huge masses of rock fallen from the cliffs above. Along a part of the chasm the railway is cut in the side of the steep mountain, descending directly to the bed of the stream

FIG. 53.—*American Canyon.*

Where the road could not be carried round or over the spurs of the mountains it passes through tunnels, often cut through solid stone. A few miles farther the line reaches the city of Ogden, in the state of Utah, the territory of the Mormons. This territory contains upwards of 65,000 square miles, and though the land is not naturally productive, it has, by irrigation, been brought into a high state of cultivation, and it abounds in valuable minerals, so that it now supports a population of 80,000 persons.

We have now arrived at Ogden, where the western portion of the great railway line connecting the two oceans unites to the Union Pacific we have just described. This western portion is known as the Central Pacific Railroad, and it stretches from Ogden to San Francisco, a distance of 882 miles.

The portion of the line which unites Sacramento to Ogden, 743 miles, was commenced in 1863 and finished in 1869, but nearly half of the entire length was constructed in 1868, and about 50 miles west of Ogden, the remarkable engineering feat of laying 10 miles of railway in one day was performed. It was thus accomplished : when the waggon loaded with the rails arrived at the end of the track, the two outer rails were seized, hauled forward off the car, and laid upon the sleepers by four men, who attended to this duty only. The waggon was pushed forwards over these rails, and the process of putting down the rails was repeated, while behind the waggon came a little army of men, who drove in the spikes and screwed on the fish-plates, and, lastly, a large number of Chinese workmen, with pickaxes and spades, who ballasted the line. The average rate at which these operations proceeded was about 240 ft. of track in $77\frac{1}{2}$ seconds, and in these 10 miles of railway there were 2,585,000 cross-ties, 3,520 iron rails, 55,000 spikes, 7,040 fish-plates, and 14,080 bolts with screws, the whole weighing 4,362,000 lbs.! Four thousand men and hundreds of waggons were required, but in the 10 miles all the rails were laid by the same eight men, each of whom is said to have that day walked 10 miles and lifted 1,000 tons of iron rails. Nothing but the practice acquired during the four previous years and the most excellent arrangement and discipline could have made the performance of such a feat possible as the laying of eight miles of the track in six hours, which was the victory achieved by these stalwart navvies before dinner.

The line crosses the great American desert, distinguished for its desolate aspect and barren soil, and so thickly strewn with alkaline dust that it appears almost like a snow-covered plain. The alkali is caustic, and where it abounds no vegetation can exist, most of the surface of this waste being fine, hard grey sand, mixed with the fragments of marine shells and beds of alkali.

The third great mountain range of the North American continent is crossed by this line, at an elevation of 7,043 ft. above the sea level. The Sierra Nevada, as the name implies, is a range of rugged wild broken mountain-tops, always covered with snow. The more exposed portions of the road are covered with snow-sheds, solidly constructed of pine wood posts, 16 in. or 20 in. across: the total length of snow-sheds on the Sierra Nevada may equal 50 miles. These sheds sometimes take fire; but the company have a locomotive at the Summit Station, ready to start at a moment's notice with cars carrying tanks of water. The snow falls there sometimes to a depth of 20 ft. in one winter; and in spring, when it falls into the valleys in avalanches, sweeping down the mountain-sides, they pass harmlessly over the sloping roofs of the snow-sheds. Where the line passes along the steep flank of a mountain, the roofs of these snow-sheds abut against the mountain-side, so that the masses of snow, gliding down from its heights, continue their slide without injury to line, or sheds, or trains. Where, however, the line lies on level ground, or in a ridge, the snow-sheds are built with a strong roof of double slope, in order to support or throw off the snow. From Summit (7,017 ft.) the line descends continuously to Sacramento, which is only 30 ft. above the sea level, and 104 miles from Summit. About 36 miles from Summit, the great American Canyon, one of the wildest gorges in the Sierra Nevada range, is passed. Here the American River is confined for a length of two miles between precipitous walls of rock, 2,000 ft. in height, and so steep that no human foot has ever yet followed the stream through this tremendous gorge (Fig. 53). A few miles beyond this the line is carried, by a daring feat of engineering, along the side of a mountain, overhanging a stream 2,500 ft. below. This mountain is known as "Cape Horn," and is a place to try the nerves of timid people. When this portion of the line was commenced, the workmen were lowered and held by ropes, until they had hewn out a standing-place on

FIG. 54.—" *Cape Horn.*"

the shelving sides of the precipice, along whose dizzy height, where even the agile Indian was unable to plant his foot, the science of the white man thus made for his iron horse a secure and direct road. (Fig. 54.)

These lines of railway, connecting Omaha with Sacramento, are remarkable evidences of the energy and spirit which characterize the Anglo-Saxon race in America. The men who conceived the design of the Central Pacific Railroad, and actually carried it into effect, were not persons experienced in railway construction; but five middle-aged traders of Sacramento, two of whom where drapers, one a wholesale grocer, and the others ironmongers, believing that such a railway should be made, and finding no one ready to undertake it, united together, projected the railway, got it completed, and now manage it. These gentlemen were associated with an engineer

FIG. 55.—*Snow Plough.*

named Judah, who was a sanguine advocate of the scheme, and made the
preliminary surveys, if he did not plan the line. The line is considered
one of the best appointed and best managed in the States; yet the project
was at first ridiculed and pronounced impracticable by engineers of high
repute, opposed by capitalists, and denounced by politicians. An eminent
banker, who personally regarded the scheme with hopefulness, would not
venture, however, to take any stock, lest the credit of his bank should be
shaken, were he known to be connected with so wild a scheme. And,
indeed, the difficulties appeared great. Except wood, all the materials
required, the iron rails, the pickaxes and spades, the waggons, the locomotives, and the machinery had to be sent by sea from New York, round
Cape Horn, a long and perilous voyage of nine months duration, and
transhipped at San Francisco for another voyage of 120 miles before they
could reach Sacramento. Add to this that workmen were so scarce in
California, and wages so high, that to carry on the work it was necessary
to obtain men from New York; and during its progress 10,000 Chinamen
were brought across the Pacific, to work as labourers. Subscriptions came
in very slowly, and before 30 miles of the line had been constructed, the
price of iron rose in a very short time to nearly three times its former
amount. At this critical juncture, the five merchants decided to defray,
out of their own private fortunes, the cost of keeping 800 men at work on
the line for a whole year. We cannot but admire the unswerving confidence
in their enterprise displayed by these five country merchants, unskilled in

railway making, unaided by public support, and even discouraged in their project by their own friends. The financial and legal obstacles they successfully surmounted were not the only difficulties to be overcome. They had the engineering difficulties of carrying their line over the steep Sierra, a work of four years; long tunnels had to be bored; one spring when snow 60 ft. in depth covered the track, it had to be removed by the shovel for 7 miles along the road; saw-mills had to be erected in the mountains, to prepare the sleepers and other timber work; wood and water had to be carried 40 miles across alkali plains, and locomotives and rails dragged over the mountains by teams of oxen. The chief engineer, who organized the force of labourers, laid out the road, designed the necessary structures, and successfully grappled with the novel problem of running trains over such a line in all seasons, was Mr. S. S. Montague. The requirements of the traffic necessitate not only solidly constructed iron-covered snow-sheds, but massive snow-ploughs to throw off the track the deep snow which could in no other way be prevented from interrupting the working of the line. These snow-ploughs are sometimes urged forward with the united power of eight heavy locomotives. Fig. 55 represents one of these ploughs cleaning the line, by throwing off the snow on to the sides of the track. The cutting apparatus varies in its arrangements, some forms being designed to push the snow off on one side, some on the other, and to fling it down the precipices; and others, like the one represented, are intended merely to throw it off the track.

FIG. 56.— *The first Steam Railroad Train in America.*

Sacramento is 1,775 miles from Omaha, and is connected with San Francisco by a line 139 miles long. At San Francisco, or rather at Oakland, 1,911 miles from Omaha and 3,212 miles from New York, is the terminus of the great system of lines connecting the opposite shores of the vast North American continent. San Francisco, situated on the western shore of a bay, is connected with Oakland by a ferry; but the railway company have recently constructed a pier, which carries the trains out into the bay for 2¼ miles. This pier is strongly built, and is provided with a double set of rails and a carriage-road, and with slips at which ships land and embark passengers, so that ships trading to China, Japan, and Australia can load and unload directly into the trains, which may pass without change from the shores of the Pacific to those of the Atlantic Ocean. San Francisco is a marvellous example of rapid increase, for the population now numbers 170,000, yet a quarter of a century ago 500 white settlers could not be found in as many miles around its site. The first house was erected in 1846, and in 1847 not a ship visited the bay, but now forty large steamships ply regularly, carrying mails to China, Japan, Panama, South America, Australia, &c., and there are, of course, hundreds of other steamers and ships.

The descriptions we have given of only two lines of railway may suffice to show that the modern engineer is deterred by no obstacles, but boldly drives his lines through places apparently the most impracticable. He shrinks from no operations however difficult, nor hesitates to undertake works the mere magnitude of which would have made our forefathers stand aghast. Not in England or America alone, but in almost every part of the world, the railways have extended with wonderful rapidity ; the continent of Europe is embraced by a network of lines ; the distant colonies of Australia and New Zealand have thousands of miles of lines laid down, and many more in progress ; the map of India shows that peninsula traversed in all directions by the iron roads ; and in the far distant East we hear of Japan having several lines in successful operation, and the design of laying down more. In connection with such works, at home and abroad, many constructions of great size and daring have been designed and erected ; many navigable rivers have been bridged, and not seldom has an arm of the sea itself been spanned ; hundreds of miles of embankments and viaducts have been raised ; hills have been pierced with innumerable cuttings and tunnels, and all these great works have been accomplished within the experience of a single generation of men, and have sprung from one single successful achievement of Stephenson's — the Liverpool and Manchester Railway, completed and opened in 1830. We in England should also have pride in remembering that the growth of the railways here is due to the enterprise, industry, and energy of private persons ; for the State has furnished no funds, but individuals, by combining their own resources, have executed the works, and manage the lines for their common interest and the public good. It is said that the amount of money which has been spent on railways in Great Britain is not far short of 500 millions of pounds sterling. The greatest railway company in the United Kingdom is the London and North-Western, which draws in annual receipts about seven millions of pounds ; and the total receipts of all the railway companies would nearly equal half the revenue of the State.

F IG. 57.—*Railway Embankment near Bath.*

PLATE VI.

MOUNT WASHINGTON INCLINED TRACK.

INCLINED RAILWAYS.

THE construction of railways over lofty ranges of mountains will be found illustrated by the brief notices in other pages of the Union Pacific line in the United States, and of the St. Gothard railway over the Alps. In such cases, the track has been to a great extent carried over the spurs or along the sides of the mountains, so that such inclines might be obtained as the ordinary locomotive was capable of ascending. The expensive operation of tunnelling was resorted to only where sinuous deviations from the more direct route involved a still greater expenditure of initial cost, or a continual waste of time and energy in the actual working of the line. Sometimes winding tracks, almost returning by snake-like loops on their own route, as projected on the map, were required in order that the ascent could be made with an incline practicable for the ordinary locomotive. In the earlier development of railways, there were to be met with cable inclines, where the traction of the locomotive had to be superseded or supplemented by that of a rope or chain wound round a drum actuated by a stationary steam-engine. The more powerful locomotives of the present day are able to mount grades of such inclination that the employment of cable traction is no longer requisite, except in but a few cases. Railways had carried passengers about in all parts of the world for many years before the engineer addressed himself to the problem of easily and quickly taking people up heights of steep and toilsome ascent, sought generally for the sake of the prospect, etc. Such, at least, has been the object of most of the inclined railways already constructed, but to this their utility is by no means limited, and as their safety and stability has been proved by many years of use, they may find wider applications than the gratification of the tourist and pleasure-seeker.

The toothed rail or rack which was formerly supposed necessary to obtain power of traction on rails has been already mentioned (p. 101), and as early as 1812 such a contrivance appears to have been in use in England, near Leeds, the invention of a Mr. Blenkinsop. This mode of traction received no development or improvement worthy of notice until Mr. S. Marsh constructed, in 1866, a railway ladder—for so it may be called—for the ascent of Mount Washington in the United States. In this case there was a centre rail formed of iron, angle iron laid between and parallel to the metals on which ran the wheels of the carriages. In this centre rail angle irons were connected by round bars of wrought iron, which the teeth of a pinion of the locomotive engaged, so that a climbing action, resembling somewhat that of a wheel entering on the successive rounds of a ladder, was produced, and in this way an ascensive power was obtained sufficient to overcome gravity, the gradient not much exceeding a rise of one foot in three at any point (12 vertical to 32 horizontal). This railway was completed in 1869, and for more than a quarter of a century it has carried thousands of tourists to the summit of Mount Washington without a single fatal accident. This system of ascending mountains was soon adopted in Europe with certain improvements, for in 1870 an inclined railway was constructed to the summit of the Rigi, in which a system of involute gearing was substituted for the ladder-like rounds of Mr. Marsh. A certain vibratory action, due to the successive engagements of the teeth in the central rack, which was some-

what disagreeable for passengers, was soon afterwards obviated in the
Abt system, in which two racks are used, with the teeth of one opposite
the spaces of the other, and a double pinion provided, so that greater
uniformity in the acting power is obtained. With certain modifications
in detail, such as horizontal instead of vertical pinions, this system has
been largely adopted wherever cables have been dispensed with. In the
inclined railway by which Mount Pilatus, near Lucerne, is now ascended,
horizontal teeth project from both sides of a centre rail, and these are
engaged by horizontal pinions. The incline here is very steep, being in
places nearly 30 degrees ; teeth perpendicular to the plane of the incline

FIG. 57a.—*Train Ascending the Rigi.*

would have offered a less margin of safety than those on the plan actually
adopted. In some places, as among the Alps, and more particularly in
South America, there are railways in which the ordinary mode of traction
and that with the rack are combined ; that is, where the gradient exceeds
the ordinary limit, a central rack-rail is laid down, on approaching which
the engineer slackens his speed, and allows pinions, moved by the locomotive, to become engaged in the double rack, by which he slowly climbs
the steep ascent until a level tract is reached which permits of the
ordinary traction being resumed.
    Instead of climbing the inclines by rack-work rails, there is another
system which offers great advantages for economy in working, and one
generally resorted to where the incline can be made in one vertical plane.
This is the balanced cable, in which the gravitation force of a descending
car or train is utilised to draw up, or assist to draw up, the ascending car
or train. These cars are attached to the ends of a cable which passes
round a drum at the top of the incline, and means are provided, according to circumstances, so that the drum may be turned, or its revolutions

controlled by brakes. When there is a water supply at the upper end of the incline, a simple and economical mode of working the cable is available; for all that is necessary is to provide each car with a water-tank capable of being rapidly filled and emptied. The upper car is made the heavier when required, by filling its tank with water, when it raises the lower car, and on itself arriving at the bottom, the water is discharged before the load to be taken up is received.

Many inclined railways are now in operation in various parts of the world, as at Mount Vesuvius, where two of the slopes have a combined length of 10,500 feet ; at Mount Supurga and at Mount San Salvatore there are others. At Burgen-stock in Switzerland there is one having a

FIG. 57*b*.—*At the Summit of the Rigi.*

slope 57 feet vertical to 100 feet horizontal. These are cable inclines ; but a rack is also used with a pinion regulated by a friction-brake to avoid accident, in case of the cable parting. The largest inclined railway in America is at the Catskill Mountains, where an ascent of 1,600 feet is made in a horizontal distance of 6,780 feet. In this a novel plan has been adopted for compensating the varying weight that has to be moved, for it is obvious that at the commencement the load at the top of the incline has to raise not only that at the bottom, but the whole weight of the cable also, equal to 35,000 pounds of wire rope, and again after the middle point has been passed, the descending power is constantly increasing, while the load being raised is diminishing. Now, in order that the engine may work with more uniform effect, the engineer has not made the incline a straight line, but with the slope lightest at the bottom and gradually increased towards the top, so that the line is really a curve in the vertical plane, and has at every point just the inclination required for balancing the weight of the wire cable, as this shifts from the one track to the other. Instead of a rack pinion and brake to control a too rapid descent from any accident, the cars are provided with clutches, which are automatically thrown out on wooden guard-rails, when a safe speed is exceeded. Inclined railways have also been constructed to the summit of Snowdon, in North Wales, and to that of the Jungfrau, in Switzerland.

PLATE VII.

PIKE'S PEAK RAILWAY, ROCKY MOUNTAINS.

FIG. 58.—*The Great Eastern at Anchor.*

STEAM NAVIGATION.

THE first practically successful steamboat was constructed by Symington, and used on the Forth and Clyde Canal, in 1802. A few years afterwards Fulton established steam navigation in American waters, where a number of steamboats plied regularly for some years before the invention had received a corresponding development in England, for it was not until 1814 that a steam-packet ran for hire in the Thames. From that time, however, the principle was quickly and extensively applied, and steamers made their appearance on the chief rivers of Great Britain, and soon began also to make regular passages from one sea-port to another, until at length, in 1819, a steamer made the voyage from New York to Liverpool. It does not appear, however, that such ocean steam voyages became at once common, for we read that in 1825 the captain of the first steam-ship which made the voyage to India was rewarded by a large sum of money. It was not until 1838 that regular steam communication with America was commenced by the dispatch of the *Great Western* from Bristol. Other large steamers were soon built expressly for the passage of the Atlantic, and a new era in steam navigation was reached when, in 1845, the *Great Britain* made her first voyage to New York in fourteen days. This ship was of immense size, compared with her predecessors, her length being 320 ft., and she was

9

moreover made of iron, while instead of paddles, she was provided with a screw-propeller, both circumstances at that time novelties in passenger ships. Fulton appears to have made trial in America of various forms of mechanism for propelling ships through the water. Among other plans he tried the screw, but finally decided in favour of paddle-wheels, and for a long time these were universally adopted. Many ships of war were built with paddle-wheels, but the advantages of the screw-propeller were at length perceived. The paddle-wheels could easily be disabled by an enemy's shot, and the large paddle-boxes encumbered the decks and obstructed the operations of naval warfare. Another circumstance perhaps had a greater share in the general adoption of the screw, which had long before been proposed as a means of applying steam power to the propulsion of vessels. This was the introduction of a new method of placing the screw, so that its powers were used to greater advantage. Mr. J. P. Smith obtained a patent in 1836 for placing the propeller in that part of the vessel technically called the *dead-wood*, which is above the keel and immediately in front of the rudder. When the means of propulsion in a ship of war is so placed, this vital part is secure from injury by hostile projectiles, and the decks are clear for training guns and other operations. Thus placed, the screw has been proved to possess many advantages over paddle-wheels, so that at the present time it has largely superseded paddle-wheels in vessels of every class, except perhaps in those intended to ply on rivers and lakes. Many fine paddle-wheel vessels are still afloat, but sea-going steamers are nearly always now built with screw-propellers. In the application of the steam engine to navigation the machine has received many modifications in the form and arrangement of the parts, but in principle the marine engine is identical with the condensing engine already described. The engines in steam-ships are often remarkable for the great diameter given to the cylinders, which may be 8 ft. or 9 ft. or more. Of course other parts of the machinery are of corresponding dimensions. Such large cylinders require the exercise of great skill in their construction, for they must be cast in one piece and without flaws. The engraving, Fig. 59, depicts the scene presented at the works of Messrs. Penn during the casting of one of these large cylinders, the weight of which may amount to perhaps 30 tons. Only the top of the mould is visible, and the molten iron is being poured in from huge ladles, moved by powerful cranes. In paddle vessels the great wrought iron shaft which carries the paddle-wheels crosses the vessel from side to side. This shaft has two cranks, placed at right angles to each, and each one is turned by an engine, which is very commonly of the kind known as the side-lever engine. In this engine, instead of a beam being placed above the cylinder, two beams are used, one being set on each side of the cylinder, as low down as possible. The top of the piston-rod is attached to a cross-head, from each end of which hangs a great rod, which is hinged to the end of the side-beam. The other ends of the two beams are united by a cross-bar, to which is attached the connecting-rod that gives motion to the crank. Another favourite form of engine for steam-ships is that with oscillating cylinders. The paddle-wheels are constructed with an iron frame-work, to which flat boards, or floats, are attached, placed usually in a radial direction. But when thus fixed, each float enters the water obliquely, and in fact its surface is perpendicular to the direction of the vessel's course only at the instant the float is vertically under the axis of the wheel. In order to avoid the loss of power consequent upon this oblique movement of the floats, they are sometimes hung upon centres, and are so moved by suitable

PLATE VIII.

THE "CLERMONT," FROM A CONTEMPORARY DRAWING.

FIG. 59.—*Casting Cylinder of a Marine Steam Engine.*

mechanism that they are always in a nearly vertical position when passing through the water. Paddle-wheels constructed in this manner are termed *feathering* wheels. They do not appear, however, to possess any great advantage over those of the ordinary construction, except when the paddles are deeply immersed in the water, and this result may be better understood when we reflect that the actual path of the floats through the water is not circular, as it would be if the vessel itself did not move; for all points of the wheel describe peculiar curves called *cycloids*, which result from the combination of the circular with the onward movement.

The next figure, 60, exhibits a very common form of the screw propeller, and shows the position which it occupies in the ship. The reader may not at once understand how a comparatively small two-armed wheel revolving in a plane perpendicular to the direction of the vessel's motion is able to propel the vessel forward. In order to understand the action of the propeller, he should recall to mind the manner in which a screw-nail in a piece of wood advances by a distance equal to its pitch at every turn. If he will conceive a gigantic screw-nail to be attached to the vessel extending along the keel, —and suppose for a moment that the water surrounding this screw is not able to flow away from it, but that the screw works through the water as the nail does in the wood,—he will have no difficulty in understanding that, under such circumstances, if the screw were made to revolve, it would advance and carry the vessel with it. The reader may now form an accurate notion of the actual propeller by supposing the imaginary screw-nail to have the thread so deeply cut that but little solid core is left in the centre, and supposing also that only a very short piece of the screw is used—say the length of one revolution—and that this is placed in the dead-wood. Such was the construction of the earlier screw-propellers, but now a still shorter portion of the screw is used ; for instead of a complete turn of the thread, less than one-sixth is now the common construction. Such a strip or segment of the screw-thread forms a *blade*, and two, three, four, or more blades are attached radially to one common axis. The blades spring when there are two from opposite points in the axis, and in other cases from points on the same circle. The

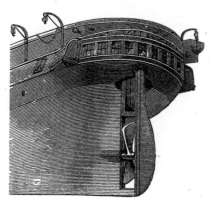

blades of the propeller are cut and carved into every variety of shape according to the ideas of the designer, but the fundamental principle is the same in all the forms. It need hardly be said that the particles of the water are by no means fixed like those of the wood in which a screw advances. But as the water is not put in motion by the screw without offering some resistance by reason of its inertia, this resistance reacting on the screw operates in the same manner, but not to the same extent, as the wood in the other case. When we know the pitch of the screw, we can calculate what distance the screw would be moved forward in a given number of revolutions if it were working through a solid. This distance is usually greater than the actual distance the ship is propelled, but in some cases the vessel is urged through the water with a greater velocity than if the screw were working in a solid nut. The shaft which carries the screw extends from the stem to the centre of the ship where the engines are placed, and it passes outward through a bearing lined with wood, of which *lignum vitæ* is found to be the best kind, the lubricant for this bearing being not oil but water. The screw would not have met with the success it has attained but for this simple contrivance; for it was found that with brass bearings a violent thumping action was soon produced by the rapid rotation of the screw. The wearing action between the wood and the iron is very slight, whereas brass bearings in this position quickly wear and their adjustments become impaired. The screw-shaft is very massive and is made in several lengths, which are supported in appropriate bearings ; there is also a special arrangement for receiving the thrust of the shaft, for it is by this thrust received from the screw that the vessel is propelled, and the strain must be distributed to some strong part of the ship's frame. There is usually also an arrangement by which the screw-shaft can, when required, be disconnected from the engine, in order to allow the screw to turn freely by the action of the water when the vessel is under sail alone.

FIG. 60.—*Screw-Propeller.*

A screw-propeller has one important advantage over paddle-wheels in the following particular : whereas the paddle-wheels act with the best effect when the wheel is immersed in the water to the depth of the lowest float, the efficiency of the screw when properly placed is not practically altered by the depth of immersion. As the coals with which a steamer starts for a long voyage are consumed, the immersion is decreased—hence the paddle-wheels of such a steamer can never be immersed to the proper extent *throughout* the voyage; they will be acting at a disadvantage during the greater part of the voyage. Again, even when the immersion of the vessel is such as to give the best advantage to the paddle-wheels, that advantage is lost whenever a side-wind inclines the ship to one side, or whenever by the action of the waves the immersion of the paddles is changed by excess or defect. From all such causes of inefficiency arising from the position of the vessel the screw-propeller is free. The reader will now understand why paddle-wheel steamers are at the present day constructed for inland waters only.

A great impulse was given to steam navigation, by the substitution of iron for wood in the construction of ships. The weight of an iron ship is only two-thirds that of a wooden ship of the same size. It must be remembered that, though iron is many times heavier than wood, bulk for bulk, the required strength is obtained by a much less quantity of the former. A young reader might, perhaps, think that a wooden ship must float better than an iron one; but the law of floating bodies is, that the part of the floating body which is below the level of the water, takes up the space of exactly so much water as would have the same weight as the floating body, or in fewer words, a floating body displaces its own *weight* of water. Thus we see that an iron ship, being lighter than a wooden one, must have more buoyancy. The use of iron in ship-building was strenuously advocated by the late Sir W. Fairbairn, and his practical knowledge of the material gave great authority to his opinion. He pointed out that the strains to which ships are exposed are of such a nature, that vessels should be made on much the same principles as the built-up iron beams or girders of railway bridges. How successfully these principles have been applied will be noticed in the case of the *Great Eastern.* This ship, by far the largest vessel ever built, was designed by Mr. Brunel, and was intended to carry mails and passengers to India by the long sea route. The expectations of the promoters were disappointed in regard to the speed of the vessel, which did not exceed 15 miles an hour; and no sooner had she gone to sea than she met with a series of accidents, which appear, for a time, to have destroyed public confidence in the vessel as a sea-going passenger ship. Some damage and much consternation were produced on board by the explosion of a steam jacket a few days after the launch. Then the huge ship encountered a strong gale in Holyhead Harbour, and afterwards was disabled by a hurricane in the Atlantic, in which her rudder and paddles were so damaged, that she rolled about for several days at the mercy of the waves. At New York she ran upon a rock, and the outer iron plates were stripped off the bottom of the ship for a length of 80 ft. She was repaired and came home safely; but the companies which owned her found themselves in financial difficulties, and the big ship, which had cost half a million sterling, was sold for only £25,000, or only about one-third of her value as old materials.

The misfortunes of the *Great Eastern,* and its failure as a commercial speculation in the hands of its first proprietors, have been quoted as an illustration of the ill luck, if it might be so called, which seems to have attended several of the great works designed by the Brunels—for the Thames Tunnel was, commercially, a failure; the Great Western Railway, with its magnificent embankments, cuttings, and tunnels, has reverted to

the narrow gauge, and therefore the extra expense of the large scale has
been financially thrown away; the Box Tunnel, a more timid engineer
would have avoided; and then there is the *Great Eastern*. It is, however,
equally remarkable that all these have been glorious and successful achievements as engineering works, and the scientific merit of their designers
remains unimpaired by the merely accidental circumstance of their not
bringing large dividends to their shareholders.   Nor is their value to the
world diminished by this circumstance, for the Brunels showed mankind
the way to accomplish designs which, perhaps, less gifted engineers woulk
never have had the boldness to propose.   The Box Tunnel led the way to
other longer and longer tunnels, culminating in that of Mont Cenis ; but
for the Thames Tunnel—once ranked as the eighth wonder of the world—
we should probably not have heard of the English Channel Tunnel—a
scheme which appears less audacious now than the other did then ; if no

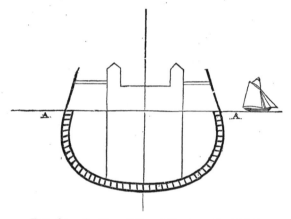

FIG. 61.—*Section of Great Eastern Amidships.*

Great Eastern had existed, we should not now have had an Atlantic Telegraph.   Possibly this huge ship is but the precursor of other still larger,
and it is undoubtedly true that since its construction the ideas of naval
architects have been greatly enlarged, and the tendency is towards increased
size and speed in our steam-ships, whether for peace or war.
      The accidents which had happened to the ship had not, however, materi-
ally damaged either the hull or the machinery; and the *Great Eastern* was
refitted, and afterwards employed in a service for which she had not been
designed, but which no other vessel could have attempted.   This was the
work of carrying and laying the whole length of the Atlantic Telegraph
Cable of 1865, of which 2,600 miles were shipped on board in enormous
tanks, that with the contents weighed upwards of 5,000 tons.   The ship has
since been constantly engaged in similar operations.*   The *Great Eastern*
is six times the size of our largest line-of-battle ships, and about seven
times as large as the splendid steamers of the Cunard line, which run
between Liverpool and New York.   She has three times the steam power
of the largest of these Atlantic steamers, and could carry twenty times as

* She was broken up for old iron, 1889.

FIG. 62.—*The Great Eastern in course of Construction.*

many passengers, with coal for forty days' consumption instead of fifteen.
Her length is 692 ft.; width, 83 ft.; depth, 60 ft.; tonnage, 24,000 tons;
draught of water when unloaded, 20 ft.; when loaded, 30 ft.; and a promenade round her decks would be a walk of more than a quarter of a mile.
The vessel is built on the cellular plan to 3 ft. above the water-line; that
is, there is an inner and an outer hull, each of iron plates $\frac{3}{4}$ in. thick, placed
2 ft. 10 in. apart, with ribs every 6 ft., and united by transverse plates, so
that in place of the ribs of wooden ships, the hull is, as it were, built up of
curved cellular beams of wrought iron. The ship is divided longitudinally by
two vertical partitions or bulkheads of wrought iron, $\frac{1}{2}$ in. thick. These are
350 ft. long and 60 ft. high, and are crossed at intervals by transverse bulkheads, in such a manner that the ship is divided into nineteen compartments, of which twelve are completely water-tight, and the rest nearly so.
The diagram (Fig. 61) represents a transverse section, and shows the cellular
construction below the water-line. The strength and safety of the vessel
are thus amply provided for. The latter quality was proved in the accident
to the ship at New York; and the former was shown at the launch, for
when the vessel stuck, and for two months could not be moved, it was found
that, although one-quarter of the ship's length was unsupported, it exhibited
no deflection, or rather the amount of deflection was imperceptible. Fig. 62
is from a photograph taken during the building of the ship, and Fig. 63
shows the hull when completed and nearly ready for launching, while the
vignette at the head of the chapter exhibits the big ship at anchor when
completely equipped. The paddle-wheels are 56 ft. in diameter, and are

FIG. 63.—*The Great Eastern ready for Launching.*

turned by four steam engines, each having a cylinder 6 ft. 2 in. in diameter, and 14 ft. in length. The vessel is also provided with a four-bladed screw-propeller of 24 ft. diameter, driven by another engine having four cylinders, six boilers, and seventy-two furnaces. The total actual power of the engines is more than that of 8,000 horses, and the vessel could carry coals enough to take her round the world—a capability which was the object of her enormous size. The vessel as originally constructed contained accommodation for 800 first-class passengers, 2,000 second class, and 1,200 third class—that is, for 4,000 passengers in all. The principal saloon was 100 ft. long, 36 ft. wide, and 13 ft. high. Each of her ten boilers weighs 50 tons, and when all are in action, 12 tons of coal are burnt every hour, and the total displacement of the vessel laden with coal is 22,500 tons.

The use of steam power in navigation has increased at an amazing rate. Between 1850 and 1860 the tonnage of the steam shipping entering the port of London increased threefold, and every reader knows that there are many fleets of fine steamers plying to ports of the United Kingdom. There are, for example, the splendid Atlantic steamers, some of which almost daily enter or leave Liverpool, and the well-appointed ships belonging to the Peninsular and Oriental Company. The steamers on the Holyhead and Kingston line may be taken as good examples of first-class passenger ships. These are paddle-wheel boats, and are constructed entirely of iron, with the exception of the deck and cabin fittings. Taking one of these as a type of the rest, we may note the following particulars : the vessel is 334 ft. long, the diameter of the paddle-wheels is 31 ft., and each has fourteen floats, which are 12 ft. long and 4 ft. 4 in. wide. The cylinders of the engines are 8 ft. 2 in. in diameter, and 6 ft. 6 in. long. The ship cost about £75,000. The average passage between the two ports—a distance of 65½ miles—occupies 3 hours 52 minutes, and at the measured mile the vessel attained the speed of 20·811 miles per hour. As an example of the magnificent vessels owned by the Cunard Company, we shall give now a few figures relating to one of their largest steam-ships, the *Persia*, launched in 1858, and built by Mr. N. Napier, of Glasgow, for the company, to carry mails and passengers between Liverpool and New York. Her length is 389 ft., and her breadth 45 ft. She is a paddle-wheel steamer, with engines of 850 horse-power, having cylinders 100 in. in diameter with a stroke of 10 ft. The paddle-wheels are 38 ft. 6 in. in diameter, and each has twenty-eight floats, 10 ft. 8 in. long and 2 ft. wide. The *Persia* carries 1,200 tons of coal, and displaces about 5,400 tons of water.

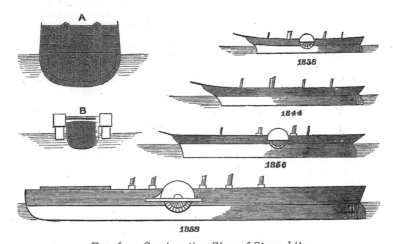

FIG. 64.—*Comparative Sizes of Steamships.*

1838, *Great Western;* 1844, *Great Britain;* 1856, *Persia;* 1858, *Great Eastern.*
A, Section amidships of *Great Eastern;* B, The same of *Great Western.* Both on the same scale, but on a larger one than their profiles.

A velocity of twenty-six miles per hour appears to be about the highest yet attained by a steamer.* This is probably near the limit beyond which the speed cannot be increased to any useful purpose. The resistance offered by water to a vessel moving through it increases more rapidly than the velocity. Thus, if a vessel were made to move through the water by being pulled with a rope, there would be a certain strain upon the rope when the vessel was dragged, say, at the rate of five miles an hour. If we desired the vessel to move at double the speed, the strain on the rope must be increased fourfold. To increase the velocity to fifteen miles per hour, we should have to pull the vessel with nine times the original force. This is expressed by saying that the resistance varies as the square of the velocity. Hence, to double the speed, the impelling force must be quadrupled, and as that force is exerted through twice the distance in the same time, an engine would be required of eight times the power—or, in other words, the power

* This has now (1895) been far surpassed.—*Vide infra.*

of the engine must be increased in proportion to the *cube* of the velocity; so that to propel a boat at the rate of 15 miles an hour would require engines twenty-seven times more powerful than those which would suffice to propel it at the rate of five miles an hour.

The actual speed attained by steam-ships with engines of a given power and a given section amidships will depend greatly upon the shape of the vessel. When the bow is sharp, the water displaced is more gradually and slowly moved aside, and therefore does not offer nearly so much resistance as in the opposite case ; but the greater part of the power required to urge the vessel forward is employed in overcoming a resistance which in some degree resembles friction between the bottom of the vessel and the water.

The wonderful progress which has, in a comparatively short time, taken place in the power and size of steam-vessels, cannot be better brought home to the reader than by a glance at Fig. 64, which gives the profiles of four steamships, drawn on one and the same scale, thus showing the relative lengths and depths of those vessels, each of which was the largest ship afloat at the date which is marked below it, and the whole period includes only the brief space of twenty years !—for this, surely, is a brief space in the history of such an art as navigation. All these ships have been named in the course of this article, but in the following table a few particulars concerning each are brought together for the sake of comparing the figures :

Date.	Name.	Propulsion.	Length.	Breadth.
1838	*Great Western* ...	Paddles	236 ft.	36 ft.
1844	*Great Britain* ...	Screw	322 ,,	51 ,,
1856	*Persia*..............	Paddles	390 ,,	45 ,,
1858	*Great Eastern* ...	Screw and paddles	690 ,,	83 ,,

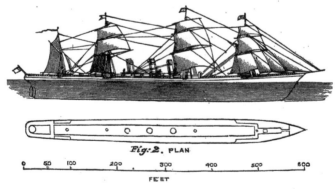

Fig. 2. PLAN

FIG. 65.—*The s.s. City of Rome.*

Several passenger ocean-going steamships have been built since the *Persia*, of still greater dimensions, and of higher engine power. These have generally been surpassed in late years by some splendid Atlantic liners, such as the sister vessels owned by the International Navigation Co., and now named respectively the *New York* and the *Paris*. The *City of Rome*, launched in 1881 by the Barrow Steamship Co., is little inferior in length to the *Great Eastern*, although the tonnage is only about one-third. The *City of Rome* is 560 ft. long, 52 ft. wide, and 37 ft. deep. Her engines are capable of working up to 10,000 indicated horse-power. Fig. 65 is a sketch of this ship, and shows that she carries four masts and three funnels. The main shaft measures more than 2 ft. across, and the screw-propeller is 24 ft. in diameter. She has accommodation for 1,500 passengers, and is fitted with all the conveniences and luxuries of a well-appointed hotel. The International Navigation Co.'s ship *Paris*, has made the passage across the Atlantic in less than six days, and appears to be the fastest vessel in the transatlantic service. In August, 1889, she made the run from shore to shore in 5 days, 22 hours, 38 minutes.

The extraordinary increase in the speed of steamships that has been effected within the last few years depends mainly upon the improvements that have latterly been made in the marine engine—a machine of which we have been unable to give an account, because its details are too numerous and complicated to be followed out by the general reader. Suffice it to say, that the use of higher steam pressures with compound expansion (p. 18), condensers which admit of the same fresh water being used in the boilers over and over again, and better furnace arrangements, are among the more important of these improvements. But not only have the limits of practicable speed been enlarged, but a greater economy of fuel for the work done has been attained; the result being that ocean carriage is now cheaper than ever. The outcome of this will not cease with simply a greatly extended steam navigation, but appears destined ultimately to produce effects on the world at large comparable in range and magnitude with those that may be traced to the use of the steam engine itself since its first invention.

Among the curiosities of steamboat construction may be mentioned a remarkable ship which was built a few years ago for carrying passengers across the English Channel without the unpleasant rolling experienced in the ordinary steamboats. The vessel, which received the name of the *Castalia*, was designed by Captain Dicey, who formerly held an official position at the Port of Calcutta. His Indian experience furnished him with the first suggestion of the new ship in the device which is adopted there for steadying boats in the heavy surf. The plan is to attach a log of timber to the ends of two outriggers, which project some distance from the side of the vessel; or sometimes two canoes, a certain distance apart, are connected together. Some of these Indian boats will ride steadily in a swell that will cause large steamers to roll heavily. Improving on this hint, Captain Dicey built a vessel with two hulls, each of which acted as an outrigger to the other. Or, perhaps, the *Castalia* may be described as a flat-bottomed vessel with the middle part of the bottom raised out of the water throughout the entire length, so that the section amidships had a form like this—The two hulls were connected by what we may term "girders," which extended completely across their sections, forming transverse partitions or bulkheads, and these girders were strongly framed together, so as to form rigid triangles. These united the two hulls so completely, that there was not any danger of the vessel being strained in a sea-way.   The decks were also formed of iron, although covered with wood, so that the whole vessel really formed a box girder of enormous section.

The reason why the steamers which until lately ran between Dover and Calais, Folkestone and Boulogne, and other Channel ports, were so small, was because the harbours on either side could not receive vessels with such a draught as the fine steamers, for example, which run on the Holyhead and Kingston line. Now, the *Castalia* drew only 6 ft. of water, or 1 ft. 6 in. less than the small Channel steamers, and would, therefore, be able to enter the French ports at all states of the tide. Yet the extent of the deck space

FIG. 66.—*The Castalia in Dover Harbour.*

was equalled in few passenger ships afloat, except the *Great Eastern* and some of the Atlantic steamers.   The vessel was 290 ft. in length, with an extreme breadth of 60 ft.   The four spacious and elegantly-fitted saloons —two of which were 60 ft. by 36 ft., and two 28 ft. by 26 ft.—and the roomy cabins, retiring rooms, and lavatories, were the greatest possible contrast to the "cribbed, cabined, and confined" accommodation of the ordinary Channel steamers.   There were also a kitchen and all requisites for sup- plying dinners, luncheons, etc., on board.   But besides the above-named saloons and cabins, there was a grand saloon, which was 160 ft. long and 60 ft. wide ; and the roof of this formed a magnificent promenade 14 ft. above the level of the sea.   There was comfortable accommodation in the vessel for more than 1,000 passengers.

The inner sides of the hulls were not curved like the outside, but were straight, with a space between them of 35 ft. wide, and the hulls were each 20 ft. in breadth, and somewhat more in depth.   There were two paddle-wheels, placed abreast of each other in the water-way between the two hulls, and each of these contained boilers and powerful engines. The designers of this vessel calculated that she would attain a speed of $14\frac{3}{4}$ knots per hour, but this result failed to be realized. Probably there were no data for the effect of paddles working in a confined water-space. The position of the paddles is otherwise an advantage, as it leaves the sides of the vessel free and unobstructed. The ship had the same form at each end, so it could move equally well in either direction. There were rudders at both ends, and the steering qualities of the ship were good. Although

FIG. 67.—*The Castalia in Dover Harbour—End View.*

the speed of the *Castalia* was below that intended, the vessel was a success as regards steadiness, for the rolling and pitching were very greatly reduced, and the miseries and inconveniences of the Channel passage obviated.

The *Castalia* is represented in Figs. 66 and 67. She was constructed by the Thames Iron Shipbuilding Co., and launched in June, 1874, but after she had been tried at sea, it was found necessary to fit her with improved boilers, and this caused a delay in placing the vessel on her station.

The *Castalia* proved a failure in point of speed, and she was soon replaced by another and more powerful vessel constructed on the same general plan, and named the *Calais-Douvres*. But this twin-ship again failed to answer expectations, and as the harbour on the French shore was meanwhile deepened and improved, new and very fine paddle-wheel boats, named the *Invicta, Victoria,* and *Empress* have been placed on the service. As the latter boat, at least, has steamed from Dover to Calais, nearly twenty-six miles, under the hour, there is nothing more to be desired in point of speed. A fourth vessel is to take the place of the twin-ship, *Calais-Douvres,* and will receive the same name.

Another very novel and curious invention connected with steam navigation was the steamer which Mr. Bessemer built at Hull in 1874. This invention also was to abolish all the unpleasant sensations which landsmen are apt to experience in a sea voyage, by effectually removing the cause of the distressing *mal de mer.* The ship was built for plying between the shores of France and England, and the method by which he purposed to carry passengers over the restless sea which' separates us

FIG. 68.—*Bessemer Steamer.*

from our Gallic neighbours was bold and ingenious. He designed a spacious saloon, which, instead of partaking of the rolling and tossing of the ship, was to be maintained in an absolutely level position. The saloon was suspended on pivots, much in the same way as a mariner's compass is suspended ; and by an application of hydraulic power it was intended to counteract the motion of the ship· and maintain the swinging saloon perfectly horizontal. It was originally proposed that the movements should be regulated by a man stationed for that purpose, where he could work the levers for bringing the machinery into action, so as to preserve the saloon in the required position. This plan was, however, improved upon, and the adjustments made automatic. It may be well to mention that it is a mistake to suppose that anything freely suspended, like a pendulum, on board a ship rolling with the waves, will hang vertically. If, however, we cause a heavy disc to spin very rapidly, say in a horizontal plane, the disc cannot be moved out of the horizontal plane without the application of some force. A very well-made disc may be made to rotate for hours, and would, by preserving its original plane of rotation, even show the effect of the earth's diurnal motion. Mr. Bessemer designed such a gyroscope to move the valves of his hydraulic apparatus, and so to keep his swinging saloon as persistently horizontal as the gyroscope itself. Mr. Bessemer's ship was 350 ft. long, and each end, for a distance of 48 ft., was only about 4 ft. from the line of floating. Above the low ends a breastwork was raised, about 8 ft. high, and 254 ft. long. In the centre, and occupying the space of 90 ft., was the swinging saloon intended for first-class passengers. At either end of this apartment were the engines and boilers. The engines were oscillating and expansive, working up to 4,600 horse-power, which could be increased to 5,000. There were two pairs of engines, one set at either end of the ship, and each having two cylinders of 80 in. in diameter, and a stroke of 5 ft., working with steam of 30 lbs. pressure per square inch, supplied from four box-shaped boilers, each boiler having four large furnaces. The paddle-wheels, of which there were a pair on either side of the vessel, were 27 ft. 10 in. in diameter outside the outer ring, and each wheel has twelve feathering floats. The leading pair of wheels, when working at full speed, were to make thirty-two revolutions per minute, and the following pair of wheels move faster.

Entrance into the Bessemer saloon was gained by two broad staircases leading to one landing, and a flexible passage from this point to the saloon. The saloon rested on four steel gudgeons, one at each end, and two close together near the middle. These were not only to support the saloon, but also to convey the water to the hydraulic engines, by which the saloon was to be kept steady. For this purpose the after one was made hollow, and connected with the water mains from powerful engines, and also with a supply-pipe leading to a central valve-box, by means of which the two hydraulic cylinders on either side were supplied with water. Between the two middle gudgeons, a gyroscope, worked by a small turbine, filled with water from one of the gudgeons, enabled Mr. Bessemer to dispense with the services of a man, and thus completed his scheme of a steady saloon, by making the machinery completely automatic. The saloon was 70 ft. long, 35 ft. wide, and 20 ft. high. The Bessemer ship proved to be a total failure, and never went to sea as a passenger boat.

On board of some modern warships where speed is essential, and where the engines are driven at a very great number of revolutions per minute, as in the case of torpedo-boat catchers, the vibration throughout the whole of the vessel becomes extremely trying, not only for the nerves of the crew, but for the security of the structure itself. The cause of this vibration and consequent strain and loss of power is not far to seek. The cylinders of marine engines are always of a large diameter, 6 feet, 8 feet, or even more sometimes, and the pistons and piston-rods are necessarily of great strength and corresponding weight. Now, at every half revolution of the engines, this heavy mass of piston and piston-rod, though moving at an exceedingly high speed in the middle of the stroke, has to be brought to a stand-still, and an equal velocity in the opposite direction imparted to it. A large portion of the power is therefore uselessly expended in stopping a great moving mass, and reversing its motion. All the force required to do this reacts on the vessel's frame. Many attempts have been made to construct rotatory steam-engines, and some hundreds of patents taken out for such inventions, which in general have a piston revolving about a shaft; but the great friction, and consequent liability to wear out, have prevented their practical use.

Lately, a method of using steam on the principle embodied in the water turbine has been developed, and within the last six or seven years has found successful application in propelling electro-dynamos at very high speeds. In the steam turbine there are no pistons, piston-rods, or other reciprocating parts, the effect depending on the same kind of reaction that is taken advantage of in the water turbine (which has a high efficiency in giving out a large proportion of energy), and the power is applied with smoothness and an entire absence of the oscillations that would shake to pieces any vessel that an ordinary steam-engine could propel at the same rate.

The advantages of the steam turbine have been proved by the performances of a small experimental vessel lately built at Newcastle, and appropriately named the *Turbinia*. She is only 100 feet in length, and 9 feet in breadth, with a displacement of some 44 tons. Now the highest record speed for any vessel of that size is 24 knots per hour; but the *Turbinia*, in a heavy sea, showed, at a measured mile, the speed of 32¾ knots, which is believed to be greater than that of any craft now afloat, being nearly 37¾ miles an hour, or equal to that of an ordinary railway train. Besides that, it has been found by experiment, that an arrangement of the blades of the screw propeller more suitable to high velocities will enable a still greater speed to be obtained. The weight of the turbine engines of this vessel is only 3 tons, 13 cwts., and the whole weight of the machinery, including boilers and condensers, is only 22 tons, with an indicated H.P. of 1576, and a steam consumption of but 16 lbs. per hour. The weight of the turbine is only one-fifth of that of marine engines of equal power; the space occupied is smaller; the initial cost is less; not so much superintendence is required; the charges of maintenance are diminished; reduced dimensions of propeller and shaft suffice; vibration is eliminated; speed is increased; and greater economy of fuel is secured.

THE RIVER AND LAKE STEAM-BOATS OF AMERICA.

THE chapter on "Steam Navigation," in the foregoing pages, has dealt mainly with the progress of the ocean-going steam-ship, from the establishment of regular transatlantic services down to the building of the splendid liners, the *New York* and the *Paris*, and we have recorded, in addition, the performances of the pair of hitherto unsurpassed sister ships, the *Campania* and the *Lucania*. The importance and interest attaching to steam navigation is, however, by no means confined to ocean-going vessels, and the chapter demands a supplementary notice of the great developments of the steam-ship in other parts of the world than Britain, more particularly where great rivers, navigable for hundreds of miles, and lakes, spreading their waters over vast areas, present conditions of traffic and opportunities for adaptation to an extent that could not be required within the range of Britain or British oceanic lines.

PLATE IX.

THE "MARY POWELL."

If the reader will cast his eye on the map of the United States, he will see towards the northern boundary a great fresh-water system, comprising five enormous lakes, the least of which is nearly two hundred, and the largest nearly three hundred miles in length, in all presenting a total area greater by far than that of England and Scotland together thrice told. This lake system has a line of coast to be reckoned only by thousands of miles, and for a long time an enormous traffic has been carried across its waters by sailing vessels of all kinds, two or three-masted schooners, brigs, and other craft, carrying wood, stone, lime, and other commodities. On the map, the position of the Detroit River, which leads from the southern extremity of Lake Huron to Lake Erie, will readily be recognised, and this strait, which is in the only line of transport from the three great upper lakes, formerly presented all the picturesqueness that crowds of boats of every build could impart. Especially was this the case at Amherstburg, its southern extremity, where sometimes a northern wind would make the passage impracticable for several days in succession, and a fleet of a hundred or two hundred sailing vessels would collect to await the opportunity of a favouring breeze in order to carry them against the current to Port Huron. Then, taking advantage of the right moment, they would set their sails, and in a compact body move slowly up the strait. This was not quick enough to meet the traffic, and, before long, larger vessels were built, which were towed up and down the Detroit by steam-tugs. The next step of replacing sailing ships by steam-vessels was not long in following, and though there still exist fine specimens of sailing craft on the lakes, their day may be said to be over. The navigation of these lakes, before the extensive development of the railway systems near their shores, comprised a large passenger traffic, which was carried on by big paddle-wheel steamers, and at the time of the great westward set of emigration to Michigan, Wisconsin, and Minnesota, these steamers were crowded to their utmost capacity. The great improvement which in recent years has become possible for passenger steamers in speed, cabin accommodation, and other particulars, above all, the growth of great cities on the shores, the progress of the territories adjoining the lake system, and other circumstances, are now combining to renew the passenger traffic on a larger scale than ever. " Fifteen millions of people," says Mr. H. A. Griffin, the Secretary of the Cleveland Board of Control (*Engineering Magazine*, iv., 819), " now live upon the shore lines of the lakes, or within six hours' travel by rail, and nearly all of that population is south of the United States boundary line. The territory directly tributary to the lakes, north and south of the line, is capable of easily maintaining a population of 100,000,000. . . . It does not require a very lively imagination to foresee the Great Lakes surrounded by the most prosperous and progressive people on earth, and crossed and recrossed by scores of lines of passenger steam-ships, in addition to a still greater number of freight lines." The number of first-class passenger steamers already launched or on the stocks is an indication that the revival of passenger traffic will not lag or be delayed.

The unique conditions and requirements of this lacustrine traffic were bound to lead to types of vessels differing in many respects from the steam-ships to be seen in the harbours of Great Britain. The introduction of iron shipbuilding gave a great impetus to the construction of the lake steamers, for vessels of more than 3,000 tons could be built with a comparatively shallow draught of water (15½ feet), which was one of the necessities of the situation. As far back as 1872, iron shipbuilding had been fully established at Cleveland and Detroit, and at the latter place scores of splendid steel steam-ships have been turned out. The Cleveland builders have not been far behind, and Buffalo, Milwaukee, Chicago, and other places, have followed suit. At the beginning of 1893, there were on the lakes more than fifty vessels of over 2,000 tons each, while the total number of steam vessels of all kinds was considerably over 1,600, and sailing vessels with steam-tugs counted over 2,000. ➤The tonnage of the ships on the lakes has been estimated at about 36 per cent. of the whole mercantile marine of the United States, and it is said that 40,000 men are employed upon the vessels. The total freight passing Detroit in 1892 was calculated to exceed 34,000,000 tons, an amount greater than the whole foreign and coasting trade of the port of London. There are more than thirty shipbuilding concerns on the lakes, and some of them possess large dry docks of their own ; but there are also independent companies owning dry docks of great size. Some of these shipbuilding establishments have turned out steel ocean-going tugs, paddle and screw passenger steamers, cargo-carrying boats, vessels for carrying railway trains across the Detroit river, etc., etc.

FIG. 68a.—*A Whaleback Steamer, No. 85, Built at West Superior, Wisconsin.*

The extent and importance which steam navigation has attained in a definite region have been indicated in the preceding paragraphs ; but an attempt to show by illustration and description the several characteristic forms the steam-ship has now assumed in these lacustrine waters would carry us far beyond our allotted limits. The steam vessels now on the lakes are almost exclusively actuated by screw-propellers, whether they are passenger or freight boats. The boilers and engines are near the stern, and the hulls are usually of great length ; in fact, some of these steamboats will compare in dimensions with the *Persia*, which was the transatlantic marvel about the year 1857. (See p. 137.) Such is the *Mariposa*, launched in 1892, which is 350 feet long and 45 feet broad, carrying 3,800 net tons, with a draught of only 15½ feet. There are others, 380 feet long, with engines of 7,000 horse-power, steaming at 20 miles an hour, and providing ample accommodation for 600 passengers. The newest and most novel type of steam-ship on the lakes is the "whaleback." The celerity with which ships of this kind have been constructed on occasion is perfectly marvellous. One of them, named the *Christopher Columbus*, designed to carry passengers to and from the World's Fair at Chicago in 1893, was launched in fifty-six days after the keel had been laid, yet it was a ship intended to carry 5,000 passengers, having a length over all of 362 feet, breadth 42 feet, depth 24 feet. The "whaleback" steamers are designed to give the greatest carrying capacity with a given draught of water, and all the structures usually fitted to the upper deck of a steamer are in them replaced by the plain curved and closed deck, over which, when the vessel is in a storm, waves may sweep harmlessly, thus avoiding the shocks received by ships with high sides.

The river steam-boat was, as we have seen, nearly coeval with the nineteenth century, and although its practicability was first demonstrated in British waters, regular steam navigation was not established until a few years afterwards, when, in 1807, Robert Fulton placed on the River Hudson its first steam-boat. To this others were soon added, so that in 1813 there were six steam-boats regularly plying on the Hudson before a single one ran for hire on the Thames. An article by Mr. Samuel Ward Stanton, in a recent number of *The Engineering Magazine*, gives a very full account of the Hudson River steam-boats from the beginning down to 1894, and to this article we are mainly indebted for the details we are about to give.

The Hudson River washes the western shore of Manhattan Island, on which stands by far the greater part of the city of New York, with its vast population. The river is here straight, and has a nearly uniform width of one mile; at New York it is commonly called the *North River*, because of the direction of its course, for it descends from almost the due north. It is not one of the great rivers of the United States as regards length or extent of navigation; not, *e.g.*, like the Mississippi and the Missouri, which are ascended by steam-boats to thousands of miles above their mouths; but it has one of the world's great capitals on its shores, and at the quays, which occupy both its banks to the number of eighty or more, may be seen in multitudes some of the finest ocean-going steam-ships, trading to every considerable port in the world. The North River separates New York from what are practically the populous suburbs of Jersey City and Hoboken, though these are controlled by their own municipalities.

It was on the River Hudson that steam navigation was inaugurated by Fulton with a vessel which was 133 feet long, 18 feet broad, and 7 feet deep, and was named the *Clermont*. The speed attained was but five miles an hour. The first trip was made on the 7th August, 1807, to Albany, 150 miles up the river from New York, with twenty-four passengers on board, and the new kind of locomotion was so well patronised that during the following winter, when the Hudson navigation had to be suspended on account of the ice, it was considered expedient to enlarge the capacity of the boat by adding both to her length and width; at the same time her name was changed to *The North River*, and she plied regularly for several seasons afterwards. Her speed down the river with the current was evidently greater than that of the first trip up the river, for on 9th November, 1809, the New York *Evening Post* announced that

"The North River steam-boat arrived this afternoon in twenty-seven and a half hours from Albany, with sixty passengers."

The paddle-wheels were of a primitive form, and as they were unprovided with paddle-boxes, the arrangement had the appearance of a great under-shot mill-wheel on each side of the boat, above the deck of which was placed the steam-engine, a position it has retained in all these river-boats, in which a huge, rhombus-shaped beam, oscillating high above the deck, is a conspicuous feature. Another boat of much larger dimensions was built the following year, having a tonnage of nearly 300, and from that time there has been a more or less regular increase in the sizes of the vessels, until in 1866 a tonnage of nearly 3,000 was reached. In 1817 a vessel called the *Livingstone* was launched, which was able to go up to Albany in eighteen hours. In 1823 was launched the *James Kent*, a novel feature in which vessel was the boiler made of copper, and weighing upwards of 30 tons. It was so planned that if it happened to burst, the hot water would be carried through the bottom of the vessel by tubes or hollow pillars. From this it appears that considerable apprehension existed as to the liability of the boilers exploding. We are told that the cost of the copper boiler was in this case nearly one-third of that of the whole vessel. The cabins are described as having been very handsomely fitted up, and the speed was such that fourteen hours sufficed for the trip up river to Albany. Many fine boats were placed on the river during the twenty following years, and these were marked by various improvements, as when, in 1840, anthracite coal was for the first time substituted for wood as the fuel for the furnaces, with the effect of reducing the cost of this item to one-half. Then, in 1844, iron began to be used for construct-ing the hulls, and a few years afterwards, steamers having a speed of twenty miles an hour and over, became quite common. In 1865, and again in the eighties, some four screw-propeller boats were built; but this type does not appear to have found much favour on the Hudson, for the large paddle-wheels and the single or double beam, working high above the deck, have continued the almost universal form of construction. A very popular and famous boat was placed on the Hudson in 1861. This was the *Mary Powell*, called the "Queen of the Hudson," which, although a boat of moderate tonnage (983), was able on occasion to steam at the rate of twenty-five miles an hour. This vessel was placed on the line between New York and Rondont, and was still running in 1894.

One of the most modern and most elegant boats on the Hudson is the *New York*, launched in 1887, and declared by Mr. Stanton to be one of the finest river steam-boats in the world, well arranged, and beautifully finished and furnished. She is built on fine lines, is 311 feet long, 40 feet broad, and with a tonnage of 1,552, draws only $12\frac{1}{4}$ feet of water. She can steam at twenty miles an hour, and is placed on one of the New York and Albany lines. Throughout the summer there are both day and night boats for Albany, and the latter especially are of great size, three stories high, and provided with saloons, state-rooms, and, in fact, all the accommodation of a luxurious first-class hotel. The vessels named in this notice include but a few of the splendid boats that ply on the River Hudson, and, in respect of their numbers, speed, and comfort, it may safely be asserted that they cannot be equalled on any other river in the world.

PLATE X.

THE " NEW YORK."

FIG. 69.—*H.M.S. Devastation in Queenstown Harbour.*

SHIPS OF WAR.

——◆——

" TAKE it all in all, a ship of the line is the most honourable thing that man, as a gregarious animal, has ever produced. By himself, unhelped, he can do better things than ships of the line; he can make poems, and pictures, and other such concentrations of what is best in him. But as a being living in flocks, and hammering out with alternate strokes and mutual agreement, what is necessary for him in those flocks to get or produce, the ship of the line is his first work. Into that he has put as much of his human patience, common sense, forethought, experimental philosophy, self-control, habits of order and obedience, thoroughly wrought hand-work, defiance of brute elements, careless courage, careful patriotism, and calm expectation of the judgment of God, as can well be put into a space of 300 ft. long by 80 ft. broad. And I am thankful to have lived in an age when I could see this thing so done." So wrote Mr. Ruskin about forty years ago, referring, of course, to the old wooden line-of-battle ships. It may be doubted whether he would have written thus enthusiastically about so unpicturesque an object as the *Glatton*, just as it may be doubted whether the armour-plated steamers will attain the same celebrity in romance and in

149

verse as the old frigates with their "wooden walls." Certain it is that the
patience, forethought, experimental philosophy, thoroughly wrought hand-
work, careful patriotism, and other good qualities which Mr. Ruskin saw
in the wooden frigates, are not the less displayed in the new ironclads.

Floating batteries, plated with iron, were employed in the Crimean War
at the instigation of the French Emperor. About the same time the question of protecting ships of war by some kind of defensive armour was
forced upon the attention of maritime powers, by the great strides with
which the improvements in artillery were advancing; for the new guns
could hurl projectiles capable of penetrating, with the greatest ease, any
wooden ship afloat. The French Government took the initiative by constructing *La Gloire*, a timber-framed ship, covered with an armour of rolled
iron plates, 4½ in. thick. The British Admiralty quickly followed with the
Warrior, a frigate similar in shape to the wooden frigates, but built on
an iron frame, with armour composed of plates 4½ in. thick, backed by 18 in.
of solid teak-wood, and provided with an inner skin of iron. The *Warrior*
was 380 ft. long, but only 213 ft. of this length was armoured. The defensive
armour carried by the *Warrior*, and the ironclads constructed immediately
afterwards, was quite capable of resisting the impact of the 68 lb. shot,
which was at that time the heaviest projectile that could be thrown by
naval guns. But to the increasing power of the new artillery it soon became
necessary to oppose increased thickness of iron plates. The earlier ironclads carried a considerable number of guns, which could, however, deliver
only a broadside fire, that is, the shots could, for the most part, be sent
only in a direction at right angles to the ship's length, or nearly so. But
in the more recently built ironclads there are very few guns, which are,
however, six times the weight of the old sixty-eight pounders, and are capable of hurling projectiles of enormous weight. The ships built after the
Warrior were completely protected by iron plates, and the thickness of the
plates has been increased from time to time, with a view of resisting the
increased power which has been progressively given to naval guns. A
contest, not yet terminated, has been going on between the artillerist and
the ship-builder; the one endeavouring to make his guns capable of penetrating with their shot the strongest defensive armour of the ships, the
other adding inch after inch to the thickness of his plates, in order, if possible, to render his ship invulnerable.

One of the finest of the large ironclads is the *Hercules*, of which a section
amidships is presented on the next page. This ship is 325 ft. in length,
and 59 ft. in breadth, and is fitted with very powerful engines which will
work up to 8,529 indicated horse-power. The tonnage is 5,226; weight of
hull, 4,022 tons; weight of the armour and its backing, 1,690 tons; weight
of engines, boilers, and coals, 1,826 tons; total with equipment and armament, 8,676 tons. Although the *Hercules* carries this enormous weight of
armour and armament, her speed is very great, excelling, in fact, that of
any merchant steamer afloat, for she can steam at the rate of nearly 17
miles an hour. She also possesses, in a remarkable degree, the property
which naval men call *handiness;* that is, she can be quickly turned round
in a comparatively small space. The handiness of a steamer is tested by
causing her to steam at full speed with the helm hard over, when the vessel
will describe a circle. The smaller the diameter of that circle, and the
shorter the time required to complete it, the better will the vessel execute
the movements required in naval tactics. Comparing the performances of
the *Warrior* and the *Hercules,* we find that the smallest circle the former

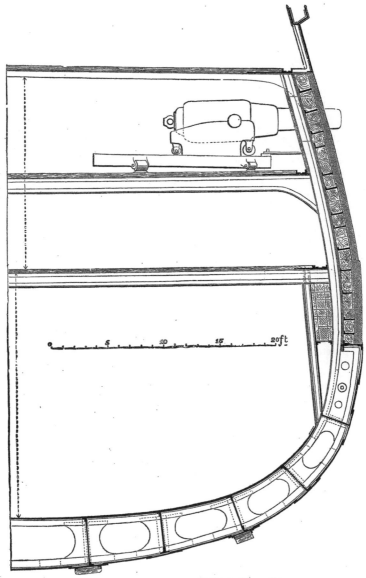

FIG. 70.—Section of H.M.S. Hercules.

can describe is 1,050 yards in diameter, and requires nine minutes for its completion, whereas the latter can steam round a circle of only 560 yards diameter in four minutes. The section (Fig. 70) shows that, like the *Great Eastern*, the *Hercules* is constructed with a double hull, so that she would be safe, even in the event of such an accident as actually occurred to the *Great Eastern*, when a hole was made by the stripping off of her bottom plates, 80 ft. long and 5 ft. wide. The defensive armour of the *Hercules* is, it will be observed, greatly strengthened near the water-line, where damage to the ship's side would be most fatal. The outer iron plates are here 9 in. thick, while in other parts the thickness is 8 in., and in the less important positions 6 in. The whole of the hull is, however, completely protected above the water-line, and the iron plates are backed up by solid teak-wood for a thickness of from 10 in. to 12 in. The teak is placed between girders, which are attached to another iron plating 1½ in. thick, supported by girders 2 ft. apart. The spaces between these girders are also filled with teak, and the whole is lined with an inner skin of iron plating, ¾ in. thick. The belt along the water-line has thus altogether 11¼ in. of iron, of which 9 in. are in one thickness, and this part is, moreover, backed by additional layers of teak, as shown in the section; so that, besides the 11¼ in. of iron, the ship's side has here 3 ft. 8 in. total thickness of solid teak-wood. The deck is also covered with iron plates, to protect the vessel from vertical fire. The *Hercules* carries eight 18-ton guns as her central battery, and two 12-ton guns in her bow and stern: these guns are rifled, and each of the larger ones is capable of throwing a shot weighing 400 lbs. The guns can be trained so as to fire within 15° of the direction of the keel; for near the ends of the central battery the ports are indented, and the guns are mounted on Scott's carriages, in such a manner that any gun-slide can be run on to a small turn-table, and shunted to another port, just as a railway-carriage is shunted from one line to another. Targets for artillery practice were built so as to represent the construction of the side of the *Hercules*, and it was found, as the result of many experiments, that the vessel could not be penetrated by the 600 lb. shot from an Armstrong gun, fired at a distance of 700 yds. The production of such iron plates, and those of even greater thickness which have since been used, forms a striking example of the skill with which iron is worked. These plates are made by rolling, and it will be understood that the machinery used in their formation must be of the most powerful kind, when it is stated that plates from 9 in. to 15 in. thick are formed with a length of 16 ft. and a breadth of 4 ft. The plates are bent, while red hot, by enormous hydraulic pressure, applied to certain blocks, upon which the plates are laid, the block having a height adjusted according to the curve required. The operation requires great care, as it must be accomplished without straining the parts in a manner injurious to the strength of the plate.

Fig. 71 on the next page is the section of another ship of war, the *Inconstant*, which has not, like the *Hercules*, been designed to withstand the impact of heavy projectiles, but has been built mainly with a view to speed. The *Inconstant* has only a thin covering of iron plating, except in that portion of the side which is above water, where there is a certain thickness of iron diminishing from the water-line upwards, but not enough to entitle the *Inconstant* to be classed as an armoured vessel. This ship, however, may be a truly formidable antagonist, for she carries a considerable number of heavy guns, which her speed would enable her to use with great effect against an adversary incapable of manœuvring so rapidly. She could give

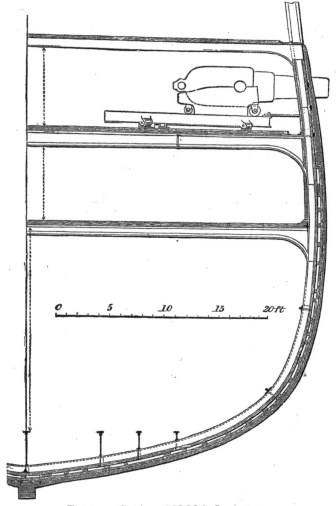

FIG. 71.—*Section of H.M.S. Inconstant.*

chase, or could run in and deliver her fire, escaping by her speed from hostile pursuit in cases where the slower movements of a ponderous iron-clad would be much less effective. The *Inconstant* carries ten 12-ton guns of 9 in. calibre, and six 6-ton 7 in. guns, all rifled muzzle-loaders, mounted on improved iron carriages, which give great facilities for handling them.

The ship is a frigate 338 ft. long and 50 ft. broad, with a depth in the hold of 17 ft. 6 in.   She is divided by bulkheads into eleven water-tight com-partments.   The engines are of 6,500 indicated horse-power, and the vessel attains an average speed of more than 18½ miles per hour.

  A new system of mounting very heavy naval guns was proposed by

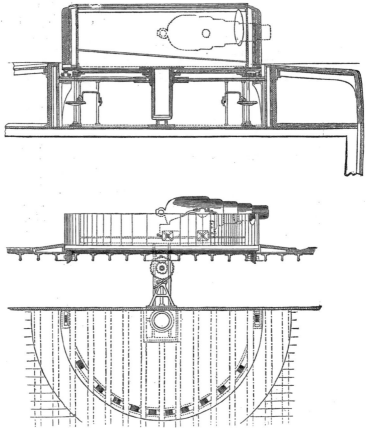

FIG. 72.—*Section, Elevation, and Plan of Turret of H.M.S. Captain.*

Captain Coles about 1861. This plan consists in carrying one or two very heavy guns in a low circular tower or turret, which can be made to revolve horizontally by proper machinery.   The turret itself is heavily armoured, so as to be proof against all shot, and is carried on the deck of the ship, which is so arranged that the guns in the turret can be fired at small angles with the keel. The British Admiralty having approved of Captain

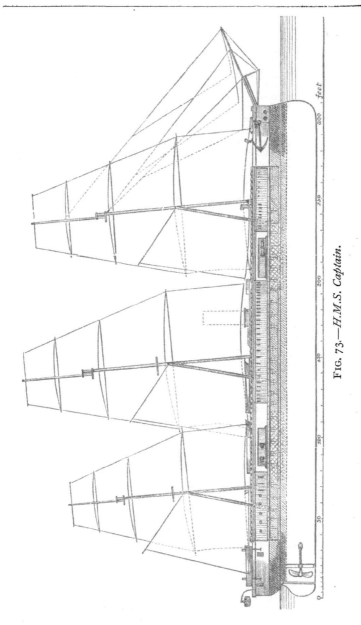

FIG. 73.—*H.M.S. Captain.*

Colès' plans, two first-class vessels were ordered to be built on the turret system. These were the *Monarch* and the *Captain*—the latter of which we select for description on account of the melancholy interest which at taches to her. On page 155 a diagram is given representing the profile of the *Captain*, in which some of the peculiarities of the ship are indicated—the turrets with the muzzles of two guns projecting from each being easily recognized. The *Captain* was 320 ft. long and 53 ft. wide. She was covered with armour plates down to 5 ft. below the water-line, as repre sented by the dark shading in the diagram. The outer plating was 8 in. thick opposite the turrets, and 7 in. thick in other parts. It was backed up by 12 in. of teak ; there were two inner skins of iron each $\frac{3}{4}$ in. thick, then a framework with longitudinal girders 10 in. deep. The deck was plated in the spaces opposite the turrets with iron $1\frac{1}{2}$ in. thick. The *Captain* was fitted with twin screws—that is, instead of having a single screw, one was placed on each side, their shafts being, of course, parallel with the vessel's length. The object of having two screws was not greater power, for it is probable that a single screw would be more effectual in propelling the ship ; but this arrangement was adopted because it was considered that, had only one screw been fixed, the ship might easily be disabled by the breaking of a blade or shaft ; whereas in the case of such an accident to one of the twin screws, the other would still be available. The twin screws could also be used for steering, and the vessel could be controlled without the rudder, as the engines were quite independent of each other, each screw having a separate pair. The diameter of the screws was 17 ft. The erections which are shown on the deck between the turrets afforded spacious quarters for the officers and men. These structures were about half the width of the deck, and tapered off to a point towards the turrets, so as leave an unimpeded space for training the guns, which could be fired at so small an angle as 6° with the length of the vessel. Above these erections, and quite over the turrets, was another deck, 26 ft. wide, called the "hurricane deck." The ship was fully rigged and carried a large spread of canvas. But the special features are the revolving turrets, and one of these is represented in detail in Fig. 72, which gives a section, part elevation, and plan. Of the construction of the turret, and of the mode in which it was made to revolve, these drawings convey an idea sufficiently clear to obviate the necessity of a minute description. Each turret had an outside diameter of 27 ft., but inside the diameter was only 22 ft. 6 in., the walls being, therefore, 2 ft. 3 in. thick—nearly half this thickness consisting of iron plating. Separate engines were provided for turning the turrets, and they could also be turned by men working at the handles shown in the figures. Each turret carried two 25-ton Armstrong guns, capable of receiving a charge of 70 lbs. of gunpowder, and of throwing a 600 lb. shot.

After some preliminary trials the *Captain* was sent to sea, and behaved so well, that Captain Coles and Messrs. Laird, her designer and contrac-tors, were perfectly satisfied with her qualities as a sea-going ship. She was then sent in the autumn of 1870 on a cruise with the fleet, and all went well until a little after midnight between the 6th and 7th September, 1870, when she suddenly foundered at sea off Cape Finisterre. The news of this disaster created a profound sensation throughout Great Britain, for, with the exception of nineteen persons, the whole crew of five hundred persons went down with the ship. Captain Coles, the inventor of the turrets, was in the ill-fated vessel and perished with the rest, as did also Captain Burgoyne, the gallant commander, and the many other distinguished naval officers who had been appointed to the ship ; among the rest was a son of Mr. Childers, then First Lord of the Admiralty. Although the night on which this unfortunate ship went down was squally, with rain, and a heavy sea running, the case was not that of an ordinary shipwreck in which a vessel is overwhelmed by a raging storm. It might be said, indeed, of the loss of the *Captain* as of that of the *Royal George :*

> " It was not in the battle ;
> No tempest gave the shock ;
> She sprang no fatal leak ;
> She ran upon no rock."

One of the survivors, Mr. James May, a gunner, related that, shortly after midnight he was roused from his sleep by a noise, and feeling the ship uneasy, he dressed, took a light, and went into the after turret, to see if the guns were all right. He found everything secure in the turret, but that moment he felt the ship heel steadily over, and a heavy sea having struck her on the weather side, the water flowed into the turret, and he got out through the hole in the top of the turret by which the guns were pointed, only to find himself in the water. He swam to the steam-pinnace, which he saw floating bottom upwards, and there he was joined by Captain Burgoyne and a few others. He saw the ship turn bottom up, and sink stern first, the whole time from her turning over to sinking not being more than a few minutes. Seeing the launch drifting within a few yards, he called out, " Jump, men ! it is your last chance." He jumped, and with three others reached a launch, in which were fifteen persons, all belonging to the watch on deck, who had found means of getting into this boat. One of these had got a footing on the hull of the ship as she was turning over, and he actually walked over the bottom of the vessel, but was washed off by a wave and rescued by those who in the meantime had got into the launch. It appears that Captain Burgoyne either remained on the pinnace or failed to reach the launch. Those who were in that boat, finding the captain had not reached them, made an effort to turn their boat back to pick him up, but the boat was nearly swamped by the heavy seas, and they were obliged to let her drift. One man was at this time washed out of the boat and lost, after having but the moment before exclaimed, " Now, lads, I think we are all right." After twelve hours' hard rowing, without food or water, the survivors, numbering sixteen men and petty officers and three boys, reached Cape Finisterre, where they received help and attention. On their arrival in England, a court-martial was, according to the rules of the service, formally held on the survivors, but in reality it was occupied in investigating the cause of the catastrophe. The reader may probably be able to understand what the cause was by giving his attention to some general considerations, which apply to all ships whatever, and by a careful examination of the diagrams, Figs. 74 and 75, which are copied from diagrams that were placed in the hands of the members of the court-martial. The letters B and G and the arrows are, however, added, to serve in illustration of a part of the explanation. The vessel is represented as heeled over in smooth water, and the gradations on the semicircle in Fig. 74 will enable the reader to understand how the heel is measured by angles. If the ship were upright, the centre line would coincide with the upright line, marked O on the semicircle, and drawn from its centre. Suppose a level line drawn through the centre of the semicircle, and let the circumference between the point where the last line cuts it and the point O be divided into ninety equal parts, and let these parts be numbered, and straight lines

drawn from the centre to each point of division. In the figure the lines
are drawn at every fifth division, and the centre line of the ship coincides
with that drawn through the forty-fifth division. In this case the vessel is
said to be inclined, or heeled, at an angle of forty-five degrees, which is
usually written 45°. In a position half-way between this and the upright
the angle of heel would be 22½°, and so on. The reader no doubt perceives
that a ship, like any other body, must be supported, and he is probably
aware that the support is afforded by the upward pressure of the water.
He may also be familiar with the fact that the weight of every body acts
upon it as if the whole weight were concentrated at one certain point, and
that this point is called the centre of gravity of the body. Whatever may
be the position of the body itself, its centre of gravity remains always at

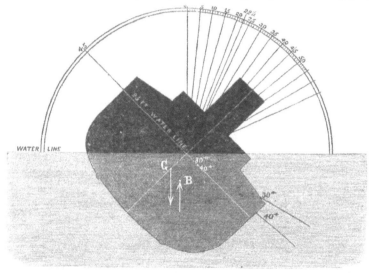

FIG. 74.

the same point with reference to the body. When the centre of gravity
happens to be within the solid substance of a body, there is no difficulty in
thinking of the force of gravitation acting as a downward pull applied at
the centre of gravity. But this point is by no means always within the
substance of bodies: as often as not it is in the air outside of the body.
Thus the centre of gravity of a uniform ring or hoop is in the centre, where,
of course, it has no material connection with the hoop; but in whatever
position the hoop may be placed, the earth's attraction pulls it *as if* this central point were rigidly connected with the hoop, and a string were attached
to the point and constantly pulled downwards. This explanation of the
meaning of centre of gravity may not be altogether superfluous, for, when
the causes of the loss of the *Captain* were discussed in the newspapers, it
became evident that such terms as " centre of gravity " convey to the minds
of many but very vague notions. One writer in a newspaper enjoying a

large circulation seriously attributed the disaster to the circumstance of the ship having lost her *centre of gravity !* The upward pressure of water which supports a ship is the same upward pressure which supported the water before the ship was there—that is, supported the mass of water which the ship displaces, and which was in size and shape the exact counterpart of the immersed part of the ship. Now, this mass of water, considered as a whole, had itself a centre of gravity through which its weight acted downwards, and through which it is obvious that an equal upward pressure also acted. This centre of gravity of the displaced water is usually termed the " centre of buoyancy," and, unlike the centre of gravity, it changes its position with regard to the ship when the latter is inclined, because then the immersed part becomes of a shape different for each inclination of the ship. Now, recalling for an instant the fundamental law of floating bodies— namely, that the weight of the water displaced is equal to the weight of the

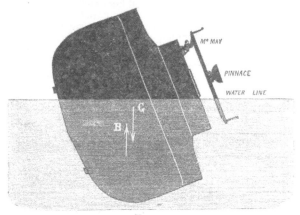

FIG. 75.

floating body— we perceive that in the case of a ship there are two equal forces acting vertically, viz., the weight of the ship or downward pull of gravitation acting at G, Fig. 74, the centre of gravity of the ship, and an equal upward push acting through B, the centre of buoyancy. It is obvious that the action of these forces concur to turn a ship placed as in Fig. 74 into the upright position. It is by no means necessary for this effect that the centre of gravity should be below the centre of buoyancy. All that is requisite for the stability of a ship is, that when the ship is placed out of the upright position, these forces should act to bring her back, which condition is secured so long as the centre of buoyancy is nearer to the side towards which the vessel is inclined than the centre of gravity is. When there is no other force acting on a ship or other floating body, these two points are always in the same vertical line. The two equal forces thus applied in parallel directions constitute what is called in mechanics a " couple," and the effect of this in turning the ship back into the upright position is the same as if a force equal to its weight were applied at the end of a lever equal in length to the horizontal distance between the lines through B and

G. The righting force, then, increases in proportion to the horizontal distance between the two points, and it is measured by multiplying the weight of the ship in tons by the number of feet between the verticals through G and B, the product being expressed in statical foot-tons, and representing the weight in tons which would have to be applied to the end of a lever 1 ft. long, in order to produce the same turning effect. When a ship is kept steadily heeled over by a side wind, the pressure of the wind and the resistance of the water through which the vessel moves constitute another couple exactly balancing the righting couple. The moment of the righting couple, or the righting force, or statical stability as it is also called, is determined by calculation and experiment from the design of the ship, and from her behaviour when a known weight is placed in her at a known distance from the centre. Such calculations and experiments were made in the case of the *Captain*, but do not appear to have been conducted with sufficient care and completeness to exhibit her deficiency in stability. After the loss of the ship, however, elaborate computations on these points were made from the plans and other data. The following table gives some of the results, with the corresponding particulars concerning the *Monarch* for the sake of comparison :

	Monarch.	*Captain.*
I. Angle at which the edge of the deck is immersed	28°	14°
II. Statical righting force in foot-tons at the angle at which the deck is immersed	12,542	5,700
III. Angle of greatest stability	40°	21°
IV. Greatest righting force in foot-tons	15,615	7,100
V. Angle at which the righting force ceases	59°	54°
VI. Reserve of dynamical stability at an angle of 14° in *dynamical* foot-tons	6,500	410

From No. V. in the above table we learn that if the *Captain* had been heeled to 54°, the centre of gravity would have overtaken the centre of buoyancy—that is, the two would have been in one vertical line. Any further heeling would have brought the points into the position shown in Fig. 75, where it is obvious that the action of the forces is now to turn the vessel still more on its side, and the result is an upsetting couple instead of a righting couple.

These figures and considerations refer to the case of the vessel floating in smooth water, but the case of a vessel floating on a wave is not different in principle. The reader may picture to himself the diagrams inclined so that the water-line may represent a portion of the wave's surface ; then he must remember that the very action which heaves up the water in a sloping surface is so compounded with gravity, that the forces acting through G and B retain nearly the same position relatively to the surface as before.

No. VI. in the foregoing table requires some explanation. To heel a ship over to a certain angle a certain amount of *work* must be done, and in the scientific sense *work* is done only when something is moved through a space against a resistance. When the weight of a ton is raised 1 ft. high, one foot ton of work is said to be done; if 2 tons were raised 1 ft., or 1 ton were raised 2 ft., then two foot-tons of work would be done, and so on. The same would be the case if a pressure equal to those weights were applied so as to move a thing in any direction through the same distances. It should be carefully noticed that the foot-ton is quite a different unit in this case from what it is as the moment of a couple. If we heel a ship over by applying a pressure on the masts, it is plain that the pressure must act through a certain space, and the same heel could be caused either by means of a smaller pressure or a greater, according as we apply it higher up or lower down ; but the space through which it must act would vary, so that the product of the pressure and space would, however, be always the same. No. VI. shows the amount of work that would have to be done in order completely to upset each of the vessels when already steadily heeled over to 14°. The amounts in the two cases are so different that we can easily understand how a squall which would not endanger the *Monarch* might throw the *Captain* over. A squall suddenly springing up would do more than heel a vessel over to the angle at which it is able to maintain it : it would swing it beyond that position by reason of the work done on the sails as they are moving over with the vessel, and the latter would come to a steady angle of heel only after a series of oscillations. Squalls, again, which, although suddenly springing up in this manner, could not heel the ship over beyond the angle where the stability vanishes, might yet do so if they were intermittent and should happen to coincide in time with the oscillations of the ship—just as a series of very small impulses, coinciding with the time of the vibrations of a heavy pendulum, may accumulate so as to increase the range of vibration to any extent. It is believed that in the case of the *Captain* the pressure of the wind on the under-side of the hurricane assisted in upsetting the vessel. This, however, could only have exerted a very small effect compared to that produced by the sails. The instability of the *Captain* does not appear to have been discovered by such calculations as were made before the vessel went to sea. It was observed, however, that the ship when afloat was 1 ft. 6 in. deeper in the water than she should have been—in other words, the freeboard, or side of the ship out of the water, instead of being 8 ft. high as intended, was only 6 ft. 6 in., and such a difference would have a great effect on the stability.

The turret system has been applied to other ships on quite a different plan. Of these the *Glatton* is one of the most remarkable. Her appearance is very singular, and totally unlike that which we look for in a ship as may be seen by an inspection of Fig. 76, page 162. The *Glatton*, which was launched in 1871, is of the *Monitor* class, and was designed by Mr. E J. Reed, who has sought to give the ship the most complete protection With this view the hull is covered with iron plates below the water-line and the deck also is cased with 3 in. iron plates, to resist shot or shell fall ing vertically. The base of the turret is shielded by a massive breastwork which is a peculiarity of this ship. The large quantity of iron required for all these extra defences has, of course, the effect of increasing the immersion of the vessel, and therefore of diminishing her speed. The freeboard when the ship is in ordinary trim is only 3 ft. high, and means are provided for admitting water to the lowest compartment, so as to increase the immersion by 1 ft., thus reducing the freeboard to only 2 ft. when the vessel is in fighting trim, leaving only that small portion of the hull above water as a mark for the enemy. The water ballast can be pumped out when no longer needed. The *Glatton* is 245 ft. long and 54 ft. broad, and she draws 19 ft. of water with the freeboard of 3 ft., displacing 4,865 tons of water, while, with the 2 ft. freeboard, the displacement is 5,179 tons. This ship

11

cost £210,000. Mr. Reed
wished to construct a vessel
of much larger size on the
same plan — a proposal to
which, however, the Ad-
miralty did not then con-
sent. The *Glatton* is, nevertheless, one of the most
powerful ships of war ever
built, and may be considered
as an impregnable floating
fortress. Above the water-
line the hull is covered with
armour plates 12 in. thick,
supported by 20 in. of teak
backing, and an inner layer
of iron 1 in. thick. Below
the water-line the iron is 8
in. thick, and the teak 10 in.
The revolving turret carries
two 25-ton guns, firing each
a 600 lb. shot, and is covered
by a massive plating of iron
14 in. in thickness. Besides
this the base of the turret is
protected by a breastwork
rising 6 ft. above the hull.
This breastwork is formed
of plates 12 in. thick, fastened on 18 in. of teak. The
turret rises 7 ft. above the
breastwork, and therefore
the latter in no way impedes
the working of the guns.
The *Glatton* has a great
advantage over all the other
turret ships in having a perfectly unimpeded fore range
for her guns, for there is no
mast or other object to prevent the guns being fired
directly over the bow. There
are no sails, the mast being
intended only for flying signals and hoisting up boats,
&c. The hull is divided by
vertical partitions into nine
water-tight compartments,
and also into three horizontal flats—the lowest being
air-tight, and having ar-
rangements for the admis-
sion and removal of water,

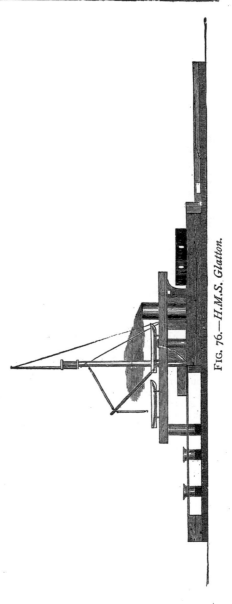

FIG. 76.—*H.M.S. Glatton.*

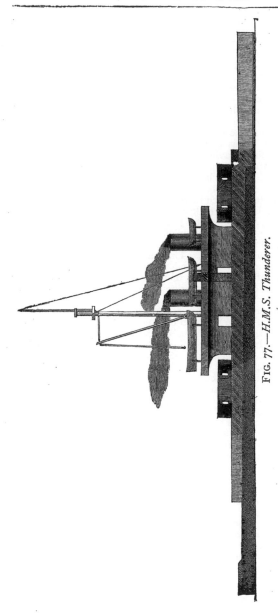

FIG. 77.—*H.M.S. Thunderer.*

as already mention-
ed. The stem of the
ship is protruded
forwards below the
water for about 8 ft.,
thus forming a huge
ram which would it-
self render the *Glat-
ton* a truly formid-
able antagonist at
close quarters even
if her guns were not
used. The engines
are capable of being
worked up to 3,000
horse-power, giving
the ship a speed of
9½ knots per hour,
and means are pro-
vided for turning
the turret by steam
power. The turret
can be rotated by
manual labour, re-
quiring about three
minutes for its com-
plete revolution, but
by steam power the
operation can be
effected in half a mi-
nute. The comman-
der communicates
his orders from the
pilot-house on the
hurricane deck to
the engine-room,
steering-house, and
turret, by means of
speaking-tubes and
electric telegraphs.
The *Glatton* was
not designed to be
ocean-going, but is
intended for coast
defence.

The British navy
contains two pow-
erful turret-ships
constructed on the
same general plan
as the *Glatton*, but
larger, and capable
of steaming at a

greater speed, and of carrying coal for a long voyage. These sister ships are named the *Devastation*, Fig. 69, and the *Thunderer*, Fig. 77. The *Thunderer* has two turrets and a freeboard of 4 ft. 6 in. Space is provided for a store of 1,800 tons of coal, of which the *Glatton* can carry only 500 tons. The vessel is fitted with twin screws, turned by two pairs of independent engines, capable of working up to 5,600 horse-power, and she can steam at the rate of 12 knots, or nearly 14 miles, an hour. With the large supply of coal she can carry, the *Thunderer* could make a voyage of 3,000 miles without re-coaling. Though the freeboard of the heavily-plated hull is only 4 ft. 6 in., a lighter iron superstructure, indicated in the figure by the light shading, rises from the deck to the height of 7 ft., making the real freeboard nearly 12 ft. This gives the ship much greater stability, and prevents her from rolling heavily when at sea. The length is 285 ft. and the width 58 ft., and the draught 26 ft. The hull is double, the distance between the outer and inner skins of the bottom being 4 ft. 6 in. The framing is very strong and on the longitudinal principle, and the keel is formed of Bessemer steel. Each turret is 24 ft. 3 in. in internal diameter, and is built with five layers of teak and iron. Beginning at the inside, there is a lining of 2⅝ in. iron plates; then 6 in. of teak in iron frames, arranged horizontally; 6 in. of armour plates; 9 in. of teak, placed vertically; outside of all, 8 in. armour plates. Each turret carries two Fraser 35-ton guns, rifled muzzle-loaders. The turrets revolve by hand or by steam-power. There are no sails, and thus a clear range for the guns is afforded fore and aft. The bases of the turrets are protected by the armoured breastwork, of which a portion is seen in the figure in advance of the fore turret.

Another very powerful ship of war, which possesses some special features, is represented in the diagram on page 165, Fig. 78. This vessel, named the *König Wilhelm*, was built at Blackwall for the Prussian Government by the Thames Ironworks and Steam Shipbuilding Company, from designs by Mr. Reed. Her length is 365 ft., width 60 ft.; burthen, 6,000 tons; displacement, 8,500 tons. She is framed longitudinally, that is, girders pass from end to end, about 7 ft. apart, and the stem projects into a pointed ram. In this case also the hull is double; there is, in fact, one hull within another, with a space of 4½ ft. between them. The armour plates are 8 in. in thickness, with 10 in. of teak backing; but on the less important parts the thickness of the iron is reduced to 6 in., and in some places to 4 in. This ship has a broadside battery, and there are no turrets, but on the deck there are, fore and aft, two semicircular shields, formed of iron plates and teak, pierced with port-holes for cannon, and also with loop-holes for muskets. From these a fore-and-aft fire may be kept up. The ship is fully rigged, and has also steam engines of 7,000 horse-power, by Maudslay and Co. Her armament consists of four three-hundred-pounders, capable of delivering fore-and-aft as well as broadside fire, and twenty-three other guns of the same size between decks. These guns are all Krupp's steel breech-loaders.

The great contest of armour plates *versus* guns has already been alluded to, and to the remarks then made it may be added that, while on the one hand, guns weighing 110 tons are mounted in turrets, ships are already designed with 18 in. and even 20 in. of steel armour plates. It would be very difficult to predict which side will sooner reach the limit beyond which increase of size and power cannot go. The gradual increase of thickness of plating, attended by increased weight of guns, projectiles, and charges

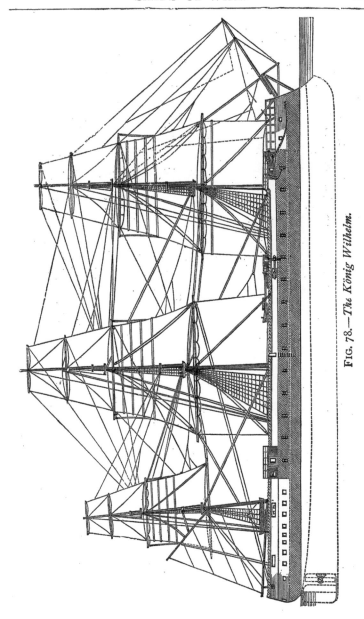

FIG. 78.—The König Wilhelm.

FIG. 78a.—*The " Victoria" leaving Newcastle-on-Tyne.*

of powder, may be illustrated by stating in a condensed form a few
details of some ships, as regards the thickness of armour, and its resisting
power, which is nearly in proportion to the square of its thickness ; and
also some particulars respecting the guns originally carried by those ships.

	Warrior.	Hercules.	Glatton.	Thunderer.	Victoria.
Date when completed...	1861	1868	1872	1877	1889
Thickness of iron plating in inches	4½	9	12	14	18
Relative resisting power of plating	20	81	144	196	324
Guns carried	Cast iron, smooth bore	Wrought iron, rifled.	Wrought iron, rifled.	Wrought iron, rifled.	Steel, Rifled.
Weight of guns in tons	4¾	18	25	35	111
Charge of powder in lbs.	16	60	70	120	960
Weight of projectiles in lbs....	68	400	600	700	1,800
Destructive power of projectiles in foot-tons	452	3,863	5,165	8,404	56,000

One of the latest additions out of the thirty or forty armoured ships that have been added to the British Navy since the preceding pages were written is included in the above table for the sake of comparison. Our ironclad fleet now includes vessels protected and armed in many different ways. Some have the protective armour extended continuously along the water-line, others have it for only a greater or less part of their length. The armaments are also very diverse as to the size of the guns and the way in which they are mounted. A few carry one or two of the huge 110-ton gun mounted in massive revolving turrets; others have their guns in central batteries, or in *barbettes*, and others again are arranged as broadside ships; while these plans are also variously combined so as to form a great number of different types. In the ships built within the last 15 years, steel has been almost invariably used instead of iron for the armour-plating. A great increase of speed has been obtained in late years. The largest British armoured ships yet launched have displacements between 10,000 and 12,000 tons, but another class of first-rate line-of-battle ships of still greater size is in process of construction, and of these it is estimated that four will be completed in 1893. They are all of the same design and armament, and will have a displacement of 14,150 tons, a length of 380 feet, and a breadth of 75 feet. The armour plates at the sides will be 18 inches thick. Each ship will carry four 67-ton breech loading rifled guns, ten 6-inch quick firing guns, and 18 other smaller guns, also quick firing. These vessels are expected to realize a speed of about 20 miles per hour; but this is somewhat less than a few of the heavy ironclads now afloat have given by actual trial, a rate equal to $21\frac{1}{2}$ miles an hour having been attained by some. Several of our rapid unarmoured cruisers are able to steam at 25 miles an hour.

Before the close of 1894, the British navy possessed no fewer than eight of the largest armoured line of battle-ships mentioned in the foregoing paragraph, each being of 14,150 tons displacement, and having engines of 13,000 horse-power. At the same period there were in course of construction four ships surpassing even these in tonnage, though of somewhat less engine-power. Two were building at Portsmouth, to be called the *Majestic* and the *Royal George*, whilst the *Jupiter* was in progress at Glasgow and the *Mars* at Birkenhead. All these are very heavily armoured vessels, each displacing 14,900 tons, provided with engines of 12,000 horse-power, and a very effective armament of guns. Among the powerful ships of the navy may now also be noted the *Blake*, the *Blenheim*, which, although the displacement is only 500 tons greater than that of *König Wilhelm*, have engines of nearly three times the power, namely, of 20,000 horse-power. Of large armoured ships, namely, those of *9,000 tons and upwards*, Great Britain now has afloat at least fifty; and the advance that has taken place in the size and power of war-ships during the last twenty years may be inferred by reference to the foregoing paragraphs giving the dimensions, &c., of the *Glatton* and the *Thunderer*, which paragraphs are, for the sake of comparison, allowed to appear as they did in the first edition (1876) of this book. Besides these very large armoured vessels, of which the smallest is nearly twice as big as the largest of twenty-five years ago, the British navy comprises ships of every size and for every purpose, and so many of them that their names and classifications would occupy many pages.

Two recent additions representing new type of ships claim notice before this article is concluded. These are first the *Terrible*, with a sister ship the *Powerful*. The former, of which a representation* is given in Plate V., is pronounced, for its size, armour, armament, and speed taken together, to be the most powerful cruiser in the world. The length is 538 ft., breadth 71 ft., depth 43 ft., and the displacement is 14,250 tons. A special object in the design of this vessel was high speed, and she is provided with twin-screws and two engines, the combined effort of which is equal to 25,000 horse-power. There are forty-eight boilers and four funnels, the ship being capable of carrying 3,000 tons of coal. The vessel is built on the lines of the great Atlantic steamers, and the engines, guns, and magazines are protected by a thick curved armour deck. The vessel has a speed of 22 knots, or 25¼ miles per hour. Her armament consists of two 22-ton guns, twelve 6-in. quick-firing, and many other smaller machine guns, and she carries besides four submerged torpedo tubes. A second ship to be noted is amongst those designed mainly to exceed all other craft in speed, and ranging in tonnage from 3,800 to 4,500. The *Janus*, a torpedo-boat destroyer of this class, was found, at a recent trial over a measured mile, to attain the then unexampled speed of 28 knots, or 32¼ miles per hour. But even this has been beaten by a new torpedo-boat destroyer, built by Messrs. Yarrow at Poplar for the Russian Government, and launched in August, 1895. This vessel, within a few hours after leaving the stocks, cut through the water at the rate of 30·285 knots, or nearly 35 miles, per hour.

A sad fate befell the *Victoria*, which was one of the heaviest armed of British ships (*vide* page 129), when taking part in some naval manœuvres off Tripoli, on the Syrian coast, where she was the flag-ship of Admiral Tryon, commander-in-chief of the squadron. On the 22nd June, 1893, in consequence of an inconsiderate order given by the admiral himself, the *Victoria* was struck by the formidable ram of the *Camperdown* (10,600), and in fifteen minutes turned over and sank in sight of the whole fleet, carrying down with her the admiral, 30 officers, and 320 men, out of a crew of 600. [1895.]

FIG. 78*b.*—*Firing at a floating battery.*

* From *Graphic*, 1st June, 1895.

PLATE XI.

H.M.S. "TERRIBLE."

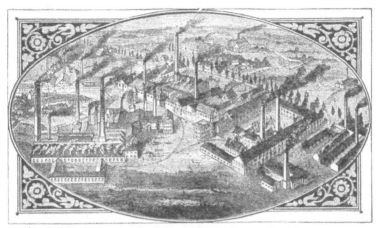

FIG. 79.—*Krupp's Works, at Essen, Prussia.*

FIRE-ARMS.

THE invention of gunpowder—or rather its use in war—appears at first sight a device little calculated to promote the general progress of mankind. But it has been pointed out by some historians that the introduction of gunpowder into Europe brought about the downfall of the feudal system with its attendant evils. In those days every man was practically a soldier: the bow or the sword he inherited from his father made him ready for the fray. But when cannons, muskets, and mines began to be used, the art of war became more difficult. The simple possession of arms did not render men soldiers, but a long special training was required. The greater cost of the new arms also contributed to change the arrangements of society. Standing armies were established, and war became the calling of only a small part of the inhabitants of a country, while the majority were left free to devote themselves to civil employments. Then the useful arts of life received more attention, inventions were multiplied, commerce began to be considered as honourable an avocation as war, letters were cultivated, and other foundations laid for modern science. If such have really been the indirect results of the invention of gunpowder, we shall hardly share the regret of the fine gentleman in "Henry IV." :

> "That it was great pity, so it was,
> That villanous saltpetre should be digged
> Out of the bowels of the harmless earth,
> Which many a good tall fellow had destroyed
> So cowardly."

We often hear people regretting that so much attention and ingenuity as are shown by the weapons of the present day should have been expended upon implements of destruction. It would not perhaps be difficult to show that if we must have wars, the more effective the implements of destruction, the shorter and more decisive will be the struggles, and the less the total loss of life, though occurring in a shorter time. Then, again, the exasperated and savage feelings evoked by the hand-to-hand fighting under the old system have less opportunity for their exercise in modern warfare, which more resembles a game of skill. But the wise and the good have in all ages looked forward to a time when sword and spear shall be everywhere finally superseded by the ploughshare and the reaping-hook, and the whole human race shall dwell together in amity. Until that happy time arrives—

> " Till the war-drum throbs no longer, and the battle flags are furl'd
> In the Parliament of man, the Federation of the world—
> When the common sense of most shall hold a fretful realm in awe,
> And the kindly earth shall slumber, lapt in universal law,"—

we may consider that the more costly and ingenious and complicated the implements of war become, the more certain will be the extension and the permanence of civilization. The great cost of such appliances as those we are about to describe, the ingenuity needed for their contrivance, the elaborate machinery required for their production, and the skill implied in their use, are such that these weapons can never be the arms of other than wealthy and intelligent nations. We know that in ancient times opulent and civilized communities could hardly defend themselves against poor and barbarous races. But the world cannot again witness such a spectacle as Rome presented when the savage hordes of Alaric swarmed through her gates, and the mighty civilization of centuries fell under the assaults of the northern barbarians. In our day it is the poor and barbarous tribes who are everywhere at the mercy of the wealthy and cultivated nations. The present age has been so remarkably fertile in warlike inventions, that it may truthfully be said that the progress made in fire-arms and war-ships within the second half of the nineteenth century surpasses everything that had been previously accomplished from the time gunpowder came into use. Englishmen have good reason to be proud of the position taken by their country, and may feel assured that her armaments will enable her to hold her own among the most advanced nations of the world.

The subject of fire-arms embraces a very wide ground, as will appear if we consider the many different forms in which these weapons are constructed in order best to serve particular purposes. Pertaining to this subject, attention must also be directed to the modern projectiles and to the newer explosives that have largely taken the place of ordinary gunpowder. The shot gun, fowling-piece, and sporting rifle properly come under the head of fire-arms, and in the march of improvement these forms have most commonly been in advance of military muskets and rifles, the ingenuity bestowed on all their details being worthy of admiration. Nevertheless it is to the implements of war that general interest attaches ; for on them depends so much the fate of battles and the destiny of nations, that whenever any country is engaged in war the question of arms becomes one of surpassing importance, enlisting the patriotic instincts of every citizen. Hence in the following pages our space will be devoted mainly to weapons of war, and more particularly to those that have been adopted by our own country.

Everyone of course is aware that guns, cannon, and gunpowder are by no means inventions of the nineteenth century ; but there are fewer acquainted with the fact that rifling, breech-loading, machine guns, and revolvers were all invented and tried hundreds of years before. The devices by which some of these ideas were sought to be realised in past ages appear to us in some instances very primitive, not to say childish, when compared with modern work : but it must be remembered that nearly all the appliances required for producing such weapons had themselves to wait for their invention until the nineteenth century ; such, for instance, as the steam-hammer, powerful and accurate tools, refined measuring implements, material entirely reliable such as the new steel, and also scientific investigations of all the conditions involved. The military fire-arms are of so many different forms and patterns that we can deal here with but a selection from the various services. If a rough classification had to be made, the most obvious distinction would be between the weapons the soldier carries in his hands (small-arms) and those which are mounted on some kind of carriage and discharge projectiles of much greater weight (ordnance). Ordnance again includes guns mounted on forts, carried in ships, or taken with an army into the field, in each case coming into action under different conditions. Partaking somewhat of the nature of both field-guns and of small-arms are the machine guns, of which the French mitrailleur was the first example, afterwards developing into much more effective weapons in the hands of Gatling, Gardner, Nordenfelt, Maxim, and Hotchkiss.

As much will have to be said about *rifling* the bores of muskets and cannon, we may here explain the nature and object of this device. The projectiles used in all guns down to comparatively recent times were almost invariably of spherical form, and could indeed scarcely be otherwise with smooth-bore weapons. As the diameter of the shot would necessarily be something less than that of the bore of the barrel, a considerable loss of power would result from the escape of the powder gases between the shot and the barrel, which escape is known as *windage*. Another disadvantage of the spherical projectile is that for the same weight of metal the air offers a greater resistance to its passage, and consequently checks its speed more quickly than that of any other circular form ; for the air resistance is proportional to the square of the diameter, and therefore if we take a ball of 1 in. diameter and a cylinder of 1 in. in length, each having the same weight of metal, the diameter of the cylindrical shot will be a little more than four-fifths of an inch, and the air resistance to the ball will be exactly half as much again as to the cylinder, that is, in the proportion of 3 to 2. Again, the passage of the spherical shot within the barrel of the gun will not be in a straight line, but in a series of rebounds from side to side, and its direction on leaving the muzzle will depend upon which part of the bore it just before impinges on, as from that it will also take a rotatory "twist" that will in part determine its path through the air.

Now if an elongated projectile were fired from a smooth-bore gun, its course through the air would be erratic to a degree impossible to the spherical shot, for it would turn end over end with deviations that would make aiming impracticable. But if the elongated projectile is made to spin rapidly enough about its longitudinal axis, it flies through the air quite steadily, the axis of rotation remaining parallel to that of the gun throughout the whole flight. The steadiness due to rapid rotation has familiar examples in spinning tops, in gyroscopic tops, in the way arrows are feathered so that the air may cause them to revolve axially, and so on. The axial rotation of the projectile is effected by ploughing out in the cylindrical barrel of the gun a number of spiral or twisting grooves, which the projectile is compelled to follow as it travels along the barrel, either by means of corresponding protuberances formed upon its surface in the first instance, as in Jacob's bullets, or by studs let into it, as in the studded shots and shells for ordnance which constituted at one time the regulation plan adopted by the British Government; or otherwise by making the force of the explosion expand some portion of the projectile in such a manner that this portion shall completely fill up the grooves, thus preventing windage, and causing the projectile to follow the twist of the grooves. This is the more general method, especially since the adoption of breech-loading. The Lancaster rifling, and that advocated by Whitworth, are the same in principle, but differ in appearance, from the section of the barrel being made in the one case oval, in the other hexagonal or polygonal, but with the twist necessary to produce rotation.

Incident to the discharge of all fire-arms, great and small, is a phenomenon of which we have to speak, because it is one which in the mounting of heavy ordnance especially has to be taken into account. And as it also illustrates in a very direct way one of the most general laws of nature, while people often have very vague and erroneous ideas of its cause and operation, it deserves the reader's attention. In gunnery it is called the *recoil*, and is familiar to anyone who has ever fired a pistol, fowling-piece, or rifle, in the kick backwards felt at the moment of the discharge. This law is in operation whenever the condition of a body in respect to its rest or motion is changing. That is, whenever a body at rest has motion given to, or if when already moving it is made to go faster or slower, or to stop, or when the direction of the motion is changed from that in a straight line. Now although these changes or actions are frequently occurring before our eyes, the operation in them of Newton's third law of motion does not generally present itself to common observation. This third law was stated by Sir Isaac Newton thus :—"To every action there is always an opposite and equal reaction." Now the expanding gases due to the gunpowder explosion press the bullet forwards and the barrel (with its attachments) backwards, with the same pressure in both cases, but at the end of the bullet's passage along the bore the same velocity is not imparted to the two bodies, because the same pressure acting for the same time on bodies of unequal *mass* always produces velocities that are inversely proportional to the *masses*. The reader should try to acquire this conception of *mass*, remarking that it is a something quite distinct from that of *weight*. A given lump of metal, for instance, would have exactly the same *mass* in any part of the universe, whereas its weight would depend upon its position; as, for instance, at the distance from the earth of the moon's orbit, it would *weigh* only $\frac{1}{3600}$th part of its weight at the earth's surface, and if it could be carried to the very centre of the earth it would there have no weight at all. Though the lump of metal will have different weights at different parts of the earth's surface, it has been found (by experiment) that the weights of bodies at any one place are proportional to their masses. Therefore the same numbers that express the weights of bodies might also express their masses; but for certain good reasons these quantities are referred to different units. In England a piece of metal weighing 32 lbs. under standard conditions is said to have mass = 1 ; and so on. As with the *same pressure acting for the same time*, the velocities imparted are inversely proportional to the masses, it follows that the number expressing the velocity multiplied by that representing the mass in each such case of action and reaction will give the same product, or in other words the *amount of motion* (momentum) will be the same. This is what Newton meant by saying the reaction is *equal* to the action. We may now by way of illustration calculate the velocity of recoil of a rifle under conditions similar to those that might occur in practice. Let us suppose that the rifle, including the stock and all attachments, weighs 10 lbs., and that from it is fired a bullet weighing one-sixteenth of a pound, with a velocity at the muzzle of 1,200 ft. per second. To obtain the amount of motion or the momentum, we should here multiply the number expressing the *mass* of the bullet by 1,200, but for our present purpose the weight numbers may be used for the sake of simplicity; therefore $\frac{1}{16} \times 1,200 = 75$ will represent (proportionately) the forward momentum of the bullet, and according to Newton's law the backward momentum of the rifle will be, on the same scale, 75 also. We must therefore find the number which multiplied by 10 will give 75, and this obviously is 7·5. That is as much as to say that at the instant the bullet is leaving the muzzle, the rifle itself, *if free to move*, would be moving backward at the speed of $7\frac{1}{2}$ ft. per second. Observe that this result would be the same if the rifle were fired where weight is non-existent; nor is the recoil due, as sometimes is erroneously supposed, to the resistance of the air to the passage of the bullet along the barrel, for even if the air were abolished, the recoil, so far as due to the masses and velocities, would remain the same, as indeed may be seen from the fact of our calculation taking no account of the bore of the rifle or of the shape of the bullet, circumstances of the utmost importance where atmospheric resistance is concerned.

The foregoing calculation however involves an assumption not in exact conformity with actual conditions, by taking for granted that the *centre of gravity* of the rifle is in the line of the axis of the barrel, while in fact this centre is almost always lower, and therefore the kick of the recoil acts in part as a turning-over push, tending to tilt up the muzzle of the gun, and for that reason the firer must hold the weapon very firmly or he will miss his aim. When such a rifle as we have supposed is fired, say from the shoulder, it would follow from the above calculation that the backward kick of the recoil is equivalent to a blow from a 10-lb. weight moving at the speed of $7\frac{1}{2}$ ft. per second. This would certainly be a very uncomfortable experience, but the backward momentum must be met somehow. We have supposed that the gun is free to move, but we know the firer presses it firmly against the muscles of his shoulder, and the stock of the gun is spread out and provided with a smooth hollow heel plate, so that any pressure from it is felt as little as possible, especially as the muscle against which it is applied acts as an elastic pad. With the rifle thus firmly held we may regard the marksman and his rifle as forming only *one mass*, and the centre of gravity of this being now much below the axis of the barrel, the effect of the recoil tends to overthrow the man backwards; but he learns to resist this by standing firmly, so that the elasticity of his whole frame comes into play; and besides this, the mass factor of the momentum being now so large, the velocity factor becomes comparatively insignificant.

Although the momenta of gun and projectile are, according to Newton's law, *equal* and opposite, the case is very different with regard to their *energies*, or powers of doing work, for the measure of these is jointly mass *and the square of the velocity.* The *energy* (*vis viva*) of a body of weight in pounds = W, moving with the velocity of v feet per second is always $\dfrac{Wv^2}{64\cdot4}$, that is, it will do this number of foot-lbs. units of work before it comes to rest. It would require too much space to demonstrate and fully explain here what this means, but the reader may refer to our index under the entries "Energy" and "Work," or to any modern elementary treatise on dynamics. If the calculation be made of the energies of the ball and of the rifle due to our calculated velocities of recoil, it will be found that that of the ball is 160 times greater than that of the other, and the ball possesses this energy in a much smaller compass.

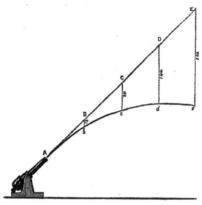

FIG. 80.—*Trajectory of a Projectile.*

The course or track of a projectile through the air after it leaves the gun is called the *trajectory*, and this has been studied both experimentally and theoretically, with interesting results. Assuming that the shot passed through empty space, or that the air offered no resistance to its passage, it would be very easy to trace the path of a projectile. Let us suppose that Fig. 80 represents a gun elevated at a high angle. The moment the projectile leaves the muzzle, gravity begins to act upon it, causing it to move vertically downwards with ever-increasing velocity until it finally reaches the ground; the onward uniform movement parallel to the axis of the piece being continued all the time. We could find the position of the projectile at the end of successive equal periods of time by drawing a straight line AC, a prolongation of the axis of the piece, or a line of the same inclination; on this we mark off equal distances representing by scale the velocity of the projectile per second, the points B, C, D, E being the positions the projectile would be in at the end of each successive second if gravity did not act. In order to bring the diagram within moderate compass, we suppose the projectile to have only the small velocity of 115 ft. per second. At the end of the first second it would be at B, but now suppose that gravity is allowed to act for one second, it would at the end of that time have fallen 16 ft. vertically below B and have arrived at *b*. Similarly we may set off by scale on verticals through C, D, and E distances representing 64 ft., 144 ft., and 256 ft. respectively. Because, for instance, the ball, without gravity acting, would at the end of 3 seconds be at D, where we may suppose its course arrested and gravity then allowed to act for 3 seconds to pull the ball down from its position of rest at D; at the end of this period, gravity alone acting, its position would be 144 ft. vertically below D, because gravity pulls a body that distance in 3 seconds, and the actual position 3 seconds after the ball had left the muzzle would be at *d*, after it had described the curved path A, *b, c, d*. Supposing *d* to be the highest point of the trajectory, another 3 seconds would bring the ball along a downward curve, and at the end of 6 seconds from the discharge it would be at a point on the same level as A. Now the complete curve would be symmetrical on each side of a vertical line through its highest point, and it would be in fact a regular *parabola* with its vertex at *d*.

The foregoing presupposes that the air offers no resistance to the passage of the projectile through it. The fact however is quite otherwise, for no sooner does the projectile begin its flight than its velocity is constantly diminished by the air's resistance. Now this resistance is complex, depending upon a number of different conditions, the effect of which can be taken into account only by extremely complex calculations. Obviously it will vary according to the area of the section presented by the projectile to the line of its flight, and again by the shape of its front, for a pointed shot will cleave the air with less resistance than one with a flat front. Then the density of the air at the time will also enter into the calculation. The mass of the projectile and also its velocity, upon which depend its *vis viva*, energy, or power of overcoming resistance in doing work, will also have to be considered. Most complex of all is the law, or rather laws (*i.e.* relations), which connect the air resistance with the velocity; for this relation no definite expression has been found. It is a function of the velocity (known only by experiment under defined conditions), and varying with the velocity itself. Thus for velocities up to 790 ft. per second, it is a function (determined experimentally) of the second power or square of the velocity; between 790 ft. per second and 990 ft. per second the law of resistance is changed and becomes a function of the third power of the velocity; between 990 ft. and 1,120 ft. velocity the law again changes and is related to the sixth power of the velocity; between 1,120 ft. and 1,330 ft. the resistance is again related to the third power of the velocity; and with higher speeds than that last named it is again more nearly related to the square of the velocity. It will be seen that to calculate the path of a projectile is really a very difficult mathematical problem, and indeed one which can be solved only approximately when all the known data are supplied.

The air resistance to the motion of a projectile is much greater than before trial would be supposed. Let us take an experiment that has actually been recorded, in which a bullet three-quarters of an inch in diameter, weighing one-twelfth of a pound, was found to have a velocity of 1,670 ft. per second at a distance of 25 ft. from the gun, and this 50 ft. farther was reduced to 1,550 ft. per second. Now if the reader will calculate, according to the formula we have given above, the *energy* due to the bullet's velocity at these points, he will find it must have done 500 foot-lbs. units of work in traversing the 50 ft., and as this could have been expended only in overcoming the resistance of the air, we learn that this last must have been equivalent to a mean or average pressure of 10 lbs. thrusting the bullet backwards.

It will be interesting to compare the difference in the trajectory of a projectile under defined conditions, worked out with the air resistance taken into account, compared with the trajectory when the air is supposed to be non-existent. We find an example of the former problem fully worked out by many elaborate mathematical formulæ in Messrs. Lloyd and Hadcock's treatise on Artillery. The problem is thus stated :—" An 11-in. breech-loading howitzer" (a howitzer is a piece of ordnance used for firing at high angles) "fires a 600-lb. projectile with an initial velocity of 1,120 foot-sec. at an elevation of 20°. Find the range, time of flight, and angle of descent." We shall calculate these points on the suppositions adopted with regard to Fig. 80, and with no higher mathematics than common multiplication and division.

It will have been observed that we supposed two motions that really take place simultaneously to take place successively and independently : one in the direction of the line of fire, due to the initial velocity ; the other vertically downwards, due to the action of gravity, the final result being the same. This affords an excellent illustration of another of Newton's laws of motion, and should be considered by the reader in this connection. The law itself admits of being stated in various ways, as thus :—" Whenever a force acts on a body, it produces upon it exactly the same change of motion in its own direction, whether the body be originally at rest or in motion in any direction with any velocity whatever—whether it be at the same time acted on by other forces or not." Or again : "When two forces act in any direction whatever on a body free to move, they impress upon it a motion which is the *superposition* (or compounding) of those that it would receive if each force acted separately." The law is given also in the following form (Thomson and Tait) :—" When any forces act on a body, then, whether the body be originally at rest or moving with any velocity and in any direction, each force produces in the body the exact change of motion which it would have had had it acted singly on the body originally at rest." In all of these expressions the word "forces" is used, and a very convenient word it is, but it may be noted in passing, nothing but a word ; for it stands for no real self-existing things, since, apart from observed changes of motion in bodies, forces for us have no existence. Nevertheless, it is useful for the sake of abbreviating statements about changes of motion, to regard these actions as produced by imaginary agents—imagined for the time and for this purpose, and therefore vainly to be sought for in the realm of reality.

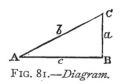

FIG. 81.—*Diagram.*

In dealing with the trajectory of the howitzer's projectile through airless space we have no concern with its diameter nor with its weight. We use the little diagram, Fig. 81, to represent the motions,—*c* being a horizontal line, *a*, a vertical one, the angle at B is therefore a right angle, and we assume that at A to be 20°. Now, the most elementary geometry teaches us that every triangle having these angles will have the lengths of its sides in the same invariable proportions one to another whatever may be the size of the triangle itself, and it has been found convenient to calculate these proportions once for all, not merely for angle 20°, but for every angle up to 90°. Besides this, distinct names have been given to the proportions of every side of the triangle to each of the other two sides. Thus in the triangle before us, if we take *a*, *b*, and *c* to represent the numbers expressing the lengths of the sides against which they are

placed, *a* divided by *b*, that is $a \div b$, or $\frac{a}{b}$, is called the *sine* of angle 20°,

while $\frac{c}{b}$ is named the *cosine* of that angle, etc. These therefore are

numbers which are given in mathematical tables, and we find by these that *sine* 20°=0·3420201, and *cosine* 20°=0·9396926, and these with the initial velocity give us all the data we require. We may first find the *time* the projectile would take to reach the ground level, or strictly that of the muzzle of the gun at B. Taking *t* to stand for this time, we know that AC $=$ 1,120 $\times$ *t*, but CB will be the distance that a body would fall from rest at C by the influence of gravity in that same time, *t*, and it is known by experiment that this distance is 16·1 feet multiplied by the *square* of the time from rest in seconds. We have now therefore the length of the line CB, and put $\frac{a}{b} = \frac{CB}{AC} = \frac{16\cdot1 \times t^2}{1,120 \times t} = sine\ 20° = \cdot3420201$, and dividing numerator and denominator by *t* and multiplying the above 3rd and 5th expressions by 1,120, we have

$$16\cdot1 \times t = 1,120 \times \cdot3420201$$

and therefore $t = \frac{1,120 \times \cdot3420201}{16\cdot1} = 23\cdot7927$ secs.

Having obtained the time, it will be easy to work out the lengths *b* and *a* as 26,648 ft. and 9114·1 ft. respectively; and as $\frac{c}{b} = cosine\ 20°$, we have $c = 26,648 \times \cdot9396926 = 25040\cdot8$ ft., which is the *range*. The trajectory will be a curve (parabola) symmetrical on each side of a vertical line half-way between A and B, and the length of this line within the triangle will be equal to half of *a*, and in half of 23·7927 seconds the projectile, supposed to move only along the line AC, would reach the point where this vertical axis intersects AC. If during this half-time it had been falling from rest at the same intersection, it would have reached a point below by a space just one quarter of CB (the spaces fallen through being as the *squares* of the times), and therefore at this its highest point its distance above AB would also be one quarter the length of *a* $=$ 2278·525 ft., which distance is called the *height* of the trajectory; and the descending curve being in every respect symmetrical to the ascending branch, the angle at which this would be inclined to AB would be 20°, but in the opposite direction to BAC, while the velocity would be the same as at A. We may now compare these results with those calculated when the air resistance is taken into account :—

	Without air resistance.	With air resistance.	Difference.
Time of descent..........	23·7927 secs.	22·61 sec.	−1·18 sec.
Angle of descent	20°	23° 49′	+3° 49′
Velocity of descent ...	1120 foot-secs.	868·8 foot-secs.	−251·2 f.-s.
Range	25040·8 ft.	20,622 ft.	−4418·8 ft.
Height of trajectory ...	2278·5 ft.	1989 ft.	−288·5 ft.

With the air resistance the trajectory will no longer be a symmetrical curve : its highest point, instead of being on the vertical line midway between A and B, will be on one 1,050 ft. nearer to B than to A, and the descending branch will be steeper than the ascending. The total time, it will be observed, is less, although the final, and therefore the mean, velocity, is also less; but this shortening of the time is due to the trajectory itself being much less in length. The range of the projectile is decreased

12

by 4,418 ft., or 1,473 yards, or more than four-fifths of a mile. The loss of velocity at the descent is very notable, and the reader will find it interesting to calculate the corresponding loss of energy by the formula already given.

The reader should now easily understand that the projectile from a rifle or gun discharged horizontally through airless space at the height of 16·1 ft. above a level plain would strike the ground in one second at a range or distance from the gun exactly equal to the initial velocity, or if the gun were on a tower and its axis 64·4 ft. above the plain, the range would then be 2v. It will be seen therefore that, corresponding to the range intended, there must be in general a certain inclination given to the axis of the piece in aiming, and this is done by means of the *sights*, one of which near the muzzle is usually fixed, while that next the breech is adjustable by sliding along an upright bar, which is graduated so that the proper elevation may be given for any required range. These graduations are made from experiments, and of course have reference only to some standard quantity and quality of ammunition and a standard of weight, shape, and material in the projectile. Sometimes large pieces of ordnance are laid by elevation in degrees, etc., marked on their mounting, the angles being taken from a table prepared for that particular gun and ammunition, from experiments at different ranges.

After these generalities about fire-arms we may enter upon certain particulars about the construction of some varieties, beginning with

THE MILITARY RIFLE.

IN Fig. 82 are represented the muzzle-loading musket and muzzle-loading rifles which formed the regulation weapons of the British infantry from the beginning of the century up to the year 1864. Somewhat slow in its earlier stages was the development of the modern military rifle from the old smooth bore musket with its flint-lock, which was the ordinary weapon of the British and other armies up to nearly the middle of this century. Partly, perhaps, owing to the inherent conservatism of government departments, and partly to the very serious outlay involved in arming all the troops of a nation with a new weapon, it has happened that many improvements in small arms were in use as applied to sporting guns, long before they were adopted in the regulation weapons of armies. The advance towards the modern arm of precision has been made along all the several directions that converge in the latest product, and it may be said that the most obvious of these are spiral rifling, breech-loading, and improved ammunition. The improvements in any one of these particulars would have been of little advantage unless the others had been kept in line with it. How long antiquated systems may continue in use may be illustrated by the case of the flint-lock, which was retained in the British army from the time it superseded the old match-lock, in the latter part of the seventeenth century, down to almost the middle of this present nineteenth. It is quite possible that not a few readers still in their fifties may never have seen a flint-lock outside of a museum, yet this was the firing apparatus of the weapon that used to be affectionately known to our soldiers as "Brown Bess," and that for a century and a half continued the regulation arm of British troops helping Wellington

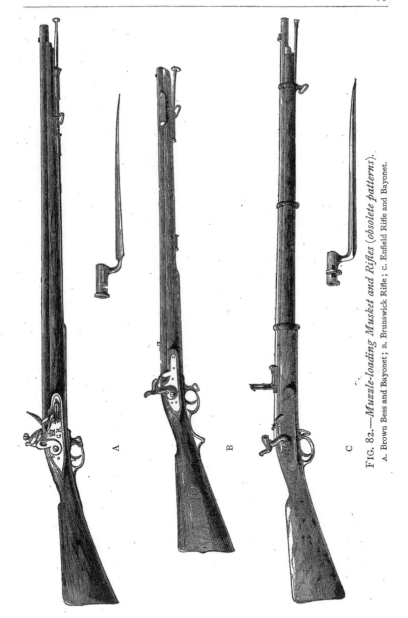

Fig. 82.—*Muzzle-loading Musket and Rifles (obsolete patterns).*
A. Brown Bess and Bayonet; B. Brunswick Rifle; C. Enfield Rifle and Bayonet.

to win his victories, and superseded by the percussion musket only in 1842. The "Brown Bess" of the earlier part of the century had a smooth-bore barrel of three-quarters of an inch diameter (0·753 inch), and 39 inches long ; this musket weighed, with its bayonet, 11 lbs., 2 oz. The bullet was spherical, and made of lead, in weight a little over one ounce. The diameter of the bullet was slightly smaller than the bore of the barrel, because a closely fitting ball could not be used, on account of the great force required to push it home with a ram-rod. The bullet was therefore wrapped in loosely fitting material, called a "patch," and this made the gun easy to load, even when the barrel was "fouled" by the solid residues that always remain after the explosion of gunpowder. "Brown Bess" was credited with a range of 200 yards, but its want of accuracy was such that the soldier was directed not to fire until he could see the whites of the enemies' eyes. But in 1800 one or two British regiments were armed with the muzzle-loading rifle known as Baker's, and again in 1835 these were provided with the *Brunswick rifle.* These regiments afterwards became known as the Rifle Brigade. The bullets in both cases were spherical, and as the earlier pattern had a seven-grooved barrel, there was so much difficulty in introducing the bullets into the muzzles that mallets had to be used. The bullet of the Brunswick rifle was encircled by a projecting band, which fitted into two rather deep grooves diametrically opposite to each other in the barrel. This bullet, wrapped in some slightly greased material, could be readily dropped into the muzzle, and rammed home without difficulty. Moreover, whereas in Baker's rifle the grooves made only a quarter of a turn in the length of the barrel, the grooves of the Brunswick rifle made more than one complete turn. This was so much an improvement on "Brown Bess" that the effective range was more than doubled. For the rank and file of the infantry regiments the flint-lock smooth-bore musket was, however, the regulation weapon until 1842, when it was superseded by the percussion musket. The percussion-cap is now comparatively little used, as, since the introduction of cartridges containing their own means of ignition, it is rapidly becoming a thing of the past. The copper percussion-cap, in the form it still retains, was invented about 1816, and was universally adopted for sporting-guns a long time before it was used for the military weapon. In 1842 the percussion musket was definitely adopted as the weapon of the British army, but up to that date the flint-locks still continued to be made at Birmingham.

The barrel of the percussion musket then issued was shortly afterwards rifled, when about the year 1852 the Minié system was adopted, and the Government awarded to M. Minié, a Frenchman, the sum of £20,000 for the bullet he had invented. What the meaning of this improvement was may now be explained, and we must begin by mentioning the various forms of grooving, or, at least, such forms as found some approval during the present century, for grooved barrels had been tried long before. At first the grooves appear to have been intended merely to receive the fouling, and these were often made without any twist or spiral, but parallel to the axis of the barrel. The grooves are hollow channels of greater or less depth, and of various forms ; square, triangular, rounded, or of such a form that the inner line of a section of the barrel would present the form of a ratchet wheel. The numbers of the grooves made use of have varied between two and twelve, or more, and different rates of twist, or numbers of turns of the spiral in the length of the barrel have been resorted to, these ranging from half a turn to twelve turns. The Brunswick rifle had been found wanting in accuracy, when at length in 1846 General Jacobs proposed the adoption of the conical bullet with projecting spiral ridges which fitted into grooves cut in the rifle barrel. The difficulty in using muzzle-loading rifles consisted in the force required to ram down the bullet, which had to adapt itself to the grooves, and fill them up so that the gases due to the explosion of the powder should not escape. If the bullet simply dropped into the bore of the rifle easily, it did not effectually fill the grooves, which then became channels of this *windage*, and if, on the other hand, the leaden bullet was made to fill the grooves from the muzzle, great force was required, and the time and effort expended in ramming the missile home, detracted enormously from the efficiency of the rifle as a military weapon. Mr. Lancaster produced rifles having a slightly oval, instead of a circular, bore, making, of course, the necessary twist within the barrel. A bullet of the corresponding section, but nearly globular, much as if the projecting belt of the Brunswick bullet had been laterally extended to its opposite poles, could be easily dropped in at the muzzle, without force being required to make it take grooves, the barrel being internally smooth throughout. It was, however, soon found that this easy-fitting ball allowed a considerable amount of *windage*, and the Minié system was definitely adopted, in which advantage was taken of a fact observed some years before by a French artillery officer, who found that an elongated leaden bullet, if hollowed out at the base, was so expanded by the pressure of the powder gases that the material was forced into the grooves of a rifle. Minié made his bullet elongated, pointed in front, and hollowed out part of its length by a conical space, the widest part of which was at the base, and was covered by a small iron cup, that, when driven inwards by the pressure of the gases, caused an expansion of the bullet by which the lead was forced into the grooves of the rifling. But the forces operating on the base of the bullet would at times cause the iron cup to cut the bullet in two, and propel the anterior portion only, leaving the base in the form of a ring clinging to the rifling. The military authorities had many comparative trials carried out between the smooth-bore percussion musket and the Minié rifle. The greater accuracy of the latter may be inferred from the results of practice made by men firing at a target 6 feet high and 20 feet broad;

FIG. 83.—
The Minié Bullet.

when at 100 yards distance, 74 hits out of 100 shots were made from the musket, against 94 from the rifle ; and the superiority of the latter, at longer ranges, was increasingly marked. Thus at 260, 300, and 400 yards the respective percentages of hits were for the musket 42, 16, 4½, but for the same ranges the rifle gave 80, 55, 52.

Curiously enough, the principle of the expanding bullet had been brought forward by the late Mr. W. Greener seventeen years before the government prize was awarded to M. Minié. Mr. Greener's bullets were of an oval form, being half as long again as their diameter, with one end flattened where the lead was excavated in a narrowing hollow nearly through the bullet. In this opening was inserted the end of a tapering plug of hard metal, and when the rifle was fired this plug was driven home, and the lead thus expanded took the grooves, so preventing windage, and giving range and accuracy; while allowing the piece to be loaded with as much ease as the smooth-bore musket. The invention, though favourably reported on by the military authorities at the time, did not receive the attention it would seem to have deserved. However, in 1857, Mr. Greener's claim of priority for the first suggestion of the expanding bullet was acknowledged by a government award of £1,000.

FIG. 84.—*Greener's Expanding Bullet.*

Sir Joseph Whitworth, having been invited by the British military authorities to institute experiments with a view to producing the best type of rifle, with the help of the most perfect machinery, constructed the barrels with a polygonal bore, a plan which he had before adopted for large guns. The barrels were accurately bored out to a hexagonal section, and experiments were made to find what number of turns in the twist would give the projectile a sufficient rapidity of rotation to maintain it during its flight parallel to its axis. It was found that one turn in 20 inches was sufficient, and the projectile was made by machinery to fit accurately but easily into the rifled bore, so that it dropped into its place, and the loading could be expeditiously performed. The bullet was long, compared with the bore, which was made smaller than before, and it was found that the explosion caused it to expand sufficiently to fill up the corners of the hexagon, so that there was no loss from windage. The accuracy of aim of the Whitworth rifle was superior to that of any weapon of the kind that had, up to that time, been produced. When officially tried against the Enfield, its mean deviation at 500 yards range was only 4½ inches, while that of the Enfield at the same range was 27 inches. Mr. Whitworth had proved the advantages of using a small bore, an elongated bullet, and a sharp twist in the rifling ; and it was acknowledged that as a military weapon his rifle was superior to all other arms of similar calibre that had before been produced. Some doubt appears to have been entertained, however, as to whether the mechanical perfection of the trial rifles could be maintained if they came to be manufactured on the large scale; and also as to whether an adequate supply of the polygonal ammunition would be procurable when required. The Whitworth rifle was never adopted into the government service, and soon after these trials in 1857, the adoption of another type of weapon became imperative, as the results obtained by the Germans with their needle-gun, demonstrated the enormous advantages of a breech-loading rifle.

The French then adopted the Chassepot rifle (so called after its inventor), which embodied the same principle as the needle-gun, but with improvements. This arm has a rifled barrel, with a breech mechanism of great simplicity, which is represented in section in Fig. 85. The piece marked B corresponds to what is called the "hammer" in the old lock used with percussion-caps, and the first operation in charging the rifle consists in drawing out B, as shown in the cut, until, by the spring, C, connected with the trigger, A, falling into a notch, the hammer, if we may so term it, is retained in that position. The effect of this movement is to draw out also a small rod attached to the hammer, and terminated in front by a needle, about $\frac{1}{2}$ in. long, at the same time that a spiral spring surrounding the rod is compressed, the spring being fastened to the front end of the rod, and abutting against a screw-plug, which closes the hinder end of F, and permits only the rod to pass through it. The piece F, which is also movable, has projecting from its front end a little hollow cylinder, through the centre of which the needle passes, and this little cylinder has a collar, serving to retain its position, an india-rubber ring surrounding a portion of the cylinder, and forming a plug to effectually close the rear end of the barrel. It will be noticed that the cylinder is continued by a smaller projection, which forms a sheath for the point of the needle. The movable breech-piece, F, is provided with a short lever, E, by which it is worked. The second movement performed by the person who is charging the piece

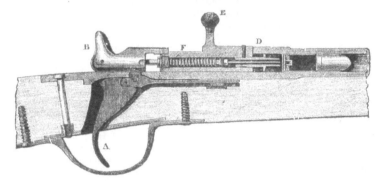

FIG. 85.—*The Chassepot Rifle.—Section of the Breech.*

is to turn this lever from a horizontal to a vertical position, which thus causes the piece F to turn 90° about its axis, and then by drawing the lever towards him he removes the piece F from the end of the barrel, which, thus exposed, is ready to receive the cartridge. The cartridge contains the powder and the bullet in one case, the posterior portion containing also a charge of *fulminate* in the centre, and it is by the needle penetrating the case of the cartridge and detonating this fulminate that the charge is exploded. When the cartridge has been placed in the barrel, the piece F is pushed forward, the metallic collar and india-rubber ring stop up the rear of the barrel, and on turning the lever, E, into a horizontal position, the breech is entirely closed. If now the trigger be drawn, the hammer is released, and the spring carries it forward, at the same time impelling the needle through the base of the cartridge-case, where it immediately causes the explosion of the fulminate. The bullet is conical, and its base having a slight enlargement, the latter moulds itself to the grooves with which the barrel is rifled. When the piece has not to be fired immediately, the lever is not placed horizontally, but in an inclined position, in which the hammer cannot move forward, even if the trigger be drawn. The Chassepot has an effective range of 1,093 yards, and the projectile leaves the piece with a velocity of 1,345 ft. per second, the trajectory being such that at 230 yards the bullet is only 18 in. above the straight line. The piece can be charged and fired by the soldier in any position, and it was found that it could be discharged from seven to ten times per minute, even when aim was taken through the sights with which it is furnished, and fourteen or fifteen times per minute without sighting. The ordinary rifled musket, which this arm superseded, could only be fired twice in a minute, and could only be loaded when the soldier was standing up.

Other nations followed either by adopting as their infantry arm some form of breech-loader, or by converting their muzzle-loaders into breech-loaders as a temporary expedient, pending the selection of some more perfect type. When in 1864 a committee which had been appointed to investigate the question of proper arms for our infantry, recommended that that branch of the service should be supplied with breech-loaders, our Government, considering that no form of breech-loader had up to that time been invented which would unequivocally meet all the requirements of the case, wisely determined that, pending the selection of a suitable arm, the service muzzle-loaders should meanwhile be converted into breech-loaders. The problem of how this was to be done was solved by the gunmaker Snider, and in the "Converted Enfield" or "Snider" the British army was provided for a time with an arm satisfying the requirements of that period. This change of weapon was effected at a comparatively small outlay, for the conversion cost less than twenty shillings an arm. The breech action in the Snider consisted of a solid piece of metal which closed the breech end of the barrel, and, being hinged on the right-hand side parallel to the barrel, could be turned aside, making room for the insertion into the conically widened bore of a metallic cartridge case, invented by Colonel Boxer, which contained the projectile, the powder charge, and the means of ignition in itself. A short backward movement of the breech-lock caused a claw acting on the base of the spent cartridge case to withdraw it from the barrel, and then the reaction of a spring brought the breech-block back into position, after insertion of a new cartridge. This cartridge proved very effective in increasing the range and accuracy of the weapon. It should be mentioned that all the breech-loading mechanisms are provided with arrangements by which the metallic cases of the spent cartridges are automatically extracted from the barrel. The authorities having, in 1866, offered gunmakers and others handsome prizes for the production of rifles best fulfilling certain conditions, nine weapons were selected out of 104 as worthy to compete. No first prize was awarded, but the second was given to Mr. Henry, while Mr. Martini was seventh on the list. In order to obtain a weapon fulfilling all the requirements, a vast number of experiments were made by the committee appointed for that purpose, as to best construction of barrel, size of bore, system of rifling, kind of cartridge, and other particulars, and assistance was rendered by several eminent gunsmiths and engineers.

After a severe competition it appeared that the best weapon would be produced by combining Henry's system of rifling with Martini's mechanism for breech-loading. The parts constituting the lock and the mechanism for working the breech, shown in Fig. 86, are contained in a metal case, to which is attached the woodwork of the stock, now constructed in two parts. To this case is attached the butt of the rifle by a strong metal bolt 6 in. in length, A, which is inserted through a hole in the heel-plate. The part that closes the breech—termed the "block"—is marked B. It turns loosely on a pin, C, passing through its rear end and fixed into the case at a level somewhat higher than the axis of the barrel. The end of the block is rounded off so as to form with the rear end of the case, D, which is hollowed out to receive it in a perfect knuckle joint. Let it be observed that this rounded surface, which is the width of the block, receives the whole force of the recoil, no strain being put on the pin, C, on which the block turns. In the experiments a leaden pin was substituted, and the action of the mechanism was not in the least impaired. This arrangement serves greatly to diminish the wear and the possibility of damage from the recoil. As the pin on which the block turns is slightly above the axis at the barrel, it follows that the block, when not supported, immediately drops down below the barrel. Behind the trigger-guard is a lever, E, working on a pin, F, fitted into the lower part of the case. To this lever is attached a much shorter piece

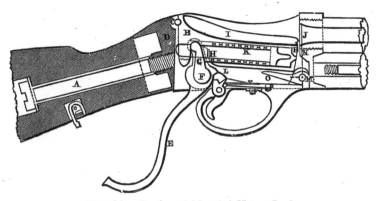

FIG. 86.—*Section of Martini-Henry Lock.*

called the "tumbler," which projects into the case, G. It is this tumbler which acts as a support for the block, and raises it into its firing position or lowers it according as the lever, E, is drawn toward a firer or pushed forward. How this is accomplished will be readily understood by observing the form of the notch, H, in which the upper end of the tumbler moves. It will be noticed that the piece being in the position for firing, if the lever be pushed back, G slides away from the shallower part of the notch into the deeper, and the block accordingly falls into the position shown in the figure ; and if again the lever is drawn backward, G acting on H will raise the block to its former position. The block or breech-piece is hollowed out on its upper surface, I, so as to permit the cartridge to be readily inserted into the exploding chamber, J. The centre of the block is bored out, and contains within the vital mechanism for exploding the cartridge, namely, a spiral spring, of which the little marks at K are the coils in section. These coils pass round a piece of metal called the "striker," which is armed with a point, capable of passing through a hole in the front

face of the block exactly behind the percussion-cap of the cartridge when
the block is in the firing position. When the lever handle is moved *forward*, it causes the tumbler, which works on the same pin, to revolve, and
one of its arms draws back the striker, compressing the spring in so doing,
so that as the block drops down the point of the striker is drawn inwards.
In this position the piece receives the cartridge into the chamber. The
lever, E, being now drawn backward, the piece is forced into the notch, H,
and the block is kept firmly in its place ; besides this, there is a further
compression of the spring by the tumbler, and in this position the spring
is retained by the rest-piece, L, which is pushed into a bend in the tumbler.
By pulling the trigger this piece is released, so that the tumbler can revolve
freely, and relieve the pent-up spring, whose elasticity impels the striker
forward, so that this enters the carriage directly. A very important and
ingenious part of this arrangement is the contrivance for extracting the
case of the exploded cartridge. The extractor turns on the pin, M, and
has two arms pointing upwards, N, which are pressed by the rim of the
cartridge pushed home into two grooves cut in the sides of the barrel. It
has another arm, O, bent only slightly upwards and pointing towards the

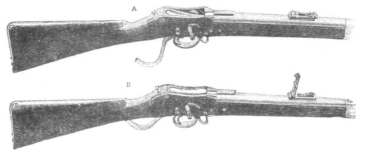

FIG. 87.—*The Martini-Henry Rifle.*
A, ready for loading ; B, loaded and ready for firing.

centre of the case, and forming an angle of about 80° with the above-
mentioned upright arm ; when, by pushing forward the lever, its short
arm drops into the recess, the block, no longer supported, falls, and hits
the point of the bent arm of extractor, so causing the two upright arms to
extract the cartridge-case a little way.

The barrel is of steel; the calibre is 0·451 in. It is rifled on Mr. Henry's
patent system. The section of the bore may be generally described as a
heptagon with re-entering angles at the junctions of the planes, so that
there are fourteen points of contact for the bullet, viz., one in the middle
of each plane, and one at each of the re-entering angles. The twist of
rifling is one turn in 22 in. The charge consists of 85 grains of powder,
and a bullet weighing 480 grains, of a form designed by Mr. Henry. The
cartridge is of the same general construction as the " Boxer " cartridge,
used in the Snider rifle, but it is bottle-shaped, the diameter being enlarged from a short distance in rear of the bullet, in order to admit of its
being made shorter, and consequently stronger, than would be otherwise
possible. A wad of bees'-wax is placed between the bullet and powder, by

which the barrel is lubricated at each discharge. The sword-bayonet to be used with this rifle is of a pattern proposed by Lord Elcho. It is a short sword, broad towards the point, and furnished on a portion of the back with a double row of teeth, so as to form a stout saw. It is so balanced as to form a powerful chopping implement, so that, in addition to its primary use as a bayonet, it will be useful for cutting and sawing brushwood, small trees, &c.

The following are the principal particulars of weight, dimensions, &c., of the Martini-Henry rifle:

Length of rifle	Without bayonet .	4 ft. 1 in.
	With bayonet fixed	5 „ 8 „
	Of barrel . .	2 „ 9·22 „
Calibre		0·451 „
Rifling . . .	Grooves . .	7
	Twist . . .	1 turn in 22 in.
Weight . . .	Without bayonet .	8 lbs. 7 oz.
	With bayonet .	10 „ 4 „
Bayonet . . .	Length . .	2 ft. 1½ in.
	Weight without scabbard	1 lb. 8 oz.
Charge of powder		85 grains.
Weight of bullet		480 „

The rifle is sighted to 1,400 yards.

As an evidence of the accuracy of fire in this rifle, it may be stated that of twenty shots fired at 1,200 yards, the mean absolute deflection of the hits from the centre of the group was 2·28 ft. The highest point in the trajectory at 500 yards is rather over 8 ft., so that the bullet would not pass over a cavalry soldier's head within that distance. The trajectory of the Snider at the same range rises to nearly 12 ft. The bullet will pass through from thirteen to seventeen ½ in. elm planks placed 1 in. apart at 20 yards distance; the number pierced by the Snider under similar circumstances being from seven to nine. As regards rapidity of fire, twenty rounds have been fired in 53 seconds; and one arm which had been exposed to rain and water artificially applied for seven days and nights, and had during that time fired 400 rounds, was then fired, without cleaning, twenty rounds in 1 minute 3 seconds.

Rifles of the Martini-Henry and Chassepot type were soon superseded, for as early as 1876 Switzerland had armed her troops with a magazine rifle of a smaller calibre than any then in use, and this weapon was found so effective that in a few years after every European nation had followed suit, as also had the United States and Japan, each country adopting some particular pattern of a weapon with certain modifications. Of these the Mannlicher and the Mauser are much used. A magazine rifle is one that can be fired several times successively without reloading. Like revolvers, the magazine arms repeat their fire, but instead of having several distinct firing chambers, they have but one, from which the empty cartridge cases are automatically extracted by the breech mechanism, for the magazine rifle is necessarily a breech-loader. The magazine rifle carries a supply of cartridges, which one after another are brought into the firing chamber by the simple action of the breech mechanism, so that the soldier is enabled to discharge several rounds in any position without reloading. The several varieties of the magazine rifle may be classed according to the position of the magazine. This may be: First, in the stock; second, under the barrel; third, in a box under the breech; or fourth, in a box above the breech. In the first and second variety the cartridges are in line in a tube, out of which they are moved on by a spiral spring, and this was the earlier form of the weapon. The box above or below the breech is the later development, and has the advantage of holding the cartridges lying side by side, and thus in a position in which they are not so liable to injure each other as in the tubular arrangements. Then, again, the movement of the cartridge in the breechbox in arriving at the firing chamber is much less than in the linear magazines, and the centre of gravity of the whole changes but little when the supply is exhausted. With any of the varieties of magazine a suitable modification of the mechanism may be adopted, so that the weapon can at will be used as a single firing rifle, but changeable in an instant to the magazine form. Again, the box magazine may be made as a fixture on the rifle, or it may be detachable. Commissions of military authorities had for several years been deliberating upon the best models for their respective nations, while Professor Hebler was working out his researches as to the best calibre for military rifles. Hebler published a work showing the great advantages of a bore one-third less in diameter than that commonly in use, which was about 0.45 inch, as in our Martini-Henry. The small-calibre rifle shoots straighter and hits harder than the large bore one, and the recoil is less, and so is the weight of the weapon. Lead is found to be too soft a material for the bullet of the small-bore rifles, as it does not keep in the rifling, which has a sharper turn than that in the older weapon; hence the bullets are now cased in steel or nickel. These bullets have remarkable power of penetration. Some will go through a steel plate $1\frac{1}{4}$ inch thick, making a clean hole in it, and the Lebel bullet penetrates 15 inches of solid oak, at a distance of 220 yards. Such a missile would, therefore, be capable of going completely through the bodies of several men or horses.

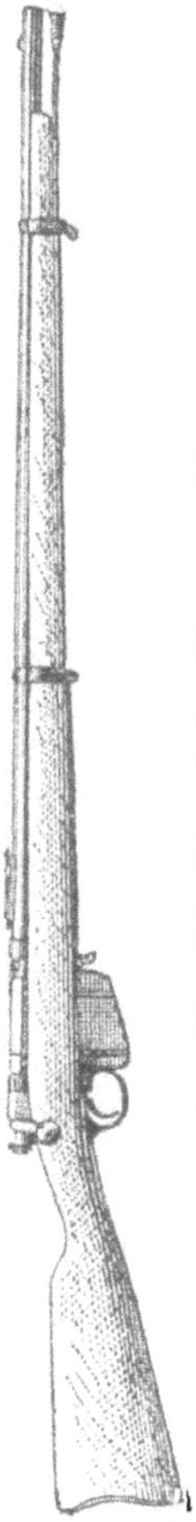

- The Germans, about 1888, adopted a magazine rifle known as the Mauser. It had a fixed tubular magazine for eight rounds below the barrel, and a breech mechanism of the Remington-Keene type. The French followed suit with their famous Lebel gun, the construction of which was long kept secret. It also has a fixed under barrel tubular magazine, and the cartridges used with it contain smokeless powder. It is said that a new gun of practically the same pattern has been adopted by Russia, but with a detachable magazine to contain five rounds. The Russian gun will also use smokeless powder. In England, a small-bore rifle of 0.303 inches calibre is now issued to all troops. It has an under breechbox magazine, modified from the Lee rifle. The box is detachable, so that the weapon could normally be used as a single loader, and the magazine attached only when required. But the British authorities have decided that the

FIG. 88.—*The Mannlicher Magazine Rifle.*

magazine box is to be attached to the weapon by a chain. The first
issue of this pattern of rifle to British soldiers took place early in 1890.
The Austrians are adopting the Mannlicher pattern, in which the
magazine idea is embodied in a complete and practical form. This rifle
has a fixed box magazine below the breech. From this box, in which
the cartridges—five in number—lie side by side, they are fed up by
springs as they are disposed of by the movement of the breech mechanism.
The magazine is recharged by placing in it a tin case containing five
cartridges, and the case drops out when all the cartridges have been fired.
In this form there is of course no necessity for providing any mechanism
for holding the magazine in reserve while the rifle is used as a single loader.
As to calibre, the Austrian authorities follow other countries in adopting
a small bore, namely, 0·315 inch. Italy has converted her single-fire
Vetterli rifle into a magazine arm, with a box something like the Mannlicher, and Belgium has adopted a gun of the same type. The rate of
fire from charged magazines of such guns as the "Lee," "Mannlicher"
and "Vetterli," worked with the right hand without bringing the piece
down from the shoulder is, for all of them, about one shot per second ;
but the time that is required to recharge the magazines varies much

FIG. 89.—*The Magazine and Breech of the Mannlicher Rifle.*

according to the contrivance used. The number of rounds the magazine
of a rifle is capable of containing when fully charged is from 5 to 12, or
more, according to the difference of system. It is considered that in the
detachable Lee or the quick recharging Mannlicher five rounds are
ample for use at a critical moment.

The calibre of the military rifle has been decreased with almost every
new pattern adopted. Thus, while the *old* "Brown Bess" had a calibre
of 0·75 inch, in the last issue of it the bore was reduced to 0·693 inch ; the
Enfield (1852) had a bore of 0·577 inch ; the Martini-Henry, 0·451 inch,
which, in a newer pattern adopted in 1887, was reduced to 0·400 inch ;
and, finally, in the Lee-Metford, the calibre is only 0·303 inch. A similar
consecutive reduction of bores has taken place in the rifles adopted by
other countries, and one of the latest type, issued for the use of the
United States Navy, has a bore of only 0·236 inch, and it is even expected
that a still smaller one will become general. The advantage of the narrow
and lighter projectile is that while it has a higher initial velocity with a
given charge, its flight is less checked by the resistance of the atmosphere,
the section it presents being so much less. Thus the bullet of 0·236 inch
diameter has a section little more than one-fourth that of the 0·45 inch

bullet. The difference is well shown in the comparative heights of the *trajectory* (or path of the bullet) of the Martini-Henry 0·450 inch bullet, and that of the 0·303 inch Lee-Metford (the latter with *cordite* ammunition); for at a range of 1,000 yards the former reaches to 48 feet above the line of sight, while the latter rises to only 25 feet.

Some form of repeating or magazine rifle has now been adopted by all the most important nations of the world. The number of shots contained in the magazines varies from 5 to 12. In the British detachable box magazine there are ten charges. The calibres of the barrels range in the infantry patterns of different nations from 0·256 inch to 0·315 inch ; the explosive used in every case is some kind of smokeless powder, and this, in the cartridge for the Lee-Metford, is *cordite*. The bullets are not made simply of lead, but of lead coated with a harder metal or alloy such as steel, cupro-nickel, nickel steel, or they consist entirely of some of these alloys.

Although the magazine rifle is now the regulation weapon of the infantry of all great armies, it is not improbable that at no distant future it may be superseded by one in which, as in certain machine guns, the force of the recoil will be used for actuating the breech and lock movements. Many patents have already been taken out for rifles on this principle, and several patterns have actually been constructed, in which a merely momentary contact of the breech-piece with the end of the barrel is sufficient; the recoil of the barrel with the reaction of a spring performs all the requisite movements with such rapidity that an amazing speed of firing has been obtained. It is said that such an automatic gun can send forth bullets at a perfectly amazing rate. Of course the mechanism of such a gun is somewhat intricate, and it is impossible to explain its construction and action without a great number of diagrams and much description.

RIFLED CANNON.

HAVING briefly sketched in the foregoing section the development of the military rifle from such weapons as our own "Brown Bess," down to the repeating or magazine rifle, we now purpose to adopt a similar course with regard to ordnance, giving also some particulars of the methods of manufacture, etc., and following in general the order of history.

Naturally there is nothing that accelerates progress in war-like inventions so much as the exigences of war itself. This is well exemplified in circumstances attending the Crimean War, which was waged in 1854 by England and France in alliance against Russia. The desire of having ships that could run the gauntlet of the heavy guns mounted on Russian forts led to the construction of *La Gloire* and other armour-plated vessels, as we have already seen, and a suggestion of the French Emperor, as to improving metal for guns, made to Mr. Bessemer, led incidentally but ultimately to the great revolution in the manufacture of steel, although it is true that Krupp of Essen had begun to produce small cast-steel ordnance as early as 1847. But what determined the necessity for rifled ordnance was more particularly the greater comparative effects obtained by the muzzle-loading rifles over the field artillery then in use in the several engagements that took place in the Crimea, especially in the battle of Inkerman (1854). The rifles so much surpassed in accuracy at long ranges the smooth-bore field-pieces firing spherical projectiles, that field artillery was on the point of losing its relative importance, and even in the matter of range the latter lost so much by windage that the men serving the artillery could sometimes be leisurely picked off by the rifle sharp-shooters. Inventors were soon at work on devising methods of increasing the accuracy of ordnance fire with both light and heavy pieces, and before the end of the war some cast-iron guns rifled on Lancaster's plan had been mounted on forts and in ships, without proving very successful except in regard to increase of range when elongated pointed projectiles were used with them.

Now let us see of what kind was the ordnance used for some years after the middle of the century, in order that we may be the better able to appreciate the progress that has since been made. Ordnance is, as already noticed, of several species, as guns mounted on fortresses, naval guns, siege guns, field-guns, etc., and the size of the pieces under each of those heads is distinguished sometimes in one, sometimes in another of

FIG 90.—32-*pounder*, 1807.

three different ways. We may name it by the weight of the gun itself in tons or hundredweights, as "a 35-ton gun," etc. ; or by the weight of its projectile, as "a 68-pounder," etc. ; or by its calibre, that is the diameter of its bore, as "a 4-inch gun," etc. We may take the naval guns with which Nelson won his battles (Trafalgar, 1807) as representative of all except field ordnance up to about 1856. They were all made of cast iron, threw spherical projectiles, and were very rudely mounted. The gun most commonly mounted on board our ships of war was the 32-pounder, weighing 32 cwt., shown with its carriage in Fig. 90. The carriage was of wood, and consisted of two side pieces joined back and front by two transverse pieces and carried by four low wooden wheels. The trunnions of the gun fitted into bearings at the top of the side-pieces, and were secured by iron plates that passed over them in a semi-cylindrical form and were bolted down to the wood. The position of the trunnions on the gun was always such that the breech end of the gun preponderated, being supported on an adjustable wooden wedge ; and when the muzzle of the gun had to be lowered, this was done by raising the breech end with handspikes and pushing in the wedge so far as to prevent the breech from dropping down again. There was a vent or narrow passage to contain a train of powder from the touch-hole at the upper part of the breech to the rear of the charge. When the gun was fired, with its muzzle protruding a little way out of the port-hole, the recoil would trundle it inwards about its own length, when its course would be stayed by a thick rope attached to the sides of the vessel ; and by other tackle it would be kept in position until loaded, when it would be allowed to roll back, or would be drawn by ropes and pulleys out to the port-hole, and by the same means such lateral inclination as might be required could be given. This last adjustment was called *training* the gun. A 32-pounder required the services altogether of a dozen or fourteen men, but these by virtue of constant drill would learn to handle the clumsy machine with a certain amount of expedition. If we except a notch on the highest point of the muzzle, the pieces were devoid of anything of the nature of sights, though sometimes marks were made on the adjustable wedge under the breech to correspond with certain elevations. Nor were sights required ; for the mode of fighting then was to get quite close to the adversary's ship and pour in a *broadside* by firing simultaneously all guns on the enemy's side when they had been *trained* (by rough methods), so as to concentrate their effect as much as possible on one point of the antagonist. Nelson's famous ship the *Victory* carried a few larger guns than the 32-pounders, namely, two 68-pounders, called *carronades* (from having first been cast at Carron in Scotland), and some 42-pounders. The 32- and 42-pounders numbered together thirty, and there were also as many 24-pounders, with forty 12-pounders. These were all simply cast of the required dimensions, and were not made with the one single improvement which after two centuries' use of cast-iron guns had been introduced into France about fifty years before, namely, the *boring* of the chase out of a solid casting.

On the outbreak of the Crimean War (1854) the minds of many inventors were occupied by the problem of ordnance construction, and this also engaged the attention among others of two of the most eminent British mechanical engineers of the day. These were Sir W. Armstrong and Sir J. Whitworth, who, with others, were invited by our War Department to submit the best models of field and heavy guns their skill was severally able to produce. Two years afterward, Sir W. Armstrong had, after many experiments, completed a gun of 1·88 in. calibre. This had a forged steel barrel 6 feet in length ; but it was only after eight such forgings had been bored and rejected on account of flaws revealed only by the boring that a sound barrel was at length obtained. This barrel was strengthened on the outside by *jackets* made from coils of wrought iron bars welded into a piece and shrunk on while hot (of which process we shall have something more to say presently) ; the barrel was rifled with many shallow grooves, and the pointed projectile, 3 calibres long, was made of lead, for which afterwards iron coated with lead was substituted. This gun was a breech-loader, the breech being closed by a block let into a slot after loading, and then pressed against the barrel by some turns of a screw which advanced parallel to the axis of the piece, and was made hollow for loading through, before the closing block was put in. In a trial of the various pieces ready in 1857, it was found that the Armstrong gun made as just described had an accuracy and range immensely greater than any weapon that had ever been tested, and the Government authorities approved of the system of construction, except that they preferred muzzle-loading pieces to breech-loading, as being simpler in action, more easily kept in repair, and cheaper in original cost and ammunition.

When Sir Joseph Whitworth's gun was, in 1863, submitted to a competitive trial against the Armstrong, as to their endurance and mode of ultimate failure when fired with ever-increasing charges of powder and shot, at the forty-second round the Armstrong breech-loader split, and at the sixtieth the Armstrong muzzle-loader had one of its coils cracked; while it was not until the ninety-second round that the Whitworth gun burst violently into eleven pieces. These competing guns were 12-pounder field-guns weighing 8 cwts., and from each 2,800 regulation rounds had been fired before they were subjected to the bursting proofs. The result of these trials being that the authorities considered that steel was not then sufficiently reliable, and they decided to adopt the system of building up rifled guns with iron jackets over an inner tube of steel. Sir Joseph Whitworth made his guns entirely of steel, and they were striking examples of beautiful and accurate workmanship. His system of rifling consisted in forming the bore of the gun so that its section is a regular

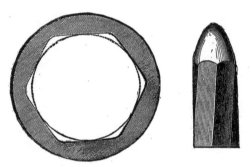

FIG. 91.—*Whitworth Rifling and Projectile.*

hexagon, and the projectile is an elongated bolt with sides exactly fitting the barrel of the gun : the projectile is, in fact, a twisted hexagonal prism. Fig. 91 shows at the left-hand side the section of the barrel, and on the right we see the form of the projectile on a smaller scale, this last representing, in fact, the exact size and shape of the bullet of the Whitworth *rifle* mentioned on another page. Sir Joseph's guns were muzzle-loaders, and they were remarkable for their long range and accuracy of fire. One of these guns, with a charge of 50 lbs. of gunpowder, threw a 250-lb. shot a distance of nearly six miles, and on another occasion a 310-lb. shell was hurled through the air, and first touched the ground at a distance of more than six and a quarter miles from the gun. These distances are greater than any to which shot or shell had previously been thrown.

As the material of these Whitworth guns was very costly, and very perfect workmanship was required in the formation of the barrel and the shots, the expense attending their manufacture and use was much greater than that incurred in the case of the Armstrong guns. Sir W. Armstrong's estimate for a 35-ton gun was £3,500, and Sir J. Whitworth's, £6,000. The gun, as constructed at Woolwich on Mr. Fraser's plan, was estimated to cost £2,500. The first cost of a gun is a matter for consideration, since each piece, even the strongest, is able only to fire a limited number of rounds before it becomes unsafe or useless. It appears that no cannon has yet been constructed capable of withstanding without alteration the tremendous shocks given by the explosion of the gunpowder, and these alterations, however small at any one discharge, are summed up and ultimately bring to an end what may be termed the "life of the piece."

About the year 1858 Sir William Armstrong (afterwards Lord Armstrong) established at Elswick, Newcastle-on-Tyne, a manufactory of ordnance, which has since developed into the great arsenal now so well known all over the world. Here all the resources of science have been applied to the problems of artillery, and experiments carried on with a prodigality of cost and promptness of execution impossible at a govern-

FIG. 92.—600-*pounder Muzzle-loading Armstrong Gun.*

ment establishment trammelled with official regulations. Here, and also at Woolwich, our national ordnance factory, guns have always since been constructed on the building-up plan advocated by Sir W. Armstrong, whose principle consists in disposing of the fibre of the iron so as best to resist the strains in the several parts of the gun. Wrought iron being fibrous in its texture has, like wood, much more strength in the direction of the grain than across it. The direction of the fibre in a bar of wrought iron is parallel to its length, and in that direction the iron is nearly twice as strong as it is transversely. A gun may give way either by the bursting of the barrel or by the blowing out of the breech. The force which tends to produce the first effect acts transversely to the axis of the gun ; hence the best way to resist it is to wrap the iron round the barrel, so that the fibres of the metal encircle it like the hoops of a cask. The force which tends to blow out the breech is best resisted by disposing the fibres of the iron so as to be parallel to the axis of the gun ; hence Sir W. Armstrong makes the breech-piece from a solid forging with the fibre in the required direction. But the

Elswick building-up principle involves much more than the direction of the fibres of the iron, for each coil or jacket, after having its spires welded together, was bored out on a lathe, and the exterior of the part of the gun on which it was to be placed was also turned with the utmost exactness, so that when the enveloping piece was heated to a certain temperature and in this state brought into position, it would in cooling compress the parts it encircled just to that degree which careful calculations showed would best strengthen the gun without unduly straining the metal at any part. The Elswick guns being built up of several superimposed jackets of calculated lengths and thicknesses, the means was afforded of distributing the tensions throughout the whole mass of metal to the best advantage. In the simpler form, arranged by Mr. Fraser, and for the sake of economy adopted by the authorities at Woolwich in 1867, the greater part of the benefit derivable from adjustment of tension was no doubt sacrificed to cheapness of manufacture. These, and also the forms of Armstrong guns that have not yet been described, ceased to be made after 1880, by which time steel had replaced iron in every part of the construction and fittings of guns, and muzzle-loading had been definitely abandoned in favour of breech-loading

FIG. 93.—*35-ton Fraser Gun.*

Now, in 1874, when the first edition of the present work was in preparation, the Fraser-Woolwich guns were in full vogue, being spoken of by the public press as the *ne plus ultra* of artillery construction in size, efficiency, and economy. When, accordingly, the author had been privileged to visit the arsenal and witness the production of these guns in every stage of their manufacture, he wrote a description of it which is here retained as printed at the time, seeing that it may not be without historical interest, particularly since great numbers of these guns must still be extant, mounted on our forts in various parts of the world, and seeing also that the description of the simpler formations may render more easily to be understood future references to similar operations in gun-making as have been retained in the later developments. Of course, the following description was written in the *present tense*, and therefore in perusing it the reader must constantly bear in mind that the guns with which our ships of war have since been equipped are in *every respect entirely different* from

The Fraser-Woolwich Guns, 1867–1880.

Until the year 1867 the guns made at Woolwich were constructed according to the original plan proposed by Sir W. Armstrong, and on this system one of the large guns consisted of as many as thirteen separate pieces. These guns, though unexceptionable as to strength and efficiency, were necessarily so very costly that it became a question whether anything could be done to lessen the expense by a simpler mode of construction or by greater economy in the material. The problem was solved in the most satisfactory manner by Mr. Fraser, of the Royal Gun Factory, who proposed an important modification of the original plan, and the adoption of a kind of iron cheaper than had been previously employed, yet perfectly

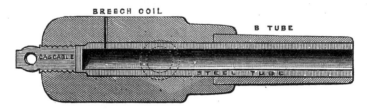

FIG. 94.—*Section of 9 in. Fraser Gun.*

suited for the purpose. Mr. Fraser's modification consisted in building up the guns from only a few coils, instead of several, the coils being longer than Sir W. Armstrong's, and the iron coiled upon itself two or even three times : a plan which enabled him to supersede the breech-piece, formerly made in one large forging, by a piece formed of coils. In order to perceive the increased simplicity of construction introduced by Mr. Fraser, we need but glance at the section of a 9 in. gun constructed according to his system, Fig. 94, and remember that a piece of the same size made after the original plan had ten distinct parts, whereas the Fraser is seen to have but four. Compare also Figs. 92 and 93. We shall now describe the process of making the Fraser 9 in. gun. The parts of the gun as shown in the section, Fig. 94, are : 1, the steel barrel; 2, the B tube; 3, the breech-coil; 4, the cascable screw. The inner steel barrel is made from a solid cylinder of steel, which is supplied by Messrs. Firth, of Sheffield. This steel is forged from a cast block, the casting being necessary in order to obtain a uniform mass, while the subsequent forging imparts to it greater solidity and elasticity. After the cylinder has been examined, and the suitable character of the steel tested by trials with portions cut from it, the block is roughly turned and bored, and is then ready for the toughening process. This consists in heating the tube several hours to a certain temperature in an upright furnace, and then suddenly plunging it into oil, in which it is allowed to remain for a day. By this treatment the tenacity of the metal is marvellously increased.

A bar of the steel 1 in. square previous to this process, if subjected to a pull equal to the weight of 13 tons, begins to stretch and will not again recover its original form when the tension is removed, and when a force of 31 tons is applied it breaks. But the forces required to affect the toughened steel in a similar manner are 31 tons and 50 tons respectively. The process, unfortunately, is not without some disadvantages, for the barrel is liable to become slightly distorted and even superficially cracked. Such cracks are removed by again turning and boring; the hardness the steel acquires by the toughening process being shown by the fact that in the first boring $8\frac{1}{4}$ in. diameter of *solid* steel is cut out in 56 hours, yet for this slight boring, in which merely a thin layer is peeled off, 25 hours are required; and lest there should be any fissures in the metal, which, though not visible to the eye, might make the barrel unsound, it is filled with water, which is subjected to a pressure of 8,000 lbs. per square inch. If under this enormous pressure no water is forced outwards, the barrel is considered safe. It is now ready to have the B tube shrunk on it.

The B tube, like certain other portions of these guns, is constructed from coiled iron bars, and this constitutes one great peculiarity of Sir W. Armstrong's system. It has the immense advantage of disposing the metal so that its fibres encircle the piece, thus applying the strength of the iron in the most effective way. The bars from which the coils are prepared are made from "scrap" iron, such as old nails, horse-shoes, &c. A pile of such fragments, built up on a wooden framework, is placed in a furnace and intensely heated. When withdrawn the scraps have by semi-fusion become coherent, and under the steam hammer are soon welded into a compact mass of wrought iron, roughly shaped as a square prism. The glowing mass is now introduced into the rolling-mill, and in a few minutes it is rolled out, as if it were so much dough, into a long bar of iron. In order to form this into a coil it is placed in a very long furnace, where it can be heated its entire length. When sufficiently heated, one end of the bar is seized and attached to an iron core of the required diameter, and the core being made to revolve by a steam engine, the bar is drawn out of the furnace, winding round the core in a close spiral, so that the turns are in contact. The coil is again intensely heated, and in this condition a few strokes of the steam hammer in the direction of its axis suffice to combine the spires of the coil into one mass, thus forming a hollow cylinder.

The B tube for the 9 in. gun is formed of two double coils. When the two portions have been completely welded together under the steam hammer, the tube, after cooling, is roughly turned and bored. It is again fine bored to the required diameter, and a register of the diameter every few inches down the bore is made. These measurements are taken for the purpose of adapting most accurately the dimensions of the steel barrel to the bore of the B tube, as it is found that perfect exactness is more easily obtained in turning than in boring. The steel barrel is therefore again turned to a size slightly *larger* than the bore of the B tube, and is then placed muzzle end upwards, and so arranged that a stream of water, to keep it cool, shall pass into it and out again at the muzzle, by means of a syphon, while the B tube, which has been heated until it is sufficiently expanded, is passed over it and gradually cooled.

If now the B tube were allowed to cool spontaneously, its ends would, by cooling more rapidly than the central part, contract upon the steel barrel and grip it firmly at points which the subsequent cooling would tend to draw nearer together longitudinally, and thus the barrel would be subjected to injurious strains. In order to prevent this, the B tube is made to cool progressively from the breech end, by means of jets of water made to fall upon it, and gradually raised towards the muzzle end, which has in the meanwhile been prevented from shrinking by having circles of gas-flames playing upon it.

The breech-coil, or jacket, is formed of three pieces welded together. First, there is a triple coil made of bars 4 in. square, the middle one being coiled in the reverse direction to the other two. After having been intensely heated in a furnace for ten hours, a few blows on its end from a powerful steam hammer welds its coils perpendicularly, and when a solid core has been introduced, and the mass has been well hammered on the sides, it becomes a compact cylinder of wrought iron, with the fibres all running round it. When cold it is placed in the lathe, and the muzzle end is turned down, leaving a shoulder to receive the trunnion-ring. The c coil is double, welded in a similar manner to the B coil, and it has a portion turned off, so that it may be enclosed by the trunnion-ring.

The trunnion-ring is made by punching a hole in a slab of heated iron first by a small conical mandrel, and then enlarging by repeating the process with larger and larger mandrels. The iron is heated for each operation, and the trunnions are at the same time hammered on and roughly shaped —or, rather, only one has to be hammered on—for a portion of the bar which serves to hold the mass forms the other. The trunnion-ring is then bored out, and after having been heated to redness, is dropped on to the triple breech-coil which is placed muzzle end up, and the turned end of the c coil (of course, not heated) is then immediately placed within the upper part of the trunnion-ring. The latter in cooling contracts so forcibly as to bind the ends of the coils together, and the whole can thus be placed in a furnace and heated to a high temperature, so that when removed and put under the steam hammer, its parts are readily wielded into one mass. The breech-coil in this state weighs about 16 tons ; but so much metal is removed by the subsequent turnings and borings, that it is reduced to nearly half that weight in the gun. It is then turned in a lathe of the most massive construction, which weighs more than 100 tons. Fig. 34, page 95, is from a drawing taken at Woolwich, and shows one of the large guns in the lathe. No one who witnesses this operation can fail to be struck with the apparent ease with which this powerful tool removes thick flakes of metal as if it were so much cheese. The projections of the trunnions prevent the part in which they are situated from being finished in this lathe, and the gun has to be placed in another machine, where the superfluous metal of the trunnion-ring is pared off by a tool moving parallel to the axis of the piece. Another machine accomplishes the turning of the trunnions, the "jacket" being made to revolve about their axis. The jacket is then accurately bored out with an enlargement or socket to receive the end of the B tube, and a hollow screw is cut at the breech end for the cascable.

The portion of the gun, consisting of the steel barrel with the B tube shrunk on it, having been placed upright with the muzzle downwards, the breech-piece, strongly heated, is brought over it by a travelling crane, and slips over the steel barrel, while the recess in it receives the end of the B tube. Cold water is forced up into the inside of the barrel in order to keep it cool. As the breech cools, which it is allowed to do spontaneously, it contracts and grips the barrel and B tube with great force. The cascable requires to be very carefully fitted. It is this piece which plays so important a part in resisting the force tending to blow out the end of the barrel. The cascable is a solid screw formed of the very best iron, and its inner end is wrought by scraping and filing, so that when screwed in there may be perfect contact between its face and the end of the steel barrel. A small annular space is left at the circumference of the inner end, communicating through a small opening with the outside. The object of this is, that in case of rupture of the steel barrel, the gases escaping through it may give timely warning of the state of the piece.

Besides minor operations, there remain the important processes of finishing the boring, and of rifling. The boring is effected in two operations, and after that the interior is gauged in every part, and "lapping" is resorted to where required, in order to obtain the perfect form. Lapping consists in wearing down the steel by friction against fine emery powder and oil, spread on a leaden surface. The piece is then ready for rifling. The machinery by which the rifling is performed cannot be surpassed for its admirable ingenuity and simplicity.

In this operation the gun is fixed horizontally, its axis coinciding with that of the bar, which carries the grooving tools. This bar is capable of two independent movements, one backwards and forwards in a straight line in the direction of the length of the bar, and the other a rotation round its axis. The former is communicated by a screw parallel to the bar, and working in a nut attached to the end of it. For the rotatory movement the bar carries a pinion, which is engaged by a rack placed horizontally and perpendicularly to the bar, and partaking of its backward and forward movement, but arranged so that its end must move along another bar placed at an angle with the former. It is this angle which determines the pitch of the rifling, and by substituting a curved guide-bar for the straight one, an increasing twist may be obtained in the grooves.

The projectile used with these guns is of a cylindrical form, but pointed at the head, and the moulds in which these shots are cast are so arranged that the head of the shot is moulded in iron, while the body is surrounded with sand. The rapid cooling induced by the contact of the cold metal causes the head of the shot to solidify very quickly, so that the carbon in the iron is not separated as in ordinary casting. In consequence of this treatment, the head of the shot possesses the hardness of steel, and is therefore well adapted for penetrating iron plates or other structures. The projectiles are turned in a lathe to the exact size, and then shallow circular cavities are bored in them, and into these cavities brass studs, which are simply short cylinders of a diameter slightly larger than the cavities, are forced by pressure. The projecting studs are then turned so as accurately to fit the spiral grooves of the guns. Thus the projectile in traversing the bore of the piece is forced to make a revolution, or part of a revolution, about its axis, and the rapid rotation thus imparted has the effect of keeping the axis of the missile always parallel to its original direction. Thus vastly increased accuracy of firing is obtained.

Shells are also used with the Woolwich rifled guns. The shells are of the same shape as the solid shots, from which they differ in being cast hollow, and having their interior filled with gunpowder. Such shells when used against iron structures require no fuse ; they explode in coming into collision with their object. In other cases, however, the shells are provided with fuses, which cause the explosion when the shot strikes. Fig. 93, page 195, represents one of the 35-ton guns, made on the plan introduced by Mr. Fraser. This piece of ordnance is 16 ft. long, 4 ft. 8 in. in diameter at the breech, and 1 ft. 9 in. at the muzzle. The bore is about 1 ft. Each gun can throw a shot or bolt 700 lbs. in weight, with a charge of 120 lbs. of powder. It is stated that the shot, if fired at a short range, would penetrate a plate of iron 14 in. thick, and that at a distance of 2,000 yards it would retain sufficient energy to go through a plate 12 in. thick. The effect

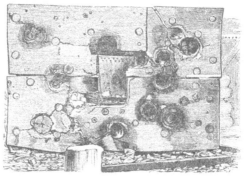

FIG. 95.—*Millwall Shield after being battered with Heavy Shot.—Front View.*

of these ponderous missiles upon thick iron plates is very remarkable. Targets or shields have been constructed with plates and timber backing, girders, &c., put together in the strongest possible manner, in order to test

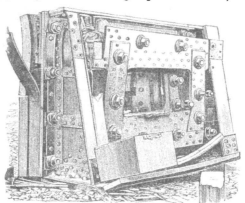

FIG. 96.—*Rear View of the Millwall Shield.*

the resisting power of the armour plating and other constructions of our iron-clad ships. The above two cuts, Figs. 95 and 96, are representations of the appearance of the front and back of a very strong shield of this description, after having been struck with a few 600 lb. shots fired from the 25-ton gun. The shots with chilled heads, already referred to, sometimes were found to penetrate completely through the 8 in. front plate, and the 6 in. of solid teak, and the 6 in. of plating at the back. The shield, though strongly constructed with massive plates of iron, only served to prove the relative superiority of the artillery of that day, which was at the time when our century had yet about thirty years to run. Up to 1876 no confidence was placed in steel as a resisting material, a circumstance perhaps not much to be wondered at, as its capabilities had not then been developed by the newer processes of manufacture, described in our article on Iron ; nor had mechanicians acquired the power of operating with large masses of the metal. Since then it has come about that only steel is relied upon for efficiently resisting the penetration of projectiles, iron being held of no account except as a backing. There has always been a rivalry between the artillerist and the naval constructor, and this contest between the attacking and the defending agencies is well illustrated in the table on page 166, where the parallel advance in the destructive power of guns and in the resisting power of our war-ships is exhibited in a numerical form.

The 35-ton Fraser guns were at the time of their production humorously called in the newspapers " Woolwich infants " ; but it was not long before they might in another sense be called infants in comparison with a still larger gun of 81 tons weight constructed at Woolwich shortly before iron-coiling and muzzle-loading were set aside. Fig. 97 shows the relative

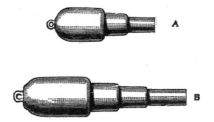

FIG. 97.—*Comparative Sizes of* 35 *and* 81 *ton Guns.*

A, 35-ton ; B, 81-ton.

dimensions of the 35-ton and 81-ton guns : the latter was built up in the same way as the 9-inch gun described above, but the coils were necessarily longer and the chase was formed in three parts instead of two. The total length of this gun was 27 feet, and the bore was about 24 feet long and 14 in. in diameter, and the weight of the shot about 1000 lbs., with sufficient energy to penetrate at a considerable distance an iron plate 20 in. in thickness. It was for the manufacture of these very large guns that the great steam hammer, represented in Plate III., was erected at Woolwich.

The 81-ton gun was the largest muzzle-loader ever made in the national gun factory at the time when such huge weapons were in request ; but in 1876 its dimensions were surpassed by those of a few 100-ton guns built at Elswick to the order of the Italian Government for mounting on their most formidable ironclads. These guns have a calibre of 17·72 inches, and are provided with a chamber of somewhat larger bore to receive the charge of powder. They are built up on the Armstrong shrinkage principle, and comprise as many as twenty different tubes, jackets, hoops, screws, etc., and are undoubtedly the most powerful muzzle-loading weapons ever constructed. It happened, just as these guns were completed, that the British Government, apprehensive at the time of a war with Russia, exercised its rights of purchasing two of them, one to be mounted at Gibraltar, the other at Malta.

The Elswick establishment soon afterwards surpassed all its former achievements in building great guns, by designing and constructing the huge breech-loaders, one of which forms the subject of our Plate XII. These are known as the Armstrong 110-ton guns; they are formed of solid steel throughout, and their weight is accurately 247,795 lbs., or 110 tons 12 cwts. 51 lbs. The total length of the gun is 43 ft. 8 in., and of this 40 ft. 7 in. is occupied by the bore, along which the rifling extends 33 ft. 1 in. The calibre of the rifled part is 16¼ in., and the diameter of the powder chamber is somewhat greater. The regulation charge of powder weighs 960 lbs., although the guns are tested with still greater charges. The weight of the projectile is 1,800 lbs., and it leaves the muzzle with a velocity of 2,128 ft. per second, which is equivalent to a dynamical energy of 56,520 foot-tons. What this means will perhaps be better understood, not by describing experiments such as those on the Millwall Shield, the results of which are depicted in Figs. 95 and 96, but by stating that if the shot from the 110-ton gun encountered a solid wall of wrought iron a yard thick, it would pass through it. The Elswick 110-ton gun is, in fact, the most powerful piece of ordnance that has ever been constructed. There are no trunnions to these great guns, but they are encircled by massive rings of metal, between which pass strong steel bands that tie the gun to its carriage, or, rather, to the heavy steel frame on which it is mounted, and which slides on a couple of girders. The force of the recoil acts on a hydraulic ram that passes through the lower part of the supporting frame. The whole working of the gun is done by hydraulic power, and, indeed, the same method has been applied by the Elswick firm to the handling of all heavy guns. By hydraulic power, maintained automatically by a pumping engine exercising a pressure of from 800 lbs. to 1,000 lbs. per square inch, are operated the whole of the movements required for bringing the cartridge and the projectile from the magazine ; for unscrewing the breech block, withdrawing it, and moving it aside ; for pushing home the shot and the cartridge to their places in the bore ; for closing the breech and screwing up the block ; for rotating the turret within which the gun is mounted, or in other cases for ramming the piece in or out, and for elevating or depressing it. It is, indeed, obvious that such ponderous masses of metal as form the barrels and projectiles of these 110-ton and other guns of the larger sizes could not be handled to advantage by any of the ordinary mechanical appliances. But by the application of the hydraulic principle, a very few men are able to work the largest guns with the greatest ease, for their personal labour is thus reduced to the mere manipulation of levers. On board ship the power required for working large guns has lately been sometimes supplied by a system of shafting driven by a steam engine and provided with drums and pulleys, exactly as in an engineer's workshop. Great care has also been bestowed upon the mounting of the smaller guns, which are so nicely

PLATE XII.

THE 110-TON ARMSTRONG GUN.

poised on their bearings and provided with such accurately fitted racks, pinions, etc., that a steel gun of 10 ft. in length can easily be pointed in any direction by the touch of a child's hand. The mechanical arrangements are now so admirably adapted for facility of working that, unless in the rude shocks of actual warfare the nicely adjusted machinery is found to be liable to be thrown out of gear, these applications of the engineer's skill may be considered as having done all that was required to bring our modern weapons to perfection.

With the construction of the 110-ton we arrive at a period when commences a new era in guns—and especially in the armament of war-ships—necessitated by various circumstances, amongst which may be named the invention of torpedoes and the building of swiftly moving torpedo-boats, and of still swifter "torpedo-boat catchers or destroyers"; so that guns that could be worked only at comparatively long intervals were at a great disadvantage. Again, about 1880, were published the records of a most elaborate and important series of researches conducted by Captain Noble and Sir F. Abel, the chemist of our War Department. They had investigated all the conditions attending the combustion of gunpowder in confined spaces, the nature and quantities of the products, the temperature and pressures of the confined gases, etc. The information thus afforded was extremely valuable; but besides this, direct experiments made with actual guns were carried out, more particularly at Elswick, in which the speed of the projectiles at every few inches of their travel along the bore of the piece was ascertained, and also the pressures of the powder gases at any point. The way in which this is done we shall explain on another page. (*See article on Recording Instruments.*)

So long as muzzle-loading was in use, guns were necessarily made short, for had they not to be run in from the port-holes and embrasures of forts in order to be loaded? Now there was an obvious disadvantage in this, for the projectile left the gun before the expansive force of the gases had been spent that could have imparted additional velocity. When however muzzle-loading was abandoned, and especially when strong and trustworthy steel became available for the construction of the gun throughout, there was no reason to waste in this way the power of the charge, so that barrels were made lighter, much longer in proportion to the calibre, and every part accurately adapted in strength to the strain to be resisted. For instances of increasing length, take the 38-ton 12-inch guns built up at Woolwich (of only seven pieces) for H.M.S. *Thunderer* (see Fig. 93), on Mr. Fraser's plan. These had a bore equal to only 16 times their calibre, while in the Armstrong 100-ton guns the bore is 21 calibres long; and in the 110-ton guns the total length of the chase is 31 times the diameter of the rifled part. It has since been the practice to make the bore of guns from 30 to 40 calibres in length.

The effect of a longer chase used with an appropriate charge is very clearly and instructively shown by the diagram Fig. 98, which is by permission copied from the very comprehensive work by Messrs. E. W. Lloyd and A. G. Hadcock, entitled *Artillery : its Progress and Present Position.* The reader should not pass over this diagram until he has thoroughly understood it, for it is an excellent example of the graphic method of presenting the results of scientific investigations. At the lower part of the diagram there are drawn to scale half-sections of a long and of a short gun. The horizontal line above is marked in equal parts representing feet numbered from the base of the projectiles. The upright line on the left numbered at every fourth division is the scale for the pressures in tons per square inch on the base of the projectile, and these are represented by the height of the plain curves above the horizontal line at each point in the travel of the shot. The dotted lines represent in the same way, but *not* on the same scale as the former, the velocity with which the base of the projectile passes every point in the chase. The figures 2, 4, and 6 on the upright line at the right-hand side refer only to pressures : the velocities scale is such that the point where the dotted meets the right-hand one is 2,680 units above the horizontal line, as the middle upright in the same way is 1,561 high, and the heights of the dotted lines represent each on the same scale the velocities of the bases of the projectiles at the corresponding parts of the chases. The shorter gun has the rifled part of the chase 15·4 calibres long ; the corresponding part of the longer is nearly 33 calibres. The short 7-inch gun has a charge of 30 lbs. of gunpowder, and its projectile weighed 115 lbs. The longer 6-inch gun was not charged with gunpowder, but with the more powerful modern explosive *cordite* (see Index), of which there was 19·5 lbs., and its projectile weighed 100 lbs. The charges were so adjusted that the shots had the same initial maximum pressure of 20 tons per square inch applied to them. Now the cordite, though much more powerful than gunpowder (that is, a given weight will produce far more gas), is slower in its ignition, continuing longer to supply gas. The maximum pressure, 20 tons in both cases, is suddenly attained by the gunpowder gases, when the shot has hardly moved 6 inches onward, and the pressure declines rapidly as the moving shot leaves more space for the gas ; while the cordite gases produce their greatest pressure more gradually at a part where the shot is already about 20 inches on its way, and not only do their highest pressures continue for a greater distance,— but the decline is far less rapid than in the other case. It will be observed by the intersection of the dotted lines, that when the shots in each case have moved about 2 ft. their velocities are equal. They finally leave the muzzles with the velocities marked on the diagram, and if the reader will apply the formula given on page 174 he will obtain their respective energies in *foot-lbs.* ; but for large amounts like these it is more usual to state the energy in *foot-tons*, which of course will be arrived at by dividing the *foot-lbs*. numbers by 2,240, and these will work out in the one case to 4978·9 ft.-tons, and in the other to 1942·5 ft.-tons. The shot from the long gun will therefore have more than 2½ times the destructive power of the other.

The operations required in constructing guns are multiform, and have to be very carefully conducted so that the workmanship shall be of the best quality. The finest ores are selected for reduction, and the steel is obtained by the Siemens-Martin process already described. It must be free from sulphur and phosphorus, and contain such proportions of carbon, silicon, and manganese as experience has shown to be best, and its composition is ascertained by careful chemical analysis before it is used. The fluid steel is run into large ladles lined with fire-brick, and provided with an opening in the bottom from which the metal can be allowed to run out into the *ingot* moulds, the size and proportions of these being in accordance with the object required ; some admitting of as much as 80 tons at one operation. When a barrel or hoop is required of not less than 6 inches internal diameter the ingot is cut to the required length and roughly bored. The ingot is then heated, a long cylindrical steel bar

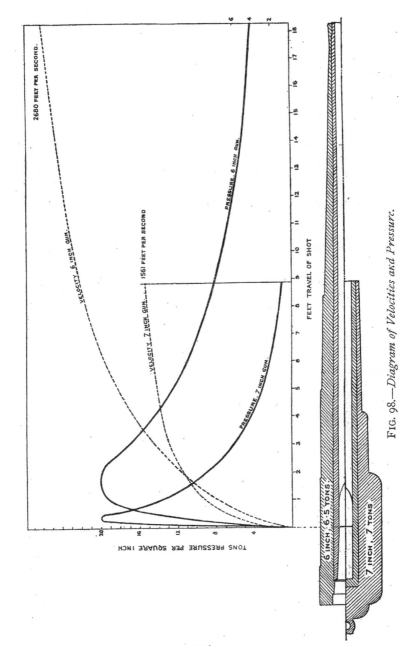

FIG. 98.—*Diagram of Velocities and Pressure.*

205

is put through the hole, and under a hydraulic press the hot metal is
squeezed into greater length and less diameter. The hole first bored
through the ingot is of somewhat greater, and the steel bar (called a
mandril) of less, diameter than required in the finished piece. Portions
are cut from each end of what is now called the *forging* and subjected to
mechanical tests : if these are satisfactory, the forging is rough bored and
turned on the outside. It is then annealed, by being heated and allowed
to cool very slowly. The next operation is to *harden* the metal by raising
it to a certain temperature, at which it is immersed in rape oil until cold.
Then the piece is again annealed, and fine-turned and bored. All these
operations have to be performed not only on the barrel, but also on each
hoop, before the hoops are shrunk on, and the greatest nicety of measurement is required in each piece. Then the gun has to be turned on the
outside, the screw for the breech piece cut, the bore rifled, etc. The
object of the annealing is to relieve the metal from internal strains. It
will not be wondered at that months are required for the construction of
the larger kind of guns. Thus at Elswick a 6 in. quick firing gun, upon
which men are employed night and day, cannot be completed in less
than five months, and sixteen months are required for making a 67-ton
gun.

We may take as an illustration of the progress of modern artillery one
of the products of the Elswick factory which has just been referred to,
and for which the demand from all quarters has been unprecedentedly
great, namely, the 4·7 inch gun. This weapon is mounted in various
manners according to the position it has to occupy, whether for a land
defence, or on ship-board between decks, or on the upper deck. The
arrangement shown in Fig. 99, which is reproduced from Messrs. Lloyd
& Hadcock's work, is known as the centre pivot mounting, and is
suitable for such a position as the upper deck of a ship. The reader
should compare the proportions and mounting of this weapon with those
of the old 32-pounder sketched in Fig. 90, observing the very much greater
comparative length of the modern weapon, and the mechanism for elevating and training it (which, however, the scale of the drawing crowds into
too small a space to show as it deserves). C is a projection from the
breech, to which is attached the piston of the recoil press ; at T is the
handle for training, which actuates a worm at V ; the elevation is regulated by the turning of the four-armed wheel. The long chase of the
gun projects in front ; but the mounting and the breech machinery are
protected by shields of thick steel, of which the sections of two plates are
denoted by the dark upright parts in front. These are fixed ; but a
movable plate above the gun can be raised or lowered into an inclined
position, for better taking sights. In the figure this is shown as open
and in a horizontal position. This gun is provided with sights by which
it can be aimed at night ; that is, the sights can be illuminated by small
electric lamps suitably placed ; the wires connecting these with voltaic
battery cells carried on the mounting are indicated. The figure represents the gun as constructed about 1893, but the improvements that
are continually being made have brought about some modifications in
the details.

Very notable among the productions of the great Elswick factory are
the *quick firing* guns. These were at first confined to guns of small
calibre, such as the 6-pounders. They are, of course, all breech-loaders,
and the powder and shot are both contained in a single metallic cartridge

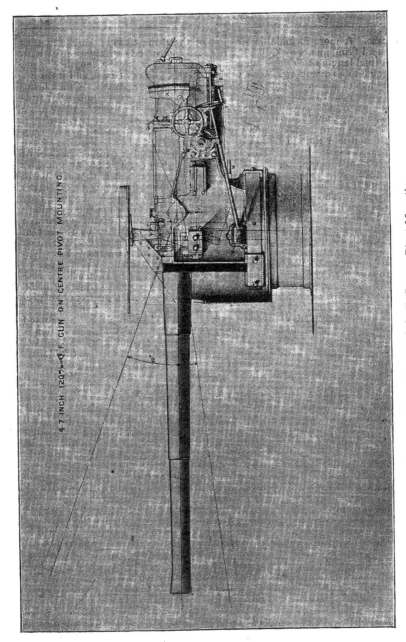

4·7 INCH 120°× Q.F. GUN ON CENTRE PIVOT MOUNTING.

FIG. 99.—*Elswick 4·7 inch Q. F. Gun on Pivot Mounting.*

case. A more formidable weapon of the same class is the 45-pounder rapid firing gun, which, like the rest, is constructed entirely of steel, with a total length of 16 ft. 2½ in., a calibre of 4,724 in., and a length of bore equal to 40 diameters. The weight of this gun is 41 cwt., and it throws a shell of 45 lbs. weight with a 12-lb. charge of gunpowder. Quick firing guns having a calibre of 6 in. are now also made in great numbers for arming our ironclads. The breech block in the quick firing guns turns aside on a hinge, and after the introduction of the cartridge it is closed and screwed up to its place by a slight turn of a handle. The piece is then pointed and trained by aid of mechanical gearing as in the case of the heavier guns. But Mr. Hotchkiss has introduced a simpler method of elevating and training his 3-pounder and 6-pounder quick firing guns, by attaching to the rear, and unaffected by the recoil, a shoulder piece against which the marksman can lean, and move the weapon as he takes his aim. Though these guns weigh respectively 4½ cwt. and 7 cwt., they can thus be pointed with the greatest ease. The firing is done by pulling a trigger in what seems like the stock of a pistol. The empty cartridge case is automatically extracted from the firing chamber by the act of opening the breech, and it drops to the ground. Ten or twelve rounds per minute can be fired from these guns, and Lord Armstrong has advocated the use of a number of them for naval armament in preference to that of a few ordinary breech-loaders of more unwieldy dimensions. He has calculated that in a given time a far greater weight of metal can be projected from a vessel armed with quick firing guns than from one provided only with the heavier class of cannons.

The breech pieces in the Elswick guns are closed on the "interrupted screw" system—that is, a very large screw thread of V-shaped section is cut in the barrel at the breech end, and a corresponding thread on the principal part of the breech block, which is, of course, capable of rotating about the axis. The screw threads, however, are not continuous, segments parallel to the axis being cut away, the spaces in the outer thread corresponding with the projecting parts in the inner, and *vice versâ*, so that when the block is pushed home, one very small part of a turn suffices to engage all the threads. The screw is also made conical, and is so cut into steps, as it were, that great resisting power is brought into play. The Elswick guns are provided with hydraulic buffers for checking the recoil, and the principle is applied in various modified forms. In some cases the pistons allow for the water a passage, which towards the end gradually diminishes. This is the arrangement for the 3-pounder rapid firing Hotchkiss gun, and the force of the recoil is made at the same time to compress two springs, which serve to return the gun to the firing position. This very handy gun is said to be able to fire twenty rounds per minute. In Mr. Vavasseur's plan of mounting, the recoil is checked by ports, or openings, in the piston of a hydraulic cylinder being gradually closed, which is easily arranged by making a spiral groove within the cylinder, which gives a small axial motion to part of the piston.

An extremely effective plan for the defence of coasts and harbours was originated by Colonel Moncrieff, when about 1863 he contrived a method of mounting large guns on the disappearing system, by which almost complete protection against hostile fire is given to both gun and gunners. He utilizes the recoil as a means of bringing the gun down into a protected position the moment it has been fired, and retains this energy by a simple arrangement until the piece has been reloaded, when it is allowed to expend itself by again raising the gun above the parapet into the original firing position. The configuration and action of Colonel

FIG. 100.—*The Moncrieff Gun raised and ready for firing.*

FIG. 101.—*Moncrieff Gun lowered for loading.*

Moncrieff's gun-carriage will be understood by an inspection of the annexed illustrations, where in Fig. 100 is shown the gun raised above

14

the parapet and ready for firing. When the discharge takes place, the gun, if free, would move backwards with a certain speed, but the disposition of the mounting is such that this initial velocity receives no *sudden* check, the force being expended in raising a heavy counterpoise, and at the same time the gun is permitted to descend, while maintaining a direction parallel to its firing position. At the end of the descent, which, it must be understood, is caused by the force of the recoil, and not by the counterpoise, for this more than balances the weight of the gun, the latter is retained as shown in Fig. 101 until it has been reloaded; and when it has again to be fired, it is released so as to allow the descent of the counterpoise to raise it once more into position. The great advantage of this invention is the protection afforded to the artillerymen and gun while loading; and even the aiming can be accomplished by mirrors, so that the men are exposed to no danger, except from "vertical fire," which involves but little risk.

Colonel Moncrieff took out a patent for his invention in 1864, but committed the practical working out of his idea to the firm of Sir W. G. Armstrong & Co., in whose hands the design was ultimately transformed from the original somewhat cumbersome arrangement of the mounting into the compact and manageable form shown in Fig. 102, which represents a 13·9 inch 68-ton breech-loading disappearing gun on the Elswick hydro-pneumatic mounting. The principle of hydraulic power is fully explained in our article on that subject, and an example of its application to cranes as devised by Sir W. Armstrong is there described. When guns began to be made very large, and projectiles weighing several cwts. had to be dealt with, the application of power in some form became essential for loading, running out, elevating, training, etc.; and though steam-power naturally was first used, hydraulic power was adopted at Elswick, and has been there applied to the mountings of large guns with the greatest success by Mr. G. W. Rendel. To mention the various arrangements in which this power is applied, or to attempt any description of the elaborate machinery by which it is regulated, would carry us far beyond our limits. But the powerful weapon depicted in Fig. 102 is designed to be worked only by the manual effort of a few men. In this mounting the pressure of condensed air sustains the gun in the firing position; that pressure, acting upon the water in the recoil presses, having previously forced up their rams so as to turn into a nearly vertical position the strong brackets or beams on which the trunnions are supported. The recoil is checked in the usual way by the forcing of the water through small ports or valves as the ram descends, but these valves are so arranged that the water is in part forced back into the air chamber, and there recompresses the air, to restore the power for again raising the gun. The pressure in the air chamber when the gun is down may be about 1,400 lbs. per square inch; when it is up this will be reduced to perhaps one half by the expansion of the air in doing work. We have here the reaction of compressed air taking the place of the gravity of the counterpoise originally designed. There are in this hydro-pneumatic mounting a number of adjusting appliances, such as forcing pumps, brakes, etc., for regulating the pressures, or quantity of liquid, as, for instance, when lowering the gun without any recoil action in operation. Then again, with any change in the weight of projectile or in the powder charge, there would be a corresponding change in the power of the recoil, and therefore the necessity for compensatory adjustments, which are made

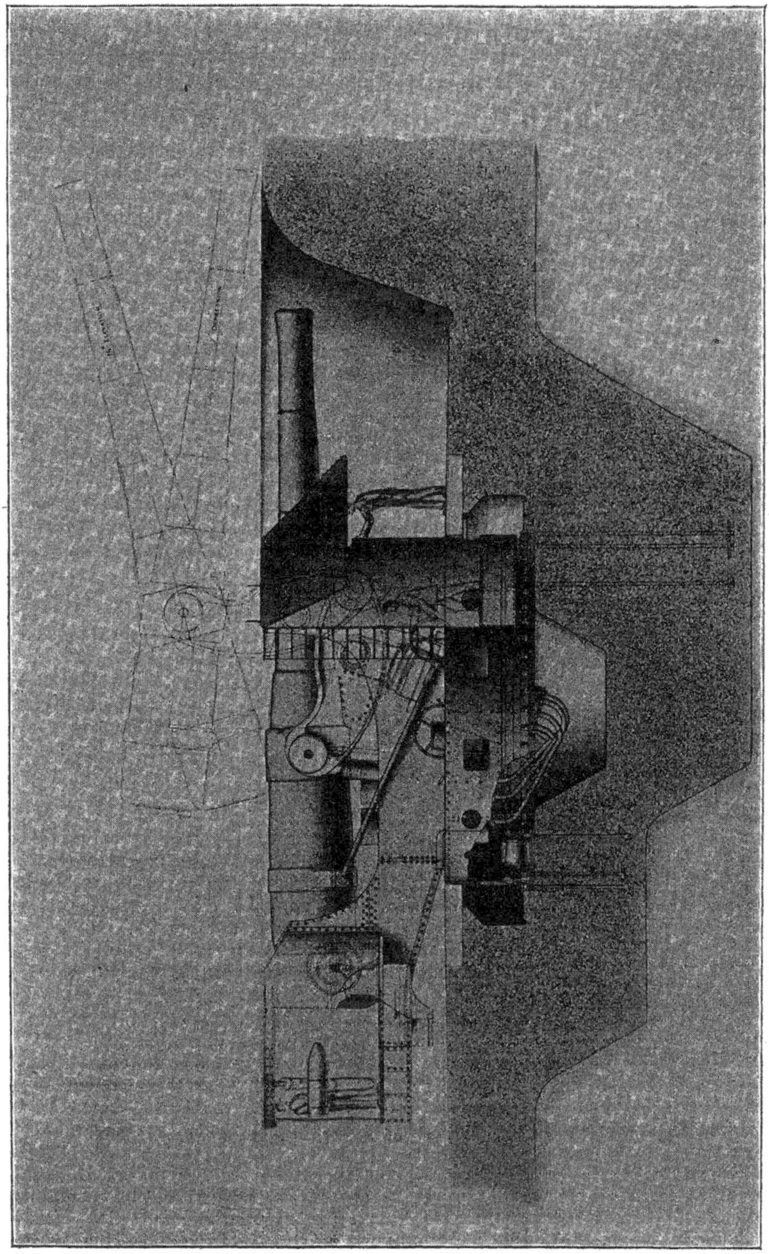

FIG. 102.—68-ton Gun on Elswick Hydro-Pneumatic Mounting.

with great readiness. The nicety with which the parts are adapted to each other in this mounting must be obvious, when we observe the magnitude of the mass to be moved with the least delay, and brought to rest, quite gently and exactly, in a new position. Details cannot here be given even of the method by which the valves in the recoil cylinders are automatically controlled for this purpose. Means are also supplied for setting the gun, while still in its protected position, to the required angle of elevation or depression. The adjustment is made by the long rods attached near the breech and set at their lower ends to the position giving the intended angle to the raised gun. The varied and powerful strains to which the parts of this mechanism are subject, and which have had to be calculated and provided for, may be inferred from the enormous recoil energy of the gun, which under ordinary conditions amounts to no less than 730 foot-tons. The gun is provided with ordinary, and also with reflecting, sights, so that no one need be exposed to the enemy's fire. Protective armour above the gun is not required, as the pit itself being usually on some elevation is imperceptible to the enemy, and the gun is visible but for a few seconds, forming a quite inconspicuous object. The pit in which the gun is mounted is commonly lined with concrete. Italy, England, Norway, Japan and other countries have appreciated the advantages of the disappearing system in providing the most powerful coast defences yet devised, and a great many guns have been mounted on this principle.

An extraordinary piece of ordnance is represented in Fig. 103. It is one of two huge mortars, the idea of which presented itself to Mr. Mallet during the Crimean War, the intention being to throw into the Russian lines spherical shells a yard in diameter, which would, in fact, have constituted powerful mines, rendering it impossible for the fortifications to continue tenable. Mallet's original design was to project these shells from mortars of no less than 40 tons weight. When it was pointed out that the transport of so heavy a mass would be impracticable, the design was changed to admit of the mortar being made in pieces not exceeding eleven tons in weight, and built up where required. During the most active period in the siege of Sebastopol this plan was submitted to Lord Palmerston, who at once ordered two of these apparently formidable pieces to be constructed, without waiting for official examinations of the scheme, and the usual reports of experts,—promptness in this case being considered of the utmost importance. A contract was made with a private firm, who undertook to deliver them in ten weeks. But the difficulties attending such constructions not being understood at the time, delays arose, the contractors failed, and two years elapsed before the mortars were completed. In the meantime peace had been concluded, and the mortars were never fired against any hostile works ; but experiments were made with one of them at Woolwich. The heaviest of the shells it was intended to project weighed 2,940 lbs., and for this it was proposed to use a charge of 80 lbs. of gunpowder. In the experiments the charges first used were low, but gradually increased : when it was found that after every few rounds repairs became necessary in consequence of the weak points in the construction, and after the nineteenth round the mortar was so much damaged that the trials were definitely discontinued. The other mortar, though mounted, was never fired, but remains at Woolwich, an object of some interest to artillerists, especially since there has been some talk of reverting to this very old-fashioned form of ordnance as a means of attacking ironclads in their most vulnerable direction by the so-called vertical fire. In one of the rounds of the Mallet mortar tried at Woolwich, a shell weighing 2,400 lbs. was thrown by a charge of 70 lbs. of gunpowder a distance of more than a mile and a half, and it buried itself in the soil to a very great depth.

FIG. 103.—*Mallet's Mortar.*

For high-elevation firing, howitzers will more probably be the form of ordnance most in use. The range of the howitzer is determined by the angle at which it is elevated, whereas with the mortar it is chiefly by variation of the powder charge that the aim is adjusted. Many of the old short 9 in muzzle-loaders have already been converted into 11 in.

rifled howitzers, and these are likely to prove of great service in defending our harbours and channels against war vessels.

Some account has been given in a preceding article of the great steel works of Krupp & Co. at Essen, and the place has been noted as one of the greatest gun factories in the world during the second half of our century. The process there practised of casting crucible steel ingots, and already described, is precisely that used in the first stage of gun-making. The steel for guns put into the crucibles is a carefully adjusted mixture of one quality of iron puddled into steel and subjected to certain treatment; the other portion is made from a different quality of iron from which all the carbon has been puddled out. The cast ingot is forged under a great steam hammer, bored, turned, and steel hoops shrunk upon it, in several layers; and other operations are performed upon it like those which have already been mentioned. A 14 in. gun is said to require sixteen months for its manufacture, and its cost to be about £20,000.

Artillerists had long carried on a warm controversy as to the relative merits of wrought iron and steel in gun construction, the latter material

FIG. 104.—*32-pounder Krupp Siege Gun, with Breech-piece open.*

being regarded with shyness on account of its want of uniformity as formerly produced. Krupp however began as early as 1847 to make guns of his excellent crucible steel, and through bad report and good report confined himself to this material until, it is asserted, by 1878 he had supplied over 17,000 steel guns of all calibres. He began by making a 3-pounder gun, but soon produced pieces of larger size, all of which were bored and turned out of solid masses of metal. At a later period the plan of shrinking on strengthening hoops of steel was adopted. The Krupp guns have found extensive favour, and many very heavy ones have been made, some indeed of greater weight than the 110-ton Armstrong; but the excess of weight is due to the mass of metal which the Krupp construction of the breech mechanism requires. Thus Krupp's 120-ton gun has a muzzle energy of but 45,796 foot-tons, while that of the Elswick piece is 55,105 foot-tons.

The breech arrangement in the Krupp guns consists of a lateral slot into which slides a closing block after the charge has been inserted from the rear. An obsolete form of this breech piece is seen in Fig. 104, which represents a 32-pounder gun such as was used in sieges by the Prussians in the Franco-German War. It will be observed here that the slot and breech piece are of rectangular form ; but this shape, causing the piece to be weak where most strength was required, was afterwards altered into a D-shaped section, the curved side being of course to the rear. That difficulty which baffled the earliest attempts at breech-loading is the same that has given much trouble to modern gunmakers. It consists in so closing the breech that no escape of the powder gases can take place there at the moment of discharge. When we remember that the momentary pressure of the gases in the powder chamber may amount to more than 40 tons on the square inch, we can well understand the enormous velocity with which they will rush forth from even the smallest

FIG. 105.—*The Citadel of Strasburg after the Prussian Bombardment.*

interspace between the base of the gun and the breech block, but we can hardly realise without actual inspection the mechanical action they produce in their passage : when once the escape occurs, a channel is cut in the metal as if part had been removed by an instrument, and the piece in that condition is disabled for further use. Several devices are in use obtaining perfect closure of the breech, which is technically called *obturation* (Latin, *obturare*, to close up). One of these consists in fitting closely into the circumference of the bore a ring of very elastic steel, turned up at the edges towards the powder chamber. The gas pressure forces the edge of this ring still more closely against the interior of the powder chamber, much in the same way as the Bramah collar acts in the hydraulic press (see Fig. 165). The shaded circle shown on the breech piece in Fig. 104 is an additional device for obtaining obturation. The Broadwell ring, as the above-mentioned contrivance is called, is not used in English guns, but another plan of obtaining a gas-joint has been much adopted, in which a squeezable pad is by compression forced outwards to close up the bore.

A very long range was claimed for Krupp's guns at the time of the Franco-German War, for at the siege of Paris (1870) it was said they could hurl projectiles to the distance of five miles, though probably there was some exaggeration in this statement. There is no doubt however that the Prussians had very effective and powerful artillery, as may be gathered from Fig. 105, which is taken from a photograph of part of the fortifications of Strasburg after the bombardment of that fortress. The explosive shells used by the Prussians against masses of troops were not precisely segment shells of the form already described, but the principle and effect were the same, for the interior was built up of circular rings, which broke into many pieces when the shell exploded.

Out of the very numerous forms in which modern ordnance is constructed, we have been able to select but a few examples for illustration and description. These will suffice, it is hoped, to give an idea of the progress that the century has witnessed. It would be beyond our scope to give details of the ingenious mechanical devices that have come to be applied to guns : such as the breech-closing arrangements, the various ways in which recoil is controlled and utilized, etc. A good illustration, had space permitted, of the scientific skill applied to ordnance would be found in the contrivances fitted to certain projectiles in order to determine their explosion at the proper moment. These are very different from the cap or time fuse that did duty in the first half of the century. We have indeed said little of the projectiles themselves beyond mention of the Palliser chilled shot and the obsolete studded projectiles. We have not explained how bands of copper, or other soft metal, are put round a certain part of the shot or shell, in order that, being forced into the grooves, the axial rotation may be imparted, or how windage is prevented by "gas checks" attached to the base of the projectiles. We must now be contented to conclude this section by showing the structure of two kinds of explosive shells which have been much used.

Shrapnel shell takes its name from Lieutenant Shrapnel, who was its inventor about the end of last century, but the projectile began to be used only in 1808. Fig. 105*a* is a section showing the shell as a case containing a number of spherical bullets, of which in the larger shells there are very many, the interspaces being filled with rosin, poured in when melted ; the bullets are thus prevented from moving about. The figure shows the shell without the fuse or percussion apparatus, which screws into the hollow at the front. The bursting charge of gunpowder is behind the bullets, and when it explodes they travel forward with a greater velocity than the shell, but with trajectories more or less radiating, carrying with them wide-spreading destruction and death.

A shrapnel shell may be said to be a short cannon containing its charge of powder in a thick chamber at the breech end ; the sides of the fore part of the shell are thinner than those of the chamber, and may be said to form the barrel of the cannon. This cannon is loaded up to the muzzle with round balls, which vary with the shell in size. An iron disc between the powder and the bullets represents the wad used in ordinary fowling-pieces. A false conical head is attached to the shell, so that its outward appearance is very similar to that of an ordinary cylindro-conoidal shell :

that is to say, it looks like a very large long Enfield bullet. The spinning motion which had been communicated to the shell by the rifling of the gun from which it had been fired causes the barrel filled with bullets to point in the direction of the object at which the gun has been aimed. Consequently, when the shrapnel shell is burst, or rather fired off, the bullets which it contained are streamed forward with actually greater velocity than that at which the shell had been moving; and the effect produced is similar to firing grape and canister from a smooth-bore cannon at a short range.

Segment shells were first brought into use by Lord Armstrong in 1858 in connection with his breech-loading guns. The segment shell consists of a thin casing like a huge conical-headed thimble, with a false bottom attached to it. It is filled with small pieces of iron called "segments,"

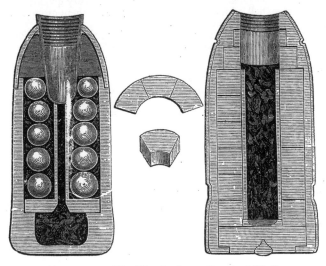

FIG. 105*a*.—*The Shrapnel and Segment Shells.*

cast into shapes which enable them to be built up inside the outer casing into two or more concentric circular walls. The internal surface of the inmost wall forms the cavity of the compound or segment shell, and contains the bursting charge. The segment shell is fitted with a percussion fuse, which causes it to explode when it strikes. In the shrapnel shell, the powder charge is situated in rear of the bullets, and consequently produces the chief effect in a forward direction. In the segment shell, the powder is contained inside the segments, and therefore produces the chief effect in a lateral direction. When the shrapnel shell is burst at the right moment, its effect is greatly superior to that of the segment shell; on the other hand, the segment shell, when employed at unknown or varying distances, is far more unlikely to explode at the proper time.

Shrapnel and segment shells can be used with field artillery, *i.e.*,

9-pounders, 12-pounders, 16-pounders ; and also with heavy rifled guns in fortresses, viz., 40-pounders, 64-pounders, 7-in. and 9-in. guns. But the conditions of their service are very different in each case. With regard to field artillery, the distance of the enemy is rarely known, and is constantly changing, and hence the men who have to adjust the fuses would probably be exposed to the fire of the enemy's artillery, and, consequently, could not be expected to prepare the fuses with the great care and nicety which are absolutely necessary to give due effect to the shells. There are, however, some occasions when the above objections would not hold good—as, for instance, when field artillery occupy a position in which they wait the attack of an enemy advancing over ground in which the distances are known.

Segment shells require no adjustment of their percussion fuse. They enable the artillerymen to hit off the proper range very quickly, since the smoke of the shell which bursts on striking tells them at once whether they are aiming too high or too low.

With regard, however, to the service of heavy rifled guns in fortresses, the conditions are quite different. In the first place, the distance of all objects in sight would be well known beforehand ; and in the second place, the fuses of the shells would be carefully cut to the required length in the bomb-proofs, where the men would be completely sheltered. The 7-in. shrapnel contains 227 bullets, and a 9-in. shrapnel would contain 500 bullets of the same size, and these shells could be burst with extraordinary accuracy upon objects 5,000, 6,000, or 7,000 yards off.

MACHINE GUNS.

THE name of machine guns has been applied to arms which may be regarded as in some respects intermediate between cannons and rifles, since in certain particulars they partake of the nature of both. Like the former, they are fired from a stand or carriage, and in some of their forms require more than one man for their working : in the calibre of their barrels and the weight of their projectiles, they are assimilated to the rifle, but they are capable of pouring forth their missiles in a very rapid succession—so rapid indeed as practically to constitute volley firing. The firing mechanism of the machine gun has always an automatic character, but the rifle has acquired this feature, so that it cannot be made a distinguishing mark : on the other hand, since machine guns have been made to discharge projectiles of such weights as 1 lb. or 3 lb. there is nothing to separate them from quick-firing ordnance unless it be the automatic firing.

The idea of combining a number of musket-barrels into one weapon, so that these barrels may be discharged simultaneously or in rapid succession, is not new. Attempts were made two hundred years ago to construct such weapons ; but they failed, from the want of good mechanical adjustments of their parts. Nor would the machine gun have become the effective weapon it is, but for the timely invention of the rigid metallic-cased cartridge. Several forms of machine guns have in turn attracted much attention. There is the Mitrailleur (or Mitrailleuse), of which so much was heard at the commencement of the Franco-German War, and of whose deadly powers the French managed to circulate terrible and mysterious reports, while the weapon itself was kept concealed.

Whether this arose from the great expectations really entertained of the destructive effects of the mitrailleur, or whether the reports were circulated merely to inspire the French troops with confidence, would be difficult to determine. Our own policy in regard to new implements of war is not to attempt to conceal their construction. Experience has shown that no secret of the least value can long be preserved within the walls of an arsenal, although the French certainly apparently succeeded in surrounding their invention with mystery for a while. The machine gun, or "battery," invented by Mr. Gatling, an American, is said by English artillerists to be free from many defects of the French mitrailleur. In 1870 a committee of English military men was appointed to examine the powers of several forms of mitrailleur, with a view to reporting upon the

FIG. 105*b.—The Gatling Gun.—Rear View.*

advisability or otherwise of introducing this arm into the British service. They recommended for certain purposes the Gatling battery gun.

In the Gatling the barrels, ten in number, are distinct and separate, being screwed into a solid revolving piece towards the breech end, and passing near their muzzles through a plate, by which they are kept parallel to each other. The whole revolves with a shaft, turning in bearings placed front and rear in an oblong fixed frame, and carrying two other pieces, which rotate with it. These are the "carrier" and the lock cylinder. Fig. 105*b* gives a rear view, and Fig. 105*c* a side view, of the Gatling battery gun. The weapon is made of three sizes, the largest one firing bullets 1 in. in diameter, weighing ½ lb., the smallest discharging bullets of ·45 in. diameter. The small Gatling is said to be effective at a range of more than a mile and a quarter, and can discharge 400 bullets or more in one minute. Mr. Gatling thus describes his invention :

The gun consists of a series of barrels in combination with a grooved carrier and lock cylinder. All these several parts are rigidly secured upon a main shaft. There are as many grooves in the carrier, and as many holes in the lock cylinder, as there are barrels. Each barrel is furnished with one lock, so that a gun with ten barrels has ten locks. The locks work in the holes formed in the lock cylinder on a line with the axis of the barrels. The lock cylinder, which contains the lock, is surrounded by a casing, which is fastened to a frame, to which trimmers are attached. There is a partition in the casing, through which there is an opening, and into which the main shaft, which carries the lock cylinder, carrier, and barrels, is journaled. The main shaft is also at its front end journaled in the front part of the frame. In front of the partition in the casing is placed a cam, provided with spiral surfaces or inclined planes.

"This cam is rigidly fastened to the casing, and is used to impart a reciprocating motion to the locks when the gun is rotated. There is also in the front part of the casing a cocking ring which surrounds the lock cylinder, is attached to the casing, and has on its rear surface an inclined plane with an abrupt shoulder. This ring and its projection are used for cocking and firing the gun. This ring, the spiral cam, and the locks make up the loading and firing mechanism.

" On the rear end of the main shaft, in rear of the partition in the casing, is located a gear-wheel, which works to a pinion on the crank-shaft. The rear of the casing is closed by the cascable plate. There is hinged to the frame in front of the breech-casing a curved plate, covering partially the grooved carrier, into which is formed a hopper or opening, through which the cartridges are fed to the gun from feed-cases. The frame which supports the gun is mounted upon the carriage used for the transportation of the gun.

" The operation of the gun is very simple. One man places a feed-case filled with cartridges into the hopper; another man turns the crank, which, by the agency of the gearing, revolves the main shaft, carrying with it the lock cylinder, carrier, barrels, and locks. As the gun is rotated, the cartridges, one by one, drop into the grooves of the carrier from the feed-cases, and instantly the lock, by its impingement on the spiral cam surfaces, moves forward to load the cartridge, and when the butt-end of the lock gets on the highest projection of the cam, the charge is fired, through the agency of the cocking device, which at this point liberates the lock, spring, and hammer, and explodes the cartridge. As soon as the charge is fired, the lock, as the gun is revolved, is drawn back by the agency of the spiral surface in the cam acting on a lug of the lock, bringing with it the shell of the cartridge after it has been fired, which is dropped on the ground. Thus, it will be seen, when the gun is rotated, the locks in rapid succession move forward to load and fire, and return to extract the cartridge-shells. In other words, the whole operation of loading, closing the breech, discharging, and expelling the empty cartridge-shells is conducted while the barrels are kept in continuous revolving movement. It must be borne in mind that while the locks revolve with the barrels, they have also, in their line of travel, a spiral reciprocating movement ; that is, each lock revolves once and moves forward and back at each revolution of the gun.

" The gun is so novel in its construction and operation that it is almost impossible to describe it minutely without the aid of drawings. Its main features may be summed up thus : 1st.—Each barrel in the gun is provided with its own independent lock or firing mechanism. 2nd.—All the locks revolve simultaneously with the barrels, carrier, and inner breech, when the gun is in operation. The locks also have, as stated, a reciprocating motion when the gun is rotated. The gun cannot be fired when either the barrels or locks are at rest.

There is a beautiful mechanical principle developed in the gun, viz., that while the gun itself is under uniform constant rotary motion, the locks rotate with the barrels and breech, and at the same time have a longitudinal reciprocating motion, performing the consecutive operations of loading, cocking, and firing without any pause whatever in the several and continuous operations.

The small Gatling is supplied with another improvement called the "drum feed." This case is divided into sixteen sections, each of which contains twenty-five cartridges, and is placed on a vertical axis on the top of the gun. As fast as one section is discharged, it rotates, and brings another section over the feed aperture, until the whole 400 charges are expended.

FIG. 105c.—*The Gatling Gun.—Front View.*

After a careful comparison of the effects of field artillery firing shrapnel, the committee concluded that the Gatling would be more destructive in the open at distances up to 1,200 yards, but that it is not comparable to artillery in effect at greater distances, or where the ground is covered by trees, brushwood, earthworks, &c. The mitrailleur, however, would soon be knocked over by artillery if exposed, and therefore will probably only be employed in situations under shelter from such fire. An English officer, who witnessed the effects of mitrailleur fire at the battle of Beaugency, looks upon the mitrailleur as representing a certain number of infantry, for whom there is not room on the ground, suddenly placed forward at the proper moment at a decisive point to bring a crushing fire upon the enemy. Many other eye-witnesses have spoken of the fearfully deadly effect of the mitrailleur in certain actions during the Franco-German War.

Mr. Gatling contends that, shot for shot, his machine is more accurate than infantry, and certainly the absence of nerves will ensure steadiness;

while so few men (four) are necessary to work the gun that the exposure of life is less. No re-sighting and re-laying are necessary between each discharge. When the gun is once sighted its carriage does not move, except at the will of the operator.; and the gun can be moved laterally when firing is going on, so as to sweep the section of a circle of 12° or more without moving the trail or changing the wheels of the carriage. The smoke of battle, therefore, does not interfere with its precision.

Whatever may be the part this new weapon is destined to play in the wars of the future, we know that every European Power has now provided itself with some machine guns. The Germans have those they took from the French, who adhere to their old pattern. The Russians have made numbers of Gatlings, each of which can send out, it is said, 1,000 shots per minute, and improvements have been effected, so as to obtain a lateral sweep for the fire.

A competitor to the Gatling presents itself in the Belgian mitrailleur, the

FIG. 105*d.—The Montigny Mitrailleur.*

Montigny, Fig. 105*d.* This gun, like the Gatling, is made of several different sizes, the smallest containing nineteen barrels and the largest thirty-seven. The barrels are all fitted into a wrought iron tube, which thus constitutes the compound barrel of the weapon. At the breech end of this barrel is the movable portion and the mechanism by which it is worked. The movable portion consists mainly of a short metallic cylinder of about the same diameter as the compound barrel, and this is pierced with a number of holes which correspond exactly with the position of the gun-barrels, of which they would form so many prolongations. In each of the holes or tubes a steel piston works freely; and when its front end is made even with the front surface of the short cylinder, a spiral spring, which is also contained in each of the tubes, is compressed. The short cylinder moves as a whole backwards and forwards in the direction of the axis of the piece, the movement being given by a lever to the shorter arm of which the movable piece is attached. When the gun is to be loaded this piece is drawn backwards by raising the lever, when the spiral springs are relieved from compression, and the heads of the pistons press lightly against a flat steel plate in front of them. The withdrawal of the breech-block gives space for a steel plate, bored with holes corresponding to the barrels, to be slid down vertically ; and this plate holds in each hole a cartridge, the head of each cartridge being, when the plate has dropped into its position, exactly opposite to the barrel, into which it is thrust, when the movable breech-block is made to advance. The anterior face of this breech-block is formed of a plate containing a number of holes again corresponding to the barrels, and in each hole is a little short rod of metal, which has in front a projecting point that can be made to protrude through a *small* aperture in the front of the plate, the said small apertures exactly agreeing in position with the centres of the barrels, and being the only perforations in the front of the plate. The back of the plate has also openings through which the heads of the pistons can pass, and by hitting the little pieces, or strikers, cause their points to pass out through the apertures in front of the plate, and enter the base of the cartridges, where *fulminate* is placed. The plate filled with cartridges has a bevelled edge, and the points of the strikers are pushed back by it as it descends. The heads of the pistons are separated until the moment of discharge from the recesses containing the strikers by the flat steel plate or shutter already mentioned. The effect, therefore, of pushing the breech-block forward is to ram the cartridges into the barrels, and at the same time the spiral springs are compressed, and the heads of the pistons press against the steel shutter which separates them from the strikers, so that the whole of the breech mechanism is thus closed up. When the piece is to be fired a handle is turned, which draws down the steel shutter and permits the pistons to leap forward one by one, and• hit the strikers, so that the points of the latter enter the cartridges and inflame the fulminate. The shutter is cut at its upper edge into steps, so that no two adjoining barrels are fired at once. The whole of the thirty-seven barrels can be fired by one and a quarter turns of the handle, which may, of course, be given almost instantly, or, by a slower movement, the barrels can be discharged at any required rate.

The barrels of the machine guns we have described do not, as is generally supposed, radiate ; on the contrary, they are arranged in a perfectly parallel direction. In consequence of this, the bullets are at short ranges directed nearly to one spot. The Gatling gun was adopted as a service weapon by the British navy, and in several minor actions it had proved effective, but in its original form it was superseded by the Gardner gun, in which the barrels are fixed horizontally side by side, and are in number five or fewer ; each barrel is able to fire 120 rounds per minute. A new system of feed was afterwards applied to the Gatling gun by Mr. Accles, by means of which this gun was greatly improved and its rate of firing was increased to more than 1,000 rounds per minute ; indeed, 80 rounds have been fired from it within 2 seconds. The Gatlings in this improved form have ten barrels, and are provided with feed drums, each containing 104 cartridges, and capable, when empty, of being almost instantly replaced by a full one. The contents of one drum can, if necessary, be discharged in about $2\frac{1}{4}$ seconds, so that in this time 104 rifle bullets would be fired ; or considerably more than the rate of 1,000 rounds per minute could easily be maintained. The weapon is so mounted, that without moving its carriage it can be pointed at any angle of elevation or depression, and through a considerable lateral range.

Mr. Nordenfelt has brought out a machine gun, which, on account of the simplicity and strength of its firing mechanism, has proved the most reliable weapon of its class, and it also has been adopted into the British service, and indeed into that of nearly every nation in the world. In this gun there are five barrels arranged as in the Gardner, but the firing is operated by a lever working backwards and forwards at the rate of 600 rounds per minute.

In the firing of all these weapons, by turning a crank, or moving a lever at one side, any attempt at exact aiming must obviously be difficult

FIG. 105e.—*A Hotchkiss Gun.*

if not impossible, from the liability of the gun to get moved. Several designs have been proposed for making the firing mechanism entirely automatic so as to require no effort on the part of the firer, whose attention can then be directed solely to pointing the piece. It would not be easy to explain in detail the way in which this is accomplished in these very ingenious guns ; for while the principle of their action is sufficiently clear, namely, that the force of the recoil is made to extract the spent cartridge, open the breech, insert a fresh cartridge, close the breech, and fire the charge, the mechanism of the reacting springs, etc., by which this is effected could scarcely, even by the aid of elaborate diagrams, be made intelligible to any other than a gunsmith. The Maxim is one of those automatic guns : it has but one barrel, and after the first discharge it will go on firing with marvellous rapidity the cartridges supplied to it in a continuous chain, and this without any deviation from such direction as may be given to it by the operator, for he has neither crank to turn nor lever to move, but merely sits behind an iron shield directing the weapon at will, which, without interference, fires hundreds of shots per minute from one barrel, so long as the long bands of cartridges are supplied to it.

Mr. Nordenfelt and Mr. Hotchkiss have also both contrived quick firing guns for 1-lb., 3-lb., and 6-lb. projectiles, and these, it has been thought, will be of great service in naval warfare as against torpedo boats.

Though the automatic mechanism, whereby the breech operations are all performed by the force of the recoil of the barrel, which is allowed to slide backwards, and is then returned to its place by a spring, is too complicated for illustration here, mention may be made of a quite recent device by which the recoil action is dispensed with, and the mechanism so far simplified that scarcely more than half the number of parts in the lock mechanism are required. Imagine a closed tube beneath the barrel, parallel to it, and communicating with it only by a small boring near the muzzle ; through this opening the expanding gases will pass, in a degree depending on its size and position, and by their action on a piston near the breech, impulses are supplied that will actuate the lock mechanism so long as cartridges are supplied, as they may be in a continuous band. A weapon of this construction has been already tried, and its discharges are so rapid that the sound of them is described as being quite deafening. This plan appears to be equally applicable to small arms, and to machine or field guns. A very effective gun of the kind, which fires ordinary rifle bullets, has been contrived by Mr. Hotchkiss, and is represented in Fig. 105e. It is capable of sending forth as many as 1,000 shots in one minute.

Modern ordnance has required certain modifications in the making of gunpowder, so that the original name of *powder* would now hardly be applicable at all. The large charges now used, if introduced in the form of fine powder, would certainly shatter the guns from the suddenness of the exploding force. Hence the material is made up into larger or smaller masses, generally rounded like small pebbles. The explosive used for the huge 110-ton guns presents itself in the form of chocolate-coloured hexagonal prisms, two or three inches long and about an inch in diameter. These are obtained by compressing the specially prepared material into moulds with a hydraulic press. The reason for this process is that, in order to obtain precision and uniformity in the effects, not only must the composition of the powder be always the same, but the size, shape, weight, and number of the several portions that make up the charge must be invariable. It has not been found possible to fire one of these monster guns many times without such signs of deterioration as

15

would suggest a short "life" for each of them. But the greatest necessity for modern fire-arms is a smokeless powder or other explosive. It is
obvious that the advantages of quick firing, whether of large or of small
fire-arms, are greatly reduced if the soldier or gunner is prevented by
smoke from taking aim. The invention of a smokeless gunpowder has
several times been announced, and great advances have, indeed, been
made towards its realization. Certain compositions, which appeared to
meet the requirement of being practically smokeless, have, however,
been found liable to chemical changes, or to corrode the bore, or to
possess other objectionable properties. In this country the explosive
coming into use as best adapted for quick firing guns, etc., presents
itself in appearance like whitish or grey strings, and has hence received
the name of *cordite*. The composition and mode of manufacture of these
new substitutes for gunpowder are not readily disclosed, each military
authority jealously guarding its own secrets. The problem of smokeless powder has, however, been almost completely solved, for at a military
review that took place on the Continent in 1889, the discharge of the
rifles (loaded with blank cartridges, of course) is said to have been
attended with no more smoke than the puff of a cigar. The new invention will cause some changes in military tactics, for the manœuvres
formerly executed under cover of the battle smoke will no longer be
possible. Some particulars as to the nature of smokeless powders will
be found in the article on "Explosives."

FIG. 106.—*Harvey's Torpedo. Working the Brakes.*

TORPEDOES.

THE notion of destroying ships or other structures by explosions of gunpowder, contained in vessels made to float on the surface of the water,
or submerged beneath it, is not of very modern origin. Two hundred and
fifty years ago the English tried "floating petards" at the siege of Rochelle.
During the American War of Independence similar contrivances were used
against the British, and from time to time since then "torpedoes," as they
were first termed by Fulton, have been employed in warfare in various
forms; but up to quite a recent period the use of torpedoes does not appear
to have been attended with any decided success, and it is probable that but
for the deplorable Civil War in the United States we should have heard
little of this invention. During that bitter fratricidal struggle, however,
when so much ingenuity was displayed in the contrivance of subsidiary
means of attack and defence, the torpedo came prominently into notice,
having been employed by the Confederates with the most marked effects.
It is said that thirty-nine Federal ships were blown up by Confederate
torpedoes, and the official reports own to twenty-five having been so
destroyed. This caused the American Government to turn their attention
to the torpedo, and they became so convinced of the importance of this
class of war engine that they built boats expressly for torpedo warfare, and
equipped six *Monitors* for the same purpose.

It has been well remarked that the torpedo plays the same part in naval
warfare as does the mine in operations by land. This exactly describes the

purpose of the torpedo where it is used defensively, but the comparison fails to suggest its capabilities as a weapon of offence. There are few occasions where a mine is made the means of attack, while the torpedo readily admits of such an employment, and, used in this way, it may become a conspicuous feature of future naval engagements. Many forms of this war engine have been invented, but all may be classified, in the first place, under two heads : viz., stationary torpedoes, and mobile or offensive torpedoes; while independent distinctions may be made according to the manner of firing the charge; or, again, according to the mode of determining the instant of the explosion. The stationary torpedo may be fixed to a pile or a raft, or attached to a weight ; the offensive torpedo may be either allowed to float or drift against the hostile ships, or it may be propelled by machinery, or attached to a spar of an ironclad or other vessel. The charge may be fired by a match, by percussion, by friction, by electricity, or by some contrivance for bringing chemicals into contact which act strongly upon each other, and thus generate sufficient heat to ignite the charge. The instant of explosion may be determined by the contact of the torpedo with the hostile structure (in which case it is said to be " self-acting "), or by clockwork, or at the will of persons directing the operations. In some cases lines attached to triggers are employed; in others electric currents are made use of.

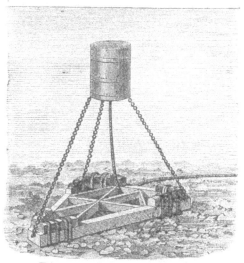

FIG. 107.—*Submerged Torpedo.*

In the American Civil War the stationary torpedoes at first laid down were self-acting, that is, they were so arranged as to explode when touched by a passing vessel. Such arrangements present the great disadvantages of being as dangerous to friendly as to hostile ships. The operation of placing them is a perilous one, and when once sunk, they can only be removed at great risk. Besides this, they cannot be relied on for certain action in time of need, as the self-acting apparatus is liable to get out of order. The superiority of the method of firing them from the shore when the proper instant arrived, became so obvious that the self-acting torpedo was soon to a great extent superseded by one so arranged that an observer could fire it at will, by means of a trigger-line or an electric current. Similar plans had often been previously employed or suggested. For example, during the war between Austria and Italy the Austrian engineers at Venice had very large electric torpedoes sunk in the channels which form the approaches to the city. They consisted of large wooden cases capable of containing 400 lbs. of gun-cotton, moored by chains to a wooden framework, to which weights were lashed that sufficed to sink the whole apparatus, Fig. 107. A cable containing insulated wires connected the torpedo with an electrical arrangement on shore, and the explosion could take place only by the operator sending a current through these wires. The torpedo was wholly submerged, so that there was nothing visible to distinguish its position. There was no need of a buoy or other mark, as in the case of self-acting torpedoes, to warn friendly vessels off the dangerous spot, and therefore nothing appeared to excite an enemy's suspicions. But it is, however, absolutely necessary that the defenders should know the precise position of each of their submarine mines, so that they might explode it at the moment the enemy's ship came within the range of its destructive action. This was accomplished at Venice in a highly ingenious manner, by erecting a camera obscura in such a position that a complete picture of the protected channels was projected on a fixed white table. While the torpedoes were being placed in their positions an observer was stationed at the table, who marked with a pencil the exact spot at which each torpedo was sunk into the water. Further, those engaged in placing the torpedoes caused a small boat to be rowed round the spot where the torpedo had been placed, so as to describe a circle the radius of which corresponded to the limit of the effective action of the torpedo. The course of the boat was traced on the picture in the camera, so that a very accurate representation of the positions of the submarine mines in the channels was obtained. Each circle traced on the table was marked by a number, and the wire in connection with the corresponding torpedo was led into the camera, and marked with the same number, so that the observer stationed in the camera could, when he saw the image of an enemy's ship enter one of the circles, close the electric circuit of the corresponding wire, and thus instantly explode the proper torpedo. The events of the war did not afford an opportunity of testing practically the efficiency of these preparations.

Another mode of exploding torpedoes from the shore has been devised by Abel and Maury. It has the advantage of being applicable by night as well as by day. The principle will be easily understood with the assistance of the diagram, Fig. 108, in which, for the sake of simplicity, the positions of only three torpedoes, 1, 2, 3, are represented.

In this arrangement two observers are required at different stations on the shore. At each station—which should not, of course, be in any conspicuous position—is a telescope, provided with a cross-wire, and capable of turning horizontally about an upright axis. The telescope carries round with it, over a circular table of non-conducting substance, a metallic pointer which presses against narrow slips of metal let into the circumference of the table. To each slip of metal a wire passing to a torpedo is attached, and another wire is connected with the axis of the pointer, so as to be put into electric contact with each of the others when the pointer touches the

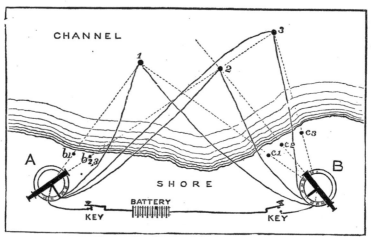

FIG. 108.—*Mode of Firing Torpedo.*

corresponding piece of metal on the rim of the table. The mode in which these wires are connected with the torpedoes, the telescopes, and the electric apparatus is shown by the lines in the diagram. At each station is a key, which interrupts the electric circuit except when it is pressed down by the operator. There are thus four different points at which contacts must be simultaneously made before the circuit can be complete or a torpedo explode. In the diagram three of these are represented as closed, and in such a condition of affairs it only remains for the observer to depress the handle of the key at station B to effect the explosion of torpedo No. 2. The observer at station A is supposed to see the approaching vessel in the line of torpedo No. 2, and recognizing this as an enemy's ship, he depresses the key at his station. The operator at B, by following the course of the vessel with his telescope, will have brought the pointer into contact with the wire leading to No. 2 torpedo, and he then causes the explosion to take place by completing the circuit by depressing his key. A modification of this plan is proposed by which the position of the torpedoes is indicated by placing marks, such as differently-coloured flags, or by night lamps with coloured glasses, throwing their light only towards the telescopes. These marks are placed in the line of direction of each torpedo from the telescope as at c_1, c_2, c_3 and b_1, $b_{2\,3}$; and if they can be put at some distance, the position of the torpedo is determined with great accuracy by the intersection of the lines of sight of the two telescopes. Electric wires connect the stations and the torpedoes in the same manner as we have before described. Such methods of firing torpedoes are no doubt the most efficient, for the destructive charge may be sunk so far below the surface that not a ripple or an eddy can excite an enemy's suspicion, or the channel appear otherwise than free and unobstructed, while friendly ships may pass and repass without risk; for the current which determines the explosion only passes when the two sentinels complete the circuit by simultaneously depressing their keys.

Attempts have often been made to convert the torpedo into an offensive weapon, by causing vessels containing explosive charges to drift by currents, or otherwise, into contact with the enemy's ships. The results have been always unsatisfactory, as there is great uncertainty of the machine coming into contact with its intended mark. Besides, it is easy to defend vessels against such attacks by placing nets, &c., to intercept the hostile visitors, especially if the attack is made by day, and by night the chance that a torpedo drifting at random would strike its object is very small indeed. One condition essential to the success of such attacks is that the approach of the insidious antagonist may be unobserved. Accordingly divers schemes

FIG. 109.—*Explosion of Whitehead's Torpedo.*

have been projected for propelling vessels wholly submerged beneath the surface of the water, so that they may approach their object unperceived, and exert their destructive effect precisely at that part of the vessel where damage is most fatal, and where an ironclad vessel is most vulnerable, namely, below the water-line. Vessels have been built, propelled by steam and so contrived that their bodies are wholly submerged, only the funnel being visible above the surface. These *quasi* submarine ships carry small crews, and are fitted with a long projecting spar in front, at the end of which is carried the torpedo.

The Federal navy sustained several disasters from torpedo-boats of this kind. For example, the commander of the United States steamer *Housatonic* reported the loss of that vessel by a rebel torpedo off Charleston on the evening of the 17th February, 1864, stating that about 8.45 p.m. the officer of the deck discovered something in the water about 100 yards from, and moving towards, his ship. It had the appearance of a plank moving in the water. It came directly towards the ship, the time from when it was first seen till it was close alongside being about two minutes; and hardly had it arrived close to the ship before it exploded, and the ship began to sink. The torpedo-boat, with its commander and crew, were lost, having, it is supposed, gone into the hole made by the explosion, and sunk with the *Housatonic.* In general, however, the performance of submarine boats has been unsatisfactory. There is the difficulty of determining accurately the

FIG. 110.—*Effect of the Explosion of Whitehead's Torpedo.*

course of the boat; there is great danger to the men manning it, as exemplified in the case above; and there is again the problem of providing a means of propulsion which shall enable such a boat to advance or retreat for, say, a mile or more, without making its presence conspicuous by smoke or otherwise. The latter condition would appear to exclude the use of steam for such purposes, as the inevitable smoke and vapour would betray the presence of the wily craft. Another power which has been proposed is air strongly compressed, and recently a still more portable agent has been suggested in solid carbonic acid, which is capable of exerting a pressure of forty atmospheres by passing into the gaseous form. A locomotive form of torpedo, invented by Mr. Whitehead, has the explosive charge, which consists of about 18 lbs. of glyoxyline, placed in the front part of a cigar-shaped vessel, the other part containing mechanism for working a screw-propeller, by means of compressed air contained in a suitable reservoir. This torpedo having been sunk a few feet below the water, the motive power may be set in action by drawing a cord attached to a detent, when the mechanical fish proceeds in a straight line under the water. It is said that this torpedo is effective at 500 yards from the ship attacked, and may even be made sufficiently powerful to travel 1,000 yards under the water. The great objection to such arrangements is the uncertainty of the missile arriving at its destination, for even supposing that the water were without currents, the least deviation from the straight course would cause the torpedo to pass wide of the mark at 1,000 yards distant. It is said that at the experimental trials more than one projector of such war engines has been startled by his machine, after pursuing a circuitous submarine course, exploding in dangerous proximity to the place whence it was sent off, the engineer narrowly escaping being "hoist with his own petard." The experiments which have been made with Whitehead's torpedo in smooth water appear, however, to have been so far successful that we may probably hear of this invention being put in practical operation in certain cases. Fig. 109 shows the upthrow of water produced by the explosion of one of these torpedoes against an old hulk. The large mass of water thus heaved up is a proof of the mechanical energy of the explosion, and the effect on the hulk is shown in Fig. 110, which exhibits the damage done to her timbers, from the effects of which, it need hardly be said, she immediately sank. In Fig. 112 we have the repre-

FIG. 111.—*Experiment made by the Royal Engineers with a Torpedo charged with* 10*lbs. of Gun-Cotton.*

sentation of the explosion of one of Whitehead's torpedoes containing 67 lbs. of gun-cotton, instead of the glyoxyline. The accurate delineation of these pyramids of water could not have been obtained but by the aid of instantaneous photography, and it constitutes a good example of the great value of such an application of that art, for the instantaneous photographs obtained in these experiments enabled the engineers to calculate accurately the volume and height of the column of water, which thus furnishes a measure of the power of the explosion.

The ordinary torpedo adopted by the British authorities for coast defence consists of a cylinder of boiler plate, 4 ft. long and 3 ft. in diameter. It is intended to contain 432 lbs. of loose gun-cotton, equivalent in explosive energy to about a ton of gunpowder. The effect of one of these torpedoes exploded 37 ft. beneath the surface of the water is depicted in Fig. 113, and in Fig. 114 is shown the effect produced when the same charge was exploded at the depth of 27 ft. below the surface. Gun-cotton appears to be the most effective explosive for torpedoes, if we may judge by the large volume of water heaved up, as witness Fig. 111, which shows the result with a small torpedo, containing only 10 lbs. of gun-cotton, exploded at a less depth than those already mentioned. The ordinary torpedoes are moored by an anchor attached to the torpedo, and floating above it is a buoy shaped like an inverted cone. This cone contains a mechanical arrangement of such a nature that when it is struck by a passing vessel, an electric circuit is closed by bringing into contact two wires connecting the torpedo with a voltaic battery on shore. While the apparatus may thus be at any moment made fatal to a hostile vessel touching it, from the control it is under by the engineer having the management of the battery contacts, friendly vessels may pass over it with impunity.

FIG. 112.—*Explosion of Whitehead's Torpedo, containing 67 lbs. of Gun-Cotton.*

The employment of torpedoes develops, as a matter of course, a system of defence against them. Nets spread across a channel will catch drifting torpedoes, and stationary ones may be caused to explode harmlessly by nets attached to spars pushed a great distance forward from the advancing ship.

Before the final adoption of Whitehead's torpedo, presently to be described, the British Government had, after various official trials, approved of a towing torpedo designed for offensive operations. It is the invention of Commander Harvey, and is worthy of a detailed description for the ingenuity of its construction.

The shape of Harvey's torpedo, as may be noticed on reference to Fig. 118, is not symmetrical, but it has some remote resemblance to a boat, though constructed with flat surfaces throughout. The outside case is formed of wood well bound with iron, all the joints being made thoroughly water-tight. The length is 5 ft. and the depth 1¾ ft., while the breadth is only 6 in. Within this wooden case is another water-tight case made of thick sheet copper, from the top of which two very short wide tubes pass upwards to what we may term the deck of the wooden case. These are the apertures through which the charge of gunpowder or other explosive material is introduced ; and when the tubes have been securely stopped with corks, brass caps are screwed on. The centre of the internal case is

FIG. 113.—*Explosion of* 432 *lbs. of Gun-Cotton in* 37 *feet of Water.*

occupied by a copper tube, *g*, Fig. 115, which passes the entire depth, and is soldered to the top and bottom of the copper case, so that the interior of the tube has no communication with the body of the torpedo, the principal charge merely surrounding it. Thus the tube forms a small and quite in-

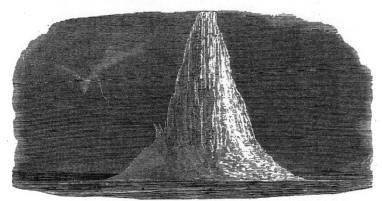

FIG. 114.—*Explosion of* 432 *lbs. of Gun-Cotton in* 27 *feet of Water.*

dependent chamber in the midst of the large one, which latter is capable of containing 80 lbs. of gunpowder. The copper tube or priming-case contains also a charge, *a*, which when exploded bursts the tube, and thus fires the torpedo in its centre. The priming charge is put in from the lower end of the tube, which is afterwards closed by a cork and brass cap, *h* ;

for the centre of the priming-case is occupied by a brass tube, *b*, closed at the bottom, but having within a pointed steel pin projecting upwards. In this tube works the exploding bolt *c d*, which requires a pressure of 30 or 40 lbs. to force it down upon the steel pin. This pressure is communicated to the bolt by the straight lever working in the slot at its head, *d*, and itself acted on at its extremity by the curved lever to which it is attached.

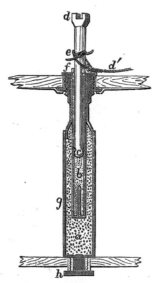

FIG. 115.—*Section of Priming-Case and Exploding Bolt.*

Thus from the mechanical advantage at which the levers act a moderate downward pressure suffices to force the exploding bolt to the bottom of the brass tube. The lower end of this bolt has a cavity containing an exploding composition sufficient in itself to fire the torpedo, even independently of the priming charge contained in the copper tube. This composition is safely retained in the end of the bolt by a metallic capsule, *c*, which, when the bolt is forced down, is pierced through by the steel pin at the bottom of the brass tube, and then the explosion takes place. The bolts are not liable to explosion by concussion or exposure to moderate heat, and they can be kept for an indefinite period without deterioration.

The mode of producing the explosion is not stated : it consists probably of an arrangement for bringing chemicals into contact. Besides the two levers already mentioned, a shorter curved lever working horizontally will be noticed. The object of this is to make a lateral pressure also effective in forcing down the bolt—a result accomplished by attaching to the short arm of the lever a greased cord, which, after passing horizontally through a fair-leader, runs through an eye (see Fig. 117) in the straight lever, and has its extremity fastened so that a horizontal movement of the short lever draws the other down. A very important part of the apparatus is the safety key, *f*, Fig. 115, a wedge which passes through a slot in the exploding bolt, and resting on the brasswork of the priming-case, retains the muzzle 1 in. above the pin. Through the eye of the safety key and round the bolts passes a piece of packthread, *e*, which being knotted is strong enough to keep the key securely in its place, but weak enough to yield when the strain is put on the line, *d'*, used for withdrawing the safety key at the proper moment. This line is attached to the eye of the key, and passes through one of the handles forming the termination of the iron straps. As represented in Fig. 117, it forms the centre one of the three coils of rope. The bottom of the torpedo is ballasted with an iron plate, to which several thicknesses of sheet lead can be screwed on as occasion requires. Fig. 117 shows the arrangement of the slings by which the torpedo is attached to the tow-rope, and it will be seen that another rope passes backwards through an eye in the stern to the spindle-shaped object behind the torpedo. This is a buoy, of which two at least are always used, although only

FIG. 116.—*Harvey's Torpedo.*

one is represented in the figure. Each buoy, in length 4½ ft., is made of solid layers of cork built up on an iron tube running through it lengthways, so that the buoys admit of being strung upon the rope.

Having thus described the construction of the torpedo, we proceed to explain how it is used. It must be understood that if the torpedo and its attached buoys are left stationary in the water, the tow-rope being quite slack, the torpedo will, from its own weight, sink several feet below the surface. But when they are *towed*, the strain upon the tow-line brings the torpedo to the surface, to dip below it again as often as the tow-line is slackened. There is another peculiarity in the behaviour of the torpedo, and that is that, when towed, it does not follow in the wake of the vessel, but diverges from the ship's track to the extent of 45°. Its shape and the mode in which it is attached to the tow-line are designed so as to obtain this divergence. But, according as the torpedo is required to diverge to the right or to the left, there must be the corresponding shape and arrangement of tow-line and levers; hence two forms of torpedo are required, the starboard and the port. The figures represent the port torpedo, or that which is launched from the left side of the torpedo-ship, and diverges to the left of its course. The efficiency of the torpedo depends upon the readiness and certainty with which it can be brought into contact with the hostile ship, and this is accomplished by duly arranging the course of the torpedo vessel, and by skilfully regulating the tow-line so as to obtain the requisite amount of divergence, and to cause the torpedo to strike at the proper depth. The tow-rope is wound on a reel, furnished with a powerful brake, the action of which will be readily understood by inspection of **Fig. 116,** which represents also a similar smaller reel for the line attached to the safety key. Leather straps, sprinkled with rosin to increase the friction, encircle the drums of the reels, and can be made to embrace them tightly by means of levers, so that the running out of the lines can be checked as quickly as may be desired. Handles are attached to the straps, so that they can be lifted off the drum when the line is being drawn in by working

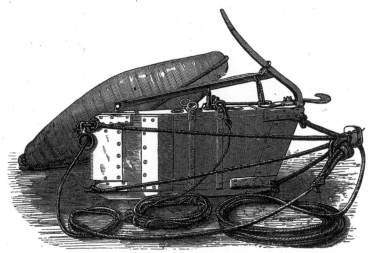

FIG. 117.—*Harvey's Torpedo.*

the handles. When the torpedo is ready for action and has been launched, a suitable length of tow-line, which is marked with knots every ten fathoms, is allowed to run off its reel, while the safety key-line is at the same time run off the small reel, care being taken to avoid fouling or such strains on the line as would prematurely withdraw the key. Fig. 106 will make clear the mode of controlling the lines, but it is not intended to represent the actual disposition in practice, where the men and the brakes would be placed under cover. On the left of the figure a starboard torpedo is about to be launched; on the right a port torpedo has been drawn under the iron-clad and is in the act of exploding, the safety key having been withdrawn by winding in its line when the torpedo came into proximity to the attacked vessel.

When the torpedo has been launched over the vessel's side, the latter being in motion, the torpedo immediately diverges clear of the ship; and when the buoys have also reached the water, the men working the reels pay out the line steadily, occasionally checking the torpedo to keep it near the surface, but avoiding a sudden strain upon the slacked tow-rope, which would cause the torpedo to dive, and in shallow water this might lead to the injury or loss of the torpedo. The torpedo can be gradually veered out to the distance required, at the same time that the safety-key is so managed that sufficient strain may be put upon it to prevent it from forming a long

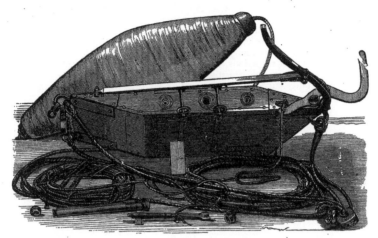

FIG. 118.—*Harvey's Torpedo.*

bight astern of the torpedo, but avoiding such a strain as would break the yarn holding the safety-key in its place. The distance to which the tow-line may be paid will depend upon the circumstances of the attack. More

FIG. 119.—*Official Trial of " Harvey's Sea Torpedo," February,* 1870.

than 50 fathoms is, however, a disadvantage, as the long bight of tow-lines makes the torpedo drag astern. The torpedo can always be made to dive several feet below the surface by suddenly letting out two or three fathoms of tow-line. The torpedo vessel should, of course, be a steamer of considerable speed—able to outstrip when necessary all her antagonists, and, as a rule, it is found best to make the attack at night. Let us imagine two ships of war at anchor, and parallel to each other at perhaps a distance of 60 fathoms; and suppose that, under cover of darkness, a hostile torpedo vessel boldly steams up between them, having launched both its starboard and port torpedoes. In such a case neither ship could fire at the torpedo vessel for fear of injuring the other, while the torpedo vessel would in all probability succeed in bringing its floating mines into contact with both its enemies.

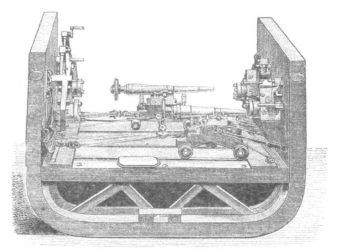

FIG. 120.—*Model of Submarine Guns.*

Another device for submarine attacks upon vessels on which much ingenuity has been expended is the submarine gun. It has been sought to propel missiles beneath the surface of the water, these missiles being usually provided with a charge which, on contact with the vessel's side, would explode, and by making a hole below the water-line, cause the certain destruction of the ship. It is obvious that such a mode of attack would reach the only vulnerable parts of a thickly-plated ironclad, and therefore the project has been recently revived in several forms. Fig. 120 is taken from the photograph of a model of an invention of this kind. The guns which are to propel the submarine projectiles, have port-holes formed by valves in such a manner that the gun when loaded can be run out without allowing water to enter; it can then be fired while the muzzle is below the surface, and again drawn in without the port being at any time so opened that water can pour into the vessel. All contrivances of this kind have hitherto been failures; indeed, it does not appear possible that they could succeed, except at very close quarters, for the resistance offered by water to a body moving rapidly in it is extremely great, and, as we have already had occasion to state, the resistance increases as the square of the velocity, and probably in even a higher degree for very great velocities. Any one who will remember the effort it requires to move one's hand quickly backwards and forwards through water will easily understand that the resistance it presents would, in a comparatively short space, check the speed of a projectile, however great that speed might be at first. A good many years ago Mr. Warner produced a great sensation by an invention which appears to have been essentially a floating torpedo. The cut below, Fig. 121, represents the result of an experiment publicly made by him off Brighton, in 1844, upon a barque, which was towed out by a steamer to a distance of a mile and a half from the shore. Mr. Warner was on board the steamer, and the barque was 300 yards astern. Five minutes after a signal had been made from the shore, the torpedo was caused to explode, striking the barque amidships, throwing up a large column of water and *débris*, shooting the mainmast clean out of the vessel, the mizen going by the board, and dividing the hull into two parts, so that she sank immediately. Yet this invention, though apparently so successful, does not seem to have ever been put in practice.

Fig. 121.—*The Warner Experiment off Brighton.*

The stationary torpedoes of the kind mostly used in the American Civil War were, as already stated, *self-acting*; that is, they exploded when touched by a passing vessel. They would now be more generally called *self-acting mines*, and are to be distinguished from that form of the weapon in which the explosion is determined by some manipulation on shore, such as the closing of an electric circuit, when the hostile vessel comes within the area of destructive action. This form receives the name of *observation mines*. Stationary mines are essentially instruments of defence, and as such are employed for the protection of rivers and harbours The self-acting varieties usually contain a charge of 70 lbs. to

80 lbs. of gun-cotton, and are commonly arranged in lines. Of course, when the occasion for which such mines have been laid down is past, they must be removed, and the operation of picking them up is one of great danger. The observation mines, on the other hand, do not require immediate removal, and they can be taken up with little risk. In the British service the observation mine contains about 500 lbs. of gun-cotton, and a line of these is sometimes moored in a water-way, from 35 feet to 50 feet below the surface. The area of destructive action in this case is a circle of about 30 feet radius, and therefore a line of seven such mines laid across a channel at intervals of 120 feet apart would ensure the almost certain destruction of any war vessel of ordinary breadth that might attempt to pass up a river of 840 feet in width. This is about the distance across the Thames near the Tower of London, but the depth of the river there being only 12 feet at low-water and 33 feet at high-water, would not suffice to give effect to the full energy of so large a charge of gun-cotton; for it has been found that, for a given charge, there is a certain depth under water at which its explosion will produce the maximum effect, and this depth will be greater with heavier charges than with light ones. The regulation "observation mine" of the British service has a cylindrical case of stout plate-iron, 32 inches in diameter and 34 inches high, with domed ends. Within this gun-cotton is contained, in a wet condition, in a number of copper envelopes, which have holes for access of water to wet the charge from time to time as occasion requires, the wet condition being the safest for the carriage of gun-cotton. The centre of the case is a tin charged with some discs of dry gun-cotton, and the detonator required to bring about the explosion of the whole charge when the electrical contact is made, fires the fuze contained within the primer. These cases are arranged to have a certain buoyancy, and are moored with wire ropes to heavy iron sinkers, the mooring ropes being of such length as to keep the explosive case at the proper depth below the surface, and of sufficient strength to resist the force of the currents in the water-way. There is also another type of submarine mine, operated by an electric current, the circuit of which is closed by contact of a passing vessel, if at the time a battery on shore is included in the circuit. In this way the mines can be made harmless or dangerous for passing vessels at the will of the operator on shore. The passing vessel is made to complete the circuit by tilting over the cylindrical case so far that some mercury contained in a small part of it is upset, and makes the requisite metallic contact. This arrangement is known as the *electro-contact* mine.

In former pages of this article on torpedoes will be found representations of the effects produced by *Whitehead's torpedo*, which, being automobile and travelling altogether under the surface of the water, was capable of being made a very formidable weapon of offence. When the earlier editions of this work were going through the press, it was understood that the Whitehead torpedo left much to be desired as regards speed, certainty of direction through the water, and perhaps in other points, the inventor being constantly engaged in effecting improvements. At that time particular pains were taken to keep secret the nature of the most important parts of the internal mechanism. The work of construction was carried on in a room with locked doors, blocked-out windows, and a military guard outside. The earlier experimental forms of this automobile torpedo were constructed in complete secrecy by the inventor himself, with the help of only one trusted, skilled mechanic and a boy, who was no other than Mr. Whitehead's own son. The history of the invention is very interesting, and exemplifies the power of skill and perseverance to overcome a multitude of difficulties, the result being a machine which is simply a marvel of ingenuity and of delicate nicety of adaptation.

The first notion of the automobile torpedo appears to have occurred to an Austrian naval officer ; but it took rather the form of a small vessel containing within itself some propelling power by which it could move along the surface of the water, its course being directed by ropes or guiding lines from the shore or from a ship. The fore part of the little vessel was to hold an explosive, to be fired automatically by the self-propelled torpedo coming into contact with the side of the hostile vessel. The propelling power, as first suggested, was clockwork, if that could be made efficient, or steam as an alternative. The Austrian authorities, however, considered that it would be impracticable to direct the course of the torpedo in the manner proposed, and that there were also great objections to each of the methods of obtaining motive power. The assistance of a thoroughly competent and skilful mechanician was then sought, and Mr. Whitehead, at that time the director of an engineering establishment at Fiume, devoted himself to solving the problem of devising a torpedo which should be able to travel beneath the surface of the water, and, when once started, should require no external guidance to keep it on its proper course. After some years of experimental labours, Mr. Whitehead produced the first form of the weapon with which his name is associated, but to this he has since added from time to time many ingenious improvements. A committee of experts having been appointed by the Austrian Government to test the capabilities of the new invention, it was made the subject of a long series of trials, after which the committee recommended its immediate adoption in the Austrian navy. The earlier form of the Whitehead torpedo had, however, the defect already mentioned, of being sometimes very erratic in its course ; its speed was small (6 knots) compared with that of the more recent patterns (30 knots), and its range of travel proportionately less. The British Admiralty having invited Mr. Whitehead to visit England with some specimens of his invention, a committee was appointed to make complete trials of the capabilities of two weapons he had brought with him. Although by this time great improvements had been made on the original design, and in particular, Mr. Whitehead had almost completely overcome the difficulty of keeping the torpedo at a uniform depth during its course, by means of delicate adjustments in what we may call the steering chamber (to be presently mentioned), much remained to be accomplished before the weapon attained the perfection of the modern patterns. Indeed, the inventor may be said to have from time to time redesigned his contrivances, as when in 1876 the speed was increased to 18 knots, and again in 1884 more powerful engines brought up the speed to 24 knots. Further improvements have been made by Mr. Whitehead, who designed a new form of the weapon in 1889, and some of the more recent patterns can now show a speed of 30 knots or more. The committee appointed by the Admiralty to conduct experiments with the first pair of torpedoes brought to England, after having tested them in various ways for a period extending over six months, reported that they believed that "any maritime nation failing to provide itself with submarine locomotive torpedoes, would be neglecting a great source of power, both for offence and defence." Upon this recommendation the Admiralty immediately purchased from Mr. Whitehead for £15,000 the secret of the internal mechanism of his invention and the rights of manufacturing it. The self-adjusting apparatus within the steering chamber, by means of which the torpedo was kept at its due depth, was then a jealously-guarded secret ; but when the arrangement with Mr. Whitehead was effected, the Government immediately set about the manufacture of these torpedoes on a large scale. The artificers employed in making the Whitehead torpedoes were now numerous, and the internal structure of these weapons could not advantageously be altogether concealed from those who had to handle them on board of the ships, so that it inevitably happened that some details of their construction leaked out, and came into the possession of other powers, whereupon all the maritime states followed the example of Great Britain by providing their navies with Whitehead or some such form of locomotive torpedo. It is no part of our plan to enter into the mechanical *minutiæ* of the Whitehead torpedo. We may, however, give the reader such an idea of the external appearance and internal arrangement of the Whitehead torpedo as will enable him to appreciate to some extent the ingenuity and skill that have been brought to bear upon its construction.

There are in existence many different patterns of the weapon—twenty-four, it is said—and this is what might be expected from the fact of its being produced at several different manufactories, each striving to effect whatever improvements its resources will supply. Some torpedoes have been made at Fiume, very many at Mr. Whitehead's works at Portland, as also at the Government establishment at Woolwich, while private enterprise in this direction is encouraged by contracts with some private firms, such as that of Messrs. Greenwood & Bately at Leeds. The greatest diameter of the large torpedo is 18 inches, but in some it is rather more, in others 14 inches or 16 inches ; and the length may vary between 14 feet and 19 feet. Many of our Whitehead torpedoes are made of polished steel, but in the later patterns phosphor-bronze is partly made use of, as being not liable to corrode. The interior of the torpedo is divided by transverse partitions into five distinct compartments. The foremost of these, called the " head," contains the explosive charge when the weapon is ready for use in actual warfare. This section, which may occupy about one-sixth of the total length, is an air-tight case made of phosphor-bronze, one-sixteenth of an inch thick, and it is kept permanently charged with slabs of wet gun-cotton, which may amount to 200 pounds weight in all, and is ready to be attached by a screw and bayonet joint to the body of the torpedo ; but this is done only at the time immediately before it is required for its destructive employment. Its place at other times, as when the torpedo is used for drill practice, and to test its running powers, is occupied by a dummy head of steel, of exactly the same shape and size, and packed with wood in such a manner that its weight and centre of gravity are like those of the explosive head when the latter is ready for action. The wet gun-cotton requires a *detonative* explosive of dry material close to it, in order to determine its own detonation. The explosive heads of the Whitehead are not fitted with the pistol and priming tube until all is ready for the discharge of the weapon, as this would render the handling of the torpedo highly dangerous. This priming apparatus is merely a metallic tube that slips into a corresponding hollow in the explosive head so far as to reach well within the wet gun-cotton charge, although still separated from the latter by a metal casing. The posterior extremity of the priming tube contains a few ounces of dry gun-cotton, and just in front of this is a copper cap containing some fulminate of mercury, which readily explodes when struck by the point of a steel rod, occupying the centre of the tube and projecting a short distance out at the "nose" of the torpedo, so as to be driven inwards by the impact of the latter on a ship's side. The explosion of the fulminate causes the detonation of the dry gun-cotton at the bottom of the priming tube, and this is taken up by the whole mass of the explosive with destructive effect. The danger of premature or accidental explosion by anything coming in contact with the projecting striker is obviated by several checks which prevent any chance blow driving the rod home against the fulminate charge. The anterior projecting end of the rod has a screw thread worked upon it, and on this turns freely a nut provided with wings like a small fan, revolving in such a manner that as the torpedo is moved through the water, the nut is spun off, and the striker is free to be driven back, except in so far as it is still retained by a small copper pin, the breaking of which requires a considerable blow. Again, the little fan above mentioned cannot begin to spin off the rod until another pin or wedge has been withdrawn, which operation is performed just before launching the weapon.

Immediately behind the exploding head of the torpedo is the air-chamber, which occupies a considerable space in the length, *i.e.*, about one-third of the whole. This part is made of the toughest steel, nearly $\frac{3}{8}$ of an inch thick, and contains the power actuating the motor, in the form of air forced into it by powerful pumps on board the ship, until the pressure reaches the enormous amount of 1,300 lbs. or more on the square inch, or, at least, this is what is made use of in the newer patterns when charged for action. In the largest size of the weapon the weight of air injected may be more than 60 lbs., and, of course, considerably detracts from the buoyancy of this part.

Behind the air-chamber comes another much shorter compartment we have called the "steering chamber," in which are contained the most ingenious and delicate parts of the apparatus, namely, the mechanism by which this extraordinary artificial fish adjusts itself, after the manner of a living thing, to the required conditions. Among other contrivances, it contains several valves controlling the action of the compressed air on the engines, etc. The enormous pressure to which the air-chamber is charged, if allowed to act unchecked, would give at first a power almost sufficient to shatter the machinery, and, in order to prevent this, a "reducing valve" is interposed so that only a moderate and uniform pressure of air is allowed to act upon the engines. Then there is the "starting valve" by which the air is admitted cr cut off from the engines, and still another valve which is contrived to delay the action of the compressed air for the short interval during which the torpedo is passing from the discharging tube until it enters the water. For during this interval the propellers not having to act against the water, but only against the resistance of the atmosphere, would be whirled round at an enormous speed, and the machinery would sustain such shocks and strains as might endanger the whole apparatus. It is to prevent this that the "delay action valve" is provided.

The automatic apparatus by which the torpedo's course is regulated is a very remarkable part of the invention, and it admits of the nicest adjustments. This was the crown of Mr. Whitehead's ingenuity, but the details were, by an arrangement between the government and the inventor, not to be made public, though necessarily communicated to certain officers in the service, and known to the chief artizans employed in their fabrication. These persons are all, we believe, required to give pledges not to divulge the arrangement of particular parts. But such details could scarcely be made intelligible, even should they be interesting, to the general reader. The principles upon which the controlling apparatus are arranged may, however, be comprehended without difficulty.

The tail of the torpedo is provided with two rudders, one in its central vertical plane, and the other in its central horizontal plane. Their action in directing the torpedo's course is exactly that which the tail supplies to a fish, or the rudder to a boat. Suppose that while the torpedo is passing through the water the vertical rudder is by any means turned towards one side, the course of the metallic fish will be diverted towards that side; or again, a turning upwards of the horizontal rudder would have the effect of directing the nose towards the surface, and would make the torpedo rise, and so on. Now the positions of the horizontal rudder are regulated from the "steering chamber," in which a heavy weight is suspended like a pendulum, so as to be capable of swinging fore and aft. This pendulous weight actuates the horizontal rudder through a system of rods and levers, so that when it hangs vertically the horizontal rudder is level, but if from any cause the nose of the torpedo were directed downwards, the pendulous weight would come to a more forward position in the steering chamber, and would raise the rudder, and thus turn the nose towards the surface until the original horizontal position were regained. In the contrary case, of course, the reverse action would take place. But the torpedo, while preserving a horizontal position, might tend to sink to too great a depth, or rise too near the surface, and this is prevented by another adjustment, namely, a piston receiving the pressure of the water, which, on the other side, is opposed by a spring. If the torpedo sinks a little the pressure increases, the piston, which moves with perfect freedom without allowing water to pass in, is forced inwards, and its movement is communicated to the same levers that connect the pendulous weight with the horizontal rudder, the latter is raised, and then the nose of the torpedo is directed upwards, and it consequently approaches the surface again. In the contrary case the spring, relieved from some of the external pressure, operates the levers in the other direction.

The compartment immediately behind the "steering chamber" contains the engines which are of the Brotherhood type, provided with *three* single acting cylinders. The threefold throw prevents any possibility of the engine getting on a "dead point." Though this compartment is the shortest in the torpedo, the engines in the larger sizes are capable of indicating as much as thirty horse power. It has for simplicity been stated above that the pendulous weight and the balanced piston act by means of rods on the horizontal rudder; this was so in the early patterns of the torpedo, but it was soon found that they did not do so with sufficient steadiness and promptitude, and the force they could apply was in the larger and swifter forms quite ineffective. Nowadays the engine compartment always contains a little piece of apparatus which is an arrangement of cylinder and piston, upon which the compressed air acts in one or the other direction according to the way its slide-valve is moved. It is this slide-valve that the rods from the "steering chamber" move, and allow the force of the compressed air to turn the rudder up or down. This auxiliary apparatus has the same relation to the torpedo rudder that the steam-steering apparatus of a large vessel has to its rudder. Although it is only about a few inches long, its power and delicacy are such that the pressure of half an ounce on its slide admits to its piston a force equal to 160 lbs., and its introduction has given the torpedo the power of steadily steering itself.

Behind the engine compartment, but completely shut off from it, is another almost empty division occupying a considerable part of the length of the torpedo, and known as the "buoyancy chamber." But it contains, attached to the bottom of it, a certain amount of ballasting, so adjusted to balance the weights of the other parts that the whole floats horizontally, and at the same time preserving the tube in one vertical position as regards its transverse diameter, *i.e.*, so that the horizontal rudder is always horizontal. The shaft from the engine passes through this compartment, as also the rod from the small motor that moves the horizontal rudder. These, of course, pass through water-tight bearings.

At the tail of the torpedo, behind the rudders, are *two* three-bladed screw propellers, of which the anterior one is mounted on a tubular shaft having a connon axis with the other, but made to revolve in the opposite direction by means of a bevel wheel mounted on each independent shaft, with a third such wheel connecting them. The object of the double screw is to obviate "slip," that is, ineffective motion of the blades through the water, and by this means the full power of the engines can be developed ; while any tendency to *deviation* to right or left, due to the rotation, is reduced to a minimum. We have spoken of one horizontal and one vertical rudder, although externally there appear to be two of each kind, right and left, above and below, on the tail of the torpedo. These pairs, however, are so connected as to be always in the same respective planes. The controlling mechanism acting in two different ways on the horizontal rudder has been already indicated, but nothing has yet been said about the vertical rudder. It is not moveable by anything within the torpedo, but is commonly fixed by clamping screws in or about the same vertical plane as the axis of the torpedo, and it performs the same function as a kind of back fin, which, in the earlier forms, extended nearly the whole length of the tube ; and that is obviating any tendency of the torpedo to roll about its axis. The vertical rudder can also be fixed at a considerable inclination to the axis should occasion require, and the effect of that would be to cause the torpedo to pursue a circular course of greater or less radius, according to the less or greater degree of inclination. Very rarely, however, would this be required, and the vertical rudder may be considered as fixed in the axial plane, or having such slight inclination as may, on trial, have been found necessary to counteract any tendency to lateral deviation.

There are several different methods for discharging the Whitehead torpedoes from ships. They may be sent from a tube below the water-line, but the arrangements for that purpose are complicated and difficult to manage, while, on the other hand, the launch of the weapon is not perceived by the enemy, and it is at the same time out of the reach of any blow from a hostile missile while yet in its discharging tube. More commonly the discharging tube is arranged above the water-level. On regular torpedo boats, the tubes are sometimes mounted on pairs upon a revolving table, provided with many nice adjustments, and even the single above-water torpedo tube, as used between decks, is an apparatus having somewhat complicated appliances. The torpedo is expelled from the tube now preferably by a small charge of *cordite*. But in the Royal Navy no fewer than some twenty different patterns of torpedo tubes have been in use for the various sizes of torpedoes. In some of these, compressed air, in others gunpowder or *cordite*, in others, again, mechanical impulse propels the torpedo into its element. It would obviously be impossible within our limits to enter into details of these various constructions, or to attempt descriptions of *all* the ingenious contrivances applied to the torpedo itself, or to give an account of the means of defence against mines and torpedoes, this last being a matter belonging to naval tactics. The adoption of the torpedo as a naval weapon has given rise to special types of boats adapted for its employment, and these again have required other boats to destroy them ("torpedo-boat destroyers" or "catchers"). Light draught and high speed were desired in these last; but in many cases the intended speed was inferior to that of the torpedo boats that were to be caught.

The following particulars about the British torpedo-boat destroyer *Daring* may be compared with those given of the cruiser *Majestic*. The *Daring* is 185 feet long, 7 broad, and she draws only 7 feet of water. Her speed is about 28½ knots per hour, with a steam pressure in the boilers of 200 lbs. per square inch, and an air pressure in the stoke-holds equivalent to 3 inches of water (forced draught.)

The importance attached to the prospective use in war of the automobile torpedo may be shown by the fact that at the end of 1890 the number of torpedo boats built or laid down for England was 206, and for France 210, while other nations followed with numbers proportionate to their means. Forty "torpedo-boat destroyers" were in building for the British Navy towards the close of the year 1896, and now (March, 1897) it is announced that the number of torpedo boats and torpedo-boat destroyers in the French Navy is to be increased by 175.

FIG. 122.—*M. Ferdinand de Lesseps.*

SHIP CANALS.

ARTIFICIAL CANALS are amongst the oldest of inventions, for, centuries ago, they have been constructed, even of very large dimensions, in various parts of the world. There is in China, for instance, a great canal, 900 miles in length and 200 feet broad, which is supposed to have been made 800 years ago. The advantages of canals did not escape the attention of the Egyptians, Greeks and Romans. We read of very early attempts to cut through isthmuses, in order to form a water communication between regions where other carriage would be long and difficult. It appears to be admitted that canals connecting the Red Sea with the Mediterranean existed some centuries before the Christian era, and to cut the Isthmus of Corinth by a waterway was a cherished project with several Roman Emperors, and now it appears that in this nineteenth century this project will shortly be realized. But as the canal-lock is but a comparatively modern invention, dating only from the fourteenth century, and first used in Holland, all the canals anterior to that period had to be designed as level cuts, a restriction which greatly increased the difficulties of the problem. Canals were in use in various parts of Europe, particularly in Holland and France, long before any were constructed in England, as, for example, the Languedoc Canal, which, by a cut of 150 miles, connects the Bay of Biscay with the Mediterranean. It is 60 feet broad, and attains, at its highest level, an elevation of 600 feet above the sea. The canal system in England was first introduced in the middle of the eighteenth century, and soon afterwards, the Duke of Bridgewater engaged the famous Brindley to construct a canal, connecting his collieries at Worsley with Manchester, about seven miles distant, and afterwards extended his scheme, so as to open up a more direct water communication between Manchester and Liverpool. Before the making of this canal, the cost of the carriage of goods between these towns had been forty shillings per ton by land, and twelve shillings by water. After that, they were conveyed with regularity for six shillings per ton. The system was soon extended, so as to connect the Trent with the Mersey, and the boldness of both the projectors and their engineer in carrying out this scheme is memorable in the history of such undertakings. Brindley was equal to the task of coping with the difficulty of carrying his canal over what had hitherto been supposed an insuperable obstacle, for he pierced Harecastle Hill with a tunnel more than a mile and a half in length—a then unheard of piece of engineering—to say nothing of several shorter tunnels, many aqueducts, and scores of locks. The Duke of Bridgewater, who at one period had been unable to raise £500 on his own bond for the prosecution of his scheme, died in 1803, in receipt of a princely income from the profits of his useful undertaking. For its creation, he had, however, denied himself the present enjoyments of his patrimonial revenue, by reducing his expenses at one period to the modest sum of £400 per annum. Before his death, the Duke, for taxation purposes, estimated his income at £110,000 per annum. Before the railway system was fully established a network of canals had united the most populous places in England, the total length of the waterways being not much less than two thousand miles. With the rise of railways the importance of canals as channels for the conveyance of merchandise declined. But, nevertheless, in consequence of the continued increase of traffic and the great cheapness with which goods can be carried by water, canals are often able to compete with railways in the carriage of bulky or heavy goods when speed of transit is not an object. The English canals have, therefore, never been disused or abandoned, notwithstanding the ubiquitous ramifications of the railway lines. Nay, the value of the Bridgewater Canal system, about to be superseded so far as concerns the communication between Liverpool and Manchester by the greater scheme we have presently to describe, is such that £1,710,000 is now required for its purchase ; and that is the value in spite of four lines of railway connecting those great towns, and all competing for the carriage of goods. In these canals, designed for inland communication only, the navigation is confined to boats or barges of very insignificant dimensions compared with the sea-going ships that some great modern canals are constructed to receive.

To the present century belongs the famous "Caledonian Canal," as the waterway is often called that extends in a straight line for more than 60 miles across Scotland, in north-east and south-west directions. The canal work here was commenced in 1802, under the direction of Telford, and though it was opened for traffic in 1822, the work as it now exists was not completed until 1847. But the length of the actual canal construction in this case did not much exceed 23 miles, for a natural waterway, navigable for ships of any burden, is formed by the series of narrow lakes that fill what is called the "Great Glen of the Highlands." This glen has many of the characteristics of a great artificial ditch : its highest point is only 90 feet above the tide level in Loch Linnhe ; a circumstance not a little remarkable in so mountainous a country. What is also remarkable is the great depth of these lakes, which in some places exceeds 900 feet. The banks also are generally very steep, and indeed at one time it was impracticable to pass along the shores of Loch Ness, the longest of the lakes. But there are now good roads along both banks. Although the ground traversed by the artificial channels of the Caledonian Canal is chiefly alluvial, the cost of the undertaking proved to be great, amounting, it is said, to about one and a quarter million pounds sterling. Indeed, had it not been for the introduction of steam navigation before the completion of the work, and the consequent increase and facility of water conveyance, it is doubtful whether the utility of this canal would have been commensurate with its cost, or its receipts have made any profit for its promoters. By the Caledonian Canal large steamers and other vessels may pass from sea to sea, and in the summer time the steamers that traverse it are crowded with tourists attracted by the magnificent scenery it presents throughout the greater part of its length.

But whatever had previously been done in canal construction was surpassed in enterprise and importance by Lesseps' great work in Egypt.

THE SUEZ CANAL.

AS we have already seen, the idea of opening a waterway between the Red Sea and the Mediterranean is by no means a product of the present century. The ancient Egyptians do not appear to have cut directly through the Isthmus, but Herodotus describes a canal made by Necho about the year 600 B.C., from Suez through the Bitter Lakes to Lake Timsah and then westward to Bubastis on the Nile. He mentions certain water gates, and states that vessels took four days in sailing through. This canal became silted up with sand ages ago, but it was cleared out again and re-opened in the seventh century of our era by the Caliph Omar, and traces of it are still visible. According to some recent discoveries in the chief archives of Venice, as early as the end of the fifteenth century, when Vasco da Gama had discovered the Cape of Good Hope, and the Portuguese took that new route to India, hitherto the exclusive property of the Venetian and Genoese merchants, a re-cutting of the Isthmus of Suez was thought of. Plans were prepared and embassies sent to Egypt for paving the way for the accomplishment of this great enterprise, which, it is said, was only foiled by the persistent opposition of some patricians, who were probably bribed by foreign gold to prevent the execution of the plan. One of our Elizabethan poets, Christopher Marlowe, appears, in the following lines, to have anticipated M. de Lesseps :—

> "Thence marched I into Egypt and Arabia,
> And here, not far from Alexandria,
> Whereat the Terrene and the Red Sea meet,
> Being distant less than full a hundred leagues
> I meant to cut a channel to them both,
> That men might quickly sail to India."

For at that period travellers going to India in the famous sailing ships, called "East Indiamen," were obliged to sail round the Cape of Good Hope and pass from the Southern to the Indian Ocean. The reader who wishes to understand the importance of the Suez Canal should look at the map of the Eastern Hemisphere, where he will have no difficulty in finding the position of the vast continent of Africa, which is washed on the north by the Mediterranean Sea, on the west by the Atlantic, on the south by the Southern Ocean, and on the east and north-east by the Indian Ocean and the Red Sea. If he now traces the waterway round Africa, on coming to the head of the Red Sea he will find the only interruption of the oceanic continuity in the narrow neck of land called the Isthmus of Suez. But for this, ships might long ago have made complete circuits round this vast, and, even as yet, but partially explored continent. The circuit would, indeed, be a great one of some 15,000 miles; but the barrier that the Isthmus presented to inter-oceanic communication between the eastern and the western worlds was a piece of physical geography which has undoubtedly been a most important factor in determining the course of history. It has been said that had there existed at Suez a strait like that of Gibraltar or that of Messina, instead of a sandy isthmus, the achievements of Diaz, Vasco da Gama, and Columbus would have lost much of their significance ; but the advantages to the world's commerce would have been incalculable, and the progress of the race might have been more rapid.

The Emperor Napoleon I. had the idea of restoring the old canal ; but it was only when steam navigation had taken its place on the seas that the scheme was looked upon as offering any chance of financial success. But General Chesney, who made some surveys for the French Government in 1830, had come to the conclusion that there was a considerable difference of level between the two seas—a difference, he calculated, of about 30 feet. The existence of such a state of things would, of course, have been very unfavourable for the undertaking; but the General's supposition was soon proved to have been erroneous.

The suggestion of carrying out the project of constructing a ship canal through the Isthmus was seriously revived by Père Enfantin, the St. Simonian, in the year 1833. He then induced M. Ferdinand Lesseps, the French vice-consul, and Mehemet Ali, the Pasha of Egypt, to take some practical measures towards its accomplishment. Surveys were made, but owing to the breaking out of a plague, and to other causes, not much more was heard of the scheme till 1845. In 1846 *La Société d'Etude du Canal de Suez* was formed, and among those who turned their attention to the subject was Robert Stephenson. His report was wholly unfavourable to the enterprise. He recommended the construction of a railway through Egypt, and a line was accordingly made between Alexandria and Suez. But, notwithstanding the opinion of Mr. Stephenson, M. Lesseps persevered with wonderful energy, believing, on the report of other engineers, that the scheme could be successfully carried out. It is right, however, to state that Mr. Stephenson did not say it was impossible to complete the Suez Canal—he merely gave it as his opinion that the cost of making the canal, and keeping it in a proper state for navigation, would be so great that the scheme would not pay. However, in 1854, the Viceroy of Egypt signed the concession, and in 1860 the work was actually commenced, but not on a plan that was advocated by the English engineers of making the canal 25 feet above the sea level.

There were also some political and financial difficulties to be overcome. The Suez Canal Company, it was said, had expended twelve millions of money in what was considered to be chiefly shifting sands.

When the Suez Canal was projected, many prophesied evil to the undertaking, from the sand of the Desert being drifted by the wind into the canal, and others were apprehensive that where the canal was cut through the sand, the bottom would be pushed up by the pressure of the banks. They imagined that the sand would behave exactly like the ooze of a soft peat-bog, through which, when a trench has been cut, the bottom of the trench soon rises, for the soft matter has virtually the properties of a liquid : it acts, in fact, exactly like very thick treacle.

Sand, however, is not possessed of liquid properties ; it has a definite angle of repose, which is not the case with thin bog. This behaviour of sand is familiarly illustrated in the sand-glass, which the diagram Fig. 123, will recall to mind. It may be observed that the sand falling in a slender stream from the upper compartment is in the lower one heaped up in a little mound, the sides of which preserve a nearly constant inclination of about 30°. In this property it is distinctly different from peat-bog or such-like material, which has no definite angle of repose. It need hardly be said that all apprehensions as to the safety of the canal from the causes here alluded to have proved unfounded.

But if some English engineers appeared to oppose the project, another eminent one, Mr. Hawkshaw, certainly helped it on at a moment when the Viceroy of Egypt was losing confidence ; and, had his opinion been adverse to the project reported

FIG. 123.—*The Sand-Glass.*

upon, the Viceroy would certainly not have taken upon himself additional liability in connection with the undertaking, and the money expended up to that date would have been represented only by some huge mounds of sand and many shiploads of artificial stone, thrown into the bottom of the sea to make the harbour of Port Said. And that M. Lesseps appreciated the good offices of Mr. Hawkshaw is shown from the fact that, when he introduced that engineer to various distinguished persons, on the occasion of the opening of the canal, he said, "This is the gentleman to whom I owe the canal." It cannot, therefore, be said of the English nation that they were jealous of the peaceful work of their French neighbours, or opposed it in any other sense but as a "non-paying" and apparently unprofitable scheme.

The Canal was opened in great state by Napoleon III.'s Empress Eugénie, in November, 1869, when a fleet of fifty vessels passed through, and the fact was thus officially announced in Paris :—"The canal has been traversed from end to end without hindrance, and the Imperial yacht, *Aigle*, after a splendid passage, now lies at her moorings in the Red Sea.

"Thus are realized the hopes which were entertained of this great undertaking—the joining of the two seas.

"The Government of the Emperor cannot but look with satisfaction upon the success of an enterprise which it has never ceased to encourage. A work like this, successfully accomplished in the face of so many obstacles, does honour to the energetic initiative of the French mind, and is a testimony to the progress of modern science."

An Imperial decree was then issued, dated the 19th of November, appointing M. de Lesseps to the rank of Grand Cross of the Legion of Honour, in consideration of his services in piercing the Isthmus of Suez.

The Suez Canal is 88 geographical, or about 100 statute miles long: its average width is 25 yards, and the minimum depth, 26 feet. At intervals of five or six miles, the canal is widened, for a short space, to 50 yards, forming thus sidings (*gares*) where only vessels can pass each other. At these, therefore, a ship has often to wait until a file of perhaps twenty steamers, coming the other way, has passed. Occasionally a ship gets across, or "touches," and then the canal is blocked for hours. So much inconvenience has been found from the restricted dimensions of the work, that in 1886 it was proposed to widen the canal, or, alternatively, to construct a second canal, and use the two like the lines of a railway, so that

FIG. 124.—*A Group of Egyptian Fellahs, and their Wives.*

vessels would never have occasion to pass each other. The amount of traffic is very large, and has been steadily increasing. Thus, in 1874, the tonnage of the vessels passing through was 5,794,400 tons ; in 1880, the tonnage was 8,183,313, and the receipts of the Company amounted to £2,309,218. In 1875, the British Government purchased, from the Khedive, £4,000,000 worth of shares.

The Suez Canal is not so much a triumph of engineering as a monument of successful enterprise and determination on the part of its great promoter, M. Lesseps, in the face of great difficulties. According to the original programme, the canal was to have been constructed by forced labour, supplied by the Viceroy. The unhappy peasantry of the country, called "fellahs," were compelled to give their labour for a miserable pittance of rice. No doubt, in ancient times, when forced labour was in use, every peasant might cheerfully work, because it was for the general benefit to bring sweet water from the Nile to other dry and thirsty places in Egypt ; but to be obliged to work at a waterway of salt, which was only to be of use to foreigners who passed through the country, could not be expected of human beings, and therefore the carrying out of the work was not unaccompanied by cruelties of the nature attending slave labour in other lands. This was one of the reasons why the late Lord Palmerston opposed the canal scheme, for the kind hearted statesman bore in mind the loss of health and life occasioned to poor Egyptians by this mode of labour, and the more so because it had been originally proposed that one of the conditions on which the French Company was to take up the project should be the execution of the work by *free labour*. In consequence, no doubt, of representations from free countries, the Porte was induced to put a veto on the employment of forced labour, and everyone thought that this would be the deathblow to the completion of the canal : but M. Lesseps did not give way to despair, and he since stated that if he had depended on the labours of the fellahs only, the difficulties of the work never could have been surmounted ; and that, in fact, the successful prosecution of the work was owing to his having turned his attention to the mechanical contrivances used for dredging on the Thames and the Clyde, from which he obtained better results in half the time and at half the cost.

FIG. 125.—*Dredges and Elevators at Work.*

The dredges used in the construction of the canal were of a new description. They were wonderful mechanical contrivances, and but for them the canal would not have been finished. They were not the contrivance of M. Lesseps, but of one of the contractors, a distinguished engineer, who received his technical education in France but his practical experience in England. The use of the dredging machines was prepared for by digging out a rough trough by spade work, and as soon as it had been dug to the depth of from six feet to twelve feet, the water was let in. After the water had been let in, the steam dredges were floated down the stream, moored along the bank, and set to work. The dredges were of two kinds. The great *couloirs* consisted of a long, broad, flat bottomed barge, on which stood a huge framework of wood, supporting an endless chain of heavy iron buckets. The chain was turned by steam, and the height of the axle was shifted from time to time, so that the empty buckets, as they revolved round and round, should always strike the bottom of the canal at a fixed angle. As they were dragged over the soil they scooped up a quantity of mud and sand and water, and as each bucket reached its highest point in the round, it discharged its contents into a long iron pipe which ran out at right angles to the barge. The further extremity of this pipe stretched for some yards beyond the bank of the canal, and therefore, when the dredging was going on, there was a constant stream of liquid mud pouring from the pipe's mouth upon the shore, and thus raising the height of the embankment. When the hollow scooped out by the buckets had reached the required depth, the dredge was moved to another place, and the same process was repeated over and over again. These stationary dredges, however, though very effective, required much time in moving, and the lighter work of the canal was chiefly effected by movable dredges of a smaller size. These machines were of the same construction as those described ; the only difference was that the mud raised by their agency was not poured directly on shore by pipes attached to the dredges, but was emptied in the first instance into large barges moored alongside the dredge. These barges were divided into compartments, each of which contained a railway truck, and when the barge was filled it steered away to the bank, where an elevator was fixed. The trucks, filled with mud were raised by a crane worked by steam power, and placed upon inclined rails, attached to the elevator, which sloped upwards at an angle of 45 degrees towards the bank. They were then drawn up the rails by an endless rope, and as each truck reached the end of the rails its side fell open, the mud was shot out upon the bank, and the empty truck returned by another set of rails to the platform on which the elevator was placed, and was thence lowered into the barge to which it belonged. As the elevator could un-load and re-load a barge much faster than the dredges could fill it with mud, each elevator

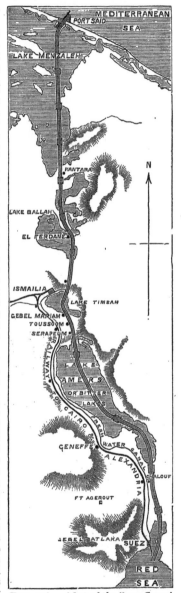

FIG. 126.—*Map of the Suez Canal.*

was fed by half a dozen dredges, and thus the mud raised from the canal by several dredges was carted away without difficulty at one and the same time. As these floating dredges were much easier to shift than those encumbered by the long *couloir* pipes, the work of excavating the bed went on much more rapidly. But in places where there was any great mass of earth or sand to be removed, the large *couloirs* could scoop out a given volume in a shorter time.

The traveller who wishes to see the canal should go to France, and, embarking at the port of Marseilles, cross the Mediterranean Sea, and steam to Port Saïd, which is about 150 miles east of the port of Alexandria, where the isthmus is crossed by the railroad, and is used by travellers to India, being known as the "overland route." And this railway conveys the mail to and from India, thus saving the great sea voyage round Africa and the Cape of Good Hope. Nevertheless, it involves two transhipments—from the steamer to the rail at Alexandria, and from the railway to the steamer at Suez.

Let us notice in order the places passed by the traveller in going from Port Saïd to Suez and the Red Sea. The arrow (Fig. 126) points in the direction of the compass, and shows that the canal runs very nearly from north to south. Port Saïd is the little town at the northern or Mediterranean en-

FIG. 127.—*Port Saïd, the Mediterranean entrance to the Suez Canal.*

trance to the canal, situated on the flat sands at the entrance of the canal, and is built chiefly of wood, with straight wide streets and houses, and although it now contains several thousand inhabitants, before the making of the canal was begun one hundred people could hardly have been got together. The town contains nothing deserving of notice, and has a striking resemblance to the newly settled cities of America. But in it reside agents who represent numerous varied interests—administrative, financial, mercantile and political. It is provided with docks, basins, quays and warehouses, and has a harbour stretching out a couple of miles or so into the sea, for to that distance two piers, or rather breakwaters, run out.

17

Fig. 128 shows these two converging breakwaters, which have been built out into the Mediterranean from the coast, the larger and more westerly one being one mile and a half long, the shorter about a mile and a quarter, and the distances between the two lighthouses erected on the extremities of the breakwater being half a mile.

The piers are made of concrete which was cast in blocks weighing 10 tons each. This composition has of late years been greatly approved by engineers where stone cannot be procured. The sea-face of the great canal in Holland is composed of a similar artificial stone, and it is found to bear the wear and tear of the waves almost, if not quite, as well as ordinary stone. It is stated that 25,000 blocks, each weighing 10 tons,

FIG. 128.—*Bird's-eye View of Port Saïd.*

were used. They were not laid with the regularity of ordinary masonry, but had been dropped from large barges, so that they presented a very rugged and uneven appearance (Fig. 129) ; but the object of throwing out these great bulwarks is for the purpose of preventing the sand brought down by the Nile silting in and closing up the canal. Along the western pier there is, from this cause, a constant settlement of sand, which was partially washed through the interstices left between the blocks of artificial stone, and might have given some trouble by forming sandbanks in the harbour ; but this was prevented by the introduction of smaller stones, which could readily be carried out in boats at the low tide.

Beginning with the Mediterranean Sea and Port Saïd, there is a run of 28 miles to Kantara, through Lake Menzaleh. Although called a lake, it is, in truth, nothing but a shallow lagoon or swamp, in which water-fowl of all kinds are very abundant, the great flocks of white pelicans and pink flamingoes being especially striking. The waters of this lagoon cover lands that once were fertile, and the salt sea-sands doubtless conceal the remains of many an ancient town.

Of all portions of the undertaking, this one, M. Lesseps states, was the most arduous and difficult, though, at the time, it attracted the least

FIG. 129.—*One of the Breakwaters at Port Saïd.*

attention. A trough had to be dredged out of the bed of the shallow lagoon, and on either side of this hollowed out space high sand-banks had to be erected, and the difficulty of making a solid foundation for these sand banks was found to be extreme. The difficulty, however, was surmounted, and such is the excellence of the work, that the water neither

FIG. 130.—*Lake Timsah and Ismaïlia.*

leaks out, nor does any of the brackish water of the lagoon infiltrate and undermine the great embankments.

At Kantara, the canal crosses the track of the highway between Cairo and Syria—a floating bridge carries the caravans across ; and near this spot is stationed an Egyptian man-of-war, which supplies the police for the proper watch and ward of the canal. From Kantara to El Fendane is a distance of 15 miles—that is to say, to the southern extremity of Lake

Ballah, where the canal still passes through sand embankments, raised within a mere. The lake is, however, almost dried up, and therefore the difficulties which had to be surmounted at Lake Menzaleh were not felt here.

The traveller may now be supposed to have arrived at Lake Timsah, where, no doubt, in the days of the Pharaohs, a lake existed. When taken in hand by M. Lesseps, it was a barren, sandy hollow, containing a few shallow pools, through which a man could easily wade, but now it is filled with the waters of the Mediterranean Sea. It is a pretty, inland, salt water lake, about three miles in width. On the northern shore stands the town, or, rather, small settlement of Ismaïlia, which is, in fact, the "half way house" where most of the officials of the Suez Canal Company resided, as they could get to either end of the canal with greater facility, or to Cairo by the railroad, which comes to this point, and continues, with the canal, to Suez.

When the canal was opened, in November, 1869, Ismaïlia was the scene

FIG. 131.—*Railway Station at Ismaïlia.*

of the most brilliant part of the opening ceremony, in which the French Empress Eugénie, the Empress of Austria, the Crown Prince of Prussia, and other distinguished personages took share. The Khedive built himself a summer palace, and M. Lesseps erected a villa, and the town was most artistically laid out, with every prospect of becoming a flourishing place. But the drainage had been so entirely overlooked, that it is said the sewage found its only outlet in the fresh water canal; and the consequence was fever broke out and so infected the town, that it was soon almost quite deserted. In 1882, Ismaïlia was once more the scene of bustle and activity, for here was the base of Sir Garnet Wolseley's operations in his brilliant campaign against Arabi. The British Navy entered the canal, and took possession of Ismaïlia, where the army and the military stores were rapidly concentrated. From this place, Sir Garnet advanced along the route of the railway and the Sweet Water Canal, and, after storming the lines of Tel-el-Kebir, occupied Cairo, without further resistance, after a campaign of only three weeks' duration.

From Lake Timsah to the Bitter Lakes the canal again passes for eight miles or so through the desert, where, by partial excavations by hand labour and subsequent flooding to admit the dredges, it was considered that a sufficiently deep channel could be made. The *couloirs* were set to work, when suddenly " a lion arose in their path" in the shape of a great rock, about 80 feet in length, and lying 12 feet only below the surface, and right in the middle of the main channel. If anything could show the indomitable energy of M. Lesseps it was his courage in dealing with this difficulty, and at a time when a few months only could elapse before the advertised day of the opening. He attacked the sunken rock with gunpowder. A large raft, or floor, supported on barges, was moored over the sunken rock, and from this men, armed with long poles shod with steel, drilled numerous holes, into which charges of gunpowder were placed, and fired in the usual manner by the electric battery. This temporary obstruction occurred opposite to the landing place at Sérápeum.

Passing by Sérápeum, the traveller arrives at a vast expanse of water called the "Bitter Lakes," because the dry sandy hollow formerly con-

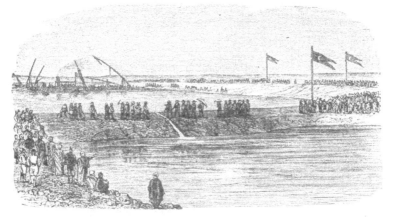

FIG. 132.—*The Viceroy of Egypt cutting the last embankment of the Reservoir of the Plain of Suez, to unite the two seas—the Mediterranean and the Red Sea.*

tained a marsh, or mere, of very brackish water. The possibility of keeping this great area filled with sea water had been denied by the opponents of the canal, who said the water would sink into the sand or be evaporated by the intense heat of the sun; but none of these prognostications have been verified, and it is now a great inland sea, far surpassing Lake Timsah, being 25 miles long and from six to seven miles wide. The only difficulty in filling this enormous natural basin arose from the rapidity and force with which the waters flowed in. This was done when the water at Suez was at low tide, and then subsequently the Red Sea was allowed to flow in. Though the expanse of water in the Bitter Lakes is great enough, the available channel is still narrow. But the steamers can proceed at full speed, as here there are no banks to be washed away.

Since the two seas have joined their waters, a strong current has set in from south to north, but there is no eddy or fall at the place where the waters meet. The tide runs up the canal with great force, and there is a difference of six or seven feet between high and low water : but the tide does not extend beyond the Bitter Lakes, where it is gradually diffused and lost. The colour of the current of water from Suez is said to be green, whilst that portion fed by the Mediterranean is blue. Since the Bitter Lakes have been filled the mean temperature of the districts on the banks has fallen 5° Centigrade. It is also stated that, although the canal swarms with sea fish they keep to their respective ends of the canal, as if the Mediterranean fish would not consort with those of the Red Sea, or, rather, make themselves at home in strange waters. There is also, perhaps, another cause, and that is the very bitter nature of the water at the northern end of the Bitter Lakes, which acts as a natural barrier, through which the fish may decline to pass.

The bed of the Bitter Lakes is the only portion of the canal's course in which it was not necessary to make a cutting. Buoys are laid down to mark the best channel, but such is the width and depth of the water that vessels need not exactly keep within them. Quitting the Bitter Lakes we again enter the canal proper. In order to reach the vast docks which the Suez Canal Company has constructed on the western coast of the Red Sea, the canal is now quitted, and the vessel crosses the neck of the Red Sea. The Cairo and Alexandria Railway has been extended two miles, and is carried through the sea on an embankment, which lands the train close to the docks and quays of the canal, so that passengers by the overland route are able to embark from the train on board the steamer, and thus escape the troublesome transhipment of themselves and luggage.

THE MANCHESTER SHIP CANAL.

THE project of constructing a ship canal to connect Manchester with the sea appears to have been started just before the railway era, but it was then abandoned, as the opening of the Liverpool and Manchester Canal brought about an immediate reduction in the rates of carriage. Perhaps it was the success of the Suez Canal which caused the revival of this scheme, in 1880, combined with the depression of the cotton trade at that period, when the Liverpool dock dues and the comparatively high railway rates proved a heavier tax than usual on the great Lancashire industry. The first definite steps were taken two years afterwards, when two plans were submitted for the selection of a committee. One scheme proposed to construct the canal without any locks ; but, as Manchester is 60 feet above the sea level, there would, it was felt, be certain inconveniences in loading or unloading ships in a deep depression. The other plan was submitted by Mr. Leader Williams, a well known canal engineer, who proposed to take the canal from Runcorn, a distance of 20 miles, and making use of locks. When Parliament was applied to for powers authorizing the prosecution of the enterprise, there was, of course, much opposition offered by the various interests involved, and the inquires before the Committees of each House of Parliament were unusually protracted, for they extended in all to 175 days, and the cost to the promoters is said to have amounted to £150,000. Then, when the Bill had passed, it was found that the capital (£8,000,000) could not be raised owing to the financial depression, and partly also to some want of confidence in

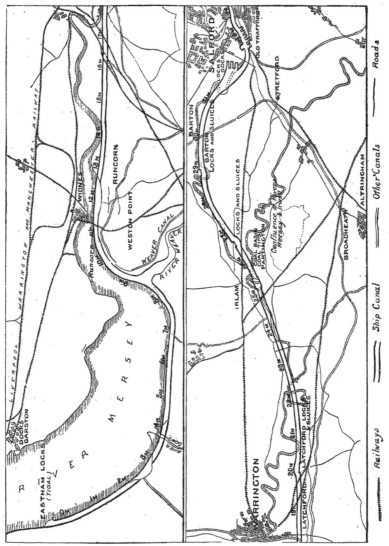

FIG. 133.—*Western Portion.*　　FIG. 134.—*Eastern Portion.*
FIGS. 133 AND 134.—*Map of the Manchester Ship Canal.*

the soundness of the undertaking on the part of the Lancashire capitalists. But the promoters submitted the whole scheme to a representative committee, who should consider any possible objections. This committee reported (after sitting almost daily for five weeks) upon every point, and were unanimous in pronouncing the undertaking to be perfectly practicable and commercially sound. After this there was no difficulty in raising the required capital, which was subscribed by corporate bodies as well as private persons. The contract was let for £5,750,000, and the work was commenced in November, 1887, the contractor undertaking to have the canal completed and ready for traffic by January 1st, 1892.

The Manchester Docks of this canal will cove⁻ an area of nearly 200 acres at the south-western suburb of that city, a ¹ from there the canal traverses the Valley of the Irwell, following, indee, the general course of the river, but not its windings, so that the bed o⁴ ...e river is, in the distance of eight miles, or down to its junction with the Mersey, repeatedly crossed by the line of the canal. From the confluence of the rivers, the canal traverses the Valley of the Mersey, for this is the name retained by the combined streams. The course of the river, in its progress towards the sea, now makes wider bends, but the canal proceeds, by a slight and nearly uniform curve, to Latchford, near Warrington, passing to the south of which last named place it follows a straight line to Runcorn, which is at a distance of 23 miles from Manchester. Here it reaches what is now the estuary of the Mersey, but the embankments are continued along the southern shore to Eastham, where the terminal locks are placed. In this part of the canal, the engineer had difficulties to overcome of a different nature from those encountered in the upper part, where it was chiefly a matter of cutting across the ground intervening between the bends of the river, so as to form for its waters a new and direct channel everywhere of the requisite breadth and depth. But when Runcorn has been passed, and Weston Point rounded, there is the mouth of the River Weaver to be crossed, and this is marked by a great expanse of loose and shifting mud. Other affluents of the Mersey are dealt with by means of sluices, and in one instance the waters of a river are actually carried beneath the course of the canal by conduits of 12 feet in diameter. The total length of the canal from Manchester to the tidal locks at Eastham is 35 miles.

The minimum width of the canal at the bottom is 120 feet, its depth 26 feet. But for several miles below Manchester this width will be increased, so that ships may be moored along the sides, and yet sufficient space left for the up and down lines of traffic in the middle. In this way, works and manufactories on the banks will be able to load and unload their cargoes at their own doors, and it may be expected that the advantages so offered will cause the banks of the canal to be much in request for the sites of works of all kinds. At the several places where the locks are placed there will be a smaller and a larger one, side by side, so that water shall not be needlessly used in passing a moderate sized vessel through the greater locks. As these last are 550 feet long and 60 feet wide, they are capable of receiving the largest ships, whilst the smaller locks are 300 feet long and 40 feet wide. Again, both the larger and the smaller are provided with gates in the middle, so that only half their length may be used when that is found sufficient. Coming down the canal from Manchester, the first set of locks will be at Barton, about three miles distance, just below the place where the Bridgewater Canal is carried across the Irwell, which

FIG. 135.—*A Cutting for the Manchester Ship Canal.*

is now to become the ship canal, by means of the aqueduct of 1760, by
which Brindley became so famous. There is a story told about Brindley
being desirous of satisfying the duke about the practicability of his plan,
and requesting the confirmatory opinion of another engineer. When,
however, this gentleman was taken to the place where it was proposed to
construct the aqueduct, he shook his head, and said that he had often
heard of castles in the air, but had never before been shown where any ot
them were to be erected. This aqueduct is about 600 feet long, and the
central one of its three arches spans the river at a height of nearly 40 feet
above the water. But the Manchester Ship Canal requires a clear headway of 75 feet, and Mr. Williams is going to replace the fixed stone

FIG. 136.—*Blasting Rocks for the Manchester Ship Canal.*

structure by a swinging aqueduct, or trough of iron, which can be turned
round, so as to give a clear passage for ships in his canal. This trough,
or great iron box, will have gates at each end, and gates will be provided
in the aqueduct at each side, so that no water will be lost when the water
bridge is turned aside. But more than this ; hydraulic lifts have been
designed, so that, in a few minutes, vessels can be lowered from the
Bridgewater Canal into the Manchester Canal, or raised from the latter
into the former while still floating in water. The supply of water for the
canal will be ample, as it has the rivers Irwell, Mersey and Bollin, with
their tributary streams, to draw from. It should be mentioned that the
terminal locks at Eastham will be of somewhat larger dimensions than
those already referred to, and will be three in number. The largest,

which is on the south or landward side, will be 600 feet by 80 feet, the middle one 350 feet by 50 feet, and the smallest one 150 feet by 30 feet. These three locks will be separated by concrete piers 30 feet wide, on which will be placed the hydraulic machinery for opening and closing the gates. Besides the ordinary gates, there will be provided for each lock at Eastham an outer pair of storm-gates that will be closed only in rough weather. These gates will shut from the outside against the lock sills, and, by resisting the force of wind and waves, will protect the ordinary tidal gates from being forced open. The lock gates throughout will be made of a wood obtained from British Guiana, and known as *greenheart.* This timber is the product of a large tree (*Nectandra Rodiæi*) belonging

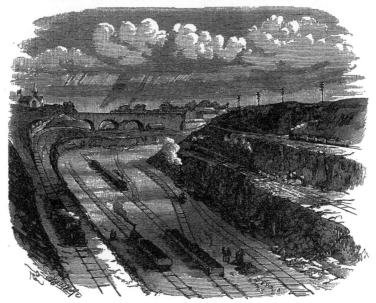

FIG. 137.—*Manchester Ship Canal Works, Runcorn.*

to the laurel family. It is a very heavy and close grained wood, the strength and endurance of which have been proved many years ago by its use in ship-building, etc., and some of the logs imported for the canal are remarkably fine specimens, being 22 inches square and 60 feet long. A pair of the largest gates weigh about 500 tons. The gates of the tidal locks at Eastham will all be open for half the time of each tide, when there will be a depth of water, above the sills, greater by 11 feet than that of any dock in Liverpool or Birkenhead.

The way in which the difficulty is overcome of crossing the several busy lines of railway that intersect the course of the new canal, so that their traffic shall not be impeded, is one of special interest in this bold scheme. The London and North Western Railway crosses the Mersey at Runcorn by a bridge that leaves a clear headway of 75 feet at high water, and it was determined that this headway should be maintained in the bridges over the canal. The use of swing bridges on lines of railway over which trains are constantly passing being out of the question, it is necessary that the railways be carried over the canal at the required height. It is accordingly laid down in the Act of Parliament that before the Canal Company can cut the existing lines of railway it shall construct permanent bridges, and carry over them lines rising by gradients not exceeding 1 in 135, and not only so, but these deviation lines must be previously given up to the several railway companies for six months to be tried experimentally in that period for goods traffic. The cost of constructing these deviation lines, which, in all, will not be far short of

FIG. 137*a.*—*The French Steam Navvy.*

12 miles of new railway, will not be much less than £500,000. The traffic of the canal will probably have great feeders at certain points in the other canals and the railway lines that reach it. For instance, the Bridgewater Canal, now incorporated with the greater undertaking, will bring traffic from the Staffordshire potteries, the river Weaver brings salt laden barges from Cheshire, and at other points the railways will bring the produce of the excellent coal fields of South Yorkshire and South Lancashire, which will be automatically transferred from the waggons into ocean going steamships.

Though the general notion of the construction of the canal as a deep, wide trench, or cutting following the course shown on the map, is sufficiently simple, the operation of carrying this into practice involves the exercise of great skill and ingenuity in dealing with mechanical obstacles. Man's operations in the world consist but in changing the position of masses of matter; and the properties of matter—its inertia, cohesion, gravitation, etc., are the forces that oppose his efforts. The quantity of matter to be shifted in excavating this trench of thirty-five miles long across the country was no less than sixty millions of tons. The number of "navvies" employed at one time has been 15,000; but even this army of workmen would have made but slow progress with a cutting of this magnitude, had not the "strong shouldered steam" been also called into operation for scooping out the soil. The illustrations (Figs. 137a and 137b) will show the arrangement of two forms of "steam

FIG. 137b.—*The English Steam Navvy.*

navvies" that were much used on the works. One (Fig. 137a) is similar to the dredgers used for clearing mud out of rivers and canals: it consists of a series of scoops, or buckets, mounted on an endless chain, so as to scrape the material from an inclined embankment and tip it into waggons for removal. The other (Fig. 137b) may be compared to a gigantic ladle made to scrape against the face of a cutting in rising, and filling each time its bucket with nearly a ton of the material. It is most interesting to witness the perfect control which the man at the levers exercises over this machine, the movements of which he directs with as much precision as if he were handling a spoon. One of these steam navvies is able to fill 600 waggons or more—that is, to remove 3,000 tons of material—in one day ; and as many as eighty of them have been simultaneously used on the Canal works. The value of the plant employed by the contractor is estimated at £700,000, and the length of temporary railway lines (see Fig. 137), for transport of the " spoil," etc., is said to exceed 200 miles. There is a main line running through from one end of the canal to the other, and known to the workmen as the " Overland Route." From this diverge numerous branches, some to the bottom of the excavations in progress, others to embankments down which is tipped out the " spoil," as the dug out material is called ; while others connecting with brickfields and quarries, or with existing canals and railway lines, serve to bring supplies of the materials used in the constructions. Some 150 locomotives are constantly at work on these temporary lines, and the coal consumed by them, and by the steam navvies, steam cranes, pumping engines, etc., is equivalent to about two train loads every day.

Though the Manchester Ship Canal is to be nearly twice as wide as the Suez Canal, its width for some miles below Manchester will be still greater, for there the banks will form long continuous wharves for the accommodation of the works and factories that are certain to be attracted to the spot. Indeed, so obvious are the advantages of ocean shipment, and so extensive the industries of South Lancashire, that it is not improbable the whole course of the canal may, in process of time, be lined with wharves, and the two great cities of Manchester and Liverpool may be united by a continuous track of dense population. Be that as it may, there seems every reason to believe that the undertaking will be a financial success. Calculation has shown that if the cotton alone that enters and leaves Manchester were carried by the canal at half the rates charged by the railways, there would result not only an annual saving of £456,000 to the cotton trade, but a clear profit to the canal company sufficient to pay more than 3 per cent. interest on its own capital. And, again, the railway and other local interests that have hitherto been opposed to this great enterprise can hardly fail to be in the long run benefited by the enlarged prosperity and increased general trade and manufactures it will develop. So that it will presently be found that there is room enough and work enough for both canal and railways.

The Manchester Ship Canal, so far from having been ready for traffic on the 1st January, 1892, was not completed until the end of 1893, and it was only on the 16th December, 1893, that the directors and their friends made the trial trip throughout its entire length, accomplishing the distance of 35½ miles in 5½ hours. The total cost of the canal was greatly in excess of the estimates, which placed it at eight million pounds, as fifteen millions is the sum actually expended upon it. With such a vast capital expenditure, it may be some time before the ordinary shareholders can look for dividends, especially as there has not been any sudden rush of traffic, such as many sanguine people expected. On the other hand, traffic is continuously and steadily increasing, and there is reason to believe that this great work will ultimately prove a commercial, as it has an engineering, success.

THE NORTH SEA CANAL.

L IKE several other canals for sea-going ships this last addition to the achievements of modern engineering is but the realisation of a project conceived at a long past period. The idea of a canal to connect the Baltic and the North Sea dates back into the Middle Ages, and indeed a short canal was constructed in 1389, which by uniting two secondary streams of the peninsula really did provide a waterway between the two seas. The inefficiency of this means of communication may be inferred from the fact of there having been proposed since that period no fewer than sixteen schemes of canalisation between these two seas, of which the recently completed North Sea Canal is the sixteenth, and it need hardly be said the greatest, so that in comparison with it the rest vanish into insignificance. The canal was commenced in 1887, and on the 20th of June, 1895, it was opened by the reigning Emperor of Germany, William II., with a very imposing naval pageant in which nearly a hundred ships of war from the great navies of the world took part. A glance at the accompanying sketch-map will show the great importance of this

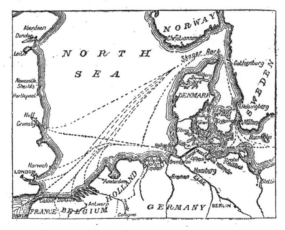

FIG. 137c.—*Sketch Map of The North Sea Canal.*

canal as a highway of commerce. The entrance to the Baltic has hitherto been round the peninsula of Denmark and through the narrow "belts" and "sounds" that divide the Danish Islands, a course beset with imminent perils to navigators, for the channels abound in rocks and dangerous reefs, to say nothing about the frequent storms and the impediments of ice floes. Yet as many as 35,000 vessels have lately had to take that course annually; these representing a total tonnage of no less than 20,000,000 tons. The figures speak for the magnitude of the Baltic shipping intercourse with the rest of the world; while the losses incurred in traversing these forbidding waters may be gathered from the statement that since 1858, nearly 3000 ships have been wrecked in them, and a greater number much damaged. Indeed, for large vessels, there is hardly a more dangerous piece of navigation in all Europe. The importance of this canal must not therefore be estimated solely by the saving of length in ships' course, though that is great, as the map shows.

The North Sea Canal is 61 miles long, 200 ft. wide at the surface, 85 ft. wide at the bottom, and it will admit of vessels of 10,000 tons register passing through, the average time of transit being about twelve hours. The estimated cost of this undertaking was nearly eight and a quarter million pounds sterling, and about one-third of this sum was contributed by Germany, for whom the canal is of the greatest strategic importance in case of war, for her fighting ships need not then traverse foreign waters. The construction was therefore pushed forward with unusual energy, as many as 8,600 men having been engaged on the works at one time. An important naval station already exists at Kiel, the Baltic end of the canal, where there is a splendid harbour. The engineer and designer of this water-way is Herr Otto Baensch, who has devised much ingenious machinery in connection with the immense tidal locks at the extremities of the canal, and the swing bridges by which several lines of railway are carried across it. In the construction of this canal there were no vast engineering difficulties to be overcome, and hence striking feats of mountain excavation or valley bridging are not to be met with in its course, though in places there are some deep cuttings. The methods of excavating and of steam dredging that were made use of have already been illustrated in relation to the other works described in this article. The country through which the canal passes does not present any unusually picturesque features.

THE PANAMA AND NICARAGUA CANAL PROJECTS.

THE several undertakings described in our chapter on Ship Canals are now all completed and in active operation, and but for financial mismanagement and dishonest speculations, the same might probably have been said of another great project, the name of which was on everyone's lips a short time ago, but in which public interest has lately waned ; perhaps from a mistaken impression that the construction itself is involved in a common ruin with the fortunes of so many of its promoters, or that the scheme was frustrated by some unforeseen and insurmountable engineering difficulties. These assumptions have so little justification that it is quite probable that Lesseps' last great project may yet be completed under more favourable auspices, and the Panama Canal unite the Atlantic and Pacific Oceans. The Panama Canal Company still exists, and possesses not only a very large part of the work almost quite finished, but all the extensive plant in perfect condition for resuming operations. The original scheme provided for a tidal water-way between the two oceans, without the intervention of a single lock. The canal was to be nearly 47 miles in length, 100 feet wide at the surface of the water, 72 feet wide at the bottom, and 29 feet deep. The entrances are at Colon on the Atlantic side, and at Panama on the Pacific. The latter is the eastern extremity, and the western one is on the Atlantic side, owing to the configuration of the isthmus which curves round the Panama Gulf that opens to the south. A railway crosses the isthmus between the points already named, and the route of the canal is laid down almost parallel with this railway, from which it is nowhere far distant. For the first 20 miles from the Atlantic side the land is only at a very moderate elevation above the sea-level, say 25 or 30 feet, but the next 11 miles is more hilly, the elevations reaching at some points 150 to 170 feet, but these are only for short distances. A few miles farther on, they rise still higher, until at Culebra the highest point is met with, about 323 feet above the sea-level, and a cut of this depth, 1,000 feet long, would be required. Through this highest part it has been proposed to drive a tunnel, but the total extent of the deep cutting at this part of the canal would be nearly 2 miles in length. This would no doubt be a work of the most formidable magnitude, for it has been calculated that no less than 24,000,000 cubic yards of material, consisting for the most part of solid rock, would have to be removed. It is not supposed, however, to offer any great difficulty in an engineering point of view. Doubtless it would be costly, and would take some time to accomplish. Another heavy piece of work would consist in constructions for controlling a mountain torrent called the Rio Chagres, through the valley of which the canal passes. This stream is very variable in the quantity of water it discharges, rising in the rainy season 45 feet above its ordinary level, and sending down forty times as much water as it does in the dry season.

Mr. Saabye, an American engineer, who examined unofficially the works of the Panama Canal in 1894, considers that about one half of the total excavation has already been done, and one half of the total length of the canal almost finished, and remaining in comparatively good condition. At both ends, including 15 miles on the Atlantic side, there is water 18 to 24 feet deep. "Besides the work already done, the Canal Company has on hand, distributed at both terminals, and at convenient points along the canal route, an immense stock of machinery, tools, dredges, barges, steamers, tug-boats, and materials for continued construction. At Panama, La Boca, and Colon, as well as along the canal, are numerous buildings—large and small—for offices, workshops, storehouses, and warehouses, and for lodging and boarding the men who were employed on the work. The finished work, as well as all the machinery, tools, materials, buildings, etc., are well taken care of and looked after. The Canal Company employs one hundred uniformed policemen, besides numerous watchmen, machinists, and others, whose sole duty consists in watching the canal and looking after needed repairs of plant and care of materials. In fact, the work and the whole plant is in such a condition, so far as I could ascertain, that renewed construction could be taken up and carried to a finish at any time it is desired to do so, after the Company's finances will permit."

An enormous amount of money has already been expended on the Panama Canal, and much of it lavishly and unnecessarily. A reorganised

18

company may probably be able to form such estimates of the probable cost of completing the work under careful and efficient management, that financial confidence in it may be restored. The canal not only already possesses the requisite plant, but the route has the special advantages or assistance in transport from the railway everywhere at but a short distance from it, and fine commodious harbours for its ocean mouths. If it were finished as originally designed, vessels could pass through it with one tide, say in about six hours. It is understood that before the Panama enterprise is again proceeded with, the Company think that a sum of about £25,000 should be expended in a complete survey and re-study of all the conditions, and the results submitted to the most eminent engineers.

A rival scheme for carrying a ship canal across the isthmus that divides the Atlantic and Pacific Oceans is that known as the Nicaragua Canal, as the proposed route is to cross Lake Nicaragua, an extensive sheet of water situated some 400 or 500 miles north-west of the Panama Canal. The lake is 110 miles long and 45 miles broad, and is on its western side separated from the Pacific by a strip of land only 12 miles wide, having at one point an elevation not exceeding 154 feet, which is probably the lowest on the isthmus. The lake drains into the Caribbean Sea on the east, by the San Juan river, a fine wide stream, 120 miles in length, which is navigable for river boats from the Caribbean Sea up to the lake, except near its upper part, where some rapids at certain times prevent the passage of the boats. This canal project first took definite form in 1850, when a survey was made and routes reported on. The scheme attracted some attention in the United States, and in 1872, and again in 1885, further surveys and estimates were made at the instance of the States Government. The earlier schemes provided for the rise and fall between sea and lake—108 feet, a considerable number of locks— eleven on each side, making the total length from sea to sea 181 miles. The report of the latter advocated the canalization of the San Juan by a very bold measure, namely, the construction of an immense dam, by which the waters were to be retained in the valley for many miles at the level of the lake. A company was formed to promote the project, and again in 1890 there were more surveys and estimates made. This company actually expended a considerable sum of money in attempting to improve the harbour at Greytown, which would have formed the eastern terminus, but had become silted up. But it was found afterwards that it would be better to recommend the formation of an artificial harbour at another point, by constructing two long piers running out into the sea, although this change would involve the abandonment of a few hundred yards of canal already excavated by the company near Greytown. The company has also laid down about 12 miles of railway along the proposed route, with wooden and iron sheds as workshops, offices, etc., and, more- over, had dredges and other appliances at work. At this stage it was proposed that the United States Government should guarantee the bonds of the Nicaragua Canal Company to the extent of more than twenty million pounds sterling. By an Act of Congress passed in March, 1895, a commission of engineers was appointed for the purpose of ascertaining the feasibility, permanence, and cost of construction and completion of the Nicaragua Canal by the route contemplated. The report of this commis- sion is an elaborate and exhaustive review of the whole scheme based upon a personal examination of the route, and on the plans, surveys, and esti- mates made for the company, whose records, however, are stated in the report to be deficient in the supply of many important data. The Canal Company's project provided for the improvement of Greytown harbour, as already stated, and from that place the canal was to proceed westward at the sea-level to the range of high ground on the eastern side of the isthmus, which elevation was to be ascended by three locks of unusual depth, and a deep cut more than 3 miles in length, through rock to a maximum depth of 324 feet. After passing this enormous cut, the route provides for a series of deep basins, in which the water is confined by numerous dams or embankments, the canal excavations being confined to short sections through higher ground separating these basins. The total length of these embankments will be about 6 miles, and their heights will vary from a few feet to more than seventy. About 31 miles from Greytown the canal reaches the San Juan river, which, however, by means of an enormous dam across the valley at a place called Ochoa, 69 miles below the point at which it receives the waters of Lake Nicaragua, is there practically converted into an arm of the lake. This dam, which would raise the water of the river 60 feet above its present level, and would, of course, flood the valley back to the lake, is the most notable feature of the project. Its maximum height would be about 105 feet, and the weirs on its crest, to discharge the surplus water, would require a total length of nearly a quarter of a mile. Twenty-three smaller embankments would also be needed for retaining the waters; the river would have to be deepened in the upper part, and a channel dredged out in the soft mud of the lake for 14 miles beyond the river. The big Ochoa dam is said to have no precedent in engineering construction, on account of its great height and the enormous volume of the waters it is intended to retain. No doubt its construction and safe maintenance are within the range of engineering skill, when a thoroughly exhaustive survey of the site has been made, and the necessary funds are forthcoming. From the western shore of the lake its level would also be extended by another great dam crossing the valleys of the Tola and the Rio Grande, with a length of 2,000 feet and a height of 90 feet. The canal would then be carried to the sea-level by a series of locks. The length of the canal from sea to sea would be 170 miles, but of this only 40 miles of channel would require to be excavated. The total cost of the work, as estimated by the Nicaragua Canal Company, would be about fifteen million pounds sterling, but the State Commission of Engineers thinks about double that amount would be a safer calculation, and taking into account the imperfection of the data, even this might be exceeded in certain contingencies. The Government of the United States has been urged to expend a few thousand pounds on another engineering commission, to make complete surveys, and consider all the practical problems involved, including the final selection of a route.

FIG. 138.—*Britannia Bridge, Menai Straits.*

IRON BRIDGES.

THE credit of having invented the arch is almost universally assigned to the ancient Romans, though the period of its introduction and the date of its first application to bridge building are unknown. That some centuries before the Christian era, the timber bridges of Rome had not been superseded by those of more permanent construction is implied in the legend of the defence of the gate by Horatius Cocles—a tale which has stirred the heart of many a schoolboy, and is known to everybody by Macaulay's spirited verses, in which

"Still is the story told,
How well Horatius kept the bridge,
In the brave days of old."

Some of the arched bridges built by the Romans remain in use to this day to attest the skill of their architects. The Ponte Molo at Rome, for example, was erected 100 B.C. ; and at various places in Italy and Spain many of the ancient arches still exist, as at Narni, where an arch of 150 ft.

span yet remains entire. Until the close of the last century the stone or brick arch was the only mode of constructing substantial and permanent bridges. And in the present century many fine bridges have been built with stone arches. The London and Waterloo Bridges across the Thames are well-known instances, each having several arches of wide span, attaining in the respective cases 152 ft. and 120 ft. The widest arch in England, and one probably unsurpassed anywhere in its magnificent stride of 200 ft., is the bridge across the Dee at Chester. built by Harrisson in 1820. At the end of last century *cast* iron began to be used for the construction of bridges, a notable example being the bridge over the Wear at Sunderland, of which the span is 240 ft. But with the subsequent introduction of *wrought* iron into bridge building a new era commenced, and some of the great results obtained by the use of this material will be described in the present article. In order that the reader may understand how the properties of wrought iron have been taken advantage of in the construction of bridges, a few words of explanation will be necessary regarding the strains to which the materials of such structures are exposed.

Such strains may be first mentioned as act most directly on the materials of any structure or machine, and these are two in number, namely, extension and compression. When a rope is used to suspend a weight, the force exerted by the latter tends to stretch the rope, and if the weight be made sufficiently great, the rope will break by being pulled asunder. The weight which just suffices to do this is the measure of the *tenacity* of the rope. Again, when a brick supports a weight laid upon it, the force tends to compress the parts of the brick or to push them closer together, and if the force were great enough, the brick would yield to it by being crushed. Now, a brick offers so great a resistance to a crushing pressure, that a single ordinary red brick may be capable of supporting a weight of 18 tons, or 40,320 lbs.—that is, about 1,000 lbs. on each square inch of its surface. Thus the bricks at the base of a tall factory chimney are in no danger of being crushed by the superincumbent weight, although that is often very great. The *tenacity* of the brick. however, presents the greatest possible contrast to its strength in resisting pressure, for it would give way to a pull of only a few pounds. Cast iron resembles a brick to a certain extent in opposing great resistance to being crushed compared to that which it offers to being pulled asunder, while wrought iron far excels the cast metal in tenacity, but is inferior to it in resistance to compression.

The following table expresses the forces in tons which must be applied for each square inch in the section of the metals, in order that they may be torn apart or crushed:

	Tenacity per square inch, in tons.	Crushing pressure per square inch, in tons.
Cast iron	8	50
Wrought iron	30	17
Iron wire	40	...

Besides the direct strains which tend to simply elongate or compress the materials of a structure or of a machine, there are modes of applying forces which give rise to transverse strains, tending to twist or wrench the pieces, or to bend them, or rupture them by causing one part of a solid to slide away from the rest. Strains of this kind no doubt come into play in certain subordinate parts of bridges of any kind; but if we divide bridges according to the nature of the strains to which the essential parts of the structure are subject, we may place in a class where the materials are exposed to crushing forces only, all bridges formed with stone and brick arches; and in a second class, where the material is subjected to extension only, we can range all suspension bridges; while the third class is made up of bridges in which the material has to resist both compression and extension. This last includes all the various forms of girder bridges, whether trussed, lattice, or tubular. The only remark that need be here made on arched bridges is, that when cast iron was applied to the construction of bridges, the chief strength of the material lying in its resistance to pressure, the principle of construction adopted was mainly the same as that which governs the formation of the arch; but as cast iron has also some tenacity, this permitted certain modifications in the adjustment of the equilibrium, which are quite out of the question in structures of brick and stone.

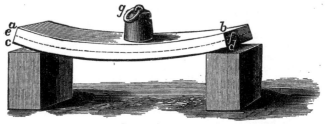

FIG. 139.

The general principle of the construction of girder bridges is easily explained by considering a simple case, which is almost within everybody's experience. Let us suppose we have a plank supported as in Fig. 139. The plank will by its own weight sink down in the centre, becoming curved in the manner shown; or if the curvature be not sufficiently obvious, it may always be increased by placing weights on the centre, as at g. If the length of the plank had been accurately measured when it was laid flat upon the ground, it would have been found that the upper or concave surface, $a\,b$, had become shorter, and the lower or convex surface, $c\,d$, longer when the plank is supported only at the ends—a result sufficiently obvious from the figure it assumes. It is plain, then, that the parts of the wood near the upper surface are squeezed together, while near the lower surface the wood is stretched out. Thus, the portions in the vicinity of the upper and lower surfaces are in opposite conditions of strain; for in the one the tenacity of the material comes into play, and in the other its power of resisting compression. There is an intermediate layer of wood, however, which, being neither extended or compressed, receives no strain. The position of this is indicated by the line $e\,f$, called the *neutral line*. If the plank, instead of being laid flat, is put upon its edge, as in Fig. 140, the deflection caused by its weight will hardly be perceptible, and it will in this position support a weight which in its former one would have broken it down. There is in this case a neutral line, *e f*, as before; but as the part which is most compressed or extended is now situated at a greater distance from the neutral line, the resistance of the material acts, as it were, at a greater leverage. Again the portions near the neutral line are under no strain; they do not, therefore, add to the strength, although they

FIG. 140.

increase the weight to be supported, and they may, for that reason, be re-moved with advantage, leaving only sufficient wood to connect the upper and lower portions rigidly together. The form of cast iron beams, Fig. 141, which were used for many purposes, depends upon these principles. The

FIG. 141

FIG. 142.

sectional area of the lower flange, which is subjected to tension, is six times that of the upper one, which has to resist compression, because the strength of cast iron to resist pressure is about six times greater than its power of resisting a pull. If the upper flange were made thicker, the girder would be weaker, because the increased weight would simply add to the tension of the lower one, where, therefore, the girder would be more ready to give way than before. If we suppose the vertical web divided into separate vertical portions, and disposed as at Fig. 142, the strength of the girder, and the principle on which that strength depends, will be in no way changed, and we at once obtain the box girder, which on a large scale, and arranged so that the roadway passes through it, forms the tubular bridge. It is only necessary that the upper part should have strength enough to resist the compressing force, and the lower the extending force, to which the girder may be subject; and wrought iron, properly arranged, is found to have the requisite strength in both ways, without undue weight. The various forms of trussed girders, the trellis and the lattice girders, now so much used for railway bridges, all depend upon the same general principles, as does also the Warren girder, in which the iron bars are joined so as to form a series of triangles, as in Fig. 143.

FIG. 143.

Girders have been made of wrought iron up to 500 ft. in length, but the cost of such very long girders is so great, that for spans of this width other modes of construction are usually adopted.

GIRDER BRIDGES.

THE Britannia Bridge, which carries the Chester and Holyhead Railway across the Menai Straits, is perhaps the most celebrated example of an iron bridge on the girder principle. It was designed by Stephenson, but the late Sir W. Fairbairn contributed largely by his knowledge of iron to the success of the undertaking, if he did not, in fact, propose the actual form of the tubes. Stephenson fixed upon a site about a mile south of Telford's great suspension bridge, because there occurred at this point a rock in the centre of the stream, well adapted for the foundation of a tower. This rock, which rises 10 ft. above the low-water level, is covered at high water to about the same depth. On this is built the central tower of the bridge, 460 ft. from the shore on either side, where rises another tower, and at a distance from each of these of 230 ft. is a continuous embankment of stone, 176 ft. long. The towers and abutments are built with slightly sloping sides, the base of the central or Britannia tower being 62 ft. by 52 ft., the width at the level where the tubes pass through it, a height of 102 ft., being reduced by the tapering form to 55 ft. The total height of the central tower is 230 ft. from its rock foundation. The parapet walls of the abutments are terminated with pedestals, the summits of which are decorated by huge lions, looking landwards. As each line of rails has a separate tube, there are four tubes 460 ft. long for the central spans, and four 230 ft. long for the shorter spans at each end of the bridge. Each line of rails, in fact, traverses a continuous tube 1,513 ft. in length, supported at intervals

FIG. 144.—*Section of a Tube of the Britannia Bridge.*

by the towers and abutments. The four longer tubes were built up on the shore, and were floated on pontoons to their positions between the towers, and raised to the required elevation by powerful hydraulic machinery. The external height of each tube at the central tower is 30 ft., but the bottom line forms a parabolic curve, and the other extremities of the tubes are reduced to a height of 22¾ ft. The width outside is 14 ft. 8 in. Fig. 144 shows the construction of the tube, and it will be observed that the top and bottom are cellular, each of the top cells, or tubes, being 1 ft. 9 in. wide, and each of the bottom ones 2 ft. 4 in. The vertical framing of the tube consists essentially of bars of T-iron, which are bent at the top and bottom, and run along the top and bottom cells for about 2 ft. The covering of the tubes is formed of plates of wrought iron, rivetted to T- and L-shaped ribs. The thickness of the plates is varied in different parts from ⅓ in. to ¾ in. The plates vary also in their length and width in the different parts of the tubes, some being 6 ft. by 1¾ ft., and others 12 ft. by 2 ft. 4 in. The joints are not made by overlapping the plates, but are all what are termed *butt* joints, that is, the plates meet edge to edge, and along the juncture a bar of T-iron is rivetted on each side, thus: ⊥. The cells are also formed of iron plates, bolted together by L-shaped iron bars at the angles.   The rails rest on longitudinal timber sleepers, which are well secured by angle-iron to the T-ribs of the framing forming the lower cells.   More than two millions of rivets were used in the work, and all the holes for them, of which there are seven millions, were punched by special machinery.   The rivets being inserted while red hot, and hammered up, the contraction which took place as they cooled drew all the plates and ribs very firmly together.   In the construction of the tubes no less than 83 miles of angle-iron were employed, and the number of separate bars and plates is said to be about 186,000.   The expansion and contraction which take place in all materials by change of temperature had also to be provided for in the mode of supporting the tubes themselves.   This was accomplished by causing the tubes, where they pass through the towers, to rest upon a series of rollers, 6 in. in diameter, and these were arranged in sets of twenty-two, one set being required for each side of each tube, so that in all thirty-two sets were needed.   There are other ingenious arrangements for the same purpose at the ends of the tubes resting on the abutments, which are supported on balls of gun-metal, 6 in. in diameter, so that they may be free to move in any manner which the contractions and expansions of the huge tubes may require.   Each of the tubes, from end to end of the bridge, contains 5,250 tons of iron.   The mode in which these ponderous masses were raised into their elevated position is described in the article on " Hydraulic Power," as it furnishes a very striking illustration of the utility and convenience of that contrivance.   The foundation-stone of the central tower was laid in May, 1846, and the bridge was opened in October, 1850. The tubes have some very curious acoustic properties: for example, the sound of a pistol-shot is repeated about half a dozen times by the echoes, and the tubular cells, which extend from one end of the bridge to the other, were used by the workmen engaged in the erection as speaking-tubes.   It is said that a conversation may thus be carried on with a person at the other end of the bridge, a distance of a quarter of a mile.   The rigidity of the great tubes is truly wonderful.   A very heavy train, or the strongest gale, produces deflections in the centre, vertical and horizontal respectively, of less than one inch.   But when ten or a dozen men are placed so that they can press against the sides of the tube, they are able, by timing their efforts so as to agree with the period of oscillation proper to the tube, to cause it to swing through a distance of $1\frac{1}{4}$ in.—an illustration of facts of great importance in mechanics, showing that even the most strongly-built iron structure has its own proper period of oscillation as much as the most slender stretched wire, and that comparatively small impulses can, by being isochronous with the period of oscillation, accumulate, as it were, and produce powerful effects.   Bridges are often tried by causing soldiers to march over them, and such regulated movements form the severest test of the freedom of the structures from dangerous oscillation.   The main tubes of the Britannia Bridge make sixty-seven vibrations per minute.   The expansion and contraction occurring each day show a range of from $\frac{1}{2}$ in. to 3 in.   The total cost of the structure was £601,865.

A stupendous tubular bridge has also been built over the St. Lawrence at Montreal, and the special difficulties which attended its construction render it perhaps unsurpassed as a specimen of engineering skill.   The magnitude of the undertaking may be judged of from the following dimensions : Total length of the Victoria Bridge, Montreal, 9,144 ft., or $1\frac{3}{4}$ miles; length of tubes, 6,592 ft., or $1\frac{1}{4}$ miles : weight of iron in the tubes, 9,044

FIG. 145.—*Albert Bridge, Saltash.*

tons ; area of the surface of the ironwork, 32 acres ; number of piers, 24, with 25 spans between the piers, each from 242 ft. to 247 ft. wide.

Another singular modification of the girder principle occurs in the bridge built by Brunel across a tidal river at Saltash, Fig. 145. Here only a single line of rails is carried over the stream, which is, however, 900 ft. wide, and is crossed by two spans of about 434 ft. wide. A pier is erected in the very centre of the stream, in spite of the obstacles presented by the depth of the water, here 70 ft., and by the fact that below this lay a stratum of mud 20 ft. in depth before a sound foundation could be reached. This work was accomplished by sinking a huge wrought iron cylinder, 37 ft. in diameter and 100 ft. in height, over the spot where the foundation was to be laid. The cylinder descended by its own weight through the mud, and when the water had been pumped out from its interior, the workmen proceeded to clear away the mud and gravel, till the rock beneath was reached. On this was then built, within the cylinder, a solid pillar of granite up to the high-water level, and on it were placed four columns of iron 100 ft. high, each weighing 150 tons. The two wide spans are crossed by girders of the kind known as "bow-string" girders, each having a curved elliptical tube, the ends of which are connected by a series of iron rods, forming a catenary curve like that of a suspension bridge. To these chains, and also to the curved tubes, the platform bearing the rails is suspended by vertical suspension bars, and the whole is connected by struts and ties so nicely adjusted as to distribute the strains produced by the load with the most beautiful precision. When the bridge was tested, a train formed wholly of locomotives, placed upon the entire length of the span, produced a deflection in the centre of 7 in. only. This bridge has sometimes been called a suspension bridge because of the flexible chords which connect the ends of the bows; but this circumstance does not in reality bring the bridge as a whole under the suspension principle. The section of the bow-shaped tube is an ellipse, of which the horizontal diameter is 16 ft. 10 in. and the vertical diameter 12 ft., and the rise in the centre about 30 ft. Beside the two fine spans which overleap the river, the bridge is prolonged on each side by a number of piers, on which rest ordinary girders, making its total length 2,240 ft., or nearly half a mile; 2,700 tons of iron were used in the construction. As in the case of the Britannia Bridge, the tubes were floated to the piers, and then raised by hydraulic pressure to their position 150 ft. above the level of the water. The bridge was opened by the late Prince Consort in 1860, and has received the name of the Albert Bridge.

SUSPENSION BRIDGES.

THE general principle of the suspension bridge is exemplified in a chain hanging between two fixed points on the same level. If two chains were placed parallel to each other, a roadway for a bridge might be formed by laying planks across the chains, but there would necessarily be a steep descent to the centre and a steep ascent on the other side. And it would be quite impossible by any amount of force to stretch the chains into a straight line, for their weight would always produce a considerable deflection. Indeed, even a short piece of thin cord cannot be stretched horizontally into a perfectly straight line. It was, therefore, a happy thought which occurred to some one, to hang a roadway from the chains, so that it might be quite level, although they preserved the necessary curve. In designing such bridges, the engineer considers the platform or roadway as itself constituting part of the chain, and adjusts the loads in such a manner that the whole shall be in equilibrium, so that if the platform were cut into sections, the level of the road would not be impaired.

Public attention was first strongly drawn to suspension bridges by the engineer Telford, who, in 1818, undertook to throw such a bridge across the Menai Straits, and the work was actually commenced in the following year. The Menai Straits Suspension Bridge has been so often described, that it will be unnecessary to enter here into a lengthy account of it, especially as space must be reserved for some description of other bridges of greater spans. The total length of this bridge is 1,710 ft. The piers are built of grey Anglesea marble, and rise 153 ft. above the high-water line. The distance between their centres is 579 ft. 10½ in., and the centres of the main chains which depend from them are 43 ft. below the line joining the points of suspension. The roadway is 102 ft. above the high-water level, and it has a breadth of 28 ft., divided into two carriage-ways separated by a foot-track. The chains are formed of flat wrought iron bars, 9 ft. long, 3¼ in. broad, and 1 in. thick. In the main chains, of which there are sixteen, no fewer than eighty such bars are found at any point of the cross section, for each link is formed of five bars. These bars are joined by cross-bolts 3 in. in diameter. The main chains are connected by eight

FIG. 146.—*Clifton Suspension Bridge, near Bristol.*

.transverse stays formed of cast iron tubes, through which pass wrought iron bolts, and there are also diagonal ties joining the ends of the transverse stays. The time occupied in the construction was $6\frac{1}{2}$ years, and the cost was £120,000. This bridge has always been regarded with interest for being the first example of a bridge on the suspension principle carried out on the large scale, and also for its great utility to the public, who, instead of the hazardous passage over an often stormy strait, have now the advantage of a safe and level roadway.

The Clifton Suspension Bridge over the Avon, near Bristol, is noted for having a wider span than any other bridge in Great Britain, and it is remarkable also for the great height of its roadway. The distance between the centres of the piers—that is, the distance of the points between which the chains are suspended—is more than 702 ft. Part of the ironwork for this bridge was supplied from the materials of a suspension bridge which formerly crossed the Thames at London, and was removed to make room for the structure which now carries the railway over the river to the Charing Cross terminus. Five hundred additional tons of ironwork were used in the construction of the Clifton Bridge, which is not only much longer than the old Hungerford Bridge, but has its platform of more than double the width, viz., 31 ft. wide, instead of 14 ft. A view of this bridge is given in Fig. 146, where its platform is seen stretching from one precipitous bank of the rocky Avon to the other, and the river placidly flowing more than 200 ft. below the roadway. The picturesque surroundings of this elegant structure greatly enhance its appearance, and the view looking south from the centre of the bridge itself is greatly admired, although the position may be at first a little trying to a spectator with weak nerves. The work is also of great public convenience, as it affords the inhabitants of the elevated grounds about Clifton a direct communication between Gloucestershire and Somersetshire, thus avoiding the circuitous route through Bristol, which was required before the completion of the bridge

The use of iron wire instead of wrought bars has enabled engineers to far exceed the spans of the bridges already described. The table on page 199 shows that iron wire has a tenacity nearly one-third greater than that of iron bars, and this property has been taken advantage of in the suspension bridge which M. Chaley has thrown over the valley at Fribourg, in Switzerland. This bridge has a span of no less than 880 ft., and is constructed entirely of iron wires scarcely more than $\frac{1}{10}$ in. in diameter. The main suspension cables, of which there are two on each side, are formed of 1,056 threads of wire, and have a circular section of $5\frac{1}{2}$ in. diameter. The length of each cable is 1,228 ft., and at intervals of 2 ft. the wires are firmly bound together, so as to preserve its circular form. But as the cable approaches the piers, the wires are separated, and the two cables on each side unite by the spreading out of the wires into one flat band of parallel wire, which passes over the rollers at the top of the piers, and is again divided into eight smaller cables, which are securely moored to the ground. Each of the mooring cables is 4 in. in diameter, and is composed of 528 wires. In order to obtain a secure attachment for the mooring cables, shafts were sunk in the solid rock 52 ft. deep, and the ingenious mode in which, by means of inverted arches, an anchorage in the solid rock is formed for the cables, will be understood by a reference to Fig. 147. The cables pass downwards through an opening made in each of the middle stones, and are secured at the bottom by stirrup-irons and keys. The suspension piers are built of blocks of stone, very carefully shaped and put together with cramps and ties, so as to constitute most substantial structures. These piers are embellished with columns and entablatures, forming Doric porticoes, enclosing the entrances to the bridge, which are archways 43 ft. high and 19 ft.

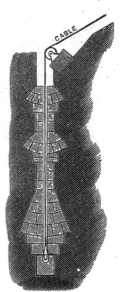

FIG. 147.

wide. The roadway is 21 ft. wide, and is supported on transverse beams, 5 ft. apart, upon which is laid longitudinal planking covered by transverse planking. The roadway beams are suspended to the main cables by vertical wire cables, 1 in. in diameter. The length of these suspension cables of course varies according to their position, the shortest being $\frac{1}{2}$ ft. and the longest 54 ft. in length. Each suspension cable is secured by the doubling back of the wires over a kind of stirrup, through which passes a plate of iron, supported by the two suspension cables, the latter being close together, and, indeed, only separated by the thickness of the suspension cables, which hang between them. The roadway has a slight rise towards the centre, its middle point being from 20 to 40 in. above the level of the ends, according to the temperature.

To test the stability of the bridge, fifteen heavy pieces of artillery, accompanied by fifty horses and 300 people, were made to traverse it at various speeds, and the results were entirely satisfactory. Indeed, a few years afterwards the people of Fribourg had another wire bridge thrown over the gorge of Gotteron, at about a mile from the former. This, though not so long (640 ft.), spans the chasm at a great height, and in this respect is probably not surpassed by any bridge in the world—certainly not by any the length of which can compare with its own. The height of the roadway above the valley is 317 ft., or about the same as that of the golden gallery of St. Paul's Cathedral above the street. The structure is very light, and the sensation experienced when, looking *vertically* downwards through the spaces between the flooring boards, you see the people below diminished to the apparent size of flies, and actually feel yourself suspended in mid-air, is very peculiar, as the writer can testify.

The Americans have, however, outspanned all the rest of the world in their wire suspension bridges. They have thrown a suspension bridge of 800 ft. span over the Niagara at a height of 260 ft. above the water, to carry not only a roadway for ordinary traffic, but a railway. Suspension bridges are not well adapted for the latter purpose, but there seemed no other solution of the problem possible under the circumstances. The bridge, however, combines to a certain extent the girder with the suspension principle. The girder which *hangs* from the main cables (for they are made of wire), carries the railway, and below this is the suspended roadway for passengers and ordinary carriages. The engineer of this work was Roebling, who also designed many other suspension bridges in America.

The spans of any European bridges are far exceeded by that of the wire suspension bridge which crosses the Ohio River at Cincinnati, with a stride of more than 1,000 ft. ; and this is, in its turn, surpassed by another bridge which has been thrown over the Niagara. This bridge, which must not be confounded with the one mentioned above, or with the Clifton Bridge in England already described, merits a detailed description from the audacity of its span, which is nearly a quarter of a mile, and entitles it to the distinction of being the longest bridge in the world of one span.

The new suspension bridge at the Niagara Falls, called the Clifton Bridge, of which a view is given in Fig. 147*a*, is intended for the use of passengers and carriages visiting the Falls, and it is also the means of more direct communication between several small towns near the banks of the river. The bridge is situated a short distance below the Falls, crossing the river at right angles to its course at a point where the rocks which form the banks are about 1,200 ft. apart. The distance between the centres of the towers is 1,268 ft. 4 in., and the bridge has by far the longest single span of any bridge in the world, the distance between the points of suspension being more than twice that of the Menai Bridge, and more than six times the span of the widest stone bridge in England. This remarkable suspension bridge was constructed by Mr. Samuel Keefer, and was opened for traffic on the 1st of January, 1869, the actual time employed in the work having been only twelve months. The cables and suspenders are made of wire, which was drawn in England at Warrington and Manchester, and the wires for the main cables were made of such a length, that each wire passed from end to end of the cable without weld or splice. The length of each of the two main cables is 1,888 ft., and of this length 1,286 ft. usually hangs between the suspending towers, the centre being about 90 ft. below the level of the points of suspension. This last distance, however,

FIG. 147a.—*Clifton Suspension Bridge, Niagara.*

varies considerably with the temperature, for in winter the contraction produced by the cold brings up the centre to 89 ft. below the level line, while in summer it may be 3 ft. lower. The centre of the bridge is about 190 ft. above the water in summer, and 193 ft. in winter. The cables are each formed of seven wire ropes, and each rope consists of seven strands, each strand containing nineteen No. 9 Birmingham gauge wires of the diameter of 0·155 in. The cables of this bridge do not hang in vertical planes, since in the centre they are only 12 ft. apart; while at the towers, where they pass over the suspension rollers, they are 42 ft. apart. The end of the platform which rests on the right bank is 5 ft. higher than the other, and if a straight line were drawn from one end to the other, the centre of the roadway would be in winter 7 ft. above it, and in summer 4 ft. From each point of suspension twelve wire ropes, called "stays," pass directly to certain points of the platform. The stays are not attached to the cables, but pass over rollers on the tops of the towers, and are anchored in the rock, independently of the cables. The longest stays are tangential to the curve formed by the main cables, and they are fixed to the platform at a point about half-way to the centre. Other stays proceed from the platform at intervals of 25 ft., between the longest and the end of the bridge. The thickness of the stays is varied according to the strain they have to bear, and they form not only a great additional support to the platform, but they also serve to stiffen the bridge and lessen the horizontal oscillations to which the platform would be liable from the shifting loads it has to bear. There are also stays which transversely connect the two cables. The wire ropes by which the platform is suspended to the main cables are $\frac{5}{8}$ths of an inch in diameter, and have such a strength that the material would only yield to a strain of 10 tons. These suspenders are placed 5 ft. apart and are 480 in number, the lengths, of course, being different according to the position. To each pair of suspenders is attached a transverse beam, $13\frac{1}{2}$ ft. long, 10 in. deep, and $2\frac{1}{2}$ in. wide. Upon these beams—which are, of course, 5 ft. apart from centre to centre—rests the flooring, formed of two layers of pine planking $1\frac{1}{2}$ in. thick; and the roadway thus formed constitutes a single track 10 ft. in width. Along each side of the platform is a truss the whole length of the bridge, formed of an upper and a lower beam, $6\frac{1}{4}$ ft. apart, united by ties and diagonal pieces. The lower chord of the truss is 2 ft. below the road, and on it rolled iron bars are bolted continuously from one end of the bridge to the other. The last arrangement contributes greatly to stiffen the platform, vertically and horizontally. In the central part of the bridge the flooring-boards are bolted up to the cables, and there are studs formed of 2 in. iron tubes, so that the platform cannot be lifted vertically without raising the cables also; and as thus 81 tons of the weight of the cables vertically rest upon the platform, great steadiness is secured, inasmuch as the central part of the cables must partake of any movement of the platform, and their weight greatly increases the inertia to be overcome. In order still further to prevent oscillations as much as possible, a number of "guys" are attached to the bridge. These are wire ropes of the same thickness as the suspenders, and they connect the platform with various points of the bank—some going horizontally to the summit of the cliffs, others vertically, but the majority obliquely. There are twenty-eight guys on the side of the bridge next the falls, and twenty-six on the other side. The thickness of the wire rope of which they are made being little more than $\frac{1}{2}$ in., they are scarcely visible, or rather appear like spider lines. About 400 ft. of the length of the bridge in the centre is without either

19

guys or stays except two small steel ropes, which, tightly strained from cliff to cliff, cross each other nearly at right angles at the centre of the bridge. The suspension towers are pyramidal in form and are built of white pine, the timbers being a foot square in section and very solidly put together, so that they are capable of bearing forty times the load which can ever be put upon them. The towers are surmounted by strong frames of cast iron, to which are fixed the rollers carrying the cables and stays to their anchorage. The weight of the bridge itself, together with the greatest load it can be required to bear, amounts to 363 tons. Its cost was £22,000, and it was constructed without a single accident of any kind.

The foam of the great falls is carried by the stream beneath the bridge, and in sunshine the spectator who places himself on the centre of its platform sees in the spray driven by the wind, not a mere fragment of a rainbow, or a semicircular arc, but the complete circle, half of which appears beneath his feet. The gorge of the Niagara is very liable to furious blasts of winds, for by its conformation it seems to gather the aërial currents into a focus, so that a gentle breeze passing over the surrounding country is here converted into a strong gale, sweeping down with great force between the precipitous banks of the river. Indeed, one would suppose that the cavern from which Æolus allows the winds to rush out, must be situated near Niagara Falls. The bridge is not disturbed by ordinary winds, although during its construction, before the stays and guys were fixed, it was subject to considerable displacement from this cause. The peculiar arrangement of the cables, by which they hang, not vertically, but widening out from the centre of the bridge, giving what has been termed the "cradle" form, has proved of the highest advantage, so that, with the aid of the guys and stays, and the plan of attaching the central part of the roadway to the cables, the bridge is believed to be capable of withstanding without damage a gale having the force of 30 lbs. per square foot, although its total pressure on the structure might then amount to more than 100 tons. The stability of the structure was severely tested soon after its erection by a furious gale from the south-west, by which the guys were severely strained; in fact, many of them gave way. In one case an enormous block of stone, 32 tons in weight, to which one of the guys was moored, was dragged up and moved 10 ft. nearer the bridge. This and some lateral distortion of the platform, which was easily remedied, was all the damage sustained by the bridge. By an increase of the strength of the guys, &c., and the addition of the two diagonal steel wire ropes mentioned above, the bridge was soon made stronger than before. Some years ago, when the Menai suspension bridge was exposed to a storm of like severity, that structure suffered great damage, the platform having been broken and some of it swept away. In the great gale which swept down upon the Niagara bridge, although the force of the wind was so great that passengers and carriages could not make headway, the vertical oscillations of the bridge never exceeded 18 in., an amount which must be considered extremely satisfactory in a bridge of the kind, having a span of nearly a quarter of a mile.*

* Notwithstanding the skill displayed in its construction, this bridge has, since the above account was written, been destroyed by a tremendous hurricane

CANTILEVER BRIDGES.

THE great Forth Bridge, now (December, 1889) approaching completion, is the first bridge on the cantilever and central girder principle that has been erected in Great Britain, and it has also the distinction of being by far the widest spanned bridge in all the world. We are told by the engineers of the bridge that the cantilever and girder principle is by no means new, for it has been adopted hundreds of years ago by comparatively rude tribes in the construction of timber bridges, to which it readily lends itself. Such bridges are described as having been erected by the natives of Hindoostan, Canada, Thibet, etc., even at remote periods. The principle of the cantilever and girder construction was well illustrated by Mr. Baker, one of the engineers of the bridge, at a lecture given by him at the Royal Institution, by means of what he termed "a living

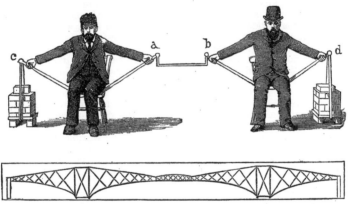

FIG. 147*b*.—*Living Model of the Cantilever Principle.*

model," of which (Fig. 147 *b*) shows the general arrangement. Two men, seated on chairs, extend their arms and hold in their hands sticks, of which the other ends butt against the chairs. The central girder is represented by a shorter stick, suspended at *a* and *b*. We have here the representation of two double cantilevers, the ropes at *c* and *d*, connected with the weights, representing the anchorages of the landward arms of the cantilevers. When a weight is placed on *a b*, which was done in the "living model," by a third man seating himself thereon, a tensile strain comes into action in the ropes and in the men's arms, while the sticks abutting on the chairs have to resist a compressing force, and the weight of the whole is borne by the legs of the chairs, also under compression. Now let the reader imagine the men's heads to be 360 feet above the ground, and about a third of a mile apart, while the distance between *a* and *b* is 350 feet, and he will have a rough but sufficiently clear idea, not only of the principle upon which the Forth Bridge is constructed, but also of the magnitude of one of its spans. To complete the comparison, Mr. Baker further invited his hearers to suppose that the pull upon each arm of the men is equal to 10,000 tons, and that the legs of each chair press on the ground with the weight of more than 100,000 tons.

The Forth Bridge spans the estuary at Queensferry nine miles north-west from Edinburgh, and its purpose is to afford uninterrupted railway communication along the eastern side of Scotland. It will, in effect, shorten the railway journey between Edinburgh and Perth, or Aberdeen, by nearly two hours. Queensferry had long been established as a usual place for crossing the Forth, and readers of Scott's "Antiquary" will remember that the first chapter describes how Monkbarns and Lovel, by some accidental delays to the coach, lost the tide, and had to wait, to sail "with the tide of ebb and the evening breeze," finding themselves, in the meanwhile, pretty comfortable over a good dinner at the "Hawes Inn." This inn still stands, its situation being close to the southern end of the great bridge. A design for the erection of a light suspension bridge at the same spot was published at the beginning of the present century, but although the spans were to be equal to those of the present bridge (17,000 feet), the different scale of the projects may be inferred from the total weight of iron to be used being estimated at 200 tons, while 50,000 tons will be required for the structure now approaching completion.

In 1873, an Act of Parliament was obtained authorizing the construction of a suspension bridge at Queensferry, to carry the railway over the estuary. The design comprised practically two bridges, each carrying a single line of rails, the bridges being braced together at intervals. The central towers were to have been 600 feet high, or about 100 feet loftier than any other erection then existing in the world. The designer was the late Sir Thomas Bouch, and preparations were made for carrying out the plans by the erection of workshops and the manufacture of bricks for the piers. But the project was knocked on the head by the terrible disaster at the Tay Bridge, in December, 1879, when several of the central piers were overturned by the force of the wind, with swift destruction to a passing train, which was precipitated into the water, and every one of about ninety persons in the train perished. Sir Thomas Bouch having been the designer of the Tay Bridge, public confidence in his plan was shaken to such an extent, that the four railway companies who were promoting the construction of the suspension bridge abandoned the project in favour of a design on the cantilever and central girder system, which was then brought forward by Mr. (now Sir John) Fowler and Mr. Baker. When the Bessemer process had made steel attainable at a cheap rate, these engineers recognized the advantages which cantilever bridges, made of that material, presented for the wide spans required for carrying railways across navigable rivers, and in 1865 they had designed such a bridge, with 1,000 feet spans for a viaduct, across the Severn, near the position of the present tunnel. It was not, however, until 1881 that the designs for the Forth Bridge were published in English and American engineering journals. These designs at once attracted attention, and scarcely a year had elapsed before a railway bridge was built for the Canadian and Pacific Railway, on the same principle, and this has been followed by others since. It is, however, absurd to allege that the engineers took their ideas from America, merely because these smaller undertakings have been completed before the great work that dwarfs them all was open for traffic. The construction of the Forth Bridge on its present design was commenced in January, 1883. Its site at Queensferry is at a point where the estuary narrows, and where, in the very middle of the channel, there is a small

·PLATE XIII.

THE FORTH BRIDGE,

rocky island, called Inchgarvie, that furnishes a solid foundation for the great central pier. On each side of this island the channels are about one-third of a mile wide, and more than 200 feet deep, and through them the tide rushes with great velocity. The impossibility of building up any intermediate piers, under such circumstances, is sufficiently obvious—the currents must be crossed at one span, if a railway bridge had to be made. The formation of the piers for such a work presented many novel problems, and much of the work had to be commenced in deep water ; that is, the ground of rock or hard clay had to be prepared, in some parts, as far as 90 feet below high water. Each pier stands on four caissons, which are great tubes or drums of iron and steel, filled up with concrete. Each weighed, when empty, about 400 tons, but when filled up with concrete, the weight would be about 3,000 tons. The diameter of each is 70 feet, and the deepest one is sunk 89 feet below the water, and it was with no little labour that some of them were put in their places. Each caisson has an outer and an inner tube, is 70 feet in diameter at the base, and 60 feet at the top. Seven feet from the bottom, an air-tight partition formed a chamber in the lower part of the caisson, about 70 feet in diameter, by 7 feet high, and shafts sufficiently large to admit the passage of men and tools led from the top. Air was forced into this chamber, when the caisson had been sunk, expelling the water, and then men descended through the shafts and locks, in which a high pressure of air was also maintained, and excavated the material at the bottom, until the caisson had, by its own weight, sunk to the depth required. The work in this air chamber was carried on by means of electric lights, and ten or twelve weeks were occupied in sinking each caisson. The pressure of the air in the working chamber was sometimes as high as 35 pounds per square inch, or sufficient to maintain the mercurial column in a barometer 72 inches high, instead of the ordinary 29 or 30 inches. It was found that the labour in the compressed air chamber could not be done by our home workmen, as they were quite unaccustomed to the high air pressures required to keep out the water ; but arrangements were made for the assistance of a staff of French workmen, inured to the conditions by long working under water in the construction of the docks at Antwerp.

The stores, offices and workshops, situated on a slight eminence near the south end of the bridge, are very extensive, occupying, it is said, an area of 50 acres. Here are great furnaces, cranes and machinery for shaping and fitting the steel plates and bars ready for taking their appointed places in the vast structure. An hydraulic crane may, for instance, be seen lifting a ton weight flat steel plate that has been heated to redness in a regenerative gas furnace, and transfering it to an hydraulic press, where it is quickly and quietly bent to the required shape. The plate is then cooled, and, when the edges have been planed, it is placed in position with the adjoining plates, and the rivet holes are drilled by an ingenious machine, specially designed by Mr. Arrol, the contractor, for that purpose. It works upon 8-feet lengths of the tubes, and simultaneously cuts ten rivet holes at different points in the circumference. All the different parts of the structure are temporarily fitted together to ascertain that every piece is properly adjusted. They are then marked according to the position they are to take, and are laid aside until they are wanted. Thus the work at the bridge has proceeded without any awkward hitches arising from ill adjusted sections being brought together. At times, 1,800 tons of finished steelwork has been turned out of these shops in a month, and this material, which was supplied by the Steel Company of Scotland, has been found thoroughly trustworthy in every respect. Its strength is one-half greater than that of the best wrought-iron, and the plates have thrice the ductility of iron plates. The steel plates for the great tubes are supplied in lengths of 16 feet, and of different thicknesses, between ⅜ths of an inch and 1¼ inch.

The sketch, Fig. 147 c, shows the general dimensions of the bridge proper, or that part of the viaduct which will actually span the estuary. Of the three great piers that support the cantilevers, it will be observed that the central one, which rests on Inchgarvie, is wider than the other two. Each consists mainly of four tubes, 12 feet in diameter, made of plates of steel 1¼ inch in thickness, and these rise to the highest part of the bridge, which is 361 feet above the water, so that the structure is as lofty as St. Paul's Cathedral. These great tubes are not placed vertically, but incline inwards towards the top, so that while the "straddle legs" of each pair are 120 feet apart at the base, they are only 33 feet apart at the top. These lofty columns are also braced together diagonally by other steel

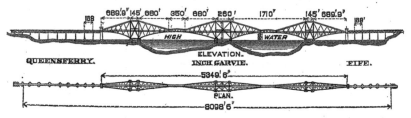

FIG. 147c.—*Principal Dimensions of the Forth Bridge.*

tubes—that is, a tube passes from the foot of every column to each of the other three. At the base of each column, the lowest spanning member springs also (which appears like an arch, but is not so), as a tube of 12 feet diameter. Thus abutting or resting on enormously thick plates of steel that cap the masonry of each pier, are five tubular steel limbs, three of which are 12 feet in diameter, and two are 8 feet; and, besides these five, girder members diverge from nearly the same centre. One of the large tubular members is the first strut that rises obliquely to support the upper structure. From the point where this strut meets the upper member, a stay passes downwards with an opposite inclination to the lower member, from its point of junction with which another strut rises, and so on. All the struts, as being subject to compressing force, are made of steel tubes; the straight upper members and the stays are lattice braced girders of rectangular section. The apparent curve of the lower member —for it is really made up of sections of straight tubes—may suggest the notion of an arch; but the reader must remember that the principle of this bridge has no relation to that of the arch. The cantilevers do not unite the long arms they stretch, but each is an independent structure with its own perfect stability, and it will not be clutched on or locked up to its neighbours by the central girders. The weight of one of these 1,700 feet spans is about 16,000 tons, and the heaviest train loads might be two coal trains, weighing together, say 800 tons, or only one-twentieth of the dead weight of the structure. But, what would not generally be supposed, the pressure of the wind is an element of much more importance in considering the stability of the bridge than the weight of the rolling load. It is to resist the wind pressure that the lofty columns that are only 33 feet apart at the top across the bridge, plant their bases 120 feet asunder. The estimated lateral pressure of the wind on one of the cantilevers, assuming it as equal to 56 lbs. per square foot, would amount to 2,000 tons. These strains are so fully provided for that the engineers are confident that a hurricane of such a force as would desolate the country would leave the Forth Bridge intact, even if the wind blew in opposite directions on the two arms of the cantilever. To rend asunder the top ties, a pull equivalent to the weight of 45,000 tons would be required, whilst the utmost strain that passing trains could possibly bring upon these ties would be less than 2,000 tons. A striking illustration of the strength of these huge brackets was lately given by Mr. Baker himself, when in a public lecture he assured his audience that half a dozen of our ponderous modern ironclads might be hung from the cantilevers. Everyone knows that a bracket requires to be strongest nearest the base, and the lower steel arms that stretch out 680 feet each diminish in diameter until at the end it has decreased to five feet, and the pairs approach each until, from being 120 feet apart at the base, they are only 33 feet apart at the ends. The central girders will each weigh about 1,000 tons, and only one end of each will be attached to a cantilever, the other ends will simply rest on what are called " rocking columns," so that there may be freedom of motion to allow play for the changes of position that will be induced by changes of temperature expanding or contracting the huge masses of metal.

The reader can hardly have failed to observe that the chief element in the stability of the structure depends upon balancing a great mass of metal on the one side of a pier by an equal mass on the other side. But while each end of the central cantilever bears half the weight of a central girder, the two shoreward cantilevers have this load at their inner ends only. How is their balance maintained? In this way: the shoreward arms are made about 10 feet longer than those that stretch over the water and their extremities are also loaded with about 1,000 tons of iron, built up within the shore piers.

The lofty columns of the piers were erected without any external staging, from a temporary platform surrounding the piers and supporting the necessary machinery. The weight of this platform with the machinery on it was about 400 tons, and as the work proceeded it was raised as required by hydraulic machines placed within the vertical columns. As the height of these increased, the men and materials had to be conveyed to the platform by cages moving between guide ropes and worked by steam engines. From this platform were constructed not only the main columns, but the great diagonal tubes, the bracing girders, and the viaduct girder. The cantilevers were also put together without scaffolding. When the first few feet of the lower member had been built out from the base, a movable platform was hung round it, and on this platform were the cranes for putting the plates into position, the furnace for heating the rivets, and the hydraulic riveter of specially designed construction, without noise or hammering, the riveting being completed by the application of a pressure equal to 3 tons per square inch. The building up of the cantilever arms on either side of each pier always proceeded at the same rate, so that the balance was constantly maintained. This building out from each side of the pier, without the necessity of relying upon any temporary scaffolding from below, is one great advantage of the cantilever system, as it is both easier and safer than a system which relies upon the temporary scaffolding raised from below. The Forth is for the time the longest spanned bridge in the world; but it may not retain that honour long, for the legislature of the United States has already authorized the construction of a cantilever bridge, the spans of which are to be 2,480 feet. Still more gigantic is the project lately put forward by some competent French engineers of bridging the English Channel from Folkestone to Cape Grisnez in 70 spans on the cantilever system. The designs have been completed and the calculations made, and no one doubts of the engineering practicability of the scheme. But the cost is estimated at about 34 million pounds sterling, or nearly six times as much as that required for constructing the proposed Channel Tunnel; so that the scale could be turned in favour of the bridge only if the political reasons that were opposed to the tunnel were held not to be applicable to the bridge. But it is difficult to conceive that the existing traffic could ever be developed to such an extent as to make an undertaking of this magnitude a commercial success.

Since the above account was written, the Forth Bridge was formally opened on the 4th March, 1890, by the Prince of Wales, in the presence of a great gathering of railway directors, eminent engineers, and other distinguished persons from all parts. A very strong gale was blowing at the time, and at this very hour the bridge was therefore subjected to another severe but undesigned test of its stability. The perfect steadiness and security of the structure impressed all who were present on that occasion, and the train crossed the bridge, exposed to a wind pressure, registered by the gauge, of 25 lbs. per square foot. At the luncheon following the opening ceremony, the Prince announced that baronetcies had been conferred upon Mr. M. W. Thompson (the chairman of the Bridge Company) and upon Sir John Fowler, and that Mr. Baker and Mr. Arrol, the contractor for the works, were to be knighted. Sir John Fowler, the engineer-in-chief, was born in 1817, and has been engaged in many other important works of railway construction in Yorkshire, in that of the London and Brighton Railway, in the Sheffield Waterworks, &c. The Metropolitan Railway in London, which also was carried out by Sir John Fowler, would alone suffice to make him famous as an engineer. Sir Benjamin Baker is a much younger man, who has had a large and varied practice in railway engineering in various parts of the world. He is in much request on the American continent, and is now engaged in carrying out a ship railway in Canada and a tunnel under the Hudson at New York. Sir William Arrol began life at nine years of age as a "piecer" in a cotton mill, but was afterwards apprenticed as an engineer. Subsequently he was employed as a foreman by engineering firms in Glasgow. In 1866, he began business on his own account at Dalmarnock, and obtained contracts at first for smaller then for larger works connected with bridge and viaduct building. He is distinguished for the energy and inventive resources he displays in carrying out his undertakings.

THE TOWER BRIDGE, LONDON.

A LITTLE more than four years after the opening of the Forth Bridge, in June 1894, another great enterprise which had been commenced eight years before, was inaugurated by the Prince and Princess of Wales. as representatives of Her Majesty the Queen. This was the Tower Bridge, which not only is one of the most important public works of the century, but one that presents features of interest and novelty that have never before been combined in any single structure. The want of an adequate communication between the shores of the Thames eastward of London Bridge had long been felt, and was for years a subject of serious consideration for the Metropolitan authorities. The congested state of the traffic across London Bridge has often furnished a spectacle for the sight-seer, and figures are not wanting to show that the number of foot-passengers alone who daily traverse that bridge, which altogether is only 54 feet wide, would be equal to the whole population of many considerable cities : for in 1882 a count showed the daily average of pedestrians to be 110,525, while the number of vehicles was 22,242. There was much difference of opinion as to the best method of providing the required means of communication ; but there was an almost universal agreement as to its position being selected just eastward of the Tower of London. The map of the districts connected by the Tower Bridge which is given in Fig. 147*d*, will show a reader who has any acquaintance with London the suitability of the site. The problem of traversing the river at this point involved complex conditions as affecting the vehicular traffic and the navigation, and many different schemes were proposed and examined, comprised under the three heads of bridges, tunnels and ferries. But a ferry is always an imperfect means of communication, liable to accidents and interruptions from fogs, and in severe weather from ice, rendering the transit impossible for sometimes many days together. A tunnel beneath the river would, of course, leave the navigation without impediment, but among its special disadvantages are the great expense of construction and maintenance, for it has been found that tunnels beneath water-ways are very costly in both respects. Besides, there would have to be long inclined approaches at each end, and the cost would be enormously increased by the amount of valuable land these would occupy. It was indeed proposed that the tunnel should be provided instead with hydraulic lifts at each end, like those often found in connection with the sub-ways at railway stations ; but such would have to be of Brobdignagian dimensions, and would daily entail heavy expense. Then, as regards the bridges, schemes of various kinds were proposed, some even bridging the whole 850 feet width of the river at a single span, but all distinguishable by these important characteristics : they either provided a high level roadway which requires long inclines to reach it, but permitted lofty-masted ships to pass under it ; or, on the other hand, the roadway was to be made at a low level with a clear headway above the water of moderate height. While avoiding the inclined approaches, this plan would either prevent fully rigged vessels passing to the wharves above the bridge, or some part of the structure would have to open or swing aside, that the ships might pass through the opening, thus completely interrupting the pedestrian and vehicular traffic for the time, with an amount of inconvenience that may be imagined when, as often happens, twenty large ships or more might pass in the course of a day, each causing a stoppage of five minutes in the road traffic. Nor would it be without risks that large vessels could pass through a comparatively narrow opening in a strong tide-way. Plans for sub-ways, for high level roadways and for low level roadways, were examined by Parliamentary Committees when powers to construct the works were successively applied for by the Metropolitan authorities, and much valuable evidence having been given, such objectionable features of each scheme as have been already referred to were duly noted. At length in 1878, Mr. Horace Jones, the late architect to the City of London, in a report on the various projects, suggested the general plan on which the present bridge is built, and this having been approved of by the Common Council, steps were taken to obtain Parliamentary powers to raise the necessary capital and to proceed with the works ; but, for various reasons, it was not until 1885 that the Act authorising the undertaking was passed. In the meantime Mr. John Wolfe Barry was appointed engineer of the structure, while Mr. Jones was to superintend the architectural details ; but after having received the honour of knighthood in 1885, he died in the same year ; and Mr. Barry, reconsidering the joint design, introduced some new features and somewhat modified the architectural expression of the structure. One striking point of originality about the Tower Bridge is that while it is essentially an iron and steel construction as much as the Forth Bridge, the heavy stiff metal-work is encased in masonry of elegant and appropriate architectural design, by which the general desire that the bridge should harmonize so far as might be, with the ancient historical fortress it adjoins, has been happily realised. Then again, by the ingenious engineering, the public have the advantage of a low level roadway, while the largest vessels may pass freely through a wide space without risk. These apparently incompatible advantages have been obtained by the adoption of what is the *bascule* principle on a hitherto unattempted scale. *Bascule* is a French engineering term, which is probably less familiar to most of our readers than the thing itself. It is applied to the platform of a draw-bridge which turns as the lid of a box does on its hinges, to afford a passage over the stream or moat when it is horizontal, and when drawn up vertically denies such passage. Smaller *bascule* bridges on exactly the same plan as in the Tower Bridge may often be seen in places having docks or canals, such as Hull, &c. In these a flap or platform is let down from each side from the vertical position, in which the water-way is open until the free edges meet together to form the roadway. These platforms turn on horizontal pivots, and are counterpoised by loads of stone or metal, so that they are without difficulty raised and lowered by a winch or handle that turns a cogged pinion engaging the teeth of a large quadrant.

The following general description of the Tower Bridge is mainly abstracted from a very full and excellent account of it drawn up in 1894 by Mr. J. E. Tuit, engineer to Sir W. Arrol & Co., the contractors, in which are embraced the whole of the technical details of the structure. The map, Fig. 147*d*, shows the site of the bridge and its approaches, of which the northern one begins close to the mint and passes along the east side of the Tower of London to the northern abutment. This approach is formed of a series of brick arches, and is nearly 1,000 feet long and 35 feet wide in the roadway, with a footpath 12½ feet wide on either side of it. The

PLATE XIV.

THE TOWER BRIDGE IN COURSE OF CONSTRUCTION.

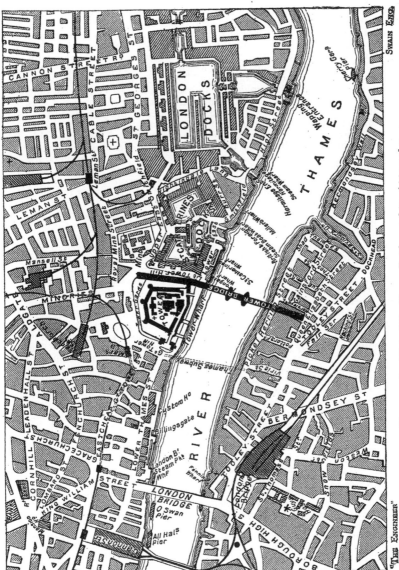

FIG. 147d.—Map of the Tower Bridge and its Approaches.

incline is only a rise of 1 in 60, but the southern approach is slightly steeper, namely, 1 in 40 leaving the street level at Tooley Street. At each abutment there are also stairs connecting the banks of the river with the roadway of the bridge. The width of the river between the two abutments is 880 feet, and this is divided, as shown in Fig. 147e, into two side spans, each 270 feet wide, and one central span of 200 feet clear, making together 740 feet, the river piers, each of which is 70 feet wide, completing the total span. The clear headway above high water, when the bascules or leaves are down, is, in the middle span, 29½ feet in the centre, but only 15 feet at the ends; but when the leaves are raised for ships to pass, it is about 143 feet. The headway at the shore sides of the piers is 27 feet, but this is lessened to 23 feet and 20 feet at the north and south abutments respectively. The roadway and footpaths are continued along the side spans of the same width as on the approaches, but over the central span the road is 32 feet, and each footway 8½ feet wide. The river piers are said to be the largest in the world of the same kind, and their great area was necessitated by the nature of the London clay on which they rest, which was found incapable of bearing a load much exceeding four tons per square foot without some risk of undue settlement.

The part of the piers below the bed of the river is formed of concrete, while the upper part is brickwork, set in cement and faced with Cornish granite. Upon each of the river piers rest four octagonal columns, built up of flat steel plates, connected together at their edges by splayed angle-bars. The columns are 120 feet high, and 5½ feet in diameter; those on each pier are securely braced together, at certain stages also by plate girders, 6 feet deep, to form a floor or landing, and the tops of the columns are similarly joined together. At the height of 143 feet above high water there are two footways, each 12 feet wide and 230 feet long, carried on girders over the central span, and supported by the columns on each pier. It must be noted that all the roadway, and, in fact, all the practical and useful structure of the bridge, depend upon the steel-work alone, which is supported mainly by the eight octagonal columns just mentioned. The architectural features, which so appropriately clothe all the steel columns, are added for æsthetic considerations, and their masonry takes no part in bearing the weights and strains of the structure. Indeed, the stone-work of the towers is carefully separated from the columns, which were covered with canvas while the masonry was built round them, and spaces were left at every point where compression of the steel-work would bring weight upon the stone-work. This investment of the metal-work by beautiful architecture is, as already mentioned, one of the most original features of the Tower Bridge. The view of the work in progress, as given in Plate VIII., which is one of the many beautiful illustrations in Mr. Tuit's book, will give the reader an opportunity of judging how much the structure gains in sightliness by the addition of the architectural features. Two hydraulic lifts are placed in each tower to convey pedestrians to and from the higher level footways, when the moving parts of the bridge are open, and stairs also are provided for the same purpose for those who prefer them to using the lifts.

The side spans are really suspension bridges, but the chains have only two links, connected at the lowest point by a pin 2½ feet in diameter, while their higher ends are supported on the columns of the piers, and on similar but shorter columns on the abutments. The horizontal pulls of the chains on the piers are made to balance each other by connecting the

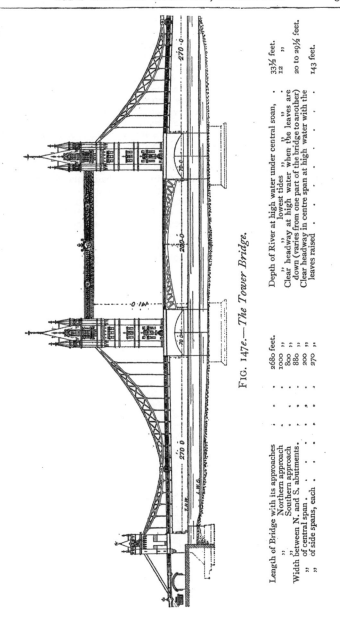

FIG. 147e.—The Tower Bridge.

Length of Bridge with its approaches . . . . 2680 feet.
"   Northern approach . . . . 1000 "
"   Southern approach . . . . 800 "
Width between N. and S. abutments. . . . 880 "
"   of central span . . . . 200 "
"   of side spans, each . . . . 270 "

Depth of River at high water under central soan, . 33½ feet.
"   "   lowest tides ". . . 12 "
Clear headway at high water when the leaves are
 down (varies from one part of the bridge to another) 20 to 29½ feet.
Clear headway in centre span at high water with the
 leaves raised . . . . 143 feet.

chains to tie bars stretching across the central span, and the landward ends of the chains, after passing over the lower columns of the abutments, are securely anchored in enormous masses of concrete.

Each of the opening parts, or *bascules*, or leaves, as they may be called, consists of four girders 18½ feet apart, rigidly braced together, and connected at the pier end with a great shaft, 48 feet long and 1 foot 9 inches in diameter, which turns in massive bearings, resting upon four fixed girders. The leaf is counterbalanced on the shore side of the pivot shaft by 350 tons of lead and iron ; the short leverage of the centre-weight and small space available for it required the greater part of this weight to be of lead, rather than of the less expensive metal. The pivot shaft passes through the centre of gravity of the whole, so that, although the total weight is nearly 1,200 tons, no very great power is required to set it in motion, as the pivot shaft rests on rollers to diminish the friction. The power for moving the leaf is applied to toothed quadrants of 42 feet radius, of which two are fixed to the outside girders of each leaf, and are geared into cogs moved by eight large hydraulic engines, with six accumulators, into which water is pumped by two engines, each of 360 horse-power.

The total length of the bridge, including the approaches, is just half a mile, and the height of the towers from the foundations is 293 feet, so that if one of them were placed beside St. Paul's Cathedral, it would compare with it in height as shown in the sketch, Fig. 147*f.*

FIG. 147*f.—Sketch.*

THE GREAT BROOKLYN BRIDGE.

THE Clifton Bridge at Niagara Falls, which for a time had the distinction of being the longest in span of any suspension bridge in the world, has been fully described in previous pages ; but more recently this bridge has been surpassed in span, and in all other respects, by a structure that immediately connects two of the most populous localities in the United States of America. The Island of Manhattan, which is occupied by the city of New York proper, has a population of nearly two millions, and a strait on its eastern side, connecting Long Island Sound with New York Harbour, alone divides it from the other great seats of population, called respectively Long Island City and Brooklyn. This channel is about ten miles long, and of a varying width, which may average three-quarters of a mile. There are many ferries between the opposite shores, and the waters are busy with steamers, sailing-boats, tugs, and craft of all kinds, engaged either in traffic with ports near at hand, or in trade with distant lands. At the southern end of this strait, near the point of its junction with New York Bay, is the narrowest part of its course, and it is here that it is crossed by the magnificent suspension bridge, known indifferently as the East River Bridge, or Brooklyn Bridge, which provides land communication between New York, with its population of two millions, and Brooklyn, the fourth city of the States in point of size, with inhabitants numbering about one million. Brooklyn is largely a residential place for persons whose daily business is in New York. It has wide, well-planned streets, many shaded by the luxuriant foliage of double rows of trees, and possesses parks, public buildings, institutes, churches, etc., on a scale commensurate with its importance.

The central span of Brooklyn Bridge, from tower to tower, is 1,595 feet, and each shore part, extending from the tower to the anchorage of the cables, is 930 feet span, while the two approaches beyond the anchorage together add 2,534 feet to the total length, which is 5,989 feet, or considerably over a mile. The centre span, it will be observed, is much greater than that of the Niagara Falls Clifton Bridge, which was less than one quarter of a mile, whereas the Brooklyn Bridge span extends to something approaching one-third of a mile, or, more exactly, a few yards longer than three-tenths. The width of the Brooklyn is another one of its remarkable features, for this is no less than 85 feet, and includes two roadways for ordinary vehicles, and two tramway tracks, on which the carriages are moved by an endless cable, worked by a stationary engine on the Brooklyn side. There is also a foot-path, 13 feet wide, for pedestrians. In this structure, as in many other suspension bridges, advantage has been taken of the great tenacity of steel wire as compared with iron bars. But here the wires are not twisted in strands like ropes, but are laid straight together, and bound into a cylindrical form, each wire being 3,572 feet long, and extending from end to end of the cables, which are four in number, each calculated to bear a strain of 12,200 tons. The number of wires in each cable is very great, for instead of about the thousand of which the stranded wire cables usually consist, there are 5,296 steel wires wrapped closely round, and forming a cylinder 15¾ inches in diameter. Each wire is galvanised, that is, coated with zinc, and then coated with oil. The towers over which the cables pass are of masonry, and rise to 272 feet above high-water; their dimensions at the water level are 140 feet by 50 feet, which offsets diminish until at the top they are 120 feet by 40 feet. At the anchor structures, the cables enter the masonry at nearly 80 feet above high-water, and pass 28 feet into the stonework for connection with the anchor chains. The anchorages are masses of masonry, measuring at the base 129 feet by 119 feet, and at the top 117 feet by 104 feet, with a height of 89 feet in front and 85 feet in the rear. The weight of each anchor-plate is 23 tons. The roadway of the bridge is suspended from the cables above the buildings and streets between the towers and the anchorages. The approaches, on the Brooklyn side 971 feet, on the New York side 1,563 feet, are carried on stonework arches, which are utilised as warehouses, but where these approaches cross streets, iron bridges are thrown over. The clear headway between the centre of the roadway over the river at high-water is 135 feet, so that there is no obstruction to navigation, and the headway at the towers is 119 feet, so that the roadway rises towards the centre about 3 feet 3 inches in 100 feet. The two towers comprise more than 85,000 cubic yards of masonry, and for various purposes 13,670 tons of concrete were used. The work was commenced in January, 1870, and the first wire was carried across on 29th May, 1877. The bridge was opened to the public on the 24th of May, 1883, and the tramway four months later. The bridge was made free for pedestrians in 1891, and in 1894 the tram-car fares were reduced to five cents (2½d.) for two journeys. In that year, 41,927,122 passengers were carried on the cars. The average number of persons daily crossing the bridge is estimated at about 115,000, although on one day (11th Feb., 1895) as many as 225,645 passengers have been carried on the cars. The cost of the work connected with this great bridge was $15,000,000 (£3,125,000).

In relation to the subject of wide-spanning bridges, the erection has been contemplated of structures which would surpass in magnitude and boldness any of those yet named. Thus, in 1894, the New York Chamber of Commerce proposed to throw across the River Hudson, which washes the western side of New York, a bridge with a clear span of 3,200 feet (six-tenths of a mile), and 500 feet clear height; and the project was declared by an eminent and experienced engineer to be quite feasible.

PLATE XV.

THE BROOKLYN BRIDGE.

FIG. 148.—*Newspaper Printing-Room, with Walter Machines.*

PRINTING MACHINES.

A VOLUME might be filled with descriptions of the machines which in every department of industry have taken the place of slow and laborious manual labour. But if even we selected only such machines as from the beautiful mechanical principles involved in their action, or from their effects in cheapening for everybody the necessaries and comforts of life, might be considered of universal interest, the limits of the space we can afford for this class of inventions would be far exceeded. The machines for spinning, for weaving fabrics, for preparing articles of food, are in themselves worthy of attention; then there is a little machine which in almost every household has superseded one of the most primitive kinds of handwork, and that is the sewing machine. But all these we must pass over, and confine our descriptions of special machines to a class in which the interest is of a still more general and higher character, since their effect in promoting the intellectual progress of mankind is universally acknowledged. We need hardly say that we allude to Printing Presses, and if we add a few lines on printing machines other than those which have given us cheap literature, it is because these other machines also have contributed to the general culture by giving us cheap decorative art, and in their general principles they are so much akin to the former that but little additional description is necessary.

LETTERPRESS PRINTING.

THE manner in which the youthful assistants of printers came to receive their technical appellation of " devils " has been the subject of many ingenious explanations. One of these is to the effect that the earlier productions of the press, having imitated the manuscript characters, the uninitiated supposed the impressions were produced by hand-copying, and in consequence of their rapid production and exact conformity with each other, it was thought that some diabolical agency must have been invoked. Another story relates that one of Caxton's first assistants was a negro boy, who of course soon became identified in the popular mind with an imp from the nether world. A very innocent explanation is put forward in another tale, relating that one of the first English printers had in his employment a boy of the name of De Ville, or Deville, which name was soon corrupted into the now familiar title, and became the inheritance of this youth's successors in the craft. Perhaps a more probable and natural explanation might be found in the personal appearance which the apprentices must have presented, with hands, and no doubt faces also, smeared over with the black

FIG. 149.—*Inking Balls.*   FIG. 150.—*Inking Roller.*

ink which it was their duty to manipulate. For the ink was formerly always laid upon large round pads or balls of leather, stuffed with wool. When these balls, Fig. 149, which were, perhaps, about 12 in. in diameter, had received a charge of ink, the apprentice dabbed the one against the other, working them with a twisting motion, and after having obtained a uniform distribution of the ink on their surfaces with many dexterous flourishes, he applied them to the face of the types with both hands, until all the letters were completely and evenly charged. The operation was very troublesome, and much practice was required before the necessary skill was obtained, while it was always a most difficult matter to keep the balls in good working condition.

The first important step towards the possibility of a printing machine was made, when for these inking balls was substituted a cylindrical roller, mounted on handles, Fig. 150. The body of the roller is of wood, but it is thickly coated with a composition which unites the qualities of elasticity, softness, and readiness to take up the ink and distribute it evenly over the types. The materials used for this composition are chiefly glue and treacle, and sometimes also tar, isinglass, or other substances. Glycerine and various other materials have also been proposed as suitable ingredients for these composition rollers, but it is doubtful whether the original compound is not as efficacious as any yet tried. The composition is not unlike india-rubber in its appearance and some of its properties. Fig. 150 represents equally the mode in which the roller is applied to the type in hand presses, and that in which it is charged with ink, by being moved backwards and forwards over a smooth table upon which the ink has been spread.

From the time of the first appearance of printing presses in Europe down to almost the beginning of the present century, a period of 350 years, no improvement in the construction appears to have been attempted. They were simply wooden presses with screws, on exactly the same plan as the cheese-presses of the period. Earl Stanhope first, in 1798, made a press entirely of iron, and he provided it with an excellent combination of levers, so that the "platen," or flat plate which overlies the paper and receives the pressure, is forced down with great power just when the paper comes in contact with the types. Such presses are capable of turning out about 250 impressions per hour, and it should be noted that the very finest book printing is still done by presses upon this principle. One reason is that in such cases, where it is desired to print with the greatest clearness and depth of colour, the ink employed is much thicker, or stiffer, and requires more thorough distribution and application to the type than a machine can effect. Stanhope's press was not of a kind to meet the desire for rapid production, to which the increasing importance of newspapers gave rise. The first practical success in this direction was achieved by König, who, in 1814, set up for Mr. Walter, the proprietor of the "Times," two machines, by which that newspaper was printed at the rate of 1,100 impressions per hour, the machinery being driven by steam power.

The "Times" of the 28th November, 1814, in the following words made its readers acquainted with the fact that they had in their hands for the first time a newspaper printed by steam power:

" Our journal of this day presents to the public the practical result of the greatest improvement connected with printing since the discovery of the art itself. The reader of this paragraph now holds in his hand one of many thousand impressions of 'The Times' newspaper, which were taken off by a mechanical apparatus. A system of machinery almost organic has been devised and arranged, which, while it relieves the human frame of its most laborious efforts in printing, far exceeds all human powers in rapidity and dispatch. That the magnitude of the invention may be justly appreciated by its effects, we shall inform the public that after the letters are placed by the compositors, and enclosed in what is called the 'form,' little more remains for man to do than to attend upon and watch this unconscious agent in its operations. The machine is then merely supplied with paper, itself places the form, inks it, adjusts the paper to the form newly inked, stamps the sheet, and gives it forth to the hands of the attendant, at the same time withdrawing the form for a fresh coat of ink, which itself again distributes, to meet the ensuing sheet now advancing for impression, and the whole of these complicated acts is performed with such a velocity and simultaneousness of movement that no less than 1,100 sheets are impressed in one hour. That the completion of an invention of this kind, not the effect of chance, but the result of mechanical combinations, methodically arranged in the mind of the artist, should be attended with many obstructions and much delay may be readily admitted. Our share in this event has, indeed, only been the application of the discovery, under an agreement with the patentees, to our own particular business; yet few can conceive, even with this limited interest, the various disappointments and deep anxiety to which we have for a long course of time been subjected. Of the person who made the discovery we have little to add. Sir Christopher Wren's noblest monument is to be found in the building which he erected: so is the best tribute of praise which we are capable of offering to the inventor of the printing machine comprised in the preceding description, which we have feebly sketched, of the powers and utility of his invention. It must suffice to say further, that he is a Saxon by birth, that his name is König, and that the invention has been executed under the direction of his friend and countryman, Bauer."

Each of the machines erected by König for the "Times" printed only one side of the sheet, so that when they had been half printed by one machine, they had then to be passed through the other, in order to be "perfected," as it is technically termed. These machines were greatly improved by Messrs. Applegath and Cowper, who contrived also a modification by which the sheets could be perfected in one and the same machine.

As the principle of these machines has been followed, with more or less

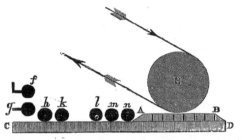

FIG. 151.—*Diagram of Cowper and Applegath's Single Machine.*

diversity of detail, in most of the printing machines at present in use, it is very desirable to lay that principle clearly before the reader. The diagram, Fig. 151, will make the action of Applegath and Cowper's single-printing machine easily understood. The type is set up on a flat form, A B, which occupies part of the horizontal table, C D, the rest of which, A C, is the inking table. E is a large cylinder, covered with woollen cloth, which forms the "blanket." The paper passes round this cylinder, and it is pressed against the form. The small black circles, *f, g, h, k, l, m, n,* represent the rollers for distributing the ink. *f* is called the *ductor* roller. This roller, which revolves slowly, is made of metal, and parallel to it is a plate of metal, having a perfectly straight edge, nearly, but not quite, touching the cylinder, and at the other side, as well as at the extremities, bent upwards, so as to form a kind of trough, to contain the ink, as a reservoir. The slow rotation of the ductor conveys the ink to the next roller, which is covered with composition, and being made to move backwards and forwards between the ductor roller and the table at certain intervals, it is termed the *vibrating* roller. The ink having thus reached the inking-table, is spread evenly thereon by the *distributing rollers, h, k,* and it is taken up from the inking table, as the latter passes under, by the *inking* rollers, *l, m, n.* The table, C D, as a whole is constantly moving right and left in a horizontal direction.

so that the form passes alternately under the impression cylinder, E, and the inking rollers, *l, m, n.* The axles of the inking and distributing rollers are made long and slender, and instead of turning in fixed bearings, they rest in slots or notches, in order that, as the form passes below them, they may be raised, so that they rest on the inking slab, and on the types, only by their own weight. They are placed not quite at right angles to the direction of the table, but a little diagonally. The sliding motion caused by this, helps very much in the uniform spreading of the ink. By these arrangements the form is evenly smeared with ink, since each inking roller passes over it *twice* before it returns to meet the paper under E.

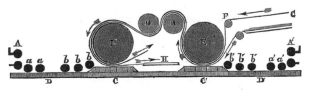

FIG. 152.—*Diagram of Applegath and Cowper's Perfecting Machine.*

Fig. 152 is a similar diagram, to show the action of the double or perfecting printing machine, in which the sheets are printed on both sides. It will be observed that the general arrangement of impression cylinder,

FIG. 153.—*Cowper's Double Cylinder Machine.*

rollers, &c., is represented in duplicate, but reversed in direction. There are also two cylinders, B B, the purpose of which, as may be gathered from an inspection of the diagram, is to reverse the sheets of paper, so that after one side has been printed under the cylinder, E', the blank surface may be turned downward, ready to receive the impression from the form, A B. Fig. 153 gives a view of the Cowper and Applegath double machine, as actually constructed. The man standing up is called the *feeder* or *layer-on.* He pushes the sheets forward, one by one, towards the tapes, which carry them down the farther side of the more distant cylinder, under which they pass, receiving the impression; and so on in the manner already indicated in the diagram, Fig. 152, until finally they reach a point where, released by the separation of the two sets of tapes, they are received by the *taker-off* (the boy who is represented seated on the stool), and are placed by him on a table. The bed or table which carries the form moves alternately right and left, impelled by a pinion acting in a rack beneath it, in such a manner that the direction of the table's motion is changed at the proper moment, while the driving pulley continues to revolve always in the same direction. The movements of the table and of the cylinders are performed in exact harmony with each other, for these pieces are so connected by trains of wheels and rackwork that the sheets of paper may always receive the impression in the proper position as regards the margins, and therefore, when the sheets are printed on both sides, the impressions will be exactly opposite to each other. This gives what is technically called " true register,"

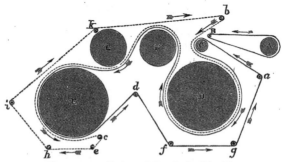

FIG. 154.—*Tapes of Cowper's Machine.*

and as this cannot be secured unless the paper travels over both cylinders at precisely the same rate, these are finished with great care by turning their surfaces in a lathe to exactly the same diameter. The action of the machine will not be fully understood without a glance at the arrangement of the endless tapes which carry the paper on its journey. The course of these may be followed in Fig. 154, and a simple inspection of the diagram will render a tedious description unnecessary.

In Fig. 155 we have a representation of a steam-power printing machine, such as is now very largely used for the ordinary printing of books, newspapers of moderate circulation, hand-bills, &c., and in all the ordinary work of the printing press. In this the table on which the form is placed has a reciprocating motion, but the large cylinder moves continuously always in the same direction. The feeder, or layer-on, places the sheet of paper against certain stops, and at the right moment the sheet is nipped by small steel fingers, and carried forwards to the cylinder, which brings it into contact with the inked type. This is done with much accuracy of register, for the impression cylinders gear in such a manner with the rest of the parts that their revolutions are synchronous. This is a perfecting machine, for the paper, after having received the impression on one side, is carried by tapes round the other cylinder, where it receives the impression on the other side, " set-off sheets " being passed through the press at the same time. The axles of the impression cylinders are mounted at the

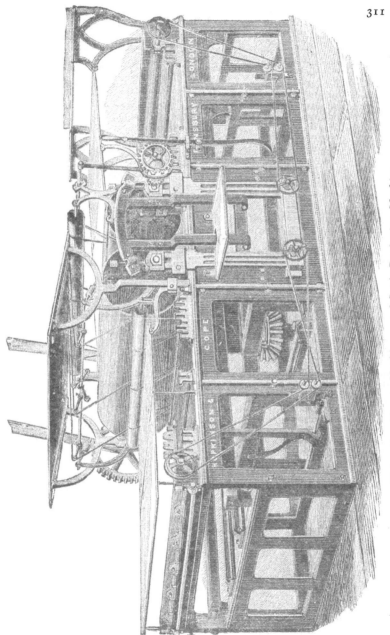

FIG. 155.—*Messrs. Hopkinson and Cope's Perfecting Machine.*

ends of short rocking beams, by small oscillations of which the cylinders are alternately brought down upon, or lifted off, the form passing below them. A machine of this kind can print 900 impressions per hour, even of good book-work, and for newspaper or other printing, where less accuracy and finish are required, it may be driven at such a rate as to produce 1,400 perfected impressions per hour.

The machines used for lithographic printing by steam power are almost identical in their general arrangement with that just described, which may be taken as a representative specimen of the modern printing machine.

To such machines as those already described the world is indebted for cheap books, cheap newspapers, and cheap literature in general. But when, with railways and telegraphs, came the desire for the very latest intelligence, the necessities of the newspaper press, as regards rapidity of printing, soon required a greater speed than could possibly be attained by any of the flat form presses; for in these the table, with the forms placed upon it, is unavoidably of a considerable weight, and this heavy mass has to be set in motion, stopped, moved in the opposite direction, and again stopped during the printing of each sheet. The shocks and strains which the machine receives in these alternate reversals of the direction of the movement impose a limit beyond which the speed cannot be advantageously increased. When Mr. Applegath was again applied to by the proprietors of the "Times" to produce a machine capable of working off a still larger number of impressions, he decided upon abandoning the plan of reciprocating movement, and substituting a continuous rotary movement of the type form. And he successfully overcame the difficulties of attaching ordinary type to a cylindrical surface. The idea of placing the type on a rotating cylinder is due to Nicholson, who long ago proposed to give the types a wedge shape, so that the pieces of metal would, like the stones of an arch, exactly fit round the cylindrical surface. The wedge-shaped types were, however, so liable to be thrown from their places by the centrifugal force, that Nicholson proposed also certain mechanical methods of locking the types together after they had been placed on the circumference of the drum. The plan he suggested for this purpose involved, however, such an expenditure of time and trouble that his idea was never carried into practice. Mr. Applegath used type of the ordinary kind, which was set up on flat surfaces, forming the sides of a prism corresponding to the circumference of his revolving type cylinder, which was very large and placed vertically. The flat surfaces which received the type were the width of the columns of the newspaper, and the type forms were firmly locked up by screwing down wedge-shaped rules between the columns at the angles of the polygon. These form the "column rules," which make the upright lines between the columns of the page, and by their shape they served to securely fix the type in its place. The diameter of the cylinder to which the form was thus attached was 5 ft. 6 in., but the type occupied only a portion of its circumference, the remainder serving as an inking table. Round the great cylinder eight impression rollers were placed, and to each impression roller was a set of inking rollers. At each turn, therefore, of the great cylinder eight sheets received the impression. These cylinders were, as already stated, placed vertically, and, as it was necessary to supply the sheets from horizontal tables, an ingenious arrangement of tapes and rollers was contrived, by which each sheet was first carried down from the table into a vertical position, with its plane directed towards the impression roller, in which position it was stopped for an instant, then moved horizontally forwards round the impression cylinder, and was finally brought out, suspended vertically, ready for a taker-off to place on his pile. This machine gave excellent results as to speed and regularity. From 10,000 to 12,000 impressions could be worked off in an hour, and the advantage was claimed for it of keeping the type much cleaner, by reason of its vertical position. The power of this machine may be judged of from one actual instance. It is stated that of copies of the " Times " in which the death of the Duke of Wellington was announced, 14th November, 1852, no less than 70,000 were printed in one day, and the machines were not once stopped, either to wash the rollers or to brush the forms. It may be mentioned, in order to give a better idea of the magnitude of the operation of printing this one newspaper, that one average day's copies weigh about ten tons, and that the paper for the week's consumption fills a train of twenty waggons.

At the " Times " office and elsewhere, the vertical machine has some years ago been superseded by others with horizontal cylinders. The fastest, perhaps, of all these printing machines is that which is now known as the " Walter Press," so called either because its principle was suggested by the proprietor of the " Times," or merely out of compliment to him. The improvements which are embodied in the Walter Press have been the subject of several patents taken out in the names of Messrs. MacDonald and Calverley, and it is to these improvements that we must now direct the attention of the reader. But we must premise that such machines as the Walter Press became possible only by the discovery of the means of rapidly producing what is called a stereotype plate from a form of type. A full account of the methods of effecting this is reserved for a subsequent article, but here it may suffice to say, that when a thick layer of moist cardboard, or rather a number of sheets of thin unsized paper pasted together and still quite moist, is forced down upon the form by powerful pressure, a sharp even mould of the type is obtained, every projection in the latter producing a corresponding depression in the *papier maché* mould. When the paper mould is dry, it may be used for forming a *cast* by pouring over it some fusible metallic alloy, having the properties of becoming liquid at a temperature which will not injure the mould, of taking the impressions sharply, and of being sufficiently hard to bear printing from. One of the improvements in connection with the Walter Press is in the mode of forming cylindrical stereotype casts from the paper mould. For this purpose the mould is placed on the *internal* surface of an iron semi-cylinder, with the face which has received the impression of the type inwards. The central part of the semi-cylinder is occupied by a cylindrical iron core, which is adjusted so as to leave a uniform space between its convex surface and the concave face of the mould. Into this space is poured the melted metal, and its pressure forces the mould closely against the concave cylindrical surface to which it is applied, so that the thickness becomes quite uniform. The iron core has a number of grooves cut round it, and these produce in the cast so many ribs, or projections, which encircle the inner surface, and serve both to strengthen the cast and afford a ready means of obtaining an exact adjustment. Not the complete cylinder, but only half its circumference, is cast at once, the axis of the casting apparatus being placed horizontally, and the liquid metal poured in one unbroken stream between the core and the mould from a vessel as long as the cylinders. Fig. 156 is a section of the casting apparatus, in which *a* is the core, *b* the *papier maché* mould, *c* the iron semi-cylinder containing it, *d* the metal which has been poured in at the widened space, *e*. When the metal has solidi fied, the core is simply lifted off, and the cast is then taken out, in the form of a semi-cylinder, the internal surface of which has exactly the diameter of the external surface of the roller of the machine on which it is to be placed, in company with another semi-cylindrical plate, so that the two together encircle half the length of the roller, and when another pair of semi-cylinders have been fixed on the other part of the roller, the whole

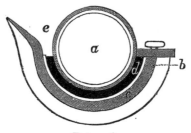

FIG. 156.

matter of one side of the newspaper sheet, usually four pages, is ready for printing. One great advantage of working from stereotype casts made in this way is that the form-bearing cylinder of the machine has no greater circumference than suffices to afford space for the matter on one side of the paper. The casts are securely fixed on the revolving cylinder by elbows, which can be firmly screwed down. The casts are usually made to contain one page each, so that four semi-cylinders, each half the length of the revolving cylinder, are fixed on the circumference of the latter. The process of casting in no way injures the paper mould, which is in fact generally employed to produce several plates.

The Walter Machine is not fed with separate sheets of paper, but takes its supply from a huge roll, and itself cuts the paper into sheets after it has impressed it on both sides. This is done by a very simple but effective plan, which consists in passing the paper between two equal-sized rollers, the circumference of which is precisely the length of the sheets to be cut. These rollers grip the paper, but only on the marginal spaces; and on the circumference of one of them, and parallel to its axis, is a slightly projecting steel blade, which fits into a corresponding recess, or groove, in the circumference of the other, and at this time the whole width of the sheet is firmly held by a projecting piece acted on by a spring. Although the Walter Machine, as actually constructed, presents to the uninitiated spectator an apparently endless and intricate series of parallel cylinders and rollers, yet it is in reality exceedingly simple in principle, as may be seen by the diagram given in Fig. 157. In this we may first direct the reader's attention to the two cylinders, F_1, F_2, which bear the stereotype casts—one of the matter belonging to one side of the sheet, the other of the matter belonging to the other side, for the Walter Press is a perfecting machine—and the web of paper having been printed by F_1, against which it is pressed by the roller, P_1, passes straight, as shown by the dotted line, to the second pair of cylinders, in order to be printed on the other side; and here, of course, the form cylinder, F_2, is below, and the impression cylinder, P_2, above, and an endless cleaning blanket is supplied to the latter to receive the *set-off.* The web of paper then passes between the cutting rollers, c, c_1, by which it is cut in sheets. But the knife has a narrow notch in the centre, and one at each end, so that the paper is not severed at those parts, narrow strips or tags being left, which maintain for a while a slight connection. But the tapes, t_1, t_2, between which the paper is now carried, are driven at a rather quicker rate than the web issues from c, c_1; and the result is, that the tags are torn, and the sheet becomes separated from the portion next following it. Thus, as a separate sheet, it arrives at the horizontal tapes,

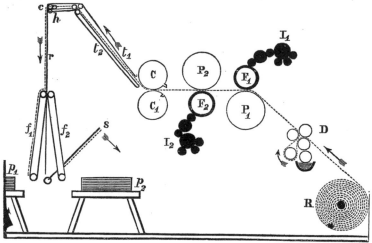

FIG. 157.—*Diagram of the Walter Press.*

h, and is brought to another set of tapes mounted on the frame, r, rocking about the centre, c, by which it is brought finally to the tapes, f_1, f_2, which by the movement of r receive the sheets alternately. A sheet-flyer, s, oscillates between the tapes, f_1, f_2; and as fast as the sheets arrive, lays them down right and left alternately, and it only remains for the piles, p_1, p_2, so formed, to be removed. The inking apparatus of each form-cylinder is indicated by the series of rollers marked I_1, I_2; and in this part of the machine there are also some improvements over former presses, for the distributing rollers are not made of composition, but of iron, turned with great exactness to a true surface, and arranged so as not quite to touch each other. At D is an apparatus for damping the paper, in which there are hollow perforated cylinders, covered by blanket, and filled with some porous material, which is kept constantly wet. These cylinders being made to rotate rapidly, the centrifugal force causes the water to find its way uniformly to the outside. Here the paper also passes between rollers intended to flatten and to stretch it. At R is the great roll of paper, from which the machine takes its supply. These rolls contain, perhaps, five miles length of paper, and at first it was a matter of some difficulty to fix them firmly on their wooden axles, so that they might be steadily unwound; but the contrivers of the Walter Press make these spindles as tight as may be required by forming them in wedge-shaped pieces, which can be made to increase the thickness of the spindle by drawing one upon another by screws.

The great speed of the Walter Machine is secured by the paper being drawn by the machine itself from a continuous web, instead of being laid on in a separate sheet, so that the machine is not dependent on the dexterity of the layers-on, who are besides necessarily highly-skilled workmen, and therefore a great economy of wages results from using a machine which does not require their services; and as the Walter Press also itself lays down the perfected sheets, the necessary attendants are as few as possible. The waste of paper and loss of time by stoppages are said to be extremely small with this machine.

Fig. 148 will give some idea of the appearance of the printing-room where one of the leading London daily papers is being printed by Walter Presses.

Another fast printing machine is the type revolving cylinder machine invented by Colonel Richard M. Hoe, and manufactured by the well-known firm of Hoe and Company, New York, with whose name the history of fast printing machines must ever be associated. In these machines the type is placed on the circumference of a cylinder which rotates about a horizontal axis, and the difficulties of securely locking up the type are successfully overcome. The machines are made with two, four, six, eight, or ten impression cylinders, and at each revolution of the great cylinder the corresponding number of impressions are produced. The engraving on the opposite page, Fig. 158, represents the two-cylinder machine, and an examination of the figure will render its general action intelligible. The form of type occupies about one-fourth of the circumference of the great cylinder, the remainder being used as an ink-distributing surface. Round this main cylinder, and parallel to it, are placed smaller impression cylinders, from two to ten in number, according to the size of the machine. When the press is in operation, the rotation of the main cylinder carries the type form to each impression cylinder in succession, and it there impresses the paper, which is made to arrive at the right time to secure true register. One person is required for each impression cylinder, to supply the sheets of paper, which have merely to be laid in a certain position, when, at the proper moment, they are seized by the " grippers," or fingers of the machine, and after having been printed, are carried out by tapes, and laid in heaps by self-acting sheet-flyers, by which the hands which are required to receive and pile the sheets in other machines are dispensed with. The ink is contained in a fountain placed beneath the main cylinder, and is conveyed by means of rollers to the distributing surface of the main cylinder. This surface, being lower than that of the type forms, passes by the impression cylinders without touching them. For each impression cylinder there are two inking rollers, receiving their supply of ink from the distributing surface of the main cylinder. These inking rollers, the bearings of which are, by springs, drawn towards the axis of the main cylinder, rise as the form passes under them, and having inked it, they again drop on to the distributing surface. Each page of the matter is locked up on a detachable segment of the large cylinder, which segment constitutes its bed and chase. The column-rules are parallel with the shaft of the cylinder, and are consequently straight, while the head, advertising, and dark rules have the form of segments of a circle. The column-rules are in the shape of a wedge,

FIG. 158.—*Messrs, Hoe's Type Revolving Cylinder Machine.*

with the thin end directed towards the axis of the cylinder, so as to bind the types securely. These wedge-shaped column-rules are held in their place by tongues projecting at intervals along their length, and sliding in grooves cut crosswise in the face of the bed. The spaces in the grooves between the column-rules are accurately fitted with sliding blocks of metal level with the surface of the bed, the ends of the blocks being cut away underneath, to receive a projection on the sides of the tongues of the column-rules. The locking up is effected by means of screws at the foot of each page, by which the type is held as securely as in the ordinary manner upon a flat bed. The main cylinder of the machine represented in Fig. 158 has a diameter of 3 ft. 9 in., and its length is, according to the size of the sheets to be printed, from 4 ft. 5 in. to 7 ft. 4 in. The whole is about 20 ft. long, 10 ft. wide, including the platforms, and a height of 9 ft. in the room in which it is placed suffices for its convenient working. The steam power required is from one to two horse-power, according to the length of the main cylinder. The speed of these machines is limited only by the ability of the feeders to supply the sheets fast enough. The ten-cylinder machine has, of course, ten impression cylinders, instead of two, and there are ten feeding-tables, arranged one above the other, five on each side. The main cylinder has a diameter of 4 ft. 9 in., and is 6 ft. 8 in. long. The machine occupies altogether a space of 31 ft. by 16 ft., and its height is 18 ft. A steam engine of eight horse-power is sufficient to drive the ten-cylinder machine, which is then capable of producing 25,000 impressions per hour. The mechanism of the larger machines is precisely similar to that of the two-cylinder machine, except such additional devices as are necessary to carry the paper to and from the main cylinder at four, six, eight, or ten points of its circumference. Much admirable contrivance is displayed in the manner of disposing feeders as closely as possible round the central cylinder.

In some machines, such as Messrs. Hoe's, Fig. 158, the sheet-flyers are interesting features, for they form an efficient contrivance for laying down and piling up, with the greatest regularity, sheet after sheet as it issues from the press. The sheet-flyer is in fact an automatic taker-off, and therefore it supersedes the services of the boy who would otherwise be required. It is simply a light wooden framework of parallel bars, turning on one of its sides as a centre; and the tapes carrying the sheet, passing down between the bars, bring the paper down upon the frame, where its progress is then stopped, the frame makes a rapid turn on its centre, lays down the sheet, and quickly rises to receive another from the tapes. One can hardly see a printing machine in action without being struck with the deftness with which the sheet-flyer does its duty; for the precision with which it receives a sheet, lays it down, and then quickly returns, to be ready for the next, suggest to the mind of the spectator rather the movements of a conscious agent than the motions of an unintelligent piece of mechanism. The sheet-flyer is seen at the left-hand side of Fig. 158, where it is in the act of laying down a sheet on the pile it has already formed.

The modern improvements in printing presses are well illustrated by the machine represented on the opposite page, Fig. 159, which has been designed by the Messrs. Hoe to work exclusively by hand. It is intended for the newspaper and job work of a country office, and it works easily, without noise or jar, by turning the handle always in the same direction, producing 800 impressions in an hour. The bed moves backwards and forwards on wheels running on rails, the reciprocating movement being

319

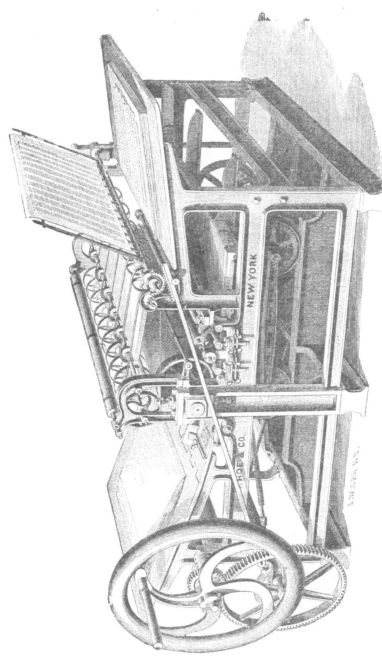

FIG. 159.—*Messrs. Hoe's "Railway" Machine.*

derived from the circular one by means of a crank. From the mode in which the table is carried backwards and forwards, the manufacturers call this the " Railway Printing Machine." The paper is fed to the underside of the cylinder, which, after an impression has been given, remains stationary while the bed is returning, and while the layer-on is adjusting his sheet of paper. The axle of the impression cylinder carries a toothed wheel working in a rack on the bed or table, the wheel having at two parts of its circumference the teeth planed off so as to permit of the return of the table without moving the impression cylinder, which is again thrown into gear with the rack by a catch, so that the same tooth of the rack always enters the same space on the toothed wheel, and thus a good register is secured. The impression cylinder remains unaltered, whatever may be the size of the type form, it being only necessary to place the forward edge of the form always on the same line of the bed. Machines of a very similar

FIG. 160.—*Napier's Platen Machine.*

construction, but driven by steam power, are used in lithographic printing; and in some of these machines advantage is elegantly taken of the fact that, when a wheel rolls along, the uppermost point of its circumference is always moving forward at exactly twice the velocity of its centre. Hence, if the table of a printing machine rests on the *circumference* of wheels, a backward and forward movement of the centres of these wheels, produced by the throw of a crank through a space of 2 ft., would produce a rectilineal reciprocating movement through a distance of 4 ft. of a table resting on the circumference of the wheels. Any reader who is interested in geometry or mechanics would do well to convince himself that the lowest point of the wheel of a railway carriage, for example, is stationary (considered while it is the lowest point), that the *centre* of the wheel is moving forwards with the velocity of the train, and that the highest point of the wheel is moving forwards with just twice the speed of the train. There is no difficulty about the rate of rectilineal motion of the centre, but the reader cannot possibly perceive the truth of the statement regarding the lowest and highest points unless he reflects on the subject, or puts it to the test of experiment. Another form of press which is used for good book printing is represented in the engraving, Fig. 160, which shows Napier's platen machine. There the action is similar to that of the ordinary hand presses as regards the mode in which the paper is pressed against the face of the type ; but the movements are all performed by steam power, applied through the driving belt, shown in the figure.

The various kinds of printing machines adapted to each description of work are too numerous to admit of even a passing mention here ; but those which have been described may fairly be considered as representing the leading principles of modern improvements. This article relates only to the mechanism by which an impression is transferred from a form to the surface of paper : the interesting and novel *processes* by which the form itself may be produced—processes which have amazingly abridged the printers' labour and extended the resources of the art—deserve a separate chapter, and will furnish matter for an article on Printing Processes, which will be the better understood by being placed after chapters wherein the scientific bases of some of these processes are discussed.

PATTERN PRINTING.

THE machines used for printing patterns are, in principle, very similar to those for letterpress printing ; but the circumstance of several different colours having frequently to go to the production of one pattern leads to the multiplication, in the present class of machines, of the apparatus for distributing the colours and impressing the materials. Pattern printing machines are most extensively used for impressing fabrics, such as calicoes, muslins, &c., and for producing the wall-papers for decorating apartments. The machines employed for calicoes and for papers are so much alike, that to describe the one is almost to describe the other.

The papers intended for paper-hangings are, in the first instance, covered with a uniform layer of the colour which is to form the ground, and this is done even in the case of papers which are to have a white ground. The colours thus laid on, and those which are applied by the machine, are composed of finely-ground colouring matters mixed with thin size or glue to a suitable consistence, and the ground-tint is given by bringing the upper surface of the paper, as it is mechanically unwound from a great roll, into contact with an endless band of cloth emerging from a trough containing a supply of the fluid colour. The paper then passes over a horizontal table, where the layer of colour is uniformly distributed over its surface by brushes moved by machinery, and the paper, after having been thoroughly dried, is ready to receive the impressions. The impressions may be given by flat blocks of wood on which the pattern is carved in relief, or from revolving cylinders on which the pattern is similarly carved. The former is the process of hand labour called " block printing," and it requires much skill and care on the part of the operator ; but with these, excellent results are obtained, as a correct adjustment of the positions of the parts of the pattern can always be secured. The latter is the mode of printing mechanically on rollers, corresponding with the type-bearing cylinders of the machines already described; but for pattern printing on paper they are made of fine-grained wood, mounted on an iron axle, and they are carved so that the design to be printed stands out in relief on their surface. One of these rollers is represented in Fig. 161, and it should be clearly understood that each colour in the pattern on a wall-paper requires a separate roller, the design cut on which corresponds only with the forms the particular colour contributes to the pattern. Such rollers being necessarily somewhat expensive, as the pattern is usually repeated many times over the cylindrical surface, the plan has been adopted of fastening a mass of hard composition in an iron axle, and when this has been turned to a truly cylindrical surface, it is made to receive plates of metal, formed of a fusible alloy of lead, tin, and nickel. These plates are simply casts

FIG. 161.—*Roller for Printing Wall-Papers.*

from a single carved wooden mould of the pattern, which has thus only once to be formed by hand. The plates are readily bent when warmed, and are thus applied to the cylindrical surface, to which they are then securely attached. It is found advantageous to cover the prominent parts of the rollers which produce the impressions with a thin layer of felt, as this substance takes up the colours much more readily than wood or metal, and leaves a cleaner impression.

The machine by which wall-papers are printed is represented in Fig. 162, where it will be observed that the impression cylinder has a very large diameter, and that a portion of its circumference forms a toothed wheel, which engages a number of equal-sized pinions placed at intervals about its periphery. Each pinion being fixed on the axle of a pattern-bearing roller, these are all made to revolve at the same rate. There is, however, some adjustment necessary before that exact correspondence of the impressions with each other is secured, which is shown on the printed pattern by each colour being precisely in its appointed place. The rollers are constantly supplied with colour by endless cloths, which receive it from the troughs that are shown in the figure, one trough being appropriated to each roller. Some of these machines can print as many as eighteen or twenty different colours at once, by having that number of rollers; and it is easy to see how, by dividing each trough into several vertical compartments, in each of which a different colour is placed, it would be possible to triple or even quadruple the number of colours printed by one machine.

The machinery by which calicoes are printed is almost identical in construction with that just described, and presents the same general appearance. There is, however, an important difference in the rollers, which in calico printing are of copper or bronze, and have the design engraved upon their polished cylindrical surface, not in relief, but in hollows. After the whole surface of the roller becomes charged with colour, there is in the machine a straight-edge, which removes the colour from the smooth sur-

FIG. 162.—*Machine for Printing Paper-Hangings.*

face, leaving only what has entered into the hollow spaces of the design, which, as the roller comes round to the cloth, yield it up to the surface of the latter. Thus, by a self-acting arrangement, the rollers are charged with colour, cleaned, and made to give up their impressions to the stuff by parting with the colour in the hollows. Rollers having patterns in relief are also used in calico printing, the mechanism being then almost identical with that of the former machine. It need hardly be said that great pains are taken in the construction of such machines to have each part very accurately adjusted, so that the impression may fall precisely upon the proper place, without any blurring or confusion of the colours, and the fact that an intricate design, having perhaps eighteen or twenty tints, can be thus mechanically reproduced millions of times speaks volumes for the accuracy and finish of the workmanship which are bestowed on such printing machines.

FIG. 163.—*Chain-Testing Machine at Messrs. Brown and Lenox's Works, Millwall.*

HYDRAULIC POWER.

IF a hollow sphere, *a*, Fig. 173, be pierced with a number of small holes at various points, and a cylinder, *b*, provided with a piston, *c*, fitted into it, when the apparatus is filled with water, and the piston is pushed inwards, the water will spout out of all the orifices equally, and not exclusively from that which is opposite to the piston and in the direction of its pressure. The jets of water so produced would not, as a matter of fact, all pursue straight paths radiating from the centre of the sphere, because gravity would act upon them; and all, except those which issued vertically, would take curved forms. But when proper allowance is made for this circum-stance, each jet is seen to be projected with equal force in the direction of a radius of the sphere. This experiment proves that when pressure is applied to any part of a liquid, that pressure is transmitted *in all directions equally*. Thus the pressure of the piston—which, in the apparatus repre-sented in the figure, is applied in the direction of the axis of the cylinder only—is carried throughout the whole mass of the liquid, and shows itself by its effect in urging the water out of the orifices in the sphere in all direc-tions; and since the force with which the water rushes out is the same at every jet, it is plain that the water must press equally against each unit of area of the inside surface of the hollow sphere, without regard to the position of the unit.

If we suppose the piston to have an area of one square inch, and to be pushed inwards with a force of 10 lbs., it cannot be doubted that the square inch of the inner surface of sphere immediately opposite the cylinder will receive also the pressure of 10 lbs. ; and since the pressures throughout the interior of the hollow globe are equal, every square inch of its area will also be pressed outwards with a force equal to 10 lbs. Hence, if the total area of the interior be 100 square inches, the whole pressure produced will amount to a hundred times 10 lbs.

That water or any other liquid would behave in the manner just described might be deduced from a property of liquids which is sufficiently obvious, namely, the freedom with which their particles move or slide upon each other. The equal transmission of pressure in all directions through liquids was first clearly expressed by the celebrated Pascal, and it is therefore known as " Pascal's principle." He said that " if a closed vessel filled with water has two openings, one of which is a hundred times as large as the other; and if each opening be provided with an exactly-fitting piston, a man pushing in the small piston could balance the efforts of a hundred men pushing in the other, and he could overcome the force of ninety-nine."

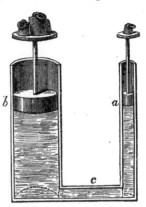

FIG. 164.—*Pascal's Principle.*

Pascal's principle—which is that of the hydraulic press—may be illustrated by Fig. 164, in which two tubes of unequal areas, a and b, communicate with each, and are supposed to be filled with a liquid—water, for example, which will, of course, stand at the same level in both branches. Let us now imagine that pistons exactly fitting the tubes, and yet quite free to move, are placed upon the columns of liquid—the larger of which, b, we shall suppose to have five times the diameter, and therefore twenty-five times the sectional area, of the smaller one. A pressure of 1 lb. applied to the smaller piston would, in such a case, produce an upward pressure on the larger piston of 25 lbs.; and in order to keep the piston at rest, we should have to place a weight of 25 lbs. upon it. Here then a certain force appears to produce a much larger one, and the extent to which the latter may be increased is limited only by the means of increasing the area of the piston. Practically, however, we should not by any such arrangement be able to prove that there is exactly the same proportion between the total pressures as between the areas, for the pistons could not be made to fit with sufficient closeness without at the same time giving rise to so much friction as to render exact comparisons impossible. We may, however, still imagine a theoretical perfection in our apparatus, and see what further consequences may be deduced, remembering always that the actual results obtained in practice would differ from these only by reason of interfering causes, which can be taken into account when required. We have supposed hitherto that the pressures of the pistons exactly balance each other. Now, so long as the system thus remains in equilibrium no *work* is done; but if the smallest additional weight were placed upon either piston, that one would descend and the other would be pushed up. As we have supposed the apparatus to act without friction, so we shall also neglect the effects due to difference in the levels of the columns of liquid when the pistons are moved ; and further, in order to fix our ideas, let us imagine the smaller tube to have a section of 1 square inch in area, and the larger one of 25 square inches.   Now, if the weight of the piston, *a,* be increased by the smallest fraction of a grain, it will descend.   When it has descended a distance of 25 in., then 25 cubic inches of water must have passed into *b,* and, to make room for this quantity of liquid, the piston with the weight of 25 lbs. upon it must have risen accordingly.   But since the area of the larger tube is 25 in., a rise of 1 in. will exactly suffice for this ; so that a weight of 1 lb. descending through a space of 25 in., raises a weight of 25 lbs. through a space of 1 in.   This is an illustration of a principle holding good in all machines, which is sometimes vaguely expressed by saying that *what is gained in power is lost in time.*   In this case we have the piston, *b,* moving through the space of 1 in. in the same time that the piston *a* moves through 25 in. ; and therefore the *velocity* of the latter is twenty-five times greater than that of the former, but the time is the same.   It would be more precise to say, that what is gained in force is lost in space ; or, that no machine, whatever may be its nature or construction, is of itself capable of doing *work.*   The " mechanical powers," as they are called, can do but the work done upon them, and their use is only to change the relative amounts of the two factors, the product of which measures the work, namely, space and force.   Pascal himself, in connection with the passage quoted above, clearly points out that in the new mechanical power suggested bʸ him in the hydraulic press, " the same rule is met with as in the old ones —such as the lever, wheel and axle, screw, &c.—which is, that the distance is increased in proportion to the force ; for it is evident that as one of the openings is a hundred times larger than the other, if the man who pushes the small piston drives it forward 1 in., he will drive backward the large piston one-hundredth part of that length only."   Though the hydraulic press was thus distinctly proposed as a machine by Pascal, a certain difficulty prevented the suggestion from becoming of any practical utility.   It was found impossible, by any ordinary plan of packing, to make the piston fit without allowing the water to escape when the pressure became considerable.   This difficulty was overcome by Bramah, who, about the end of last century, contrived a simple and elegant plan of packing the piston, and first made the hydraulic press an efficient and useful machine.   Fig. 166 is a view of an ordinary hydraulic press, in which *a* is a very strong iron cylinder, represented in the figure with a part broken off, in order to show that inside of it is an iron piston or ram, *b,* which works up and down through a water-tight collar ; and in this part is the invention by which Bramah overcame

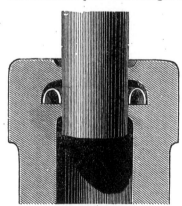

FIG. 165.—*Collar of Hydraulic Cylinder.*

the difficulties that had previously been met with in making the hydraulic press of practical use.   Bramah's contrivance is shown by the section of the cylinder, Fig. 165, where the interior of the neck is seen to have a groove surrounding it, into which fits a ring of leather bent into a shape resembling an inverted ∪. The ring is cut out of a flat piece of stout leather, well oiled and bent into the required shape. The effect of the pressure of the water is to force the leather more tightly against the ram, and as the pressure becomes greater, the tighter is the fit of the collar, so that no water escapes even with very great pressures. To the ram, *b*, Fig. 166, a strong iron table, *c*, is attached, and on this are placed the articles to be compressed. Four

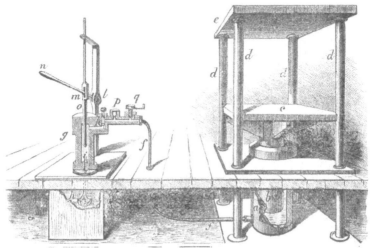

FIG. 166.—*Hydraulic Press.*

wrought iron columns, *d d d d*, support another strong plate, *e*, and maintain it in a position to resist the upward pressure of the goods when the ram rises, and they are squeezed between the two tables. The interior of the large cylinder communicates by means of the pipe, *f f*, with the suction and force-pump, *g*, in which a small plunger, *o*, works water-tight. Suppose that the cylinders and tubes are quite filled with water, and that the ram and piston are in the positions represented in the figure. When the piston of the pump, *g*, is raised, the space below it is instantly filled with water, which enters from the reservoir, *h*, through the valve, *i*, the valve *k* being closed by the pressure above it, so that no water can find its way back from the pipe, *f*, into the small cylinder. When the piston has completed its ascent, the interior of the small cylinder is therefore completely filled with water from the reservoir; and when the piston is pushed down, the valve, *i*, instantly closes, and all egress of the liquid in that direction being prevented, the greater pressure in *g* forces open the valve, *k*, and the water flows along the tube, *f*, into the large cylinder. The pressure exerted by the plunger in the small cylinder, being transmitted according to the principles already explained, produces on each portion of the area of the large plunger equal to that of the smaller an exactly equal pressure. In the smaller hydraulic presses the plunger of the forcing-pump is worked by a lever, as represented in the figure at *n;* so that with a given amount of force applied by the hand to the end of the lever, the pressure exerted by the press will depend upon the proportion of the sectional area of *b* to that of *o*, and also upon the proportion of the length *m n*, to the length *m l*. To fix our ideas, let us suppose that the distance from *m* of the point *n* where the hand is applied is ten times the distance *m l*, and that the sectional area of *b* is a hundred times that of *o*. If a force of 60 lbs. be applied at *n*, this will produce a downward pressure at *m* equal to 60×10, and then the pressure transmitted to the ram of the great cylinder will be 60×10×100=60,000 lbs. The apparatus is provided with a safety-valve at *p*, which is loaded with a weight; so that when the pressure exceeds a desired amount, the valve opens and the water escapes. There is also an arrangement at *q* for allowing the water to flow out when it is desired to relieve the pressure, and the water is then forced out by the large plunger, which slowly descends to occupy its place.   The body of the cylinder is placed beneath the floor in such presses as that represented in Fig. 166, in order to afford ready access to the table on which the articles to be compressed are placed.

The force which may, by a machine of this kind, be brought to bear upon substances submitted to its action, is limited only by the power of the materials of the press to resist the strains put upon them.   If water be continually forced into the cylinder of such a machine, then, whatever may be the resistance offered to the ascent of the plunger, it must yield, or otherwise some part of the machine itself must yield, either by rupture of the hydraulic cylinder, or by the bursting of the connecting-pipe or the forcing-pump. This result is certain, for the water refuses to be compressed, at least to any noticeable degree, and therefore, by making the area of the plunger of the force-pump sufficiently small, there is no limit to the pressure per square inch which can be produced in the hydraulic cylinder; or, to speak more correctly, the limit is reached only when the pressure in the hydraulic cylinder is equal to the cohesive strength of the material (cast or wrought iron) of which it is formed.   It has been found that when the internal pressure per square inch exceeds the cohesive or tensile strength of a rod of the metal 1 in. square (see page 207), no increase in the thickness of the metal will enable the cylinder to resist the pressure.   Professor Rankine has given the following formula for calculating the external radius, R, of a hollow cylinder of which the internal radius is *r*, the pressure per square inch which it is desired should be applied before the cylinder would yield being indicated by *p*, while *f* represents the tensile strength of the materials:

$$R = r\sqrt{\left(\frac{f+p}{f-p}\right)}$$

We may see in this formula that as the value of *p* becomes more and more nearly equal to *f*, the less does the divisor (*f—p*) become, and therefore the greater is the corresponding value of R; and when *f=p*, or *f—p*=0, the interpretation would be that no value of R would be sufficiently great to satisfy the equation.   Thus a cylinder, made of cast iron, of which the breaking strain is 8 tons per square inch, would have its inner surface ruptured by that amount of internal pressure, and the water passing into the fissures would exert its pressure with ever-increasing destructive effect.

With certain modifications in the proportions and arrangement of its parts, the hydraulic press is used for squeezing the juices from vegetable substances, such as beetroots, &c., for pressing oils from seeds, and, in fact, all purposes where a powerful, steady, and easily regulated pressure is needed. Cannons and steam boilers are tested by hydraulic pressure, by forcing water into them by means of a force-pump, just as it is forced into the cylinder of the hydraulic press described above. This mode of testing the strength has several great advantages; for not only can the pressure be regulated and its amount accurately known; but in case the cannon or steam boiler should give way, there is no danger, for it does not explode— the metal is simply ruptured, and the moment this takes place, the water flows out and the strain at once ceases.

The strength of bars, chains, cables, and anchors is also tested by hydraulic power, and the engraving at the head of this article, Fig. 163, represents the hydraulic testing machine at the works of Messrs. Brown and Lenox, the eminent chain and anchor manufacturers, of Millwall. Immediately in front of the spectator are the force-pumps, and the steam engine by which they are driven. It will be observed that four plungers are attached to an oscillating beam in such a manner that the water is continuously forced into the hydraulic cylinder. The outer pair of plungers are of much larger diameter than the inner pair, in order that the supply of water may be cut off from the former when the pressure is approaching the desired limit, and the smaller pair alone then go on pumping in the water, the pressure being thus more gradually increased. Behind the engine and forcing pump is the massive iron cylinder, where the pressure is made to act on a piston, which is forced towards that end of the cylinder seen in the drawing. The piston is attached to a very thick piston-rod, moving through a water-tight collar at the other end of the cylinder. The effect of the hydraulic pressure is, therefore, to draw the piston-rod into the cylinder, and not, as in the apparatus represented in Fig. 166, to force a plunger out. The head of the piston-rod is provided with a strong shackle, to which the chains to be tested can be attached. In a line with the axis of the cylinder is a trough, some 90 ft. long, to hold the chain, and at the farther end of the trough is another very strong shackle, to which the other end of the chain is made fast. A peculiarity of Messrs. Brown and Lenox's machine is the mode in which the tension is measured. In many cases it is deemed sufficient to ascertain by some kind of gauge the pressure of the water in the hydraulic cylinder, and from that to deduce the pull upon the chain; but the Messrs. Brown have found that every form of gauge is liable to give fallacious indications, from variations of temperature and other circumstances, and they prefer to measure the strain directly. This is accomplished by attaching the shackle at the farther extremity of the trough to the short arm of a lever, turning upon hard steel bearings, the long arm of this lever acting upon the short arm of another, and so on until the weight of 1 lb. at the end of the last lever will balance a pull on the chain of 2,240 lbs., or 1 ton. The tension is thus directly measured by a system of levers, exactly resembling those used in a common weighing machine, and this is done so accurately that even when a chain is being subjected to a strain of many tons, an additional pull, such as one can give to the shackle-link with one hand, at once shows itself in the weighing-room. The person who has charge of this part of the machine places on the end of the lever a weight of as many pounds as the number of tons strain to which the chain to be tested has to be submitted. The engineer sets the pump in action, the water is rapidly forced into the cylinder, the piston is thrust inwards, and the strain upon the chain begins; the engineer then cuts off the water supply from the larger force-pumps, and the smaller pair go on until the strain becomes sufficient to raise the weight, and then the person in the weighing-room, by pulling a wire, opens a valve in connection with the hydraulic cylinder, which allows the water to escape, and the strain is at once taken off. This testing machine, which is capable of testing cables up to 200 tons or more, was originally designed by Sir T. Brown, the late head of the firm, and not only was the first constructed in the country, but remains unsurpassed in the precision of its indications.

The testing of cables, which we have just described, is a matter of the highest importance, for the failure of cables and anchors places ships and men's life in great danger, since vessels have frequently to ride out a storm at anchor, and should the cables give way, a ship would then be almost entirely at the mercy of the winds and waves. Hence the Government have, with regard to cables and anchors, very properly made certain stringent regulations, which apply not only to the navy but to merchant shipping. The chain-cable is itself a comparatively modern application of iron, for sixty years ago our line-of-battle ships carried only huge hempen cables of some 8 in. or 9 in. diameter. Chain-cables have now almost entirely superseded ropes, though some ships carry a hempen cable, for use under peculiar circumstances. The largest chain-cables have links in which the iron has a diameter of nearly 3 in., and these cables are considered good and sound when they can bear a strain of 136 tons. Such are the cables used in the British navy for the largest ships. Of course, there are many smaller-sized cables also in use, and the strains to which these are subjected when they are tested in the Government dockyards vary according to the thickness of the iron; but it is found that nearly one out of every four cables supplied to the Admiralty proves defective in some part, which has to be replaced by a sounder piece. The chain-cables made by Messrs. Brown and Lenox for the *Great Eastern* are, as might have been expected, of the very stoutest construction; the best workmanship and the finest quality of iron having been employed in their manufacture. These cables were tested up to 148 tons, a greater strain than had ever before been applied as a test to any chain, and it was found that a pull represented by at least 172 tons was required to break them. It is difficult to believe that a teacup-full of cold water shoved down a narrow pipe is able to rend asunder the massive links which more than suffice to hold the huge ship securely to her anchors, but such is nevertheless the sober fact. The regulations of the Board of Trade require that every cable or anchor sold for use in merchant ships is to be previously tested by an authorized and licensed tester, who, if he finds it bears the proper strain, stamps upon it a certain mark.

The means which is afforded by hydraulic power of applying enormous pressures has been taken advantage of in a great many of the arts, of which, indeed, there are few that have not, directly or indirectly, benefited by this mode of modifying force. An illustration, taken at random, may be found in the machinery employed at Woolwich for making elongated rifle-bullets. The bullets are formed by forcing into dies, which give the required shape, little cylinders of solid lead, cut off by the machine itself from a continuous cylindrical rod of the metal. The rod, or rather filament, of lead is wound like a rope on large reels, from which it is fed to the machine. It is in the production of this solid leaden rope or filament that hydraulic pressure is used. About 4 cwt. of melted lead is poured into a very massive iron cylinder, the inside of which has a diameter of $7\frac{1}{2}$ in., while the external diameter is no less than 2 ft. 6 in., so that the sides of the cylinder are

actually 11¼ in. thick. When the lead has cooled so far as that it has
passed into a half solid state, a ram or plunger, accurately fitting the bore
of the cylinder, is forced down by hydraulic pressure upon the semi-fluid
metal. This plunger is provided with a round hole throughout its entire
length, and as it is urged against the half solidified metal with enormous
pressure, the lead yields, and is forced out through the hole in the plunger,
making its appearance at the top as a continuous cylindrical filament, quite
solid, but still hot. This is wound upon the large iron reels as fast as it

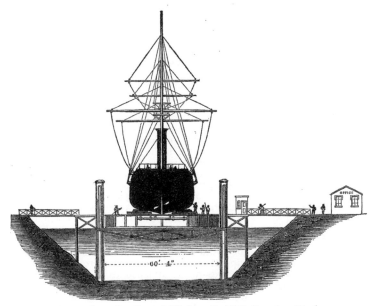

FIG. 167.—*Section of Hydraulic Lift Graving Dock.*

emerges from the opening in the plunger, and these reels are then taken to
the bullet-shaping machine, which snips off length after length of the leaden
cord, and fashions it into bullets for the Martini-Henry rifle. The leaden
pipes which are so much used for conveying water and gas in houses are
made in a similar manner, metal being forced out of an annular opening,
which is formed by putting an iron rod, having its diameter of the required
bore of the pipe, in the middle of the circular opening. The lead in escaping
between the rod and the sides of the opening takes the form of a pipe, and
is wound upon large iron reels, as in the former case.

Another interesting application of hydraulic power is to the raising of
ships vertically out of the water, in order to examine the bottoms of their
hulls, and effect any necessary repairs. The hydraulic lift graving dock, in
which this is done, is the invention of Mr. E. Clark, who, under the direction of Mr. Robert Stephenson, designed the machinery and superintended

the raising of the tubes of the Britannia Bridge, where a weight of 1,800 tons was lifted by only three presses. The suitability of the hydraulic press for such work as slowly raising a vessel was doubtless suggested to him in connection with this circumstance, and the durability, economy, and small loss of power which occurs in the action of the press, pointed it out as particularly adapted for this purpose. The ordinary dry dock is simply an excavation, lined with timber or masonry, from which the tide is excluded by a gate, which, after the vessel has entered the dock at high water, is closed ; and when the tide has ebbed, and left the vessel dry, the sluice through

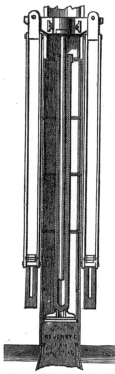

FIG. 168.—*Section of Column.*

which the water has escaped is also closed. In a tideless harbour the water has to be pumped out of the dock, and this last method is also adopted even in tidal waters, so that the docks may be independent of the state of the tides. The lift of Clark's graving dock is a direct application of the power of the hydraulic press, and we select for description the graving dock constructed at the Victoria Docks for the Thames Graving Dock Company, whose works occupy 26 acres. Fig. 167 is a transverse section of this hydraulic lift graving dock, in which there are two rows of cast iron columns, 5 ft. in diameter at the base, where they are sunk 12 ft. in the ground, and 4 ft. in diameter above the ground. The clear distance between the two rows in 60 ft., and the columns are placed 20 ft. apart from centre to centre, sixteen columns in each row, thus giving a length of 310 ft. to the platform, but vessels of 350 ft. in length may practically be lifted. The bases of the columns, one of which is represented in section in Fig. 168, are filled with concrete, on which the feet of the hydraulic cylinders rest. The outer columns support no weight, but act merely as guides for the crossheads attached to the plungers. The height of the columns is 68½ ft., and a wrought iron framed platform connects the columns at the top. In order that any inequalities in the height of the rams may be detected, a scale is painted on each column, to mark the positions of the crossheads. The hydraulic cylinders, which are within these columns, have solid rams of 10 in. diameter, with a stroke of 25 ft., and on the tops of these are fastened the crossheads, 7½ ft. long, made of wrought iron, and supporting at the ends bars of iron, to the other ends of which the girders of the platform are suspended. The girders are, therefore, sixteen in number, and together form a gridiron platform, which can be raised or lowered with the vessel upon it. The thirty-two hydraulic cylinders were tested at a pressure of more than 3 tons per square inch. The water is admitted immediately beneath the collars at the top (this being the most accessible position) by pipes of only ⅜ in. diameter, leading from the force-pumps, of which there are twelve, of 1⅞ in. diameter, directly worked by a fifty horse-power steam engine. The presses are worked in three groups—one of sixteen, and two of eight presses,—so arranged that their centres of action form a sort of tripod support, and the presses of each group are so connected that perfect uniformity of pressure is maintained. The raising of a vessel is accomplished in about twenty-five minutes, by placing below the vessel a pontoon, filled in the first instance with water, and then raising the pontoon with the vessel on it, while the water is allowed to escape from the pontoon through certain valves; then when the girders are again lowered, the pontoon, with the vessel on it, remains afloat. Thus in thirty minutes a ship drawing, say, 18 ft. of water is lifted on a shallow pontoon, drawing, perhaps, only 5 ft., and the whole is floated to a shallow dock, where, surrounded with workshops, the vessel, now high and dry, is ready to receive the necessary repairs. The number of vessels which can thus be docked is limited only by the number of pontoons, and thus the same lift serves to raise and lower any number of ships, which are floated on and off its platform by the pontoons. With a pressure in the hydraulic cylinders of about 2 tons upon each square inch, the combined action of these thirty-two presses would raise a ship weighing 5,000 tons.

Hydraulic power has been used not only for graving docks, as shown in the above figures, but also for dragging ships out of the water up an inclined plane. The machinery for this purpose was invented by Mr. Miller for hauling ships up the inclined plane of " Martin's slip," at the upper end of which the press cylinder is placed, at the same slope as the inclined plane, and the ship is attached, by means of chains, to a crosshead fixed on the plunger. Hydraulic power has also been used for launching ships, and the launch of the *Great Eastern* is a memorable instance; for the great ship stuck fast, and it was only by the application of an immense pressure, exerted by hydraulic apparatus, that she could be induced to take to the water. Water pressure is also applied to hoists for raising and lowering heavy bodies, and in such cases the pressure which is obtained by simply taking the water supply from an elevated source, or from the water-main of a town, is sometimes made use of, instead of that obtained by a forcing pump. The lift at the Albert Hall, South Kensington, by which persons may pass to and from the gallery without making use of the stairs, is worked by hydraulic pressure in the manner just mentioned. In such lifts or hoists there is a vertical cylinder, in which works a leather-packed piston, having a piston-rod passing upwards through a stuffing-box in the top of the cylinder. The upper end of the piston-rod has a pulley of 30 in. or 36 in. diameter, attached to it, and round this pulley is passed a chain, one end of which is fixed, and the other fastened to the movable cage or frame. So that the cage moves with twice the speed of the piston, and the length of the stroke of the latter is one-half of the range of the cage.

Sir William Armstrong has applied hydraulic power to cranes and other machines in combination with chains and pulleys. His hydraulic crane is represented by the diagram, Fig. 169, intended to show only the general disposition of the principal parts of this machine, which is so admirably arranged that one man can raise, lower, or swing round the heaviest load with a readiness and apparent ease marvellous to behold. Here it is proper to mention once for all, that the pressure for the hydraulic machines is obtained not only by natural heads of water, or by forcing-pumps worked by hand, but very frequently by forcing-pumps worked by steam power. It is usual to have a set of three pumps with their plungers connected respectively with three cranks on one shaft, making angles of 120° with each other. A special feature of Sir W. Armstrong's hydraulic crane is the arrangement by which the engines are made to be always storing up power by forcing water into the vessel, *a*, called the "accumulator." The accumulator—which in the diagram is not shown in its true position—may be placed in any convenient place near the crane, and consists of a large cast iron cylinder, *b*, fitted with a plunger, *c*, moving water-tight through the neck of the cylinder. To the head of the plunger is attached by iron cross-bars, *d d*, a strong iron case filled with heavy materials, so as to load the plunger, *c*, with a weight that will produce a pressure of about 600 lbs. upon each square inch of the inner surface of the cylinder. The water is pumped into the cylinder by the pumping engines through the pipe, *f*, and then the piston rises, carrying with it the loaded case, guided by the timber framework, *g*, until it reaches the top of its range, when it moves a lever that cuts off the supply of steam from the pumping engine. When the crane is working the water passes out of the cylinder, *a*, by the pipe, *h*, and exerts its pressures on the plungers of the smaller cylinders ; and the plunger of the accumulator, in beginning its descent again, moves the lever in connection with the throttle-valve of the engine, and thus again starts the pumps, which therefore at once begin to supply more water to the accumulator. The latter is, however, large enough to keep all the several smaller cylinders of the machine at work even when they are all in operation at once. Fig. 169 shows a sketch elevation and a ground plan of the crane as constructed to carry loads of 1 ton, but the size of the cylinders is somewhat exaggerated, and all details, such as pipes, guides, valves, rods, &c., are omitted. The hydraulic apparatus is entirely below the flooring—only the levers by which the valves are opened and closed appearing above the surface. The crane-post, *i*, is made of wrought iron : it is hollow and stationary ; the jib, *k*, is connected with the ties, *l*, by side-pieces, *n*, which are joined by a cross-piece at *m*, turning on a swivel and bearing the pulley, *u*. The jib and the side-pieces are attached at *o* to a piece turning round the crane-post, and provided with a friction roller, *p*, which receives the thrust of the jib against the crane-post; the same piece is carried below the flooring and is surrounded with a groove, which the links of the chain, *q*, fit. This chain serves to swing the crane round, and for this purpose the hydraulic cylinders, *r*, *r*, come into operation. The plungers of these have each a pulley, over which passes the chain *q*, having its ends fastened to the cylinders, so that when, by the pressure of the water, one plunger is forced out, the other is pushed in, and the chain passing round the groove at *s* swings the jib round. The cylinders are supplied with water by pipes—omitted in the sketch, as are also those by which the water leaves the cylinders. These pipes are connected with valves—also omitted on account of the scale of the diagram being too small to show their details—so that the movement of a lever, *t*, in one or the other direction at the same time connects one cylinder with the supply and the other with the exit-pipe. When the crane is swinging round, the sudden closing of the valves would produce an injurious shock, and to prevent this *relief-valves* are provided on both the supply and exit-pipes communicating with each cylinder. When, therefore, the valves are closed, the impetus of the jib and its load acting on the chain, and through that on the plungers, continues to move the latter, the motion is permitted to take place by the relief-valves opening, and allowing water to enter or leave the cylinders against the pressure of the water. There is also a self-acting arrangement by which, when these plungers have moved to the extent of their range in either direction, the valves are closed. The chain of the crane rests on guide pulleys, and passing over the pulley *u*, goes down the centre of the crane-post to the pulley *v*, and thence passes backwards and forwards over a series of three pulleys at *w* and two at *x*, and is fastened at its end to the cylinder, *y*. As there are thus six lines of chain, when the

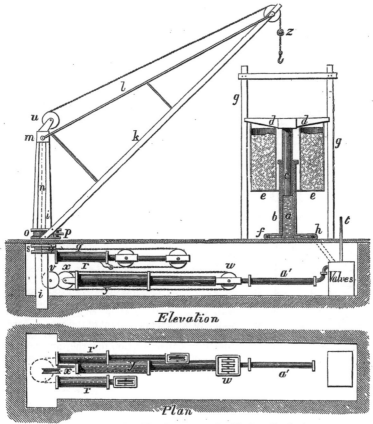

FIG. 169.—*Sir W. Armstrong's Hydraulic Crane.*

plunger of the lifting cylinder comes 1 ft. out, 6 ft. of chain pass over the guide pulley, *u*. The plunger, when near the end of its stroke in either direction. is made to move a bar—not shown—which closes the valve. When the crane is loaded, the load is lowered by simply opening the exhaust-valve, when the lift-plunger will be forced back into its cylinder by the pull on the chain. But as the chain may require to be lowered when there is no load upon it, although a bob is provided at *z* to draw the

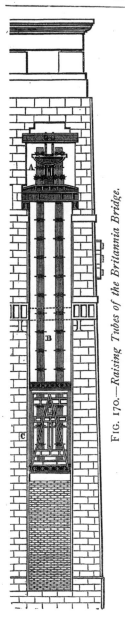

FIG. 170.—*Raising Tubes of the Britannia Bridge.*

chain down, it would be disadvantageous to increase the weight of this to the extent required for forcing back the lifting plunger. A *return* cylinder is therefore made use of, the plunger of which has but a small diameter, and is connected with the head of the lift-plunger, so that it forces the latter back when the lift-cylinder is put in communication with the exhaust-pipe. The water is admitted to the lifting cylinder from the accumulator by a valve worked by a lever, which, when moved the other way, closes the communication and opens the exhaust-pipe, and then the pressure in the return cylinder, which is constant, drives in the plunger of the lifting cylinder. The principle of the accumulator may plainly be used with great advantage even when manual labour is employed, for a less number of men will be required for working the pumps to produce the effect than if their efforts had to be applied to the machine only at the time it is in actual operation, for in the intervals they would, in the last case, be standing idle. Apparatus on the same plan has been used with advantage for opening and shutting dock gates, moving swing bridges, turn-tables, and for other purposes where a considerable power has to be occasionally applied.

A famous example of the application of hydraulic power was the raising of the great tubes of the Britannia Bridge. As already stated, the tubes were built on the shore, and were floated to the towers. This was done by introducing beneath the tubes a number of pontoons, provided with valves in the bottom, so as to admit the water to regulate the height of the tube according to the tide. The great tubes were so skilfully guided into their position that they appeared to spectators to be handled with as much ease as small boats. The mode in which they were raised by the hydraulic presses will be understood from Fig. 170, where A is one of the presses and C the tube, supported by the chains, B. The tubes were suspended in this manner at each end, and as the great tubes weighed 1,800 tons, each press had, therefore, to lift half this weight, or 900 tons. The ram or plunger of the pump was 1 ft. 8 in. in diameter, and the cylinder in which it worked was 11 in. thick. Two steam engines, each 40 horse-power, were used to force the water into the cylinders. These cylinders were themselves remarkable castings, for each contained no less than 22 tons of iron. Notwithstanding the great thickness of the metal, an unfortunate accident occurred while the plungers were making their fourth ascent, for the bottom of one of the cylinders gave way—a piece of iron weighing nearly a ton and a half having been forced out, which, after killing a man who was ascending a rope ladder to the press, fell on the top of the tube 80 ft. below, and made in it a deep indentation. The accident occasioned a considerable delay in the progress of the work, for a new cylinder had to be cast and fitted. Such an accident would assuredly have caused the destruction of the tube itself but for the foresight and prudence of the engineer in placing

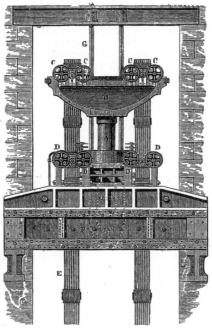

FIG. 171.—*Press for Raising the Tubes.*

beneath the ends of the vast tube as it ascended slabs of wood 1 in. thick, so that it was impossible for the tube to fall more than 1 in. It must be stated that as the tube was lifted each step, the masonry was built up from below, and then as the next lift proceeded inch by inch, a slab of wood was placed under the ends. Although by the giving way of the cylinder of the hydraulic press the end of the tube fell through no greater space than 1 in., the momentum was such that beams calculated to bear enormous weights were broken. At the time of the accident the pressure in the cylinder did not exceed that which it was calculated to bear or that which is frequently applied in hydraulic presses for other purposes. Some scientific observers attributed the failure of the cylinder to the oscillating of the tube. It had

22

been found when the similar tubes of the bridge over the Conway were
being raised, that when the engines at each end made their strokes simultaneously, a dangerous undulation was set up in the tube, and it was therefore necessary to cause the strokes of the engines to take place alternately.
The chains by which the tubes were suspended were made of flat bars 7 in.
wide and about 1 in. thick, being rolled in one piece, with expanded portions
about the " eye," through which the connecting-bolts pass.   The links of
the chain consisted of nine and eight of these bars alternately—the bars of
the eight-fold links being made a little thicker than those of the nine-fold,
so as to have the same aggregate strength.   The mode in which the
hydraulic presses were made to raise the tubes is very clearly described by
Sir William Fairbairn in his interesting work on the Conway and Britannia Bridges, and his account of the mode of raising the tubes is here given
in his own words, but with letters referring to Fig. 171 :

" Another great difficulty was to be overcome, and it was one which presented itself to my mind with great force, viz., in what manner the enor-

FIG. 172.—*Head of Link-Bar*.

mous weight of the tube was to be kept suspended when lifted to the height
of 6 ft., the proposed travel of the pump, whilst the ram was lowered and
again attached for the purpose of making another lift.   Much time was
occupied in scheming means for accomplishing this object, and after examining several projects, more or less satisfactory, it at last occurred to me
that, by a particular formation of the links (of the chain by which the
tubes were to be suspended) we might make the chains themselves support
the tube.   I proposed that the lower part of the top of each link, immediately below the eye, should be formed with square shoulders cut at right
angles to the body of the link (Fig. 172).   When the several links forming
the chain E were put together, these shoulders formed a bearing surface, or
"hold," for the crosshead B attached to the top of the ram A of the hydraulic
pump.   But the upper part of this crosshead, C C, was movable, or formed of
clips, which fitted the shoulders of the chain, and were worked by means
of right- and left-handed screws, and could be made either to clip the chain
immediately under the shoulders when the ram of the pump was down and
a lift about to be made, or be withdrawn at pleasure.   Attached to the large
girders F were a corresponding set of clips, D D, which were so placed and adjusted as to height that when the ram of the pump was at the top there was
distance between the two sets of clips equal to twice the length of the travel
of the pump, or the length of the two sets of the links of the chain.   To

render the action of the apparatus more clear, suppose the tube resting on the shelf of masonry in the position that it was left in after the operation of floating was completed, and the chains attached, and everything ready for the first lift, the ram of the pump being necessarily down. The upper set of clips attached to the crosshead are forced under the shoulders of the links, and the lower set of clips attached to the frames resting upon the girders are drawn back, so as to be quite clear of the chain; the pumps are put into action simultaneously at both ends of the tube, and the whole mass is slowly raised until it has reached a height of 6 ft. from its original resting-place. The clips attached to the crosshead, B, have so far been sustaining the weight, but it will be observed that by the time the pump has ascended to its full travel, the square shoulders of another set of links have come opposite to the lower clips on the girders, D, and these clips are advanced under the shoulders of the links, and the rams being allowed to descend a little, they in their turn sustain the load and relieve the pumps. The upper clips being withdrawn, the rams are allowed to descend, and after another attachment, a further lift of 6 ft. is accomplished; and thus, by a series of lifts, any height may be attained. The fitness of this apparatus for its work was admirable, and the action of the presses was, as Mr. Stephenson termed it, delightful."

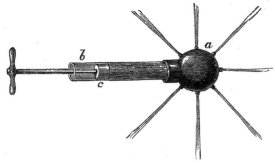

FIG. 173.—*Apparatus to prove Transmission of Pressure in all directions.*

FIG. 174.—*Pneumatic Tubes and Carriages.*

PNEUMATIC DISPATCH.

———◆———

WHEN the use of the electric telegraph became general, it was found necessary to establish in all large towns branch stations, from which messages were conveyed to the central station, or to which they were sent, either by messengers who carried the written despatch, or by telegraphing between the central and branch stations. The latter had the disadvantages of rendering the original message liable to an additional chance of incorrect transmission, and when an unusually great number of despatches had to be sent to or from a particular branch station, there was necessarily great delay in the forwarding of them. The plan of sending the written messages between the central stations by bearers was unsatisfactory on account of the time occupied. These inconveniences led to the invention of a system for propelling, by the pressure of air, the papers upon which the messages were written through tubes connecting the stations. This was first carried into practice by the Electric and International Telegraph Company, who, in this way, connected their central station in London with their City branch stations. The apparatus was designed and erected by Mr. L. Clark and Mr. Varley in 1854. The first tube laid down was from Lothbury and the Stock Exchange—a distance of 220 yards. This tube had an inside diameter of only $1\frac{1}{2}$ in.; but a larger tube, having a diameter of $2\frac{1}{4}$ in. was, some years afterwards, laid between Telegraph Street and Mincing Lane—a distance of 1,340 yards—and was used successfully. In these tubes the carriers were pushed forward by the pressure of the atmosphere, a vacuum having been produced in front by pumping out the air. The plan of propelling the carrier by compressing the air behind it was also tried with good results, and, in fact, with a gain of speed; for, while a carrier occupied 60 or 70 seconds in passing from Telegraph Street to Mincing Lane when drawn by a vacuum, it accomplished its journey in 50 or 55 seconds when it was shot forwards by compressed air, the difference in pressure before and behind it being the same in each case. A great deal of trouble was occasioned when the vacuum system was used, by water being drawn in at the joints of the pipes. This water sometimes accumulated to such a degree, especially after wet weather, that it completely overcame the power of the vacuum to draw the air through it, by lodging in the vertical portions of the tube, where they passed to the upper floors of the central station. This was remedied by improving the construction of the joints, and by arranging a syphon for drawing off any water which might be present. The best construction of the carrier was another matter which required some experience to discover. It was found that gutta-percha, or papier maché covered with felt, was the most efficient material. The tubes found by Mr. Varley to give the best results were formed of lead covered externally with iron pipes. The joints were made perfectly smooth in the inside by means of a heated steel mandrel, on which they were formed, so that the tube was of one perfectly uniform bore throughout. An ingenious arrangement was also adopted by which the air itself was made to do the work of opening and closing the valves, and even that of removing the carrier from the tube: when, by a telegraphic bell, rung from the distant station, it was announced that a carrier was dispatched, the attendant at the receiving station had only to touch for a second a knob marked " receive," which put the tube in communication with the vacuum, in which condition it remained until the arrival of the carrier, which, by striking against a pad of india-rubber, released the detent, and thus cut off the vacuum. The carrier then fell out of the receiver and dropped into a box placed to catch it. When a carrier was sent, it was placed in the tube, and a button marked " send" was touched, by which a communication was opened with a vessel of compressed air and the end of the tube behind the carrier was immediately closed by a slide, the movements being all performed by the air itself. On the arrival of the carrier, the boy at the receiving station rang an electric bell to signal its reception ; and the sender then touched another knob marked " cut off," which caused the supply of compressed air to be cut off, and the slide to be withdrawn from the end of the tube, which was then ready either to receive or send carriers. By this arrangement there was no waste of power, for the reservoirs of compressed air or of vacuum were only drawn upon when the work was actually required to be done.

The tubes laid down by the Telegraph Company are still in active operation ; but at the new Central Telegraph Station the automatic valves of Messrs. Clark and Varley appear to be dispensed with, and the attendants perform the work of closing the tube, shutting off the compressed air, &c., by a few simple movements.

In December, 1869, Messrs. Siemens were commissioned by the Postmaster-General to lay tubes on their system from the General Post Office to the Central Telegraph Station ; and the work having been accomplished in February, 1870, and proving perfectly satisfactory after six weeks' trial, it was decided to connect in the same manner Fleet Street and the West

Strand office at Charing Cross with the Central Station. The system
proposed by the Messrs. Siemens consisted in forming a circuit of tubes,
through which the carriers might be continually passing in one direction.
The diagram, Fig. 175, will give an idea of the manner in which it was
designed to arrange the tubes between the Central Telegraph Station and
Charing Cross. The arrows indicate the direction in which the air rushes
through the tubes; A is the piston in the cylinder, and valves are so
arranged as to pump air out of the chamber V, and compress it into the
chamber P. This plan has been departed from, so far as regards the
Charing Cross Station, for want of space there prevented the tube being
curved with a·radius large enough to convey the carriers without their
being liable to stick, and consequently, these are not carried round in the

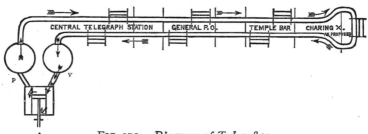

A   FIG. 175.—*Diagram of Tubes, &c*

tube. The passage of carriers being stopped here, there are, in point of
fact, two tubes : an "up" tube and a "down" tube. But these are connected by a sharp bend, so that though the tube is continuous as regards
the air current, it is interrupted as regards the circulation of the carriers.
The tubes are of iron, 3 in. internal diameter, made in lengths of about 19 ft. ;
and for the turns and bends, pieces are curved with a radius of 12 ft. Both
lines are laid side by side in a trench at about a foot depth below the streets.
The ends of the adjacent lengths form butt joints, so that the internal surface is interrupted as little as possible, and there is a double collar to fasten
the lengths together. Arrangements are also made for removing from the
inside of the tubes water or dirt, or matter which may in any manner have
got in.

One special feature of Messrs. Siemens' invention is the plan by which
the carriers are introduced into and removed from the tube at any required
station without the circulation of the air being interfered with. The simple
yet ingenious mechanism by which this is effected will be understood from
the sections shown in Figs. 176 and 177. The figures represent the position of the apparatus when placed to receive a carrier ; A' is the receptacle
into which the carrier is shot by the air rushing from A towards A". This
receptacle is ∪-shaped, the curve of the D corresponding with that of the
tube, and the upper flat part admitting of a piece of plate glass being inserted, through which the attendant may perceive when a carrier arrives.
The progress of the carrier is arrested by a perforated plate, B, which
allows the air to pass. The ends of this receptacle are fixed in two parallel
plates, F F', which also receive the ends of the plain cylinder, having precisely the same diameter as the tube, A. These plates are connected also

by cross-pieces, D E, the whole forming a sort of frame, which turns upon E as a centre; and according as it is put in the position shown by the plain line in Fig. 176, or in that indicated by the dotted lines, causes the receiving tube or the hollow cylinder to form part of the main tube, the cross-piece, D, serving as a handle for moving the apparatus. It should be remarked that the plates are made to fit the space cut out of the main tube with great nicety, otherwise much loss of power would result from leakage. When

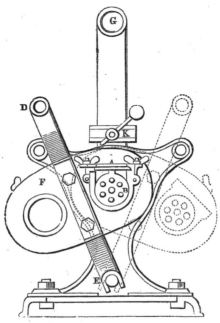

FIG. 176.—*Sending and Receiving Apparatus.*—*Transverse Section.*

the hollow cylinder is in a line with the main tube, it is plain that the carrier will not be stopped, as the tube is then continuous and uninterrupted. In this hollow cylinder also the carrier to be sent is deposited after the rocking frame has been placed on it, Fig. 177; then, on drawing the handle, the hollow cylinder is brought into the circuit, and the carrier at once shoots off. To stop a carrier, the receiving-tube is put in by another movement of the handle, and when the carrier arrives, it is removed by bringing the open cylinder, or *through tube*, into the circuit, and thus making the receiver ready for having the carrier pushed out of it by a rod which is made to slide out by moving a handle. In order to avoid the obstruction to the movement of the air which would be caused by the carrier while in the receiving-tube, a pipe, G, is provided, through which the air chiefly passes when the perforations of the plate, B, are closed by the presence of a carrier.   In this pipe at H is a throttle-valve, which is opened by tappets, K, on the rocking frames when the receiver is in circuit, and again closed when the open tube is substituted.   The current thus suffers no interruption by the action of the apparatus.

The carriers are small cylinders of gutta-percha, or papier maché, closed at one end, and provided with a lid at the other.   They are covered with felt or leather, and at the front they are furnished with a thick disc of drugget or leather, like the leathers of a common water-pump, but fitting quite loosely in the tube.   Such a carrier, being placed in the tube at the Central Station, Fig. 175, will be carried by the current in the direction of the arrows to the Charing Cross Station, where its progress will be interrupted; but according to the original plan it would continue its journey until

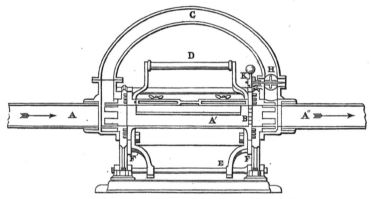

FIG. 177.—*Receiving Apparatus.—Longitudinal Section.*

it again reached the Central Station, where it would be intercepted by the diaphragm, Fig. 175.   But the carrier is stopped, if at any station the receiving-tube is placed in circuit, and this is done when an electric signal indicates to the station that a carrier intended for it has been dispatched. The tubes are worked on the " block system," that is, each section is known to be clear before a carrier is allowed to enter it, and a bell is provided, which is struck by a little lever, moved by each carrier in its passage through, so that the attendant at each station knows when a carrier has shot along the " through tube " of the station.   This mode of working the tubes renders the liability to accidents much less, but their carrying power might be increased by dispatching carriers at regular and very short intervals of time, when the limit would be only in the ability of the attendants to receive a carrier and open the circuit in sufficient time to allow the next following one to proceed without stoppage.   The length of the lines of tube laid down on this system, with the times required for the carriers to traverse them, are stated below, the pressure and the vacuum being respectively equal to the absolute pressures of 22 lbs. and $5\frac{1}{4}$ lbs. on each square inch of the reservoirs during the experiments:

	Yards.	M.	S.
Telegraph Station to General Post Office......	852	1	54
General Post Office to Temple Bar	1,206	2	28
Temple Bar to General Post Office	1,206	2	10
General Post Office to Telegraph Station......	852	1	13
	4,116	7	45

When the air was not compressed, but the vacuum only was used, the air being allowed to enter the other end of the tube at the ordinary atmospheric pressure, the time required for the carrier to traverse the circuit was 10 minutes 23 seconds. In this case the vacuum was maintained, so that the air was constantly in movement; but when the experiment was tried by allowing the air in the tube to become stationary, placing a carrier at one end, and then opening communication with the vacuum reservoir at the other, the carrier required 13½ minutes to complete the journey. This is explained by the fact of the greater part of the air having to be exhausted from the tube before the carrier could be set in motion.

The utility and advantage of the pneumatic system is well seen when its powers are compared with the wires. Thus, a single carrier, which may contain, say, twenty-seven messages, can be sent every eight minutes; and since not more than one message per minute could be transmitted by telegraph wire, even by the smartest clerks, the real average being about two minutes for each message, it follows that only four messages could be sent in the time required for a single carrier to traverse the up tube, and to do the work which could be done by the tube seven wires and fourteen clerks would be required.

Mr. R. S. Culley, the official telegraph engineer, states as his experience of the relative wear and tear of the carriers in these iron tubes and in the smooth lead tubes, that it had been found necessary to renew the felt covering of eighty-two dozen of the carriers used for three months in the iron tubes, while in the same period only thirty-eight dozen of those used in the lead tubes required to be re-covered. The numbers of carriers sent and received by the pneumatic tubes on the 21st of November, 1871, between 11 a.m. and 4 p.m., were :

Iron tubes... 135
2¼ in. lead tubes .... 1,170 } 1,697
1½ in.   „   „   .... 527 }

The mileage of the carriers sent was much greater in the lead than in the iron pipes, although the total lengths of each kind were respectively 5,974 yards and 6,826 yards. The result is remarkable, as showing the effect of apparently slight differences when their operation is summed up by numerous repetitions.

The circuit at Charing Cross having been divided on account of the difficulty mentioned above, the tubes act as separate pipes—one for "up" traffic (*i.e.*, to Central Telegraph Station), the other for "down" (*i.e.*, from the Central Station). The air, however, still accomplishes a circuit, being exhausted at one end and compressed at the other. A very noticeable and curious difference is found between the times required by the carriers to perform the "up" and the "down" journeys :

An "up" carrier requires .... 6·5 minutes
A "down" carrier requires .... 12·5   „

Together ...... 19·0   „

When two pipes were separated at Charing Cross so that the air no longer circulated from one to the other, but both were left open to the atmosphere, while the "up" pipe was worked by a vacuum only and the "down" pipe by pressure only, the times were for

An "up" carrier .... 8·5 minutes
A "down" carrier .... 11·3   „

Together ...... 19·8   „

The time, therefore, for the whole circuit was practically the same—whether the tubes were worked by a continuous current of air or separated, and one worked by the vacuum and the other by pressure. It was also seen that when the tubes were connected so that the air current was continuous, and the pump producing a vacuum at one end and a compression at the other, the neutral point where the pressure was equal to that of the atmosphere was not found midway between the two extremities—that is, at Charing Cross Station—but much nearer the vacuum end. When the tubes were disconnected, it appeared, as already shown by the figures given above, that there was a gain of speed on the down journey, and a loss of speed on the up journey ; and as the requirements of the traffic happened to require greater dispatch for the down journeys, the tubes have been worked in this manner.

It has been proposed to convey letters by pneumatic dispatch between the General and Suburban Post Offices, and the Post Office authorities have even consulted engineers on the practicability of sending the Irish mails from London to Holyhead by this system. It was calculated, however, that although the scheme could be carried out, the proportion of expense for great speeds and long distances would be enormously increased. A speed of 130 miles per hour was considered attainable, but the wear and tear of the carriers would be extremely great at this high velocity, and it was considered doubtful whether this circumstance might not operate seriously against the practical carrying out of the plan. The prime cost would be very great, for the steam power alone which would be requisite would amount to 390 horse-power for every four miles. We thus see that very high velocities would introduce a new order of difficulties in the practical working. The case as regards the velocity with which electric signals can be sent round the world is very different.

An amusing hoax appears to have been perpetrated by some waggish telegraph clerk on an American gentleman at Glasgow, with regard to the pneumatic system of sending messages; for the gentleman sent to the "Boston Transcript" a letter, in which he relates that having sent a telegraphic message from Glasgow to London, he received in a few minutes a reply which indicated a mistake somewhere, and then he went to the Glasgow telegraph office, and asked to see his message.

"The clerk said, 'We can't show it to you, as we have sent it to London.' 'But,' I replied, 'you must have my original paper here. I wish to see that.' He again said, 'No, we have not got it: it is in the post office at London.' 'What do you mean?' I asked. 'Pray, let me see the paper I left here half an hour ago.' 'Well,' said he, 'if you must see it, we will get it back in a few minutes, but it is now in London.' He rang a bell, and in five minutes or so produced my message, rolled up in pasteboard. . . . I inquired if I might see a message sent. 'Oh, yes; come round here.' He slipped a number of messages into the pasteboard scroll, popped it into the tube, and made a signal. I put my ear to the tube and heard a slight rumbling noise for seventeen seconds, when a bell rang beside me, indicating that the scroll had arrived at the General Post Office, 400 miles off. It almost took my breath away to think of it."

In the journal called "Engineering," into which this curious letter was copied, it is pointed out that to travel from London to Glasgow, a distance of 405 miles, in seventeen seconds, the carrier must have moved at the rate of 24 miles per second, or 5 miles a second faster than the earth moves in its orbit, and the carrier would have in such a case become red hot by its friction against the tube before it had travelled a single second.

A plan of conveying, not telegraph messages, but parcels, was proposed and carried into effect some time ago, and more recently has been applied to lines of tubes in connection with the General Post Office. These tubes pass from Euston Station down Drummond Street, Hampstead Road, Tottenham Court Road, to Broad Street, St. Giles's, whence, with a sharp bend, they proceed to the Engine Station at Holborn, and then to the Post Office. The tube is formed chiefly of cast iron pipes of a ⌂-shaped section, 4 ft. 6 in. wide and 4 ft. high, in 9 ft. lengths. There are curves with radii of 70 ft. and upwards, and at these parts the tube is made of brickwork and not of iron. The carriages run on four wheels, and are so constructed that the ends fit the tubes nearly, and the interval left is partly closed by a projecting sheet of india-rubber all round. The carriages are usually sent through the tube in trains of two or three, and the trains are drawn forward by an exhausting apparatus formed by a fan, 22 ft. in diameter, worked by two horizontal steam engines having cylinders 24 in. in diameter and a stroke of 20 in. The air rushes by centrifugal force from the circumference of the fan, and is drawn in at the centre, where the exhaust effect is produced. The tubes which convey the air from the main tube open into the latter at some distance from its extremities, which are closed by doors, so that after the carriage passes the entrance of the suction tube, its momentum is checked by the air included between it and the doors, which air is, of course, compressed by the forward movement of the carriage. At the proper moment the doors are opened by a self-acting arrangement, and the carriage emerges from the tube. There are two lines of tube—an "up" and a "down" line—and means are provided for rapidly transferring the carriages from one to the other at the termini. The time occupied in the transit is about 12 minutes. Some of the inclines have as much slope as 1 in 14, yet loads of 10 or 12 tons weight are drawn up these gradients without difficulty. The mails are sent between Euston Station and the Post Office by means of these tubes. Passengers have also made the journey as an experiment by lying down in the carriages. Fig. 174 shows one of the carriages and the entrance to the tubes.

Great expectations have been formed by some persons of the applications of pneumatic force. Some have suggested its use for moving the trains in the proposed tunnel between England and France. But calculations show that for long distances and large areas such modes of imparting motion are enormously wasteful of power. Thus, in the tunnel alluded to it must be remembered that not only the train, but the whole mass of air in the tunnel would have to be be drawn or pushed forward. The drawing of a train through by exhausting the air would be very similar to drawing it through by a rope; in fact, the mass of air may be regarded as a very elastic rope, but by no means a very light one, or one that could be drawn through without some opposing force which has a certain resemblance to friction coming into operation. Indeed, it has been calculated that in the case named, only five per cent. of the total power exerted by the engines in exhausting the air could possibly produce a useful effect in moving the train.

Air has also been made the medium for conveying intelligence in another manner than by shooting written messages through tubes, for its property of transmitting pressure has been applied to produce at a distance signals like those made use of in the electric telegraph system. A few years ago, an apparatus for this object was contrived by Signor Guattari, whose invention is known as the " Guattari Atmospheric Telegraph." In this there is a vessel charged with compressed air by a compression-pump, and the pressure is maintained by the same means, while the reservoir is being drawn upon. A valve is so arranged that the manipulator can readily admit the compressed air to a tube extending to the station where the signals are received, at which the pressure is made to move a piston as often as the sender opens the valve. This movement is made to convey intelligence when a duly regulated succession of impulses is sent into the tube—the receiving apparatus being arranged either to give visible or audible signals, or to print them on slips of paper, according to any of the methods in use with the electric telegraph. Certain advantages over the electric system are claimed for this pneumatic telegraph—as, for example, greater simplicity and less liability to derangement. The tubes, which are merely leaden piping of small bore, are also exempt from the inconvenient interruptions which electric communication sometimes suffers from electrical disturbances in the atmosphere. The pneumatic system is easily arranged, and from its great simplicity any person can in a few hours learn to use the whole apparatus, while it is calculated that the expense of construction and working would not be above half of that incurred for the electric system. For telegraphs in houses, ships, warehouses, and short lines, this invention will doubtless prove very serviceable; but for long lines a much greater force of compression would be required, and the time needed for the production of an impulse at the distant ends of the tubes would be considerably increased. [1875].

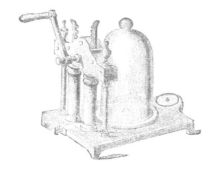

FIG. 178.—*The Sommeiller Boring Machines.*

ROCK BORING.

ALLUSION has already been made to one great characteristic of our age, namely, the replacement, in every department of industry, of manual labour by machines. A brief notice of even the main features of the various contrivances which have been made to take the place of men's hands would more than occupy this volume. Accordingly, we must omit all reference to many branches of manufacture, although the products may be of very great utility, and the processes of very high interest ; and in taking one example here and another there, we must be guided mainly by the extent and depth of the influence which the new invention appears destined to exert. This consideration has, with scarcely an exception, decided the selection of the topics already discussed, and it has also determined the introduction of the present article, which relates to machines of no less general importance than the rest, although at first sight it might seem to enter upon the details of merely a special branch of industry. But so general are the interests connected with the subject we are about to lay before our readers, that we are not sure it would not have been more logical to have placed the present article before all the rest. For whence comes the iron of which our steam engines, tools, rails, ships, cannon, bridges, and printing presses are made ?—whence comes the fuel which supplies force to the engines ?—whence come, in fine, the substances which form the *matériel* of every art ? Plainly from the earth—the nurse and the mother of all, and in most cases from the bowels of the earth, for her treasures are hidden far below the surface—the coal, and the ores of iron and other metals, are not ready to our hand, exposed to the light of day. The railways also, and the canals, can be made only on condition that we cut roads through the solid rocks, and pierce with tunnels the towering mountains. Hence the tools which enable us to penetrate into the sub-stance of the earth present the highest general interest from a practical point of view, and this interest is enhanced by the knowledge of the struc-ture and past history of our planet acquired in such operations.

The operations by which solid rocks are penetrated in the sinking of shafts for mines, or in the driving of tunnels, drifts, headings, galleries, or cuttings for railways, mines, or other works, are easily understood. In the first place a number of holes—perhaps 3 ft. or 4 ft. deep and 2 in. or 3 in. in diameter—are formed in the rock. The holes are then charged with gun-powder or other explosive materials, a slow-burning match is adjusted, the miners retire to a safe distance, the explosion takes place—detaching, shattering, and loosening masses of the rock more or less considerable ; and then gangs of workmen clear away the stones and *débris* which have been detached by the explosion, and the same series of operations is renewed. The holes for the blasting charges are formed by giving repeated blows on the rock with a kind of chisel called a *jumper*—the end of which is formed of very hard steel, so that the rock is in reality chipped away. The *débris* resulting from this operation is cleared away from time to time by a kind of auger or some similar contrivance. But for many purposes it is necessary to drill holes in rocks to great depths, hundreds of feet perhaps, as for example, in order to ascertain the nature of underlying strata, or to verify the presence of coal or other minerals before the expense of sinking a shaft is incurred. These bore-holes were commonly formed in exactly the same manner as the blast-holes already mentioned, by repeated blows of a chisel or jumper, which was attached to the end of a rod ; and as the hole deepened, additional lengths of rod were joined on, and the rods were withdrawn from time to time to admit of the removal of the *débris* by augers, or by cylinders having a valve at the bottom. The reciprocating movement is given to the chisels and rods either by hand or by steam or water power. When the length of the rods becomes considerable, of course the difficulty of giving the requisite blows in rapid succession is greatly increased, for the whole length of rods has to be lifted each time, and if allowed to fall with too much violence, the breaking of the chisel or the rods is the inevitable result. The time requisite for drawing out the rods, removing the fragments chipped out, and again attaching the rods and lowering, also increases very much as the bore gets deeper. Messrs. Mather and Platt, the Manchester engineers, have, in order to obviate these difficulties, constructed machines in which the chipping or cutting is done by the fall of a tool suspended from a rope, the great advantage re-sulting from the arrangement being the facility and rapidity with which the tools used for the cutting and for the removal of the *débris* are lowered to their work and drawn up. It is necessary in using the jumper, whether in cutting blast-holes or bore-holes, to give the tool a slight turn after each blow, in order that the rock may be chipped off all round, and the action of the tool equalized. Many attempts have been made to drill rocks after the fashion in which iron is drilled—that is, by drilling properly so called, in which the tool has a rapid rotary motion. But even in comparatively soft rock, it is found that no steel can sufficiently withstand the abrading action of the rock, for the tool becomes quickly worn, and makes extremely slow progress. We shall have presently to return to the subject of bore-holes ; but now let us turn our attention to an example which will illustrate the nature and advantages of the machinery which has in recent times been applied to work the jumpers by which the holes for blasting are formed.

THE MONT CENIS TUNNEL.

THE successful construction, by the direction of Napoleon, of a broad and easy highway from Switzerland into Italy, crossing the lofty Alps amid the snows and glaciers of the Simplon, has justly been considered a feat of skill redounding to the glory of its designers. But we have recently witnessed a greater feat of engineering skill, for we have seen the Alps conquered by the stupendous work known as the Mont Cenis Tunnel. This tunnel is $7\frac{1}{2}$ English miles in length; but it is not the mere length which has made the undertaking remarkable. The mountain which is pierced by the tunnel is formed entirely of hard rock, and what added still more to the apparently impracticable character of the proposal when first announced was the circumstance that it was quite impossible to sink vertical shafts, so that the work could not, as in the usual process, be carried on at several points simultaneously, but must necessarily be continued from the two extremities only, a restriction which would occasion a vast loss of time and much expense, to say nothing of the difficulties of ventilating galleries of more than three miles in length. The reader must bear in mind that the importance of this question of ventilation depends not simply on the renewing of the air contaminated by the respiration of the workmen, but on the quick removal of the noxious gases produced in the explosions of the blasting charges. A work surrounded by such difficulties would probably have never been attempted had not Messrs. Sommeiller and Co. invited the attention of engineers to an engine of their invention, worked by compressed air, and capable of automatically working "jumpers" which could penetrate the hardest rock. These rock-boring machines, having been examined by competent authorities in the year 1857, were pronounced so efficient that the execution of the long-spoken-of Alpine tunnel was at once resolved upon, and before the close of that year the work had actually been commenced, after a skilful and accurate survey of the proposed locality had been made, and the direction of the tunnel set out. The tunnel does not pass through Mont Cenis, although the post road from St. Michel to Susa passes over part of Mont Cenis, which gives its name to the pass. The mountain really pierced by the tunnel is known as the Grand Vallon, and the tunnel passes almost exactly below its summit, but at a depth the perpendicular distance of which is as nearly as possible one mile. The northern end of the tunnel is near a village named Fourneaux.

Pending the construction of the Sommeiller machines, and other machinery which was to supply the motive force, the work of excavation was commenced at both ends, in 1857, in the ordinary manner, that is, by hand labour, and in 1858 surveys of the greatest possible accuracy were meanwhile made, in order that the two tunnels might be directed so that they would meet each other in the heart of the mountain. The reader will at once perceive that the smallest error in fixing on the direction of the two straight lines which ought to meet each other would entail very serious consequences. The difficulties of doing this may be conceived when we remember that the stations were nearly 8 miles apart, separated by rugged mountains, in a region of snows, mists, clouds, and winds, over which the levels had to be taken, and a very precise triangulation effected. So successfully were these difficulties overcome, and so accurately were the measurements and calculations made, that the junction of the centre lines of the completed tunnel failed by only a *few inches*, a length utterly insignificant under the conditions.

The work was carried on by manual labour only, until the beginning of 1861, for it was found, on practically testing the machinery, that many important modifications had to be made before it could be successfully employed in the great work for which it was designed. After the machinery had been set to work, at the Bardonnêche end, breakages and imperfections of various parts of the apparatus, or the contrivances for driving it, caused delay and trouble, so that during the whole of 1861 the machines were in actual operation for only 209 days, and the progress made averaged only 18 in. per day, an advance much less than could have been effected by manual labour. The engineers, not disheartened or deterred by these difficulties and disappointments, encountered them by making improvement after improvement in the machinery as experience accumulated, so that a wonderful difference in the rate of progress showed itself in 1862, when the working days numbered 325, and the average rate of advance was *three feet nine inches per day*.

At the Fourneaux extremity more time was required for the preparation of the air-compressing machinery, and the machines had been at work in the other extremity, with more or less interruption, for nearly two years before the preparations at Fourneaux were completed.

The illustration at the head of this article, Fig. 178, represents the Sommeiller machines at work, the motive power being compressed air, conveyed by tubes from receivers, into which it is forced until the pressure becomes equal to that of six atmospheres, or 90 lbs. per square inch. The compression was effected by taking advantage of the natural heads of water, which were made to act directly in compressing the air ; the pressure due to a column of water 160 ft. high being made to act upwards, to compress air, and force it through valves into the receivers; then the supply of water was cut off, and that which had risen up into the vessel previously containing air was allowed to flow out, drawing in after it through another valve a fresh supply of air; and then the operations were repeated by the water being again permitted to compress the air, and so on, the whole of the movements being performed by the machinery itself. The compressed air, after doing its work in the cylinders of the boring tools, escaped into the atmosphere, and in its outrush became greatly cooled, a circumstance of the greatest possible advantage to the workmen, for otherwise, from the internal warmth of the earth, and that produced by the burning of lights, explosions of gunpowder, and respiration, the heat would have been intolerable. At the same time, the escaping air afforded a perfect ventilation of the workings while the machines were in action. At other times, as after the explosion of the charges, it was found desirable to allow a jet of air to stream out, in order that the smoke and carbonic acid gas should be quickly cleared away. Even had the work been done by manual labour alone, a plentiful supply of compressed air would have been required merely for ventilation, so that there was manifest advantage in utilizing it as the motive power of the machines.

FIG. 179.—*Transit by Diligence over Mont Cenis.*

The experience gained in the progress of the work suggested from time to time many improvements in the machinery and appliances, which finally proved so effectual that the progress was accelerated beyond expectation. At the end of 1864, when the machines had been in work about four years, it was calculated that the opening of the tunnel might be looked for in the course of the year 1875. But in point of fact it happened that on the 25th December, 1870, perforator No. 45 bored a hole from Italy into France, by piercing the wall of rock, about 4 yards thick, which then separated the workings from each other. The centre lines of the two workings, as set out from the different sides of the mountain, failed to coincide by only a foot, that set out on the Fourneaux side being this much higher than the other, but their horizontal directions exactly agreeing. The actual length of the tunnel was found to be some 15 yards longer than the calculated length, the calculation having given 7·5932 miles for the length, whereas by actual measurement it was found to be 7·6017 miles. The heights above the sea-level of the principal points are these:

	Feet.
Fourneaux, or northern entrance	3,801
Bardonnêche, or southern entrance	4,236
Summit of tunnel	4,246
Highest point of mountain vertically over the tunnel	9,527

23

The tunnel is lined with excellent brick and stone arching, and it is connected with the railways on either side by inclined lines, which are in part
tunnelled out of the mountain, so that the extremities of the tunnel referred
to above are not really entered by the trains at all; but these lateral tunnels
join the other and increase the total distance traversed underground to
very nearly 8 miles, or more accurately, 7·9806 miles. The time required
by a train to pass from one side to the other is about 25 minutes. What a
contrast is this to the old transit over the Mont Cenis pass by "diligence"!
We have the scene depicted in Fig. 179, where we perceive, sliding down
or toiling up the steep zigzag ascents, a series of curious vehicles drawn by
horses with perpetually jingling bells.

The cost of the Mont Cenis Tunnel was about £3,000,000 sterling, or
upwards of £200 per yard; but as a result of the experience gained in this
gigantic work, engineers consider that a similar undertaking could now be
carried out for half this cost. It is supposed that the profit to the contractors for the Mont Cenis Tunnel was not much less than £100 per yard.
The greatest number of men directly employed on the tunnel at one time
was 4,000, and the total horse-power of the machinery amounted to 860.
From 1857 to 1860, by hand labour alone, 1,646 metres were excavated;
from 1861 to 1870 the remaining 10,587 metres were completed by the
machines. The most rapid progress made was in May, 1865, in which
month the tunnel was driven forward at one end the length of 400 feet.
When the workings were being carried through quartz, a very hard rock,
the speed was greatly reduced—as, for example, during the month of April,
1866, when the machines could not accomplish more than 35 ft.

The perforators used in the Mont Cenis Tunnel were worked by compressed air, conveyed to a small cylinder, in which it works a piston, to the
rod of which the jumper is directly attached. The air, being admitted behind the piston, impels the jumper against the rock, and the tool is then
immediately brought back by the opening of a valve, which admits compressed air in front of the piston, at the same time that the air which has
driven it forward is allowed to escape, communication with the reservoir
of compressed air having previously been closed behind it. The whole
of these movements are automatic, and they are effected in the most
rapid manner, four or five blows being struck in every second, or between
two and three hundred in one minute. Water was constantly forced into
the holes, so as to remove the *débris* as quickly as it was formed. A number
of these machines were mounted on one frame, supported on wheels, running on the tramway which was laid along the gallery. The perforators had
no connection with each other, for each one had its own tube for the conveyance of compressed air, and its own tube to carry the water used for
clearing out the hole, and the cylinders were so fixed on the frames that
the jumpers could be directed in any desired manner against any selected
portion of the rock. They were driven to an average depth of about 2½ ft.,
and the process occupied from forty to fifty minutes. When a set of holes
had thus been formed, the cylinders were shifted and another series commenced, until about eighty holes had been bored, the formation of the
whole number occupying about six or seven hours, and the holes being so
arranged that the next operation would detach the rock to the required
extent. The flexible tubes, which conveyed the air and water to the machines from the entrances, were then removed from the machines and
stowed away, the frame bearing the perforators was drawn back along the
tramway, workmen advanced whose duty it was to wipe out the holes,

charge them with powder, and fix the fuses ready for the explosion. When the slow-burning match was ignited, all retired behind strong wooden barricades, at a safe distance, until the explosion had taken place; and after the compressed air had been allowed to stream into the working, so as to clear away all the smoke and gas generated by the explosion, the workmen ran up on a special tramway the waggons which were to carry away all the detached stones; and when this had been done, the floor was levelled, the tramways were lengthened, and the frame bearing the drilling machines was brought up to begin a fresh series of operations, which were usually repeated about twice in the course of every twenty-four hours. A great part of the rock consists of very hard calcareous schist, interspersed with veins of quartz, one of the hardest of all rocks, which severely tries the temper of the steel tools, for a few blows on quartz will not unfrequently cause the point of a jumper to snap off.

ROCK-DRILLING MACHINES.

SEVERAL forms of rock-drills, or perforators, have been constructed on the same principle as that used in the Mont Cenis Tunnel, and a description of one of them will give a good notion of the general principle of all. We select a form devised by Mr. C. Burleigh, and much used in America, where it has been very successfully employed in driving the Hoosac Tunnel, effecting a saving in the cost of the drilling amounting to one-third of the expense of that operation, and effecting also a still greater saving of time, for the tunnel, which is 5 miles in length, is to be completed in four years, instead of twelve, as the machines make an advance of 150 ft. per month, whereas the rate by hand labour was only 49 ft. per month. These machines are known as the " Burleigh Rock Drills," and have been patented in England for certain improvements by Mr. T. Brown, who has kindly supplied us with the following particulars:

The Burleigh perforator acts by repeated blows, like Bartlett and Sommeiller's, but its construction is more simple, and the machine is lighter and not half the size, while its action is even superior in rapidity and force. The Burleigh machines are composed of a single cylinder, the compressed air or steam acting directly on the piston, without the necessity of flywheel, gearing, or shafting. The regular rotation of the drills is obtained by means of a remarkably simple mechanical contrivance. This consists of two grooves, one rectilinear, the other in the form of a spiral cut into the piston-rod. In each of these channels, or grooves, is a pin, which works freely in their interior : these pins are respectively fixed to a concentric ring on the piston-rod. A ratchet wheel holds the ring, and the pin slides into the curve, causing it to turn always in the same direction, without being able to go back. By this eminently simple piece of mechanism, the regular rotation of the drill-holder is secured. The slide-valve is put into motion by the action of a projection, or ball-headed piston-rod, on a double curved momentum-piece, or trigger, which is attached to the slide-rod or spindle by a fork, thus opening and shutting the valve in the ascent and descent of the piston. Fig. 180 represents one of the machines attached in this instance by a clamp to the frame of a tripod. The principal parts of the machine are the cylinder, with its piston, and the cradle with guide-ways, in which the cylinder travels. The action of the piston is similar to that of the ordinary steam hammer, with this difference, that. in addition

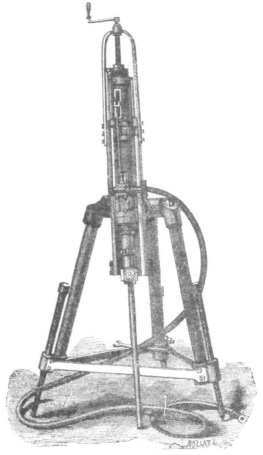

FIG. 180.—*Burleigh Rock Drill on Tripod.*

to the reciprocating, it has also a rotary, motion. The drill-point is held in a slip-socket, or clamp, at the end of the piston-rod, by means of bolts and nuts. The drill-point rotates regularly at each stroke of the piston, making a complete revolution in every eighteen strokes. For hard rocks it is generally made with four cutting edges. in the form of a St. Andrew's cross, thus striking the rock in seventy-two places in one revolution, each cutting edge chipping off a little of the stone at each stroke in advance of the one preceding. The jumper makes, on an average, 300 blows per minute, and such is the construction of the machine, that the blows are of an elastic, and not of a rigid, nature, thus preventing the drill-point from being soon blunted. It has been found in practice, that a drill-point used in the Burleigh machine can bore on an average 20 ft. of Aberdeen granite without re-sharpening. As the drill pierces the rock, the machine is fed down the guide-ways of the cradle by means of the feed-screw (see Fig. 180), according to the nature of the rock and the progress made. When the cylinder has been fed down the entire length of the feed-screw, and if a greater depth of hole is required, the cylinder is run back, and a longer drill is inserted in the socket at the end of the piston-rod. The universal clamp may be attached to any form of tripod, carriage, or frame, according to the requirements of the work to be done; it enables the machines to work vertically, horizontally, or at any angle.

The following advantages are claimed for this machine : Any labourer can work it; it combines strength, lightness, and compactness in a remarkable degree, is easily handled, and is not liable to get out of order. No part of the mechanism is exposed ; it is all enclosed within the cylinder, so there is no risk of its being broken. It is applicable to every form of rockwork, such as tunnelling, mining, quarrying, open cutting, shaft-sinking, or submarine drilling ; and in hard rock, like granite, gneiss, ironstone, or quartz, the machine will, according to size, progress at the incredible rate of *four inches* to *twelve inches per minute*, and bore holes from ¾ in. up to 5 in. diameter. It will, on an average, go through 120 ft. of rock per day, making forty holes, each from 2 ft. to 3 ft. deep, and it can be used at any angle and in any direction, and will drill and clear itself to any depth up to 20 ft.

The following extract from the "Times," September 24th, 1873, gives an account of some experiments with the machine, made at the meeting of the British Association in that year, before the members of the Section of Mechanical Science :

"Yesterday, considerable interest was taken in this section, as it had been announced that a 'Burleigh Rock Drilling Machine' would be working during the reading of a paper by Mr. John Plant. The machine was not, however, in the room, but was placed in the grounds outside, where it was closely examined by the members after the adjournment, and seen in full operation, boring into an enormous block of granite. The aspect of the machine cannot be called formidable in any respect, for it looks like a big garden syringe, supported upon a splendid tripod; but when at work, under about 80 lbs. pressure of compressed air, it would be deemed a very revolutionary agent indeed, against whose future power the advocates for manual labour in the open quarry, the tunnel, and even the deep mine, may well look aghast. Placed upon a block of granite a yard deep, the machine was handled and its parts moved by the fair hands of many of the lady associates of scientific proclivities ; but once the source of power was turned on, the drill began its poundings, eating holes 2 in. in diameter in the block of granite, and making a honeycomb of it as easily as a schoolboy would demolish a sponge cake. It pounds away at the rate of 300 strokes, and progresses forward about 12 in., in the minute, making a complete revolution of the drill in eighteen strokes, and keeping the hole free of the pounded rock. The machine was fixed to work at any angle, almost as

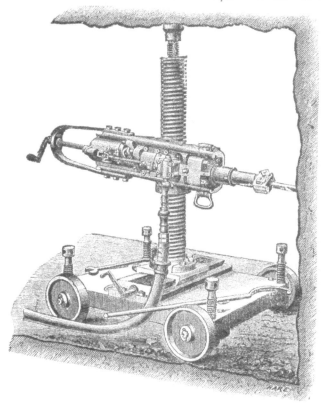

FIG. 181.—*Burleigh Rock Drill on Movable Column.*

readily as a fireman can work his hose ; and its adaptation to a wide range
of stone-getting, by drilling for blasting, and cutting large blocks for building and engineering, with a saving of capital and labour, was admitted by
many members of the section.   The tool is called the 'Burleigh Rock
Drill,' invented by Mr. Charles Burleigh, a gentleman hailing from Massa-
chusetts, United States.   The patent is the property of Messrs. T. Brown
and Co., of London.   The principal feature of this new machine is, that it
imitates in every way the action of the quarryman in boring a hole in the
rock."

Many forms of carriages and supports have, from time to time, been
made to suit the work for which the 'Burleigh' machines have been
required.   The machine is attached to these carriages, or supports, by
means of the universal clamp, by which it can be worked in any direction
and at any angle.   Of these carriages we select for notice only two forms,
one of which is shown in Fig. 181.   This carriage can be used to great

advantage in adits and drifts.   It consists of an upright column, with a screw clamp-nut for holding and raising or lowering the machine, the whole being mounted on a platform which can slide right across the carriage, and thus the machine can be brought to work on any point of a heading.   It is secured in position by means of a jack-screw in the top of the column ; and as the carriage is mounted on wheels, it is easily moved to permit of blasting.   Fig. 182 represents a carriage which is the result of many years' experience with mining machinery, and it is considered a very perfect appliance.   It is constructed of wood and iron, and it runs on wheels.

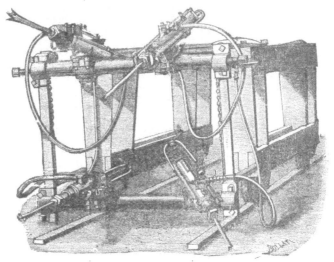

FIG. 182.—*Burleigh Rock Drills mounted on a Carriage.*

The supports for the machines, four of which may be mounted at once, are two horizontal bars, the lower of which can be raised or lowered, as may be necessary.   The two parallel sides of the carriage are joined only at the upper side, and there is nothing to prevent it from being run into the heading, though the way between the rails may be heaped up with broken rock, if only the rails are clear.   Drilling, and the removal of the broken rock, may then proceed simultaneously ; for, by means of a narrow gauge inside the carriage rails, small cars may be taken right up to the *débris*.   It is made in different sizes, to suit the dimensions of the tunnel required.   To give the carriage steadiness in working, it is raised from the wheels by jack-screws, and held in position by screws in a similar manner to the carriage represented in Fig. 181.

An extremely interesting system of drilling rocks—totally different from that on which the machines we have just described are constructed—has, within the last few years, been introduced by Messrs. Beaumont and Appleby.   What does the reader think of boring holes in rocks with diamonds ?   It has long been a matter of common knowledge that the diamond is the hardest of all substances, and that. it will scratch and wear down any other substances, while it cannot itself be scratched or worn by anything but diamond. In respect to wearing down or abrading hard stones, the diamond, according to experiments recently made by Major Beaumont, occupies a position over all other gems and minerals to a degree far beyond that which has been generally attributed to it ; for in these experiments it was found that on applying a diamond, or rather a piece of the "carbonate" about to be described. fixed in a suitable holder, to a grindstone in rapid rotation, the grindstone was quickly worn down ; but on repeating a similar experiment with sapphires and with corundum, it was these which were worn down by the grindstone. Without, on the present occasion, entering into the natural history of the diamond, we may say that there are, besides the pure colourless transparent crystals so highly prized as gems, several varieties of diamond, and that those which are tinged with pink, blue, or yellow, are far from having the same value for the jeweller. Then there is another impure variety called *boort,* which appears to be employed only to furnish a powder by which the brilliants are ground ánd polished. In the diamond gravels of Brazil, from which we derive our regular supply of these gems, there was discovered in 1842 a curious variety of dark-coloured diamond, in which the crystalline cleavage, or tendency to split in certain directions (which belongs to the ordinary stones), appears to be almost absent ; and the substance might be regarded as a transition form between the diamond and graphite but for its hardness. This substance was until lately used for the same purposes as *boort,* which is a nearer relative of the pure crystal. and like it, splits along certain planes. It received from the miners the name of "*carbonado,*" and with regard to the application we are considering. it has turned out to be a sort of Cinderella among diamonds ; for its unostentatious appearance is more than compensated for by its surpassing all its more brilliant sisters in the useful property to which reference has been made. This Brazilian term is doubtless the origin of the English name by which the substance in question is known among the English diamond merchants, who call it "carbonate"—an unfortunate

FIG. 183.

word, for it is used in chemistry with an entirely different signification. "Carbonate" it is, however, which supplies the requirements of the rock-drill, and the selected stones are set in a crown, or short tube, of steel, represented by *c* in Fig. 183. In this they are secured as follows : holes are drilled in the rim of the tube, and each hole is then cut so that a piece of the diamond exactly fits it, and when this piece has been inserted, the metal is drawn round by punches, so as almost to cover the stone, leaving only a point projecting, *b b.* The portions of the crown between the stones are somewhat hollowed out, as at *a,* for a purpose which will presently be mentioned. The crown thus set with the boring gems is attached to the end of a steel tube, by which it is made to rotate with a speed of about 250 revolutions per minute while pressed against the rock to be bored. Water is forced through the steel tube, and passing out between the rock and the crown, especially under the hollows, *c c,* makes its escape between the outside of the boring-tube and the rock, thus washing away all the *débris* and keeping the drill cool. The pressure with which the crown is forced forward depends, of course, on the nature of the rock to be cut, and varies from 400 lbs. to 800 lbs. In this way the hardest rocks are quickly penetrated—sometimes, for example, at the rate of 4 in. per minute, compact limestone at 3 in., emery at 2 in., and quartz at the rate of 1 in. per minute. It is found that, even after boring through hundreds of feet of such materials, the diamonds are not in the least worn, but as fit for work as before : they are damaged only when by accident one of the stones gets knocked out of its setting; and this machine surpasses all in the rapidity with which it eats its way through the firmest rocks. This, it must be observed, is the special privilege of the diamond drill—that, since the begemmed steel crown and the boring-rods are alike tubular, the rock is worn away in an annular space only, and a solid cylinder of stone is detached from the mass, which cylinder passes up with the hollow rods, where, by means of certain sliding wedges, it is held fast, and is drawn away with the rods.

When the diamond drill is used merely for driving the holes for blasting, this cylinder of rock is not an important matter; but there is an application of the drill where this cylinder is of the greatest value, furnishing as it does a perfect, complete, and easily preserved section of the whole series of strata through which the drill may pass when a bore-hole is sunk in the operation of searching for minerals (which is so significantly called in the United States " prospecting," a phrase which seems to be making its way in England in mining connections); for the core is uniformly cylindrical, the surface is quite smooth, and any fossils which may be present come up uninjured, so far as they are contained in the solid core, and thus the strata are readily recognized. Contrast this with the old method, where the bore-hole in prospecting is made by the reciprocating action imparted to a steel tool, and merely the *pounded* material is obtained, usually in very small fragments, by augers or sludge-pumps : the fossils, which might afford the most valuable indications, crushed and perhaps incapable of being recognized; and instead of the beautifully definite and continuous cylinder, a mere mass of *débris* is brought up. In the prospecting-bores the diameter of the hole is from 2 in. to 7 in. The size adopted depends on the nature of the strata to be penetrated, and on the depth to which it is proposed to carry the boring. When the strata are soft, the operation is commenced with a bore of 7 in., and when this has been carried to an expedient depth, the danger of the sides of the hole falling in is avoided by putting down tubes, and then the diamond drill, fixed to tubes of a somewhat smaller diameter, will be again inserted, and the boring recommenced; or the hole can be widened, so as to receive the lining-tubes. Of course, in boring through hard rocks, such as compact limestones, sandstone, &c., no lining-tubes are necessary.

In a very interesting paper, read before the members of the Midland Institute of Mining Engineers, by Mr. J. K. Gulland, the engineer of the Diamond Rock-Boring Company, who have the exclusive right of working the patents for this remarkable invention, that gentleman concludes by remarking that " the leading feature of the diamond drill is that it works without percussion, thus enabling the holing of rocks to be effected by a far simpler class of machinery than any which has to strike blows. Every mechanical engineer knows, often enough to his cost, that he enters upon a new class of difficulties when he has to recognize it as a normal state of things with any machinery he is designing that portions of it are brought violently to rest. These difficulties increase very much when the power, as in the case of deep bore-holes, has to be conveyed for a considerable distance. Where steel is used a percussive action is necessitated, as, if a scraping action is used, the drill wears quicker than the rock. The extraordinary hardness of the diamond places a new tool in our hands, as its hardness, compared with ordinary rock, say granite, is practically beyond comparison. Putting breakages on one side, a piece of "carbonate" would wear away thousands of times its own bulk of granite. Irrespective of the private and commercial success which this invention has attained, it is a boon to a country such as ours, where minerals constitute in a great measure our national wealth and greatness."

The advantages of the diamond drill may be illustrated by the case of what is termed the Sub-Wealden Exploration. From certain geological considerations, which need not be entered upon here, several eminent British and continental geologists have arrived at the conclusion that it is probable that coal underlies the Wealden strata of Kent and Sussex, and that it may be perhaps met with at a workable depth. If such should really prove to be the case, the industrial advantages to the south of England would be very great, for the existence of coal so comparatively near to the metropolis would prove not only highly lucrative to the owners of the coal, but confer a direct benefit upon thousands by cheapening the cost of fuel. A number of property owners and scientific men, having resolved that the matter should be tested by a bore, raised funds for the purpose, and a 9 in. bore had been carried down to a depth of 313 ft. in the ordinary manner, when a contract was entered into with the Diamond Rock-Boring Company for a 3 in. bore extracting a cylinder of rock 2 in. in diameter. The company, as a precautionary measure, lined the old hole with a 5 in. steel tube; and in spite of some delay caused by accidents, they increased the depth of the hole to 1,000 ft. in the interval from 2nd February, 1874, to 18th June, 1874—the progress of the work being regarded with the greatest interest by the scientific world. Unfortunately, the further progress of the work has been prevented by an untoward event, namely, the breaking of the boring-rod, or rather tube; and, although the company is prepared with suitable tackle for extracting the tubes in case of accidents of this kind, and generally succeeds in lifting them by a taper tap, which, entering the hollow of the tube, lays hold of it by a few turns—yet, in this instance, where there have been special difficulties, the extraction of so great a length of tubes is, as the reader may imagine, by no means an easy task. Six attempts have been made to remove the boring-rods which have dropped down; but so difficult has this operation proved, that all these efforts having failed, it has been decided to abandon the old work and commence a new boring on an adjacent spot. A contract has been entered into with the Diamond Boring Company, who have undertaken to complete the first 1,000 ft. for £600, which is only £200 more than it would have cost to completely line the old bore-holes with iron tubes—an operation which was contemplated by the committee in charge of the exploration. The terms agreed to by the company are very favourable to the promoters of the Sub-Wealden Exploration, although the cost of the second 1,000 ft. will be £3,000 more; and the committee are relying upon the public for contributions to enable them to carry on their enterprise. It is most probable that funds will be forthcoming, and should the boring result in the finding of coal measures beneath the Wealden strata, all the nation will be the richer and participate in the advantages resulting from an undertaking carried on by private persons. Already a totally unexpected source of

FIG. 184.— *The Diamond Drill Machinery for deep Bores.*

wealth has been met with by the old bore showing the existence of considerable beds of gypsum in these strata, and the deposits of gypsum are
about to be worked. Whether coal be found or not found, there is no doubt
that a bore-hole going down 2,000 ft. will greatly increase our geological
knowledge, and may reveal facts of which we have at present no conception.

The boring-tubes, it may be remarked, are made in 6 ft. lengths, and are
so contrived that the joints are nearly flush—that is, there is no projection

at the junctions of the tubes. Fig. 184 is engraved from a photograph of the machinery used for working the diamond drill when boring a hole for " prospecting." This looks at first sight a very complicated machine, but in reality each part is quite simple in its action, and is easily understood when its special purpose has been pointed out. We cannot, however, do more than indicate briefly the general nature of the mechanism. The reader will on reflection perceive that, although the idea of causing a rod to rotate in a vertical hole may be simple, yet in practically carrying it out a number of different movements and actions have to be provided for in the machinery. The weight of the rods cannot be thrown on the cutters, nor borne by the moving parts of the machine—hence the movable disc-shaped weights attached to the chains are to balance the weight of the boring-rods as the length of the latter is increased. There must also be a certain amount of *feed* given to the cutters, regulated and adjusting itself to avoid injurious excess : hence a nut which feeds the drill is encircled by a friction-strap in which it merely slips round without advancing the cutter when the proper pressure is exceeded. There must be means of throwing this into or out of gear, or advancing the tool in the work and of withdrawing it—hence the handles seen attached to the brake-straps. Water must be drawn from some convenient source, and caused to pass down the drill-tube—hence the force-pump seen in the lowest part of the figure. The rods must be raised by steam power and lowered by mechanism under perfect control—hence suitable gearing is provided for that purpose.

The reader may be interested in learning what is the cost of " prospecting" with this unique machinery. The company usually undertake to bore the first 100 ft. for £40, but the next 100 ft. cost £80—that is, for 200 ft. £120 would be charged ; the third 100 ft. would cost £120—that is to say, the first 300 ft. would cost £240, and so on—each lower 100 ft. costing £40 more than the 100 ft. above it. Some of the holes bored have been of very great depth, and have been executed in a marvellously short space of time. Thus, in 54 days, a depth of 902 ft. was reached at Girrick in a boring for ironstone ; another for coal at Beeston reached 1,008 ft. ; and at Walluff in Sweden 304½ ft. were put down in one week !

These machines are peculiarly suitable for submarine boring, for they work as well under water as in the air ; and they will no doubt be put into requisition in the preliminary experiments about to be made for that great project which bids fair to become a sober fact—the Channel Tunnel between England and France; and as, by the time these pages will be before the public, the work of the greatest and boldest rock-boring yet attempted will have commenced, and the scheme itself will be the theme of every tongue, the Author feels that the present article would be incomplete without some particulars of the great enterprise. [1875.]

THE CHANNEL TUNNEL.

THE notion of connecting England and France by a submarine line of railways is not of the latest novelty, but has been from time to time mooted by the engineers of both countries. The most carefully prepared scheme, however, is embodied in the joint propositions of Sir J. Hawkshaw and Messrs. Brunlees and Low among English engineers; and those of M. Gamond on the French side, which these gentlemen have prepared at the invitation of the promoters of the scheme, give the clearest and most authentic account of the considerations on which this gigantic enterprise will be based, and from this document we draw the following passages :

The undersigned engineers, some of whom have been engaged for a series of years, in investigating the subject of a tunnel between France and England, having attentively considered those investigations and the facts which they have developed, beg to report thereon jointly for the informa-· tion of the committee.

These investigations supported the theory that the Straits of Dover were not opened by a sudden disruption of the earth at that point, but had been produced naturally and slowly by the gradual washing away of the upper chalk; that the geological formations beneath the Straits remained in the original order of their deposit, and were identical with the formations of the two shores, and were, in fact, the continuation of those formations.

Mr. Low proposed to dispense entirely with shafts in the sea, and to commence the work bv sinking pits on each shore, driving thence, in the first place, two small parallel driftways or galleries from each country, connected at intervals by transverse driftways. By this means the air could be made to circulate as in ordinary coal-mines, and the ventilation be kept perfect at the face of the workings.

Mr. Low laid his plans before the Emperor of the French in April, 1867, and in accordance with the desire of his Majesty, a committee of French and English gentlemen was formed in furtherance of the project.

For some years past Mr. Hawkshaw's attention has been directed to this subject, and ultimately he was led to test the question, and to ascertain by elaborate investigations whether a submarine tunnel to unite the railways of Great Britain with those of France and the Continent of Europe was practicable.

Accordingly, at the beginning of the year 1866, a boring was commenced at St. Margaret's Bay, near the South Foreland; and in March, 1866, another boring was commenced on the French coast, at a point about three miles westward of Calais; and simultaneously with these borings an examination was carried on of that portion of the bottom of the Channel lying between the chalk cliffs on each shore.

The principal practical and useful results that the borings have determined are that on the proposed line of the tunnel the depth of the chalk on the English coast is 470 ft. below high water, consisting of 175 ft. of upper or white chalk and 295 ft. of lower or grey chalk; and that on the French coast the depth of the chalk is 750 ft. below high water, consisting of 270 ft. of upper or white chalk and 480 ft. of lower or grey chalk; and that the position of the chalk on the bed of the Channel, ascertained from the examination, nearly corresponds with that which the geological inquiry elicited.

In respect to the execution of the work itself, we consider it proper to drive preliminary driftways or headings under the Channel, the ventilation of which would be accomplished by some of the usual modes adopted in the best coal-mines.

As respects the work itself, the tunnel might be of the ordinary form, and sufficiently large for two lines of railway, and to admit of being worked by locomotive engines, and artificial ventilation could be applied; or it might be deemed advisable, on subsequent consideration, to adopt two single lines of tunnel. The desirability of adopting other modes of traction may be left for future consideration.

Such are the essential passages of the report which, in 1868, was submitted to the Government of the Emperor Louis Napoleon, and was made the subject of a special commission appointed by the Emperor to inquire into the subject in all its bearings. The commission presented its report in 1869, and these are the chief conclusions contained in it ·

I. The commission, after having considered the documents relative to the geology of the Straits, which agree in establishing the continuity, homogeneity, and regularity of level of the *grey chalk* between the two shores of the Channel,
Are of opinion that driving a submarine tunnel in the lower part of this chalk is an undertaking which presents reasonable chances of success.

Nevertheless they would not hide from themselves the fact that its execution is subject to contingencies which may render success impossible.

II. These contingencies may be included under two heads : either in meeting with ground particularly treacherous—a circumstance which the known character of the grey chalk renders improbable; or in an influx of water in a quantity too great to be mastered, and which might find its way in either by infiltration along the plane of the beds, or through cracks crossing the body of the chalk.

Apart from these contingencies, the work of excavation in a soft rock like grey chalk appears to be relatively easy and rapid; and the execution of a tunnel, under the conditions of the project, is but a matter of time and money.

III. In the actual state of things, and the preparatory investigations being too incomplete to serve as a basis of calculation, the commision will not fix on any figure of expense or the probable time which the execution of the permanent works would require.

The chart, Fig. 185, and the section, Fig. 186, will give an idea of the course of the proposed tunnel, which will connect the two countries almost at the nearest points. The depth of the water in the Channel along the proposed line nowhere exceeds 180 ft.—little more than half the height of St. Paul's Cathedral, which building would, therefore, if sunk in the midst of the Channel, stiil form a conspicuous object rising far above the waves. But the tunnel will pass through strata at least 200 ft. below the bottom of the Channel, rising towards each end with a moderate gradient; and from the lower points of these inclines the tunnel will rise slightly with a slope of 1 in 2,640 to the centre, or just sufficient for the purposes of drainage. On the completion of the tunnel a double line of rails will be laid down in it, and trains will run direct from Dover to Calais. Companies have already been formed in England under the presidency of Lord Richard Grosvenor, and in France under that of M. Michel Chevalier, and the legislation of each country has sanctioned the enterprise. Verily the real magician of our times is the engineer, who, by virtually abolishing space, time, and tide, is able to transport us hither and thither, not merely one or two— almost like the magicians we read of in the " Arabian Nights," with their enchanted horses or wonderful carpets—but by hundreds and by tens of hundreds.

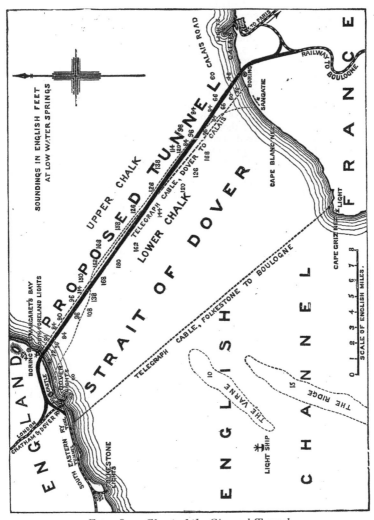

FIG. 185.—*Chart of the Channel Tunnel.*

The "Daily News" of January 22nd, 1875, in presenting its readers with a chart of the proposed tunnel, offered also the following sensible and interesting comment on the subject :

"This long-debated project has at length emerged from the region of

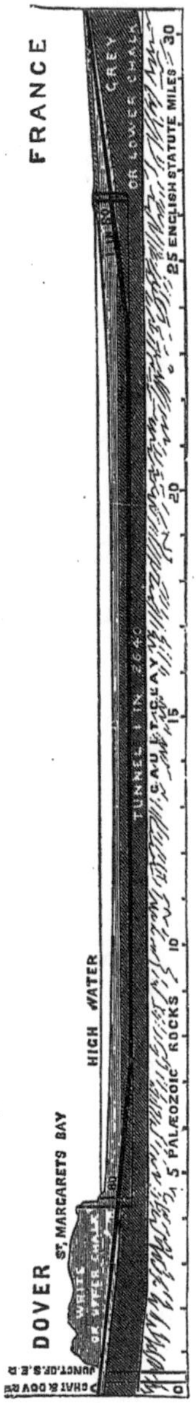

FIG. 186.—*Section of the Channel Tunnel.*

speculation, and is entering the stage of practical experiment. On this side the Channel a company has been formed to carry out the work, and on the other side the French Minister of Public Works has presented to the Assembly a Bill authorizing a French company to co-operate with the English engineers. The enterprise is one worthy of the nations which have in the present generation joined the two shores of the Atlantic by an electric cable, and cut a ship canal through the Isthmus of Suez, and of the age which has obliterated the old barrier of the Alps. All these gigantic undertakings seemed almost as bold in conception and as difficult of execution as the great work now about to commence. Those twenty miles of sea have long been crossed by telegraph lines ; they will soon be bridged, as it were, by splendid steamers ; but even our own generation, accustomed as it is to gigantic engineering works, has scarcely regarded the construction of a railway underneath the waves as within the reach of possibility. M. Thomé de Gamond, who first made the suggestion five and thirty years ago, was long regarded as an over-sanguine person, who did not recognize the inevitable limits of human skill and power. A tunnel under twenty miles of stormy sea seemed very much like an engineer's dream, and it is only within the last few years that it has been regarded as a feasible project. Of its possibility, however, there seems now to be no manner of doubt. It is merely a stream of sea-water, and not a fissure in the earth, which divides us from the Continent. Prince Metternich was right in speaking of it as a ditch. The depth is nowhere greater than one hundred and eighty feet ; and so far as careful soundings can ascertain the condition of the soil underneath the water, it consists of a smooth unbroken bed of chalk. The success of the experiment depends on this bed of chalk being continuous and whole. Should any very deep fissure exist, which is extremely improbable, the tunnel may probably not be driven through it. But given, what every indication shows to exist, a homogeneous chalk bed some hundreds of feet in thickness, the driving of a huge bore for twenty miles through it is a mere question of time, money, and organization, and as the engineers have these resources at their command, they are sanguine, and we may even say confident, of success.

"The method by which it is proposed that the excavation shall be made is in some respects similar to that which was successfully employed in tunnelling the Alps. Mont Cenis was pierced by machinery adapted to the cutting of hard rock ; the chalk strata under the Channel are to be bored

FIG. 187.—*View of Dover.*

by an engine, invented by Mr. Dickenson Brunton, which works in the
comparatively soft strata like a carpenter's auger. A beginning will be
made simultaneously on both sides of the Channel, and the effort will at
first be limited to what we may describe as making a clear hole through
from end to end. This small bore, or driftway as it is called, will be some
seven or nine feet in diameter. If such a communication can be successfully made, the enlargement will be comparatively easy. Mr. Brunton's
machine is said to cut through the chalk at the rate of a yard an hour. We
believe that those which were used in the Mont Cenis Tunnel cut less than
a yard a day of the hard rock of the mountain. Two years, therefore,
ought to be sufficient to allow the workers from one end to shake hands
with those from the other side. The enlargement of the driftway into the
completed tunnel would take four years' more labour and as many millions
of money. The millions, however, will easily be raised if the driftway is
made, since the victory will be won as soon as the two headways meet
under the sea. One of the great difficulties of the work is shared with the
Mont Cenis Tunnel, the other is peculiar to the present undertaking. The
Alps above the one, and the sea above the other, necessarily prevent the
use of shafts. The work must be carried on from each end ; and all the
débris excavated must be brought back the whole length of the boring, and
all the air to be breathed by the workmen must be forced in. The provision of a fit atmosphere is a mere matter of detail. In the great Italian
tunnel the machines were moved by compressed air, which, being liberated
when it had done its work, supplied the lungs of the workers with fresh

24

oxygen. The Alpine engineers, however, started from the level of the earth : the main difficulty of the Submarine Tunnel seems to be that it must have as its starting-point at each end the bottom of a huge well more than a hundred yards in depth. The Thames Tunnel, it will be remembered, was approached, in the days when it was a show place, by a similar shaft, though of comparatively insignificant depth. This enterprise may indeed be said to bear something like the relation to the engineering and mechanical skill of the present day which Brunel's great undertaking bore to the powers of an age which looked on the Thames Tunnel as the eighth wonder of the world. Probably the danger which will be incurred in realizing the larger scheme is less than that which Brunel's workmen faced.

" It is, of course, impossible for any estimate to be formed of the risks of this enormous work. They have been reduced to a minimum by the mechanical appliances now at our disposal, but they are necessarily considerable. The tunnel is to run, as we understand, in the lower chalk, and there will be, as M. de Lesseps told the French Academy, some fifty yards of soil—a solid bed of chalk, it is hoped—between the sea-water and the crown of the arch. Moreover, an experimental half-mile is to be undertaken on each side before the work is finally begun ; the engineers, in fact, will not start on the journey till they have made a fair trial of the way. Altogether the beginning seems to us to be about to be made with a combination of caution and boldness which deserves success, even though it should be unable to command it. Unforeseen difficulties may arise to thwart the plans, but the enterprise, so far, is full of promise. The opening of such a communication between this country and the Continent will be a pure gain to the commercial and social interests on both sides. It obliterates the Channel so far as it hinders direct communication, yet keeps it intact for all those advantages of severance from the political complications of the Continent, which no generation has more thoroughly appreciated than our own. The commercial advantages of the communication must necessarily be beyond all calculation. A link between the two chief capitals of Western Europe, which should annex our railway system to the whole of the railways of the Continent, would practically widen the world to pleasure and travel and every kind of enterprise. The 300,000 travellers who cross the Channel every year would probably become three millions if the sea were practically taken out of the way by a safe and quick communication under it. The journey to Paris would be very little more than that from London to Liverpool. It is, however, quite needless to enlarge on these advantages. The Channel Tunnel is the crowning enterprise of an age of vast engineering works. Its accomplishment is to be desired from every point of view, and, should it be successful, it will be as beneficent in its results as the other great triumphs of the science of our time."

The Channel Tunnel is not yet a *fait accompli*, although the preliminary trial works have been made at both ends. Drift-ways of some ten feet diameter have been cut beneath the waters of the strait, and instead of the experimental half mile mentioned in the foregoing paragraph, the works have been pushed forward on the English side for about a mile and a quarter with complete success. As was anticipated, no physical difficulties were met with, for the machines did their work with the greatest ease, and the drift has now remained for some years practically free from any infiltration of water. These results indicate that the scheme might be completed with speed and safety. Parliament, how-

ever, has refused to allow the undertaking to proceed, being moved to
this course by the opinions of military authorities, who see dangers to
England in the completion of this enterprise, or at least such a disturbance of the British complacency at the notion that our island might be
reached otherwise than "by the inviolate sea," that the whole land would
be liable to terrors and alarms from invasion by stratagem. It is represented that huge fortresses and a special army for that purpose would
become necessary to guard the mouth of the tunnel were it made. This
is, perhaps, the kind of objection which such an enterprise could not fail
to raise. But it can hardly be expected that all the commercial and
international advantages which the realization of the scheme would
undoubtedly secure are for ever to stand in abeyance for such opinions
as have, for the present. caused the operations to be suspended. It has
been pointed out that there are many ways of instantly rendering such
a tunnel impracticable in case of a sudden alarm. But the necessity
could only arise after a supposed paralysis or destruction of such army
and navy as Britain could bring together to defend her land. Perhaps
military skill will presently devise less costly methods of defence than
those authorities now suppose the tunnel would require ; or, even if such
armaments were really necessary for our sense of insular security, the
expense might be no unprofitable outlay for the advantages to be gained.
It is satisfactory to know that the promoters of the scheme are sanguine
of the subsidence of the military and political prejudices, which are now
the only obstacles to its accomplishment. A somewhat unexpected
result from the operations in connection with the experimental drift-ways
has been the discovery, on the Kentish coast, of seams of coal underlying
the chalk at a workable depth.

THE ST. GOTHARD RAILWAY.

SINCE the completion of the Mont Cenis Tunnel, a still greater piece
of rock boring has been begun and finished in the great tunnel of the
St. Gothard Railway. The construction of a railway to connect Italy
with Switzerland, was a project conceived as far back as 1838, when the
first railway company in the latter country was constructed. The route
of the proposed line was a matter of much debate, not alone on account
of difference of engineering opinions, but also by reason of the various
competing interests that would have to be reconciled and induced to
co-operate in the work. The St. Gothard route was only one of the
several schemes that were advocated, and the first decisive step appears
to have been taken at Lucerne, where, in 1853, a meeting was called by
the authorities of the canton to consider the merits of the project ; the
result being that the Lucerne Government addressed to the Federal
Council a representation of the advantages this route would afford. More
discussion ensued, and it was only when Switzerland appeared likely to
have no share in the traffic between the Milan district and the more
northern parts of Europe that, in 1861, the partizans of the St. Gothard
route appointed a provisional committee to take action in the matter.
This committee had plans prepared, and sent a deputation to obtain the
assent of the Italian Government. The canton of Tessin, through which

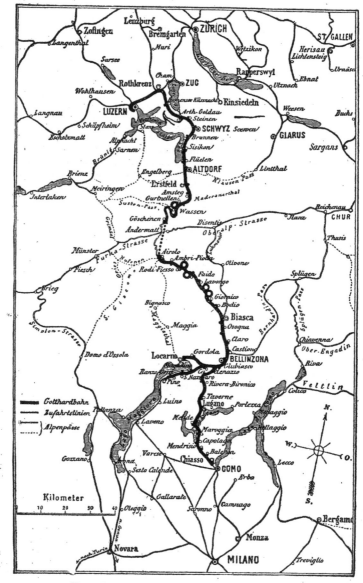

FIG. 187*a*.—*Map of the St. Gothard Railway.*

the projected line, or its then surviving rival, was designed to pass, became a lively scene in the game of speculation, for promoters rushed in to secure, if possible, concessions which they might sell at a very advanced price to the winning party. For this purpose came to that poor Swiss canton Jews and Christians from every land. The St. Gothard route gained the day, and a Union was, in 1863, formed by the concurrence of the two principal Swiss railways and fifteen of the cantons most interested in the scheme. Difficulties and delays were, however, encountered before the necessary compacts could be concluded with the neighbouring states—and then there came the war of 1867. So that it was not until the latter part of 1872 that the construction of the line was actually entered upon. Before the great work of piercing the St. Gothard had been completed, the undertaking was embarrassed by financial difficulties arising from the fact of the lines on the Italian side costing more than double the estimated amount. The Swiss Government, however, voted a special subsidy, and the work, which had been suspended for a while, was proceeded with ; much attention being paid to its economical prosecution. In 1881, when the line was opened, the mails were carried between Zurich and Bellinzona in seven hours, instead of in thirty hours as previously required for transit by the excellently appointed mail carriages under the Federal Administration.

Besides the great tunnel, the St. Gothard line has some unique devices in railway construction which cannot fail to interest the reader. Several of the passes over the Alps have been made use of from time immemorial. We know that Hannibal led his Carthaginian hosts over one of them, and that they have been traversed by Roman legions, as well as by Germanic hordes. But, although the St. Gothard is the most direct of all the routes, it never afforded a passage to armies or migratory tribes. The road through this pass was not formed by the use of any elaborate appliances for overcoming the natural obstacles: it was rather the work of simple peasants and mountain shepherds, with such rough constructions in wood as might give a sufficiently secure passage across the torrents and gorges. The old road keeps beside the Reuss from the head of the lake of Lucerne until it reaches the highest level of the pass, where the water-shed occurs. It then descends steeply, with many twists and windings, to the banks of the Ticino, and it follows the course of this river to its embouchure at Tresa, on Lago Maggiore. The railway follows the same course, except that it cuts off the higher part of the pass by the great tunnel piercing the mountain. The scenery throughout could, perhaps, be nowhere equalled for the variety of its wild grandeur.

The great tunnel of the St. Gothard passes from Gœschenen, on the Reuss, beneath the col of the pass, and emerges close to the village of Airolo, on the banks of the Ticino. The length of this tunnel is rather more than nine and a quarter miles, so that it is about one and a half miles longer than the Mont Cenis Tunnel. Its northern end is 3,638 feet above the sea level ; its southern end is higher, namely, 3,756 feet ; but there is an intermediate point in the tunnel higher than either—3,786 feet—and from this there is a uniform incline in each direction. The tunnel is 300 yards beneath the lowest part of the valley of Andermatt, and the summits of the mountains it traverses are at least a mile above it. The motive power by which the rock-drilling machines used in driving the tunnel were actuated was, as in the case of the Mont Cenis Tunnel, compressed air; and the power used for compressing the air was, in this case also, a head of water,—but this was not applied in the same way. The waters of the Reuss at the northern side, and those of the Tremola and of the Ticino at the southern side, were taken at a considerable height in very large cast-iron pipes, and were made to act upon powerful turbines that gave motion to the compressing machines. These were capable of compressing the air so that its volume was reduced to one-twentieth, and the pressure it then exercised would, of course, be equal to that of twenty atmospheres, or about 300 lbs. on the square inch,—or more than three times as much as was made use of in the Mont Cenis Tunnel. The compressed air, carried through pipes to the head of the workings in the rock, was there allowed to exert its force on the pistons of the perforators in the manner already described. There was, in fact, a continual repetition of exactly the same cycle of operations of boring, charging, firing, etc., that are mentioned on page 355. A large quantity of the compressed air was always allowed to rush into the work immediately after each blasting, in order that the smoke and other products might be driven out and the atmosphere rendered fit for respiration. In attacking the mountain simultaneously from each side it was, of course, essential that the tunnels should be driven in precisely the same direction, and therefore the positions of the points of departure had to be determined by very careful surveys. At Gœschenen, the gorge of the Reuss did not naturally admit of a sufficient distance of vision to fix the direction with the required accuracy, and it became necessary to pierce a thick mass of rock with a special tunnel for the purpose of taking a sight sufficiently far back. At Airolo, again, the tunnel had to enter the valley by curving towards the village; and here a provisional gallery had to be driven in the straight line.

Several contractors competed for the work of constructing this great tunnel, and it was at first supposed that an Italian company, which was managed by some of the principal engineers engaged on the Mont Cenis, would be almost certain to obtain the contract. The promoters, however, intrusted the work to a private individual, M. Louis Favre, of Geneva. This gentleman undertook to complete the tunnel in eight years, at the price of 2,800 francs per mètre for the work of excavation merely, exclusive of masonry, etc. This cost would be not far from £101 per English yard. The contract was signed on August 7th, 1872, and on September 12th of the same year M. Favre commenced operations at the southern end, and the work at the northern end was begun on October 9th following. The operations were carried on with great energy, and even during the period of the Company's financial difficulties there was no stoppage of the works between Gœschenen and Airolo. It has been suggested that it was largely due to the regular and successful progress of this great piece of rock boring that the Company were enabled to re-establish themselves on a basis that ensured the completion of the whole undertaking. The contractor, on his part, did not fail to encounter many physical difficulties. At the southern end much trouble was caused by torrents of water gushing from the soil, many of these being of great volume and force; in fact, the work was here carried on for nearly a whole year in the midst of water—for the ground for the first mile consisted of glacial and other deposits, which were intersected by subterranean watercourses. Reaching the solid rock was here a relief. But at Gœschenen little of loose formation was met with; but the rock encountered was of extreme hardness—consisting, indeed, of almost pure

PLATE XVI.

THE NORTH MOUTH OF THE GREAT TUNNEL, ST. GOTHARD RAILWAY.

quartz, which had the effect of quickly blunting the points of even the best tempered tools. But another kind of difficulty had to be overcome when the workings got beneath the vale of Urseren. Here, at several places, layers of argillaceous matter were found between the masses of hard rock. These layers were easy enough to pierce through, but on account of the pressure of the rocks in which they were interspersed, they were squeezed out and gradually protruded within the tunnel, which would soon have become entirely obstructed. At first a very massive lining of timber was tried, but it was soon found that this must be replaced by a solid vaulting of stone. The first vault failed to sustain the pressure, and so did the second, although the thickness of the material was more than a yard. In some places these operations had to be several times repeated, and from this cause the cost of parts of the tunnel has been nearly £1,000 per yard.

The instances above mentioned may be taken as mere specimens of the physical difficulties attending a work of this kind. There are often others arising from the unusual circumstances under which the workmen are placed, and others again from accidental causes alone. M. Favre experienced some of these, as, for example, when one year a fire destroyed the greater part of the village of Airolo ; another year there was a strike on the part of the workmen. The high temperature in the workings was, especially towards the

FIG. 187b.—*The Uppermost Bridge over the Maienreuss.*

end, a source of great trouble. The cause of the heat is no doubt the same as that which is held to support the theory of the earth's central heat. Numberless observations have established the fact that the temperature of the earth's crust increases as we go deeper. The increase appears not to be uniform in different places—at least there is much discrepancy in the estimates that have been made. But as a sufficient approximation to a general statement, it may be taken as proved that for every seventy feet or so that you go below the surface of the ground, there is an increase of the temperature of the strata equal to 1° Fahrenheit. Now, the workmen in the two sections of the tunnel had, at last, to carry on their labour in a temperature of more than 100° Fahrenheit. This, perhaps, might have been one cause of some unprecedented kinds of malady that appeared amongst the tunnel labourers. M. Favre himself was not destined to witness the completion of his great undertaking, for, on July 19th, 1879, as he was returning from an inspection of the tunnel, he fell into the arms of his companions, struck down by a fatal attack of apoplexy. On February 29th, 1880, the last fuse required to blast down the rock separating the two tunnels was fired by one of the few workmen who had been engaged in the operations during the whole period from their commencement. It was found that the two tunnels met exactly and coincided in direction.

The construction of such a line of railway as the St. Gothard tries the skill of the engineer, and taxes all the resources of his art. The problems presented by the nature of the route, and the requirements of the iron road, have in this case been successfully solved by bold expedients—by new and ingenious devices. The reader will readily understand that the ordinary cart road may wander about, so to speak, of its own will ; it is not confined to the limited gradient of the line ; or obliged to make its turns and curves of at least a certain radius. Now, there are portions of the valley where the general slope is too steep for the railway to follow, and where it was necessary to form it in zig-zags, so that certain sections of the gorge or valley may be several times traversed by the line returning upon itself. Fig. 187*d* is a view showing an incident of this kind, and one of the most interesting spots on the route. The dark line on the spectator's right is the track of the railway ; the white trace, which in the lower part of the view is seen on the other side of the Reuss, is the ordinary road. If this last be followed up the valley, it will be seen to cross first the Reuss, and then a tributary stream (the *Maïenreuss*) descending through a gorge on the right, after which it zig-zags up a hill to the village of *Wasen* (the church of which village is seen crowning the eminence in the centre of our view), and then it continues its course up the valley, passing through a small village, and disappearing over the shoulder of a hill on the right bank of the river. Let us now carefully follow the railway from where the train at the bottom of the picture is seen ascending the gradient. The line presently passes under a bridge, and then enters the tunnel, near to the entrance of which a small building will be noticed. The course of the tunnel is shown by the *curve* marked in dots, for this tunnel makes a round within the rock, and the railway emerges to day again at a point lower down in the course of the valley than at the entrance to the tunnel, but at a higher level. It is seen in the figure appearing from behind the rocks in the right-hand lower corner, passing under a short tunnel and continuing along the mountain side. The curved tunnel resembles in direction part of the turn of a

FIG. 187c.— *The Bridges over the Maïenreuss near Wasen.*

FIG. 187d.—*Windings of the Line near Wasen.*

corkscrew ; it is one of a series of *helicoidal* tunnels of which there are several examples on the line. The entrance to this tunnel is 2,539 feet, the exit 2,654 feet, above the sea level. It is known as *Pfaffensprung* (Monk's Leap) Tunnel. The line again enters' a short tunnel, and immediately , crosses the deep gorge of the Maïenreuss, to plunge again into another tunnel at the base of this hill on which Wasen stands. Higher up it crosses the Reuss and enters the helicoidal tunnel of *Wattingen* (dotted line). On emerging from this, the line re-crosses the Reuss, and may now be traced *down* the valley, but higher up on the mountain side, coming in the reverse direction, and after passing *Wasen* on the other side, re-crossing the Maïenreuss gorge by a second bridge. Then turning back again through another helicoidal tunnel (*Leggistein*) the line crosses the Maïenreuss for the third time, and continues its course up the valley. Fig. 187c gives us a near view of *Wasen*, and a glimpse up the gorge of the Maïenreuss from its junction with the Reuss. The bridge with the large single arch is that which carries the ordinary road, and higher up we see the three iron bridges that carry the railway backwards and forwards in its doublings. We can well imagine the perplexity of anyone ascending the valley in the train for the first time, and ignorant of the peculiarities of this extraordinary railway. In crossing the first, or lowest bridge, over the Maïenreuss, he would catch a glimpse of the church of Wasen, perched on its hill, high above him, and on his right. After being carried through more tunnels, and over more bridges, he would some minutes afterwards be disposed to think that his eyes were deceiving him, for there, still on his right, he would see the same church, but now on about the same level as the train. Again, after more tunnels and bridges, the church would once more appear, transferred to the left of the line, and sunk very far down. These several apparitions of the same building in different positions, after the train has seemed to have been pursuing its onward course the while,—which course would not be judged by any impressions the traveller would usually receive to be other than rectilinear,—are indeed a regular bewilderment to the inexperienced traveller. He is then obliged finally to resign himself passively to be carried he knows not whither or how, for his sense of direction is completely at fault ;—the train comes out of tunnels which seem turned the wrong way ; the river, which he expected to find on the left hand, he sees on the right ; and the Reuss appears to have reversed the direction of its flow.

It is understood that the St. Gothard line has been a great commercial success, for the number of passengers entering and leaving Italy by that route has been enormous, and still shows a large annual increase. Indeed, the prosperity of the line has been so great that the project has been revived of carrying another railway over the Alps to connect Italy and Switzerland by way of the Simplon. If this scheme should be carried out, the mountains will be pierced by a tunnel of a length double that of the St. Gothard.

FIG. 188.

LIGHT.

THE foregoing pages have been devoted to the description of inventions or operations in which mechanical actions are the most obvious features. Some of the contrivances described have for their end and object the communication of motion to certain bodies, others the arrangement of materials in some definite form, and all are essentially associated with the idea of what is called *matter*. But we are now about to enter on another region—a region of marvels where all is enchanted ground—a region in which we seem to leave far behind us our grosser conceptions of matter, and to attain to a sphere of more refined and subtile existence. For we are about to show some results of those beautiful investigations in which modern science has penetrated the secrets of Nature by unfolding the laws of light—

"Light
Ethereal, first of things, quintessence pure."

The diversity and magnificence of the spectacles which, by day as well as by night, are revealed to us by the agency of light, have been the theme of the poet in every age and in every country. It cannot fail to arrest the attention to find Science declaring that all the loveliness of the landscape, the fresh green tints of early summer and the golden glow of autumn, the brilliant dyes of flowers, of insects, of birds, the soft blue of the cloudless sky, the rosy hues of sunset and of dawn, the chromatic splendour of rubies, emeralds, and other gems, the beauties of the million-coloured rainbow,—are all due to light—to light alone, and are not qualities of the bodies themselves, which merely *seem* to possess the colours. The following quaint stanzas, in which a poet of the seventeenth century addresses "Light" have a literal correspondence with scientific truth:

"All the world's bravery, that delights our eyes,
Is but thy several liveries;
Thou the rich dye on them bestowest,
Thy nimble pencil paints this landscape as thou goest.

"A crimson garment in the rose thou wearest:
A crown of studded gold thou bearest;
The virgin lilies, in their white,
Are clad but with the lawn of almost naked light.

"The violet, Spring's little infant, stands
Girt in thy purple swaddling-bands:
On the fair tulip thou dost dote;
Thou clothest it in a gay and parti-coloured coat."

All these beauties are indeed derived from the imponderable and *invisible* agent, light; and the variety and changefulness of the effects we may constantly observe show that light possesses the power of impressing our visual organs in a thousand different ways, modified by the surrounding circumstances, as witness that ever-shifting transformation scene—the sky. In the skies of such a climate as that of England there are ceaseless changes and ever-beautiful effects, producing everywhere more perfect and diversified pictures than the richest galleries can show. In the night how changed is the spectacle, when the sun's more powerful rays are succeeded by the soft light of the moon, sailing through the azure star-bestudded vault! What limitless scope for the artist is afforded by these innumerable modifications of a single subtile agent, in light and shade, brightness and obscurity, in the contrasts and harmonies of colours, and in the countless hues resulting from their mixtures and blendings!

It will be necessary, before attempting to explain the discoveries and inventions which prove how successfully science, aided by the powerful mathematical analysis of modern times, has acquired a knowledge of the ways of light, to discuss such of the ordinary phenomena as have a direct bearing upon the subjects to be considered.

FIG. 189.—*Rays.*

SOME PHENOMENA OF LIGHT.

IT may be considered as a matter of common experience that light is
able to pass through certain bodies, such as air and gases, pure water,
glass, and a number of other liquids and solids, which, by virtue of this
passage of light, we term *transparent*, in opposition to another class of
bodies, called *opaque*, through which light does not pass. That light traverses a vacuum may be held as proved by the light of the sun and stars
reaching us across the interplanetary spaces; but it may also be made the
subject of direct experiment by an apparatus described below, Fig. 190.
Another fact, very obvious from common observation, is that light usually
travels in straight lines. Some familiar experiences may be appealed to
for establishing this fact. For example, every one has observed that the
beams of sunlight which penetrate an apartment through any small opening pursue their course in perfectly straight lines across the atmosphere, in

which their path is rendered visible by the floating particles of dust. It is by reason of the straightness with which rays of light pursue their course that the joiner, by looking along the edge of a plank, can judge of its truth, and that the engineer or surveyor is able by his theodolite and staff to set out the work for rectilinear roads or railways. On a grander scale than in the sunbeam traversing a room, we witness the same fact in the effect represented in Fig. 189, where the sun, concealed from direct observation, is seen to send through openings in the clouds, beams that reveal their paths by lighting up the particles of haze or mist contained in the atmosphere. It is not the air itself which is rendered visible; but whenever a beam of sunlight, or of any other brilliant light, is allowed to pass through an apartment which is otherwise kept dark, the track of the beam is always distinctly visible, and, especially if the light be concentrated by a lens or concave mirror, the fact is revealed that the air, which under ordinary circumstances appears so pure and transparent, is in reality loaded with floating particles, requiring only to be properly lighted up to show themselves.

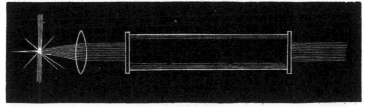

FIG. 190.

Professor Tyndall, in the course of some remarkable researches on the decomposition of vapours by light, wished to have such a glass tube as that represented in Fig. 190, filled with air perfectly free from these floating particles. When the beam of the electric lamp passed through the exhausted tube, no trace of the existence of anything within the tube was revealed, for it appeared merely like a black gap cut out of the visible rays that traversed the air; thus proving that light, although the agent which makes all things become visible, *is itself invisible*—that, in fact, we see not light, but only illuminated substances. When, however, air was admitted to the tube, even after passing through sulphuric acid, the beam of the light became clearly revealed within the tube, and it was only by allowing the air to stream very slowly into the exhausted glass tube through platinum pipes, packed with platinum gauze and intensely heated, that Professor Tyndall succeeded in obtaining air "optically empty," that is, air in which no floating particles revealed the track of the beams. The destruction of the floating matter by the incandescent metal proves the particles to be organic; but a more convenient method of obtaining air free from all suspended matter was found by Professor Tyndall to be the passing of the air through a *filter of cotton wool*. It must not be supposed that it is only occasionally, or in dusty rooms, laboratories, or lecture-halls, that the air is charged with organic and other particles—

"As thick as motes in the sunbeams."

"The air of our London rooms," says Tyndall, "is loaded with this organic dust, nor is the country air free from its pollution. However ordinary daylight may permit it to disguise itself, a sufficiently powerful beam causes the air in which the dust is suspended to appear as a semi-solid, rather than as a gas. Nobody could, in the first instance, without repugnance, place the mouth at the illuminated focus of the electric beam and inhale the dust revealed there. Nor is this disgust abolished by the reflection that, although we do not see the nastiness, we are drawing it in our lungs every hour and minute of our lives. There is no respite to this contact with dirt ; and the wonder is, not that we should from time to time suffer from its presence, but that so small a portion of it would appear to be deadly to man." The Professor then goes on to develop a very remarkable theory, which attributes such diseases as cholera, scarlet fever, small pox, and the like, to the inhalation of organic *germs* which may form part of the floating particles. But we must return to our immediate subject by a few words on the

VELOCITY OF LIGHT.

FIG. 191.—*Telescopic appearance of Jupiter and Satellites.*

IT may be stated at once, that this velocity has the amazing magnitude of 185,000 miles in one second of time, and that the fact of light requiring time to travel was first discovered, and the speed with which it does travel was first estimated, about 200 years ago, by a Danish astronomer, named Roemer, by observations on the eclipses of the satellites of Jupiter. The satellites of Jupiter are four in number, and as they revolve nearly in plane of the planet's orbit, they are subject to frequent eclipses by entering the shadow cast by the planet; in fact, the three inner satellites at every revolution. Fig. 191 represents the telescopic appearance of the planet, from a drawing by Mr. De La Rue, and in this we see the well-known "belts," and two of the satellites, one of which is passing across the face of the planet, on which its shadow falls, and is distinctly seen as a round black spot, while the other may be noticed at the lower right-hand corner of the cut. The satellite next the planet (Io) revolves round its primary in about 42½ hours, and consequently it is eclipsed by plunging into the shadow of Jupiter at intervals of 42½ hours, an occurrence which must take place with the greatest regularity as regards the duration of the intervals, and which can be calculated by known laws when the distance of the satellite from the planet has been determined. Nevertheless, Roemer observed that the actual intervals between the successive immersions of Io in the shadow of Jupiter did not agree with the calculated period of rotation when the distance between Jupiter and the earth was changing, in consequence chiefly of the movement of the latter (for Jupiter requires nearly twelve years to complete his revolution, and may, therefore, be regarded as stationary as compared for a short time with the earth). Roemer saw also, that when this distance *was increasing*, the observed intervals between the successive eclipses were a little greater, and that when the distance *was decreasing* they were a little less, than the calculated period. And he found that, supposing the earth, being at the point of its orbit nearest to Jupiter, to recede from that planet, the *sum of all the retardations* of the eclipses which occur while the earth is travelling to the farthest point of its orbit, amounts to 16½ minutes, as does also the *sum of the deficiencies* in the period when the earth, approaching Jupiter, is passing from the farthest to the nearest point of her orbit. While, however, the earth is near the points in her orbit farthest from, or nearest to Jupiter, the distance between the two planets is not materially *changing* between successive eclipses, and *then* the observed intervals of the eclipses coincide with the period of the satellite's rotation. The reader will, after a little reflection, have no difficulty in perceiving that the 16½ minutes represent the time which is required by the light to traverse the diameter of the earth's orbit; or, if he should have any difficulty, it may be removed by comparing the case with the following.

Let us suppose that from a railway terminus trains are dispatched every quarter of an hour, and that the trains proceed with a common and uniform velocity of, say, one mile per minute. Now, a person who remains stationary, at any point on the railway, observes the trains passing at regular intervals of fifteen minutes, no matter at what part of the line he may be placed. But now, let us imagine that a train having that very instant passed him, he begins to walk along the line towards the place from which the trains are dispatched: it is plain that he will meet the next train before fifteen minutes—he would, in fact, meet it a mile higher up the line than the point from which he began his walk fourteen minutes before; but the train, taking a minute to pass over this mile, would pass his point of departure just fifteen minutes after its predecessor. And our imaginary pedestrian, supposing him to continue his journey at the same rate, would meet train after train at intervals of fourteen minutes. Similarly, if he walked away from the approaching trains, they would overtake him at intervals of sixteen minutes. And again, it would be easy for him to calculate the speed of the trains, knowing that they passed over each point of the line every fifteen minutes. Thus, suppose him to pass *down* the line a distance known to be, say, a quarter of a mile; suppose he leaves his station at noon, the moment a train has passed, and that he takes, say an hour, to arrive at his new station a quarter of a mile lower; here, observing a train to pass at fifteen seconds after one o'clock, and knowing that it passed his original station at one, he has a direct measure of the speed of the trains. Here we have been explaining a discovery two centuries old; but our purpose is to prepare the reader for an account of how the velocity of light has been recently measured in a direct manner, and it certainly appears a marvellous achievement that means have been found to measure a velocity so astounding, not in the spaces of the solar system, or along the diameter of the earth's orbit, but within the narrow limits of an ordinary room! The reliance with which the results of these direct measures will be received, will be greatly increased by the knowledge of the astronomical facts with which they show an entire concordance. In

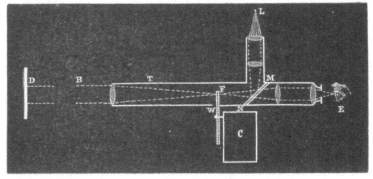

FIG. 192.

taking leave of Roemer, we may mention that his discovery, like many others, and like some inventions which have been described in this book, did not for some time find favour with even the scientific world, nor was the truth generally accepted, until Bradley's discovery of the aberration of light completely confirmed it.

To two gifted and ingenious Frenchmen we are indebted for independent measurements of the velocity of light by two different methods. The general arrangement of M. Fizeau's method is represented in Fig. 192, in which the rays from a lamp, L, after passing through a system of lenses, fall upon a small mirror, M N, formed of unsilvered plate-glass inclined at an angle of 45° to the direction of the rays; from this they are reflected along the axis of a telescope, T, by the lens of which being rendered parallel, they become a cylindrical beam, B, which passes in a straight line to a station, D, at a distance of some miles (in the actual experiment the lamp was at Suresnes and the other station at Montmartre, 5½ miles distant) whence the beam is reflected along the same path, and returns to the little plate of glass at M N, passing through which it reaches the eye of the observer at E. At W is a toothed wheel, the teeth of which pass through the point F, where the rays from the lamp come to a focus ; and as each tooth passes, the light is stopped from issuing to the distant station. This wheel is capable of receiving a regular and very rapid rotation from clockwork in the case, C, provided with a register for recording the number of its revolutions. If the wheel turns with such a speed that the light permitted to pass through one of the spaces travels to the mirror and back in exactly the same time that the wheel moves and brings the next space into the tube, or the second space, or the third, or any *space*, the reflected light will reach the spectator's eye just as if the wheel were stationary ; but if the speed be such that a *tooth* is in the centre of the tube when the light returns from the mirror, then it will be prevented from reaching the spectator's eye at all, so long as this particular speed is maintained, but either a decrease or an increase of velocity would cause the luminous image to reappear. Speeds between those by which the light is seen, and those by which it entirely disappears, cause it to appear with merely diminished brilliancy. It is only necessary to observe the speed of the wheel when the light is at its brightest, and when it suffers complete eclipse, for then the time is known which is required for space and tooth respectively to take the place of another space—and hence the time required for the light to pass to the mirror and back is found.

M. Foucault's method is similar in principle to that used by Wheatstone in the measurement of the velocity of electricity. He used a mirror which was made to revolve at the rate of 700 or 800 turns *per second*, and the arrangement of the apparatus was such as to admit of the measurement of the time taken by light to pass over the short space of about four yards ! More recently, however, he has modified and improved his apparatus by adopting a most ingenious plan of maintaining the speed of the mirror at a determined rate, which he now prefers should be 400 turns per second, while the light is reflected backwards and forwards several times, so that it traverses a path of above 20 yards in length. The time taken by the light to travel this short distance is, of course, extremely small, but it is accurately measured by the clockwork mechanism, and found to be about the $\frac{1}{150000000}$th of a second ! The results of these experiments of Foucault's make the velocity of light several thousand miles per second less than that deduced from the astronomical observation of Roemer and Bradley, in which the distance of the earth from the sun formed the basis of the calculations ; and hence arose a surmise that this distance had been over-estimated. That such had, indeed, been the case was confirmed almost immediately afterwards by a discussion among the astronomers as to the correctness of the accepted distance, the result of which has been that the mean distance, which was formerly estimated at 95 millions of miles, has, by careful astronomical observations and strict deductions, been now estimated at between 91 and 92 millions of miles. The famous transit of Venus December 9th, 1873—to observe which the Governments of all the chief nations of the world sent out expeditions—derived its astronomical and scientific importance from its furnishing the means of calculating, with greater correctness than had yet been attained, the distance of the earth from the sun.

REFLECTION OF LIGHT.

L ONG before plate glass backed by brilliant quicksilver ever reflected
the luxurious appointments of a drawing-room ; long before looking-
glass ever formed the mediæval image of "ladye fair" ; long before the
haughty dames of imperial Rome were aided in their toilettes by *specula;*
long before the dark-browed beauties of Egypt peered into their brazen
mirrors ; long, in fact, before men knew how to make glass or to polish
metals, their attention and admiration must have often been riveted by
those perfect and inverted pictures of the landscape, with its rocks, trees,
and skies, which every quiet lake and every silent pool presents. Enjoyment
of the spectacle probably prompted its imitation by the formation artificially of smooth flat reflecting surfaces ; and no doubt great skill in the pro=

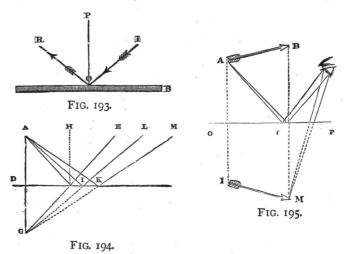

FIG. 193.

FIG. 195.

FIG. 194.

duction of these, and their application to purposes of utility, coquetry, and
luxury, preceded by many ages any attempt to discover the laws by which
light is reflected. The most fundamental of these laws are very simple, and
for the purpose we have in view, it is necessary that they should be borne
in mind. Let A B, Fig. 193, be a *plane reflecting surface,* such as the surface
of pure quicksilver or still water, or a polished surface of glass or metal,
and let a ray of light fall upon it in the direction, I O, meeting the surface at
O. it will be reflected along a line, O R,—such that if at the point O we draw
a line, O P, perpendicular to the surface, the incident ray, I O, and the reflected ray, O R, will form equal angles with the perpendicular—in other
words, the angle of incidence will be equal to the angle of reflection, and
the perpendicular, the incident ray, and the reflected ray, will all be in
one plane perpendicular to the reflecting plane. It would be quite easy to

prove from this law that the luminous rays from any object falling on a plane reflecting surface are thrown back just *as if* they came from an object placed behind the reflecting surface symmetrically to the real object. The diagrams in Figs. 194 and 195 will render this clear. In the second diagram, Fig. 195, it will be noticed that only the portion of the mirror between Q and P takes any part in the action, and therefore it is not necessary, in order to see objects in a plane mirror, that the mirror should be exactly opposite to them ; thus the portion O Q might be removed without the eye losing any part of the image of the object A B.

There are many very interesting and important scientific instruments in which the laws of reflection from plane surfaces are made use of—such, for example, as the *sextant* and the *goniometer;* but passing over all these, we

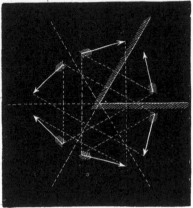

FIG. 196.

may say a word about the formation of several images from one object by using two mirrors. It has already been explained that the action of a plane mirror is equivalent to the placing of objects behind it symmetrically disposed to the real object. The reflections, or *virtual images* in the mirror, behave optically exactly *as if* they were themselves real objects, and are reflected by other mirrors in precisely the same manner. From this it follows that two planes inclined to each other at an angle of 90° give three images of an object placed between them, the images and the object apparently placed at the four angles of a rectangle. When the mirrors are inclined to each other at an angle of 60°, five images are produced, which, with the original object, show an hexagonal arrangement. The formation of these by the principle of symmetry is indicated in Fig. 196. It was these symmetrically disposed images which suggested to Sir David Brewster the construction of the instrument so well known as the *kaleidoscope*, in which two—or, still better, three—mirrors of black glass, or of glass blackened on one side, are placed in a pasteboard tube inclined to each other at 60° : one end of the tube is closed by two parallel plates of glass ; the outer one ground, but the inner transparent, leaving between them an interval, in which are placed fragments of variously-coloured glass, which every movement of

FIG. 197.—*Polemoscope.*

the instrument arranges in new combinations. At the other end of the
tube is a small opening—on applying the eye to which one sees directly
the fragments of glass, with their images so reflected that beautifully
symmetrical patterns are produced; and this with endless variety. When
this instrument was first made in the cheap form in which it is now so
familiarly known, it obtained a popularity which has perhaps never been
equalled by any scientific toy, for it is said that no fewer than 200,000
kaleidoscopes were sold in London and Paris in one month.

By way of contrast to the mirrors of the kaleidoscope harmlessly producing beautiful designs, by symmetrical images of fragments of coloured
glass, we show the reader, in Fig. 197, mirrors which are reflecting quite
other scenes, for here is seen the manner in which even the plane mirror
has been pressed into the service of the stern art of war. The mirrors are
employed, not like those of Archimedes, to send back the sunbeams from
every side, and by their concentration at one spot to set on fire the enemy's
works, but to enable the artillerymen in a battery to observe the effect of
their shot, and the movement of their adversaries, without exposing themselves to fire by looking over the parapet of their works. The contrivance
has received the appropriate name of *Polemoscope* (πολεμος, *war*, and σκοπεω,

FIG. 198.—*Apparatus for Ghost Illusion.*

to view), and it consists simply, as shown in the figure, of two plane mirrors so inclined and directed, that in the lower one is seen by reflection the localities which it is desired to observe.

We return once more to the arts of peace, in noticing the advantage which has been lately taken of plane mirrors for the production of spectral and other illusions, in exhibitions and theatrical entertainments, the improvement in the manufacture of plate-glass having permitted the production of enormous sheets of that substance. Among the most popular exhibitions of this class was that known as "Pepper's Ghost," the arrangement of the mirrors having been the subject of a patent taken out by Mr. Pepper and Mr. Dircks jointly. The principle on which the production of the illusion depends, may be explained by the familiar experience of everybody who has noticed that, in the twilight, the glass of a window presents to a person inside of a room the images of the light or bright objects in the apartment, while the objects outside are also visible through the glass. As, by night coming on, the reflections increase in brilliancy, the darkness outside is almost equivalent to a coat of black paint on the exterior surface of the glass; but, on the contrary, in the daylight no reflection of the interior of the room is visible to the spectator inside, on looking towards the window. The reflections are present, nevertheless, in the day-time as well as at night, only they are overpowered and lost when the rays which reach the eye *through* the glass are relatively much more powerful. Even in the day-time the image of a lighted candle is usually visible, in the absence of

direct sunshine, against a dark portion of the exterior objects as a back-
ground. The visibility, or otherwise, of the internal objects by reflection, and
of the external objects seen through the glass, depends entirely on the relative intensities of the illumination, for the more illuminated side overpowers
and conceals the other, just as the rising sun causes the stars " to pale their
ineffectual fires." Hence, on looking through the window on a dark night,
we cannot see objects out of doors unless we screen off the reflection of the
illuminated objects in the room. If the rays transmitted through the glass,
and those which are reflected, have intensities not very different, we see
then the reflected images mixed up in the most curious manner with the
real objects. It is exactly in this way that the ghosts are made to appear
in the illusion of which we are speaking. The real actors are seen through
a large plate of colourless and transparent glass, and from the front surface
of this glass rays are reflected which apparently proceed from a phantom
taking a part in the scene among the real actors. The arrangement is
shown in Fig. 198, where E G is the stage, separated from the auditorium,
H, by a large plate of transparent glass, E F, placed in an inclined position,
and not visible to the spectators, for the lights in front are turned down,
and the stage is also kept comparatively dark. Parallel to the large plate
of glass is a silvered mirror, C D, placed out of the spectators' sight, and
receiving the rays from a person at A, also out of sight of the spectators,
and strongly illuminated by an oxy-hydrogen lime-light at B. The manner
in which the rays are reflected from the silvered mirror to the plate-glass,
and hence reflected so as to reach the spectators and give them the impression of a figure standing on the stage at G, is sufficiently indicated by
the lines drawn in the diagram. The apparitional and unsubstantial character of the image is derived from its seeming transparency, and from
the manner in which it may be made to melt away, by diminishing the
brightness of the light which falls on the real person. The introduction of
the second mirror was a great improvement, for by this the phantom is
made to appear erect, while its original stands in a natural attitude. Whereas, with only the plate-glass, E F, the *ghost* could not be made to appear
upright, unless, indeed, as was sometimes done, the plate was inclined at
an angle of 45°, and the actor of the ghost lay horizontally beneath it. A
scene of the kind produced by the improved apparatus, is represented in
Fig. 198a.

   Another illusion is produced by the help of a large silvered mirror,
placed at an inclination of 45°, sloping backwards from the floor, and,
in consequence, presenting to the spectators the image of the ceiling,
which appears to them the back of the scene. The mirror is perforated
near the centre by an opening, through which a person passes his head,
and, all his body being concealed by the mirror, the effect produced is that
of a head floating in the air. Means are provided of withdrawing the
mirror, when necessary, while the curtain is down, and then the real back
of the scene appears, which, of course, is exactly similar to the false one
painted on the ceiling. Fig. 199 represents a scene produced at the Polytechnic by a somewhat similar arrangement of mirrors, under the management of Mr. Pepper. Plane mirrors were employed in another piece of
natural magic which this gentleman exhibited to the public, who were
shown a kind of large box, or cabinet, raised from the floor, and placed
in the middle of the stage, so that the spectators might see under it and
all round it. Inside of the box were two silvered mirrors the full height of
it, and these were hinged to the farther angles, so that each one being

Fig. 198a.—*The Ghost Illusion.*

FIG. 199.—*Illusion produced by Mirrors.*

folded with its face against a side of the box, their backs formed the apparent sides, and were painted exactly the same as the real interior of the box. When the performer enters the box, the door is closed for an instant, while he, stepping to the back, turns the mirrors on their hinges until their front edges meet, where an upright post in the middle of the box conceals their line of junction. The performer thus places himself behind the mirrors in the triangular space between them and the back of the box, while the mirrors, now inclined at angles of 45° to the sides, reflect images of these to the spectators when the door is opened, and the spectators see then the box apparently empty, for the reflection of the sides appears to them as the back of the cabinet. The entertainment was sometimes varied by a skeleton appearing, on the door being opened, in the place of the person who entered the cabinet. It is hardly necessary to say that the skeleton was previously placed in the angle between the mirrors where the performer conceals himself.

FIG. 200.—*A Stage Illusion.*

To the same inventive gentleman, whose ingenious use of plane mirrors has thus largely increased the resources of the public entertainer, is due another stage illusion, the effect of which is represented in Fig. 200 ; and, although it does not depend on reflection, it may be introduced here as showing how the perfection of the manufacture of plate-glass, which makes it available for the ghost exhibition, can be applied in another way in dramatic spectacles. The female form, here supposed to be seen in a dream by the sleeper, is not a reflection, although she appears floating in mid-air, strangely detached from all supports, but the real actress. This is accomplished by making use of the transparency of plate-glass, a material strong enough to afford the necessary support, and yet invisible under the circumstances of the exhibition.

But it is not behind the turned-down footlights, or in the exhibitions of the showman, that we find the most beautiful illustrations of the laws of reflection. In the quiet mountain mere, amid the sweet freshness of nature, we may often see tree, and crag, and cliff, so faithfully reproduced, that it needs an effort of the understanding to determine where substance leaves off and shadow begins, a condition of the liquid surface indicated in two lines by Wordsworth :

> "The swan, on still St. Mary's Lake,
> Floats double, swan and shadow."

The landscape painter is always gratified if he can introduce into his picture some piece of water, and it can hardly be doubted that much of the charm of lakes and rivers is due to their power of reflecting.   Look on Fig. 201, a view of some buildings at Venice; and, in order to see how much of its beauty is owing to the quivering reflections, imagine the impression it would produce were the place of the water occupied by asphalte pavement, or a grass lawn.   The condition of the reflections here represented is perhaps even more pleasing than that produced by perfect repose: they are in movement, and yet not broken and confused :

> "In bright uncertainty they lie,
> Like future joys to Fancy's eye."

FIG. 201.—*View of Venice—Reflections.*

REFRACTION.

THAT light moves in straight lines is a statement which is true only when the media through which it passes are uniform; for it is easily proved that when light passes from one medium to another, a change of direction takes place at the common surface of the media in all rays that meet this surface otherwise than perpendicularly. As a consequence of this, it really is possible to see round a corner, as the reader may convince himself by performing the following easy experiment. Having procured a cup or basin, Fig. 202, let him, by means of a little bees'-wax or tallow, attach to the bottom of the vessel, at R, a small coin. If he now places the cup so that its edge just conceals the coin from view, and maintains his eye steadily in the same position as at I, he will, when water is poured into the cup, perceive the coin apparently above the edge of the vessel in the direction I R′, that is, the bottom of the cup will appear to have risen higher. Since it is known that in each medium the rays pass in straight lines, the bending which renders the coin visible can therefore only take place at the common junction of the media, or, in other words, the ray, R O, passing from the object in a straight line through the water, is bent abruptly aside as it passes out at the surface of the water, A B, and enters the air, in which it again pursues a straight course, reaching the eye at I, where it

FIG. 202.

gives the spectator an impression of an object at R′. This experiment is also an illustration of the cause of the well-known tendency we have to under-estimate the depth of water when we can see the bottom. The broken appearance presented by an oar plunged into clear water is due to precisely the same cause. The curious exaggerated sizes and distorted shapes of the gold-fish seen in a transparent globe have their origin in the same bending aside of the rays. This deviation which light undergoes in passing obliquely from one medium into another is known by the name of *refraction,* and it is essential for the understanding of the sequel that the reader should be acquainted with some of the laws of this phenomenon, although their discovery by Snell dates two centuries and a half anterior to the present time. Let T O, Fig. 203, be a ray of light which falls obliquely upon a plane surface, A B, common to two different media, one of which is represented by the shaded portion of the figure, A B C D, of which C D represents another plane surface, parallel to the former. If the ray, T O, suffered no refraction, it would pursue its course in a straight line to *r′*; but as a matter of fact it is found that such a ray is always bent aside at O, if the medium A B C D is more or less dense than the other. If, for example, A B′C D is water, and the medium above it glass, then the ray entering at O will take the course O *r*; but if A B C D is a plate of glass with water above and below it, the ray will take the course T O, O R, R B, suffering refraction on entering the glass, and again on leaving it, so that R B will emerge from the glass parallel to its original direction at T O. If through the point of incidence, O, we suppose a line, O P, to be drawn perpendicular to the surface, A B, then we may say that the ray in passing from the rarer medium (water, air, &c.) into the denser medium (glass, &c.) is bent towards the perpendicular, or normal, as at O; but that on leaving the denser to enter the rarer medium, as at R, it is bent away from the perpendicular.   In other words, the angle *b* O *a* is less than the angle *m* O T, and O R forms a less angle with R P′ than R B′ does.   It is also a law of *ordinary* refraction that the normal, O P, at the point of incidence, the

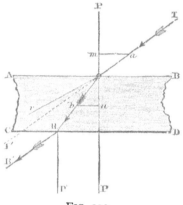

FIG. 203.

incident ray, T O, and the refracted ray, O R, are all in the same plane. Besides, there is the important and interesting law discovered by Snell and by Descartes, which may thus be explained with reference to Fig. 203.   On the incident and refracted rays, T O and O R, let us suppose that any equal distances, O *d* and O *b*, are measured off from O, and that from each of the points *a* and *b*, perpendiculars, *a m* and *b n*, are drawn to the normal, P P, which passes through O; then it is found that, whatever may be the angle of incidence, T O P, or however it is made to vary, the length of the line *a m* bears always the same proportion to the line *b n* for the same two media. Thus, if A B C D be water, and T O enters it out of the air, the length of the line *a m* divided by the length of the line *a b* will always (whatever slope T O may have) give the quotient 1·33.   This number is, therefore, a constant quantity for air and water, and is called the index of refraction for air into water.   The law just explained is expressed by the language of mathematics thus : For two given media the ratio of the sines of the angles of incidence and of refraction is constant.

It is an axiom in optical science that a ray of light when sent in the opposite direction will pursue the same path.   Thus in Fig. 203 the direction of the light is represented as from T towards B′; but if we suppose B′ R to be an incident ray, it would pursue the path B′ R, R O, O T, and in passing out of the denser medium, A B C D at O, its direction is farther from the normal, P P, or, as the law of sines, *a m* will be always longer than *n b*, and will bear a constant ratio to it.   Suppose the angle R O P to increase, then P O B will become a right angle; that is, the emergent ray, O T, will just graze the surface, A B, when the angle R O P has some definite value. If this last angle be further increased, *no light at all will pass out of the medium* A B C D, but the ray R O will be totally reflected at O back into the medium, A B C D, according to the laws of reflection. The angle which R O forms with O P when O T just skims the surface, A B, is termed the *limiting. angle*, or the *critical angle*, and its value varies with the media. The reader may easily see the total reflection in an aquarium, or even in a tumbler of water, when he looks up through the glass at the surface of the water, which has then all the properties of a perfect mirror.

The power of lenses to form images of objects is entirely due to these laws of refraction. The ordinary double-convex lens, for example, having its surfaces formed of portions of spheres, refracts the rays so that *all* the rays which from *one luminous point* fall upon the lens, meet together again at a point on the other side, the said point being termed their *focus*. It is thus that *images* of luminous bodies are formed by lenses. An explanation of the construction and theory of lenses cannot, however, be entered into in this place.

One important remark remains to be made—namely, that in the above statement of the laws of reflection and refraction, certain limitations and conditions under which they are true and perfectly general have not been expressed; for the mention of a number of particulars, which the reader would probably not be in a condition to understand, would only tend to confuse, and the explanation of them would lead us beyond our limits. Some of these conditions belong to the phenomena we have to describe, and are named in connection with them, and others, which are not in immediate relation to our subject, we leave the reader to find for himself in any good treatise on optics.

DOUBLE REFRACTION AND POLARIZATION.

ABOUT two hundred years ago, a traveller, returning from Iceland, brought to Copenhagen some crystals, which he had obtained from the Bay of Roërford, in that island. These crystals, which are remarkable for their size and transparency, were sent by the traveller to his friend, Erasmus Bartholinus, a medical man of great learning, who examined them with great interest, and was much surprised by finding that all objects viewed through them appeared double. He published an account of this singular circumstance in 1669, and by the discovery of this property of Iceland spar, it became evident that the theory of refraction, the laws of which had been studied by Snell and by Huyghens a few years before, required some modification, for these laws required only one refracted ray, and Iceland spar gave two. Huyghens studied the subject afresh, and was able, by a geometrical conception, to bring the new phenomena within the general theory of light. Iceland spar is chemically carbonate of lime (calcium carbonate), and hence is also called calc spar, and, from the shape of the crystals, it has also been termed rhombohedral spar. The form in which the crystals actually present themselves is seen in Fig. 204, which also represents the phenomenon of double refraction. Iceland spar splits up very readily, but only along certain definite directions, and from such a piece as that represented in Fig. 204 a perfect rhombohedron, such as that shown in Fig. 206, is readily obtained by cleavage; and then we have a solid having six lozenge-shaped sides, each lozenge or side having two obtuse angles of 101° 55′, and two acute angles of 78° 5′.   Of the eight solid corners, such as A B C, &c., six are produced by the meeting of one obtuse and two acute angles, and *the remaining two solid corners are formed by the meeting of three obtuse angles.*   Let us imagine that a line is drawn from one of these

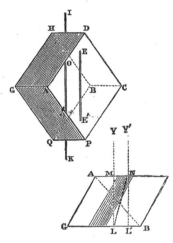

FIG. 204.   FIG. 205.

angles to the other: the diagonal so drawn forms the *optic axis* of the crystal, and a plane passing through the optic axis, A B, Fig. 205, and through the bisectors of the angles, E A D and F B G, marks a certain definite direction in the crystal, to which also belong all planes parallel to that just indicated.   Any one of such planes forms what is termed a "principal section," to which we shall presently refer.

It will be observed that in Fig. 204 the white circle on a black ground seen through the crystal is doubled; but that, instead of being white as the circle really is, the images appear grey, except where they overlap, and there the full whiteness is seen.   If we place the crystal upon a dot made on a sheet of paper, or having made a small hole with a pin in a piece of cardboard, hold this up to the light, and place the crystal against it, we see apparently two dots or two holes.   The two images will, if the dot or hole be sufficiently small, appear entirely detached from each other.   Now, if, keeping the face of the crystal against tne cardboard or paper, the observer turn the crystal round, he will see one of the images revolve in a circle round the other, which remains stationary.   The latter is called the *ordinary* image, and the former the *extraordinary* image.   Let us place the crystal upon a straight black line ruled on a horizontal sheet of paper, Fig. 205, and let us suppose, in order to better define the appearance, that we place it so that the *optic axis,* A B, is in a plane perpendicular to the paper, A being one of the two corners where the three obtuse angles meet, and B the other, and the face, A B C D, parallel to E G H B, which touches the paper. Then, according to the laws of ordinary refraction, if we look *straight* down upon the crystal, we should see through it the line I·K, unchanged in position— that is, the ray would pass perpendicularly through the crystal as shown by L M—and, in fact, a part of the ray does this, and gives us the *ordinary* image, O O′ ; but another part of the ray departs from the laws of Snell and Descartes, and, following the course L N Y′, enters the eye in the direction N Y′, producing the impression of another line at L′, which is the *extraordinary* ray, E E′. If the crystal be turned round on the paper, E E′ will gradually approach O O′, and the two images will coincide when the *principal section* is parallel to the line I K; but the coincidence is only apparent, and results from the superposition of the two images—for a mark placed on the line drawn on the paper will show two images, one of which will follow the rotation of the crystal, and show itself to the right or left of the *ordinary* image, according as C is to the right or left of A. So that there are really in every portion of the crystal two images on the line, one of which turns round the other, and the coalescence of the two images twice in each revolution is only apparent, for the different parts of the lengths of

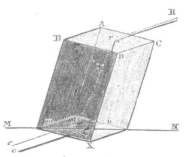

FIG. 206.

the images do not coincide. On continuing the revolution of the crystal after they apparently coincide, the images are again seen to separate, the *extraordinary* one being now displaced on the other side, or always towards the point, C. Thus, then, the ray, on entering the crystal, bifurcates, one branch passing through the crystal and out of it in the same straight line, just as it would in passing through a piece of glass, while the other is refracted at its entrance into the crystal, although falling perpendicularly upon its face, and again at its exit. And again, when a beam of light, R *r*, Fig. 206, falls obliquely on a crystal of Iceland spar, it divides at the face of the crystal into two rays, *r* O, and *r* E ; the former, which is the ordinary ray, follows the laws of ordinary refraction—it lies in the plane of incidence, and obeys the law of sines, just as if it passed through a piece of plate-glass. The *extraordinary* ray, on the other hand, departs from the plane of incidence, except when the latter is parallel to the *principal section,* and the ratio of the sines of the angles of incidence and refraction varies with the incidence. The reader who is desirous of studying these curious phenomena of *double* refraction, and those of polarization, is strongly recommended to procure some fragments of Iceland spar, which he can very easily cleave into rhombohedra, and with these, which need not exceed half an inch square, or cost more than a few pence, he can demonstrate for himself the phenomena, and become familiar with their laws. He will find very convenient the simple plan recommended by the Rev. Baden Powell, of fixing one of the crystals to the inside of the lid of a pill-box, through which a small hole has been made, and through the hole and the crystal view a pin-hole in the bottom of the box, turning the lid, and the crystal with it, to observe the rotation of the image. The same arrangement will serve, by merely attaching another rhomb of spar within the box, to study the very interesting facts of the polarization to which we are about to claim the reader's attention.

The curious phenomena which have just been described, although in themselves by no means recent discoveries, have led to some of the most interesting and beautiful results in the whole range of physical science. The examination and discussion of them by such able investigators as Huyghens, Descartes, Newton, Fresnel, Malus, and Hamilton, have largely conduced to the establishment of the undulatory hypothesis—that comprehensive theory of light, which brings the whole subject within the reach of a few simple mechanical conceptions.

It was at first supposed that it was only one of the rays which are produced in double refraction that departed from the ordinary laws, and Iceland spar was almost the only crystal known to have the property in question. At the present day, however, the substances which are known to produce double refraction are far more numerous than those which do not possess this property, for, by a more refined mode of examination than the production of double images, Arago has been able to infer the existence of a similar effect on light in a vast number of bodies. Crystals have also been found which split up a ray of light entering them into two rays, neither of which obeys the laws of Descartes. It may, in fact, be said that, with the exception of water, and most other liquids, of gelatine and other colloidal substances, and of well-annealed glass, there are few bodies which do not exercise similar power on light.

On examining the two rays which emerge from a rhomb of Iceland spar, on which only one ray of ordinary light has been allowed to fall, we find that these emergent rays have acquired new and striking properties, of which the incident ray afforded no trace ; for, if we allow the two rays emerging from a rhomb of the spar to fall upon a second rhomb, we shall find, on viewing the images produced, that their intensity varies with the position into which its second crystal is turned. Thus, if we place a rhomb of the spar upon a dot made on a sheet of white paper, we shall have, as already pointed out, two images of equal darkness. But, in placing a second rhomb of the spar upon the first, in such a manner that their *principal sections* coincide, and the faces of one rhomb are also parallel to the faces of the other, we shall still see *two* equally intense images of the dot, only the images will be more widely separated than before, and no difference will be produced by separating the crystals if the parallelism of the planes of their respective principal sections be preserved. Here, then, is at once a notable difference between a ray of ordinary light and one that emerges from a rhomb of Iceland spar ; for, in the case of rays of ordinary light, we have seen that the second rhomb would divide each ray into two, whereas it is incapable (in the position of crystals under consideration) of dividing either the ordinary or the extraordinary ray which emerges from the first rhomb. If, still keeping the second rhomb above the other, we make the former rotate in a horizontal plane, we may observe that, as we turn the upper crystal so that the planes of the *principal sections* form a small angle with each, each image will be doubled, and, as the upper crystal is turned, each pair of images exhibits a varying difference of intensity. The ordinary ray in entering the second crystal is divided by it into a second ordinary ray and a second extraordinary ray, the intensities of which vary according to the angle between the principal sections. When the two principal sections are parallel to one plane, that is, when the angle between them is either 0° or 180°, the extraordinary image disappears, and only the ordinary one is seen, and with its greatest intensity. When the two *principal sections* are perpendicular to each other, that is, when the second crystal has been turned through either 90° or 270°, the extraordinary has, on the contrary, its greatest intensity, and the ordinary one disappears. When the principal section of the second crystal has been turned into any intermediate position, such as through 45° and 135°, or any odd multiple of 45°, both images are visible and have equal intensities. This experiment shows that the two rays which emerge from the first crystal have acquired new properties, that each is affected differently by the second crystal, according as the crystal is presented to it in different directions round the ray as an axis.

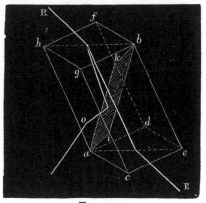

FIG. 207.

The ray of light is no longer uniform in its properties all round, but appears to have acquired different sides, as it were, in passing through the rhomb of Iceland spar. This condition is indicated by saying that the ray is *polarized*, and the first rhomb of spar is termed the *polarizer*, while the second rhomb, by which we recognize the fact that both the *ordinary* and the *extraordinary* rays emerge having different sides, has received the name of *analyser*. But, in order to study conveniently all the phenomena in Iceland spar, we should have crystals of a considerable size, otherwise the two rays do not become sufficiently separated so as to make it an easy matter to intercept one of them while we examine the other. A very ingenious mode of getting rid of one of the rays was devised by Nicol, and as his apparatus is much used for experiments on polarized light, we shall state the mode of constructing *Nicol's Prism*. It is made from a rhomb of Iceland spar, Fig. 207, in which *a* and *b* are the corners where the three obtuse angles meet, all equal. If we draw through *a* and *b* lines bisecting the angles *d a c* and *f h g*, and join *a b*, these lines will all be in one plane, which is a principal section of the crystal, and contains the axis, *a b*. Now suppose another plane, passing through *a b*, to be turned so that it is at right angles to the plane containing *a b* and the bisectors: this plane would cut the sides of the crystal in the lines *a i, i h*, *b k, k a*; and in making the Nicol prism, the crystal is cut into two along this plane, and the two pieces are then cemented together by *Canada balsam.* A ray of light, R, entering the prism, undergoes double refraction; but the ordinary ray, meeting the surface of the Canada balsam at a certain angle greater than the limiting angle, is totally reflected, and passes out of the crystal at O; while the extraordinary ray, meeting the layer of balsam at a less angle than *its* limiting angle, does not undergo total reflection, but passes through the balsam, and emerges in the direction of E, completely polarized, so that the ray is unable to penetrate another Nicol's prism of which the principal section is placed at right angles to that of the first.

Among other crystals which possess the property of doubly refracting, and therefore of polarizing, is the mineral called *tourmaline*, which is a semi-transparent substance, different specimens having different tints. In Fig. 208, A, B, represent the prismatic crystals of tourmaline, and C shows a crystal which has been cut, by means of a lapidary's wheel, into four pieces, the planes of division being parallel to the axis of the prism. The two inner portions form slices, having a uniform thickness of about $\frac{1}{20}$ in., and when the faces of these have

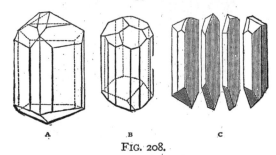

A   B   C

FIG. 208.

been polished, the plates form a convenient polarizer and analyser. Let us imagine one of the plates placed perpendicularly between the eye and a lighted candle. The light will be seen distinctly through it, partaking, however, of the colour of the tourmaline; and if the plate be turned round so that the direction of the axis of the crystal takes all possible positions with regard to the horizon, while the plane of the plate is always perpendicular to the line between the eye and the candle, *no change whatever will be seen in the appearance of the flame.* But if we fix the plate of crystal in a given position, let us say with the axial direction vertical, and place between it and the eye the second plate of tourmaline, the appearances become very curious indeed, and *the candle is visible or invisible according to the position of this second plate.* When the axis of the second is, like that of the first, vertical, the candle is distinctly seen; but when the axis of the second plate is horizontal, no rays from the candle can reach the eye. If the second plate be slowly turned in its own plane, the candle becomes visible or invisible at each quarter of a revolution, the image passing through all degrees of brightness. Thus the luminous rays which pass through the first plate are polarized like those which emerge from a crystal of Iceland spar. It is not necessary that the plates used should be cut from the same crystal of tourmaline, for any two plates will answer equally well which have been cut parallel to the axes of the crystals which furnished them. In the case of tourmaline the extraordinary ray possesses the power of penetrating the substance of the crystal much more freely than the ordinary ray, which a small thickness suffices to absorb altogether. It may be noted that in the simple experiment we have just described, the plate of tourmaline next the candle forms the *polarizer*, and that next the eye the *analyser;* and that until the latter was employed, the eye was quite incapable of detecting the change which the light had undergone in passing through the first plate, for the unassisted eye had no means of recognizing that the rays emerged with sides. The usual manner of examining light, to find whether it is polarized, is to look through a plate of tourmaline or a Nicol's prism, and observe whether any change in brightness takes place as the prism or plate is rotated. Now, it so happened that in 1808 a very eminent French man of science, named Malus, was looking through a crystal of Iceland spar, and seeing in the glass panes of the windows of the Luxembourg Palace, which was opposite his house, the image of the setting sun, he turned the crystal towards the windows, and instead of the two bright images he expected to see, he perceived only one ; and on turning the crystal a quarter of a revolution, this one vanished as the other image appeared. It was, indeed, by a careful analysis of this phenomenon that Malus founded a new branch of science, namely, that which treats of polarized light ; and his views soon led to other discoveries, which, with their theoretical investigations, constitute one of the most interesting departments of optical science, as remarkable for the grasp it gives of the theory of light as for the number of practical applications to which it has led.

The accidental observation of Malus led to the discovery that when a ray of ordinary light falls obliquely on a mirror—not of metal, but of any other polished surface, such as glass, wood, ivory, marble, or leather—it acquires by reflection at the surface the same properties that it would acquire by passing through a Nicol's prism or a plate of tourmaline : in a word, it is polarized. Thus, if a ray of light is allowed to fall upon a mirror of black glass at an angle of incidence of 54° 35′, the reflected ray will be found to be polarized in the plane of reflection—that is, it will pass freely through a Nicol's prism when the principal section is parallel to the plane of reflection ; but when it is at right angles to the latter, the reflected ray will be completely extinguished by the prism—that is, it is completely polarized. If the angle of the incident ray is different from 54° 35′, then the reflected ray is not completely intercepted by the prism—it is not completely but only partially polarized. The angle at which maximum polarization takes place varies with the reflecting substance ; thus, for water it is 53°, for diamond 68°, for air 45°. A simple law was discovered by Sir David Brewster by which the polarizing angle of every substance is connected with its refractive index, so that when one is known, the other may be deduced. It may be expressed by saying that the polarizing angle is that angle of incidence which makes the reflected and the refracted rays perpendicular to each other. The refracted rays are also found to be polarized in a plane perpendicular to that of reflection.

Instruments of various forms have been devised for examining the phenomena of polarized light. They all consist essentially of a polarizer and an analyser, which may be two mirrors of black glass placed at the polarizing angle, or two bundles of thin glass plates, or two Nicol's prisms, or two plates of tourmaline, or any pair formed by two of these. Fig. 209 represents a polariscope, this instrument being designed to permit any desired combination of polarizer and analyser, and having graduations for measuring the angles, and a stage upon which may be placed various sub-

LIGHT.

stances in order to observe the effects of polarized light when transmitted through them. It is found that thin slices of crystals placed between the polarizer and analyser exhibit varied and beautiful effects of colour, and by such

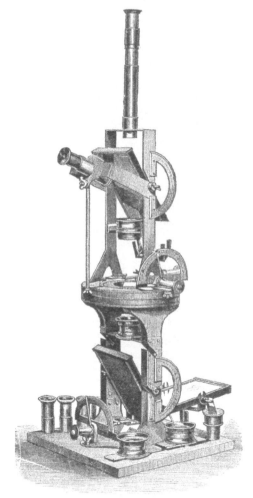

FIG. 209.—*Polariscope.*

effects the doubly refracting power of substances can be recognized, where the observation of the production of double images would, on account of their small separation, be impossible. And the polariscope is of great service in revealing structures in bodies which with ordinary light appear entirely devoid of it—such, for example, as quill, horn, whalebone, &c. Except liquids, well-annealed glass, and gelatinous substances, there are, in fact, few bodies in which polarized light does not show us the existence of some kind of structure. A very interesting experiment can be made by placing

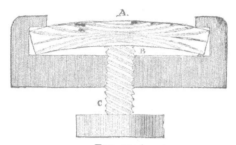

FIG. 210.

in the apparatus, shown in Fig. 210, a square bar of well-annealed glass; on examining it by polarized light, it will be found that before any pressure from the screw C is applied to the glass, it allows the light to pass equally through every part of it; but when by turning the screw the particles have been thrown into a state of strain, as shown in the figure, distinct bands will make their appearance, arranged somewhat in the manner represented; but the shapes of the figures thus produced vary with every change in the strain and in the mode of applying the pressure.

FIG. 211.—*Iceland Spar showing Double Refraction.*

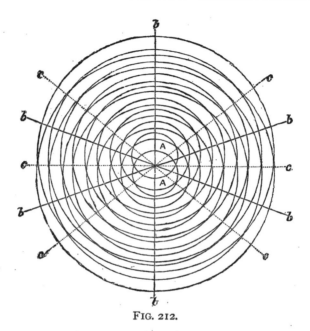

FIG. 212.

CAUSE OF LIGHT AND COLOUR.

WE have hitherto limited ourselves to a description of some of the phenomena of light, without entering into any explanation of their presumed causes, or without making any statements concerning the nature of the agent which produces the phenomena. Whatever this cause or agent may be, we know already that light requires time for its propagation, and two principal theories have been proposed to explain and connect the facts. The first supposes light to consist of very subtile matter shot off from luminous bodies with the observed velocity of light; and the second theory, which has received its great development during the present century, regards luminous effects as being due to movements of the particles of a subtile fluid to which the name of "ether" has been given. Of the existence of this ether there is no proof: it is imagined; and properties are assigned to it for no other reason than that if it did exist and possess these properties, most of the phenomena of light could be easily explained. This theory requires us to suppose that a subtile imponderable fluid pervades all space, and even interpenetrates bodies—gaseous, liquid, and solid; that this fluid is enormously elastic, for that it resists compression with a force almost beyond calculation. The particles of luminous bodies, themselves in rapid vibratory motion, are supposed to communicate movement to the particles of the ether, which are displaced from a position of equilibrium, to which they return, executing backwards and forwards movements, like the stalks of corn in a field over which a gust of wind passes. While an ethereal particle is performing a complete oscillation, a series of others, to which it has communicated its motion, are also performing oscillations in various phases—the adjacent particle being a little behind the first, the next a little behind the second, and so on, until, in the file of particles, we come to one which is in the same phase of its oscillation as the first one. The distance of this from the first is called the " length of the luminous wave." But the ether particles do not, like the ears of corn, sway backwards and forwards merely in the direction in which the wave itself advances : they perform their movements in a direction perpendicular to that in which the wave moves. This kind of movement may be exemplified by the undulation into which a long cord laid on the ground may be thrown when one end is violently jerked up and down, when a wave will be seen to travel along the cord, but each part of the latter only moves perpendicularly to the length. The same kind of undulation is produced on the surface of water when a stone is thrown into a quiet pool. In each of these cases the parts of the rope or of the water do not travel along with the wave, but each particle oscillates up and down. Now, it may sometimes be observed, when the waves are spreading out on the surface of a pool from the point where a stone has been dropped in, that another set of waves of equal height originating at another point may so meet the first set, that the crests of one set correspond with the hollows of the other, and thus strips of nearly smooth water are produced by the superposition of the two sets of waves. Let Fig. 212 represent two systems of such waves propagated from the two points A A, the lines representing the crests of the waves. Along the lines, *b b*, the crests of one set of waves are just over the hollows of the other set ; so that along these lines the surface would be smooth, while along C C the crests would have double the height. Now, if light be due to undulation, it should be possible to obtain a similar effect— that is, to make two sets of luminous undulations destroy each other's effects and produce darkness : in other words, we should be able, *by adding light to light, to produce darkness!* Now, this is precisely what is done in a celebrated experiment devised by Fresnel, which not only proves that darkness may be produced by the meeting of rays of light, but actually enables us to measure the lengths of the undulations which produce the rays.

In Fig. 213 is a diagram representing the experiment of the two mirrors, devised by Fresnel. We are supposed to be looking down upon the arrangement : the two plane mirrors, which are placed vertically, being seen edgeways, in the lines, M O, O N, and it will be observed that the mirrors are placed *nearly* in the same upright plane, or, in other words, they form an angle with each other, which is nearly 180°. At L is a very narrow upright slit, formed by metallic straight-edges, placed very close together, and allowing a direct beam of sunlight to pass into the apartment, this being the only light which is permitted to enter. From what has been already said on reflection from plane mirrors, it will readily be understood that these mirrors will reflect the beams from the slit in such a manner as to produce the same effect, in every way, as if there were a real slit placed behind each mirror in the symmetrical positions, A and B. Each virtual image of the slit may, therefore, be regarded as a real source of light at A and at B ; thus, for example, it will be observed that the actual lengths of the paths traversed by the beams which enter at L, and are reflected from the mirrors, are precisely the same as if they came from A and B respectively.

The virtual images may be made to approach as near to each other as may be required, by increasing the angle between the two mirrors, for, when this becomes 180°, that is, when the two mirrors are in one plane, the two images will coincide. If, now, a screen be placed as at F G, a very remarkable effect will be seen; for, instead of simply the images of the two slits, there will be visible a number of vertical coloured bands, like portions of very narrow rainbows, and these coloured bands are due to the two sources of light, A and B; for, if we cover or remove one of the mirrors, the bands will disappear and the simple image of the slit will be seen. If, however, we place in front of L a piece of coloured glass, say red, we shall no longer see rainbow-like bands on the screen, but in their place we shall find a number of strips of red light and dark spaces alternately, and, as before, these are found to depend upon the *two* luminous sources, A and B. We must, therefore, come to the conclusion that the two rays exercise a mutual effect, and that, by their superposition, they produce darkness at some

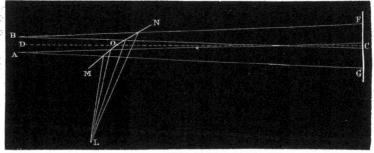

FIG. 213.

points and light at others. These alternate dark and light bands are formed on the screen at all distances, and the spaces between them are greater as the two images, A and B, are nearer together. Further, with the same disposition of the apparatus, it is found that when yellow light is used instead of red, the bands are closer together; when green glass is substituted for yellow, blue for green, and violet for blue, that the bands become closer and closer with each colour successively. Hence, the effect of coloured bands, which is produced when pure sunlight is allowed to enter at L, is due to the superposition of the various coloured rays from the white light. Let us return to the case of the red glass, and suppose that the distance apart of the two images, A and B, has been measured, by observing the angle which they subtend at C, and by measuring the distance, C O D, or rather, the distance C O L. Now, the distances of A and B from the centre of each dark band, and of each light band, can easily be calculated, and it is found that the *difference between the two distances* is always the same for the same band, however the screen or the mirrors may be changed. On comparing the *differences* of the distances of A and B in case of bright bands, with those in the case of dark ones, it was found that the former could be expressed by the even multiples of a very small distance, which we will call *d*, thus :

$$0, 2d, 4d, 6d, 8d, \ldots$$

while the differences for the dark bands followed the odd multiples of the same quantity, *d*, thus :

$$d, 3\,d, 5\,d, 7\,d, 9\,d, \ldots \ldots$$

These results are perfectly explained on the supposition that light is a kind of wave motion, and that the distance, *d*, corresponds to *half the length of a wave.* We have the waves entering L, and pursuing different lengths of path to reach the screen at F G, and, if they arrive in opposite phases of undulation, the superposition of two will produce darkness. The undulations will plainly be in opposite phases when the lengths of paths differ by an *odd* number of *half-wave* lengths, but in the same phase when they differ by an *even* number. Hence, the length of the wave may be deduced from the measurement of the distances of A and B from each dark and light band, and it is found to differ with the colour of the light. It is also plain that, as we know the velocity of light, and also the length of the waves, we have only to divide the length that light passes over in one second, by the lengths of the waves, in order to find how many undulations must take place in one second. The following table gives the wave-lengths, and the number of undulations for each colour :

Colour.	Number of Waves in one inch.	Number of Oscillations in one second.
Red	40,960	514,000,000,000,000
Orange ...	43,560	557,000,000,000,000
Yellow......	46,090	578,000,000,000,000
Green	49,600	621,000,000,000,000
Blue.........	53,470	670,000,000,000,000
Indigo......	56,560	709,000,000,000,000
Violet	60,040	750,000,000,000,000

These are the results, then, of such experiments as that of Fresnel's, and although such numbers as those given in the table above are apt to be considered as representing rather the exercise of scientific imagination than as real magnitudes actually measured, yet the reader need only go carefully over the account of the experiment, and over that of the measurement of the velocity of light, to become convinced that by these experiments *something* concerned in the phenomena of light has really been measured, and has the dimensions assigned to it, even if it be not actually the distance from crest to crest of ether waves—even, indeed, if the ether and its waves have no existence. But by picturing to ourselves light as produced by the swaying backwards and forwards of particles of ether, we are better able to think upon the subject, and we can represent to ourselves the whole of the phenomena by a few simple and comparatively familiar conceptions.

As an example of the facility with which the ether theory lends itself to aiding our notions of the phenomena of light, take the explanation of polarization. Let us suppose that we are looking at a ray of light along its direction, and that we can see the particles of ether. We should, in such a case, see them vibrating in planes having every direction, and their paths, as so seen, would be represented by an indefinite number of the diameters of a circle. Now, suppose we make the ray first pass through a rhomb of Iceland spar : we should, if we could see the vibrating particles in the emergent ordinary and extraordinary rays, perceive them swaying backwards and forwards across the direction of the rays in two planes only, as represented by the lines, B D and A C, in the two circles, O o and E e, Fig. 214—that is, half the particles would be vibrating in the direction B D, and the other half in the direction A C ; and further, the two directions would be at right angles to each other—the vibrations forming the extraordinary ray being performed in a plane at right angles to that in which the vibrations producing the ordinary ray take place. If—these planes being in the position indicated in 1, Fig. 214—we turn the crystal round through 90°, they would rotate with it, and would come severally into the position shown in 2, Fig. 214.

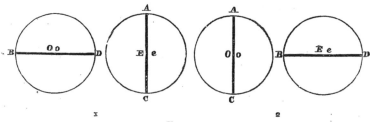

FIG. 214.

It was at one time objected to the theory which represents light as due to wave-like movements that, just as the vibrations which constitute sound spread in all directions, and go round intercepting bodies, enabling us, for example, to hear the sound of a bell even when a building intervenes, so if vibrations really produce light, these would extend within the shadows, and we ought to perceive light within the shadows, bending, as it were, round the edge of the shadow-casting body. This objection, which at one time presented a great difficulty for the wave theory, was triumphantly removed by the discovery that the luminous vibrations do extend into the shadow, and that this is in reality never completely dark. It is true that, although we can hear round a corner, we are in general unable to see round it ; but it should be noticed that in the case of hearing, the sound is much weakened by intervening objects, and that there are what may be termed *sound shadows.* A ray of light produces sensible effects only in the direction of its propagation ; but it can be shown that the successive portions of the waves advancing along it are centres of lateral disturbances producing new or secondary waves in all directions, which, however, interfere with and destroy each other. When an opaque screen intercepts a portion of the principal wave, it also stops a number of oblique or secondary waves, which would interfere more or less with the rest. Under ordinary circumstances, the remaining oblique or secondary rays are quite insensible in the presence of the direct light. But, with an apparatus which will cost but the two or three minutes' time required to construct it, the reader may see for himself that light is able to *pass round an obstacle,* and he may witness directly phenomena of the same order as those presented in the experiment of Fresnel's mirrors, which require costly apparatus for their production. He has only to take two fragments of common window-glass, and having made a piece of tinfoil adhere to one surface of each piece of glass, cut, with a sharp penknife, the finest possible slit in each piece of tinfoil, making the slit from ⅓ in. to 1 in. in length. If he will then hold one piece of glass about 2 ft. from his eye, so that it may be in the line between his eye and the sun (or other luminous body), and hold the other piece close to his eye with its slit parallel to that in the first piece, he will see the latter not simply as a line of light, but parallel to it a number of brilliantly-coloured rainbow-like bands will be seen on either side. If, instead of receiving the light from the sun, or from a candle-flame, the light given off by a spirit-lamp, with a piece of salt on its wick, be used, bright yellow stripes will be seen with dark spaces between them. Or, if the piece of glass next the sun be red-coloured, instead of plain glass, no rainbow-like bands will be visible, but a number of bright red stripes alternating with dark bands will be seen. The reader will have probably now little difficulty in perceiving that these can be easily explained as the results of interferences of a kind quite analogous to those of the waves of water represented in the diagram, Fig. 212. The rainbow-like stripes are due to the different wave-lengths of the different colours, as a consequence of which the bright and dark bands would be formed at different positions. Our limits do not admit of a full explanation of these beautiful effects, but the reader requiring further information would peruse with the greatest advantage portions of Sir John Herschel's " Familiar Lectures on Scientific Subjects."

The undulatory theory gives also an easy explanation of colours ; they being, according to the theory, only the effects, as already stated, of the different rates of vibrations of the ether. If the ether particles perform 514,000,000,000,000 oscillations in a second, we receive the impression we call red colour; if they execute 750,000,000,000,000 vibrations, the impression produced on our organ of sight is different—we call it violet; and so on. Thus science teaches us that visual impressions so different as red, green, blue, violet, and other distinct colours, are, in reality, all due to movements of one and the same——something ; and that the different sensations of colour we experience, arise merely from different rates of recurrence in these movements. In the subsequent article we shall have occasion to show that ordinary light, such as that of the sun, or of a candle, contains rays of every imaginable colour, mixed together in such proportions, that when this light falls upon a piece of paper, or upon snow, we have, in looking at these objects, the sensation of *whiteness.* But, if the light falls upon any substance which is able, in some way, to absorb or destroy some of the vibrations, the admixture of which makes up " *white light*," as it is called, then that object sending back to our eyes the rays formed of the remaining group of vibrations, gives us the sensation of *colour.* Suppose, for example, a substance to be so constituted that it is capable of absorbing, or quenching in some way, all the vibrations of the ether which occur at a quicker rate than 520,000,000,000,000 in a second : such a substance would send back to our eyes only the vibrations which constitute red light (see table, page 411), and we should say the substance in question had a *red* colour. Similarly, if the substance gave back only the vibrations which have the quickest rates, we should call the substance of a *violet* character. The agent which produces in our visual organs the impression of colour is, therefore, not in the objects, but in the light which falls upon them. The rose is red, not because it has redness in itself, but because the light which falls upon it contains some rays in which there are movements that occur just the number of times per second that gives us the impression we call redness ; in short, the colour comes not from the flower but from the light. " But," the reader might say, " the rose is always red by whatever light I see it, and therefore the colour must be in the flower. Whether I·view it by sunlight, or moonlight, or candlelight, or gaslight, I invariably see that *it is red.*" Now, it is precisely this circumstance—the seemingly invariable association of the object with a certain impression—in this case, redness—that leads our judgment astray, and makes us believe that the colour is in the object. Most people live out their lives without anything occurring to them which would give them the least idea that the colours of the objects they see around them are not in these objects themselves, but are derived from the light that falls upon the objects. And it required the comparison of many observations and experiments, and some clear reasoning, to establish a truth so unlike the most settled convictions of ordinary minds.

The point in question is fortunately one extremely easy of experiment, since we have simple means of producing light in which the vibrations corresponding to only one colour are present. The reader is strongly recommended to try the following experiment for himself. Let him procure a spirit-lamp, and place on the wick a piece of common salt about as large as a pea. Let the lamp be lighted·in a room from which all other light is completely excluded, and bring near the flame a red rose or a scarlet geranium. The flower will be seen with *all its redness gone*—it will appear of an ashy grey or leaden colour. A ball of bright scarlet wool, such as ladies use to work brilliant patterns for cushions, &c., held near this flame, is apparently transformed into a ball of the homely grey worsted with which, about a century ago, old ladies might be seen industriously darning stockings. The experiment is, perhaps, even more striking when, a little distance from the spirit-lamp, is placed a feeble light of the ordinary kind, a rushlight for example. The ball of wool, held near the latter, shows vivid scarlet, but, brought near the spirit-lamp with the salted wick, is pale, ashy grey. Moving thus the ball of worsted, first to one light then to the other, gives a most convincing and striking proof of the entire illusion we are under as to colour being an inherent quality of substances. Similar experiments may be multiplied indefinitely. A bouquet, viewed by the rushlight, shows the so-called *natural* colours of the flowers ; viewed by the salted flame, roses, verbenas, violets, larkspurs, and leaves, all appear of the uniform ashy grey, and only *yellow* flowers come out in their *natural* colours. A picture, say a chromo-lithograph after one of the most gorgeous landscapes that Turner ever painted, appears a work in monochrome, and gives exactly the effect of a sepia or indian-ink drawing. The most blooming complexion vanishes, and the countenance assumes a cadaverous aspect very startling to persons of weak nerves ; the lips especially, which might have rivalled pink coral by ordinary light, take a repulsive livid hue. All these effects may be seen to greater advantage by using the gas-flame of a Bunsen's burner, having a lump of salt placed in the flame ; or by means of a piece of *fine* wire gauze, about six inches square, supported about two or three inches above an ordinary gas-burner, from which the gas is allowed to issue without being lighted, but when to the top of the wire gauze, which is strewed with small fragments of salt, a light is applied, the gas will ignite only above the gauze, without the flame passing down to the burner below.

A fuller explanation of these strange appearances may be gathered from the subsequent article ; but it may suffice now to state that spirit, or gas burned in the way we have indicated, gives off little or no light of any kind.

If, however, common salt be introduced into the flame, then light—but light of only one particular colour—is given off, and that colour is yellow. There are no red, or green, or blue, or violet vibrations given off; and as the objects on which the light falls cannot supply these, it follows that with this light no impression corresponding to these colours can be produced on the eye, whatever may be the objects upon which it falls. Such experiments, not simply read about but actually performed, cannot fail to convince an intelligent person that the colours come from the light and not from the object. Of course, it is not denied that there is in each substance something that determines which are the rays absorbed, and which are the rays reflected to the eye—something that can destroy certain waves, but is powerless over others that rebound from the substance, and reaching the eye, there produce their characteristic impressions. And it is but this power of sending back only certain rays among the multitude which a sunbeam furnishes, that can be attributed to objects when we say that they have such or such a colour. In this sense, then, we may properly say that *the rose is red*, but it is also at the same time undeniably true that *the redness is not in the rose*.

Let it not be supposed that such scientific conclusions as those we have arrived at tend in any way to rob Nature of her beauty, or that our sense of the loveliness of colour is in any danger of being blunted by thus tracing out, as far as may be, the causes and sources of our sensations. The poets have occasionally said harsh things of science —indeed, one goes so far as to stigmatize the man of science as one who would "untwist the rainbow" and "botanize upon his mother's grave;" and another thus laments dispelled illusions :

" When Science from Creation's face
Enchantment's veil withdraws,
What lovely visions yield their place
To cold material laws ! "

Now, in the case we have been considering, the scientific view is surely as beautiful as the ordinary one. We can, it is true, no longer regard the objects as having in themselves the colours which common observation attributes to them, but we look upon the material world as being, so to speak, the neutral canvas upon which Light, the great painter, spreads his varied tints, although, unlike the real canvas of an artist, which is not only neutral, but receives indifferently whatever hues are laid upon it, the objects around us exercise a selective effect—as if the picture of Nature were produced by each part of the canvas refusing all the tints save one, but itself supplying none. The tendency of the study of science to increase our interest in the great spectacle of Nature, and to enhance our appreciation of her charms, has been more justly indicated by another poet — thus :

" Nor ever yet
The melting rainbow's vernal tinctured hues
To me have shone so pleasing, as when first
The hand of Science pointed out the path
In which the sunbeams gleaming from the west
Fall on the watery cloud, whose darksome veil
Involves the orient."

FIG. 215.—*Portrait of Professor Kirchhoff.*

THE SPECTROSCOPE.

MANY of the modern discoveries and inventions already described in
these pages have been instances of practical applications of science
to the every-day wants of mankind; but the chief interest of the subject
we now enter upon flows mainly from other sources than direct applications
of its principles in useful arts, although these applications are already neither
few nor unimportant. But that which, in the highest degree, claims our attention and excites our admiration in the revelations of the spectroscope is the
wonderful and wholly unexpected extent to which this instrument has enlarged our knowledge of the universe, and the apparently inadequate means
by which this has been accomplished. A little triangular piece of glass
gives us power to rob the stars of their secrets, and tells more about those
distant orbs than the wildest imagination could have deemed attainable to
human knowledge. One of the most acute philosophers of the present
century, a profound thinker who devoted his mind to the consideration of
the mutual relations of the sciences, declared emphatically, not very many
years ago, that all we could know of the heavenly bodies must ever be confined to an acquaintance with their motions, and to such a limited acquaintance with their features as the telescope reveals in the less distant ones.

A knowledge of their composition, he expressly asserted, could never be attained, for we could have no means of chemically examining the matter of which they are constituted. Such was the deliberate utterance of a man by no means disposed to underrate the power of the human mind in the pursuit of truth. And such might still have been the opinion of the learned and of the unlearned, but for the remarkable train of discoveries which has led us to the construction of instruments revealing to us the nature of the substances entering into the constitution of the heavenly bodies. We have now, for example, the same certainty about the existence of iron in our sun, that we have about its existence in the poker and tongs on the hearth. The last few years have seen the dawn of a new science; and two branches of knowledge which formerly seemed far as the poles asunder—namely, astronomy and chemistry—have their interests united in this new science of celestial chemistry. The progress which has been made in this department of spectroscopic research is so rapid, and the field is so promising, that the well-instructed juvenile of the future, instead of idly repeating the simple lay of *our* childhood :

> "Twinkle, twinkle, little star,
> How I *wonder* what you are !"

will probably only have to direct his sidereal spectroscope to the object of his admiration in order to obtain exact information as to what the star is, chemically and physically.

The results which have already been obtained in celestial chemistry, and other branches of spectroscopic science, are so surprising, and apparently so remote from the range of ordinary experience, that the reader can only appreciate these wonderful discoveries by tracing the steps by which they have been reached. A few fundamental phenomena of light have already been spoken of in the foregoing article ; and an acquaintance with these will have prepared the reader's mind for a consideration of the new facts we are about to describe. In discussing, in the foregoing pages, the subject of refraction, we have, in order that the reader's attention might not be distracted, omitted all mention of a circumstance attending it, when a beam of ordinary light falls upon a refracting surface, such as that represented in Fig. 203. The laws there explained apply, in fact, to elementary rays, and not to ordinary white light, which is a mixture of a vast multitude of elementary rays, red, yellow, green, &c. When such a beam falls obliquely upon a piece of glass, the ray is, at its entrance, broken up into its elements, for these, being refracted in different degrees by the glass, each pursues a different

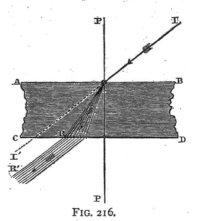

FIG. 216.

path in that medium, as represented by Fig. 216. Each elementary ray obeys the laws which have been explained, and therefore each emerges from the second surface of the plate parallel to the incident ray, and, in

27

consequence of this, the separation is not perceptible under ordinary circumstances with plates of glass having parallel surfaces. But, if the second surface be inclined so as to form such an angle with the first that the rays are rendered still more divergent in their exit, then the separation of the light into its elementary coloured rays becomes quite obvious. Such is the arrangement of the surfaces in a prism, and in the triangular pieces of glass which are used in lustres.

For the fundamental experimental fact of our subject, we must go back two centuries, when we shall find Sir Isaac Newton making his celebrated analysis of light by means of the glass prism. We shall describe Newton's experiment, for, although it was peformed so long ago, and is generally well known, it will render our view of the present subject more complete; and it will also serve to impress on the reader an additional instance of the world's indebtedness to that great mind, when we thus trace the grand results of modern discovery from their source. " It is well," is the remark of a clear thinker and eloquent writer, " to turn aside from the fretful din of the present, and to dwell with gratitude and respect upon the services of ' those mighty men of old, who have gone down to the grave with their weapons of war,' but who, while they lived, won splendid victories over ignorance."

The experiment of Sir Isaac Newton will be readily understood from Fig. 217, where C is the prism, and A C represents the path of a beam of sunlight allowed to enter into a dark apartment through a small *round*

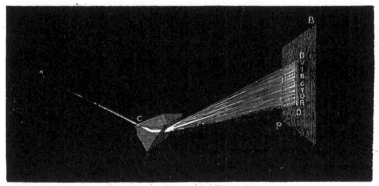

FIG. 217.—*Newton's Experiment.*

hole in a shutter, all other light being excluded from the apartment. In this position of the prism, the rays into which the sunbeam is broken at its entrance into the glass were bent upwards, and at their emergence from the glass they were again bent upwards, still more separated, so that when a white screen was placed in their path, instead of a white circular image of the sun appearing, as would have been the case had the light been merely refracted and not split up, Newton saw on the screen the variously-coloured band, D D D, which he termed the *spectrum.* The letters in the figure indicate the relative positions of the various colours, red, orange, yellow, green, blue, &c., by their initial letters. The spectrum, or prolonged coloured image of the sun, is red at the end, R, where the rays are least refracted, and violet at the other extremity, where the refraction is greatest, while, in the intermediate spaces, yellow, green, and blue pass by insensible gradations into each other.   Newton varied his experiment in many ways, as, for example, by trying the effect of refraction through a second prism on the differently coloured rays.   He found that the second prism did not divide the yellow rays, for instance, into any other colour, but merely bent them out of the straight course, to form on the second screen a somewhat broader band of yellow, and similarly with regard to the others.   From these, and a number of other experiments described in his "Opticks," (A. D. 1675), Newton concludes, "that if the sun's light consisted of but one sort of rays, there would be but one colour in the whole world, nor would it be possible to produce any new colour by reflections and re-fractions, and, by consequence, the variety of colours depends upon the composition of light." . . . . . "And if, at any time, I speak of light and rays, or coloured, or endued with colours, I would be understood to speak not philosophically and properly, but grossly, and accordingly to such con-ceptions as vulgar people in seeing all these experiments would be apt to frame.   For the rays, to speak properly, are not coloured.   In them there is nothing else than a certain power and disposition to stir up a sensation of this or that colour.   For, as sound in a bell, a musical string, or other sounding body, is nothing but a trembling motion, and in the air nothing but that motion propagated from the object, and in the sensorium 't is a sense of that motion under the form of a sound ; so colours in the object are nothing but a disposition to reflect this or that sort of rays more copi-ously than the rest : in the rays they are nothing but their dispositions to propagate this or that motion into the sensorium, and in the sensorium they are sensations of these motions under the form of colours."

These memorable investigations of Newton's have been the admiration of succeeding philosophers, and even poets have caught inspiration from this theme :

> " Nor could the darting beam of speed immense
> Escape his swift pursuit and measuring eye.
> E'en Light itself, which everything displays,
> Shone undiscovered, till his brighter mind
> Untwisted all the shining robe of day ;
> And, from the whitening undistinguished blaze,
> Collecting every ray into his kind,
> To the charmed eye educed the gorgeous train
> Of parent colours.   First the flaming red
> Sprung vivid forth ;   the tawny orange next ;
> And next delicious yellow—by whose side
> Fell the kind beams of all-refreshing green ;
> Then the pure blue, that swells autumnal skies,
> Ethereal played ;   and then, of sadder hue,
> Emerged the deepened indigo, as when
> The heavy-skirted evening droops with frost,
> While the last gleamings of refracted light
> Died in the fainting violet away.
> These, when the clouds distil the rosy show,
> Shine out distinct adown the watery bow ;
> While o'er our heads the dewy vision bends
> Delightful—melting on the fields beneath.
> Myriads of mingling dyes from these result,
> And myriads still remain.—Infinite source
> Of beauty ! ever blushing—ever new !
> Did ever poet image aught so fair,
> Dreaming in whispering groves, by the hoarse brook,
> Or prophet, to whose rapture Heaven descends ? "

The spectra which Newton obtained by admitting the solar beams through a circular aperture, were, however, not simple spectra. The circular beam may be considered as built up of flat and very thin bands of light, parallel to the edges of the prism, and a simple ray would be formed by one of these flat bands; as the round opening would allow an indefinite number of such rays to enter, each would produce its own spectrum on the screen, and the actual image would be formed of a number of spectra overlapping each other. When the aperture by which the light is admitted consists merely of a narrow slit, or line, parallel to the edges of the prism, we obtain what is termed a *pure spectrum*. When the prism is properly placed, an eye, viewing the fine slit through it, sees a spectrum formed, as it were, of a succession of virtual images of the slit in all the elementary coloured rays.

The person who first examined the solar spectrum in this manner was the English chemist Wollaston, who, in 1802, found that the spectrum thus observed was not continuous, but that it was crossed at intervals by dark lines. Wollaston saw them by placing his eye directly behind the prism. Twelve years later, namely, in 1814, the German optician Fraunhofer devised a much better mode of viewing the spectrum; for, instead of looking through the prism with the naked eye, he used a telescope, placing the prism and the telescope at a distance of 24 ft. from the slit, the virtual image of which was thus considerably magnified. The prism was so placed that the incident and refracted rays formed nearly equal angles with its faces, in which circumstance the ray is least deflected from its direction, and the position is therefore spoken of as being that of *minimum deviation*. It can be shown that this position is the only one in which the refracted rays can produce clear and sharp virtual images of the slit, and therefore it is necessary in all instruments to have the prism so adjusted. Fraunhofer then saw that the dark lines were very numerous, and he found that they always kept the same relative positions with regard to the coloured spaces they crossed; that these positions did not change when the material of which the prism was made was changed; and that a variation in the refracting angle of the prism did not affect them. He then made a very careful map, laying down upon it the position of 354 of the lines out of about 600 which he counted, and indicated their relative intensities, for some are finer and less dark than others. The most conspicuous lines he distinguished by letters of the alphabet, and these are still so indicated; and the dark lines in the solar spectrum are called " Fraunhofer's Lines." These lines, as will appear in the sequel, are of great importance in our subject. A few of the more obvious ones are shown in No. 1, Plate XVII. Fraunhofer found that these lines were always produced by sunlight, whether direct, or diffused, or reflected from the moon and planets; but that the light from the fixed stars formed spectra having different lines from those in the sun—although he recognized in some of the spectra a few of the same lines he found in the solar spectrum. The fact of these differences in the spectra of the sun and fixed stars proved that the cause of the dark lines, whatever it might be, must exist in the light of these self-luminous bodies, and not in our atmosphere. It was, however, some years afterwards ascertained that the passage of the sun's light through the atmosphere does give rise to some dark bands in the spectrum; for it was found that certain lines make their appearance only when the sun is near the horizon, and its rays consequently pass through a much greater thickness of air.

Sir D. Brewster first noticed in 1832 that certain coloured gases have the power of absorbing some of the sun's rays, so that the spectrum, when the rays are made to pass through such a gas before falling on the prism, is crossed by a series of dark lines—altogether different from Fraunhofer's lines, though these are also present. The gas in which this property was first noticed is that called "nitric peroxide"—a brownish-red gas, of which even a thin stratum produces a well-marked series of dark lines. The same property was soon discovered in the vapours of bromine, iodine, and a certain compound of chlorine and oxygen. Each substance furnishes a system of lines peculiar to itself: thus the vapour of bromine, although it has almost exactly the same colour as nitric peroxide, gives a totally different set of lines. These, therefore, do not depend on the mere colour of the gas or vapour, and this is conclusively proved by the fact of many coloured vapours producing no dark lines whatever: the vapour of tungsten chloride, for example, although in colour so exactly like bromine vapour that the two cannot be distinguished by the eye, yields no lines whatever.

In Fig. 218 is represented a lamp for burning coal-gas, which is constantly used by chemists as a source of heat. It is known as "Bunsen's burner," from its inventor the celebrated German chemist. It consists of a metal tube, 3 in. or 4 in. long, and ⅓ in. in diameter, at the bottom of which the gas is admitted by a small jet communicating with the elastic tube which brings the gas to the apparatus. A little below the level of the jet there are two lateral openings which admit air to the tube. The gas, therefore, becomes mixed with air within the tube, and this inflammable mixture streams from the top of the tube and readily ignites on the approach of a flame, the mixture burning with a pale bluish flame of a very high temperature. This little apparatus is not only the most useful pieces of chemical apparatus ever devised, but it furnishes highly instructive illustrations of several points in chemical and physical science; and to some of these we invite the reader's attention, as they have an immediate bearing on our present subject. Coal-gas is a mixture of various compounds of

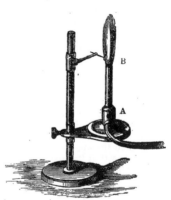

FIG. 218.—*Bunsen's Burner on a stand.*

the two elementary bodies, hydrogen and carbon; and when the gas burns, these substances are respectively uniting with the oxygen of the air, producing water and carbonic acid gas. Now, when coal-gas is burnt in the ordinary manner as a source of light, the supply of oxygen is too small to admit of the complete combustion of all its constituents; and as the oxygen more eagerly seizes upon the hydrogen than upon the carbon, a large proportion of the latter thus set free from its hydrogen compound is deposited in the flame in the solid form, and is there intensely heated. The presence of solid carbon in an ordinary gas flame is easily proved by holding in it a cold fragment of porcelain, or a piece of metal, which will become covered with soot. In the flame of the Bunsen burner there is no soot, because the increased supply of oxygen, afforded by previously mixing the gas with air, enables the whole of the constituents of the gas to be completely burnt; and this is of the greatest advantage to the chemist, who always desires to have the vessels he heats free from soot, in order that he may observe what is taking place within them. The flame of Bunsen's lamp becomes that of an ordinary sooty gas flame, when the two orifices which admit the air at the bottom of the tube are closed up, and then, of course, the temperature cannot be so high as when the whole constituents of the gas are completely burnt, but the flame becomes highly luminous; whereas when the orifices are open it gives so little light, that in a dark room one cannot see a finger held 20 in. from the lamp. Plainly the cause of this difference is connected with the presence or absence of the heated particles of solid carbon. The non-luminous flame contains no solid particles; the bright part of the other flame is full of them. To these heated particles of solid carbon we are, then, indebted for the light which burning coal-gas supplies. And, since we are able by such artificial illumination to distinguish colours, the white-hot carbon must give off rays of all degrees of refrangibility, and we should expect to find in the spectrum produced by such a flame, the red, yellow, green, and other coloured rays. And such is indeed the spectrum which these incandescent carbon particles produce: it resembles the solar spectrum, but *there is an entire absence of dark lines,* so that the appearance is that represented in No. 1, Plate XVII., if we suppose the Fraunhofer lines removed. If the pale blue flame of the Bunsen's burner be similarly examined, the spectrum, No. 14, Plate XVII., shows that only a few rays of certain refrangibilities are emitted, forming bright lines here and there, but of little intensity, while the whole of the other rays are absent. This shows that while the highly heated solid gives off all rays from red to violet without interruption, the still more highly heated gases give off only a few selected rays.

It has long been known that some substances impart certain colours to flames, and such substances have been long employed to produce coloured effects in fireworks, &c. But coloured flames do not appear to have been examined by the prism until 1822, when Sir John Herschel described the spectra of strontium, copper, and of some other substances, remarking that " The colours thus communicated by the different bases to flame afford in many cases a ready and neat way of detecting extremely minute quantities of them." A few years later, Fox Talbot described the method of obtaining a monochromatic flame, by using in a spirit-lamp diluted alcohol in which a little salt has been dissolved. The paper in which he describes this and other observations concludes thus: " If this opinion should be correct and applicable to the other definite rays, a glance at the prismatic spectrum of flame may show it to contain substances which it would otherwise require a laborious chemical analysis to detect." Here we have the first hint of that spectrum analysis which has provided the chemist with a method of surpassing delicacy for the detection of metallic elements. The spectra of coloured flames were also subsequently examined and described by Professor W. A. Miller, but the most complete investigation into the subject was made by Professors Kirchhoff and Bunsen, who also contrived a convenient instrument, or *spectroscope,* for the examination and comparison of different spectra. The instrument has received many improvements and modifications, but the essential parts are one or more prisms; a slit, through which the light to be examined is allowed to enter; a tube, having at the other end a lens to render parallel the rays from the slit; a telescope, through which the spectrum is viewed; and usually some apparatus by which the positions of the different lines may be identified.

A very elegant instrument, made by Mr. John Browning, of the Strand, is

SPECTRA.

represented in Fig. 219. It has a single prism, made of glass, of great power in dispersing the rays. The prism is supported on a little stage, placed in the middle of a horizontal circular brass table about 6 in. in diameter. On the left is seen a tube, about 15 in. long, at the outer extremity of which is the slit, formed of pieces of metal very accurately shaped. One of these pieces

FIG. 219.—*Spectroscope with one Prism.*

slides in a direction at right angles to the slit, and, by means of a spring and a fine screw, can be very nicely adjusted, so that an opening of any degree of fineness can be readily obtained. In front of the slit is a small glass prism, with its edges parallel to the slit, but only half its height. The bases of this prism are formed of two sides of a square and its diagonal, and, as shown in the figure, one side is parallel to the face of the slit, and the other to the axis of the tube. Rays of light coming from a source on the left of the slit (as seen in the figure) will, therefore, enter this little prism, and be totally reflected (see page 399) by the diagonal surface, down the axis of the tube through the lower half only of the slit. This is the only office of this prism, which has nothing to do with the dispersion of the rays: the use to which it is put will be seen presently. It is fixed in such a manner that, when required, it can be turned aside with the touch of a finger, and the *whole* length of the slit exposed. A peculiarity in these instruments of Mr. Browning's is the admirable arrangement for determining the position of any line in a spectrum. For this purpose, the eye-piece of the telescope is provided with a pair of cross-wires, and the telescope itself, which is about 18 in. in length, moves in a horizontal plane round the axis of the circular brass table, from which an arm projects, carrying a ring into which the telescope screws. This arm carries a *vernier* along the limb of the circular table, which is very accurately divided into thirds of degrees, so that with the aid of the vernier the angular position of the telescope can be read off to a minute, that is, to $\frac{1}{60}$th of a degree. The arm carrying the telescope is provided with a screw for clamping it in any desired position while the readings are taken. On placing in front of the slit the flame of a Bunsen's burner, the spectrum produced by any substance in this flame will, when the instrument is in proper adjustment, be seen on looking through the telescope, and the cross-wires being also in view, the point of their intersection may be brought into coincidence with any line of the spectrum, and the telescope being clamped in this position, the angular reading thus taken determines the position of the line. Thus, for example, the angular positions in which the principal Fraunhofer's lines are seen having been observed and recorded, the angular position of any line in another spectrum will at once determine its position among the Fraunhofer lines; or the spectrum may be mapped by laying down the angular readings of the lines by means of a scale of equal parts. And, again, in the little prism in front of the slit we have the means of bringing two spectra in view at once, one being from a light directly in front, and the other from a light at the side. The two spectra are seen one above the other, and the coincidence or difference of their lines may be directly observed. When the instrument is in use, the prism and the ends of the tube are covered with a black cloth, loosely thrown over them, by which all stray light is shut out. The author has had in use for several years one of these instruments, and he cannot forbear expressing his perfect satisfaction with its powers, which he finds amply sufficient for all ordinary chemical purposes, while the accuracy of the workmanship is really wonderful, considering the very moderate price of the instrument.

The substances the spectra of which are most conveniently examined are the metals of the alkalies and alkaline earths. Small quantities of the salts of these metals, placed in a loop of fine platinum wire, impart characteristic colours to the flame of a Bunsen burner or to that of a spirit-lamp. For the examination of the spectra the former is to be preferred, as the lines come out much more vividly. Indeed, at temperatures higher than that of the Bunsen's burner, such as in the flame of pure hydrogen, or in the voltaic arc, some substances give out additional lines. In Plate XVII., Nos. 2 to 9, is shown the appearance of the spectra produced by the Bunsen's burner when salts of the metals are held in the flame in the manner already mentioned, and the spectra are examined with the instrument just described. One of the simplest of these spectra is that produced by sodium compounds, such as common salt. The smallest particle of this substance imparts an intense yellow colour to the flame, and the spectrum is found to take the form of a single bright yellow line—No. 3. It has been estimated that the presence of the $\frac{1}{1000000000}th$ *part of a grain* of sodium can be detected by the production of this line. Indeed, the very delicacy of this sodium reaction renders it almost impossible to get rid of this line, for sodium is found to be present in almost everything,—a fact the earlier observers of spectra were not aware of, for they attributed this yellow line to water, which was the only substance they knew to be so generally diffused. If a platinum wire be heated in the flame of the Bunsen burner until all the sodium indications have disappeared, it suffices to remove the wire, and, without allowing it to come into contact with anything, to leave it exposed to the air for a few minutes, to cause it again to give the characteristic yellow colour when again plunged into the flame. This is due to the fact that the element is contained in all the floating particles which pervade the atmosphere. The spectroscope is not required to show the presence of the sodium on the platinum which has been exposed to the air, the colour imparted to the flame being plainly visible to the eye, and it needs only the Bunsen burner and 2 in. of platinum wire to prove the fact, and also to show that mere contact with the fingers is enough to highly charge the wire with sodium compounds. Any volatile compound of potassium gives the spectrum represented by No. 2, the principal lines being a red line and one in the extreme violet, the latter being somewhat difficult to observe. There is also a third rather ill-defined red line, and a portion of a faint continuous spectrum. Salts of strontium impart a bright red colour to the flame. and the spectrum they produce is shown by No. 6, in which are seen several bright red lines and a fainter blue one. Calcium, which also gives a reddish colour to flame, furnishes an entirely different set of lines (No. 5). and barium salt another, containing numerous lines, especially some very vivid green ones.

In all the cases we have named, and whenever bright-lined spectra are furnished by substances placed in the flame of a lamp, or in burning hydrogen gas, or in the intensely hot voltaic arc, there is evidence that the substances are converted into vapour or gas. We have already seen how hot solid carbon gives a continuous spectrum, while carbon in the state of gaseous combination gives most of the bright lines seen in the spectrum of coal-gas (No. 14). It is observed also that the more readily volatized are the salts, the more vivid are the bright lines they produce when heated in a flame. It must be understood that each element gives it own characteristic lines, that these are always in precisely the same position in the spectrum, that no substance produces a line in exactly the same position as another, however near two lines due to different substances may, in some cases, appear; and also, that however the salts of the different metals are mixed together, each produces its own lines, and each ingredient may be recognized. And this is done in an instant by an experienced observer— a mere glance at the superposed spectra of, perhaps, half a dozen metals, suffices to inform him which are present. There is also a peculiarity in this optical mode of recognizing the presence of bodies which gives the subject the highest interest, namely, the circumstance that the spectrum is produced and the bodies recognized, however far from the observer the luminous gas may be placed, the only condition required being that the rays reach the instrument.

Until Kirchhoff and Bunsen's spectroscopic investigations, lithium was supposed to be a rare metal, occurring only in a few minerals. It happens that this substance yields a remarkable spectrum (No. 4), for it gives an extremely vivid line of a splendid red colour, accompanied by only one other, a feeble yellow line ; and the reaction is of very great delicacy, for $\frac{1}{600000}$th of a grain can easily be detected, and an eye which has once seen the red line readily recognizes it again. A single drop of a mineral water containing lithium has been found to distinctly produce the red line, in cases where the quantity contained in a quart of the water would have escaped ordinary chemical analysis. The spectroscope has shown that lithium, so far from occurring in only four or five minerals, is a substance very widely diffused in nature. In the waters of the ocean, in mineral and river waters, in most plants, in wines, tea, coffee, milk, blood, and muscle, this metal has been found. Dr. Roscoe states that the ash of a cigar, when moistened with hydrochloric acid, and held in a platinum wire in the flame of the Bunsen's burner, at once shows the principal lines of sodium, potassium, calcium, and lithium. Salts of lithium and of strontium both impart a rich crimson tint to flames, and it is hardly possible to detect any difference in these colours with the naked eye; but, as the reader may see on comparing spectra No. 4 and No. 6, the prism makes a wide distinction.

Matter for a very interesting chapter in the history of prismatic analysis has been furnished by the discovery of four new elements by means of the

spectroscope. In 1860 Bunsen observed that the residue, after evaporation, of a certain mineral water, yielded spectra with bright lines which
he had not seen before. He concluded that they were due to some unknown elements, and, in order to separate these, he evaporated many tons
of the water, and was rewarded by the discovery of two alkaline metals,
cæsium and *rubidium*. The delicacy of the spectrum reaction may be
inferred from the fact of a ton weight of the water containing only three
grains of the salts of each of these substances. Rubidium gives a splendid
spectrum, containing red, yellow, and green lines, and also two characteristic violet lines; while cæsium has orange, yellow, and green lines, and
two very beautiful blue lines, by which it is easily recognized.

About the same time, Mr. W. Crookes discovered, in a mineral from the
Hartz, another elementary body, the existence of which was first indicated
to him by the characteristic spectrum it produces, namely, a single splendid
green line (No. 8 spectrum). In 1864 two German chemists discovered,
also in the Hartz, a fourth new element, which was detected by two well-
defined lines in the more refrangible end of the spectrum—(see spectrum

FIG. 220.—*Miniature Spectroscope.*

No. 9, in the plate). This metal was named Indium, in reference to the
colour of its lines, and the names of the other three—cæsium, rubidium,
and thallium, are also derived from the colours of their characteristic lines.

Although the reader may, from such representations of the spectra as
those given in Plate XVII., form some idea of their appearance, he would
find his knowledge of the subject much clearer if he had the opportunity
of examining for himself the actual phenomena. We have already recommended the performance of certain easy experiments involving no outlay,
but, in the matter of spectroscopes, carefully finished optical and mechanical work is absolutely necessary in the appliances. It fortunately happens
that one eminent optician, at least, has made it his study to produce good
spectroscopic apparatus at the lowest possible cost, and if the reader be
interested in this subject, and desirious of trying experiments himself, he
can, for a very moderate sum, be equipped with all the appliances for
examining the phenomena we have described. He has only to procure, in
the first place, a small direct-vision spectroscope, such as that represented
of its actual size in Fig. 220, which is sold by Mr. Browning for twenty-
two shillings; secondly, a Bunsen's burner, a few feet of india-rubber
tubing, two inches of platinum wire, and a few grains of the salts of lithium,
strontium, thallium, &c. The whole expense will probably be covered by
adding four shillings to the cost of the spectroscope, and the reader will
then be in a position to see for himself the principal Fraunhofer lines, the
spectra of the metals already referred to, and the absorption bands of the

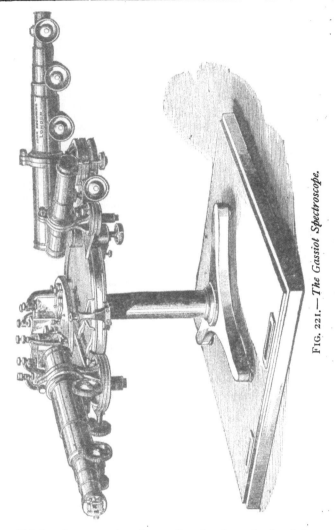

FIG. 221.—*The Gassiot Spectroscope.*

gases which have been mentioned, as well as the absorption bands in liquids which will be spoken of in the sequel.

The splitting up of a beam of light into its elements—which it is the office of the prism to produce—is accomplished by a single prism to a certain degree only. It separates the red from the green, for example ; but the colours pass into each by insensible gradations through orange, yellow, and greenish yellow. If we allow the rays to fall upon a second prism after emerging from the first, the separation is carried further; the red, for instance, is spread out into different kinds of red, and so on with the rest. And the greater the number of prisms, the greater is the extension which is given to the spectrum. Now, just as by increasing the power of the telescope, new stars become visible, whose light was before too faint, and nebulæ, or stars which before seemed single, are resolved into clusters of individual stars—so, by increasing the power of the spectroscope by employing two, four, or more prisms, lines which appear single by the less powerful instruments are, in some instances, resolved into groups of lines, and new lines come into view, which before were too faint to show themselves. For example, if we view the Fraunhofer lines through a spectroscope like that in Fig. 220, but having two prisms instead of one, we shall see that the D line is not really a single line, but is formed of two lines close together. If we use greater dispersive power by employing a greater number of prisms, we shall observe with solar light that when these two D lines are sufficiently separated, several other lines make their appearance between them. In this way the number of dark lines in sunlight, which have been carefully mapped by Kirchhoff and others, amount to upwards of 2,000; and no doubt there are many more lines waiting a still more powerful instrument. Fig. 221 is copied from a large spectroscope made by Mr. Browning for Mr. Gassiot. It has nine or more highly dispersive glass prisms; the telescope and the tube bearing the slit have focal lengths of 18 in., the lenses having a diameter of $1\frac{1}{2}$ in.; the telescope is provided with a slow motion for taking the angular position; and there is a third tube provided with a micrometer, by which the position of the lines can be measured to $\frac{1}{10000}$th of an inch.

The instruments we have mentioned, except the miniature spectroscope, show only a portion of the spectra at once, a movement of the telescope being requisite to bring each part into view. It has been already stated that the only position of the prism which will make the lines clear and well defined is that in which the *deviation* is the least. In using trains of prisms it is therefore necessary to adjust each prism for the part of the spectrum which may be under observation. This is a tedious process, and it has been obviated by a useful invention of Mr. Browning's, by which the adjustment is rendered automatic—that is, the movements of the telescope are communicated to the prisms in such a manner that they place themselves into the proper position for producing clear images of the slit, whatever may be the refrangibility of the rays under examination : Fig. 222 shows the arrangement as it appears when viewed from above. The train of six prisms can be so arranged that the ray after passing through six of them shall be totally reflected by a surface of the last prism, and pursue again its path through the six prisms in the reverse direction, becoming more and more dispersed by each prism until it emerges parallel to the axis of the telescope. The power of the instrument is, therefore, equivalent to that of one with twelve prisms ; but it can be used at pleasure with any dispersive power, from two to twelve prisms.

By making use of one of the Bunsen burners, the lines which are characteristic of some ten or twelve metals are readily seen when one of their more volatile salts is converted into vapour. For this purpose their chlorides are usually employed, but the reactions are common to all their salts. It is necessary that the metal should exist in the flame in the state of highly heated vapour or gas, in order that its characteristic rays should be given off. We usually introduce compounds of these metals into the flame ; but there is reason to believe that these are decomposed in the flame, and the disassociated metal takes the form of glowing gas, a small quantity of which suffices for the production of the bright lines. No doubt the other constituent of the compound, the chlorine for example, is also set free in the gaseous form ; but since the spectrum of the metal only is visible, we may infer that at the temperature of the flame, the non-metallic elements are not sufficiently luminous to produce a spectrum. When we repeat the experiments with salts of the less volatile metals, we obtain no spectra—

Fig. 222.—*Browning's Automatic Adjustment of Prisms.*

the temperature of the flame not being sufficiently high to convert these into vapour. Other methods have, therefore, to be resorted to, and advantage is taken of the fact discovered by Faraday, that an electric spark is nothing but highly heated matter. The spectroscope gives us reason to believe that this matter, which is formed of the substances between which the spark passes, is in the gaseous state; for it is found, on examining sparks passing between two pieces of each metal, that characteristic bright lines are produced. If one of the metals already named is submitted to this examination, the same lines are found which are seen in the spectra produced by the salts of the metal volatized in the flame, but in some cases additional bright lines appear in the spark spectrum. With the heavier metals the spark, or the electric arc, is, however, the only means of igniting their vapours. The usual mode of doing this is to make the discharges of a large induction coil pass between the two fine wires of the metals, placed about a quarter of an inch apart. A Leyden jar is commonly employed to condense the discharge, and thus produce a still higher temperature. Mr. Browning has contrived the neat little apparatus shown in Fig. 223, in which the jar is superseded by a more compact and convenient condenser inside of the box, so that it is only necessary to attach one terminal of the coil to the binding-screw, seen outside of the end of the box, and place the other wire from the coil in the binding-screw of one or the other of the pieces of apparatus supported by the upright rod. Of these it is the one on the right which at present engages our attention. Within a small glass cylinder are two sliding rods, terminated by screw-clips, which hold finely-pointed pieces of the metal under examination. The slit of the spectroscope

FIG. 223.—*Apparatus for Spark Spectra.*

is placed close to the glass cylinder, and when a very rapid succession of sparks is passing, the bright lines are seen continuously. The spectra of metals examined in this way are found to yield a very large number of lines. Thus the spectrum of calcium has 75 lines, and that of iron no fewer than 450 lines. Our limits will not permit of an account of many interesting particulars relating to these spectra, which include those of all the 50 metallic elements. It should, perhaps, be stated that a modified mode of producing spectra by sparks is sometimes found useful. This consists in causing sparks to pass between a solution of some salt of the metal and a piece of platinum wire. The apparatus for this purpose is that shown on the left side of the upright in Fig. 223.

It remains to describe the method of producing spectra of the gaseous non-metallic elements, such as oxygen, nitrogen, hydrogen, &c. For this purpose electricity is again made use of. It has been found that while an electric discharge cannot take place across a perfect vacuum, and air or gas, at ordinary densities, offers much resistance to the passage of electricity, on the other hand, a highly rarefied gas permits the discharge to take place through it with great facility. This is seen in Geissler's tubes, where a succession of discharges from a Ruhmkorff's coil causes the tubes to appear filled with light—due to the heating to incandescence of a very minute quantity of the gas. The eye readily recognizes difference of colour in the light given off by the different gases, and when this light is examined by the spectroscope, bright lines, characteristic of each gas, are observed. Nos. 12 and 13, in Plate XVII., are the spectra of hydrogen and of nitrogen respectively, which appear when the gases are examined in the manner just described. In this manner the spectra of chlorine, bromine, iodine, oxygen, sulphur, phosphorus, &c., may be studied. Silicon and some other solid non-metallic elements present great difficulties to the spectroscopist, for these elements cannot be volatized at any temperature we can command, and the spectra of their elements can only be inferred from those of their compounds. But unfortunately the spectra are found to vary with the nature of the compound, and thus it happens that in the case of carbon, for example, no definite spectrum can be assigned to the element. The flame of coal-gas, burning in the air, as in the Bunsen burner, gives the spectrum No. 14; but if this is compared with the spectrum of the flame of burning *cyanogen* (a compound of carbon and nitrogen), the two are found to differ greatly. The cyanogen spectrum has the two pale broad bands of violet-blue, the four blue lines, the two green lines, and the brightest of the greenish yellow which are seen in the coal-gas spectrum. But it has in addition a characteristic series of violet lines, a series of bright blue, two or three crimson and red lines, and bands in the orange, and several green lines, none of which occur in the coal-gas spectrum. These additional lines are not due to nitrogen, for, with perhaps the exception of some red lines, they do not coincide in position with any of the nitrogen lines. The spectrum of hydrogen, No. 12, should be noticed, as its three lines are very distinct, and it will be observed that they exactly coincide in their position with the three Fraunhofer lines, C, F, and G, in No. 1.

There is another branch of this extensive subject to which we have now to invite the reader's attention. The power of certain gases to absorb or stop certain rays of an otherwise continuous spectrum has already been mentioned; but this property is by no means confined to gases, for certain liquids and solids do this in a high degree. There is a remarkable metallic element, named *didymium*. It is a rare substance, and its presence cannot with certainty be detected by any ordinary tests. Its salts, however, form solutions *without colour*, or nearly so, which have the power of strongly absorbing certain rays. If we hold before the slit of the spectrum a small tube containing a solution of any one of the salts, and allow the rays from the sun, or from a luminous gas or candle-flame, to pass through it, we see the spectrum crossed by certain well-defined very dark bands. A spectrum of this kind is called an *absorption spectrum*, and the position, number, width, &c., of dark bands are found to be as peculiar to each substance as are the bright lines in the spectra of the elements. The method of observing them when produced by solutions is very simple. The liquid is contained in a small test-tube, which is placed in front of the slit; or, more conveniently, the liquid is put into a *wedge-shaped* vessel, and thus the thickness of the stratum of liquid through which the rays pass can easily be varied, so that the best results may be obtained. The absorption spectra are produced by many compound substances. A striking absorption spectrum is seen when a solution in alcohol of the green colouring matter of leaves (*chlorophyll*) is examined; for several distinct bands are seen, one in the red being especially well marked. Many other coloured bodies exhibit characteristic absorption bands, as, for example, permanganate of potash, uranic salts, madder, port wine, and magenta. The bands are so peculiar for each substance, that if so-called port wine, for example, owe its colour to colouring matter other than that of the grape, such as logwood, &c., the adulteration can be instantly detected by a glance at the absorption spectrum. As, however, the absorption bands are not, like the bright lines of metals, definite images of the slit, but rather broad portions of the spectra, it is very desirable in examining such spectra to compare them directly with those of known substances, by throwing two spectra into one field, by means of a side reflecting prism, as already described.

Perhaps one of the most interesting examples of absorption spectra is that of blood. A single drop of blood in a tea-cupful of water will show its characteristic spectrum when it is properly examined. If the blood is arterial or oxidized blood, two well-marked dark bands are visible; but if venous or deoxidized blood be used, we see, instead of the two dark bands, a single one in an intermediate position. These differences have been proved to be due to oxidization and deoxidization of a constituent of the blood, called *hæmoglobin*, and by using appropriate chemical reagents, the same specimen of blood may be made to exhibit any number of alternations of the two spectra, according as oxidants or reducing reagents are employed. It would be possible by an examination of the absorption spectrum of a drop of arterial blood to pronounce that a person had died of suffocation from the fumes of burning charcoal. In such case, the supply of oxygen being cut off, the hæmoglobin of the whole of the blood in the system becomes deoxidized.

The beautiful delicacy of these spectrum reactions has permitted the spectroscope to be applied to the microscope with signal success by Mr. Browning, working in conjunction with Mr. Sorby, who has devoted great attention to this subject. The Sorby-Browning instrument is a direct-vision spectroscope, with a slit, lens, &c., placed above the eye-piece of the microscope. By receiving the light through a single drop of an absorptive liquid placed under the object-glass of the microscope, the characteristic bands are made visible. The micro-spectroscope is also a valuable instrument for examining the absorption bands which are found in the light reflected from solid bodies, for the smallest fragment suffices to fill the field of the microscope. Mr. Sorby is able to obtain most unmistakably the dark bands peculiar to blood from a particle of the matter of a blood-stain weighing less than $\frac{1}{1000}$th part of a grain. It is plain from this that the spectroscope must sometimes prove of great service in giving evidence of crime from traces which would escape all ordinary observation.

The micro-spectroscope, in its most complete form, is represented in Fig. 224. As may be seen from the figure, the apparatus consists of several parts. The prism is contained in a small tube, which can be removed at pleasure; below the prism is an achromatic eye-piece, having an adjustable slit between the two lenses; the upper lens being furnished with a screw motion to focus the slit. A side slit, capable of adjustment, admits, when required, a second beam of light from any object whose spectrum it is desired to compare with that of the object placed on the stage of the microscope. This second beam of light strikes against a very small prism suitably placed inside the apparatus, and is reflected up through the compound prism, forming a spectrum in the same field with that obtained from the object on the stage. A is a brass tube carrying the compound direct-vision prism, and has a sliding arrangement for roughly focussing.

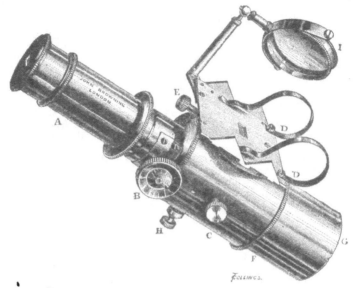

FIG. 224.—*The Sorby-Browning Micro-Spectroscope.*

B, a milled head, with screw motion to finally adjust the focus of the achromatic eye-lens.

C, milled head, with screw motion to open or shut the slit *vertically.* Another screw, H, at right angles to C, regulates the slit horizontally. This screw has a larger head, and when once recognized cannot be mistaken for the other.

D D, an apparatus for holding a small tube, that the spectrum given by its contents may be compared with that from any other object on the stage.

E, a screw, opening and shutting a slit to admit the quantity of light required to form the second spectrum. Light entering the aperture near E strikes against the right-angled prism which we have mentioned as being placed inside the apparatus, and is reflected up through the slit belonging to the compound prism. If any incandescent object is placed in a suitable position with reference to the aperture, its spectrum will be obtained, and will be seen on looking through it.

F shows the position of the field lens of the eye-piece.

G is a tube made to fit the microscope to which the instrument is applied. To use this instrument, insert G like an eye-piece in the microscope tube. Screw on to the microscope the object-glass required, and place the object whose spectrum is to be viewed on the stage. Illuminate with stage mirror if transparent, with mirror and lieberkühn and dark well if opaque, or by side reflector, bull's-eye, &c. Remove A, and open the slit by means of the milled head, H, at right angles to D D. When the slit is sufficiently open the rest of the apparatus acts like an ordinary eye-piece, and any object can be focussed in the usual way. Having focussed the object, replace A, and gradually close the slit till a good spectrum is obtained. The spectrum will be much improved by throwing the object a little out of focus.

Every part of the spectrum differs a little from adjacent parts in refrangibility, and delicate bands or lines can only be brought out by accurately focussing their own parts of the spectrum. This can be done by the milled head, B. Disappointment will occur in any attempt at delicate investigation if this direction is not *carefully attended to.* When the spectra of very small objects are to be viewed, powers of from ½ in. to 1-20th, or higher, may be employed. Blood, madder, aniline dyes, permanganate of potash solution, are convenient substances to begin experiments with. Solutions that are too strong are apt to give dark clouds instead of delicate absorption bands. Small cells or tubes should be used to hold fluids for examination.

Mr. Browning has still further improved the micro-spectroscope by the ingenious arrangement for measuring the positions of the lines, which is represented in Fig. 225, and the construction and the use of which he thus described in a paper read before the Microscopical Society :

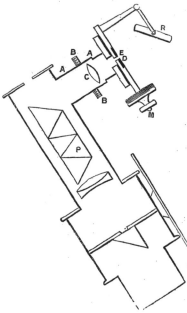

FIG. 225.—*Section of Micro-Spectroscope with Micrometer.*

Attached to the side is a small tube, A A. At the outer part of this tube is a blackened glass plate, with a fine clear white pointer in the centre of the tube. The lens, C, which is focussed by sliding the milled ring, M, produces an image of the bright pointer in the field of view by reflection from the surface of the prism nearest the eye. On turning the micrometer, M, the slide which holds the glass plate is made to travel in grooves, and the fine pointer is made to traverse the whole length of the spectrum.

It might at first sight appear as if any ordinary spider's web or parallel wire micrometer might be used instead of this contrivance. But on closer attention it will be seen that as the spectrum will not permit of magnification by the use of lenses, the line of such an ordinary micrometer could not be brought to focus and rendered visible. The bright pointer of the new arrangement possesses this great advantage—that it does not illuminate the whole field of view.

If a dark wire were used, the bright diffused light would almost obscure the faint light of the spectra, and entirely prevent the possibility of seeing, let alone measuring, the position of lines or bands in the most refrangible part of the spectrum.

To produce good effects with this apparatus the upper surface of the compound prism, P, must make an angle of exactly 45° with the sides of the tube. Under these circumstances the limits of correction for the path of the rays in their passage through the dispersing prisms are very limited and must be strictly observed. The usual method of correcting by the outer surface is inadmissible. For the sake of simplicity, some of the work of the lower part of the micro-spectroscope is omitted in the engraving. As to the method of using this contrivance : With the apparatus just described, measure the position of the principal Fraunhofer's lines in the solar spectrum. Let this be done *carefully*, in *bright* daylight. A little time given to this measurement will not be thrown away, as it will not require to be done again. Note down the numbers corresponding to the position of the lines, and draw a spectrum from a scale of equal parts. About 3 in. will be found long enough for this spectrum ; but it may be made as much longer as is thought desirable, as the measurements will not depend in any

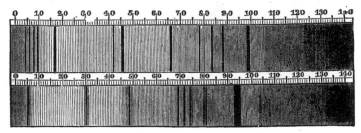

FIG. 226.

way on the distance of these lines apart, but only on the micrometric numbers attached to them. Let this scale be done on cardboard and preserved for reference. Now measure the position of the dark bands in any absorption spectra, taking care for this purpose to use lamplight, as daylight will give, of course, the Fraunhofer lines, which will tend to confuse your spectrum. If the few lines occurring in most absorption spectra be now drawn to the same scale as the solar spectrum, on placing the scales side by side, a glance will show the exact position of the bands in the spectrum relatively to the Fraunhofer lines, which thus treated form a natural and unchangeable scale (see diagram, Fig 226). But for purposes of comparison it will be found sufficient to compare the two lists of numbers representing the micrometric measures, simply exchanging copies of the scale of Fraunhofer lines, or the numbers representing them will enable observers at a distance from each other to compare their results, or even to work simultaneously on the same subject.

A simpler form of the micro-spectroscope is also made by Mr. Browning at a very modest price, and if the reader possesses a microscope, and desires to examine these interesting subjects for himself, he will do well to procure this instrument, instead of that represented in Fig. 220, as it will also answer better for other purposes. A section of the instrument is shown in Fig. 227. When used with the microscope it is slipped into the place of the eye-piece. There is an adjustable slit, a reflecting prism, by which two different spectra may be examined at once, and a train of five prisms for dispersing the rays. It can be used equally well for seeing the bright lines of metals and the Fraunhofer lines, and for viewing any two spectra simultaneously. These direct-vision spectroscopes are better adapted for general use by those who have not several different instruments, than such forms as that shown in Fig. 229, for in the direct-vision instruments the whole extent of the spectrum is visible at one view, which is by no means the case with the larger instruments.

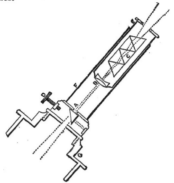

FIG. 227.—*Section of Micro-Spectroscope.*

CELESTIAL CHEMISTRY AND PHYSICS.

WE now approach that portion of our subject in which its interest culminates, for however remarkable may be some of the above-named results of this searching optical analysis, they are surpassed by those which have been obtained in the field upon·which we are about to enter. The cause of the dark lines which Fraunhofer observed in the light of the sun and of certain stars remained unexplained, he only establishing the fact that they must be due to some absorptive power existing in the sun and stars themselves, and not to anything in our atmosphere. It was reserved for Professor Kirchhoff, of the University of Heidelberg, to show the full significance of the dark lines. Fraunhofer had, on his first observation of the lines, noticed that the D lines were coincident with the bright lines in the spectrum of sodium. This interesting fact may be readily observed with any spectroscope which permits of the two spectra being simultaneously viewed. The bright line (or lines if the spectroscope be powerful) of the metal is seen as a prolongation of the dark D solar line. Even with an instrument like that shown in Fig. 220 the coincidence may be noticed. Let the observer receive into the instrument the rays in diffused daylight only, when he will still see the principal Fraunhofer lines distinctly, and let him note the exact position of the D line, while he brings in front of the slit the flame of a spirit-lamp charged with a little salt. He will then see the·bright yellow line replacing the dark D line, and by alternately removing and putting back the lamp he will be soon convinced of the perfectly identical position of the lines.

This fact remained without explanation from 1814 to 1859, when Kirchhoff accidentally found, to his surprise, that the dark D line could be produced artificially. He says : " In order to test in the most direct manner possible the frequently asserted fact of the coincidence of the sodium lines with the D lines, I obtained a tolerably bright solar spectrum, and brought a flame coloured by sodium vapour in front of the slit. I then saw the dark lines D, change into bright ones. The flame of a Bunsen's lamp threw the bright sodium lines upon the solar spectrum with unexpected brilliancy. In order to find out the extent to which the intensity of the solar spectrum could be increased without impairing the distinctness of the sodium lines, I allowed the full sunlight to shine through the sodium flame, and, to my astonishment, I saw that the *dark lines*, D, *appeared with an extraordinary* degree of clearness. I then exchanged the sunlight for the Drummond's or oxyhydrogen lime-light, which, like that of all incandescent solid or liquid bodies, gives a spectrum containing no dark lines. When this light was allowed to fall through a suitable flame, coloured by common salt, *dark* lines were seen in the spectrum in the position of the sodium lines. The same phenomenon was observed if, instead of the incandescent lime, a platinum wire was used, which, being heated in the flame, was brought to a temperature near its melting point, by passing an electric current through it. The phenomenon in question is easily explained, upon the supposition that the sodium flame absorbs rays of the same degree of refrangibility as those it emits, whilst it is perfectly transparent for all other rays." (Quoted in Roscoe's Lectures on " Spectrum Analysis.") When the light of ignited lime was similarly made to pass through flames containing the incandescent vapours of potassium, barium, strontium, &c., the bright lines which these substances would have produced had the lime-light not been present were found to be in every case changed into dark lines, occupying the very same positions in the spectrum. In such experiments the flames containing the metals in the vapourized state do all the time really give off those rays which are peculiar to each substance ; but when a more intense illumination—such as the lime-light, the electric arc, or direct sunlight—passes through them, the rays of the spectrum produced by the intense light overpower those given off by the relatively feebly coloured flames, and hence the portions of the spectrum which are occupied by these, appear black. But as the intense light would give a perfectly continuous spectrum if the incandescent metallic vapour allowed the rays corresponding to its lines to pass through it, the inference is obvious that each vapour absorbs those particular rays which it has itself the power of emitting, but allows all others to pass freely through it. Besides the experimental proofs of this fact which have been already adduced, many others might be named. The flame of a spirit-lamp with a salted wick appears opaque and smoky when we look through it at a large flame of burning hydrogen, also coloured by sodium ; for the rays emitted by the latter do not penetrate the former, which, in consequence of its feebler light, appears dark by comparison. Again, if an exhausted tube containing metallic sodium be heated so as to convert the sodium into vapour, the tube viewed by the light of a sodium flame appear to contain a black smoke, and the light from the flame will no more pass through it than through a solid object ; yet the tube appears perfectly transparent when viewed by ordinary light, and the light from a lithium or other coloured flame would also pass freely. Kirchhoff was led by purely theoretical reasoning to conclude that all luminous bodies have precisely the same power of absorbing certain rays of light as they have of emitting them at the same temperature, and he thus brought luminous rays under the same general law which had previously been established for radiant heat by Prevost, Dessains, Balfour Stewart, and others.   Here, then, a law was arrived at, and, abundantly confirmed by direct experiment as regards the more volatile metals, it was ready to supply the most satisfactory explanation of the coincidences which were everywhere discovered to exist between the Fraunhofer lines and those which belong to terrestrial substances.   For Kirchhoff also found, when mapping the very numerous lines seen in the spark spectrum of iron, that for each of the 90 bright lines of iron which he then observed, there was a dark line in the solar spectrum exactly corresponding in position.   The number of observed bright lines in the iron spectrum has been since extended to 460, and yet each is found to have its exact counterpart in a solar dark line.

So many coincidences as these made it certain that these dark lines and the bright lines of iron must have a common cause, for the chances against the supposition that the agreement was merely accidental are enormous. Kirchhoff actually calculated, by the theory of probabilities, the odds against the supposition. He found it represented by 1,000,000,000,000,000,000 to 1. The result arrived at in the case of sodium at once suggested the explanation that these lines were produced by an absorptive effect of the vapour of iron.   Now, the existence of such a vapour in our atmosphere could not be admitted, while the temperature of the sun was known to be exceedingly high, far higher, indeed, than any temperature we can produce by electricity, or any other means.   Hence, Kirchhoff concluded that his observations proved the presence of the vapour of iron in the sun's atmosphere with as much certainty as if the iron had been actually submitted to chemical tests. By the same reasoning, Kirchhoff also demonstrated the existence in the solar atmosphere of calcium, chromium, magnesium, nickel, barium, copper, and zinc.   To these, other observers have added strontium, cadmium, cobalt, manganese, lead, potassium, aluminium, titanium, uranium, and hydrogen.   It has also been demonstrated that a considerable number of the Fraunhofer lines are due to absorption in our atmosphere by its gases and aqueous vapour.   This demonstration of the existence of iron and nickel in the sun is an interesting pendent to the known composition of many meteorites which reach us from interplanetary space.

Kirchhoff was led to believe that the central part of the sun is formed of an incandescent solid or liquid, giving out rays of all refrangibility, just as white-hot carbon does; that round this there is an immense atmosphere, in which sodium, iron, aluminium, &c., exist in the state of gas, where they have the power of absorbing certain rays; that the solar atmosphere extends far beyond the sun, and forms the corona; and that the dark sunspots, which astronomers have supposed to be cavities, are a kind of cloud, floating in the vaporous atmosphere.

During total eclipses of the sun, certain red-coloured prominences have been noticed projecting from the sun's limb, and visible only when the glare of its disc is entirely intercepted by the moon.   Fig. 228 represents a total eclipse, and will give a rude notion of the appearance of the red prominences seen against the fainter light of the *corona*, which extends to a considerable distance beyond the sun's disc.   Now, two distinguished men of science simultaneously and independently made the discovery of a mode of seeing these red prominences, even when the sun was unobscured.   M. Janssen was observing a total eclipse of the sun in India, and the examination by the spectroscope of the light emitted from the red prominences showed him that they were due to immense columns of incandescent hydrogen, for he recognised the red line and blue lines which belong to the spectrum of this gas (see No. 12, Plate XVII.). Mr. Norman Lockyer at the same time also succeeded in viewing the solar prominences in London without an eclipse. He found a red line perfectly coinciding in position with Fraunhofer's C line and that of hydrogen,. another nearly coinciding with F, and a third yellow line near D. Soon after this, Dr. Huggins discovered a mode of observing the shape of the red prominences at any time, by using a powerful train of prisms and a wide slit, so that the changes in the forms of the red flames can be followed. Now, since the red prominences give off only a few rays of particular refrangibility, it is not difficult to understand that the light of the sun might be, as it were, so diluted by stretching out the spectrum, by means of a train of many prisms, that almost only the red rays, C, should enter the telescope, and occupy the field with sufficient

FIG. 228.—*Solar Eclipse,* 1869.

intensity to overpower all others, and produce an image of the object from which they originated. The nature of this action may be illustrated thus : If we hold vertically a prism, and look through it at a candle-flame, we may perceive a lengthened-out image of the flame, showing the succession of prismatic colours, and formed, as it were, of a red image of the flame close to a yellow one, and so on, but presenting no defined form. If, still viewing this spectrum, we introduce into the flame on a platinum wire a piece of common salt, we shall perceive a well-defined yellow image of the candle start out, because the rays which are emitted by the incandescent sodium, being all of one refrangibility, the prism simply refracts without dispersing them. The dispersion which weakens the light of the continuous spectrum by lengthening it out, does not sensibly detract from the brilliancy of the bright lines, as their breadth is scarcely increased—they are refracted but not dispersed. Hence, when a sufficient number of prisms is employed, the bright lines of the solar *chromosphere* may be seen in full sunshine, in spite of the greater intensity of the light emanating from the *photosphere,* which produces the continuous spectrum. The bright C line is, of course, a virtual image of the slit produced by rays of that particular refrangibility ; but by using a very high dispersive power, the slit may be opened so wide that the C rays form in the telescope a red image of the prominence from which they issue, since their light will predominate over that of any rays belonging to the continuous spectrum.

In the hands of Mr. Norman Lockyer the science of the physical and chemical constitution of the sun has made rapid progress, and new facts are continually being observed, which serve to furnish more and more definite views. Mr. Lockyer considers that, extending to a great distance around the sun is an atmosphere of comparatively cooler hydrogen, or perhaps of some still lighter substance which is unknown to us. It is this which forms what is termed the *corona,* or circle of light which is seen surrounding the sun in a total eclipse. Immersed in this, and extending to a much smaller distance from the nucleus of the sun, is another envelope, termed the *chromosphere*, consisting of incandescent hydrogen and some glowing vapours of magnesium and calcium. The brightest part of this envelope, which lies nearest the sun, is that which gives off the red rays by which the prominences may be observed without an eclipse. These prominences have been shown to be tremendous outbursts of glowing hydrogen, belched up with sometimes an enormous velocity from below, since they have been observed to spring up 90,000 miles in a few minutes. Beneath the chromosphere, and nearer to the body of the sun, are enormous quantities of the vapours of the different elements—sodium, iron, &c.—to which the dark lines of the solar spectrum are due. This stratum Mr. Lockyer calls the *reversing layer*, because it reverses (turns to dark) the lines which would otherwise have appeared bright, just as Kirchhoff's sodium vapour did in the experiment described on page 437. Beneath the reversing layer is the *photosphere*, from which emanates the light that is absorbed in part by the reversing layer, and which there is good reason to believe is either intensely heated solid or liquid matter.

In 1861 Dr. Huggins devoted himself, with an ardour which has since known no remission, to the extension of prismatic analysis to the other heavenly bodies. The difficulties of the investigations were great. There was first the small quantity of light which a star sends to the spectator ; this was obviated by the use of a telescope of large aperture, which admitted and brought to a focus many more rays from the star, and therefore the brightness of the image was proportionately increased. Not so the size of the image : the case of the fixed stars for this always remains a mere point. It was, of course, necessary to drive the telescope by clock-work, so that the light of the star might be stationary on the field of the spectroscope. As the spectrum of the image of the star formed by the object-glass would be a mere line, without sufficient breadth for an observation of the dark or light lines by which it might be crossed, it is necessary to spread out the image so that the whole of the light may be drawn out into a very narrow line, having a length no greater than will produce a spectrum broad enough for the eye to distinguish the lines in it. This is accomplished by means of a cylindrical lens placed in the focus of the object-glass, and immediately in front of the slit. Covering one-half of the slit is a right-angled prism by which the light to be compared with that of the star is reflected into the slit. The light is usually that produced by taking electric sparks between wires of the metal in the manner already described. The dispersive power of the spectroscope was furnished by two prisms of very dense glass, and the spectrum was viewed through a telescope of short focal length. Dr. Huggins's observations lead him to the conclusion that the planets Mars, Jupiter, and Saturn possess atmospheres, as does also the beautiful ring by which Saturn is surrounded ; for he noticed in the spectrum of each different dark lines not belonging to the solar spectrum.

Passing to the results obtained in the case of the fixed stars, we may remind the reader of the enormous distance of the bodies which are submitted to the new method of analysis. Sir John Herschel gives the following illustration of the remoteness of Sirius—supposed to be one of the nearest of the fixed stars : Take a globe, 2 ft. in diameter, to represent the sun, and at a distance of 215 ft. place a pea, to give the proportionate size and distance of the earth. If you wish to represent the distance of Sirius *on the same scale*, you must suppose something placed *forty thousand miles* away from the little models of sun and earth. But not only do we know with certainty some of the substances contained in Sirius, but the star spectroscope has taught us a great deal about orbs so remote, that their distance is absolutely unmeasurable. About Aldebaran we know that there are hydrogen gas and vapours of magnesium, iron, calcium, sodium, and some four or five other elements. Generally the lines indicate the presence of hydrogen in these distant suns ; but there is, at least, one

FIG. 229.—*The Planet Saturn.*

remarkable exception in *a Orionis*, the spectrum of which yields no trace of the hydrogen lines, although it is evident that magnesium, sodium, calcium, &c., are present. The spectra of celestial bodies are of several kinds. Many of the stars have, like our sun, a continuous spectrum crossed by dark lines. Such is that of Sirius, No. 10, Plate XVII. Others have, however, both dark and bright lines, and some are marked by only three bright spaces. Of the spectra of the nebulæ some have three bright lines (see No. 11, Plate XVII.), and the bodies producing them are, therefore, to be considered as masses of incandescent gas, while some give continuous spectra. One of the bright lines in the spectra of the nebulæ coincides with one of the hydrogen lines, and another—the brightest of the three— with one of the brightest nitrogen lines ; but the third does not agree with any with which it has as yet been compared. The inference from these appearances is that the nebulæ contain hydrogen and nitrogen, but the absence of the other lines of these substances has not been fully explained ; although the observation of Dr. Huggins, that when the light of incande- scent nitrogen and hydrogen is gradually obscured by interposing layers of neutral tinted glass, the lines corresponding with those in the nebular

spectra are the last to disappear, seems to suggest a probable solution of
the difficulty.

There is another very interesting line of spectroscopic research in the
power the prism gives us of estimating the velocity with which the distances
of the stars from our system are increasing or diminishing. On closely
examining the hydrogen lines of Sirius, and comparing them with the bright
lines of hydrogen rendered incandescent by electric discharges in a Geissler
tube, the spectrum of which his instrument enabled him to place side by
side with that of the star, Mr. Huggins was surprised to find that the lines
in the latter did not exactly coincide in position with those of the former,
but appeared slightly nearer the red end of the spectrum. This indicated

FIG. 230.—*Solar Prominences, No.* 1.

a longer wave-length, or increased period of vibration, according to the
theory of light, which would be accounted for by a receding motion between
Sirius and the earth, just as the crest of successive waves of the sea would
overtake a boat going in the same direction at longer intervals of time than
those at which they would pass a fixed point, while, if the boat were meeting the waves, these intervals would, on the other hand, be shorter. Hence,
if the position of the lines in the spectrum depends on the periods of vibration, that position will be shifted towards the red end when the luminous
body is receding from the earth with a velocity comparable to that of light,
and towards the violet end when the motion is one of approach. The change
in refrangibility observed by Mr. Huggins corresponded with a receding
velocity of 41·4 miles per second, and when from this was subtracted the
known speed with which the earth's motion round the sun was carrying us
from the star at the time, the remainder expressed a motion of recession

amounting to about twenty miles a second, which motion, there is reason to believe, is chiefly due to a proper movement of Sirius. These deductions from prismatic observations are of the highest value astronomically, since they will eventually enable the real motions of the stars to be determined, for ordinary observation could only show us that component of the motion which is at right angles to the visual ray, while this gives the component along the visual ray. In the same manner, it is inferred that Arc

FIG. 231.—*Solar Prominences, No. 2.*

turus, a bright star in the constellation *Boötes*, is approaching us with a velocity of fifty-five miles per second.

When the solar spots are examined with the spectroscope, the dark image of the slit produced by the hydrogen line, F, is observed to show a strange crookedness when it is formed by rays from different parts of the spot. This distortion is due to the same cause as the displacement of the stellar lines, namely, motions of approach or recession of the masses of glowing hydrogen. Mr. Norman Lockyer, to whom we are indebted for the most elaborate investigations of the solar surface, has calculated, from the position of the lines, the velocities with which masses of heated hydrogen are seen bursting upwards, and those which belong to the down-rushes of cooler gas. Velocities as great as 100 miles per second were, in this way, inferred to occur in some of the storms which agitate the solar surface. Two drawings of a solar storm, given by Mr. Lockyer, are shown in Figs. 230 and 231. These are representations of one of the so-called red prominences, the first giving its appearance at five minutes past eleven on the morning of March 14th, 1869, and the last showing the same *ten minutes after-*

wards. The enormous velocity which these rapid changes imply will be understood when it is stated that this prominence was 27,000 miles high. "This will give you some idea," says Mr. Lockyer, "of the indications which the spectroscope reveals to us, of the enormous forces at work in the sun, merely as representing the stars, for everything we have to say about the sun the prism tells us—and it was the first to tell us—we must assume to be said about the stars. I have little doubt that, as time rolls on, the spectroscope will become, in fact, almost the pocket companion of every one amongst us ; and it is utterly impossible to foresee what depths of space will not in time be gauged and completely investigated by this new method of research."

The light of comets has also been examined by the spectroscope, and many interesting results arrived at. Our limits do not, however, permit us to enter into a discussion of these interesting subjects.

Fig. 232 is a section of another of Mr. Browning's popular instruments, which is named by him the "Amateur's Star Spectroscope." It exhibits very distinctly the different spectra of the various stars, nebulæ, comets, &c.

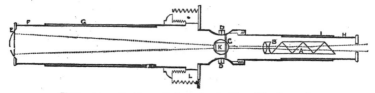

FIG. 232.—*Section of Amateur's Star Spectroscope.*

The reader who is desirous of learning more of this fascinating subject is referred to Dr. Roscoe's elegant volume, entitled, "Lectures on Spectrum Analysis." This work, which is embellished with handsome engravings and illustrated by coloured maps and spectra, gives a clear and full account of every department of the subject, and in the form of appendices, abstracts of the more important original papers are supplied, while a complete list is given of all the memoirs and publications relating to the spectroscope which have been published.

This brief account of the spectroscope and its revelations, which is all that our space permits us to give, will not fail to awaken new thoughts in the mind of a reader who has obtained even a glimpse of the nature of the subject, especially in relation to that branch of which we have last treated, for in every age and in every region the stars have attracted the gaze and excited the imagination of men. The belief in their influence over human affairs was profound, universal, and enduring ; for it survived the dawn of rising science, being among the last shades of the long night of superstition which melted away in the morning of true knowledge. Even Francis Bacon, the father of the inductive philosophy, and old Sir Thomas Browne, the exposer of "Vulgar Errors," believed in the influences of the stars ; for while recognizing the impostures practised by its professors, they still regarded astrology as a science not altogether vain. It was reserved for the mighty genius of Newton to prove that in very truth there are invisible ties connecting our earth with those remote and brilliant bodies—ties more potent than ever astrology divined ; for he showed that even the most distant orb is bound to its companions and to our planet by the same power that draws the projected stone to the ground. And now the spectroscope is revealing other lines of connection, and showing that not gravitation alone is the sympathetic bond which unites our globe to the celestial orbs, but that there exists the closer tie of a common constitution, for they are all made of the same matter, obeying the same physical and chemical laws which belong to it on the earth. We learn that hydrogen, and magnesium, and iron, and other familiar substances, exist in these inconceivably distant suns, and there exhibit the identical properties which characterize them here. We confirm, by the spectroscope, the fact partially revealed by other lines of research, that the stars which appear so fixed, are, in reality, career-ing through space, each with its proper motion. We learn also that the stars are the theatres of vast chemical and physical changes and trans-formations, the rapidity and extent of which we can hardly conceive. There is, for example, the case of that wonderful star in the constellation of the Crown, which, in 1866, suddenly blazed out, from a scarcely descernible telescopic star, to become one of the most conspicuous in the heavens, and the bright lines its beams produced in the spectroscope revealed the fact that this abrupt splendour was due to masses—who can imagine how vast?—of incandescent hydrogen. This brightness soon waned, and *τ Coronæ Borealis* reverted once more to all but telescopic invisibility. The seeming fixity of the stars is an illusion of the same nature as that which prevents a casual observer from recognizing their apparent diurnal motion, and now we have also ample evidence that permanence of physical condition, even in the stars, is impossible. Everywhere in the universe there is motion and change; there is no pause, no rest, but a continual unfolding, an endless progression.

> "Know the stars yonder,
> The stars everlasting,
> Are fugitive also,
> And emulate, vaulted,
> The lambent heat-lightning
> And fire-fly's flight."

ROENTGEN'S X RAYS.

ON page 507 reference will be made to certain remarkable effects observed by Mr. Crookes when the electric discharges from an induction coil are passed through very highly exhausted tubes. These phosphorescent and mechanical effects Mr. Crookes attributed to streams of "radiant matter" shot off from the *negative* pole with immense velocities—the matter not being that of the electrode itself, but particles of the extremely rarefied residual gas, which, being comparatively few, could mostly traverse the tube in straight lines without coming into collision with their fellows, and thus a class of phenomena, different from the striated discharges in the ordinary and less highly exhausted Geissler tubes, comes into view. The emanations from the negative pole, or *cathode*, in highly rarefied gases became known as the "cathode rays," and they began to be further examined by other observers, and more particularly in 1894 by Hittorf, and by M. Lenard, a Hungarian physicist, who found that they pass through thin plates of metal, and through wood and other substances not transparent to ordinary light. It was also observed by Lenard, and also previously by Hertz, that there are several kinds of cathode rays, which differ from each other as regards their powers of exciting phosphorescence, capability of being deflected by a magnet, and the degrees in which they are absorbed by various media.

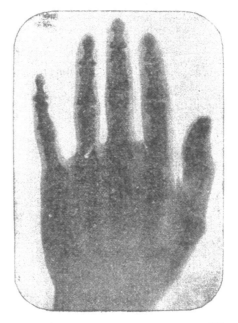

FIG. 232a.—*Living Hand. Exposure, 4 Minutes.*

But universal attention was drawn to this subject by the announcement, at the end of 1895, of certain discoveries made by Dr. W. K. Roentgen, a professor of physics at Wurzburg. He covered a highly-exhausted Crookes' tube with black cardboard, and found that when the discharge of a large induction coil was passed through the tube in a dark room, a piece of paper coated on one side with platino-cyanide of barium, and held near the covered tube, glowed with a brilliant fluorescence, no matter which side of the paper was turned towards the tube ; and even at a distance of two yards some fluorescence was still visible. On experimenting with various bodies interposed between the covered tube and the fluorescent screen, it was found that the emanations passed through nearly every substance with more or less facility. The screen lit up when placed behind a book of a thousand pages, also behind two packs of cards. A single layer of tin-foil scarcely threw a shadow, and several thicknesses were required to produce a distinct effect. Deal boards, an inch thick, offered little resistance. A very thick plate of aluminium ($\frac{6}{10}$ inch) reduced the fluorescence, but still allowed some rays to pass. The hand held before the fluorescent screen showed a dark shadow of the bones only, with but a faint outline of their fleshy investment. Copper, silver, gold, platinum, and lead, in comparatively small thicknesses, intercept these rays. Thus a plate of lead only five hundredths of an inch thick almost stops them. Increase of thickness increases the resistance to their passage in all cases ; but the comparative transparency of a body cannot be deduced from its thickness and density. Many other bodies besides platino-cyanidæ of barium become fluorescent under the influence of these rays, such as certain kinds of glass, Iceland spar, rock-salt, etc. Dr. Roentgen is convinced that these rays are not the cathode rays or any part of them ; but as the theoretical nature of the new rays has not yet been explained, he has preferred to provisionally call them the X rays, a denomination doubtless suggested by the use of the symbol x in algebra to represent unknown quantities.

The source of the X rays, Roentgen states, is at the place where the cathode rays strike the walls of the exhausted tube, and produce the most brilliant phosphorescence ; but they cannot be cathode rays which have merely passed through the glass, for, contrary to what has been observed with respect to the latter, they cannot be deflected by a magnet. Nor is glass the only substance in which they can be generated, for they were obtained from an apparatus in which the cathode rays were made to impinge upon a plate of aluminium nearly one-tenth of an inch thick. Photographic dry plates are also sensitive to the X rays, and their power to pass through wood, ebonite, etc., makes the experiments of testing the opacity, or otherwise, of various objects for them quite easy. It is necessary merely to place the object on the closed cover of the dark slide, and place the whole under the vacuum tube ; all the exposure, which is somewhat prolonged in most cases, may be made in ordinary light. But the light-tight boxes, in which photographic plates are packed, cannot, of course, be brought near the apparatus, as they are completely permeable to the X rays, and their whole contents may be rendered useless. The impression obtained on the photographic plate is not so much a photograph as a *shadow* of the interposed object—a shadow more or less dense in the positive print according to the permeability of the object, and the length of the exposure. These photographic results have sometimes been called "shadowgrams," "radiograms," "radiographs," "skiagraphs," etc. The word *skiagraph* appears the most appropriate designation. That the emanations from the phosphorescing substance on which the cathode rays impinge are entitled to be also called rays, appears from the regularity of the shadows thrown on the fluorescent screen or photographic plate ; and the fact of their propagation in straight lines was proved by Dr. Roentgen obtaining a *pin-hole* photograph of the phosphorescing part of the vacuum tube, when the latter was enveloped in black paper. Why a *pin-hole* and not a lens was used for taking this photograph will presently appear.

One of Dr. Roentgen's experiments excited the attention and interest of the general public, as well as of the scientific world, in the most extraordinary degree, and though its announcement was received in some quarters with incredulity, experimenters in all parts of the world immediately set themselves at work to test the truth of the alleged discovery. Electrical apparatus of different kinds, with various adjustments, were employed, with results that were in some cases failures, in others confirmations of the German professor's statements, and not unfrequently the variations in the conditions gave rise to increased knowledge of the phenomena generally. The experiment just alluded was one in which a dry photographic plate contained in one of the camera dark slides, now so familiar to every one, was placed (with the slide still closed by its wooden cover of nearly one quarter of an inch thick) a few inches below the Crookes' tube, and the hand of a living person being extended on the outside of the cover, a shadow of the bones of the hand, as if seen through

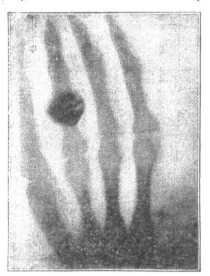

FIG. 232*b*.—*Skiagraph of a Hand, by Dr. Roentgen. The Third Finger has a Ring on it.*

the surrounding tissues, was obtained. Much popular misconception as to the powers of the " new photography " arose from want of knowledge of the process by which these strange pictures were obtained, the common notion being that these photographs were produced by some method of using a camera, and that outlines of people's bodies and skeletons could be taken instantaneously, not only through their clothes, but through doors and walls. Much nearer the mark was the allusion of a scientific writer as to the possibility of the new process realising Dickens' description of Marley's ghost : " His body was transparent, so that Scrooge, observing him, and looking through his waistcoat, could see the two buttons on his coat behind." The value of the new discovery for medical and surgical purposes was immediately recognised, and very soon its application was successfully practised.

Dr. Roentgen found that the X rays are incapable of refraction, in this respect differing from ordinary light (see pages 397 and following), and among the experiments which most impressed and astonished his auditors when he was lecturing at Potsdam on his new discovery before the Imperial Court of Germany, was one in which he showed the X rays passing in a straight line through water without undergoing refraction. The rays pass without interruption equally through the substances, whether these be coherent or in a layer of fine powder of the same thickness, and this again shows that there can be no regular reflection or refraction. Prisms and lenses, whether of glass, ebonite, or aluminium, fail to afford evidence of refractive action, hence the X rays cannot be focused like those of ordinary light, and that is why the photograph of the vacuum tube had to be taken by a pin-hole. Again, glass lenses could not be used, because this substance, so transparent to light, is particularly opaque to the X rays, and would in a great degree intercept them, while lenses of ebonite and of aluminium, which were tried, were inoperative on account of the irrefrangibility of the rays.

As to the nature of the rays themselves, Dr. Roentgen rejects the notion of their being "ultra-violet" rays, which was suggested by some. The meaning of this term is seen when it is understood that a great distance beyond the violet end of the visible spectrum there are radiations, revealed by their photographic impressions, so that the whole spectrum is really some eight times as long as the visible part. In consequence of these ultra-violet rays acting on the photographic plate, it is possible, as has long been known, to take a photograph in the dark. The eye is quite insensible to the X rays also, and although these, as we have seen, readily pass through the bodily tissues, it may be placed quite near the discharge tube, the latter being enveloped in black paper, without causing any sensation. That the new rays are in some way allied to light is the opinion held by Dr. Roentgen, and he is inclined to consider them as due to *longitudinal* vibrations in the ether ; that is, instead of the transverse waves to which light is attributed, these resemble the waves of sound, in so far that they move in the direction of propagation. This would account for the absence of any distinct refraction, or polarisation, which seems to characterise the X rays. Their connection with certain electric Maxwell-Hertz waves (see p. 541) is more problematical, as the mathematical formulæ for these admit only transverse oscillation. But on the assumption of certain conditions, due to the action of electricity, etc., on highly rarefied air, the possible existence of longitudinal vibrations has been deduced by admitting a certain variation in some of the factors of the Maxwell formulæ.

Other suggestions have been advanced in order to make the observed facts concerning the X rays fit into established theories, but so far these attempts have been unsuccessful. It would seem as if our present conceptions of light, electricity, the ether and matter, will have to be profoundly modified and enlarged in order to bring these and other recently discovered phenomena within their scope. Since the publication of Dr. Roentgen's paper, his results have received confirmation in every quarter, and many new observations have been added, some of which seem to tend not so much to elucidate the phenomena, as to prove them even more complicated than was at first supposed. Such was the announcement in June, 1896, of the discovery of several varieties of X rays.

In the meantime, various modifications have been made in the forms of the tubes and electrodes, and divers arrangements have been used for the exciting electrical apparatus. Thus it has been found that the X rays

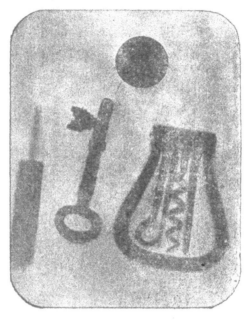

FIG. 232c.—*Metal Objects taken through Calico and Sheet of Aluminium.*

are given off from platinum more copiously than from glass, aluminium, or any other substance, and by using a tube closed by a "platinum window," on which the cathode rays impinge, Mr. Gifford has been able to reduce the time of exposure for obtaining a skiagraph of the bones on the hand to half a minute, whereas twenty times that period was formerly required. Another form of tube is advertised by Brady & Martin of Newcastle, with which, in conjunction with a new screen, a coil giving a 5-inch spark will, it is stated, yield a good skiagraph of the hand in *two seconds*, which appears to be the shortest time yet attained. Another firm of tube-makers, Newton & Co., London, state that their special form of tube, excited by a coil giving a 6-inch spark, and used with their fluorescent screen, "will work right through the human body, showing the heart, liver, spine, ribs, the movements of the heart and of the diaphragm, etc." It has recently been observed that the best results are obtained when there is a certain, but as yet undefined, relation between the degree of rarefaction of the residual gas in the tube, and the intensity or frequency of the electric discharges, and that these should be accommodated to the work required. Thus, for example, if a skiagraph of the hand be attempted with an apparatus in which these factors are carried to too high a degree, the resulting X rays will pass through the bones almost as freely as through the surrounding tissues, and their shadows will therefore not appear. If, on the other hand, the contrary conditions hold, an incomplete or maybe no result will be found. This seems to explain the failures that have sometimes occurred when tubes of appar·

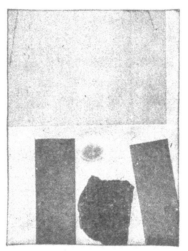

FIG. 232*d.*—*A Skiagraph of Layers of Various Substances.*

ently identical construction have been used in the hands of the same, or of different, observers. Perhaps more depends also on the time of exposure. For instance, if a short exposure be given in the case of the hand, the photograph will be merely a silhouette of that member; with a little longer exposure, this will show the nails; with still longer time, the shadow of the fleshy parts begins to grow faint and the skeleton to appear. With yet more prolonged exposure only the bones will show, in their various degrees of opacity, and the shadows of these will gradually disappear as the time of exposure is increased, until at length the image will be entirely effaced. The considerable differences as to distinctness of the various tissues, which are exhibited by the published prints of hand shadows, are thus explicable.

FIG. 233.—*Portrait of Professor Helmholtz.*

SIGHT.

THE investigations of modern science have borne rich fruit, not only by vastly extending our knowledge of the universe of things around us, but also making us acquainted with the mode in which certain agents act upon our bodily organs, and by revealing, up to a certain point, what may be termed the mechanism of that most wonderful thing—the human mind —or, at least, that part which is immediately concerned in the perceptions of an external world. Of all the physical influences which affect the human mind, those due to light are the most powerful and the most agreeable. One of the most ancient of philosophers says, in the simple words which are appropriate to the expression of an undeniable truth, " Truly the light is sweet, and a pleasant thing it is for the eyes to behold the sun." The impression produced by light alone is a source of pleasure—a cheering influence of the highest order; and there is a special character in the pleasing effects of light, from the circumstance that they do not exhaust the sense so quickly as do even pleasurable impressions on other organs—such as sweet tastes, fragrant odours, or agreeable sounds. Sight is not liable to that satiety which soon overtakes the enjoyment of sensations arising from the other senses ; it possesses, therefore, a refinement of quality of which the rest are devoid. Sight converses with its objects at a greater distance than does any other sense, and it furnishes our minds with a greater variety of ideas. Indeed, our mental imagery is most largely made up of reminiscences of visual impressions ; for when the idea of anything is brought up in our minds by a word, for example, there arises, in most cases, a more or less vivid presentation of some visible appearance. Our visual impressions are also longer retained in memory or idea than any other class of sensations.

The nature of the impressions we receive through the eye is extremely varied ; for we thus perceive not only the difference between light and darkness, but in the sensations of colour we have quite another class of effects, while the lustre and sparkle of polished and brilliant objects add new elements of beauty and variety. We find examples of the latter qualities in the verdant sheen of the smooth leaf, in the splendid reflections of burnished gold, in the bright radiance of glittering gems, and "in gloss of satin and glimmer of pearls." The eye is also the organ which conveys to our minds the impressions of visible motion, with all those pleasures of exciting spectacle which enter so largely into our enjoyment of life. It likewise discriminates the forms, sizes, and distances of objects ; but by a process long misunderstood, and dependent upon a set of perceptions which, although precisely those whence we derive our most fundamental notions of the objects around us, have been completely overlooked in that time-honoured enumeration of the senses which recognizes only five.

If such be the extent to which our minds are dependent upon the wonderful apparatus of the eye, it may easily be imagined what must be the comparative narrowness of mental development in those who have never enjoyed this precious sense, and the feeling of deprivation in those, who, having enjoyed it, have unfortunately lost it. Well may our sublime poet despairingly ask—

> Since light so necessary is to life,
> And almost life itself—if it be true
> That light is in the soul—
> The all in every part : why was the sight
> To such a tender ball as the eye confined,
> So obvious and so easy to be quenched?"

—for he himself, in his own person, experienced this deprivation, and he thus touchingly, in his great work, laments his loss :

> "Thus with the year
> Seasons return ; but not to me returns
> Day, or the sweet approach of even or morn,
> Or sight of vernal bloom, or summer's rose,
> Or flocks, or herds, or human face divine ;
> But cloud instead, and ever-during dark
> Surround me ; from the cheerful ways of men
> Cut off ; and for the book of knowledge fair
> Presented with a universal blank
> Of Nature's works—to me expunged and rased,
> And wisdom at one entrance quite shut out."

An organ which is the instrument of so many nice discriminations as is the eye must, of course, present the most delicate adjustment in its parts. So much has in recent times been learnt of the nature of its mechanism ; of the relation between the impressions made upon it and the judgments formed by the mind therefrom ; of the illusions which its very structure produces ; of the defects to which it is liable ; and of its wonderfully

refined physiological elements—that a branch of science sufficiently extensive to require a large part of a studious lifetime for its complete mastery
has grown up under the hands of modern physiologists, physicists, and
psychologists. To some of the results of their labour we would invite the
reader's attention; and in order to render the account of them intelligible,
we must, to a certain extent, describe "things new and old."

THE EYE.

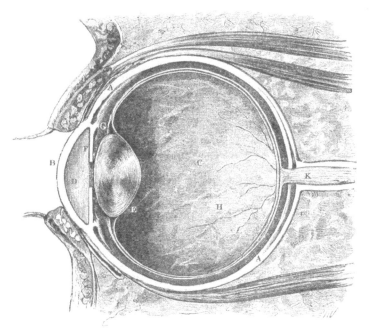

FIG. 234.—*Vertical Section of the Eye.*

THE form of the human eye and the general arrangement of its parts may
be understood by referring to Fig. 234, which is a section of the eyeball. It has a form nearly globular, and is covered on the outside by a tough
firm case, A, named the *sclerotic coat*, which is, for the most part, white
and opaque. This covering it is which forms what is commonly termed
the "white of the eye;" but in the front part of the eyeball it loses its

opacity, and merges into a transparent substance, termed the *cornea*, B. The cornea has a greater convexity than the rest of the exterior of the eyeball, so that it causes the front part of the eye to have a somewhat greater projection than would result from its general globular form. This sclerotic coat—with its continuation, the cornea—serves to support and protect the more delicate parts within, and is itself kept in shape by the *humours*, which fill the whole of the interior. The greater space is occupied by the *vitreous humour*, C ; but the space immediately behind the transparent cornea is filled with the *aqueous humour*, D. The latter is little else than pure water, and the former is like thin transparent jelly. The cavities containing these two humours are separated by the transparent double convex lens, E, called the *crystalline lens*, which, in consistence, resembles very thick jelly or soft gristle. The outward surface of this lens has a flatter curvature than the inner surface. Immediately in front of the crystalline lens is found the *iris*, F, which may be described as a curtain having in the middle a round hole. The iris is the part which varies in colour from one individual to another—being blue, brown, grey, &c. ; and the aperture in its centre is the dark circular spot termed the *pupil.*

The general disposition of the parts of the eye with regard to light will be most easily understood by comparing it with an optical instrument, to which it bears no little resemblance, namely, the *camera obscura*, so well known in connection with photography. We may picture to ourselves a still more complete resemblance, by imagining that the lens of the camera is single, that we have fixed in front of it a watch-glass, with the convex side outwards, and that we have filled with water the whole of the interior of the camera, including the space between the watch-glass and the lens. The *focussing-screen* of the camera corresponds with the inner surface of the back of the eyeball, about which we shall presently have more to say. Now, even if the camera had no lens, but were simply a box filled with water, and having in front the watch-glass, fixed in the manner just mentioned, we could obtain the images of objects on the screen, as a consequence of the curvature of the watch-glass. It would, however, in this case, be necessary to have the camera much longer, or, in other words, the rays would be brought to a focus at a greater distance than if we put in the glass lens, which would, thus placed in the water, cause the rays to converge to a focus at a much shorter distance, although its effect when surrounded by water would be less powerful than in the air. There we see the effect of the crystalline lens of the eye in bringing the rays to a focus within a much shorter distance than that which would be required had there been present only the curved cornea, and the aqueous and vitreous humours of the eye, which are but little different from pure water in their optical properties.

If we *focus* the camera by adjusting the distance between the lens and the screen so as to get a distinct image of a near object, we should find, on directing the instrument to a distant one, that the image would be blurred and indistinct, and the lens would have to be moved nearer to the screen ; or we could get the image of the distant object distinct by replacing the lens by another lens in the same position, but having some flatter curvature. It is plain that the same object would be gained if our lens could be made of some elastic material, which, on being pulled out radially at its edges, could be made to assume the required degree of flatness without losing its lenticular form. Now, it is precisely with an automatic adjustment of this kind that the *crystalline lens* of the eye is provided, for the lens is suspended by an elastic ligament, G, by the tension of which its surfaces are more flattened than they would otherwise be; but when the tension of this ligament is relaxed, by the action of certain delicate muscles which draw it down, the elasticity of the lens causes it to assume a more convex form.

These optical adjustments give, on the inner surface of the coats of the eye, a more or less perfect real image of the objects to which the eye is directed, and it is on the back part of this inner surface that the network of nerves, called the *retina*, H, is spread out. The sclerotic coat, already spoken of, is lined internally with another, named the *choroid*, which is composed of delicate blood-vessels, intermingled with a tissue of cells filled with a substance of an intensely black colour. It is upon this last layer that the delicate membrane of the retina is spread out between the choroid and the vitreous humour.

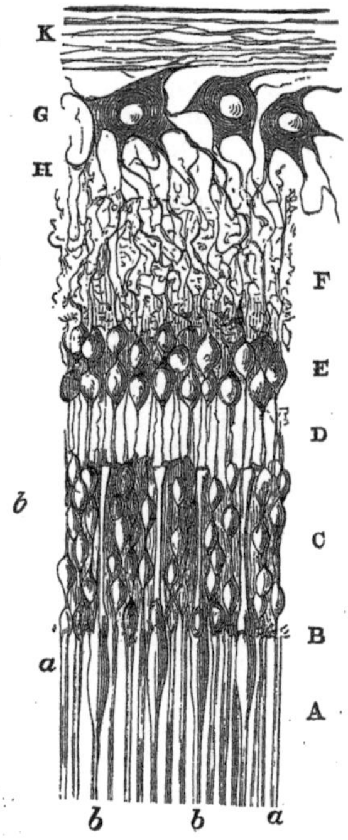

The retina is, in part, an expansion of the fibres of the optic nerve over the back part of the eyeball. If we suppose the globe of this cut vertically into two portions, and so divide the front from the back part of the eye, the retina would be seen spread out on the concave surface of the back part, and in the middle of this part, opposite the crystalline lens, would be seen a spot in which the retina assumes a yellowish colour, and in the centre of this, a little round pit or depression. The spot is called the *macula lutea*, or *yellow spot*, and the little central pit, which is of the highest importance in vision, is termed the *fovea centralis*. A little way from the yellow spot, and nearer the nose, is a point from which a number of fibres are seen to radiate, and this is, in fact, the part at which the optic nerve enters the eyeball, and from which it sends out its ramifications over the retina. This part, for a reason which will shortly appear, is called the *blind spot*.

When the minute structure of the retina is examined by the microscope, its physiological elements are found to undergo very remarkable modifications at the yellow spot. In the retina, although the total thickness does not exceed the $\frac{1}{80}$th part of an inch, no fewer than eight or ten different essential or nervous layers have been distinguished. Fig. 235 rudely represents a section. The lowest stratum, A, which is next the choroid, and forms about a quarter of the total thickness, is formed of a multitude of little rod-shaped bodies, *a*, ranged side by side, and among these are the conical or bottle-shaped bodies, *b*. This lowest stratum of the retina is called the *layer of rods and cones*. At their front extremities the rods and cones pass into very delicate fibres, which, going through an extremely fine layer of fibres, B, are connected with a series of small rounded bodies, which form

FIG. 235.
Section of Retina.

the layer of *nuclei*, C, separated by a layer of nervous fibres, D, from a granular layer, E, in front of which is a stratum of still finer granules, F, underlying a layer of ganglionic nerve-cells, G, of a larger size than any of the other elements, and these ganglionic cells send out numerous branching nerve-fibres, forming the layer H. Finally, on the front surface of the retina there is a thin stratum formed of fibres, which issue from the optic nerve, K, Fig. 234, and in fact constitute the expansion of this nerve on the inner surface of the eyeball. The terminations of some, at least, of these nerve-fibres have been traced, and have been found to form junctions with those branching from the ganglionic cells.

Of the part played by each of these delicate structures in exciting visual impressions little is yet known. How light, or the pulsations of ether, if such there be, is ultimately converted into sensation will probably for ever remain a mystery, although it is quite likely that the kind of visual impression which is conveyed by each part of the elaborate structure of the retina

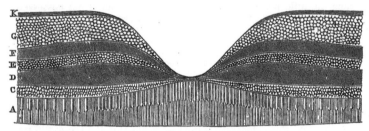

FIG. 236.

may ultimately be distinguished. One curious result of modern investigation is that *light falling directly upon fibres of the optic nerve is quite incapable of exciting any sensation whatever*. Light has no more effect on this nerve and its fibres than it would have on any other nerve of the body if exposed to its action. The apparatus of rods, cones, and other structures are absolutely essential to enable light to give that stimulus to the optic nerve which, conveyed to the brain, is converted into visual sensations. So if this apparatus were absent in our organs of vision, in vain would the optic nerve proper be spread out over the interior of the eyeball: we should be no more able to *see* with such eyes than we are able to see with our hands.

We now invite the reader's careful consideration to the diagram, Fig. 236, which is a section of the retina through the yellow spot. The upper part of the figure is the front, and the deep depression is the little pit already spoken of—the *fovea centralis*. The lowest dark line represents the basement membrane of the retina, and immediately above is seen the layer of rods and cones, and the various strata already spoken of are represented in their due order in the marginal parts of the diagram. Now observe the remarkable modifications of the nervous structures in the neighbourhood of the *fovea centralis*, some of which are visible in the diagram. In the first place, the cones are there much longer, more slender, and more closely set, so that there is a far greater number of them on a given surface; but the rods are comparatively few, and are, in fact, not found at all under the floor of the little pit. The layer of *nuclei*, into which the cones extend, is thinner, and is found almost immediately below the anterior surface, for all the other layers thin out in the fovea in a very curious manner. It is, however on the margin of the fovea that the stratum of ganglionic cells, G, Fig. 235, attains its greatest thickness, for there it is formed by the superposition of eight or ten cells, being here thicker than any other layer, while it is so thinned off towards the margin of the retina that it no longer forms even a continuous stratum. This layer, however, becomes much thinner *in the fovea*, which contains, in fact, but few superposed cells. The tint of the yellow spot is said to be derived from a colouring matter, which affects all the layers except that of the cones. The centre of the yellow spot, where the *fovea centralis* is situated, is extremely transparent, and is so delicate that it is very easily ruptured, and has frequently been taken for an aperture.

We should not have risked wearying the reader with these details concerning the little pit in the centre of the retina had it not possessed an extreme importance in the mechanism of the eye, a fact which he will at once appreciate when we say that *of the whole surface of the retina, the only spot where the image of an object can produce distinct vision is the fovea centralis.* Since this is undoubtedly true, it follows that the physiological elements which we there find are precisely those which are most essential for producing this effect. The case may be exemplified by recurring to the comparison of the eye with a photographer's camera, by supposing his screen to be of such a nature that only on one *very small spot* near its centre could a distinct image be possibly obtained of just one point of an object. Such a defect in his camera would render the photographer's art impossible, and this defect (if it may be so called) in the eye would render it almost equally useless, had not an adjustment, which more than compensates for it, been afforded in the extreme *mobility* of our organs of vision. This adjustment is so perfect that people in general do not even suspect that the image of *each point* of an object which they distinctly see must be formed on one particular spot on the retina—a spot about one-tenth of the diameter of an ordinary pin-head! We may venture, without any disrespect to the reader, to assume that the chances are that it is new to him to learn how each letter in the lines beneath his eye must successively, but momentarily, form its image in the very little pit in the centre of his retina; and the chances are at least a hundred to one that, even if aware of this, he has passively received the statement, and that he has not made the least attempt to *realize the truth for himself.* Yet nothing is easier. Let him request a friend to slowly peruse some printed page, while he meanwhile intently watches his friend's eyes. He will then perceive that before a single word can be read there is a *movement* of the eyeballs, which are, quite unconsciously to the person reading, so directed that the image of each letter (for the area of distinct vision is incapable of receiving more than this at once) shall fall upon the only parts of the retinæ from which a distinct impression can be conveyed along the optic nerve. Thus it is that the eye, without any conscious effort of the observer, is directed in succession to the various points of an object, and it is only by an effort of will in fixing the eyes upon one spot that one becomes aware of the blurred and confused forms of all the rest of the visual picture. Yet so readily do the eyeballs turn to any part of the indistinct picture on which the attention is fixed, that it is not improbable a person unversed in such experiments, wishing to verify our conclusions by looking, say, at one spot on the opposite wall, will be very apt, in thinking of the features of the rest of the picture, to direct his eyes there, and then declare that he, at least, sees no such vague forms. If such be his experience, the correction is easy. He has only to ask some one to watch closely his eyes while he repeats the experiment, and after a few trials he will succeed in maintaining the requisite immobility of the eyeballs—a condition upon which the success of many such experiments depends.

This extreme mobility of the eyeballs more than compensates for the loss of the clear and well-defined picture, for it calls into action one of the most sensitive of all the impressions of which we are capable, and one which possesses in so high a degree the power of uniting with our

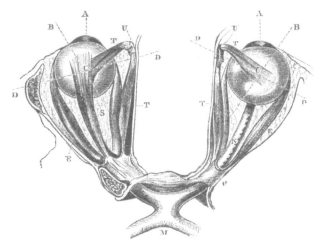

FIG. 237.—*Muscles of Eyes.*

The muscles of the eyeballs viewed from above :—B, the internal rectus; E, the external rectus; s, the superior rectus; T, the superior oblique, passing through a loop of ligament at U, and turning outwards and downwards to its insertion at c. The inferior rectus and the inferior oblique are not visible in the figure: the superior rectus is removed from the right eyeball in order to show the optic nerve N.

other sensations, that this sixth sense has been, as already stated, utterly overlooked, except by the more modern students of the nature of our sensations. It is usually termed the *muscular sense,* and to it are due some of the nicest distinctions of impressions of which we are capable. The muscles of every part of our frame take their part in producing impressions in our minds, and those of the eyeballs have a very large share in furnishing us with ideas of forms and motions. Fig. 237 is a diagram showing the general arrangement of these muscles ; and their anatomical designations, which need not much concern us at present, are given beneath the figure. The wonder is, that the sensations arising from the relative conditions of parts so few, should afford us the immense variety of notions referrible for their origin to these muscles only. We take one example in illustration. Suppose we watch the flight of a bird, at such an elevation that no part of the landscape comes into the field of view at all ; and that, again, we follow with the eye, under similar circumstances, the path of a rocket. We can unhesitatingly pronounce the motions unlike, and yet in each case there was no visual impression present but that of the object focussed upon the yellow spot. But the movement of *the muscles in one case* was different from that in the other. Nay more, we can form such a judgment of the motion as to pronounce that the object followed such and such a curve—we may recognize the parabola in one path, and the circle, perhaps, in the other. And this kind of discrimination arises from the fact, that when we have, maybe times without number, previously looked at parabolas and circles, in diagrams perhaps, the muscles of the eyeballs have performed just the same series of movements, as point after point of the line was made to form its image on the yellow spot. This is not the only class of impressions that these muscles are capable of affording ; there is, for example, little doubt that they aid us in estimating distance. But space will not permit further discussion of this subject.

Although the blurred and indefinite retinal picture may be compensated, and perhaps more than compensated, by the readiness with which the eyes move, it is, of course, possible that greater precision and delicacy of visual impression over the whole surface of the retina might be consistent with a still greater increase of our powers of perception. There are instances in which the absence of finish, as it may be termed, in all but one little spot in the picture, proves a real inconvenience and a sensible deprivation. Perhaps a friend calls our attention to the fact that a balloon is sailing through the air, or some fine morning, hearing in the fields the blithe song of the sky-lark, we look up and vainly try to bring the small image upon the place of distinct vision. Now, if an image which falls upon any other part of the retina is perceived, even indistinctly, an instant suffices to direct the eyes into the exact position requisite for clear vision—an example of the marvellous precision with which impressions are put in relation to each other by the unconscious action of the brain. But while an image on the fovea, only $\frac{1}{7000}$th of an inch diameter, produces a distinct sensation, it is found that if the image falls on the retina at a point some distance from the yellow spot, the image must be 150 times larger in order to produce any impression ; and it is in consequence of the image of balloon or bird not having the requisite size to give any impression to the less sensitive portion of the retina, that we grope blindly, as it were, until by chance the image falls near the yellow spot, when the tentative motion of the eyeballs is instantly arrested, and the image fixed. On the other hand, the field of indistinct vision which the eye takes in is extremely wide, for bright objects are thus perceived, even when their direction forms an angle laterally of nearly 90° with the axis of the eye ; and, if the object be not only bright, but in motion, its presence is noticed under such circumstances with still greater ease. Thus, an observer scanning the heavens would have a perception of a shooting star anywhere within nearly half the hemisphere. The range is, however, less than 90° in a vertical direction.

We have said that the fibres of the optic nerve, entering the back part of the eyeball, at K, Fig. 234, ramify over the anterior surface of the retina in fibres which form a layer of considerable relative thickness. The light, therefore, first encounters these nerves, and only after traversing their transparent substance does it reach the deeper seated layer of rods and cones, where it excites some action that is capable of stimulating the optic nerve. These rods and cones might naturally be supposed to be merely accessory to the fibres of the optic nerve, had we not the following conclusive evidence that the cones play a necessary part in the action, and that it is only through them that light acts upon the optic nerve :

1. The cones are more developed and more numerous in the spot where vision is most distinct.

2. The "blind spot" is full of fibres of the optic nerve, but is absolutely insensible to light, and is without rods or cones.

3. We can distinguish an image on the fovea, having only $\frac{1}{6000}$th of an inch diameter; but on the other parts of the retina the images must have larger dimensions. It is found that the size of the smallest distinguishable images agrees nearly with the diameters of the cones at the respective parts.

To some readers the fact will doubtless be new, that a considerable portion of the eye is quite insensible to light, namely, that portion already designated as the "blind spot." A simple experiment, made by help of Fig. 238, will prove this. Place the book so that the length of the figure may be parallel to the line joining the eyes, and let the right eye be exactly opposite the white cross, and at a distance from it of about 11 in. If the left eye be now closed, while with the right the cross is steadily viewed so that it is *always* clear and distinct, the white circle will completely disappear, and the ground will appear of a uniform black colour. In order to insure success, the observer must be careful not to *look at* the white circle, but at the cross, and some persons find this more difficult than others. The position of the blind spot in the eye has been already mentioned, and its significance in showing the insensibility to light of the fibres of the optic nerve has been pointed out. In the table of the dimensions of some parts of the eye, which, for convenience of reference, is given together below, it will be seen that the diameter of the blind spot is considerable compared with the size of the retina, its greatest diameter being about $\frac{8}{100}$ in.

FIG. 238.

The length on the retina of the image of a man at a distance of 6 ft. or 7 ft. is not greater than this, so that in a certain position with regard to the eye a person would, like the white circle, be quite invisible. In like manner, by looking steadily in a certain direction with one eye, the image of the full moon may be made to fall upon the blind spot, and the luminary then becomes invisible, and would be so even if its apparent diameter were eleven times greater; so that if we suppose eleven full moons ranged in a line, the whole would be quite invisible to a person looking towards a certain point of the sky at no great angular distance from them.

The following are the dimensions in English inches of some parts of the eye:

	In.
Diameter of the entrance of the optic nerve...............	0·08
Distance of centre of optic nerve from centre of yellow spot	0·138
Diameter of *fovea centralis*	0·008
Diameter of the nerve-cells of the retina	0·0005
Diameter of the *nuclei*	0·00003
Diameter of the rods...	0·00004
Diameter of the cones in yellow spot......................	0·00018
Length of rods	0·0016
Length of cones in yellow spot	0·0008
Thickness of retina at the back of the eye	0·0058

By means of an instrument to be presently described, the ophthalmoscope, it is possible to view directly the whole surface of the retina, and to observe the inverted images of the objects there depicted. It is thus observed that it is only on the parts near the yellow spot that the images are formed with clear and sharp definition. Away from this the definition is less perfect; and besides the diminished sensitiveness of the retina, this circumstance contributes to the vagueness of the visual picture, although the falling off in clearness of vision at a very little distance from the yellow spot is far more marked than the loss of definition in the image there formed.

Until within the last few years it has been most confidently asserted by many authors that the eye, considered as an *optical instrument*, is absolutely perfect, and entirely free from certain defects to which artificial instruments are liable. Thus Dr. W. B. Carpenter states, in his "Animal Physiology" (1859): "The eye is much more remarkable for its perfection as an optical instrument than we might be led to suppose from the cursory view we have hitherto taken of its functions ; for, by the peculiarities of its construction, certain faults and defects are avoided, to which all ordinary optical instruments are liable." Among the imperfections which are completely corrected in the eye, he names "spherical aberration" and "chromatic aberration"—both of which give rise to certain defects in optical instruments. But by recent careful investigations it has been conclusively shown that the eye is not free from chromatic aberration ; that it has defects analogous to spherical aberration ; and that there are, besides, certain optical imperfections in its structure, which are avoided in the artificial instruments. Professor Helmholtz, one of the most distinguished of German mathematicians, physicists, and physiologists, whose great work on "Physiological Optics" is the most complete treatise on the subject which has ever appeared, is so far from considering the eye as possessed of all optical perfections that he remarks that, should an optician send him an instrument having like *optical* defects, he would feel justified in sending it back. The defects which may be traced in the eye, *considered as an optical instrument*, do not, however, he admits, detract from the excellence of the eye *considered as the organ of vision*.

When we find that Sir Isaac Newton pointed out the chromatic aberration of the eye two centuries ago—when we find that D'Alembert, in 1767, proved that the lenses of the eye might have as great a dispersive power as glass without the want of achromatism necessarily becoming noticeable —when we find that the celebrated optician Dolland, the inventor of the achromatic lens, showed that the refractions which take place in the eye all tend to bring the violet rays towards the axis more than the red—when we find that Maskelyne the astronomer, Wollaston the physicist, Fraun-hofer the optician, and other scarcely less distinguished men of science, have made actual measurements of the distances of the *foci* in the human eye for the different rays of the spectrum—when we find how these defects have so long ago been observed, examined, and measured as to their amount —the persistence with which writer after writer has asserted the achroma-tism of the human eye appears so extraordinary, that it can only be accounted for by the prevalence of the preconceived notion that the eye is absolutely perfect—a notion not without its reason and grounds, in the fact of the exquisite adaptation of the organ of sight to the needs of humanity.

Although the want of achromatism in the eye thus escapes ordinary notice, it is, on the other hand, easy to render it evident by simple experi-ments. If, for example, we view from a certain distance the solar spectrum projected on a white screen, it will be found that, when we see the red end quite distinctly, the violet end will, at the same time, appear vague and confused, and *vice versâ*. The author believes that the following very simple experiment will at once convince any person that the fact is as stated. Procure a small piece of blue or *violet* stained glass, and another piece of *red* glass, and, having cut out of an opaque screen a rectangular opening, say ½ in. long and ¼ in. wide, place the glasses close to it, so that one-half the opening is covered by the red glass and the other half by the violet glass, the two being placed so that, on looking through the screen, a violet square and a red square are visible. The opaque screen may be made of black paper, cardboard, or tinfoil, and the edges of the opening must be cut perfectly even. On looking through this arrangement, held at a distance of about two feet from the eye, both squares may be seen dis-tinctly by a person of ordinary vision ; but, at a distance of five inches from the eye, he will find it impossible to see the squares otherwise than with vague and ill-defined edges. This is because the crystalline lens cannot adapt its curvature so as to bring the rays from the object to a focus on the retina. Now, by trial, the nearest distance at which each of the coloured squares becomes visible may be found, and it will be observed, that the violet square is first sharply defined at a less distance than the red, where-as, if the eye brought the red and violet rays to a focus at the same point, the smallest distance of distinct vision would coincide in both cases.

The reader may observe the same fact for himself, in even a still simpler manner, by turning to Fig. 238, page 461. When the white circle is viewed by one eye, at a distance of about a foot, and an opaque screen, such as a coin, is held close to the eye, so that the pupil is half covered by it, the one side of the white circle will appear bordered by a narrow fringe of blue, and the other side by a narrow fringe of orange. If the opaque screen be shifted from one side of the pupil to the other, the colours will change places, the orange appearing always on the same side of the white circle as the screen is held before the eye. The same appearances are presented in a still more marked degree when the full moon is made the subject of the experiment.

The diagram, Fig. 239, shows the course of the red and violet rays from a luminous point, A, the refraction being supposed to take place at B_1 B_2. The violet rays after refraction form the cone, B_1, E, B_2, and E is their focus; the red rays form the cone, B_1, F, B_2, and have a focus at F. The position of the retina would be intermediate between E and F, and is indicated by C_1, C_2. It will be noticed that the violet rays cross, and are received on the retina in the same circle, G G, so that the colours, then blended, would be separately imperceptible; but the point would produce a diffused circular image of the blended colours.

In viewing an object—the moon, for example—the accommodation of the eye is like that indicated in the diagram. The distinct image due to the red rays would be formed behind the retina, and that due to the violet rays would be in front of it. In the image on the retina the most intense rays—such as the orange, yellow, and green—are those which are blended by the adjustment of the eye, and the red and violet form images more out of focus (to use a common expression), and a very little larger than the more intense image. We might expect that a white disc would therefore appear with a fringe of colour, resulting from a mixture of red and violet;

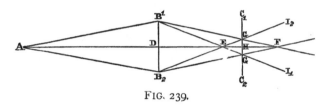

FIG. 239.

but the fringe is too narrow, and the colour itself too feeble, to become perceptible. When, however, the pupil of the eye is half covered, the red and violet images are displaced in different directions, the position of the retina being too far forward for the one, and too far back for the other. The coincidence therefore ceasing, the colours show themselves at the margins of the image.

The non-perception under ordinary circumstances of the chromatic aberration of the eye is largely due to the greater intensity of the colours which differ least in their refrangibilities. The clearness of our vision does not, therefore, practically suffer from this defect of the eye. Professor Helmholtz constructed lenses which rendered his eyes really achromatic, and looking through these when the pupil was half covered, no coloured fringes were seen at the edges of dark or light objects, or when the objects were looked at with an imperfect accommodation of the eye. He was, however, unable to detect any increase of clearness or distinctness of vision by the correction.

The eye is also subject to other aberrations and irregular refractions, which are special to itself; for example, with moderately illuminated objects the crystalline lens produces images apparently well defined, and nothing is visible to suggest the absence of uniformity in its structure. But when the light is intense, and concentrated in a small object surrounded by a dark field, the irregular structure of the crystalline lens shows itself in the most marked manner. Every one must have noticed the appearance presented by the distant street-lamps on a dark night, and by the stars. The latter we know to be for us mere points of light, and their images produced by perfect lenses would also be mere points ; instead of which we see what seem to be rays issuing from the star, an appearance which has given rise to the ordinary representation of a star as a figure having several rays. That no such rays actually do emanate from the real star may be easily proved : first, by concealing the luminous point from view, by means of a small object held up as a screen. If the rays had any existence outside of the eye, they would still be seen ; instead of which, the whole of them disappear when the luminous point, or, in the case of the street-lamp, when the flame, is covered by the screen. A second proof that the origin of the phenomenon is in the eye, and not in the object, is afforded by the fact that if, while attentively observing the rays, we incline the head, the rays turn with the eyes, so that when the head is resting on the shoulder the ray which appeared vertical becomes horizontal. The cause of these divergences from the regular image lies in the fact of the crystalline lens being built up of fibres which have refractive powers somewhat different from that of the intermediate substance. These fibres are arranged in layers parallel to the surfaces of the crystalline lens, and the direction of the fibres in each layer is generally from the centre to the circumference ; but towards the axis they form, by bending, a kind of six-rayed figure, as shown in Fig. 240, which represents the arrangement of the fibres of the external layers of the lens.

FIG. 240.

In the outermost layers the branches of the star-shaped figure are subdivided into secondary branches, which give rise to more complicated figures. When we view by night a very brilliant but small light, even these subdivisions may be traced in the radiating figure.

The light which enters the eye is partly absorbed by the black pigment of the choroid, and partly sent back by diffused reflection from the retina through the crystalline lens and pupil. The image of a luminous body as depicted on the retina of another person cannot be seen by us under ordinary circumstances, because, by the principle of reversibility already mentioned as of universal application in optics, the rays which issue from the retinal images are refracted on leaving the eye, and follow the same paths by which they entered it, so that they are sent back to the object. An observer cannot see the retinal image of a candle in another person's eye, unless he allows the rays to enter his own, and this cannot be done directly, because the head of the observer would be interposed between the candle and the eye observed, and the light would then be intercepted. By holding a piece of unsilvered plate glass vertically, we may reflect the light of a candle into the eye of another person, and then the light thrown out from the retinal image of the candle will, on again meeting the surface of the glass, be in part reflected to its source, and in part pass through the glass, on the other side of which it may be received into the eye of an observer.   The positions of the observed and observing eye may be described as exactly opposite to and near each other, while the candle is placed to one side in the plane separating the two eyes, and the glass is held so that it forms an angle of 45° with the line joining the pupils. Under these circumstances the observer may see the light at the back of the eye, but he will not be able to distinguish anything clearly, because his own eye cannot accommodate itself so as to bring to a focus the rays coming from the retina of the other, since these rays are refracted by the media through which they emerge.   But, by means of suitable lenses interposed between the two eyes, the retina and all its details may be distinctly seen and examined.   Such an arrangement of lenses and a reflecting surface con-

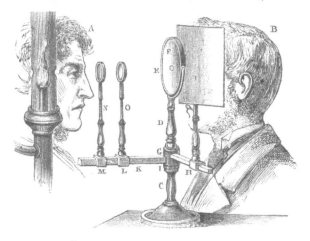

FIG. 241.—*Ruete's Ophthalmoscope.*

stitute the instrument called the *ophthalmoscope* (οφθαλμος, *the eye*) of which there are many forms, but all constructed on the principle just indicated. This principle was first pointed out by Helmholtz, who described the first ophthalmoscope in 1851.

Ruete's ophthalmoscope is represented in Fig. 241. The parts of the instrument are supported on a stand, C, and about the vertical axis of this the column, D, and the arms, H and K, can turn freely and independently; E is a concave metallic mirror, about 3 in. in diameter, and having an aperture in its centre through which the observer, B, looks.   The arm, H, merely carries a black opaque screen, which serves to shield the eye of B from the light of the lamp, and to reduce, if required, the amount of light passing through the aperture in the mirror.   The arm, K, which is about a foot in length, carries two uprights which slide along it, and in each of these slides a rod bearing a lens, which can thus be adjusted into any required position. The instrument is used in an apartment where all light but that of the lamp can be excluded. In the instrument just described an inverted image is obtained, which is sufficient for ordinary medical purposes, but this construction does not allow of the examination of retinal images, which is best performed with an instrument having a plane mirror.

The appearance presented by the back of the eye when viewed in the ophthalmoscope is represented in Fig. 242. The retina appears red, except at the place where the optic nerve enters, which is white. On the reddish ground the retinal blood-vessels can be distinguished ; A, A, A, branches of the retinal artery, have a brighter red colour, and more strongly reflect the light than the branches, B, B, B, of the retinal vein. Among these, and especially towards the margin, are seen, more or less distinctly, the broader vessels of the choroid. Above the optic nerve and a little to the right may be observed the *fovea centralis.*

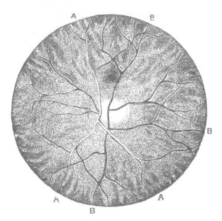

FIG. 242.

During the last twenty years the ophthalmoscope has been the chief means of extending the knowledge of oculists regarding the diseased and healthy conditions of the eye. In this way the substance of the lens and the state of the humours can be directly seen, the causes of impaired vision can be discovered, and the nature of many maladies made out with certainty. This modern invention, by which the interesting spectacle of the interior of the living eye can be observed, has therefore been far from proving a barren triumph of science. Many insidious maladies can thus be detected, and may be successfully treated before the organ has become hopelessly diseased. In some cases the ophthalmoscope gives the most certain evidence of the existence of obscure and unsuspected diseases of other parts of the body.

VISUAL IMPRESSIONS.

EVERYBODY knows that, however well the flat picture of an object may imitate the colours and forms of nature, we are never deceived into supposing that we have the real object before us. There must, therefore, be something different in the conditions ￢nder which we see real objects from those under which we view their pictures. The most favourable circumstances for receiving an illusive impression of solidity from a flat picture, is when we view it from a fixed position and with one eye. This is because one means by which we unconsciously estimate distances depends upon the changes in the perspective appearances of objects caused by changes in ꓒur point of view. In many cases these changes in the perspective are the ꓒnly means we have of judging of the relative distances of objects. But there is another circumstance which is still more intimately connected with our perception of solidity. Each eye receives a slightly different image of the objects before us (unless these be extremely remote), inasmuch as they are viewed from a different point. When the objects are very near, the two retinal images may differ considerably, as the reader may convince himself by viewing with each eye, alternately, objects immediately before him, while the other eye is closed, and the head all the while motionless. The nearer objects will plainly appear to shift their positions as seen against the background of the more distant objects ; and a somewhat more careful observation will reveal changes of perspective, or apparent form, in every one of these objects. An extreme case is presented in that of a playing card, or thin book, held in the plane which divides the eyes. The back or the face, the one side or the other, will be seen, according as the right or the left eye is opened. If we close the left eye, the displacement and change of apparent form produced by a slight movement of the head are sufficiently obvious ; a movement of the head $2\frac{1}{2}$ in. to the left causes a decided change in the relative positions of adjacent objects. It is plain, however, that it is precisely from a point $2\frac{1}{2}$ in. to the left that the left eye views these objects. and hence the perspective appearance seen by the left eye must have the difference due to this shifting of the point of view.

On the other hand, if one looks at a picture, or flat surface, placed immediately in front, no change in the relative positions of its parts is discernible by viewing it with either eye alternately. Not but that there is a difference in the retinal images in the two cases, but there is an absence of any point of comparison by which the change may be judged. If we take a photograph of a statue, it will, when viewed by one or the other eye, present the difference of the retinal images which is due to a flat surface ; the parts of the photographic image will be of slightly different proportions as seen by each eye. If, instead of the photograph we have before our eyes a statuette, each eye will see a quite different view : the right eye will see a portion which is invisible to the left eye, and *vice versâ*, and, in fact, we shall see more than half round the object. Here, then, we have certain differences of the retinal pictures when solid objects are viewed, and these differences by innumerable repetitions have, unconsciously to ourselves, become associated with notions of solidity, of something having length, breadth, and depth, or thickness. The marvellous delicacy of these perceptions will be alluded to hereafter.

Let us suppose that the lenses of two cameras are fixed in the positions occupied by the two eyes, and that a photograph is taken in each camera, the subject being, for example, a statuette. It is obvious that the differences of the two photographs would correspond with the differences of the two retinal images, and that, if a person could view with the right eye only the photograph taken in the right-hand camera, and with the left eye the left-hand photograph only, there would be formed on the retinæ of his eyes images very nearly corresponding with those which the actual object would produce, and the result would be, if these retinal pictures occupied the proper position on the eyes, that the impression of solidity would be produced, which is called the *stereoscopic effect.*

This may be done without the aid of any instrument, as almost any person may discover after some trials with nothing but a *stereoscopic slide,* if he can succeed in maintaining the optic axis of his eyes quite parallel. In such a case he will observe the stereoscopic effect by the fusing together, as it were, into one sensation, of the impression received by the right eye from the right photograph, with that received by the left eye from the left photograph. But as each eye will, at the same time, have the photograph intended for the other in the field of view, the observer will be conscious of a non-stereoscopic image on each side of the central stereoscopic one.

FIG. 243.—*Wheatstone's Reflecting Stereoscope.*

These outside images are, however, very distracting, for the moment the attention is in the least directed to them, the optic axes converge to the one side or the other, losing their parallelism, and the stereoscopic effect vanishes, because the images no longer fall in the usual positions on the retinæ. It is, in consequence, only after some practice that one succeeds in readily viewing stereoscopic slides in this manner, but the acquirement is a convenient one when a person has rapidly to inspect a number of such slides, for he can see them stereoscopically without putting them in the instrument. Many persons, however, find great difficulty in acquiring this power. In such cases it is well to begin by separating the two photographs by means of a piece of cardboard, covered with black paper on both sides. When this is held in the plane between the eyes, each eye sees only its own photograph, and the observer is not troubled with the two exterior images. After a little practice in this way, the cardboard may usually be dispensed with, and the observer will insensibly have acquired the habit of viewing the slides stereoscopically, without any aid whatever.

Instruments have, however, been contrived which enable one to obtain the desired result without effort ; and one form of these is now tolerably well known to everybody. The first stereoscope was the invention of Wheatstone. The reflecting stereoscope is represented in Fig. 243, and consists essentially of two plane metallic mirrors inclined to the front of the instrument at angles of 45°, so that in each of them the observer sees only the design which belongs to it. The rays reach the eyes as if they came from images placed in front of the observer ; and the two images having the proper differences, produce together the impression of solid objects.

Brewster's stereoscope—which is far more widely known than Wheatstone's—has two acute prisms, or, more usually, two portions of a convex lens are cut out, and placed with their margins or thin parts inwards, and they thus produce the same effect as would be obtained by combinations of a prism with a convex lens. Another very common form of the stereoscope has merely two convex lenses. The effect of the convex lenses is to increase the apparent size of the images by diminishing the divergence of the rays emitted by each point, producing the appearance of larger designs seen at a greater distance. The effect of the prism is to give the rays the direction which they would have if they proceeded from an object placed in a position immediately between the two designs, and an additional element by which we estimate distance, namely, the convergence of the optic axes, is made to aid in the illusion, when the rays proceeding from the two different pictures have approximately the inclination that they would have if they emanated from real objects at the place where the image is apparently formed. The box or case in which the lenses or lenticular prisms are placed takes various forms. One of the most common is represented in Fig.

FIG. 244.

244, but the stand on which it is mounted is not a necessary part of the instrument, although it is sometimes convenient. A handsome form is met with as a square case, enclosing a number of photographic stereoscopic views mounted on an endless chain in such a manner that they are brought successively into view by turning a knob on the outside. When an instrument of this kind is fitted up with a series of the beautiful landscape transparencies, which are produced by certain continental photographers, a more perfect reproduction of the impressions derived from nature, exclusive of colour, cannot be conceived. We seem to be present on the very spots which are so truthfully depicted by the subtile pencil of the sunbeam ; we feel that we have but to advance a foot in order to mix with the passengers in the streets of Paris or of Rome, and that a single step will bring us on the mountain-side, or place us on the slippery glacier; at our own fireside we can feel the forty centuries looking down upon us from the heights of those grand Egyptian pyramids, and find ourselves bodily confronted with the mysterious Sphinx, still asking the solution of her enigma. The truth and force with which these stereoscopic photographs reproduce the relief of buildings are such, that when one sees for the first time the real edifice of which he has once examined the stereoscopic images, it no longer strikes him as new or unknown ; for he derives from the actual scene no impression of form that he has not already received from the image.

But of all subjects of stereoscopic photography the glaciers are, perhaps, those which best show the power of the instrument as far surpassing all other resources of graphic presentation. The most careful painting fails to convey a notion of the strange glimmer of light which fills the clefts of the ice, seen through the transparent substance itself. The simple photograph commonly presents nothing but a confused mass of grey patches ; but combine in the stereoscope two such photographs, each formed of nothing but slightly different grey patches, and a surprising effect is at once produced : the masses of ice assume a palpable form, and the beautiful effects of light transmitted or reflected by the translucent solid reveal themselves. Another very beautiful class of subjects for stereoscopic slides is found in those marvellous instantaneous photographs, which seize and fix the images of the waves as they dash upon the shore. Here a scene which has tasked the power of the greatest painter is brought home to us with such force and vividness that we all but hear the wild uproar of the breakers.

But for the art of photography the stereoscope would not thus be ready

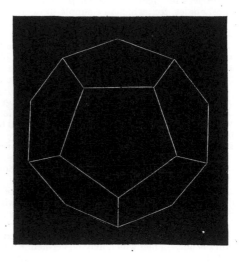

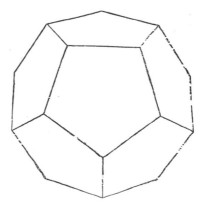

FIG. 245.

to minister to our enjoyment, for no pictures wrought by man's handiwork could approach the requisite accuracy which the two stereoscopic pictures must possess. All attempts to produce such pictures by engraving or lithography have failed, except only in the case of linear geometrical designs, such as representations of crystals. A very useful and suggestive application of the stereoscope has been made to the illustration of a treatise on *solid geometry*, where the lines representing the planes, being drawn in proper perspective, the reader by placing a simple stereoscope over the plates sees the planes stand out in relief before him, and the multitude of lines, angles, &c., which in a simple drawing might be distracting even for a practised geometrician, assume a clear and definite form.   The difference between the two retinal pictures of objects is so slight, that when the objects are at a little distance, ordinary observation fails to discover it without the aid of special instruments; and an inspection of the pair of photographs in a stereoscopic slide will convince any one that, even in these, close and careful observation is required to perceive the difference.

Some of the principles of stereoscopic drawings may be seen exemplified by the pair we give in Fig. 245.   With this figure the reader may attempt the experiment of seeing the stereoscopic effect without the stereoscope. When he has succeeded in doing this, or when he fuses the images together by placing a simple stereoscope over the page, he will find the result very singular; for he will receive the impression of a solid crystal of some dark polished substance—black lead, for instance—placed on a surface of the same material.   The edges of the solid will appear to have a certain lustre, such as one sees on the edges of a real crystal.   The reason of this impression being produced by two drawings, one of which is formed by black lines on a white ground, while the other has white lines on a black ground, is probably due to the circumstance that we very often see in nature the *lustrous* edges of an object with one eye only.   That is, one eye is in the path of the rays which are regularly reflected from the object, while the other is not,—a fact which may be verified in an instant by looking first with one eye and then with the other, at a polished pencil, or similar object, when placed in a certain position.

There is a kind of modification of the reflecting stereoscope, known under the name of the *pseudoscope*, which is highly instructive, as showing how much our notions of the solidity of objects are due to the differences of the retinal images.   In the pseudoscope the rays reach the eyes after passing through rectangular prisms in such a manner that objects on the right appear on the left, and objects on the left appear on the right; but the images agree by reason of the symmetry of the reflection, although the image of the objects that without the instrument would be formed in the right eye is, by the action of the prisms, formed in the left eye, and *vice versâ*.   The impressions produced are very curious : convex bodies appear concave—a coin, for example, seems to have the image hollowed out, a pencil appears a cylindrical cavity, a globe seems a concave hemisphere, and objects near at hand appear distant, and so on.   These illusions are, however, easily dispelled by any circumstance which brings before the mind our knowledge of the actual forms, and by a mental effort it is possible to perceive the actual forms even with the pseudoscope, and indeed to revert alternately, with the same object, from convexity to concavity. This last effect is very curious, for the object appears to abruptly change its form, becoming alternately hollow and projecting, according as the mind dwells upon the one notion or the other; but the experiment is attended with a feeling of effort, which is very fatiguing to the eyes.

Professor Helmholtz has contrived another very curious instrument, depending on the same principles as the stereoscope.   He terms it the *telestereoscope*, and while the effect of the pseudoscope is to reverse the relief of objects, the telestereoscope merely exaggerates this relief; hence this instrument is well adapted for making those objects which from their distance present no stereoscopic effect, stand out in relief. The distance between our eyes is not sufficiently great to give us sensibly different views of very distant objects, and what the telestereoscope does is virtually to separate our eyes to a greater distance. Fig. 246 is a horizontal section of the instrument. L and R represent the position of the eyes of the spectator; *a*, *b*, are two plane mirrors at 45 to his line of sight; A, B, are two larger plane mirrors, respectively nearly parallel to the former. *c d a* L and *f g b* R show the paths of rays from distant objects, and it is obvious that the right eye will obtain a view of the objects identical with that which would be presented to an eye at R', while the left eye has similarly the picture of the

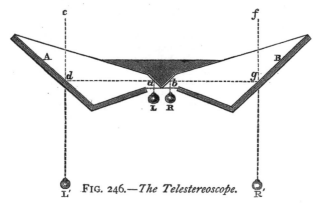

FIG. 246.—*The Telestereoscope.*

objects as seen from the point L'. The four mirrors are mounted in a box, and means are provided for adjusting the positions of the larger mirrors, as may be required. With this instrument the distant objects in a landscape—a range of mountains, for example—which present to the naked eye little or no appearance of relief, have their projections and hollows revealed in the most curious manner.

It is upon a similiar principle that stereoscopic views of some of the celestial bodies have been obtained. Admirable stereoscopic slides of the moon have been produced by photographing her at different times, when the illumination of the surface is the same, but when, in consequence of her *libration*, somewhat different views of our satellite are presented to us. Two such photographs, properly combined in the stereoscope, give not only the spherical form in full relief, but all the details of the surface: the mountains, craters, valleys, and plains are seen in their true relative projection.

The telestereoscope may be inverted, so to speak, and its effect reversed; for an arrangement of mirrors similarly disposed, but on such a scale as will permit the eyes to be respectively in the lines *c d* and *f g*, would reflect from objects in the direction L R rays which would have but little of the difference of direction to which the stereoscopic effect is due. Hence solid objects viewed with such an instrument appear exactly like flat pictures, the effect being far more marked than in simply viewing them with one eye.

An ingenious method of exhibiting a stereoscopic effect to an audience has been contrived by Rollmann. He draws on a black ground two linear stereoscopic designs—that for the left eye with red lines, that for the right eye with blue. Each individual in the audience is provided with a piece of blue glass and a piece of red: he places the red glass before the left eye, the blue glass before the right : each eye thus receives only the picture intended for it, for the blue lines cannot be seen through the red glass, or the red lines through the blue glass. The diagrams may, of course, be projected on a screen by a magic lantern, in which case the circumstances are even more favourable. Duboscq has arranged a kind of opera glass, so that a person may view appropriate designs on the large scale, and arrangements have been also contrived by which the stereoscopic effect may be seen in moving figures.

Every student of this interesting subject should examine a few stereo-scopic images produced by simple lines representing geometrical figures, or the photographs of the model of a crystal, as these exhibit in the most striking manner the conditions requisite for the production of stereoscopic effects. A person having a little skill in perspective and geometry might construct the two stereoscopic images of a body defined by straight lines, but the drawings must be executed with extreme exactitude, for the least deviation would produce the most marked effect in the stereoscopic appear-ance. The production of stereoscopic photographs now forms a consider-able branch of industrial art. At first, these photographs were made by taking the two different views with the same camera at two operations. But there were difficulties in obtaining uniformity of depth in the impres-sions, and the change in the shadows produced by the earth's rotation showed itself—although the interval between the two exposures might not exceed three or four minutes. The increased shadows in such cases show themselves in the stereoscope, like dark screens suspended in the air. It was Sir David Brewster who, in 1849, first proposed the plan now univer-sally adopted, of producing the views simultaneously by twin cameras form-ing their images on different parts of the same sensitive plate, the centres of the lenses being placed at the same distance apart as a man's eyes, that is, from $2\frac{1}{2}$ to 3 in. This is, of course, the only manner in which instan-taneous views can be secured. Helmholtz, however, advocates the photo-graphs of remote objects being taken at a much greater distance apart, for they otherwise present little appearance of relief. By selecting from an assortment of slides, two views of the Wetterhorn, taken from different points in the Grindelwald valley, and combining these in the stereoscope, he found that a far more distinct idea of the modelling of the mountain could be thus obtained than even a spectator of the actual scene would receive by viewing the mountain from any one point. Such a mode of com-bining the photographs would produce in the stereoscope the same effect as the telestereoscope would in the landscape, but the effect would be caused to a proportionately far higher degree.

The date of Wheatstone's first publication regarding the stereoscope was 1833 ; but a complete description and theory of the instrument was not published until five years afterwards. Brewster first made public, in 1843, his invention of the stereoscope with lenses, which is now sc familiar to us, and few scientific instruments have become so quickly and extensively popular ; certainly no other simple and inexpensive instrument has con-tributed so largely to the amusement and instruction of our domestic circles. And, to the philosopher who studies the nature of our perceptions, the stereoscope has been even more instructive, for, instead of vague surmises, it provided him with the solid ground of experiment on which to found his theories. The literature of this one subject—stereoscopic effect—is extensive enough to occupy a tolerably long book-shelf. It dates from 300 B.C., when Euclid touched upon the subject in his Optics ; and after a lapse of more than eighteen centuries it was taken up by Baptista Porta, in 1583 ; but the whole development of this subject belongs almost entirely to the last half-century.

The part which the muscles of the eyes take in our perceptions of form has been already alluded to, and it may be interesting to illustrate this point by a curious example or two of illusions arising from their movements. If our reader will glance at Fig. 247, he will see that the lines, *a b* and *c d*, appear to be farther apart towards the centre than at the ends, while *f g* and *h i*, on the other hand, appear nearest together in the middle. He will hardly be convinced that in each case the lines are quite parallel

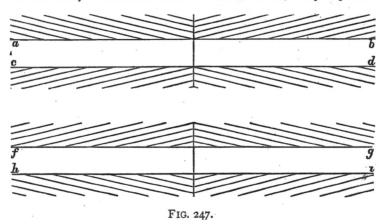

FIG. 247.

until he has actually measured the distances. A still more striking example of the same kind of illusion is shown by Fig. 248, due to Zöllner. This appears a sort of pattern, in which the broad bands are not upright, but sloping alternately to the right and left, and with the spaces between the lines wider at one end than the other. The lines in the figure are, however, strictly parallel. The illusion by which they appear divergent and convergent is still more strongly felt when the book is held so that the wider bands are inclined at an angle of 45° to the horizon. There is another illusion here with reference to the short lines, which will appear to be opposite to the white spaces on the other side of the long lines to which they are attached. That these illusions are really due to movements of the eyes may be proved by viewing the designs in any manner which entirely prevents the movement, as by fixing the gaze on one spot in the case of Fig. 247, when the illusion will vanish ; but this plan is not so easily applied to Fig. 248. A convincing proof, however, will be found in the appearance of these figures when they are viewed by the instantaneous light of the electric spark, as when a Leyden jar is discharged in a dark room. The reader viewing the figures, held near the place where the spark appears, will see them distinctly without the illusions as to the non-parallelism of the lines. In the absence of an electrical machine, or coil and jar, the reader may have an opportunity of seeing the figures by flashes of lightning at night, when the result will be the same.

FIG. 248.

There is a property of the eye which has led to the production of many amusing and curious illusions. This property in itself is no new discovery, for its presence and effects must have been noticed ages ago. The property in question is illustrated when we twirl round a stick or cord, burning with a red glow at the end. We seem to trace a *circle* of fire; but as the glowing spark cannot be in more than one point of the circle at once, it is plain that the impression produced on the eye must remain until the spark has completed its journey round the circle, and reaching each point successively renews the luminous impression. Like other subjects relating to vision, this phenomenon has been carefully examined in recent times, and its laws accurately determined.

The fact which is obvious from such an experiment, may be thus stated: Visual impressions repeated with sufficient rapidity produce the effect of objects continually present. This persistence of the visual impressions is easily made the subject of experiment by means of rapidly rotating discs; and in the common toy called a "colour top" we have a ready means of verifying some of the conclusions of science on this subject. Some very interesting results may be obtained by an apparatus as simple as this, regarding the laws of the phenomenon we are considering, and the effects of

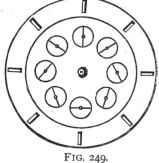

FIG. 249.

various mixtures of tints and colours. The well-known toy, the thaumatrope, depends on the same principle. In this a piece of cardboard is painted on one side, with a bird, for example, and on the other side with a cage: when the cardboard is twirled round very rapidly by means of a cord fixed at opposite points of its length, both bird and cage become visible at once, and the bird appears in the cage.

A still more ingenious application of this principle we owe to Plateau, who described it in 1833, under the title of the *phenakistiscope;* and also to Stampfer, who independently devised the same arrangement about the same time, and named it the *stroboscopic disc.* The reader may, at almost any toy-shop, purchase one of them, provided with a number of amusing figures; or he may easily construct for himself one which will exemplify the principle. He requires no other materials than a piece of cardboard, and his only tools may be a sharp penknife, a pair of compasses, and a flat ruler. Let him draw on his cardboard a circle of 8 in. diameter, and divide its circumference by eight equidistant points. From these radii should be drawn with the point of the compasses, and equal distances from the centre marked off upon them, to fix the centres of the small circles, which must all have exactly the same size (say, 1 in. in diameter) and be marked by a distinct line. In these are to be marked the

hand of a clock-face in the positions shown in Fig. 249; and finally, in the direction of the radii, narrow slips are to be cut out of the cardboard as shown. If a pin be put through the centre of the disc, attaching it thus to the flat end of a cork, so that it can freely rotate in its own plane, and the disc be turned rapidly round, as in Fig. 250, in front of a looking-glass, while the spectator looks through the slits, he will see the hand on the little dial apparently turning round, with rather a jerky movement it is true, somewhat like the dead-beat seconds-hand that is sometimes seen on clocks. The illusion is best when the slits are so narrow that only one of the several images is visible by reflection, namely, that which is adjacent to the slit. Thus, as the disc rotates,

FIG. 250.

each little circle is visible for an instant as the slit passes in front of the spectator's eye; and if the rotation be sufficiently rapid, the impression of the disc is permanent, as it is constantly being renewed by the successive circles, while, on the contrary, the hands, having different positions, pro-

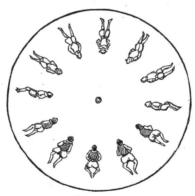

FIG. 251.

duce images in different positions, giving the appearance of a jerky rotation. The instruments sold in the shops have sometimes a thin metallic disc with the slits in it, and a series of designs printed in smaller paper discs. The paper discs may be screwed on the other disc as required, and a button on a pulley with an endless band is provided for producing the rotation more conveniently. Fig. 251 shows one of the pictures for a disc with twelve slits, and the effect produced by it is that of a dancing figure.

Another arrangement for showing the same illusion has lately become a very popular toy, and quite deservedly so, for it has the advantages of requiring no looking-glass, and of making the effect visible to a number of persons at the same time. This apparatus, which has been termed the Zoetrope, consists simply of a cylindrical box, like a drum with the upper end cut off. It is mounted on a pivot, which permits its revolving rapidly about its vertical axis when touched by the finger. The cylinder has a number of equidistant vertical slits round the upper part of its circumference. The figures which produce the illusion are printed on a slip of paper, which is placed in the lower part of the drum, and when this is in rapid rotation, and the figures are viewed through the slits, the illusion is produced in exactly the same manner as in the revolving disc.

At the end of the article on the phonograph in a subsequent page, the reader will find a remark as to the effect that might be produced by a combination of that instrument with instantaneous and simultaneous photographs of some famous speaker. This combination has now been accomplished by the great inventive genius to whom we are indebted for the phonograph. Mr. Edison has done this so effectively that he may be said to have given life to the *zoetrope* by the perfection in which the ocular illusion is produced together with the audible manifestations that keep time with it. The amount of thought and ingenuity expended on this new contrivance, which Edison has called the *kinetoscope*, will scarcely be appreciated by anyone who has not given some consideration to the many practical difficulties that have been overcome. No wonder that the announcement made at the beginning of 1892 should have been received with incredulity, for it was to the effect that Edison had contrived some happy combination of photography and electricity by which a man (presumably one who could afford to pay for luxuries) might sit in his own room and see the moving forms of the actors in an opera projected on a screen before his eyes, while at the same time he would hear their voices singing. Every movement, every change of expression, every glance of the eye, and, in fact, all that was visible to the spectator in front of the stage would appear on the screen, while not a note of vocalist, or chord of orchestra, would fail to reach the ear. And all this was to be evoked at will, and repeated as often as desired, not, therefore, of course, as a presentation of what was taking place at the time, but as a re-production of some previous performance. This wonderful result has virtually been attained by the application of delicate and ingenious machinery designed to make the phonograph and the camera work synchronously. The first part of the problem was the production of a succession of so-called instantaneous photographs at an extremely rapid rate. In the actual apparatus forty-six photographs are taken every second, a feat which would beforehand be thought impracticable. This is accomplished by making use of a band of sensitive celluloid film, which alone admits of being moved and stopped with the desired rapidity. The movement is imparted by an electric motor, and the arrangement is such that for each exposure the film is held stationary for $\frac{9}{10}$ths of $\frac{1}{46}$th of a second, during which the lens is uncovered, then for the remaining $\frac{1}{10}$th it is covered, while at the same time the film is jerked forward so as to expose a fresh surface to receive a new impression. Obviously the mass moved and stopped with this rapidity (which without the stoppages

FIG. 251a.—*Edison's Kinetographic Theatre.*

is at the rate of 26 miles an hour) must be small, and it is found that photographs about 1 in. in diameter cannot be much exceeded in view of this condition. The lens has to be entirely stopped or screened during the tenth of the short interval ($\frac{1}{460}$th of a second) in which the onward movement of the film is taking place, and it has to be practically open during the remaining $\frac{9}{10}$ths of the interval ($\frac{9}{460}$ths of one second) in which the film is held stationary in order to receive the photographic image. These alternations of movement and stoppage must take place with the utmost regularity, and Edison has used a beautifully regulated electro-motor as the active power, which also simultaneously moves a phonograph so that sights and sounds shall proceed in step, for it is thus they have to be reproduced. This is done by developing the band of film, and from it printing photographic *positives* on a similar band, whose images are successively projected on a screen by means of a lantern with a step by step movement, exactly the same as that by which the original photographs are taken, while the phonographic cylinder is so timed as to give off to a loud-speaking instrument the sounds that accompanied the photographs. A description of the ingenious mechanism by which all this is accomplished is not suitable for these pages, for it is the result, rather than the details of apparatuses, that interest the general reader. In a simpler form of kinetoscope the positive images on the band of film are viewed directly by single observers, each looking through magnifying glasses ; in this a disc with 46 slits revolves, and in its passage, as each slit momentarily permits a view of the image, an electric flash simultaneously lights it up. The same principle is, of course, used in the screen projections. From the very great number of impressions made on the eye in one second, there is none of that jerky movement that is observable in the older appliances. Mr. Edison has found it necessary to provide a special stage, or rather small theatre, in which the actors of the little dramas may be photographed with every advantage in the way of lighting, &c. Fig. 251*a* shows this kinetographic theatre with the electric camera in action. The subjects reproducible in the kinetoscope include the most rapid movements, such as quick dances, blacksmiths hammering on an anvil, &c., or incidents of ordinary life involving much gesture and change of facial expression, and nothing can be more amusing than to see all these shown to the life by the images on the screen, or by the pictures viewed through the lens, especially if at the same time the phonograph is made to emit the corresponding sounds.

FIG. 252.—*Portrait of Sir W. Thomson.*[*]

ELECTRICITY.

———◆———

ABOUT sixty years ago a popular book was published having for its theme the advantages which would flow from the general diffusion of scientific knowledge. Great prominence was, of course, given to the utility of science in its direct application to useful arts, and many scientific inventions conducing to the general well-being of society were duly enumerated. Under the head of electricity, however, the writer of that book mentioned but few cases in which this mysterious agent aided in the accomplishment of any useful end. The meagre list he gives of the instances in which he says "*even* electricity and galvanism might be rendered subservient to the operations of art," comprises only orreries and models of corn-mills and pumps turned by electricity, the designed splitting of a stone by lightning, and the suggestion of Davy that the upper sheathing of ships should be fastened with copper instead of iron nails, with a hint that the same principle might be extended in its application. At the present day the applications of electricity are so numerous and important, that even a brief account of them would more than fill the present volume. Electricity is the moving power of the most remarkable and distinguishing invention of the age—the telegraph ; it is the energy employed for ingeniously measuring small intervals of time in chronoscopes, for controlling time-pieces,

———————————————
* Now Lord Kelvin.

and for firing mines and torpedoes ; it is the handmaid of art in electroplating and in the reproduction of engraved plates, blocks, letterpress, and metal work ; it is the familiar spirit invoked by the chemist to effect marvellous transformations, combinations, and decompositions ; it is a therapeutic agent of the greatest value in the hands of the skilful physician. Such an extension of the practical applications of electricity as we have indicated implies a corresponding development of the science itself ; and, indeed, the history of electricity during the present century is a continuous record of brilliant discoveries made by men of rare and commanding genius —such as Davy, Ampère, and Faraday. To give a complete account of these discoveries would be to write a treatise on the science ; and although the subject is extremely attractive, we must pass over many discoveries which have a high scientific interest, and present to the reader so much of this recently developed science as will enable him to comprehend the principles of a few of its more striking applications.

The science of electricity presents some features which mark it with special characters as distinguished from other branches of knowledge. In mechanics and pneumatics and acoustics we have little difficulty in picturing in our minds the nature of the actions which are concerned in the phenomena. We can also extend ideas derived from ordinary experience to embrace the more recondite operations to which heat and light may be due, and, by conceptions of vibrating particles and undulatory ether, obtain a mental grasp of these subtile agents. But with regard to electricity no such conceptions have yet been framed—no hypothesis has yet been advanced which satisfactorily explains the inner nature of electrical action, or gives us a mental picture of any pulsations, rotations, or other motions of particles, material or ethereal, that may represent all the phenomena. Incapable as we are of framing a distinct conception of the real nature of electricity, there are few natural agents with whose ways we are so well acquainted as electricity. The *laws* of its action are as well known as those of gravitation, and they are far better known than those which govern chemical phenomena or the still more complex processes of organic life.

Definite as are the laws of electricity, there is no branch of natural or physical science on which the ideas of people in general are so vague. Spectators of the effects of this wonderful energy—as seen violently and destructively in the thunderstorm, and silently and harmlessly in the Aurora —knowing vaguely something of its powers in traversing the densest materials, in giving convulsive shocks, and in affecting substances of all kinds —the multitude regard electricity with a certain awe, and are always ready to attribute to its agency any effect which appears mysterious or inexplicable. The popular ignorance on this subject is largely taken advantage of by impostors and charlatans of every kind. Electric and magnetic nostrums of every form, electric elixirs, galvanic hair-washes, magnetized flannels, polarized tooth-brushes, and voltaic nightcaps appear to find a ready sale, which speaks unmistakably of the less than half-knowledge which is possessed by the public concerning even the elements of electrical science.

Electricity has also a special position with regard to its intimate connection with almost every other form of natural energy. Evolved by mechanical actions, by heat, by movements of magnets, and by chemical actions, it is capable in its turn of reproducing any of these. It plays an important, but as yet an undefined, part in the physiological actions constantly going on in the organized body, and is, in fact, all-pervading in its influence over all matter, organic and inorganic—a secret power strangely but universally concerned in all the operations of nature. We are compelled to regard electricity not as a kind of force acting upon otherwise inert matter, but rather as an affection or condition of which every kind of matter is capable, although we are still unable to form a conjecture of the precise nature of the action.

We have now to address ourselves to the task of unfolding so much of the science as will enable the reader to understand the leading principles of such important applications as electro-plating, illumination, and the telegraph ; and this will necessarily include an account of the grand discovery of the identity, or at least intimate connection, of magnetism and electricity.

ELEMENTARY PHENOMENA OF MAGNETISM AND ELECTRICITY.

THE distinctive property of a magnet is, as everybody knows, to attract pieces of iron, and this property having been observed by the ancients in a certain ore of iron which was found near the city of Magnesia, in Asia Minor, the property itself came to be called Magnetism. A bar of steel, if rubbed with the natural magnet or loadstone, acquires the same property, and if the bar be suspended horizontally or poised on a pivot, it will settle only in one definite direction, which in this country is nearly north and south. If a narrow magnetized bar be plunged into iron filings, it will be found that these are attracted chiefly by the ends of the bar, and not at all by the centre. It appears as if the magnetic power were concentrated in the extremities of the bar, and these are termed its poles, the pole at the end of the bar which points to the north is called the *north pole* of the magnet, and the other is named the *south pole*. If a north pole of one magnet be presented to the north pole of another, they will repel each other, and the same repulsion will take place between the south poles, whereas the north pole of one magnet attracts the south pole of another. In other words, poles of the same name repel each other, but poles of opposite names attract each other, or still more concisely, *like poles repel, unlike poles attract each other.*

Magnetism acts through intervening non-magnetic matter with undiminished energy. Thus, the attractions and repulsions of magnetic poles manifest themselves just as strongly when the poles are separated by a stratum of wood or stone as when merely air intervenes, and the attraction of small pieces of iron by a magnet takes place through the interposed palm of one's hand without diminution. A delicately suspended needle in even a remote apartment of a large building moves whenever a cart passes in the street. It is almost too well known to require mention here, that iron and steel are the only common substances which are capable of plainly exhibiting magnetic forces, and, indeed, there are no known substances capable of so powerful a magnetization as these. But the difference in the magnetic behaviour of iron and steel is not so well understood, and it is a point of importance for our subject, and connected with a fundamental law which governs all magnetic manifestations. A piece of pure iron is very readily cut with a file, whereas a piece of steel may be so hard that the file makes no impression upon it whatever; and hence a piece of pure iron, or rather iron holding no carbon in combination, and possessed of no steely quality, is often spoken of as *soft iron.* When a piece of soft iron is placed near the pole of a magnet, the iron becomes, for the time, a magnet. If iron filings be sprinkled over it, they will arrange themselves about the parts of the iron respectively nearest and farthest from the magnet, thus showing that the piece of soft iron has acquired magnetic poles. It will be found on examining these poles that the one nearest the magnet is of the contrary name to the pole of the magnet, and the farthest is of the same name. The conversion of the soft iron into a magnet by the influence of a magnetic pole is termed *induction.* It need hardly be said that the inductive effect is more powerful in proportion to the shortness of the distance separating the piece of soft iron from the magnetic pole, and, of course, the effect is at its maximum when there is actual contact. Induction thus explains, by aid of the law of the poles, the attraction which a magnet exercises over pieces of iron, for it is plain that the inductive influence is accompanied by attraction between the two contiguous oppositely-named poles of the magnet, and of the piece of iron. But attraction is not the only force, for the pole developed at the farthest portion of the piece of iron being of the same name as the inducing pole, these will be mutually repulsive. The attractive force will, however, be more powerful on account of the shorter distance at which it is exerted, and will predominate over the repulsive force, particularly at short distances, because then the difference will be relatively greater. At distances from the inducing pole relatively great to the distance between the two poles of the piece of iron, the difference may be so small that its effect in attracting the piece of soft iron will be imperceptible, and then the piece of iron acted on by two (nearly) equal parallel forces, will be subject to what is termed in mechanics a *couple,* the only effect of which is to turn the body into such a position that the opposing forces act along the same line. The definite direction assumed by a freely suspended needle may be explained by supposing that the earth itself is a magnet having a *south* pole in the *northern* hemisphere, and a *north* pole in the *southern* hemisphere, the line joining these poles being shorter than the axis of the earth, and not quite coinciding with it in position; and the fact of the needle being turned round but not bodily attracted is then easily accounted for, the attractive and repulsive forces being reduced to a *couple* in the manner just explained.

If the attempt be made to turn a piece of steel into a magnet, by the induction of a magnetic pole, the same results will be obtained as in the case of soft iron, but in a much feebler degree, and with this difference: the piece of steel does not lose its magnetism when the inducing magnet is withdrawn, whereas in the case of the soft iron every trace of magnetism vanishes the instant the inducing pole is removed. And if the pole of the magnet be not only put in contact with one end of the piece of steel, but rubbed on it, the piece will acquire permanent and powerful magnetism. Hence it will be noticed that a piece of soft iron can by the mere approximation of a magnetic pole be converted in an instant into a magnet, and by the removal of the magnet can as instantly be deprived of its magnetism, and made to revert into its ordinary condition; while steel is not so readily magnetized, but retains its magnetism permanently.

The elementary phenomena of electricity are extremely simple and easy of demonstration, and as the whole science rests upon inferences derived from these, the reader would do well to perform the following simple experiments for himself. Apparatus is represented in Fig. 253, but the only essential portion is a straw, B, suspended from any convenient support by a very fine filament of *white silk*. To one or both ends of the straw a little disc of gilt paper, or a small ball of elder-pith or of cork, should be attached, so that the straw may be balanced horizontally. Now rub on a piece of woollen cloth a bit of sealing-wax, or a stick of sulphur, or a piece of amber, or a penholder, paper-knife, or comb made of ebonite, and immediately present the substance to the ball at the end of the straw. It will be first attracted to the rubbed surface, but after coming into contact with it, repulsion will be manifested and the ball will separate, and may be chased round the circle by following it with the excited body. The attraction of light bodies by amber after it has been rubbed appears to be the one solitary electrical observation recorded by the ancients, but it has given its name to the science, ἐλεκτρον being the Greek name for amber. The cause, then, of this property is named *electricity*, and bodies which exhibit it are said to be *electrified*. The reader will remark that these words *explain* nothing : they are used merely to *express* a certain state of matter and the entirely unknown cause of that state. Let the pith or cork ball at the end of the straw be again charged with electricity, by bringing it into contact with a piece of sealing-wax or ebonite which has just been electrified by friction. In this condition it will, as we have just seen, be repelled by the substance which charged it, and on trial it will be found to be repelled also by all the substances we have named, after they have been excited by friction. But if, while still charged with the electricity communicated to it by contact

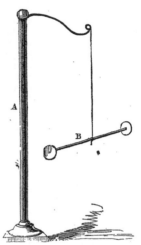

FIG. 253.—*A simple Electroscope.*

with sealing-wax, sulphur, ebonite, or amber, we present to it a warm and dry glass tube which has just been rubbed with dry silk, we shall find that the ball will be strongly attracted. After contact with the glass, repulsion will take place, and the ball will refuse again to come into contact with the excited glass. In this condition, however, it will be immediately attracted by rubbed sealing-wax or ebonite, and so on alternately : the ball when repelled by the wax is attracted by the glass, and when repelled by the glass is attracted by the wax.

These simple experiments prove that, whatever electricity may be, there are two kinds of it, or, at least, it manifests two opposite sets of forces. The electricity evolved by the friction of glass with silk was formerly called *vitreous* electricity, and that shown by excited resin, sealing-wax, amber, &c., was named *resinous* electricity. These names have now been respectively replaced by the terms *positive* and *negative*. It must be understood that these terms imply no actual excess or defect, but are purely distinguishing terms, just as we speak of the *up* and *down* line of a railway, without implying an inclination in one direction or the other. A fact of great importance in electrical theory is discovered when the substances in which electricity is developed are carefully examined: it is found that one kind is never produced without the other simultaneously appearing. Thus, the silk which has been used for rubbing the glass in the above experiments will be found to exhibit the same electricity as sealing-wax or ebonite. And, further, the *quantities* of positive and negative electricity evolved are always found to be equal, or equivalent to each other; that is, if they are put together they completely neutralize or destroy each other's effects. We have used the word " quantity," implying that electricity can be measured. No doubt, whatever electricity may be, there may be more or less of it; but can we measure an imponderable, invisible, impalpable thing, incapable of isolation? What we really measure when we say that we measure electricity is the attractive or repulsive force: we balance this against some other force (that of gravitation, for example), and we say, so much weight lifted represents so much electricity.

If we try to electrify a piece of metal by holding it in the hand and rubbing it against woollen cloth, silk, or other substance, we shall fail in the attempt: no signs of electricity will thus be shown by the metal. Hence bodies were formerly divided into two classes—those which could be electrified by friction, and those which could not. It was afterwards found, however, that there was no real ground for this division, but that, on the contrary, *no two bodies can be rubbed together, even if they are made of the same substances, without positive electricity appearing in one, and an equivalent quantity of negative electricity in the other.* The real difference between bodies which prevents the manifestation of electricity in many cases depends upon the fact that electricity is able to traverse some substances with great facility, while others prevent its passage. Thus, if we suspend horizontally a hempen cord by white silk attached to the ceiling, so that the hempen cord comes in contact with nothing but the silk, we shall find, on presenting a piece of excited ebonite to one end of the cord, that electric attraction of light bodies will be manifested at the other. If a silk cord be substituted for the hempen one, no such effect will be observed. The hemp is, therefore, said to be a *conductor*, and the silk a *non-conductor*. Again, if we substitute for one of the silk threads suspending the cord a piece of twine, or a wire, we shall fail to obtain any electric manifestations at the remote end, because the electricity will be carried off into the earth by the conducting powers of these substances. On the other hand, filaments of glass or ebonite may be used, instead of the silk, with the same effect: they do not allow the electricity to run through them to the ground, and are therefore termed, like the silk, *insulators* of electricity. The distinction of bodies into conductors on the one hand, and into non-conductors or insulators on the other, is of paramount importance in the science and in all its applications. This distinction, however, is not an absolute one: there is no substance so perfect an insulator that it will not permit any electricity to pass, and there is no conductor so perfect that it does not offer resistance to the passage. Substances may be arranged in a list which presents a gradation from the best conductor to the best insulator. The metals are by far the best conductors, but there is great relative diversity in their conductive power. Silver, copper, and gold are much the best conductors among the metals, iron offering eight times, and quicksilver fifty times, the resistance of silver. Coke, charcoal, aqueous solutions, water, vegetables, animals, and steam are all more or less conductors, while among the substances called insulators may be named, in order of increasing insulating power, india-rubber, porcelain, leather, paper, wool, silk, mica, glass, wax, sulphur, resins, amber, gum-lac, gutta-percha, and ebonite. It will now be obvious why the electricity developed by the friction of a piece of metal fails to manifest itself under ordinary circumstances, as, for instance, when held in the hand : the metal and the body being both conductors, the electricity escapes. But if the piece of metal be held by an insulating handle of glass or ebonite, the electrified condition may easily be observed.

THEORY OF ELECTRICITY.

THE few elementary facts which have been pointed out are absolutely necessary for the foundation of what is sometimes termed the theory of electricity, but which is properly no theory,—at least, not a theory in the same sense as gravitation is a theory explaining the motions cf the planets, or even in the sense in which the hypothesis of the ether and its movements explains the phenomena of light. It is absolutely necessary to have a conception of some kind which may serve to connect in our minds the various phenomena of electricity, if it were only to enable us the more easily to talk about them. In default of any supposition which will shadow forth what actually occurs in these phenomena, we have recourse to what has been aptly termed *a representative fiction :* we picture to ourselves the actions as due to *imaginary fluids*—fluids which we know *do not exist,* but are as much creations of the mind as Macbeth's air-drawn dagger; not, however, like his "false creation," proceeding from "the heat-oppressed brain," but intellectual fictions, consciously and designedly adopted for the purpose of enabling us the better to think of the facts, to readily co-ordinate them, and to express them in simple and convenient language. Non-scientific persons hearing this language usually mistake its purport, and imagine that the actual existence of an "electric fluid" is acknowledged. The accounts which appear in the newspapers of the damage done by thunder-storms are often amusing from the objectivity which the reporter attributes to the "electric fluid." It is described, perhaps, as "entering the building," "passing down the chimney," then "proceeding across the floor," "rushing down the gas-pipes," "forcing its way through a crevice, and then streaming down the wall," &c., in terms which imply the utmost confidence of belief in the existence of the "fluid." With this intimation that the hypothesis of electric fluids is merely, then, a "*façon de parler,*" the reader will not be misled by the following brief explanation of the elementary facts in the language of the theory.

In the natural state all bodies contain an indefinite quantity of an imponderable subtile matter, which may be called "neutral electric fluid." This fluid is formed by a combination of two different kinds of particles, positive and negative, which are present in equal quantities in bodies not electrified; but when there is in any body an excess of one kind of particles, that body is charged accordingly with positive or negative electricity. Both fluids traverse with the greatest rapidity certain substances termed *conductors ;* but they are retained amongst the molecules of *insulating* substances, which prevent their movement from point to point. When one body is rubbed against another, the neutral electric fluid is decomposed—the positive particles go to one body, the negative with which these positive particles were before united pass to the other body. The particles of the same name repel each other, but particles of opposite names attract each other ; and it is this attraction which is overcome when the electricities are separated by friction or in any other manner.

It will be observed that the above is nothing but the statement of the elementary facts in the language of the hypothesis. This system of the two fluids readily lends itself to the explanation of nearly all the phenomena presented in what is termed *static electricity*—that is, in those phenomena where the actions are conceivably due to a more or less permanent separation of the fluids. The grand discoveries in electricity turn, however, upon quite another condition, namely, one in which the two hypothetical fluids must be imagined as constantly combining, and here the utility of the hypothesis is less marked. Inasmuch, however, as there can be no doubt regarding the identity of the agent operating in the two sets of circumstances, the facts of *dynamical electricity* must still be expressed in the same language, with the aid of any additional conceptions which may give us more grasp of the subject.

ELECTRIC INDUCTION.

IN all electrical phenomena an inductive action occurs, which resembles that which we have already indicated with regard to magnetism. Thus, if we take an insulated metallic conductor in the uncharged state, and bring it near an electrified body, we shall find that the conductor, while still at a considerable distance, will give signs of an electrical charge. Suppose we have a cylindrical conductor, and that we present one end of it to the electrified body, but at such a distance that no spark shall pass, we shall find, if the charge on the electrified body be strong and the conductor be brought sufficiently near, that on bringing the finger near the insulated cylinder, a spark passes. While the cylinder continues in the same position with regard to the electrified body, no further sparks can be drawn from it ; but if the distance between the two bodies be increased, the insulated cylinder will be found to have another charge of electricity, which will again produce a spark. And by repeating these movements we may obtain as many sparks as we desire by these mechanical actions, without in the least drawing upon the charge on the original electrified body. The electrophorus is a device for obtaining electricity by this plan, and several rotatory electrical machines have lately been invented which yield large supplies of electricity by a similar inductive action.

It is found that in such a case as that we have above supposed, if the electrified body is charged with positive electricity, the uncharged conductor brought near it has its electricities separated—the negative attracted and held by the attraction of the positive charge in the parts of the cylinder nearest the inducing body ; while the corresponding quantity of positive electricity is driven towards the most remote parts of the insulated conductor. It is this last which gives the spark in the first case, and if it be not thus withdrawn from the conductor, it re-combines with the negative electricity when the conductor is withdrawn from the neighbourhood of the electrified body, and the conductor then reverts to the natural or unelectrified state.

But the contact of a conducting body with the conductor while it is under the influence of the electrified body withdraws only positive electricity, the negative—being held, as it were, by the attraction of the positive electricity of the charged body—is not thus removed, and in this condition it is sometimes called *disguised* or *dissimulated* electricity—a term the propriety of which is doubtful. The excess of negative "fluid" which the conductor thus acquires shows itself, however, only when the inducing body has been withdrawn. Precisely similar effects will take place, *mutatis mutandis*, if the electrified body has a negative charge. A demonstration of inductive effects is readily afforded in the action of the gold-leaf *electroscope*, Fig. 254, in which two strips of gold-leaf are suspended within a glass case from wire passing through the top, and terminated in a metal plate. This instru-

ment is often used for showing the existence of very small electric charges. Let a stick of sealing-wax be rubbed and held, say, a foot or more from the plate of the electroscope, the leaves will diverge with negative electricity. The sealing-wax being retained in the same position, touch the plate for an instant with the finger. This will remove the negative charge, but the positive electricity will be retained on the plate by the attraction of the negative of the sealing-wax. Now remove the sealing-wax, when the dissimulated charge will spread itself over the whole insulated metallic portion of the electroscope, and the leaves will diverge with a strong charge of *positive* electricity. If an excited glass tube is brought near the electroscope, the leaves will now diverge still more; if the sealing-wax is replaced in its former position, the leaves will collapse. In all these cases the electrified body parts with none of its own electricity by developing electrical effects in the neighbouring bodies.

The inductive actions we have described take place through the air, which is a non-conductor, and such actions may be made to take place through any other non-conductor. With solid non-conductors, such as glass, gutta-percha, &c., the inducing body may be brought very near to

FIG. 254.
The Gold-leaf Electro-scope.

the conductor on which it is to act; for the intervening solid substance, or *dielectric*, as it has been appropriately called, opposes a resistance to the combination of the opposite electricities, and the inductive effects are greatly intensified by the approximation. Faraday discovered that the amount of inductive action with a given charge is also dependent upon the nature of the dielectric, and that the electric forces act upon the particles of the dielectric, circumstances which are of the greatest importance, as we shall presently find, in practical telegraphy. The most familiar instance of induction is probably well known to the reader in the Leyden jar, Fig. 255, which is simply a wide-mouthed bottle of thin glass, covered internally and externally with tin-foil to within a few inches of the neck. The inner coating communicates by means of a rod and chain with a brass knob. Such a jar admits of the accumulation of a larger quantity of electricity than the conductor of a machine will retain. A very few turns of the machine will suffice usually to charge the conductor to the fullest extent; but if it be put in communication with the knob of a jar, a great many more turns will be required to attain the same charge in the conductor, and the excess of electricity represented by these additional turns will have accumulated within the jar——an effect due to the "dissimulated" electricity of its exterior.

Everybody knows the result when a metallic communication is established between the exterior and the interior of a charged Leyden jar. There is a very bright spark, a snap, and the jar is "discharged." Everybody knows, also, the sensation experienced when his body takes the place of the metallic communication, or forms part of the circuit through which the communication takes place. Everybody knows that the shock then felt may also be experienced at the same moment by any number of persons who join hands, under such conditions that they also form a part of the line of communication. Such facts irresistibly suggest the notion of something passing through the whole chain, and this notion is in perfect harmony with the

FIG. 255.—*The Leyden Jar.*

hypothesis of the " fluids," for we have only to suppose that it is one or both of these which rush through the circuit the instant the line of communication is complete. It was one of Franklin's discoveries that the electrical charges of the Leyden jar do not reside in the metallic coatings; for he made a jar with removable inside and outside coatings, which, properly taken from the glass, showed no signs of electrification, yet when replaced the jar was found to be again highly charged. This would seem to show that the charge clings to, or penetrates within, the glass.

DYNAMICAL ELECTRICITY.

L ET us take a vessel containing water, to which some sulphuric acid has been added, Fig. 256, and in the liquid plunge a plate of copper, C, and a plate of *pure* zinc, Z, keeping the plates apart from each other. As it is not easy to obtain zinc perfectly free from admixture of other metals, an artifice is commonly resorted to for obtaining a surface of pure metal, by rubbing a plate of the ordinary metal with quicksilver, which readily dissolves pure zinc, but is without action on the iron and other metals with which the zinc is contaminated, while the quicksilver is not acted upon by the diluted acid, but is merely the vehicle by which the pure zinc is presented to the liquid. Under the conditions we have described, no action will be perceived, no gas will be given off, nor will the zinc dissolve in the acid. If the electrical condition of the portion of the copper-plate which is out of the liquid be examined by means of a *delicate* electroscope, it will be found to possess a very weak charge of *positive* electricity, and a similar examination of the zinc plate will show the existence on it of a feeble charge of *negative* electricity. If the two plates be made to touch

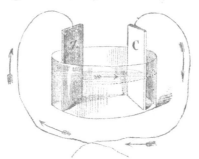

FIG. 256.—*A Voltaic Element.*

each other, or if a wire be attached to each plate, as shown in the figure, and the wires be brought into contact outside of the vessel, an action in the liquid is immediately perceptible at the surface of the *copper* plate, when a multitude of small bubbles of hydrogen gas will at once make their appearance, and the gas will be given off continuously from the copper plate so long as there is metallic contact through the wires, or otherwise, between the two plates, or until the acid is saturated with zinc—for in this action the zinc is dissolving, and, in consequence, liberating hydrogen, which strangely makes it appearance, not at the place where the chemical action really occurs, namely, at the surface of the zinc which is in contact with the acid, but at the surface of the copper which is not acted upon by the acid.

It is known that when we establish a metallic communication between two bodies charged with equivalent quantities of positive and negative electricities respectively, these combine and neutralize each other, and all signs of electricity vanish. It is obvious that the contact of the two wires has this effect, as the signs of electric charge which were before discoverable in each of the plates are no longer found while the wires are in contact. But the charges reappear the instant the contact is broken, the chemical action ceasing at the same time. If the wire connecting the two plates outside of the vessel be carefully examined, it will be found, so long as the chemical action is going on, to be endowed with new and very remarkable properties. If this wire be stretched horizontally over a freely suspended magnetic needle, and parallel to it, the needle will be deflected from its position, and, if the wire be placed very near it, will point nearly east and west, instead of north and south. Now, this effect is produced by any part whatever of the wire, and it instantly ceases if the wire be cut at any point. These facts at once suggest the idea of its being due to something flowing through the wire, so long as metallic continuity is preserved. This idea is much strengthened when we find that the action of the connecting wire upon the magnetic needle is quite definite—or, in other words, there are indications which correspond with the notion of direction. For when the wire, which we shall still suppose to be stretched horizontally *above* the needle and parallel to its direction, is so connected with the plates immersed in the acid that the portion which approaches the south-pointing pole of the needle proceeds from the copper plate, while the portion above the north pole is in connection with the zinc plate, then the north end of the needle will always be deflected towards the west—whereas, if the connections be made in the contrary manner, the deflection will be in the opposite direction ; and if the wire be below the needle, the contrary deflections will be observed with the same connections. The discovery of the action of such a wire on the magnetic needle was made by Œrsted in 1819, and it is a discovery remarkable for the wonderful extent of the field which it opened out, both in the region of pure science and in that of practical utility.

Since by such experiments as those just mentioned the notion of a *current* is arrived at, the mind recurs to the fiction of the " fluids," and pictures the " positive fluid " as rushing in one direction, and the " negative fluid " in the other, to seek a re-combination into " neutral fluid." But we must never lose sight of the fact that these ideas are consciously adopted as representative fictions to help our thoughts—just as John Doe and Richard Roe, imaginary parties to an imaginary lawsuit, used to be named in legal documents, in order to explain the nature of the proceedings. Failing, then, to

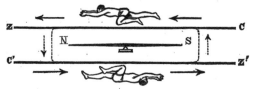

FIG. 257.—*Ampère's Rule.*

find anything really flowing along our wire, it is still absolutely necessary, seeing there is something definite in its action, to assign a direction to the supposed current ; and it has been agreed that we shall represent the current as flowing from the positively charged body to the negatively charged body—that is, in the case we have been considering, from the copper to the zinc through the wire. When this conventional representation has been adopted, the action on a magnetic needle can easily be defined and remembered by an artifice proposed by Ampère. In Fig. 257, let N S represent the magnetized needle, N being the pole which points towards the north, and S the south pole. Let C be the end of the wire connected with the copper plate, and Z that connected with the zinc. The current is therefore supposed to flow in the direction indicated by the arrows in a wire above the needle and in the wire placed below. Now, suppose that a man is swimming in the current in the same direction it is flowing, *and with his face towards the needle, then the north pole of the needle will always be deflected towards his left.* With the direction of current represented in the figure, the pole, N, will be thrown forward from the plane of the paper, or towards the spectator.

The reader who desires to study the mutual action of currents and magnets will find it necessary to fix this idea in his mind. He will now be able to see that if the wire be coiled round the needle, as shown by the lines and arrows, Fig. 257, so that the same current may circulate in reverse directions above and below the magnet, its effects in deviating the needle will everywhere concur—that is, the action of each part will be to turn the north pole towards the left. It is, therefore, plain that if the wire conveying the current be passed several times round the magnetic needle, the deflecting force will be increased; and a current, which would, by merely passing above or below the magnet, produce no marked deflection, might be made to produce a considerable effect if carried many times round it. The arrangement for this purpose is shown in Fig. 258, where it will be perceived that the needle is surrounded by a coil of wire, so that the current circulates many times about it, and the effects of each part of the circuit concur in deflecting the needle. Such an arrangement of the wire and needle constitutes what is called the *galvanometer*, an instrument used to discover the existence and direction of electric currents.

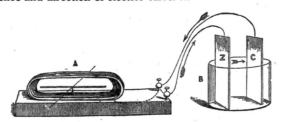

FIG. 258.—*Galvanometer.*

The arrangement of metals and acid which we have described is termed a *voltaic couple, element*, or *cell;* and a great controversy has long been carried on among men of science as to the place at which the development of electricity has its origin. Three-quarters of a century ago, the effect was attributed by Volta to the mere contact of the two dissimilar metals. In the experiment we have described this contact, supposing the wires to be of copper, would occur at the junction of the wire and the zinc plate. Now, by joining the copper plate of such a cell to the zinc plate of another cell, the copper of that to the zinc of a third, and so on, it is evident that the number of dissimilar contacts might be indefinitely increased, and the electric power should be proportionately augmented. It is found that this is really the case, but Volta's explanation has been opposed by another which regards the chemical action in the cells as the real origin of the electric manifestations. This last explanation, supported by many apparently conclusive experiments of Faraday and others, has been generally accepted. Galvanic batteries—as a series of cells joined together in a certain manner are termed—have been constructed, in which there is no contact of dissimilar metals; and no electric *current* can be obtained from an apparatus in which no chemical action takes place. The contact theory in a modified form has recently been revived by Sir W. Thomson and others. In this it is now maintained that some *separation* of electricities really does take place by contact of dissimilar *substances*, but that a *current* can be produced only when this separation is continually renewed by chemical actions. Be the true explanation what it may, the fact is undoubted that by joining cell to cell, we can really obtain vastly more powerful effects. If we take a single cell, such as that represented in Fig. 256, and connect the plates with a long and thin wire, we shall find that the current flowing through each part of the circuit is much weaker than when we connect the plates with a short and thick wire. In other words, the action in the latter case, when the wire is stretched over a magnetic needle, will be more powerful than in the former. By using a long and thin wire the current may be so weakened that it becomes necessary to surround the needle with many coils of the wire to produce a marked deflection. Again, much depends upon the material; thus a copper wire conveys a much more powerful current than a German silver one of the same dimensions. There thus appears to be a certain analogy between the flow of electricity along conductors to that of water through pipes. The longer and narrower are the pipes, the less is the quantity of water forced through them by a given head; and similarly, the resistance to the passage of a current increases with the length and narrowness of the conducting wire. When all other circumstances are the same, the *electrical resistance* of a conductor varies directly as its length and inversely as its sectional area. Hence the current flowing in the apparatus represented in Fig. 256 would be increased by making the wire thicker, and by making it shorter by bringing z and c nearer together, and by making the area they expose to the liquid larger; for in the liquid also the current flows as indicated by the arrow, a fact which may be proved by the deflection of a magnetized needle suspended above the vessel. The magnitude of the current depends, then, upon two opposing forces, namely, that which continuously separates the electricities, or drives them apart to recombine through the circuit, and that which opposes their passage. The former, which is termed the *electromotive force*, originates, according to some, from the mere contact of dissimilar materials, according to others from the chemical action. Now, we may increase the strength of the current in a given arrangement, either by increasing the electromotive force, or by diminishing the resistance. The increase of the strength of the current, produced by merely pouring more acid into the vessel, Fig. 256, is due, according to the chemical theory, to the former cause; according to the contact theory, to the latter. By multiplying the cells we increase the electromotive forces: the current receives, so to speak, an onward shove in each cell, but with each cell we introduce an additional resistance. Hence, it follows, that when the resistance of the circuit outside of the cells is extremely small, the current produced by a single cell is as powerful as that produced by a thousand. But when the external resistance is great, as when long thin wires are used, the united electromotive forces of a number of cells are needed to drive the current through the circuit. The strength of a current, c, is therefore expressible by the following simple formula, in which r stands for the internal resistance, and e for the electromotive force in each cell; n represents the number of cells in the battery, these being supposed exactly similar in every respect; R is the sum of the resistances in the circuit outside of the battery.

$$c = \frac{ne}{nr + R}$$

It is easily seen that the smaller R is made, the more nearly does the strength of the current become independent of the number of cells.

But many modifications have been made in the materials and form of the cells, by which greater power and duration of action have been attained. Our space permits a description of only two forms, and these must be described without a discussion of the principles upon which their increased efficiency depends.   Daniell's constant cell is represented in Fig. 259, where D is a battery of ten such cells, A is a cylindrical vessel of copper, C is a tube of porous earthenware, closed at the bottom, and within it is suspended the solid rod of amalgamated zinc, B.   The copper vessel and the zinc rod

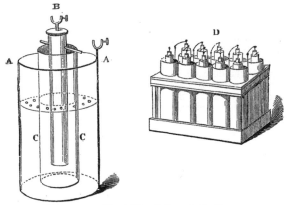

FIG. 259.—*Daniell's Cell and Battery.*

are provided with screws by which wires may be attached.   In the copper vessel is placed a saturated solution of sulphate of copper, and some crystals of the same substance are placed on the perforated shelf within the vessel. The porous tube is filled with diluted sulphuric acid.   When the battery

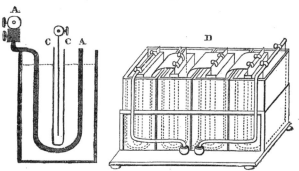

FIG. 260.—*Grove's Cell and Battery.*

is in action the zinc is dissolved by the sulphuric acid, and metallic copper is continually deposited upon the internal surface of the copper vessel. Daniell's battery, in some form or other, is much used for telegraphs and for electrotyping.   Grove's cell is shown in section in Fig. 260.   The external vessel is made of a rectangular form in glazed earthenware or glass. It contains a thick plate of amalgamated zinc, A, A, bent upwards, and between the two portions a flat porous cell, C, C, is placed, filled with strong nitric acid, in which is immersed a thin sheet of platinum. The outside vessel is charged with water, mixed with about ⅛th of sulphuric acid. D represents a battery of four such cells, in which the mode of connecting the platinum of one to the zinc of the next may be noticed. The terminal platinum and zinc form the *poles* of the battery, and to them the wires are attached which convey the current. The substitution of plates of coke for the platinum gives the form of battery known as Bunsen's, which is also sometimes made with circular cells. Gover's and Bunsen's are much more powerful arrangements than Daniell's, but the latter has the advantage as regards the duration and uniformity of its action.

When the current produced by a battery of a dozen or more such cells is conveyed by a wire, it is observed that this wire becomes sensibly hot, and, if the wire be thin enough, the heat may be sufficiently great to heat

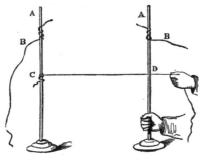

FIG. 261.—*Wire ignited by Electricity.*

the wire to redness. By stretching a piece of platinum wire between two separate rods which convey the current, as represented in Fig. 261, the length of wire through which the current passes may be adjusted so as to give any required amount of light, and the wire may even be heated to the fusing-point of platinum. This property of electricity has some interesting applications, as, for example, in firing mines and other explosive charges, and in some surgical operations. A still more interesting exhibition of heating and luminous effects is observed when the terminals of a battery of many cells are connected with two rods of coke, or gas-retort carbon. When the pointed ends of the rods are brought into contact, the current passes, and the points begin to glow with an intensely bright light, and if they are then separated from each other by an interval of $\frac{1}{10}$th of an inch or more, according to the power of the battery, a luminous arc extends between them, emitting so intense a light that the unprotected eye can hardly support it. This luminous arc is called the *voltaic arc*, and it excels all other artificial lights in brilliancy, a fact due to the extremely high temperature to which the carbon particles are heated, the temperature being, perhaps, the highest we can attain. It must not be supposed that in this brilliant light we see electricity : the light is due to the same cause as the light of a candle or gas flame, namely, incandescent particles of solid carbon. These particles are carried from one carbon point to the other, and it is found that the positive pole rapidly loses its substance, which is partly deposited on the negative pole.   But in order to obtain a steady light, it is requisite to keep the pieces of carbon at one invariable distance; and therefore the transference of the material from one pole to the other, and the loss by combustion, must be compensated by a slow movement of the carbons towards each other.   Several kinds of apparatus are used for this purpose, but they all depend upon the principle of regulating the motions by the action of an electro-magnet, formed by the current itself, which becomes weaker as the carbons are farther apart.   The movement is communicated to the apparatus by clockwork.   Duboscq's electric

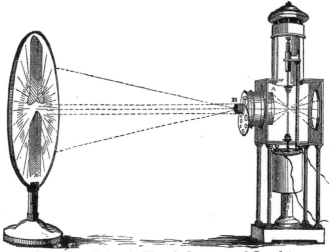

FIG. 262.—*Duboscq's Electric Lantern and Regulator.*

lantern is shown in Fig. 262, with enlarged images of the carbon points projected on a screen.   The mechanism of the regulator is contained within the cylindrical box immediately below the lantern.   The supports of both carbons are moved ; that which bears the positive carbon pole being advanced twice as fast as the other, and thus the light is maintained at the same level, for the positive carbon wears away twice as fast as the other. The light is more brilliant when charcoal is used instead of coke, but then it is necessary to operate in a vacuum, to avoid the combustion of the charcoal.   The voltaic arc has recently been applied to illuminate lighthouses, and for other purposes, and will probably soon be more widely employed, for a cheap and convenient mode of producing a uniform current of electricity has recently been discovered and will be presently described.

The current which is maintained by the chemical action taking place in the cells of the battery can also be made to do chemical work outside of the battery. When the poles of the battery are terminated by wires or plates of platinum, and these are plunged into water acidulated with sulphuric acid, bubbles of gas are seen to rise rapidly from each wire, or *electrode*, as it is termed. Fig. 263 shows an arrangement by which these gases may be collected separately, and examined, by simply placing over each electrode an inverted glass tube, filled also with the acidulated water. The gases collect at the tops of the tubes, displacing the water, and it is found that from the wire connected with the zinc end of the battery, or negative electrode, hydrogen gas is given off, while at the positive electrode oxygen gas is liberated, in volume precisely equal to half that of the hydrogen. This being the proportion in which these two substances combine to produce water, it appears that in the passage of the current a certain quantity of water is decomposed; and the quantity thus decomposed is in reality a

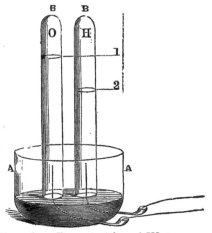

FIG. 263.—*Decomposition of Water.*

measure of the current, all the other effects of which are found to be proportional to this. When the electricity in a current is said to be measured, it is simply the power of the current to deflect a magnet, or the quantity of gas it can liberate, or some other such effect, which is in fact measured. The discharge of a Leyden jar through such an apparatus as that represented in Fig. 263 would present no perceptible decomposition of the water; yet such a discharge passed through the arms and body produces, as everybody knows, a painful shock, and is accompanied by a bright spark and a noise, while the simultaneous contact of the fingers with the positive and negative poles of the galvanic battery occasions neither shock nor spark. Thousands of discharges from large jars must be passed through acidulated water to liberate the amount of gas which a battery current of a second's duration will produce. The electricity of the jar is often spoken about as having a higher *tension* than that of the battery, but the latter sets an immensely greater quantity of electricity in motion. The idea may be illustrated thus: Suppose we have a small cistern of water placed at a great height, and that this water could fall to the ground in one mass. The fall of the small quantity from a great height would be capable of producing very marked instantaneous effects, such as smashing, as with a blow, any structure upon which it might fall. This would correspond with the small quantity of electricity which passes in the discharge of a Leyden jar. Contrast this with the case in which we allow a very large quantity of water to descend from a very small height—as when the water of a reservoir is flowing down a gently inclined channel. It is plain that a different kind of effect might be produced in this case; the current might be made, for instance, to turn a water-wheel, which the more forcible impact of the small quantity of water in the case first supposed would have broken into pieces.

It is probable that the apparent decomposition of water by the electric current is in reality a secondary effect, and that it is the sulphuric acid which is decomposed. When, instead of acidulated water, we place in the apparatus a solution of sulphate of copper, it is found that metallic copper is deposited on the negative electrode, and sulphuric acid collects at the positive electrode. The metal is deposited in a firm and coherent state, and the useful applications of this deposition of metals are of great interest and importance. For, in a similar manner, gold, silver, lead, zinc, and other metals may be made to form thin uniform layers over any properly prepared surface. The immense advantages which the arts have derived from electro-plating illustrate in a convincing manner the benefits which physical science can confer on society at large.

The process of electro-plating may be practised by the aid of apparatus of very simple character. Fig. 264 shows all that is necessary for obtaining perfect casts in copper of seals, small medals, &c. A A is a section of a common tumbler; B B is a tube, made by rolling some brown paper round a ruler, uniting the edge with sealing-wax, and closing the bottom by a plug of cork, round which the paper may be tied by a string, or in any other convenient manner. The tumbler contains a solution of sulphate of copper, and the tube is filled with water, to which about one-twentieth of its bulk of sulphuric acid has been added. A strip of *amalgamated* zinc, or a piece of thick amalgamated zinc wire, is placed in the tube, and a piece of copper bell-wire is twisted round the top of it, and has attached to its other extremity, and immersed in the copper solution, the article which is to be covered with copper. We may suppose that this is to be a cast in white wax or in plaster of one side of a medal. The cast is carefully covered with black lead by means of a soft brush, and the copper wire is inserted in such a manner as to be in contact with the black

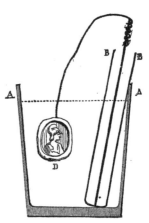

FIG. 264.—*Electro-plating.*

lead at some part. When the apparatus has been left for some hours in the position represented, a deposit of copper will be found over the black-leaded surface, and it will be a perfect impression of the wax cast.

Such a copper cast, or any article in copper having a perfectly clean surface, can be readily covered by a film of silver by means of a similar arrangement, where a solution of cyanide of potassium, in which some chloride of silver has been dissolved, is made to take the place of the sulphate of copper. Electro-plating with the precious metals has become a commercial industry of great importance; and this process has completely superseded the old plan of covering the metallic article to be plated with an amalgam of silver or of gold, and then exposing it to heat, which volatized the mercury, leaving a thin film of gold or of silver adhering to the baser metal. On the large scale a battery of several cells is used for electroplating, and the articles are immersed in the metallic solutions as the negative poles of the battery; any required thickness of deposit being given according to the length of the time they remain. At the works of Messrs. Elkington, of Birmingham, these operations are conducted on a grand scale. The liquid there employed for silvering is a solution of cyanide of silver in cyanide of potassium, and the positive pole is formed of a plate of silver, which dissolves in proportion as the metal is deposited on the negative pole. As the charging of batteries is a troublesome operation, and their action is liable to variations which affect the strength of the currents, the more uniform, more convenient, and more economical mode of producing currents by magneto-electricity, which will presently be described, has been to a great extent substituted for the voltaic battery.

The wire conveying a current not only affects a magnetic needle in the

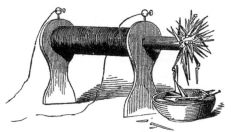

FIG. 265.—*A Current producing a Magnet.*

manner already described, but itself possesses magnetic properties, of which, indeed, its action on the needle is the result and the indication. If such a wire be plunged into iron filings, it will be found that the filings are attracted by it: they cling in a layer of uniform thickness round its whole circumference and along its whole length, and the moment the connection with the battery is broken they drop off. This experiment shows that every part of the wire conveying a current is magnetic, and it may be proved that the action is not intercepted by the interposition of any non-magnetic material. Thus the action of the wire upon the magnetic needle takes place equally well through glass, copper, lead, or wood. Consequently, if we cover the wire with a layer of gutta-percha, or over-spin it with silk or cotton, we shall obtain like results on our filings, and if we coil the covered wire round a bar of iron, while the non-conducting covering of the wire will compel the current to circulate through all the turns of the coil, it will not interfere with the magnetic action on each particle of the bar. Whenever this is done it is found that the iron is converted into a powerful magnet so long as the current passes. Fig. 265 represents in a striking manner the result when the current is made to circulate through numerous convolutions of the wire; and as each turn adds its effect to that of the rest, magnets of enormous strength may be formed by sufficiently increasing the number of the turns. The end of the iron bar is shown projecting from the axis of the coil, and below it is placed a shallow wooden bowl, containing a number of small iron nails. The instant the battery connection is completed these nails leap up to the magnetic pole, and group themselves round it in the manner shown in the cut; and again, when the current is interrupted, the iron reverts to its ordinary condition, the magnetism vanishes, and the nails drop down in an instant. These effects may be produced again and again, as often as the current flows and is broken. A magnet so produced is called an *electro-magnet,* to distinguish it from the ordinary permanent steel magnets. By coiling the conducting wire round

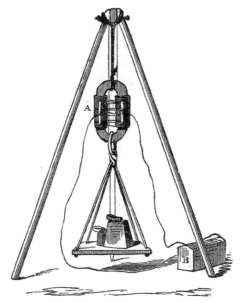

FIG. 266.—*An Electro-magnet.*

a bar of iron which has been bent into the form of a horse-shoe, very powerful magnets may be produced, and enormous weights may be supported by the force of the magnetic attraction so evoked. Fig. 266 represents the apparatus for experiments of this kind, in which weights exceeding a ton can be sustained.

Here, then, we have a striking instance of the subtile agent electricity, evoked by the contact of a few pieces of zinc with dilute acid, showing itself capable of exerting an enormous mechanical force. Engines have been constructed in which this force is turned to account to produce rotatory motion as a source of power. Such engines have certain advantages for special purposes; but the money cost for expenditure of material for power so obtained is, at least, sixty times greater than in the case of the steam engine. It is, however, in producing mechanical effects at a distance that the electric current finds the most interesting practical application of its magnetic properties. These are the actions which are so extensively utilized in the construction of telegraphic instruments, of clocks regulated by electric communication with a standard time-keeper, and of many ingenious self-registering instruments. The telegraph will be described in the next article, and we shall also have occasion in subsequent articles to describe some of the other applications of electro-magnetic and electro-chemical force.

INDUCED CURRENTS.

THESE very remarkable phenomena were discovered by the illustrious Faraday, in 1830, and this discovery, and that of magneto-electricity, may be ranked among the most memorable of his many brilliant contributions to electric science. Let two wires be stretched parallel and very near to each other, but not in contact. Let the extremities of one wire, which we shall term A, be connected with a galvanometer (page 415), so that the existence of any current through the wire may be instantly indicated. Let the two extremities of the other wire, B, be put into connection with the poles of a battery. The moment the connection is complete, and the battery current *begins* to rush through B, a deflection of the galvanometer needle will be observed, indicating a current of very short duration through A in the opposite direction to the battery current through B. This induced current, which is called the *secondary* current, does not continue to flow through A: it occurs merely at the time the *primary* or battery current is established; and though the latter continues to flow through the wire, B, no further effect is produced in the other wire. When, however, the battery connection is broken, and the primary current ceases to flow, at that instant there is set up in the wire, A, another momentary secondary current, but this one is in the *same* direction as the battery current. This is termed the *direct secondary* current, in opposition to the former, which is called the *inverse* current.

These effects are much more powerful when, instead of lengths of straight wire, or single circles of wires, we use two coils of wire, one of which, namely, that which conveys the primary currents, is placed in the axis of the other. It must be distinctly understood that the secondary currents are of momentary duration only; they are not produced at all while the battery *is flowing*, but only at the time of its commencement and cessation. If, however, we make the primary coil so that it can be slid in and out of the axis of the other, then while the primary current is continuously flowing, we can produce secondary currents in the other coil, by causing the coils to approach or recede from each other. As we bring the coils near each other, and slide the primary into the secondary, the current in the latter is *inverse;* when the one coil is receding from the other, it is *direct*. These mechanical actions are not produced without expenditure of force, for the approaching coils repel each other and the receding coils attract each other. The setting up of the battery current in the primary coil when placed within the other is equivalent to bringing it, with the current flowing, from an immense distance in an extremely small time. Similarly, when the battery current is broken, it is equivalent to an instantaneous recession. The effects, therefore, are proportionately powerful. It is found, also, and this we shall presently refer to more fully, that when, instead of the primary coil, a magnet is similarly moved into, or removed from, the axis of the secondary coil, currents in opposite directions are set up in the latter without any battery being used at all. The direction of these currents is the same as would be produced by a primary current that would form, in a piece of iron placed in the axis of the coil, an electro-magnet with poles similarly situated to those of the magnet so introduced or withdrawn. Hence, by placing a bar of soft iron in the axis of the primary coil, the secondary currents will be produced with increased force. When a long secondary coil, having the turns of its wire well insulated from each other, surrounds a primary coil provided with a core of soft iron, or still better, with a bundle of annealed iron wires, a series of powerful discharges, like those of a Leyden jar, may be obtained between the terminals of the secondary coil, when the battery contact is made and broken in rapid succession.

Such induction coils have been very carefully and skilfully constructed by

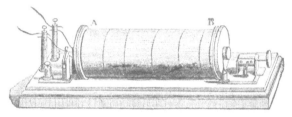

FIG. 267.—*Ruhmkorff's Coil.*

Ruhmkorff, and are therefore often called "Ruhmkorff's Coils." One of these is represented in Fig. 267. A B is the coil, and the apparatus is provided with what is termed a *condenser*, which consists of layers of tin-foil placed between sheets of thick paper, and alternately connected so that one set communicates with one extremity of the primary coil, and the other with the other. This condenser is conveniently contained in the wooden base of the instrument. Its introduction has greatly increased the intensity of the secondary current, and sparks of 18 in. or 20 in. in length have been obtained in the place of very short ones.

It should be stated that of the two secondary currents, only one has sufficient intensity to traverse the secondary circuit when there is any break in its continuity. This is the *direct secondary current*, or that which is produced on breaking the primary circuit. The reason is that the commencing current in the primary circuit induces in the spires of its own coil an inverse current, and the battery current therefore attains its full strength gradually, but still in a very short time ; while, on the cessation of the battery current, the same induction sends a wave of electricity through the primary coil in the same direction, and then the current ceases abruptly. Consequently, in the latter case, the induced electricity of the secondary coil is set in motion in much less time, and therefore possesses much greater intensity.

The magnetism of the iron core is usually made use of to break and make the current, by the attraction of a piece of iron attached to a spring, which, by moving towards the end of the core, separates from a point in connection with the battery, and, the current no longer flowing, the magnetism ceases, and the spring again brings back the iron and renews the contact.

By means of such coils many surprising effects have been produced. Perhaps one of the most beautiful experiments in the whole range of physical science is made by causing the discharges of the secondary coil to take

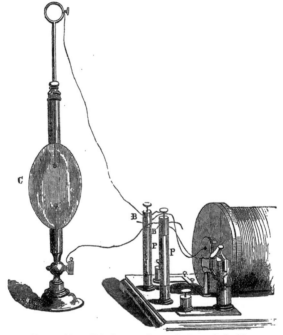

FIG. 268.—*Discharge through Rarefied Air.*

place through an exhausted vessel in the manner represented in Fig. 268. A beautiful light fills the interior of the vessel, and the terminals appear to glow with a strange radiance—one being surrounded with a kind of blue halo and another with a red. On reversing the direction of the currents, which is done by the little apparatus at the right-hand end of the coil in Fig. 267, the blue and the red radiance change places. Beautiful flashes of light may also be made to appear in the vessel, having the most marked resemblance to the streamers of the Aurora Borealis. When, instead of vessels almost free from common air, we repeat the experiment with tubes containing an extremely small residue of some other gas, such as hydrogen carbonic acid, &c., the colour of the light and other appearances change

Fɪɢ. 268a.—*Large Induction Coil at the old Polytechnic Institution, London.*

Geissler's tubes have already been spoken of in connection with the spectroscope ; but, independently of that, the various beautiful appearances which such tubes have been made to present, by the introduction of fluorescent substances and other devices, render the induction coil an instrument of the highest interest to the scientific amateur. Then there are striking physiological and other effects which the coil is capable of producing. For instance, we are able by its instrumentality to produce from atmospheric air unlimited quantities of that singular modification of oxygen which is called *ozone.* The electricity of the coil has been used for firing mines, torpedoes and cannons, and for lighting the gas-burners of large buildings.

The late Mr. Apps, who was well known as a skilful constructor of scientific apparatus, devoted much attention to improving the induction coil, and he made a very large one for the Polytechnic Institution in Regent Street, London, which Institution was at that time the home of popular science, under the direction of Mr. Pepper. This coil is represented in Fig. 268*a*, surrounded by the somewhat scenic accessories which were then supposed to be required for making science attractive to the multitude. Externally, the coil appeared as a cylinder, nearly 5 feet long and 20 inches in diameter. From each end projected smaller cylinders. All these and also the two upright pillars upon which the apparatus was supported were covered with ebonite. The large cylinder contained the primary coil, which was made of copper wire one-tenth of an inch in diameter and 3,770 yards long, covered with cotton thread, and making about 6,000 turns round the central core. This primary coil was inclosed in an ebonite tube $\frac{1}{2}$-inch thick, and outside of the tube, occupying 4 feet 2 inches of its length, was the secondary coil, containing 150 miles of silk covered wire, ·015 inch diameter, and very carefully arranged for insulation, so as to resist the tension of the electricity when the coil was in action. The condenser contained 750 square feet of tin-foil, and 40 Bunsen cells supplied the current for the primary coil. The power of this instrument was very great, for it would give a spark through the air of more than two feet in length, and the discharge could perforate a certain thickness of glass. It would charge a battery of Leyden jars having 40 square feet of tin-foil by only three breaks of contact in the primary circuit, so that the discharge would deflagrate considerable lengths of wire. The appearance of the spark, with this, as with other large induction coils, may be described as a thick line of light, surrounded by a reddish halo of less brilliancy, and this halo, unlike the line of the spark, had a sensible duration. The reddish glow might be blown aside by a current of air when a series of discharges was taking place, and partly separated from the denser looking line of light. The latter is no doubt formed by intensely heated particles of the metals between which the discharge takes place, while the former is probably due to the incandescence of the oxygen and nitrogen gases in the air. The disc shown in our illustration behind the coil was for carrying six Geissler tubes, to display the pretty experiment of the various colours of the luminous discharge in different attenuated gases. When the coil was first mounted it was provided with an ordinary contact-breaker, but as the strong sparks were found to very soon destroy the contact points, a contact-breaker was substituted on Foucault's plan. In this, the contacts are made by a platinum tipped wire dipping into mercury, that occupies the bottom of a strong glass vessel and forms part of the circuit. The vessel is filled with alcohol, which is a non-conductor, and it is therefore in the midst of this liquid that the contacts are made and broken. This apparatus is shown in the illustration, on the table at the left. A favourite experiment at the Polytechnic was to connect one of the discharging wires of the coil with the back of a large looking-glass, and bring the other wire to the front. In this case the sparks assumed a peculiar appearance, for they became thin and wiry-looking, and divided into many branches. They were very bright, and the noise of the discharges, was crackling and quite different from that produced by the blow of the flaming sparks taken through the air. Their appearance is represented in Fig. 269. The effects in this experiment were probably due to the spark taking a path on the surface of the glass determined by points of moisture or other inequalities.

Ruhmkorff's coil has been of great advantage to the electrician, for it supplies a stream of *high tension* electricity like that of the common machine, but more readily and conveniently. M. Ruhmkorff was the first person to obtain the great prize of £2,000, which the late Emperor of the French (Napoleon III.) directed, in 1852, should be awarded every five years for the most useful application of the voltaic battery. But no award had been made until 1864, when the inventor of the induction coil was properly considered worthy of it. This invention was the means of bringing into notice a new range of interesting phenomena, especially

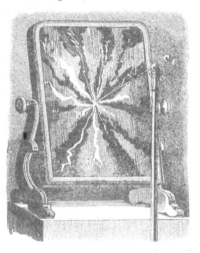

FIG. 269.—*Spark on the Looking-glass.*

those attending the discharge passed through highly exhausted vessels. Investigations into the circumstances which modify the appearances, and especially into the nature of the *stratified discharge* in which the vessels are filled with bands or flakes of light separated by dark intervals, have long engaged the attention of some of our ablest physicists. Remarkable results were obtained by Mr. Crookes with very highly exhausted vessels. These showed not only beautiful fluorescent luminous effects, but in them the discharge could produce mechanical actions, and Mr. Crookes was led to regard it as a stream of radiant matter.

MAGNETO-ELECTRICITY.

WHEN it had been shown that an electric current was capable of evoking magnetism, it seemed reasonable to expect that the reverse operation of obtaining electric currents by means of magnets should be possible. Faraday succeeded in solving this interesting problem in November, 1831, and one of his earliest, simplest, and most convincing experiments for the demonstration of the production of electricity by a magnet is represented in Fig. 270. A B is a strong horse-shoe magnet, C is a cylinder of soft iron, round which a few feet of silk-covered copper wire are wound; one end of the wire terminates in a little copper disc, and the other

FIG. 270.—*Magneto-electric Spark.*

end is bent, as shown at D, so that it is in contact with the disc, but pressing so lightly against it that any abrupt movement of the bar causes the point of the wire and the disc to separate. When the bar is allowed to fall upon the poles of the magnet, the separation occurs, and again when it is suddenly pulled off; and on each occasion a very small but brilliant spark is observed where the contact of the wire and disc is broken. It was in allusion to this experiment that a contributor to "Blackwood's Magazine" wrote:

> Around the magnet, Faraday
> Is sure that Volta's lightnings play;
>   But how to draw them from the wire?
> He took a lesson from the heart :
> 'T is when we meet, 't is when we part,
>   Breaks forth the electric fire.

If a coil of fine insulated wire be passed many times round a hollow cylinder, open at the ends, and the extremities of the wire connected with a galvanometer at some distance, then if into the axis of the coil, A B, Fig. 271, a steel magnet be suddenly introduced, an immediate deflection of the needle takes place; but after a few oscillations it returns to its former position. When the magnet is quickly withdrawn, the needle receives a momentary impulse in the opposite direction. The magnetization and demagnetization of the iron core in the induction coil would, therefore, of itself cause the induced currents already described, for these actions are equivalent to sudden insertion and withdrawal of a magnet. If we suppose

C, in Fig. 271, to represent, not a magnet, but a piece of soft iron—the reader will remember that this soft iron can be, as often as required, mag-

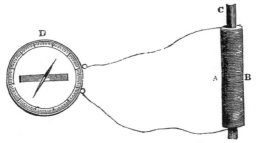

FIG. 271.—*A Magnet producing a Current.*

demagnetized by simply bringing near one end of it the pole ent magnet (see page 484). Upon this principle many ingenious

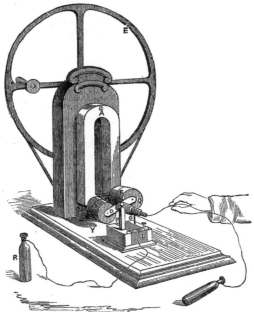

FIG. 272.—*Clarke's Magneto-electric Machine.*

machines have been constructed for producing electric currents by the relative motions of magnets and of soft iron cores surrounded by wires,

Clarke's machine is shown in Fig. 272. A is a powerful steel magnet
fixed to the upright. A brass spindle passing between the poles can be
made to rotate very rapidly by the multiplying-wheel, E, on which a handle
is fixed. There are two short cylinders of soft iron parallel to the spindle,
united together by the transverse piece of iron, D, which turns with the
spindle. Each bar is surrounded by a great length of insulated copper
wire, and the ends of the wires are so connected with springs which press
against a portion of the spindle, which is here partly formed of a non-
conducting material, that the currents generated in the coils, although in
different directions as they approach a pole and recede from it, are nevertheless made to flow in one direction in the external circuit. R R in the
figure represent two brass handles, which are grasped by a person wishing to experience the shocks the machine can give when the wheel is
turned. When the terminals of the coil are provided with insulating
handles and connected with pointed pencils of charcoal, the electric light

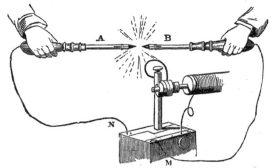

FIG. 273.—*Magneto-electric Light.*

can readily be produced by expenditure of mechanical effort-in turning the
handle. The arrangement of the points for this purpose is shown in Fig.
273, and we shall presently see what advantage has been drawn from this
experiment on a great scale as a source of light.

It will be observed that during the revolution of the armatures, as the
wire-covered iron cores are termed, there are two maximum and two
minimum points at which the currents are strongest and weakest. These
variations may be lessened by increasing the number of armatures and
of magnets, and Mr. Holmes arranged a machine with eighty-eight coils
and sixty-six magnets, and the connections were so contrived that the
currents always flowed in the same direction in the external circuit. This
machine required $1\frac{1}{4}$ horse-power to drive it when the currents were flowing,
but much less when the circuit was interrupted, and it was designed for,
and successfully applied to, the production of the electric light for lighthouse illumination. Instead of steel magnets which gradually lose their
strength, it is obvious that electro-magnets might be employed, but this
source of electricity is costly, troublesome, and inconstant. Mr. Wilde hit
upon the idea of using a small magneto-electric machine with permanent
steel magnets, to generate the current for exciting a larger electro-magnet,
and the current from this produced a still more powerful electro-magnet,

from which a magneto-electric current could be collected and applied. The same idea was subsequently applied in other forms, as by shunting off a portion of the current produced from the mere residual magnetism of an electro-magnet, to pass through its own coils and evoke a stronger magnetism, which again reacts by producing a more powerful current, and so on continually ; the limit being dependent only on the mechanical force employed, and on the power of the wires to convey the electricity, for they become very hot, and, unless artificially cooled, the insulating material would be destroyed. The armatures used in Wilde's, Ladd's, and other machines of this kind, are quite different in arrangement from those of Clarke's machine, and are far superior. They are formed of a long bar of soft iron, of a section like this, ⊢⊣, and the wire is wound longitudinally between the flanges from end to end of the bar, up one side and down the other. This armature rotates about its longitudinal axis between the pairs of the poles of a file of horse-shoe magnets, either permanent, or electro-magnets excited by the magneto-electric currents. In this case opposite poles are induced along the edges of the bar, and these poles are reversed at each half-turn. The intensity of the induced currents increases with the velocity with which the armature is made to revolve up to a certain point; but because the magnetization of the soft iron requires a sensible time to be effected, and the poles are reversed at every half-turn, it is found that a speed increasing beyond the limit is attended by decrease of the intensity of the current. The intensity in such machines has, therefore, a definite limit. But in a modification of the magneto-electric machine, which has quite recently been invented by M. Gramme, the limit is vastly extended by the ingenious disposition of the iron core and armatures, and his machines appear to solve the problem of the cheap production of steady and powerful electric currents, so that electricity will soon be applied in processes of manufacture where the cost of electrical power has hitherto placed it out of the question. We shall now endeavour to explain the principle on which the Gramme machine depends, and describe some forms in which it is constructed.

THE GRAMME MAGNETO-ELECTRIC MACHINE.

L ET x, Fig. 274, be a coil of covered wire ; then while a bar magnet, B A, is advancing towards it and passing through it, as at M, a current will

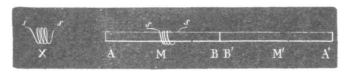

FIG. 274.

flow through the coil and along a wire connecting its ends, *s s*. The current will change its direction as the centre of the magnet is leaving the coil to advance in the direction, B A. If A A′ be a bar of soft iron, with the coil fixed upon it, we can still excite currents in the coil by magnetizing the bar inductively. If the pole of a permanent magnet be carried along from A′ to M in a direction parallel to the bar, but not touching it, the part of the bar immediately opposite will be a pole of opposite name, and the advance of this induced pole towards M will be attended with a current in the coil, and its recession by an opposite current. It need hardly be mentioned that the same result is attained if the magnetic pole is stationary, and the bar with the coil upon it moved in proximity to it. Now imagine that the

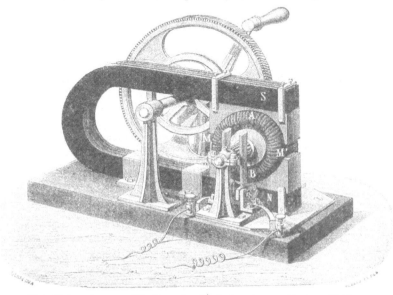

FIG. 275.—*Gramme Machine for the Laboratory or Lecture Table.*

bar is bent into a ring, the ends, A A′, being united. If the ring be made to turn round its centre in its own plane, and near a magnetic pole, it is plain that when the coil is approaching this pole a current will be produced in it, and when it is receding, an opposite current. Let the number of coils be increased, and each coil in turn will be the seat of a current, or of the electrical state which tends to produce a current. In Fig. 275 the reader may see how this disposition is realized. The figure shows a form of the Gramme Machine adapted for the lecture-table or laboratory. A M′ B M is the soft iron ring, covered with a series of separate coils placed radially, O is a compound horse-shoe steel magnet, S its south pole, N its north pole, each pole being armed with a block of soft iron hollowed into the segment of a circle and almost completely embracing the circle of coils. The magnetism of each pole is strongly developed in the interior faces of these armatures. The inductive action tends to produce two equal and opposite currents, which, like the currents of two similar voltaic batteries joined by their like poles, neutralize each other in the connected coils, but flow together through an external circuit. Fig. 276 will make clear the manner in which the coils, B B, are placed on the ring, A. The length of wire in each coil is the same, and the extremities are attached to strips of copper, R R, which are fixed on the spindle of the machine. The two ends of each wire are connected with two consecutive strips, while the coils are insulated from each other, and thus each coil, like the element of a battery, contributes to the aggregate current. The currents are drawn off, as it were, from these axial conductors at two opposite points of the ring, by springs very lightly touching them on each side of the

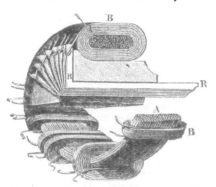

FIG. 276.—*Insulated Coils surrounding an Annulus of Iron Wires.*

spindle, as may be seen in Fig. 275. In Fig. 277 is another arrangement of

FIG. 277.—*Hand Gramme Machine, with Jamin's Magnet.*

the table apparatus with the magnet vertical, and formed according to the new plan suggested by M. Jamin, who finds the best magnets are made by tying together thin strips of steel.

33

But the importance of this invention consists in the facility which it affords for cheaply producing electricity on a scale adapted for industrial operations, for the deposition of metals, for artificial light, and for chemical purposes. The great importance of a cheap electric light for lighthouses prompted the British Government to permit the inventor to exhibit the light thus produced from the Clock Tower of the Houses of Parliament; for the signal light during the sittings of the House had previously been produced by a gas-light. This electric light was produced by a powerful Gramme machine, such as that shown in Fig. 278, driven by a small steam engine in the vaults of the Houses of Parliament, and the ordinary carbon points, reflectors, &c., were used in the Clock Tower, where the light was exhibited; copper wire ½ inch diameter being used to convey the current from the machine to the carbons. The result of these experiments may be gathered from the following extract from an official report made by the engineers of the Trinity House:

"Pursuant to the instructions received from the Deputy Master to furnish you with my opinion on the relative merits of the electric and gas lights under trial at the Clock Tower, Westminster, I beg to submit the following report:—On the evening of the 1st ultimo I was accompanied by Sir F. Arrow (who kindly undertook to check my observations by his experience) to the Westminster Palace, where we met Captain Galton, R.E., Dr. Percy, and some gentlemen connected with the electric and gas apparatus under trial. I was informed that the stipulations under which the lights were arranged were, that they be fixed white to illuminate a sector of the town surface of 180°, having a radius of three miles. I first examined the Gramme magneto-electric machine, in use for producing the currents of electricity. This machine we found attached by a leather driving-belt to the steam engine belonging to the establishment. We then proceeded to the Clock Tower, where we found the electric lamp, at an elevation of 250 ft. The Wigham gas apparatus was placed at the same elevation, within a semi-lantern of twelve sides, about 8½ ft. in diameter, and 10 ft. 3 in. high in the glazing. Near the centre of the lantern were three large Wigham burners, each composed of 108 jets. After the examination of the apparatus, we proceeded to Primrose Hill, for the purpose of comparing the electric and gas lights at a distance of three miles. The evening, which was wet and rather misty, was admirably suited to our purpose, ordinary gas-lights being barely visible at a distance of one mile"

The results of a photometric comparison of the electric and gas lights were as under, the machine making 389 revolutions per minute, and absorbing 2·66 horse-power; the illuminating power of the gas used being 25 candles, and the quantity consumed 300 cubic ft. per hour.

	Electric Light.	Wigham Gas Burner. 108 jets.
Relative intensity of lights	945·56	370·56
Or as	100	39·19
Illuminating power in standard sperm candles as units	3,066	1,199

" *Electric Light.*—Total cost per session £174 5s. 0d., being equal to 5s. 7d. per hour of exhibition of the light. Details shown in the full report. *Gas Light.*—Total cost per session of one burner of 108 jets, £159 15s. 3d., equal to 5s. 1·4d. per hour of exhibition of light, and £296 3s. 4d., equal to 9s. 5·9d. per hour of exhibition of the light, when using three burners of 108 jets each. Details shown in the full report. It will be observed from the photometric measurements, before referred to, of the electric light and 108-jet gas burner, that in the case of the electric light we have at our disposal for distribution over the required area an illuminant radiating freely in space equal to 3,066 candles ; with the gas light we have an illuminant radiating freely in space equal to 1,199 candles. It is to be remembered that in dealing with the small electric spark as the focus of a dioptric apparatus for distribution over the required area, the light can be more perfectly utilized than with the large gas flame of the Wigham burner, owing to its very small dimensions as compared with the latter. The relative cost and efficiency of the three modes of illumination may be summed up as follows :

	ELECTRIC LIGHT.	GAS.	
		One 108-jet Burner.	Three 108-jet Burners.
Cost of light per hour in pence ...	67	61·4	113·9
Or as	100	91·6	170 .
Cost of light per candle per hour in pence	·0219	·0512	·0317
Or as	100	233·8	144·7
Cost of light from a dioptric apparatus for fixed light per standard candle per hour expressed in pence	·00118	·00310	·00275
Or as	100	262·7	233·1

" Thus by adopting the electric light as a standard of intensity and cost, there is shown a superiority over the gas in intensity of 65·2 per cent. when using one 108-jet burner, and 27·1 per cent. when using three 108-jet burners. There is also shown a saving in cost per candle or unit of light per hour of 162·7 per cent. when using one 108-jet burner, and 133·1 per cent. when using three of these burners, forming a triform gas-light. It is further to be remembered that the triform gas-light actually represents the maximum power obtainable at present by gas ; but no reference has been made to the power of increase capable in the electric light by the adoption of two magneto-electric machines. By having the machine and lamp in duplicate, as estimated, and which I consider a necessity to insure perfect confidence in the regular exhibition of the electric light, this light can be doubled in intensity during such evenings as the atmosphere is found to be so thick as to impair its efficiency. This double power would be obtained at the trifling additional cost of coals and carbons consumed during the time this increased power may be found to be necessary ; this additional cost I estimate at 4d. per hour. With the arrangement proposed

FIG. 278.—*Gramme Machine, with Eight Vertical Electro-magnets.*

for the electric light, I consider this powerful illuminant, if manipulated by careful attendants, perfectly reliable : in proof of this I may state that the electric light at the Souter Point Lighthouse, on the coast of Durham, has now been exhibited two years and a half, and the light has never been known to fail for one minute."

Fig. 278 represents one of the light-producing machines. The electro-magnets are excited by a portion of the currents they themselves produce, they retaining sufficient residual magnetism to develop the currents.

FIG. 279.—*Gramme Machine, with Horizontal Electro-magnets.*

There is a pair of current-collectors on each side. This machine weighs 1,540 lbs., its height is 3 ft., and width 2 ft. It will produce a light having the intensity of 500 Carcel lamps, which may be doubled by increasing the speed. Fig. 279 is another form which is also adapted for illuminating pur-poses, and, when made with fewer coils, for electrotyping purposes also. There are in this also two sets of current-collectors, and by means of a connecting cylinder (seen at the base of the machine) the currents can be combined for quantity and for tension as may be required. This machine is only about 2 ft. square, and it produces a light equal to 200 burners ; but this may be increased, as the following table shows :

Number of revolutions per minute.	Intensity of light in Carcel Lamps.	Remarks.
650	77	No heating and no sparks.
850	125	do. do.
880	150	do. do.
900	200	do. do.
935	250	A little heat, no sparks.
1,025	290	Heat and sparks.

The value of M. Gramme's invention for electro-plating is proved by the fact of its adoption by Messrs. Christofle of Paris, whose electro-plating establishment is one of the largest in the world. This firm has no fewer than fourteen of these machines at work, and each is capable of depositing 74 ozs. of silver per hour. There is little doubt that the electric current will now soon be employed for reducing metals. Thus fine copper, which is worth 3*s*. or 4*s*. per lb., may perhaps be obtained at about the cost of ordinary copper ; potassium, sodium, and aluminium at less than half their present price ; and magnesium, calcium, and other rare metals at prices which will bring them into commercial use. The machine shown in Fig. 280 is intended for electro-plating and for general purposes : it supplies the means of readily and cheaply plating with copper, or with any other metal, such articles as steam pipes, boiler tubes, ship plates, guns, bolts, nails, marine engines, machinery, culinary vessels, cisterns, &c. The advantage of protecting iron or other material from corroding agents is obvious ; and as iron coated with copper is available not only for useful, but also for artistic, purposes, as a cheap substitute for bronze, this invention will doubtless lead to a greatly extended application of bronzed iron in buildings and ornamental structures.

The machine well illustrates how mechanical work may be changed into electricity, and electricity caused to do work. The power required to drive the machine at a given speed is much less when no current is being drawn from it, than when the current is flowing. If the current from one machine is sent through the armature of another, the latter revolves, and may be made to do work. Thus *power* may be conveyed to a distance by electricity, with only the loss caused by the resistance of the conducting wires. If, when two machines are thus connected, the direction of rotation in the first one be suddenly reversed, the armature of the second will almost immediately stop, and then resume its motion in the opposite direction. A very interesting experiment can be performed when the circuit connecting the two machines is made to include a certain length of platinum wire. When both machines are in motion, the platinum exhibits no heating effects ; but if the second machine be stopped by an assistant while the rotation of the first is continued, the wire is raised to a red heat. In this way it is shown that motion, electricity, and heat are related to each other, and are mutually convertible ; for on the stopping of the second machine, the electricity being no longer used up, so to speak, in producing motion, has its power transformed into heat.

The Gramme machine has also been ingeniously employed for railway brakes on some of the Belgian lines ; and it is applicable to telegraphy, where the cost of zinc, acids, batteries, &c., is a considerable item. It is impossible to predict the many applications for manufacturing purposes which will be made of electricity, now a cheap, reliable, and convenient mode has been discovered of producing currents of any required strength. Though by no means the first or only machine by which mechanical force can be converted into dynamical electricity, it shows an immense advance on any former one in the regularity of the action, and in the capability

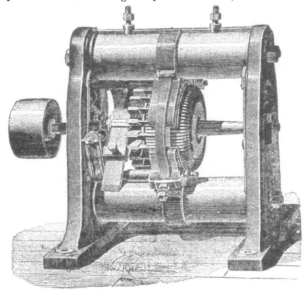

FIG. 280.

of being driven at a very high rate of speed without the inconvenient accompaniments of the heating of the conductors and destructive sparks at the movable contacts. There can be no doubt of the importance of this machine for use in lighthouses, and for metallurgical and chemical purposes, and the inventor believes the time will come when all large ocean-going vessels will carry an electric light at the masthead. The light would be sufficiently powerful to show rocks or land five or six miles ahead, and an additional safeguard of incalculable value would be thus provided for those " that go down to the sea in ships, that do business in great waters."

ELECTRIC LIGHTING AND ELECTRIC POWER.

IT was mentioned in the last section that the introduction of so convenient and reliable a means of producing electrical currents as the Gramme machine, would cause electricity to be largely applied for illuminating and other purposes. The Gramme machine was first made in 1870, and it attracted much attention, as the principle of combining the currents was quite different from that used in previous magneto-electric machines. In fact, the Gramme machine yielded quite unexpected results, and the principle employed in it opened a new field. The development that has taken place in the applications of electricity within the twenty years since 1870 has been truly marvellous. The electric light appears to have been first used in lighthouses about 1862, and the machines by which the current was produced were, in principle, combinations of a great number of Clarke's machines (see page 509). One such machine was invented by Mr. Holmes, and was used for the illumination of the South Foreland Lighthouse in 1862. Another similar form of still earlier invention had been set up in Paris as early as 1855,—not, indeed, for the purposes of illumination, but for a project which failed. Its arrangement had been originally suggested by a Belgian physicist in 1849 ; and the

FIG. 280a.—*The Alliance Machine.*

machine of 1855, having received certain improvements, afterwards became very well known by the name of the *Alliance Company's* machine, or simply the *Alliance* machine. It is represented in its improved form in Fig. 280a. Here ranges of steel horseshoe magnets will be observed, each magnet weighing about 40 lbs. and made of six plates of tempered steel, held together with screws. Each of the eight rows of magnets contains seven, and thus sixteen poles are presented at uniform distances, arranged in circles. Carried on the central axle are six discs, which revolve between the circles of sixteen poles, and on the circumference of each disc are sixteen equidistant bobbins or coils of insulated wire, so that the whole of the sixteen coils are opposite to the sixteen poles at the same moment. The extremities of the wires at the coils are connected with proper adjustments for gathering up the currents, and by means of these the coils may be arranged either for tension or for quantity, like the elements of a battery (page 494).

Wilde's machine, which has been mentioned in page 511, is shown in

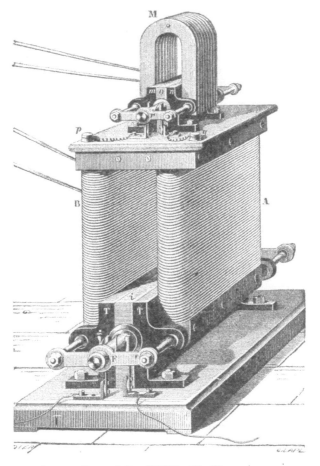

FIG. 280*b.*—*Wilde's Machine.*

fig 280*b.* It will be observed that this consists of a small machine, M, with permanent steel magnets, and the current from these circulates through the coils of the electro magnets, A B. The arrangement of the armatures, bobbin, commutators, etc., is the same in both cases. But as a speed of 2,500 revolutions per minute was needed, it was necessary to keep the bearings, T T, from heating by causing cold water to circulate through them. Mr. Ladd arranged a machine on the same principle as Wilde's, by suppressing the permanent magnets, but availing himself of the *residual* magnetism of the iron core to bring about the induction. A machine of this kind was shown at the Paris Exhibition of 1867, and people were quite astonished to see electrical power capable of producing a brilliant light developed by a small machine 2 ft. long, 1 ft. wide, and 9 in. high. But the great velocity of rotation, and the consequent heating of the bearings, left much to be desired before a really practical machine could be produced.

In the newest Siemens' machine, represented in fig 280*c*, the Gramme principle is made use of, as the revolving coil is of large diameter, and

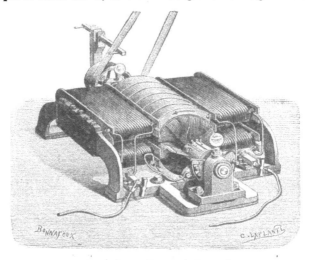

FIG. 280*c*.—*Siemens' Dynamo.*

it consists of a copper cylinder, on which are wound a number of juxtaposed coils like those of a galvanometer. The revolving cylinder is surrounded by the poles of a system of electro-magnets excited by the whole of the induced current being passed through their coils. In a paper describing this machine, Siemens first made use of the term "dynamo-electric machine," and this expression, contracted to the single word DYNAMO, has since been universally employed to designate machines of this kind. The modifications in the forms and arrangements of the different dynamos that have been invented in late years are endless, and every week patents are granted for further improvements and fresh combinations of the parts. It would be quite beyond the scope of this work to enumerate all the forms of the dynamo that have been favourably spoken of; but we shall content ourselves by adding a drawing of the Brush dynamo (Fig. 280*d*), which has been so largely used for electric lighting in the United States. In this dynamo we have a Gramme ring, but the number of coils on it is reduced to eight, the intervals being filled up with pieces of iron, and the ring revolves in a vertical plane between the poles of two double oblong electro magnets, which are arranged with poles of the same name opposite to each other. The commutators shown in the nearer part convert the alter-. nately reversed currents generated in the coils into a direct continuous one. They are formed with bundles of wires, as in the Gramme machine.

But the providing of a cheap and efficient source of current electricity, although an absolutely necessary step, would not have been capable of bringing about the present development of electric lighting, unless the appliances by which the current is made to manifest itself as light had not also been brought nearly to perfection. The conditions required to maintain a steady light from a current of electricity passing between carbon points have been already explained on page 497, and a representation of Dubosc's electric lantern and regulator is shown. The regulator

FIG. 280d.—*The Brush Dynamo.*

systems that have been invented since it became obvious that the light of the electric arc admitted of practical application on the large scale are very numerous. The earlier forms of regulator, which were used only for scientific purposes—such as lantern projections on screens, experiments on light, etc.—were complicated in their arrangements and uncertain in their action, for great variations in the light sometimes took place, and occasionally it would, indeed, be extinguished, and then again shine out as brightly as before. Nearly all the regulators that have come into use depend upon movements controlled by electro-magnetic actions produced automatically as the distance between the carbon changes. It would, however, lead us too far into the technicalities of the subject to explain minutely the mechanism of any particular form of the mechanical regulators, and the results depend so often upon the minute details, that it would be difficult to trace the action without a set of large and complete drawings. Perhaps the regulators that have been most used are those of Serrin, Siemens, Brush, Thomson, Houston and Edison. But nearly every inventor has produced different forms of his apparatus ; Siemens, for instance, has patented eight or ten regulators. Fig. 280e shows the mechanism of one of the last named inventor's regulators, in which the two actions required for the separation and approach of the carbons are determined respectively by the vibrations of the rocking lever, M Y L, actuated by the electro-magnet, E, and the simple weight of the upper carbon-holder, A A. When the lamp is not in circuit, the lever, L, is thrown back by a spring, the tension of which is regulated by the screw, R, so that the catch, Q, is disengaged from the wheel, I. The train of wheels is then free to revolve by action of the rack, A, supporting the weight of the upper carbon, until the motion stops by the carbons touching each other. Now let the lamp be connected up, and the current will pass from C, through the electro-magnet, the mass of the apparatus, and return by the wire connecting the lower carbon-holder with Z. The carbon points will glow, but the magnet then attracting M moves the lever, L, the piece, Q, engages the wheel I, pushing it one tooth forward. But this movement of the lever establishes a contact at X, so that the current abandons the electro-magnet, to pass the shorter way, and M being no longer attracted, the lever is pushed back by the spring, the contact at X is broken, and the magnet being again excited the lever turns as before, and Q pushes I round the space of another tooth. These alternating actions succeed each other with great rapidity, and effect the separation of the carbons through the train of wheels acting on the racks. These movements continue until, in a second or two, the separation of the carbons has become so great, that the current passing through the electro-magnet is no longer able to operate against the weight of the upper carbon-holder, and this happens when an arc of proper size is produced, this required result being brought about by proper adjustment of the parts of

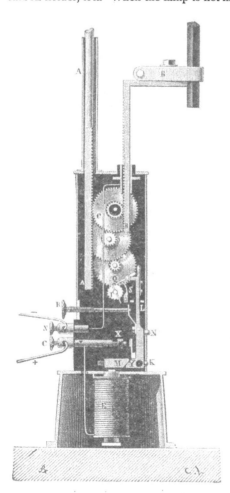

FIG. 280*e.*—*Siemens' Regulator.*

the apparatus, marked by the letters R, K and X. But as the carbons are consumed, the increase of the length of this arc further weakens the current, until the spring attached to the lever, L, prevails over the attractive force of the electro-magnet on M, and thus withdraws the catch, Q, altogether, when the wheels being free to turn, the weight operates to bring the carbons nearer together, until, with the lessened resistance, the energy of the current is restored, and Q again comes into play to arrest the approximating movement. It may be seen, from the above explanation, that this lamp is automatic ; in other words, when it has once been properly adjusted, it is lighted by merely completing the circuit. For fixing the carbons properly in their holders there are, of course, other regulating screws. How very nearly perfection the automatic regulation of the arc electric lamp has been brought by such contrivances as these, will be obvious to all who have noticed the steadiness that has been attained in all the modern installations.

An ingenious plan was devised by Jablochkoff for dispensing with all mechanism for regulating the distance of the carbons. This invention is known as the *electric candle*, and is of great interest from the fact that it was with this arrangement that the electric light was, for the first time, practically employed for street and theatre illumination. This was in 1878, when visitors to Paris, during the Exhibition, were astonished by the splendid displays in the Avenue de l'Opéra, at the shops of the Louvre, and at some of the theatres. Then it was shown, for the first time, that electric lighting was not merely a scientific curiosity, but a new and formidable rival to gas. The Jablochkoff candles were also subsequently used in the electric lamps on the Thames Embankment. The principle of the contrivance will be understood from fig. 28of. Two carbons, C and D, are placed parallel at a little distance apart, and the space between them is filled up with plaster of Paris, kaolin, or some similar material, through which the current will not pass, but which burns, fuses, volatilises, or crumbles away by the heat produced by the passage of the current between the two carbons. These carbons are, of course, fixed in insulated holders, and to start the candle a small tip of carbon paste is made to connect the carbons at the top. The Jablochkoff candles must be used with currents rapidly alternating in direction. The reason for this is, that otherwise one of the carbons (the positive one) would be consumed quicker than the other, and that would cause the dis- FIG.28of. tance between them to increase, until it became so great that *Jabloch-* the current would cease to pass, and the light would go out. In *koff Can-* order to obtain such alternating currents with the Gramme *dle:* machine, a special apparatus had to be devised to change its direct into alternately reversed currents ; but, dynamos intended to supply electric lights are now made without commutators, and they supply rapidly succeeding currents in opposite directions. In certain types of dynamos, again, the armature coils are stationary, and it is the field magnets that are made to revolve, and in these cases, not even a sliding contact is required, but the end of the armature coils are directly and permanently connected with the main circuit. But as these dynamos are self-exciting, the electricity induced in a few of the armature coils is collected apart from the main circuit, and passed through the eletcro-magnets of the machine itself, after the alternate currents have, by means of a commutator, been converted into one direct continuous current.

The arc electric light, as used for the illumination of streets and public places, is too intense and concentrated to be pleasant to the eye, and therefore it has been found necessary to surround it by globes of enamelled glass, or of porcelain, or of ground glass, or of frosted glass. By these expedients for diffusing and softening the light, it is rendered much more acceptable, but this advantage is gained at the cost of a considerable loss of the whole illuminating power, a loss which is, probably, never less than 10 per cent., but is usually much greater. The globes used in Paris, with the Jablochkoff candles, were of enamelled glass, and the apparatus was arranged, as shown in Fig. 280g, where it is partly represented in section, and with a part of the globe broken off, in order to show one of the candles placed in the holder which connects it with the circuit. In each lamp several candles were mounted, in some cases four; but the lamps in the Place de l'Opéra held twelve. At first there were mechanical arrangements, automatic and otherwise, by which, when the candle was burned down the current could be turned on to another. But M. Jablochkoff afterwards discovered that there was really no need for such a mechanism. For when the whole of the candles are simultaneously and equally connected with the circuit conductors, it is found that one of them will

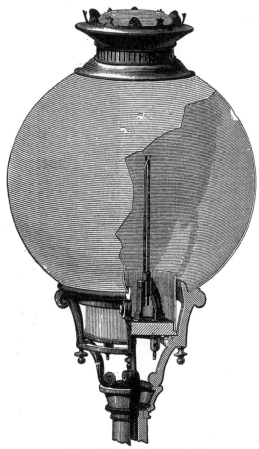

FIG. 280g.—*Electric Lamp.*

more easily transmit the current than any one of the rest, and when that particular one has once been lighted by the heat developed, the current will pass almost entirely through the arc, any loss through the connecting strip of carbon, at the tops of the other candles, being quite insignificant. When the first of the candles has burnt down completely, until the insulating porcelain holder separates the carbons, the current will at once re-establish itself at the top of one of the remaining carbons, and so on, while one is left.

The arc electric light has not been brought to its present position without the expenditure of much care and ingenuity in the preparation of the carbons used for its production. When Davy first produced the voltaic arc, the electrodes he used were simply sticks of charcoal. These were very quickly consumed, and a more durable form of carbon was sought for. This was found by Foucault, who made use of rods sawn out of the carbonaceous residue left in the retorts in the process of making coal-gas. This substance was, however, by no means uniform or sufficiently pure, and the light obtained was consequently unsteady. Many experiments were made in preparing special carbons. Pounded coke, coke and charcoal, were mixed with syrup or tar into a paste, which was moulded and compressed, and then the sticks were kept in covered vessels at a high temperature for many hours. Acids were used for purification, and also alkalis, to remove silica. At the present time there are several manufacturers of electric light carbons who carry on extensive operations by processes which probably are very similar one to another, and which may well be represented by M. Carré's, whose carbons have the highest reputation. M. Carré prefers a mixture of powdered coke, calcined lamp-black, and a syrup made of sugar and gum. The whole is well mixed and incorporated, water being added from time to time to make up for loss by evaporation, and to give the paste the proper degree of consistence. The paste is then subjected to compression, by which it is forced through draw-holes, and the carbons, having been piled up in covered crucibles, are exposed for a certain time to a high temperature.

As a practical illuminant for lighthouses, the arc electric light came into use many years ago (1862) as we have already seen. This was when the generator of the current was the magneto-electro machine; but, now, when this generator has developed into the modern dynamo, the cost of the electric supply has been enormously reduced, so that, power for power, electric lights may be worked at half the former cost, and with greater convenience and certainty. Light for light, electrical illumination is said to be far cheaper than gas. Again, the arc electric light has properties which have caused it to be employed, not only in every important light-house in England, France, Russia, America, and elsewhere, but most ships of war are provided with means of projecting a beam of electric light in any direction, in order that the presence of torpedo boats, etc., may be discovered at night, or harbours entered and signals made under circumstances when such operations would be otherwise impossible. It was by the use of the electric light that, in 1886, one of the Peninsular and Oriental Company's steamers passed safely through the Suez Canal, at night, and the experiment was so satisfactory, that the canal authorities placed beacons and light-buoys to guide such vessels, as, being provided with electric apparatus, were enabled to hold their proper course between its banks. The use of projected beams for watching the movements of enemies, and for signalling to great distances in time of war, has been recognized by all the great military powers. The advantage of the electrical light in some mines, in subterranean and submarine operations and generally, in work that has to be carried on at night by large bodies of men, is constantly finding illustration. Few readers are unacquainted with the brilliant effect of the arc lamps in exhibitions, parks, &c. ; at out of door fêtes, or applied to the illumination of fountains, such as those at the Paris Exhibition of 1889.

The arc lamps are used in series ; that is, where there are a certain number of lamps to be supplied, the same electrical current circulates through the whole of them, and this, of course, must have force enough to overcome the resistance of the whole circuit. Thus, at each lamp, the intensity of the illumination must necessarily be very great. A solution was long sought to the problem of so dividing the current energy, that it might be made to produce lights, of moderate intensity, at a greater number of points. When Mr. Edison, shortly after having invented the phonograph, announced that he had solved the problem of the electric light division, there was a great panic amongst the holders of shares in gas companies, and a heavy fall in this kind of stock immediately occurred. As it turned out, the alarm was unnecessary, for gas was not to be superseded, immediately and definitely, by electricity. Nevertheless, it is by virtue of the principle that was contained in Edison's invention, that electric lighting has assumed the wide-spread importance it has at the present day, and that it is now actually ousting gas as an illuminant in the business and domestic premises of our large towns, and in theatres, libraries, and other places of resort. The principle which has brought about this great development of electric illumination is that shown in a simple form in Fig. 261. It appears, however, that as early as 1841, a platinum wire, made incandescent by a battery current, was proposed as a source of light, and in 1845, carbon was used in the form of slender rods, by King, and also by J. W. Starr, in the United States. Both inventors inclosed their carbons in glass tubes, from which the air was exhausted, so that the carbon might not burn away. In the following year, Greener and Staite turned their attention to lamps of this kind, and, again, in 1849, Petrie worked on the same subject. After that, the problem ceased to engage attention, until, in 1873, a Russian man of science, named Lodyguine, took the matter up and patented a carbon incandescent lamp, which did not, however, prove a practical success, and although the idea was worked out in various ways by Konn, Reynier, Trouvé, and others, the apparatus they designed was, in every case, lacking in simplicity, and certainty of action. The Edison incandescent lamp, the announcement of the discovery of which so fluttered the gas companies, about 1878, was a reversion to the plan of an incandescent metallic wire. This wire was made of an alloy of platinum and iridium, which was adopted by Edison on account of the very high temperature required for its fusion. And in order to prevent the temperature from quite reaching that point, the wire was arranged in a spiral within which was a rod of metal that, by its dilatation with a certain temperature, caused a contact to be made which diverted part of the current through a shorter circuit, and thus lowered the temperature of the spiral to within the assigned limits. But the advantages presented by carbon over metallic conductors led Edison to attempt the formation of filaments by charring first slips of paper, afterwards slips of bamboo. About the same time Mr. J. W. Swan, of Newcastle-on-Tyne, was experimenting in the same direction, and, in the latter part of the year 1880, he exhibited the first incandescent lamps shown in England. Swan's carbon filaments were prepared from cotton threads which had previously been steeped in dilute sulphuric acid, washed, and passed through draw holes to give them an uniform section. They are thus made perfectly homogeneous throughout, and, after having been wound on pieces of earthenware to the required shape, they are carbonized by packing in powdered charcoal and heating. These filaments are very thin, but solid and elastic. The arrangement of the lamp (see Fig. 280*h*) is extremely simple : the filament of carbon bent into a horse-shoe form, or turned so as to form a loop, is inclosed in a glass bulb of a globular or egg shape, about two inches in diameter. The extremities of the filament are connected in an ingenious manner to two platinum wires that pass outward through the glass into which they are fused, and terminate either in binding screws or in two small loops. The bulb is exhausted first by an ordinary air-pump, and then by a Sprengel mercurial pump, the current of electricity being sent through the filament during the last stages of the process, and finally the bulb is hermetically sealed. The light yielded by these lamps is mild and steady, and its intensity depends on the electric current sent through them ; but this may,

FIG. 280*h*.

it is said, be carried as high as to make the light equal to that of twenty candles. Each horse power of force expended on the dynamo suffices to maintain ten of these lamps. At the Exhibition of Electrical Apparatus at Paris in 1881, the Swan lamp received the gold medal as being the best system in its class. The Swan and the Edison patents are now worked together by one Company, and the productions of this Company are very largely used, although there are several more or less modified systems of glow lamps prepared by other manufacturers.

The great advantages offered by electric glow lamps over gas-lights caused them to be speedily adopted by the most enterprising managers of theatres and places of amusement. Mr. D'Oyly Carte had the Savoy Theatre, in London, completely fitted up with these lamps in 1881. The light was soft and agreeable, it did away with the risks of fire both for the audience and the performers : for the foot-lights and scene-lights were also electric glow lamps, and the coolness of the house and greater purity of the air were at once appreciated. Several other London theatres have since adopted the incandescent electric lamps, and it is obvious that the system will become universal. In all ocean-going passenger steamers, electric lighting of the saloons and cabins is now the rule. No mode of illumination so readily adapts itself to the production of artistic and decorative effects as the glow lamps : for the covering glasses may be tinted of any required shade, and the lights may be placed in any position. Small glow lamps are occasionally used as personal adornments, when placed, for instance, as part of a lady's head-dress amidst diamonds, a novel effect of great brilliancy is produced. It need hardly be said that in this application the wearer is not required to carry a dynamo about with her, for the electricity is supplied in a manner much more convenient for this purpose by a device presently to be described. For several years electric incandescent lamps, supplied by the like means, have been in

34

action every night in the carriages of the trains running between London and Brighton, and more recently the Company have had electric reading lamps of five candle-power fitted up in the carriages of the main line trains. They are placed at the backs of the seats just above the passengers' head. When anyone wishes to make use of one of these lamps, he places a penny in a slot, and then, on pressing a knob, the light appears, and at the end of half an hour it is automatically extinguished ; but, of course, it can again be made to appear by another penny dropped in the slot, and so on every half-hour as long as may be required.

To maintain the electric light (whether arc or incandescent) quite steady, the greatest uniformity in the speed of the dynamo is essential ; and if the prime mover by which it is worked, whether steam-engine, gas-engine, water-wheel, or turbine, is not perfectly regular in its action, the lights will fluctuate in brightness, and thus produce an effect which is very unpleasant. This is entirely obviated by the adjunct we have now to describe, which not only is most efficient as a regulator, but is, moreover, of still more importance by also providing the means of storing up the electrical energy in a portable form. The reader will have understood that in a voltaic cell the production of an electric current is the concomitant of a chemical union of substances within the cell (p. 493). Now, in the experiment shown in Fig. 263 (p. 498), it is the reverse of combination —namely, the decomposition of the water that is supposed to be effected under the influence of the current from a galvanic battery, and the poles are so connected that the direction of the current in the liquid while the decomposition is proceeding is from the wire in the O tube to that in the H tube. If the experiment be interrupted by removing the battery, and then putting a galvanometer (Fig. 258) in its place, the galvanometer will immediately indicate a current passing through the apparatus in a direction the *reverse* of the former one—that is, in the liquid it goes from H to O, and the volumes of the gases will slowly diminish while water is reproduced by imperceptible and gradual re-combination. Batteries can be made by joining up a series of arrangements like Fig. 263, consisting of nothing but strips of platinum surrounded by hydrogen and oxygen gases and the intervening acidified water. Analogous results are obtainable by cells containing other compounds with suitable metallic poles, for when decomposition has been effected through a series of such cells by a sufficiently powerful current from a *primary* battery, the series of cells will constitute, on removal of the primary battery, a *secondary* battery, for when the terminals of this are joined, the current will flow in the reversed direction while the separated parts of the original compounds are re-combining within the cells. These *secondary* batteries are called also *polarisation* batteries. A form of secondary battery was contrived some years ago (1859) by M. Gaston Planté, in which the current of the primary battery was made to act on plates of lead immersed in dilute sulphuric acid. The effect was to coat one of the lead plates of each pair with lead oxide ; and in the action of the secondary battery this was reversed, and the plates gradually returned to their original condition, when, of course, the current ceased. Some improvements were made in the Planté battery by Faure, who coated one of a pair of very thin lead plates at once with a film of red oxide of lead, and used a layer of felt to separate it from the other plate. Such arrangements have been called "accumulators" ; another term applied to them is "storage batteries" : but it is not to be supposed that in them electricity is stored or, so to speak, bottled up. They consist merely of such an arrangement of materials as that when a current (*direct*, not alternating) from a dynamo is passing, certain substances are placed in a position of chemical separation in such a manner that in re-combining an equable current of electricity is produced in the conductor externally uniting them. We need not notice some slight modifications of the Faure cells that have been lately introduced, as no new principle is involved. The light of incandescent lamps worked by the Faure accumulator is perfectly free from the fluctuations which may usually be noticed when the lamps are directly connected with the dynamo only. Even if the engine should stop altogether, the light may be maintained for hours. The accumulator has also the advantage of giving out the electric energy that may have been imparted to it days before ; so that when a house is fitted up with an independent electric light installation, there is no necessity for running the dynamo all the time the lamps are in use, as two or three days weekly may suffice to charge all the accumulators. Then there is the portability of the accumulator, which permits electrical energy to be made use of in situations where dynamos and prime movers would be impossible. It is said that a large Faure cell weighing about 140 lbs. can receive and give out energy equal to one horse power for one hour. In the arrangement for the reading lamps in railway carriages referred to above, accumulators are placed under the seats ; and it need hardly be said that when the electric light has been seen in a *coiffure*, a small Faure cell concealed about the wearer's person has supplied the current. A very interesting and useful application of the accumulator is the portable electric light lamp for miners made by the Edison-Swan Company. It is simply an incandescent lamp protected by a strong glass cover attached to the side of a cylindrical case containing a four-celled accumulator. This lamp is provided with an ingenious contrivance by which the circuit would be interrupted, if by accident the outer glass cover of the lamp were broken. Let us now see what another new development of the applications of electricity gains by the use of accumulators by turning our attention to the *electro-motor*.

At the Vienna Exhibition of 1873, the Gramme Company showed two of their machines, and it is said that when one of these machines was at rest, a workman connected the ends of two covered copper wires with the other machine, thinking that these were placed to carry the current from that machine when in movement. Everybody was surprised when, without any power from the machinery, the ring was soon in rapid rotation. These wires were in fact joined up to the other Gramme machine which was already in action, and it was the current from this that set the former in motion. There is no reason why this story should not be perfectly true, although there are good reasons for believing that the *electro-motor* was the result of no such accidental circumstance. The attractions and repulsions between the poles of electro-magnets was soon seen to supply an available source of motive power, and the subject has been already mentioned on page 518. Professor Jacobi, of St. Petersburg, seems to have been the first who constructed an electro-magnetic engine, the exciting power being the current supplied by a voltaic battery. This was in 1834, and in a few years afterwards the Professor applied his engine to a small paddle-wheel boat, 28 feet long, which was electrically propelled for several days, but at a slow speed. The engine in this case was virtually a magneto-electric machine worked backwards, that is, instead of applying power to turn the machine and so produce a current of electricity, the current was supplied by the battery and produced power. In 1850, an electro-motor of five horse power was shown by an American, Mr. Page, the principle of which may be illustrated by supposing a reversal of the action represented in Fig. 271, thus : if, instead of producing currents by moving the magnet, c, in an J out of the coil, A B, we substitute a battery for D, we can, by alterating the direction of the current through the coil, cause a reciprocating motion of the magnet, c, and this again may be described as a magneto-electric machine worked backward. It was soon recognized that no practical electro-motor was adequate to the production of such high powers as the steam engine supplies, and that the cost must necessarily many times exceed that of steam power. But certain advantages, nevertheless, pertained to the electro-motor in certain positions, as instance in safety, and where a small force only was occasionally required. Now, when the Gramme machine was invented to supply currents of electricity under conditions much more favourable than the magneto-electric machines it superseded, and at a cost vastly less than that of any voltaic battery, it is highly improbable that the relation of the new current generator to the production of electro-motive power would long be overlooked.

The electro-motor may, therefore, be considered simply as a dynamo worked backward, and almost any form of dynamo may in this way be used as an electro-motor, that is, a current being supplied either from a battery or from a dynamo, the motor converts the electrical energy into mechanical energy. Any dynamo that supplies a direct and continuous current can thus be used ; but there are certain conditions which make it desirable to somewhat modify the proportions and arrangement of the several parts when the machine is for motor purposes.

In general, any source of current may be used, but in the applications of the electro-motor there are chiefly two methods in practice of supplying the current. The one takes the current from a dynamo in motion, the other from an accumulator which has previously been " charged" by a dynamo.

Both of these methods are used in the familiar and interesting application of the electro-motor to the propulsion of carriages on tramways and railways. For the latter, indeed, an attempt was made half-a-century ago on the Edinburgh and Glasgow railway, to employ the force of an electro-magnetic machine actuated by a battery. This was in 1842, and although this electric locomotive was fitted up completely, it did not attain a speed of more than four miles an hour. The weight with the batteries, carriage, etc., exceeded five tons. But in the recent inventions which have been in practical operation in many places, it is found quite easy to dispense with any current producer on the electric locomotive itself, for the electricity is supplied by a fixed dynamo and the current is transmitted along the line by a conductor from which a sliding contact conveys it to the electro-motor, which is attached to the framework of the carriage and acts on the driving axles of the wheels directly or by toothed gear. In such cases the return current is carried either by another conductor or by the rails themselves. In another arrangement one rail conveys the current to the locomotive and the other returns it. When the rails are so used they have, of course, to be insulated from the ground and laid with special electrical contact pieces joining their consecutive lengths, and all the carriage wheels have to be insulated, so that the currents shall flow only through the coils of the electro-motor. A railway on this system has been worked at Berlin for some time, and a short tramway on the same plan has lately been opened at Brighton. The Bessborough and Newry Electric Railway (Ireland) uses a single separate conductor three miles long, and the power is supplied at a very small cost from a dynamo station near the middle of the line, where water power is taken advantage of to drive a large turbine. Quite recently electric propulsion has been

FIG. 280*i.*—*Poles with Single Arms for Suburban Roads.*—*The Ontario Beach Railway, Rochester, N.Y.*

adopted on some of the short tunnel lines in London, and it is quite probable that ultimately the system will be adopted throughout the whole course of the underground railways, with the view of obtaining a purer and more agreeable atmosphere.

A very light electric railway has been designed, in which the cars run along rails attached to posts at such a height above the ground as may be required to make the line level, or with only slight gradients.   The rails also serve as conductors.   This is known as the telepherage system, and it is found to be well adapted for light loads in an undulating country.

The other plan which makes use of accumulators commends itself for application to ordinary tramway carriages, because no conductors are required along the line, and each car can move independently.   The chief objection is the great weight of the accumulators and the space they occupy, although they are usually placed under the seats without much inconvenience.   There are at present (January, 1890) six electric tram-cars running in London, and the accumulator system would no doubt have been applied largely as the motive power for the ordinary street omnibus, but for the difficulty of controlling them under the momentum of the great mass of the accumulators, etc.   The same objection lies against the use of the accumulators and motors for propelling tricycles, although such

FIG. 280j.—*The Glynde Telepherage Line, on the system of the late Fleeming Jenkin.*

machines have really been used.   But accidents such as occasionally happen to such vehicles would be attended with additional risks of injury from the acids of the secondary battery, etc.   But there is one mode of using electric propulsion, that is free from every objection and, indeed, offers great advantages.   Only two years ago the first electric boat on the Thames was tried experimentally between Richmond and Henley, and the result was entirely in favour of the electric over the steam launch.   The Faure battery, or so-called "storage cells," are arranged beneath the floor of the boat for most of its length in the smaller boats, and the electro-motor is directly coupled with the screw shaft.   The electric launch has these advantages : perfect safety, freedom from dirt and smoke, no thumping or vibrating, no noise of steam discharge, or smell of hot oil, no engineer or stoker is required, and much larger space available for passengers.   One of these electric launches, not going full speed, is able to travel sixty miles without having the accumulators re-charged.   A considerable number of these launches are already in use, and many more are in course of construction.   They are made of all sizes, from the smallest to those that will carry quite a large company, and may be used for excursion parties on the river. The description of one of these last states that she is 65 feet in length, and 10 feet across the beam. She can carry sixty passengers, and twenty can dine in the saloon at one time. There are lavatories, pantries, dressing rooms, etc., and a brass railed upper deck, with an awning. At night this boat is lighted up with electric glow-lamps, the current for these also being supplied by the accumulators. The Electric Launch Company has stations with Gramme machines at work to charge cells ready to replace exhausted ones at several places, namely Hampton, Staines, Maidenhead, Boulter's Lock, Henley, Reading and Oxford. There is every prospect of a general extension of the electric propulsion of boats, and visitors to the Electrical Exhibition at Edinburgh, in 1890, will find electric launches taking holiday makers as far as Linlithgow. The boats will be like those on the Thames, fitted with the Immisch motor. Some electricians are now sanguine enough to believe that even for large vessels electricity will yet be able to compete with steam in special cases.

The modes of using electric propulsion that we have just noticed furnish a very interesting chain of conversions of one form of force into another, with a reversal of the order of transformation at a certain point. Let us begin with the carbonic acid gas that existed in the atmosphere of the carboniferous geological period. The solar emanations were absorbed, and used by the leaves of the plants to separate the two elements of the gas,—the plant retaining the one in its substance and returning the other to the air. The plant becomes coal ; and ages afterwards the particles of the two separated elements are ready to re-unite and give out in the form of heat all the energy that was absorbed by their separation. This heat is in the steam-engine converted into the energy of mechanical power. This mechanical power is in the dynamo expended in moving copper wires through a magnetic field. Every schoolboy who has played with a common steel magnet—and what boy has not?—knows that the space immediately round the magnet is the seat of strange attractive and repulsive force, for he has felt their pulls and pushes on pieces of iron or steel. This mysterious space is the magnetic field, and although a person would not be able to perceive that mechanical force is expended when he moves a single copper ring across such a field, he will readily become conscious of the fact when he moves a number at once that form a closed circuit ; and he should not omit the opportunity of feeling this for himself if he is allowed to turn the handle of such a machine as that represented in Figs. 275 or 277. The mechanical power is absorbed in the dynamo because the movement induces an electric current that would of itself produce motion in the machine in the opposite direction. However, the electricity induced by magnetism and motion is made to pass through the Faure cell· or accumulator, when it does chemical work by separating oxide of lead from sulphuric acid, leaving these substances in a position to unite together again, when this action produces a reverse current of electricity through an external metal circuit. The coils of the electro-motor form this circuit ; the electricity induces magnetism, and the magnetism gives rise to visible motion and mechanical power.

From what has been already said, it will be obvious that a pair of covered copper wires connecting a dynamo with an electro-motor becomes a very convenient means of *carrying power* from one place to another. There are situations in which shafts, belts, or any other mechanical expedients are troublesome or impossible to use for this purpose.   For instance, a dynamo working at the mouth of a tunnel or coal-pit may be made to drive any machinery within with nothing between but the motionless wires.   Or a single dynamo will supply moderate power to a number of small workshops, provided each has an electro-motor, with no other connection than a pair of copper wires.   This arrangement is found very advantageous for light work and where power is required occasionally, as in watch-making, the manufacture of philosophical instruments, etc.   Such moderate power is occasionally in demand also in private houses, to drive sewing machines, lathes, etc. ; and it is obtainable from the same source as the current for lighting.   Private installations for lighting purposes usually have a dynamo driven by a gas engine, and working into a set of accumulators.   It seems not a little remarkable that if the gas were burnt in the ordinary way instead of being used in the gas engine, it would give only a fraction of the ámount of the light it causes to be given out by the electric light lamps.   But at the present time, houses and business premises are supplied with electricity by companies who carry electric mains through the streets.   In England these electric mains, which are thick insulated copper wires, are inclosed in iron pipes and laid beneath the pavement, like the gas mains.   In the United States, where electric illumination is much used, the conductors have been usually carried overhead like telegraph wires, but not a few fatal accidents have occurred from these conductors falling into the streets.   There is no reason to doubt but that in a short time it will be as common for households to draw upon such electric mains for their supply of light and power as it now is to draw gas and water from common mains.   The electric supply companies have central stations in suitable positions, where very large and powerful dynamos are regularly driven by steam power.   These stations are provided with appliances for measuring the currents and for duly controlling the energy sent out.   What will appear very extraordinary when we remember that electricity is in itself unknown, is that the quantity supplied to each house or establishment can be actually measured, and is paid for by meter as in the case of gas.   As already said (page 498) electricity can only be measured by its effects, and it is the chemical effect which it is found convenient to use for the purpose we are speaking of.   The plan is simply this : two plates of zinc dip into a solution of sulphate of zinc, and from the one to the other there is sent through the solution one-thousandth part of the current to be measured.   While the current passes, zinc is deposited on the plate towards which the current goes in the solution, and if this plate is periodically weighed this furnishes the measure of the total current.   But how is just one-thousandth of the whole current taken off from the rest and made to circulate through the measuring apparatus ?   This is very easily done by taking advantage of the law of derived circuits, which for our present purpose may be stated thus : when a current of electricity finds two different circuits along which it can pass, it will divide and circulate through both of them, but the greater part will pass through the circuit of less resistance (if there be any inequality), and by adjusting the resistances of the circuits we can divide the current between the two partial or derived circuits in any required proportions.   Electric resistances, it may be mentioned, depend upon the length, section, and nature of the conductor, and are very easily measured and adjusted.

While the method just explained serves very well to measure the quantity of electricity that has passed through a conductor in a given period, provided that the current has always been in the same direction, it will be sufficiently obvious that it would fail altogether in the case of alternating currents. And, in fact, even in the case supposed this mode of measurement does not take account of the real energy set in motion. A reference to page 498, where the differences of electric currents are men-' tioned that are commonly spoken of—*tension* and *quantity*—will show that electric effects depend upon more than the *quantity* of electricity passing. Forms of apparatus have been devised for recording the total energy supplied ; but their construction and principles are too complex to be here explained. In some cases high tension currents are required, in others it is quantity and not tension that is sought for ; and there are ways of transforming the qualities of currents so that the same source shall supply electricity of either class. An example of this may have been noticed in the action of the Ruhmkorff coil, where the mere interruption of the primary or battery circuit, which possesses so little tension that of itself it could not give rise to a spark, nevertheless produces a wave of electricity in the secondary circuit of a tension so high that sparks several feet long may be produced by it.

A somewhat recent application of the electric current of the dynamo may be just mentioned here. It is what is known as electrical welding, and depends upon the heat developed by currents being proportioned to the electrical resistance for each part of the circuit. The heat thus generated, where the current passes between two surfaces of metal, even of considerable dimensions, is sufficient to bring them to a semi-fluid condition, so that when simply pressed together they coalesce into one mass. In this way pieces of iron work can be welded together in situations where it would be either inconvenient or impossible to heat them by furnaces.

The reader who has followed the last article will probably be prepared to admit that "the magnetic field" is one of the most wonderful things in the whole realm of inorganic nature, as all the powerful effects we have been describing are the results of merely moving wires through it. A wire conveying an electrical current so modifies the space surrounding it, or so acts upon the unknown pervading medium, that conductors moved in it, have other currents generated in them. An intermittent current, like that in the primary circuit of the induction coil, is equivalent to a movement of the magnetic field in regard to the secondary coil, so that the general principle in the coil and the dynamo is fundamentally the same. Quite recently, Professor Elihu Thomson has shown some very novel mechanical effects of repulsions and rotations of conductors placed near the poles of a coil through which rapidly alternating currents are passing. [1890.]

We already hear of natural forces which have hitherto in a manner run to waste being now utilised in man's service by the advantage taken of the capability of a slender wire to convey power. A notable instance is in the case of the famous Falls of Niagara. Here the head of water is used to drive turbines ; our readers must not run away with any notion of huge water-wheels being placed below the falls. But from the high level of the water above the falls a tunnel has been cut which brings the water into pipes $7\frac{1}{2}$ feet in diameter, and these deliver it into three turbines, in passing through which it develops a force of 5,000 horse power, and this force is communicated to a steel shaft $2\frac{1}{2}$ feet in diameter, connected with the revolving parts of the dynamo. Mr. G. Forbes, the engineer, states that the company who have undertaken this enterprise are supplying, with a handsome profit to themselves, electrical current or power at $\frac{1}{4}$th of a penny per unit, for which English companies charge sixpence. That is, Niagara supplies power at $\frac{1}{48}$th of the price it can be obtained from coal.

The fact that mechanical power can be brought from a distance to everyone's door by a slender wire, and at small cost, suggests the possibility of great social and industrial changes being effected in the future by that one condition. Think of the abolition of factory chimneys and smoke, nay, even of the abolition of the factory system itself, for cheap power transmission seems to promise much in that direction, and there is a shadowing forth of still more in

THE NEW ELECTRICITY.

THE Leyden jar and a few of its most obvious and common effects have been touched upon already (page 490); but the phenomena which are revealed by a careful study of its charge and discharge show that these are by no means of the simple kind that has generally been supposed. Thus, for instance, if the magnetising effects of what is called current electricity be borne in mind, especially the *definiteness* of this action as regards the *direction* of the current (*cf.* Fig. 257), it would follow that if instead of the iron bar in Fig. 265 we should place within the coil some unmagnetised steel needles we should find after passing a current or discharge that these have become converted into permanent magnets, and that their north poles are always towards the left of the supposed current. Years ago experiments were made to ascertain whether the discharges of a Leyden jar repeatedly passed through a coil would magnetise needles in the same way, because it had been assumed that the discharge is simply a current of extremely short duration and of quite definite direction. As far back as 1824 it had, however, been observed that the needles were magnetised sometimes in the wrong direction, yet no attempt was made to explain this—it was sometimes merely mentioned in the books as "anomalous magnetisation." Dr. Henry of Washington, U.S.A., experimented on the subject, and in 1842 referred this action to a condition of the discharge which had never before been suggested. He says "we must admit *the existence of a principal discharge in one direction, and then several reflex actions backward and forward, each more feeble than the preceding, until the equilibrium is obtained.*" Some five years afterwards Helmholtz had independently arrived at the same conclusion, and from the fact that when a *succession* of Leyden jar discharges are sent through the voltameter (Fig. 263) the water is indeed decomposed, but *both* oxygen and hydrogen are evolved at *each* electrode. Sir William Thomson (now Lord Kelvin) examined the question from a theoretical point of view, and in a masterly mathematical paper published by him in 1853 not only showed that the discharge must be of an oscillating character, but gave the form of equation by which the rate of oscillation is determined.

Faraday proved, as has already been stated, that the matter of the dielectric takes part in such condensing actions as that of the Leyden jar.

The electrical charge enters into the glass, the particles of which are thrown into a certain state of strain or tension (which Faraday called polarisation), and the discharge of the jar is their release from that tension. So that it appears that whatever electricity may be, it can in some way become bound up with the particles of ordinary matter like glass and other dielectrics, and exert force upon them, which force acts always in two opposite directions. It is the opposition of the form or direction in which the electrical effect is manifested that gave rise to the conception of the two "fluids"—the "positive" and the "negative." If these "fluids" really existed it would surely have been possible to give to an insulated body an absolute charge of either of them. But this can never be done ; f, for instance, you have in the middle of a room a metallic sphere charged with positive electricity, the necessary condition is that on the walls of the apartment or on surrounding objects there is an exactly equivalent quantity or negative electricity.

The number of oscillations or alternate momentary currents in a single discharge of a Leyden jar is enormous. Theory shows that under ordinary circumstances they must be enumerated by hundreds of thousands, if not by millions ; that is, the apparently instantaneous spark is really made up of say a million surgings to and fro of the electric influence. But theory also shows that the frequency of these oscillations can be controlled or adjusted through an indefinite range. A general notion of the requisite conditions may be obtained by the analogy of sound, and for this we may take the familiar case of the strings of a musical instrument, say the violin, or the harp. Everybody knows that when a stretched string or wire is pulled a little aside it is in a state of lateral strain, striving by its elastic force to return to its position of rest, and if it is suddenly let go it not only rapidly regains that position, but by the inertia of its motion is carried beyond it against its elastic force, which, however, again brings it back, and the movement is continued nearly up to the point at which it was originally released, this swinging movement persisting for an indefinite period, during which the vibrations, which have an ascertainable and perfectly regular frequency, are communicated to the sounding-board of the instrument and from that to the air, by which they are conveyed to the ear and affect the auditor as a musical note, which note is higher as the number of vibrations per second is greater. Everybody will have observed that in the violin the note yielded by each open string is higher as the tension becomes greater by turning the peg to tighten it ; that the same string will, without any change in its tension, yield higher notes as shorter lengths of it are employed. Another circumstance upon which the pitch of the note depends may also be illustrated in the violin, in which it will be noted that the G string, which gives the lowest notes, is loaded with wire wound spirally round it. Here, then, are three circumstances that collectively determine the pitch or number of vibrations of a string—tension, length, weight ; and if you give the measures of these to a mathematician he can tell you the note the string will emit, for the number of vibrations is given (when the measures are expressed in the proper units) by the formula

$$n = \frac{\sqrt{t}}{2l \sqrt{w}}$$

This shows that we have only to adjust suitably the tension, length, and weight of a string in order to make it vibrate at any rate we please. Now in the oscillation of currents in the Leyden jar discharge there are conditions which correspond, by analogy at least, with those that determine the vibrations of a stretched string. These conditions are of course electrical, and they are definable in terms of electric units, which need not be discussed here. As we are leading the reader to the modern view of electricity, which sets aside the fluid theories and regards electricity as having no separate existence, but as being merely the manifestation of some condition of a universally pervading medium, the same, in fact, as the luminiferous ether, it is curious to remark that these electrical oscillations would seem to attribute to the incompressible and imponderable ether something very much like the characteristic property of matter we call inertia, by virtue of which the released cord flies past its position of equilibrium to the other side. Or may this quality be dependent on the matter of the dielectric in which the ether is, as it were, entangled?

The oscillatory character of the Leyden jar discharge was elegantly demonstrated before a large audience in a lecture given by Professor O. Lodge at the Royal Institution a few years ago. Clearly it is impossible to render perceptible to the senses the millions of periodic discharges that take place in the marvellously short space of time taken up by a spark, but by doing what is analogous to slackening the tension of the stretched string or increasing its length, that is by increasing the *static capacity*, which means using a large number of jars combined into a battery, and at the same time causing the discharge to pass through coils (the effect of these is to increase the *self-induction* of the circuit—called also *impedance*), an arrangement corresponding with loading the string, Dr. Lodge was able to bring down the rate of oscillation to 5,000 per second, when, instead of the crack of the ordinary discharge, a very shrill continuous sound was heard. The addition of another coil gave another load, and when the rate was thus reduced to about 500, the note emitted was that of the C above the middle A of the piano. With the rate of oscillation thus reduced, it became easy to render the discontinuity of the discharge visible by means of revolving mirrors, as in the well-known acoustical demonstrations.

Professor Lodge has devised an experiment which again shows the analogy of electrical oscillations with those by which sound is produced. It is well known that a vibrating tuning-fork will set another fork of the same pitch to vibrate also by mere approximation. A and B (Fig. 280*k*) are

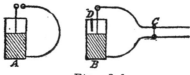

FIG. 280*k*.

two exactly similar Leyden jars, the inner and outer coatings of each being connected by a wire enclosing a considerable area in its circuit, which in the case of A contains an air gap across which sparks pass when the coatings are connected with the poles of an electrical machine. The circuit of B is provided with an adjustable sliding piece C, and the coatings are almost connected with each other by a strip of tinfoil hanging over the rim but not quite reaching to the outer coating. When the jars are placed so that their wire circuits are parallel, and sparks are passing across the air interval of A's circuit, a position of the slider on the other can be found when sparks also pass between the tin-foil and the outer coating. But if the slider be moved from this position, the two circuits will no longer be in unison, and the sparks in B will cease. This response of the oscillations in one jar to those set up in another of the same vibratory period is called *electrical resonance.*

Dr. Hertz, a professor in the University of Bonn, has opened out new paths to investigators by a brilliant series of researches which have shown that in the dielectric surrounding an electrical system executing very rapid oscillations there are waves of electro-motive and magnetic force. These researches are not capable of any condensed description here, and the reasoning is of a kind that appears mainly to the expert physicist. One of his modes of investigation required oscillations of extreme rapidity, and he obtained them by attaching to each pole of an induction coil a metal plate, and between these plates, which were in the same vertical plane, passed a stout wire interrupted by an air gap in its centre provided with small brass balls. The rate of oscillation of this arrangement was calculated as the hundred-millionth part of 1·4 second. In conjunction with this system Hertz made use of a very simple apparatus he called a resonator, which consisted merely of a piece of copper wire bent into a circle of about 28 inches diameter. The ends of the wire did not, however, meet, but were fitted with two balls, or with a ball and a point, and an arrangement by which the air gap between them could be very finely adjusted and measured. This resonator was, of course, prepared as to be in electrical tune with the original vibrator, and with it Hertz was able to examine the condition of the surrounding space. When held in the hand near the vibrator he found that sparks crossed the air space in the resonator, and that the length of the air space across which the sparks would pass varied with the position of the resonator. When the plane of the resonator was parallel with the metal planes of the vibrator and its axis in the horizontal line drawn perpendicularly through the vibrator's air space, the sparks passed readily when the air space of the resonator was at the same time vertically above or below its centre, but they ceased entirely when it was level with the centre. He obtained these sparks when the resonator was held—in free space, be it understood—in the above-mentioned position even at a distance from the vibrator of 13 yds., the length of the apartment. By examining the results with other positions of his resonator and by other and varied experiments, Hertz was able to prove the existence of definite waves of electro-magnetic and electro-motive forces, to measure their lengths, and to show that they are capable of reflection, refraction, and even polarization by the same laws that hold with the extremely short but enormously rapid vibrations constituting light. It may here be mentioned that the existence of currents in the resonator can be shown by a Geissler tube being made to take the place of the air space, which tube is thus lighted up without any metallic or visible connection with any electrical apparatus whatever, the only requisite conditions being that its circuit be tuned to the vibrator, and in a certain position in relation to the axis of the spark space of the latter. Hertz has also shown that electro-magnetic disturbances (transversal waves) are propagated in space with a determinate velocity akin to that of light, and in short the outcome of his investigations, as well as of those undertaken by others, has been a vindication of Clerk Maxwell's splendid theory by which light is regarded as an electro-magnetic action. Professor Righi of Bologna, having succeeded in obtaining shorter electrical waves than anyone before —namely, $\frac{4}{10}$ths of an inch instead of about 20 inches—was able with them to repeat all the phenomena of optics such as reflection, refraction, circular polarization, interference, &c. It appears then almost certain that light and electro-magnetic waves or radiations are but one and the same affection of a pervading medium we call the ether.

By following up in certain directions lines of research suggested by the investigations of Maxwell, Lodge, Hertz and others, and by an unreserved acceptance of the ether theory of light, electricity and magnetism, some wonderful practical results have recently been obtained by M. Nikola Tesla, an electrical engineer now resident in New York. The experiments shown by Tesla in his public lectures have excited great interest in scientific circles, and have by many persons been witnessed with something like astonishment.

One of the first objects of M. Tesla was to obtain alternating currents of high tension and great frequency. It may be seen from Fig. 272 that the movement of coils of wire in a magnetic field generates currents, and it has been stated that these currents are in alternately opposite directions as the coils approach or recede from the magnetic poles. In the machine represented in Fig. 280a, each revolution would produce 16 reversals of current. Tesla constructed a rotatory machine which gave 20,000 alternations of current in one second, because it had 400 poles and could be rotated at a very high speed. But of course the number of poles and the speed of the machine could not be increased beyond certain practical limits. By a happy application of the known principle of harmonic oscillations, in which all the rotatory movements of fly-wheels, coils and poles could be dispensed with, Tesla simplified the alternate current generator, reducing the moving parts to the minimum at the same time that he obtained a greater number of alternations and almost perfect regularity in their periodicity. The way in which this has been accomplished may be gathered from a careful inspection of Fig. 280l compared with the following explanation.

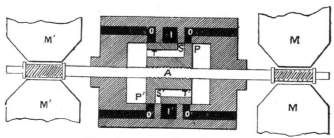

FIG. 280l.—*The Tesla Oscillator.*

pared with the following explanation. This illustration, it should be understood, is merely a diagram in which details of mechanism are altogether omitted, and only so much shown as will serve to explain the principle. We shall take the mechanical part first, and direct the reader's attention to the means by which an iron rod is made to perform very rapid to-and-fro movements in the direction of its length, and to do that

FIG. 280m.—M. Nikola Tesla.

with perfectly isochronous periods, which may be made longer or shorter at will, and which are quite independent of very considerable variations in the motive power. The diagram represents the apparatus in section, and the central part of it marked by letters P and P' is a piston through which passes what may be called a piston-rod A, which projects some distance out of the cylinder at both ends. The piston is shown in the diagram in its central position, where the impelling power has no action to move it as will presently be seen. This moving power we may assume to be the compressed air applied through the ports I I'. Just to the right of the upper one of these on the diagram will be observed in the piston a slot S opening into a hollow T, which communicates directly with the space on the left of the piston. The same arrangement, with directions reversed, is seen on the other side of the piston. If now the piston were pushed a little to the left of the position shown in the diagram, the compressed air rushing from I through the slot into the opening S T would impel the piston towards the right, and it would be carried onward by its inertia beyond the position shown in the figure towards the right, but in doing this the access of the compressed air on the left would be cut off, and the slot communicating with the space on the right hand would allow the compressed air to act in the space P, checking the further advance of the piston to the right, acting like a spring or elastic cushion, and again driving the piston to the left, during which movement the air that has done its work is allowed to escape at the outlet O O. The same cycle of operations will be rapidly repeated, but the rate of oscillation admits of control, for the larger the air chamber in which the air is compressed by the momentum of the piston and rod, the less will it be compressed and the less powerfully it will resist, while with a smaller capacity of air-chamber the more powerful will be the back spring of the imprisoned air. On the other hand, the mass that is moved may be increased ; that is the weight of the rod, &c., may be increased. In any case the oscillations will be perfectly regular, because the force which tends to bring the piston to its position of equilibrium will be always proportionate to its distance from that point. So that we have here a rod shooting in and out shuttle-wise with the utmost regularity and with almost any desired rapidity, controllable under precisely the like conditions as the stretched string already mentioned, for as the tension of the string is the measure of the force with which it strives to regain its position of equilibrium, so the compression of air in the chamber behind the piston ; and as the loaded string vibrates slower, so will the loaded piston. So much for the mechanical part of this machine, for we may omit all details of valves, &c. The electrical arrangement is very simple and of the greatest efficiency. On each projecting end of the piston are wound coils of insulated copper wire, which being shot in and out across a powerful magnetic field between the jaws of very large electro-magnets M M cut the "lines of force" to the best advantage, and from these coils alternating currents of high tension and frequency are gathered up. The vibrating rod is steadied by working in bearings (not shown). The electro-magnets are actuated as usual by coils of insulated wire surrounding their iron cores. In the motion of the moving coils there are electrical forces called into play which in mechanical effect control the movement in the same way as the air-springs, and as these electrical forces admit of certain adjustments and have calculable effects, the *mechanical period* of the machine and the *electrical* one can be made to accord, and thus to, as it were, sustain each other, and assure a perfectly isochronous periodicity, even with considerable variations of the impelling force. Though we have supposed compressed air as the actuating agent, steam has been applied in some slightly modified forms of the machine, and sometimes at the high pressure of 350 lbs. per square inch. Such is Tesla's alternating current producer, or the *Tesla Oscillator*, as it has been called. This, of course, is a very different thing from the vibrator of disruptive discharge already mentioned in connection with the experiments of Professor Hertz. Tesla also uses the disruptive discharge, and what with the high frequency and the great tension of his currents, he obtains electric oscillations of hitherto unequalled rapidity, calculable at thousands of millions per second. He claims, indeed, to be able to agitate the ether at rates of undulation comparable with those of light itself (500 billions per second). Some of the experiments he has shown certainly lend support to such an explanation. The lighting of electric lamps with but one metallic connection, and that held in a person's hand, and causing Geissler tubes to light up without any metallic connections whatever, and making gas at ordinary pressures luminous, a lump of charcoal contained in a closed glass vessel to become red-hot while the vessel is merely held in the hand, are certainly phenomena that cannot be explained on the old lines. The space between two large surfaces of metal 15 feet apart, and forming the poles of an oscillatory system, is shown to be full of light-forming influences, as when phosphorescent substances contained in closed glass vessels glow intensely, the glass being apparently no obstacle. According to Tesla, you make space and matter equally permeable to ethereal undulations when these are tuned, so to speak, to the proper frequency.

Many of the strange effects Tesla has shown are referable to the principle of electric resonance ; such are the powers of a coil with no metallic connections with any other apparatus and removed, by a distance of many feet, from any current conveying wires. Tesla's workshop was an apartment 40 feet long and 20 wide, and the wires connecting the poles of his oscillator were carried round the walls, while in the centre of the workshop stood a very large but entirely insulated coil, between the terminals of which an ordinary incandescent lamp was placed. This lamp was brilliantly illuminated when the oscillator was in action. The electric qualities of this coil were so adjusted that its currents came into tune with the ethereal vibrations propagated from the conductor round the room. But further, a single hoop of copper wire of the proper diameter and thickness could be brought into unison with the coil, and when held in the hand over the latter, even at a considerable distance, incandescent lamps attached to it were lighted up by the induced currents. Many other novel experiments have been shown by M. Tesla, but they need not here be described, as they have yet to be connected with the logical study of the entire class of phenomena. M. Tesla speaks somewhat sanguinely of being ultimately able to convey signals, and even power, to a distance, not merely with one wire but with no wires at all ! Another thing he looks forward to is to set the electricity, or rather the ether that interpenetrates the matter of the whole earth, into a state of agitation. This seems what is commercially termed " a large order ; " but we have seen that every Leyden jar, every coil, and in fact every electrical system, has its own period, and if by any possibility we could discover, or by chance hit upon the earth's electric vibration period, it is not antecedently impossible that even the com-

35

paratively small efforts of such oscillatory vibrations as we could produce, would by their accumulation agitate the earth's ether. It is well known that very small impulses, so tuned as to correspond with the natural period of a considerable mass, will produce striking mechanical effects. Thus, a troop of soldiers passing over a bridge have often been known to break down a structure that would have supported their mere weight many times over, because they were all marching together and with a step corresponding in time with the oscillatory period of the bridge. It is now always enjoined in the military orders that troops in crossing a bridge must "break step." Another familiar illustration of the accumulation of small synchronous impulses is the experiment of singing into a glass goblet the note corresponding with its vibration period. The singer merely by sustaining this note for a short time often succeeds in shivering the glass into fragments. M. Tesla believes that he has already succeeded in agitating the earth's ether to some extent ; he does at least obtain flaming purple streamers passing into the air from one end of a coil, while the other is connected with the earth.

These discoveries and theories appear likely to lead to many unforeseen results, valuable for both science and its applications, and such as may far surpass the expectations of those who take less enthusiastic views of the matter than M. Tesla and his friends do. The theoretical properties of the ether and the conditions of it, which are held capable of making it the scene and the medium of all the hitherto so-called ponderable and imponderable forces, have not been completely worked out. The experiments that have been already made show that disturbances of very different kinds may be propagated in the ether by undulations of any length from less than $\frac{1}{80000}$th part of an inch, as in the case of violet light already spoken of, to the 1.200 miles attributed to certain electrical conditions.

The foregoing sentences, describing the discoveries of Hertz and others, had not long been penned before it had become possible to announce that they had borne fruit in as extraordinary an invention as could have distinguished the close of an extraordinary century. It is the realization of what the most accomplished electrician would not long before have pronounced a dream—namely, *wireless telegraphy.* The general principle of it should not be obscure after the account of the "Hertzian waves" ; but our space does not permit a description of details of its working out in a practical form by a young Italian electrician, Signor Marconi. We have already seen that a Geissler tube, when its circuit is properly attuned, can be lighted up by the magneto-electric disturbance propagated without material contacts, and this itself would constitute a method of signalling to a distance. On the same principle, a discharge may be determined by the "wave" between conductors in certain adjustable conditions of electric tension, and in this way local circuits may be brought into play, and ordinary telegraphic effects produced, as described in the following article. The actual apparatus to receive the ethereal impulses is extremely simple—merely a little fine metallic dust (nickel and silver) in a glass tube included in the resonator circuit by a wire at each end, touching the dust. This gathers together, or coheres (hence the apparatus is called the *coherer*), under the magneto-electric influence, a local battery discharge then passes, completing a circuit, and the dust has to be shaken loose again by a mechanical agitation. Marconi has been able to signal over a distance of forty-three miles.

FIG. 281.—*Portrait of Professor Morse.*

THE ELECTRIC TELEGRAPH.

———◆———

MORE than two centuries ago a learned Italian Jesuit, named Strada, gave a fanciful account of a method by which he supposed two persons might communicate with each other, however far they might be separated. He conceived two needles magnetized by a loadstone of such virtue, that the needles balanced on separate pivots ever afterwards pointed in parallel directions ; and if one were turned to any point, the other also sympathetically moved in complete accordance with it. The happy possessors of these sympathetic needles, each having his needle mounted on a dial marked with the same letters and words similarly inscribed, would be able to communicate their thoughts to each other at preconcerted hours, by movements and pauses of the wonderful needles. The poet Akenside, when describing, in his " Pleasures of the Imagination," the effect of association in bringing ideas before our minds, illustrates his point by a happy allusion to Strada's conceit. Here is the passage :

547

" For when the different images of things,
By chance combined, have struck the attentive soul
With deeper impulse, or, connected long,
Have drawn her frequent eye ; howe'er distinct
The external scenes, yet oft the ideas gain
From that conjunction an eternal tie
And sympathy unbroken. Let the mind
Recall one partner of the various league—
Immediate, lo! the firm confederates rise.
'T was thus, if ancient fame the truth unfold,
Two faithful needles, from the informing touch
Of the same parent stone, together drew
Its mystic virtue, and at first conspired
With fatal impulse quivering to the pole.
Then—though disjoined by kingdoms, though the main
Rolled its broad surge betwixt, and different stars
Beheld their wakeful motions—yet preserved
The former friendship, and remembered still
The alliance of their birth. Whate'er the line
Which one possessed, nor pause nor quiet knew
The sure associate, ere, with trembling speed,
He found its path, and fixed unerring there."

In our own day this fancy of Strada's has been literally and completely realized in all save the convenient portability of the sympathetic dials ; but this and the other forms of apparatus which are now so familiar in electric telegraphy were produced by no sudden inspiration occurring to a single individual. Great inventions are ever the outcome not of the labours of one but of a hundred minds, and the progress of the electric telegraph might be traced, step by step, from the first suggestions, made more than a century ago, of employing, for the communication of intelligence at a distance, the imperfect electric means then known. The men who then attempted to utilize the mysterious agency of electricity failed to produce a practical telegraph, because the conditions of electrical excitation known at that time gave no scope for the realization of their project. Not the less do they deserve our grateful remembrance for the faith and energy with which they strove to overcome the difficulties of their task. Voltaic electricity was first proposed as the means of conveying signals to a distance in 1808, immediately after the discovery of the power of the pile to decompose water; and the method of communicating the signals was based upon this property. Sömmering proposed to arrange thirty-five pairs of electrodes, formed by gold pins passed through the bottom of a glass vessel containing acidulated water. Each pair of pins was marked by a letter of the alphabet or a numeral, and attached to distinct wires, which could be put into connection with a pile at the sending station. The signals were made by the gas evolved from these electrodes indicating the letter intended. The number of wires required and the slowness of working were great objections, and this system never came into practical use, although it was afterwards proposed to diminish the number of the wires from thirty-five to two—by so varying the amounts of gas given off and the periods of time as to form an intelligible system of signals. Ten or twelve years after, Mr. Ronalds, of Hammersmith, invented an ingenious system by which letters on a dial could be pointed out at a distance by frictional electricity. Two dials, on which the letters, &c., were marked, were each placed behind a screen having an aperture, which permitted only one letter to be seen at once ; and the dial was mounted on the seconds arbor of a clock with a dead-beat escapement. A pair of pith balls hung in front, insulated and connected by means of an insulated wire with the similar pair at the other end of the line, where the other clock and dial were placed. The clocks were regulated to go as nearly as possible at the same rate, so that at each end of the line the same letters were simultaneously displayed. It was easy, however, at any time to start the clocks together at the same letter by a signal previously agreed upon, and all that was really required was a synchronous motion of the discs during the time the signals were being sent. The insulated wire received from a small electrical machine a charge, which caused the pith balls at both ends to diverge; and the moment the wire was discharged, the balls collapsed suddenly and simultaneously, and this discharge was effected by the sender of the message at the instant that the letter he wished to indicate appeared at the opening in front of his dial. Since the same letter was at the same instant visible at the other end also, it was indicated to the receiver of the message by the collapse of the pith balls. Ronalds worked this telegraph experimentally with a wire 525 ft. long, but it was never adopted practically. On communicating to the Admiralty the power of his invention, he was informed that " *telegraphs of any kind were wholly unnecessary, and no other than the one in use would be adopted.*"

The memorable discovery of electro-magnetism by Œrsted in 1819 was soon followed by attempts to apply it to the production of signals at a distance. Ampère first pointed out the possibility of making an electric telegraph with needles surrounded by wires ; but he proposed to have a separate needle and wire for each signal to be transmitted. If Ampère had but thought of producing signals by different combinations of two movements, as Schweigger had before suggested for Sömmering's telegraph, thus making two wires and two needles suffice, the practical introduction of the electric telegraph would have dated some twenty years earlier than it actually did. In 1835 Baron Schilling exhibited an electric telegraph with five magnetic needles, and he afterwards improved upon it so far as to reduce the number of needles and conductors to one—for to him the happy thought seems first to have occurred that one needle could be made to produce many signals by different combinations of its movements—sometimes to the right, sometimes to the left. Thus two movements to the left might stand for A, three for B, four for C, one to left followed by one to left for D, and so on. Schilling's apparatus does not appear to have had the requisite qualities for practical working on the large scale. From this time, however, telegraphic inventions succeeded each other rapidly, and we meet with the names of Gauss, Weber, Steinheil, and others, as inventors and discoverers in the region of practical science which was now fairly opened. The first two used the magneto-electric máchine to give motion to the needle ; and the thought of using the metals of the railway line as conductors having occurred to Gauss, he found, on making the attempt, that the insulation was imperfect, but he perceived that the great apparent conductibility of the earth would allow of its being substituted for one of the metallic communicators.

But the first who succeeded, after long and persevering effort, in giving a practical character to the electric telegraph, was undoubtedly Professor Wheatstone. He had for some years been engaged in electrical researches before, in 1837—a memorable year for telegraphic inventions—he took out a patent in conjunction with Mr. W. Fothergill Cooke. In their telegraph there were five magnetic needles, arranged in a horizontal row, each needle being in a vertical position, and the various letters of the alphabet were indicated by the convergence of the needles towards the point at which the letter was marked on the dial. The first electric telegraph constructed in England was made on this system on the London and Blackwall Railway. In 1838, Messrs. Wheatstone and Cooke had reduced the number of needles to two, and many other improvements were effected in the apparatus for signalling, it being made possible for any number of intermediate stations to receive the messages. Several great railway companies erected lines with five lines of wire, but the expense of so many conductors was found to be considerable, and Messrs. Cooke and Wheatstone, after reducing the number of needles and conductors to two, ultimately (1845) patented an instrument with a single needle. It was about this time that an incident occurred which strongly drew the attention of the general public to the electric telegraph, which had, up to that time, been considered as the more immediate concern of the railway companies. A foul crime had been committed at Salthill, by the murder of a woman named Hart; and Tawell, the suspected murderer, was traced to Slough station, and there it was found he had taken the train to London; a description of his person was telegraphed, with instructions to the police to watch his movements on his arrival at Paddington. He was accordingly followed, apprehended, tried, convicted, and executed. This incident has been graphically and circumstantially described by Sir Francis B. Head, in connection with an anecdote recording a curiously expressed recognition of the value of the telegraph in furthering the ends of justice. We give the passage in full :

"Whatever may have been his fears, his hopes, his fancies, or his thoughts, there suddenly flashed along the wires of the electric telegraph, which were stretched close beside him, the following words: ' A murder has just been committed at Salthill, and the suspected murderer was seen to take a first-class ticket for London by the train which left Slough at 7.42 p.m. He is in the garb of a Quaker, with a brown great-coat on, which reaches nearly down to his feet. He is in the last compartment of the second first-class carriage.' And yet, fast as these words flew like lightning past him, the information they contained, with all its details, as well as every secret thought that had preceded them, had already consecutively flown millions of times faster ; indeed, at the very instant that, within the walls of the little cottage at Slough, there had been uttered that dreadful scream, it had simultaneously reached the judgment-seat of Heaven! On arriving at the Paddington Station, after mingling for some moments with the crowd, he got into an omnibus, and as it rumbled along he probably felt that his identity was every minute becoming confounded and confused by the exchange of fellow-passengers for strangers, that was constantly taking place. But all the time he was thinking, the cad of the omnibus—a policeman in disguise—knew that he held his victim like a rat in a cage. Without, however, apparently taking the slightest notice of him, he took one sixpence, gave change for a shilling, handed out this lady, stuffed in that one, until, arriving at the Bank, the guilty man, stooping as he walked towards the carriage door, descended the steps, paid his fare, crossed over to the Duke of Wellington's statute, where, pausing for a few moments, anxiously to gaze around him, he proceeded to the Jerusalem Coffee-house, thence over London Bridge to the Leopard Coffee-house in the Borough, and, finally, to a lodging-house in Scott's Yard, Cannon Street. He probably fancied that, by making so many turns and doubles, he had not only effectually puzzled all pursuit, but that his appearance at so many coffee-houses would assist him, if necessary, in proving an *alibi;* but, whatever may have been his motives or his thoughts, he had scarcely entered the lodging when the policeman—who, like a wolf, had followed

him every step of the way—opening his door, very calmly said to him—
the words, no doubt, were infinitely more appalling to him even than the
scream that had been haunting him—'Haven't you just come from
Slough?' The monosyllable, 'No,' confusedly uttered in reply, substan-
tiated his guilt. The policeman made him his prisoner; he was thrown
into jail, tried, found guilty of wilful murder, and hanged. A few months
afterwards, we happened to be travelling by rail from Paddington to Slough,
in a carriage filled with people all strangers to one another. Like English
travellers, they were mute. For nearly fifteen miles no one had uttered a
single word, until a short-bodied, short-necked, short-nosed, exceedingly
respectable-looking man in the corner, fixing his eyes on the apparently
fleeting posts and rails of the electric telegraph, significantly nodded to us
as he muttered aloud, 'Them's the cords that hung John Tawell!'"

So far we have followed Wheatstone and Cooke, because these gentlemen were the first who in any country made the electric telegraph a
success on the great scale. Elsewhere than in England, laboratories and
observatories had been connected by experimental lines, and models
had been exhibited to Emperors, but these two Englishmen were the first
to construct a telegraph for practical use. It must not, however, be supposed that they are entitled to be considered the exclusive inventors of the
electric telegraph, for we have already named other distinguished investigators who contributed their share to this remarkable invention. And
some years before Wheatstone and Cooke had patented their first needle
telegraph, the first ideas of a system which has largely superseded the
needles for ordinary telegraphic purposes, had presented themselves to a
mind capable of developing them into the most efficient form of telegraphic apparatus which we possess. In October, 1832, among the passengers
on board the steamship *Sully*, bound from France to the United States,
was a talented American artist who had gained some reputation in his
profession. A casual conversation with his fellow-passengers on electricity,
and the plan by which Franklin drew it from the clouds along a slender
wire, suggested to the artist the possibility of thus communicating intelligence by signals at a distance. He named his notion to a fellow-passenger,
Dr. Jackson, an American professor, who had devoted some attention to
electrical science, and this gentleman suggested several possible (and impossible) methods in which the thing might, as he thought, be accomplished. None of these suggestions, however, indicated the direction in
which the idea afterwards took practical form in Morse's hands. Jackson
had among his baggage in the hold, and therefore inaccessible on the
voyage, a galvanic battery and an electro-magnet, and these he described
to the painter by the aid of rough sketches. When, some years afterwards,
Morse had realized his ideas of electric communication, and success was
bringing him the favour of fortune, Jackson advanced a claim to a share
in the invention, and a famous law-suit, Jackson *v.* Morse, was ended by
a verdict in favour of Morse, which public and scientific opinion has
unanimously endorsed. In reference to this matter, Mr. R. Sabine, the
author of an excellent little treatise on "The History and Progress of the
Electric Telegraph," has thus placed the subject in its true light :

"Two men came together. A seed-word, sown, perhaps, by some pur-
poseless remark, took root in fertile soil. The one, profiting by that which
he had seen and read of, made suggestions, and gave explanations of
phenomena and constructions only imperfectly understood by himself, and
entirely new to the other. The theme interested both, and became a sub-

ject of daily conversation. When they parted, the one forgot or was indifferent to the matter, whilst the other, more in earnest, followed it up with diligence, toiling and scheming ways and means to realize what had only been a dream common to both. His labours brought him to the adoption of a method not discussed between them, and Morse became the acknowledged inventor of a great system. Fame and fortune smiling upon the inventor, it was natural enough that Jackson, awakening from his unfortunate indolence, should remember his share in their earlier interchange of ideas, that had, perhaps, first directed Morse's attention to the subject of telegraphy. And, although we are compelled to pronounce dishonest those attempts which Jackson made to claim the later and proper invention of Morse—that of the *electro-magnetic recorder*—and strong as is our confidence in the spotless integrity of our friend, we cannot entirely ignore Jackson—little as he has done—nor deny him an inferior place amongst those men whose names are associated with the history and progress of the electric telegraph in America."

From the time of this chance conversation with Dr. Jackson, Morse devoted his mind entirely to the subject of telegraphic communication, and although then more than forty years of age, he abandoned the profession in which he had already gained some distinction, and with the energy and elastic power of adaptibility which characterize the American mind, he gave himself up to this new pursuit to such good purpose, that a few years afterwards saw his telegraph system completely established in the United States, where the lines now exceed 20,000 miles in length. At the instigation of the late Emperor of the French, the Governments of France, Belgium, Holland, Austria, Sweden, Russia, Turkey, and the Papal States, combined to award to Professor Morse, in recognition of his services to practical science, the sum of £16,000. It was in 1836 that Morse had first brought his notions into a practical form, but his apparatus has since received many improvements at his own hands, or by the useful modifications of it which have been proposed by others. The transmitting key invented by Morse has proved a valuable piece of apparatus, and its simplicity has contributed much to the success of his invention. Telegraphs on this system were erected in America in 1837, and the Morse apparatus is now more extensively used than any other in every country.

In 1840 Professor Wheatstone had succeeded in most ingeniously applying electro-magnetism in such a manner as actually to realize Strada's sympathetic needles, by having the letters of the alphabet arranged round the circumference of a circle, and pointed at by a revolving hand. Such a dial is provided at each end of the line, and the sender of the message has only to make the index of his own dial pause for an instant at any letter; the hand of his correspondent's dial will also pause at the same letter. These dial telegraphs are particularly convenient for many purposes, as they do not require a trained telegraphist to read or send the messages. Wheatstone's plan has been greatly simplified by Breguet, of Paris, and others, and it is much used in mercantile and public establishments. From the foregoing discursive historical indication of the progress of the electric telegraph we shall now proceed to describe the systems most commonly employed in practical telegraphy, with a brief reference to some other interesting forms; and in following these descriptions, the reader will find the advantage of an acquaintance with the electrical facts discussed in the last article, with which facts we shall presume he has become to a certain extent familiar.

In every telegraphic system there are three distinct portions of the apparatus, which may be separately considered, as they may be variously combined. We have—

1°. The apparatus for producing the electricity, such as batteries, magneto-electric machines, &c.

2°. The conductors, or wires, which convey the electricity.

3°. The apparatus for sending and for receiving the messages.

Of the first we shall have little to add to what has been said in the last article; and before entering upon the description of the second, it will be better to discuss the third division.

TELEGRAPHIC INSTRUMENTS.

TELEGRAPHS may conveniently be classed according to the mode in which the actions of the sender produce their effect at the point where the message is received. A first class may include those in which the current is made to deflect magnetized needles; a second may comprise those in which the current, by magnetizing soft iron, causes an index to travel along a dial and point to the letter intended; a third may embrace those in which the same action on soft iron is made to print the despatches, either in ordinary type or in conventional signs; while in a fourth class we may put the instruments which give their indications by sounds only. It is obvious that in some of these systems signs only are used, and a special training and acquaintance with the symbols is necessary, while in the rest the ordinary alphabetic letters are shown or recorded. In the former case the apparatus is simpler, and therefore for the general business of public telegraphy it is almost exclusively employed; while for private purposes, where it is often required that the messages should be dispatched and received by persons not acquainted with the symbolic language, the dial telegraph, or that which prints the message in ordinary characters, will continue to be employed, in spite of the greater complexity and greater liability to derangement of the apparatus.

In the needle telegraphs the essential part of the apparatus is a multiplier (page 493), having its needle mounted vertically on a horizontal axis, to which is also attached an indicator, visible on the face of the instrument, and formed either of a light strip of wood, or of another magnetized needle, having its poles placed in the reverse position to those of the needle within the coil. When the current is sent through the latter, the index is deflected to the right or left, according to the direction in which the current passes. Fig. 282 represents the exterior of one of Wheatstone and Cooke's double-needle instruments, now almost entirely superseded, where needles are used at all, by the single-needle instrument. The face of the instrument is marked with letters and signs, which were supposed to aid the memory of the telegraphist, and the movements of the needles were chosen rather with that view than any other. We need not here give the code of signals, as the double instrument is now obsolete, and the code for the single-needle instrument, which was devised by Wheatstone and Cooke, has been in most cases superseded by one corresponding with the Morse code, a deflection to the right representing a dot, and a deflection to the left a dash.

The smaller case surmounting the instrument, Fig. 282, contains a bell or alarum, which serves to call the attention of the clerk at the receiving station. The first electric bell-alarum was invented by Wheatstone and

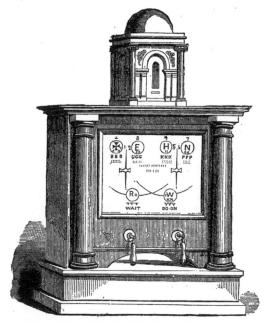

FIG. 282.—*The Double-Needle Instrument.*

Cooke. It was simply a clock alarum, put in motion by a wound-up spring. The spring was released at the proper moment by a detent, which was removed by the attraction of a soft iron armature to the core of a small electro-magnet, formed by the line wire itself; but when the current, on account of the length of the line, was too weak to produce a sufficiently strong electro-magnet, Wheatstone caused it to close the circuit of a local battery. The electric alarum has been modified in a thousand ways, and as electric alarums or bells are now coming into common use in hotels, and even private houses, we give in Fig. 283 a representation of one of the simplest forms, in which the bell is rung continuously by the electric current so long as the circuit is closed. The action is very simple: a soft iron armature, A, is attached to the steel spring, B, and prolonged into a hammer, C, which strikes the bell, D, every time the armature is attracted to the electro-magnet. The armature and the spring, E, form part of the circuit, which is continued by connectors to F, and through the coils to G. The spring, E, does not follow the armature in its motion towards the electro-magnet, and consequently the circuit is broken before the armature touches the magnet; but the hammer strikes the bell, and the elasticity of the spring,

B, brings the armature back into contact with E, the circuit is closed, and the motions are repeated, so that the bell is struck a rapid succession of blows. This *make-and-break* movement is precisely similar to that with which Ruhmkorff's coils are usually provided.

FIG. 283.—*Electro-Magnetic Bells.*

Below the dial of the instrument, in Fig. 282, may be seen two handles. Each of these is connected with an arrangement constituting the transmitting apparatus, by which the metallic contacts are varied according to the position of the handles. When the handle is vertical, all communication with the battery in connection with the instrument is cut off, but the coils are ready to receive any current from the line-wires. When the handle is turned to the right or left, the contacts are such that the battery current flows into the line, and deflects to the right or left the needles of both receiving and transmitting instruments. The single-needle instrument as now made is of a very simple and inexpensive construction, and it is the form principally used in connection with the working of lines of railway. One may see at every station in the United Kingdom the little vertical needle, mounted in the centre of a small perfectly plain green dial-plate; for the letters and signs with which it was formerly the practice to cover the dial have been found to distract the eye more than they aid the memory. A boy will after a few weeks' practice learn to read the signals and to transmit messages with considerable rapidity.

The field telegraph lines, which are used in actual warfare to enable the commander of an army to communicate with every part of his forces, require as the essential condition for their construction rapidity of erection and removal, and the greatest possible simplicity and portability in the sending and receiving instruments. The wires are fastened to trees, or other fixed supports, where such are available, but artificial supports are provided in light poles which admit of being readily planted in the ground and removed. In cases where it is inexpedient or impossible to use these, the conductor may be laid along the ground, but must then be well insulated with some non-conducting material, which is capable of withstanding the action of the weather. A kind of cable is usually employed, in which is the conductor, made of copper, protected and strengthened by hemp fibres and covered with some non-conducting material. No form of needle telegraph instrument could be simpler than that represented in Fig. 284, which has been designed for military purposes. The communicator, or

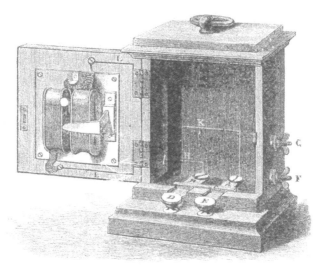

Fig. 284.—*Portable Single-Needle Instrument.*

transmitting apparatus, here shows an arrangement very compact, and not easily deranged. The springs, A B, press against the piece of metal marked C, with which good contact is insured by providing the springs with several projecting steel points. D, E are finger-keys of ebonite or ivory; underneath are two points of a metallic conductor on which the springs can be pressed down by a touch of the finger. This conductor is in communication with the binding-screw, F, from which a wire proceeds to the negative or zinc end of the battery, while the piece, C, is in metallic connection with G, to which a wire proceeding to the positive or copper end of the battery is attached. From B a wire, H, communicates through the hinge with one end of the coil, the upper end of which is connected through the upper hinge with a binding-screw not visible in the figure, and to this the end of the line conductor is attached. From A a wire K passes to another binding-screw, by which the earth connection is made. A current arriving by the line traverses the coils and passes through H and B into C, hence by A into the earth through K. When D is depressed the current from the battery passing from G through C, A, and K, into the earth, and thus to the distant station, returns through the coils of the instrument there and along the line wire, through the coils, L L, and by H B, D and F, to the negative pole of the battery. The reader will have little difficulty in tracing the course of the reverse currents, whether sent or received, which deflect the needles in the opposite direction.

The field telegraph instrument selected by the War Department of the United States Government is also extremely simple, communicating its signals, not by the deflections of a needle, but by the blows on an electromagnet of its armature. The letters are indicated by various combinations of two signals—one, a single stroke of the armature ; and the other, two blows in very rapid succession. The alphabet used is the " General Service Flag Code" of the American army and navy, and the signal numerals of this code are indicated by contacts of the transmitting key—one contact producing a single blow of the armature, implying the numeral 1, and two rapidly succeeding contacts causing two blows, which stand for the numeral 2. The signals are read merely by the sound made by the stroke of the armature. In the table below the code is given, dots being used to represent the contacts of the key in the " sending" instrument, and the blows of the armature in the " receiving" instrument—the single dots standing for one contact or sound, and the double dots for the double blows :

Letters.	Flag Code.	Telegraph Signals.	Letters.	Flag Code.	Telegraph Signals.
A.	2 2	.. ..	N	1 1	. .
B	2·1 1 2	.. . . ..	O	2 1	.. .
C	1 2 1	. .. .	P	1 2 1 2	. .. . ..
D	2 2 2	.. . ..	Q	1 2 1 1	. .. . .
E	1 2	. ..	R	2 1 1	.. . .
F	2 2 2 1	.. ..	S	2 1 2	.. . ..
G	2 2 1 1	.. .. . .	T	2	..
H	1·2 2	. .. ..	U	1 1 2	. . ..
I	1	.	V	1 2 2 2	. .. .
J	1 1 2 2,	. . .. ..	W	1 1 2 1	. . .. .
K	2 1 2 1	.. . .. .	X	2 1 2 2	.. . .
L	2 2 1	.. .. .	Y	1 1 1	. . .
M	1 2 2 1	. .. .. .	Z	2 2 2 2	.. ..

There are similar signals for the numerals and for a few often-recurring syllables.

The telegraphs we have hitherto described leave no record of the despatches sent, and hence the messages cannot be read at leisure, and errors which may occur in the transmission cannot be traced to their source. A system which registers the messages as actually received has plainly many advantages over those which merely give a visible or audible signal without leaving any trace. Hence many contrivances have been proposed for making the receiving apparatus print the message in ordinary characters. Such instruments are necessarily very much more complicated in their construction than those we have already mentioned, and by no means so simple as the system we are about to describe, namely, the Morse Telegraph, which is now so largely used, being universally adopted in America and on the continent of Europe ; and, since the telegraphic communication in Great Britain came into the hands of the Post-office authorities, here, also, the Morse is the system most approved.

The general arrangement of the transmitters, batteries, receiving instruments, &c., should be first studied in its simplest form, as represented by the diagram, Fig. 285. M represents the vertical coils of an electro-magnet upon which we are supposed to be looking down ; the armature, A, is attached to a lever, F, which, by the attraction of the electro-magnet is there-

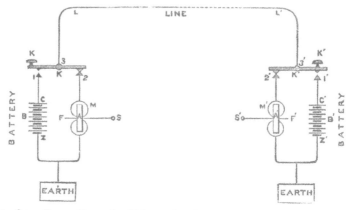

Fig. 285.—*Connections of a Telegraphic Line, with Morse Instruments.*

fore drawn down. In the position of the connections, as represented, no current is passing, but if K be pressed down so as to make connection at 1, at the same time it is broken at 2, a current will pass in from the positive pole of battery, B, into the line by 1, 3, L, L', and through 3', 2' through the coils of the electro-magnet at M' into the earth, and so back to the negative pole, Z.   The armature, A', will be attracted so long as the current continues.   Similarly, contact made at 1' and broken at 2', will affect the electro-magnet, M, from the battery at B'.   It should be noticed here that it is not a question of the reversal of currents sent from the same battery ; the key merely enables the operator to send a current in one direction, so as to affect the distant electro-magnet whenever or so long as he depresses the key.   We shall now examine the construction of the Morse receiving apparatus, one of the most complete forms of which is depicted in Fig. 286. In the present description we wish the reader to consider only the portion of the apparatus towards the left, and to suppose the absence of the electro-magnet at the right-hand side, with all the appliances immediately connected with it.   He must regard the electro-magnet, A, as corresponding with M' in Fig. 285, and remember that it is in the power of the distant operator at K to throw the current of his battery through the coils of A, by simply depressing his key.   When the current passes the armature, B, it is attracted, and the lever, C, to which it is attached, turns on its bearings at D, and the end, E, of its longer arm is pressed upwards.   At this end of the lever, in the earlier form of the instrument, was a blunt steel point which, while the armature was attracted to the electro-magnet, was

pressed into a shallow groove in a metallic roller. Between the roller and
the steel point a paper ribbon, half an inch wide, K, was unwound from
the drum, L, by the two rollers, M and N, which grip the paper between
them as they are turned by clockwork within the case, F.

   An important improvement was effected when, instead of steel points
for embossing the message, the Morse instrument was provided with an
arrangement for printing the signals in ink; since the pressure required
for embossing the paper is considerably greater than that needed merely
to bring it into contact with the edge of a little inked disc.   In the inking
arrangement the strip of paper travels just below the margin of a vertical
disc, turned by the clockwork, and having its plane parallel to the length

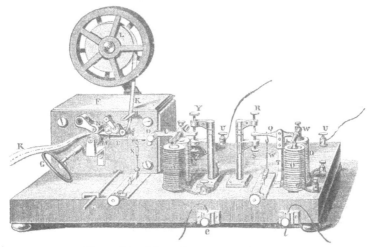

FIG. 286.—*Morse Recording Telegraph.*

of the paper strip.   The narrow edge of this disc is kept charged with
printer's ink, which it receives from a roller.   The end of the lever connected with the armature of the electro-magnet is formed of a light strip
of metal carrying a narrow projection at the end, over which the paper
passes, just beneath, but not touching, the inking disc.   When the current
passes, the little projection is lifted up, and raises the paper into contact
with the ink, printing either a dot or a dash according to the duration of
the current.   The amount of force required to raise an inch or two of the
length of the paper ribbon through a space not greater than the twen-
tieth of an inch is but small, and much less than would be required to
emboss the paper ; so that in a great many cases the part of the apparatus which is represented in Fig. 286, on the right, may be dispensed with.
In other cases it is, however, necessary; as when, from the length of the
line, the currents are too feeble to give clear indications with the printing
lever; and we shall, therefore, presently describe its arrangement and
purpose.

   The clockwork is actuated by a spring, wound by the handle G, but its action is suspended by a detent, which is released by touching the lever H. When the clockwork is in action and the current constantly circulating in the coils, a continuous line, parallel to the length of the ribbon, would be printed upon it, in consequence of the contact with the inking-disc, P, being maintained ; but when a momentary current only rushes through the coils, the armature attracted but for an instant, gives rise to merely a dot on the passing paper, while a current of a little duration will cause the paper to be marked with a short line or dash.

The dot and the dash are the elementary signs of the Morse code of signals, and these are producible according to the time the contact key is held down at the distant station. By employing various combinations of these two signs, the letters of the alphabet, numerals, &c., are indicated. In selecting the combinations Professor Morse had regard to the frequency with which the different letters recur in the English language. Thus, for the letter E, which is more frequently used than any other, the symbol chosen was a single dot ; and for T, which is the next most frequently employed, the dash was plainly the most appropriate; then the four only possible combinations of the signs in pairs fell to the next most frequent letters, and so on. The following table gives the complete Morse code. The eye of the reader will doubtless detect a kind of symmetry in the arrangement of the signs for the first five and last five numerals :

ALPHABET.

Letter.	Sign.	Letter.	Sign.	Letter.	Sign.
A	· —	J	· — — —	T	—
Ä	· — · —	K	— · —	U	· · —
B	— · · ·	L	· — · ·	Ü	· · — —
C	— · — ·	M	— —	V	· · · —
D	— · ·	N	— ·	W	· — —
E	·	O	— — —	X	— · · —
É	· · — · ·	Ö	— — — ·	Y	— · — —
F	· · — ·	P	· — — ·	Z	— — · ·
G	— — ·	Q	— — · —	Ch	— — — —
H	· · · ·	R	· — ·		
I	· ·	S	· · ·		

NUMERALS.

Numeral.	Sign.	Numeral.	Sign.
1	· — — — —	6	— · · · ·
2	· · — — —	7	— — · · ·
3	· · · — —	8	— — — · ·
4	· · · · —	9	— — — — ·
5	· · · · ·	0	— — — — —

PUNCTUATION, &c.

	Sign.		Sign.
Full stop	- - - - -	*Fraction-line	— — — — — —
Colon	— — — - - -	†Inverted com-	
Semicolon	— - — - — -	mas	- — - - — -
Comma	- — - — - —	†Parenthesis	— - — — - - —
Interrogation	- - — - - -	Italics or un-	
Exclamation	— — - - — —	derlined	- - - — - -
Hyphen	— - - - - —	New line	- — - — - -
Apostrophe	- — — — — -		

* To be placed between the numerator and demoninator of a vulgar fraction.
† To be placed before and after the words to which they refer.

OFFICIAL SIGNALS.

	Sign.		Sign.
Public mes-		Call	- — - - — - —
sage	- - -	Correction, or	
Official Tele-		rub out	- - — -
graph mes-		Interruption	- - - - - - - -
sage	- —	Conclusion	- — - — - - — -
Private mes-		Wait	- — - - -
sage	- — — -	Receipt	- — - - — - — — -

The length of a dot being taken as a unit, the length of a dash = 3 dots.
The space between the signs composing a letter = 1 dot.
   „   „   two letters of a word   = 3 dots.
   „   „   two following words   = 6 dots.

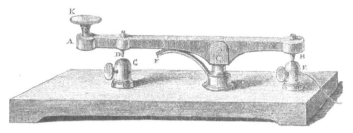

FIG. 287.—*Morse Transmitting Key.*

Fig. 287 is a view of the Morse transmitting key. A B is a brass lever, moving in bearings at C, and provided at the end of its longer arm with a large knob or button of some insulating material. Steel pins are screwed in at B and D, and they are so adjusted that while that at B is pressed against the

36

projection, E, by the action of the spring, F, when the knob, K, is pressed, contact is broken at B, and established at D.   D and E are each provided with a binding-screw, so that wires may be attached in the manner indicated in Fig. 285.   When the key is in the position shown, a current arriving by the line-wire passes from the fulcrum, C, of the lever through the con-

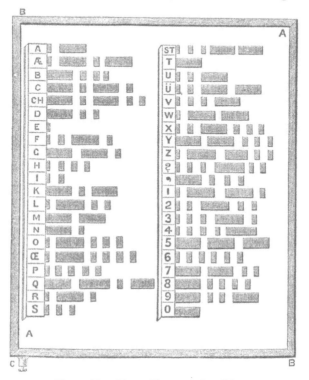

FIG. 288.—*Morse Transmitting Plate.*

tacts into the apparatus.   When the knob is pressed down the battery current enters the lever by the contact at D, and passes into the line from the fulcrum, C.   The clerks who are called upon to transmit messages usually soon learn to time the contacts very accurately in accordance with the code of signals, so as to produce the dashes and lines with accuracy.   However, with certain persons some difficulty was found in acquiring the requisite uniformity, and to obviate any objection on this score, Morse invented an arrangement for facilitating the signalling, which is represented in Fig. 288. This is a smooth tablet of a non-conducting substance, such as ivory, except the shaded portions, which are plates of metal having their surfaces even with that of the ivory, and all soldered to a plate of metal beneath the ivory, which places them all in communication with each other and with the binding-screw, C. The lengths of the strips of metal and those of the spaces between them correspond with the dots and dashes of the Morse alphabet as marked on the tablet. The battery wire is connected with the binding-screw, C, and the line-wire terminates in an elastic and flexible coil of insulated wire, which is attached to a short rod having an insulated handle and terminated by a blunt platinum point. This the transmitter takes in his hand and draws uniformly along the line of metal strips belonging to the letter which he wishes to telegraph. The circuit is closed while the point of the style is passing across the metallic strips. This arrangement appears to be but little used, but it is nevertheless admirable for its simplicity, and is described here as a good illustration of the mode in which the varied duration of the contacts is able to produce the signals of the Morse alphabet. With the ordinary transmitting key a clerk is able to telegraph, on the average, twenty or twenty-five words in a minute, but the receiving apparatus is capable of recording three times as many. Morse also invented a system of transmitting the messages automatically, by setting up the message in a kind of type, just as ordinary letters are arranged for printing. The type, if it may be so called, had simple projections like the slips of metal, corresponding with each letter in Fig. 288. The lines of the message were drawn under a contact-lever, which closed the circuit when lifted up by the projections. Thus the speed of transmission could be very greatly increased, and a single wire and apparatus had its capacity of conveying a great number of messages in a given time proportionately enlarged.

We have now to ask the reader's attention to the details of the apparatus in Fig. 286, the use of which has not already been pointed out. The electro-magnet, O O', and the parts immediately connected with it, form what is called a *relay*. The object of this may be illustrated by supposing that the instrument is at one end of a long line, such as that between Edinburgh and London. Let us suppose it is at Edinburgh: the currents sent from London by a battery of convenient size might not be powerful enough to magnetize the soft iron of A with sufficient intensity to give clearness to the signals. They are, therefore, made to circulate in the electro-magnet, O, where they act by attracting the armature, W, which has the form of a split tube of soft iron, attached to a very light lever, Q, adjusted with great delicacy, and so that it moves by little magnetic force. The end of the lever works between two adjustable screws, R and S, which are electrically insulated, except that R is in communication with one extremity of the coils of the electro-magnet, A. Q is in metallic communication through the pillar, T, and the binding-screw, U, with the zinc end of a battery at Edinburgh, which is called the local battery, the other pole of which communicates with the other ends of the coils, A, through the screw, U'. When no current from London is passing through O, Q is held down by the spring, W', and the circuit of the local battery is broken; but the instant the line-current passes, the armature, W, is attracted, and Q makes contact with R, the current from the local battery rushes through the coils, A, and the appropriate movements of the printing lever are effected by its action. X is a spring for drawing down the lever, and it is provided with a screw for adjusting its tension, and Y, Z, are screws for limiting the extent of motion of the lever; under P is the little projection by which the band of paper is pressed against the inking-disc; *l* and *e* are respectively the screws for the line and earth connections.

An extremely ingenious system of signalling, by which the speed could be greatly increased, has been devised by Sir Charles Wheatstone, and is largely adopted by the British postal authorities. In this system the message is first translated into telegraphic language by a machine, which punches certain holes in a strip of stiff paper. The apparatus originally designed for this purpose by the inventor is thus described by him in the Juror's Report, International Exhibition of 1862 :

"Long strips of paper are perforated by a machine constructed for the purpose, with apertures grouped to represent the letters of the alphabet and other signs. A strip thus prepared is placed in an instrument associated with a source of electric power, which, on being set in motion, moves it along, and causes it to act on two pins in such a manner that when one of them is elevated the current is transmitted to the telegraphic circuit in one direction, when the other is elevated it is transmitted in the reverse direction. The elevations and depressions of these pins are governed by the apertures and intervening intervals. These currents, following each other indifferently in these two opposite directions, act upon a writing instrument at a distant station in such a manner as to produce corresponding marks on a slip of paper, moved by appropriate mechanism.

"The first apparatus is a *perforator*, an instrument for piercing the slips of paper with the apertures in the order required to form the message. The slip of paper passes through a guiding groove, at the bottom of which an opening is made sufficiently large to admit of the to-and-fro motion of the upper end of a frame containing three punches, the extremities of which are in the same transverse line. Each of these punches, the middle one of which is smaller than the two external ones, may be separately elevated by the pressure of a finger-key.

"By the pressure of either finger-key, simultaneously with the elevation of its corresponding punch, in order to perforate the paper, two different movements are successively produced : first, the raising of a clip which holds the paper firmly in its position; and secondly, the advancing motion of the frame containing the three punches, by which the punch which is raised carries the slip of paper forward the proper distance. During the reaction of the key consequent on the removal of the pressure, the clip first fastens the paper, and then the frame falls back to its normal position. The two external keys and punches are employed to make the holes, which, grouped together, represent letters and other characters, and the middle punch to make holes which mark the intervals between the letters.

"The second apparatus is the *transmitter*, the object of which is to receive the slips of paper prepared by the perforator, and to transmit the currents in the order and direction corresponding to the holes perforated in the slip. This it effects by mechanism somewhat similar to that by which the perforator performs its functions. An eccentric produces and regulates the occurrence of three distinct movements : 1. The to-and-fro motion of a small frame which contains a groove fitted to receive the slip of paper, and to carry it forward by its advancing motion. 2. The elevation and depression of a spring-clip, which holds the slip of paper firmly during the receding motion, but allows it to move freely during the advancing motion. 3. The simultaneous elevation of three wires placed parallel to each other, resting at one of their ends over the axis of the eccentric, and their free ends entering corresponding holes in the grooved frame. These three wires are not fixed to the axis of the eccentric, but each end of them rests against it by the upward pressure of a spring ; so that when a light pressure is exerted on the free end of either of them, it is capable of being separately depressed. When the slip of paper is not inserted the eccentric is in action ; a pin attached to each of the external wires touches during the advancing and receding motions of the frame a different spring ; and an arrangement is adopted, by means of insulation and contacts properly applied, by which, while one of the wires is elevated, the other remains depressed ; the current passes to the telegraphic circuit in one direction, and passes in the other direction when the wire before elevated is depressed, and *vice versâ;* but while both wires are simultaneously elevated or depressed the passing of the current is interrupted. When the prepared slip of paper is inserted in the groove, and moved forward whenever the end of one of the wires enters an aperture in its corresponding row, the current passes in one direction, and when the end of the other wire enters an aperture of the other row, it passes in the other direction. By this means the currents are made to succeed each other *automatically* in their proper order and direction to give the requisite variety of signals. The middle wire only acts as a guide during the operation of the current.

" The wheel which drives the eccentric may be moved by the hand, or by the application of any motive power. Where the movement of the transmitter is effected by machinery, any number may be attended to by one or two assistants. This transmitter requires only a single telegraphic wire.

"The third apparatus is the *recording* or *printing apparatus*, which prints or impresses legible marks on a strip of paper, corresponding in their arrangement with the apertures in the perforated paper. The pens or styles are elevated or depressed by their connection with the moving parts of the electro-magnets. The pens are entirely independent of each other in their action, and are so arranged that when the current passes through the coils of the electro-magnet in one direction, one of the pens is depressed, and when it passes in the contrary direction the other is depressed ; when the currents cease, light springs restore the pens to their elevated points. The mode of supplying the pens with ink is the following : A reservoir about an eighth of an inch deep, and of any convenient length and breath, is made in a piece of metal, the interior of which may be gilt in order to avoid the corrosive action of the ink ; at the bottom of this reservoir are two holes, sufficiently small to prevent by capillary attraction the ink from flowing through them ; the ends of the pens are placed immediately above these small apertures, which they enter when the electro-magnets act upon them, carrying with them a sufficient charge of ink to make a legible mark on a ribbon of paper passing beneath them. The motion of the paper ribbon is produced and regulated by apparatus similar to those employed in other register and printing telegraphs."

The mode by which Wheatstone proposed to indicate the letters was novel, consisting in dots only, the numbers and positions of which in two lines along the paper ribbon distinguished the letters—the system of combining the symbols being still identical with the Morse code, only the dash was replaced by a dot in the lower lines :

WHEATSTONE'S DOT SIGNALS.

A	B	C	D	E	F	G	H	I	J

MORSE'S DOT AND DASH.

A	B	C	D	E	F	G	H	I	J

A single dot in the upper line stood for E, in the lower line for T; a dot in the upper line, followed by one in the lower line a little to the right, represented A ; one in the lower line, followed by another in the upper line, indicated N; and so on. By the dot printing it is said that Wheatstone would signal 700 letters per minute. There were, however, objections to the new code of signals : all the world had agreed to use the Morse alphabet, and it was perhaps less liable to incorrect reading; and for other reasons this more rapid signalling was unsuitable for submarine lines. The apparatus has therefore been modified to suit the dot and dash system of signals, and great improvements have been effected by Sir Charles on the original instruments, with a view of increasing the rapidity of transmission as much as possible. The paper as punched for the Morse signals shows a row of equidistant holes in the middle, by which the paper is guided uniformly forward, and in the outer rows are holes arranged in pairs, either exactly opposite to each other or obliquely—the former produce dots at the receiving station, the latter dashes. From 60 to 100 words can thus be sent and printed in one minute, and the automatic transmitting system can be applied to the needle, or any other form of telegraph.

After a clerk has for some time been habituated to working with the Morse instrument, he is able to read the message from the different sounds made by the armature, as dashes or dots are respectively marked, and he usually *listens* to the message, and transcribes it at once into ordinary language by the ear alone. This observation soon led to the adoption of sound alone as the means of signalling, and an instrument on this plan has already been referred to.

Among the more remarkable forms of recording telegraphs, that of Hughes may be mentioned, in which the message is printed at the receiving station in distinct Roman characters; and as only a single instantaneous current is required to be sent for each letter, the speed with which a message can be dispatched is about three times as great as with the Morse instrument. These advantages are, however, obtained only at the cost of great delicacy and complexity in the apparatus, so that it is unfit for ordinary use, although it is much employed on important lines, where competent operators and skilled mechanics and electricians are at hand to keep it duly regulated. This machine is too complicated for a full description in these pages, although it is the best form of type-printing telegraph, and possesses a special feature in the fact that the printing is done whilst the wheel carrying the types is in rapid rotation. The reader will find full and untechnical descriptions of this and of all the more important forms of telegraphic apparatus in Mr. R. Sabine's useful " History and Progress of the Electric Telegraph," or in Lardner's work as edited by Sir Charles Bright.

From the numerous forms of dial telegraphs we select two for description. All these instruments are characterized by what is called the " step-by-step" movement, and differ in their mechanical details, and in the nature of the apparatus for producing the currents, some being driven by electro-magnets and others by galvanic batteries. Their principle may be easily explained. Suppose that a ratchet-wheel, having twenty-six teeth, is mounted on an axis carrying a hand over a dial having the letters of the alphabet inscribed upon it. A simple arrangement in connection with an electro-magnet, somewhat like the escapement of a clock, will serve to advance the wheel by one tooth each time a current passes. The diagram, Fig. 289, will at once make this principle clear. E is the electro-magnet, F the armature, separated by the spring, S, from the magnet, except when the current passes, when the catch, C, draws down the tooth in which it is engaged, so that a tooth passes under the point at D ; and when the current ceases, the spring, S, brings up the catch to engage the succeeding tooth, and thus the hand moves step by step in the direction of the arrow, advancing each time the electric circuit is closed by one twenty-sixth of a revolution. In Fig. 290 is represented lecture-table models of a step-by-step indicating and transmitting instrument, as constructed by M. Froment, of Paris. These instruments are supposed to be at the extremities of a long line of wire. The left-hand figure is the manipulator, or sending instrument, in which the operator has merely to quickly turn round the index in the direction of the hands of a watch, by means of the knob, P, until it points to the desired letter, pause at the letter for an instant, and then quickly continue the movement until his index points to the cross at the top of the dial, where he pauses if the word is spelt out, and, if not,

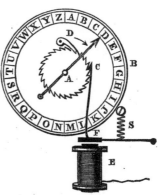

FIG. 289.—*The Step-by-step Movement.*

continues the rotation until he arrives at the next letter, and so on. All

FIG. 290.—*Froment's Dials.*

these movements and pauses the hand on the indicator will accurately repeat, and the reason of this may be seen by observing that the battery contacts are made by the projections on the metallic wheel, R, which turn with the index. The spring, N, is always in contact with the wheel, but the spring, M, has such a shape that contact is alternately made and broken as the projections and spaces pass it. It is obvious that the needle of the indicator will therefore advance over the same letters as the index of the communicator.

A very elegant dial instrument has been invented by Sir Charles Wheatstone, in which magneto-electric currents are made use of. In Fig. 291 communicator and indicator are represented mounted in one case, or small box. The larger dial is the communicator, and its circumference is divided

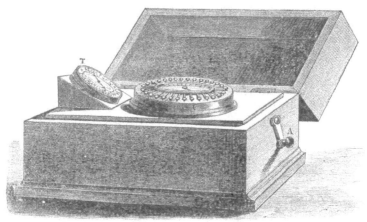

FIG. 291.—*Wheatstone's Universal Dial Telegraph.*

into thirty equal spaces, in which are the twenty-six letters of the alphabet, three punctuation marks, and a +. In an inner circle are two series of numerals and other signs. About the circumference of the dial are thirty small buttons or projecting keys, coveniently arranged, so as to be readily depressed by the touch of a finger. Inside of the box a strong permanent horse-shoe magnet is fixed, and near its poles a pair of armatures of soft iron cores with insulated wire coils revolve when the handle, A, is turned, as in the machines described in the last article. In this manner a series of waves or short currents of electricity are produced in the conductors when the circuit is complete, and the currents are alternately in opposite directions, so that fifteen revolutions of the coils will produce fifteen currents in one direction and fifteen in the other. A pinion on the same spindle as the coils works with a wheel on the axis carrying the pointer on the dial, so that the pointer makes a complete revolution as often as the handle, A, makes fifteen turns. Each of the thirty currents will pass through the indicator, I, and through the line to the distant station, where they will, by a step-by-step movement, advance the needle of the indicator. So that the hand of the dial and the needle of the indicator at the sending station, and that of the indicator at the distant station, will all simultaneously be pointing to the same letter on their respective dials; and they would continue to move round these, ever pointing to the same letter, so long as the handle, A, is turned. How, then, is the sender to cause the needle of his correspondent's instrument to pause at any desired letter? Not by stopping the revolution of the handle, A, for that could not be done so as to send just the right number of currents, inasmuch as the rotating armatures could not be instantly stopped. The mode of causing the indicators to pause at any required letter is as simple as it is ingenious. It has been already mentioned that the step-by-step movement takes place at every current which passes through the line, including the two indicators, and that thirty such currents pass at each revolution of the pointer of the communicator. But when these currents no longer flow, the indicators, of course, stop; and the stoppage of the movements is reconciled with the continuous production of the currents by having a series of little levers, each connected with one of the buttons, and so arranged that when one of these has been pushed down, the lever stops the revolution when it has come round of an arm on the same central axis as the pointer, and riding loosely on a hollow spindle, which bears the toothed wheel, driven by the pinion already spoken of. The projecting arm is provided with a spring, which falls between the teeth of the wheel, so that the arm is with certainty carried round with the wheel. But where a button has been pushed down, its lever catches the arm, lifting its spring away from the teeth of the wheel. So long as the key remains down, the arrested arm makes a short metallic circuit by its contact, and no currents pass into the line, for they take the shortest path. The key is raised only when another is depressed, and then the arm and the pointer immediately resume their revolution until they again become stationary at the letter corresponding with the key which has been pushed down. Suppose the key of +, the zero of the dial, to be down, which is the proper condition of the apparatus when a message has to be dispatched. The operator having rung a bell at the distant end, to call the attention of the person who receives his message, begins to turn the handle, A, at the rate of about two revolutions per second. In this state of affairs no current is passing into the line, and the fingers of both his communicator and indicator remain stationary, as does also that of the indicator at the distant end of the line. Now, suppose he has to spell the word "FOX." He turns the handle A continuously with his right hand the whole time he is sending the message; and, manipulating the keys with his left, he depresses that opposite to the letter F. By this action the key opposite + is raised, for the levers are pressed into notches against a watch-chain, which has just enough *slack* to allow one lever to enter a notch, and therefore the pressure of another lever always raises the key last depressed. When the operator presses down the F key, the + rises, the radial arm is instantly released, and with the index is carried on to F, where it stops; and the contacts will have, during that movement, sent six currents into the line, so that the fingers of both indicators will also point to F. When the pointer of the communicator has made just a visible pause at F, he pushes down the key of O, and all the three pointers recommence their journeys towards that letter. The operator must, of course, wait until they have reached it and paused an instant, when he depresses the button opposite X; and when the index has pointed at that, he pushes down the + key, whereby the fingers all arrest their movements at that point, indicating that the word is completed. In the case supposed the word is completed by a single revolution of the pointers; but this is, of course, not usually the case ; thus, in indicating the syllable " PON," nearly three complete revolutions would be required.

This admirable little instrument was designed for the use of private persons, and is largely used in London and elsewhere. Its great compactness and simplicity of operation render it highly suitable for this purpose. There is no battery required, and all the inconvenient attention demanded by a battery is therefore dispensed with. On the other hand, the magnets gradually lose their power, and after a time must be re-magnetized ; and the electro-motive force developed in these instruments is insufficient for lengths of line much exceeding 100 miles   For shorter lines, and for the purposes for which they are designed, these instruments are perfection.

Very interesting forms of telegraph are those in which a despatch is not merely written or printed, but actually transcribed as a *facsimile* of the writing in the original ; and in this way it is possible for a design to be drawn telegraphically at the distance of hundreds of miles.   Like the Hughes' printing telegraph, the instruments which produce these apparently marvellous results require synchronous movements at the two stations.   But although they are scientifically successful, there appears to be no public demand for these copying telegraphs. One of the best known is Bonelli's, which dispatches its messages automatically when they have been set up in raised metal types precisely similar to the Roman capitals in the type of the ordinary printer.   In Bonelli s and most other copying telegraphs the impressions are produced by chemical decompositions— effected at the receiving station on the paper prepared to receive the message.   By Bonelli's instrument it is said that when the type has been set up, messages can be sent at the extraordinary rate of 1,200 words in one minute of time !   The action of this system is such that it is proved to be possible to reproduce in a few seconds—at York, say—the very characters of a page of type the moment before set up in London.   The limits of our space will not admit of details of this invention ; but we here place before the reader a *facsimile* of the letters printed by it at the receiving stations.

BONELLI'S CHEMICAL TELEGRAPH

We have to describe two other forms of instruments for receiving telegraphic signals, both contrived with consummate skill by Sir William Thomson, and, though exhibiting no new principle in any of their parts. both fine examples of beautiful adjustment of materials for a desired end. In these forms of apparatus, the delicacy of the mechanical construction, and the accurate relations of one part to another, have produced results of the greatest practical importance. Fig. 292 represents the *mirror galvanometer,* an instrument which has not only proved of the highest value in scientific researches, but is of the first importance in submarine telegraphy. It is in principle nothing more than the single-needle telegraph, and it is exceedingly simple in construction.   A very small and light magnet, such as might be formed by a fragment of the main-spring of a watch, ⅜ths of an inch long, say, is attached to the back of a little circular mirror, made of extremely thin silvered glass, also about ⅜ths of an inch in diameter. The mirror and magnet are suspended by a single cocoon-fibre, so fine as to be almost invisible, in the centre of a coil, A, of fine silk-covered copper wire. In front of the suspended mirror, in the axis of the coil, is placed a lens of about four feet focal distance, and opposite to this is a screen having a

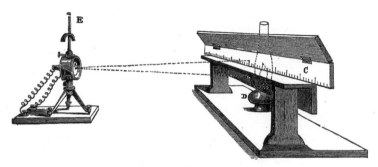

FIG. 292.—*The Mirror Galvanometer.*

slit, B, in the centre, behind which is placed a paraffin lamp, D. The screen is provided with a paper scale, C, divided into equal parts, and is placed at the distance of about two feet from the little mirror. It follows, from this arrangement, that when the light passing through the slit falls upon the mirror, it is reflected again through the lens, and an image of the slit is seen on the scale. This image is immediately above the slit when the beam falls perpendicularly upon the mirror, and this condition may be brought about by properly placing the apparatus with regard to the magnetic meridian. The directive power of the earth over the little suspended magnet is, however, *almost* annulled by properly fixing the steel magnet, E, which slides upon the upright rod, so that the suspended magnet is thus free to obey the least force impressed upon it by a current passing through the coil. And when the mirror is deflected through a certain angle, the image on the scale will be deflected to twice that angle, and thus the smallest movements of the suspended magnet are readily recognized; not only by reason of the length of the beam of light, which forms a weightless index, but because they are doubled by this increased angular deflection.

When the signals are being rapidly transmitted through a long submarine line, the currents at the receiving station are much enfeebled and retarded, and the result is that the movements of a suspended needle have by no means the decided character which is seen in the instruments connected with land lines. The signals through a submarine cable could not therefore be received by any apparatus which required a certain strength of current; but the mirror galvanometer indicates every change in the currents, and the apparently irregular motions of the spot of light can be interpreted by a skilled clerk, who, by long experience, recognizes, in quite dissimilar effects, the same signal sent by the clerk at the other end in precisely the same way. Thus a first contact, corresponding with a dot of the Morse alphabet, may cause the light to move some distance on the scale, a second contact immediately succeeding moves it but a little way farther, and a third may occasion a movement hardly perceptible.

The messages sent by the mirror galvanometer must be read as they are received; and, as a telegraphic instrument, it is wanting in the manifest advantages attending a recording instrument. Sir W. Thomson has, however, devised another receiving instrument of great delicacy, which is termed the *syphon recorder.* We cannot here describe its admirable mechanical and electrical details, but the chief feature is that the attractions and repulsions of the currents are made to produce oscillations in a syphon formed of an extremely fine glass tube, the shorter branch of which dips in a trough of ink, and the longer branch terminates opposite to, but not touching, a band of paper, which is continuously and regularly drawn along by clockwork while the message is being received. The tube is a mere hair-like hollow filament of glass, and the ink, which would not itself flow from a tube of so fine a bore, is squirted out by electrical repulsion when the insulated reservoir in which it is contained is electrified at the receiving station by an ordinary machine. The message as written by this instrument appears thus:

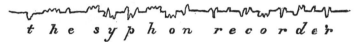

t h e   s y p h o n   r e c o r d e r

The reader, on comparing these signals with the Morse code on page 560, will have no difficulty in discovering their relation to it.

TELEGRAPHIC LINES.

IT now remains to give some account of the *line*, that is, the conductor by which the sending and receiving instruments are united, and along which the currents flow. Overhead lines are nearly always constructed with iron wires, which are usually ⅛ in. in diameter, and are coated with some substance to protect them from oxidation. Zinc is often used for this purpose, the wire being drawn through melted zinc, by which it becomes covered with a film of this metal—a process known as "galvanizing" iron. Another mode is to cover the wires with tar, or to varnish them from time to time with boiled linseed oil, and this *must* be done in populous places, where the gases in the air are liable to act upon the zinc. Sometimes *underground* wires are used, and these are often made of copper, covered with gutta-percha, and are laid in wooden troughs, or in iron pipes. They are protected by having tape or other material, saturated with tar or bitumen, wound round them. The poles employed to suspend the overhead wires are generally made of larch or fir, of such a length that when securely fixed in the ground they rise 12 ft. to 25 ft. above it, and at the top have a diameter of about 5 in. About thirty poles are required for each mile, and every tenth pole forms a "stretching-post," being made stronger than the others and provided with some appliance by which the wires can be tightened when required. The wires are attached to the posts by insulating supports; but at every pole there is always some "leakage," the amount of which depends on the form, material, and condition of the insulators. Glass is quite unsuitable, because its surface strongly attracts moisture, which thus forms a conducting film. All things considered, porcelain is found to be the best insulating material for this purpose, since moisture is not readily deposited on its surface, and even rain runs off without wetting it; and it is durable, strong, and clean. Fig. 293 shows a telegraph post, with brown salt-glazed stoneware insulators, shaped like hour-glasses, with a perspective view and section of one of them. Another form of insulator, shown in Fig. 294, has a stalk or hook of porcelain, with a notch, into which

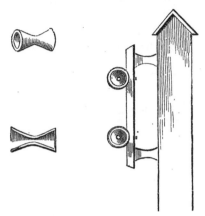

FIG. 293.—*Telegraph Post and Insulators.*

the wire is simply lifted, and is protected above by a porcelain bell. This form, or some modification of it, is that most generally used.

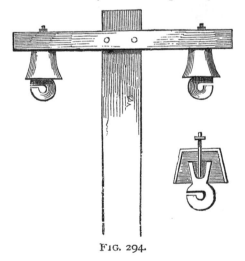

FIG. 294.

It need hardly be remarked that only a single wire is required with most of the modern instruments for communication between any two places. Each of the many wires often seen attached to the telegraph posts along a road or railway represents a distinct line of communication—that is, one wire may connect the two termini, another may join an intermediate station and a terminus, a third may belong to two intermediate stations, and so on. We have already alluded to the discovery by Steinheil of the apparent conducting power of the earth ; and if we must continue to think of complete circuits, we must regard the earth as replacing for telegraphic purposes the second or return wire, which was at first supposed essential. For instance,

FIG. 295.—*Wire Circuit.*

when a battery current had to be sent from Station A, Fig. 295, which we may suppose to be London, to Station B, which we may call Slough, it was at first thought requisite to provide a wire for the return of the current after it had traversed the coils at the receiving station. But now the connections are made as shown in Fig. 296, where the return wire is dispensed with,

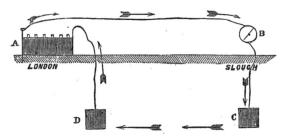

FIG. 296.—*Wire and Earth Circuit.*

except a small portion at each end, which is connected with a large plate of copper buried in the earth ; the arrows show the direction of the current, according to the commonly received notion. By this plan the current is increased in intensity, for the "earth circuit" appears to offer less resistance than the copper wire. The view, however, which regards the earth not as a conductor in the same sense as the wire, but as the great *reservoir or storehouse of electricity,* accords better with known facts.

The spread of telegraph lines, and the extent to which this mode of communication is used by the public, may be illustrated by a few particulars regarding the Central Telegraphic Office in London. The management of all the public telegraph lines in Great Britain is now in the hands of the Post Office authorities, and the arrangements at the central office in London are an admirable specimen of administrative organization. The Central Telegraph Office occupies a very large and handsome building opposite the General Post Office, St. Martin's-le-Grand. In one vast apartment in this building, containing ranges of tables, in all three-quarters of a mile long, may be seen upwards of six hundred telegraph instruments, besides a number of stations for the receipt and transmission of bundles of messages by pneumatic dispatch. The number of clerks employed in working the instruments is 1,200, and about three-fourths of these are females. The wires from each instrument are conducted below the floor of the apartment to a board where they terminate in binding-screws, marked with the number of the instrument. The same board has binding-screws, with battery connections, and others which form the terminals of the telegraph lines, and thus the requisite connections are readily made. The batteries are placed in a lower room, which contains about 23,000 cells of Daniell's construction, formed into nearly 1,000 distinct batteries, in each of which the number of cells varies according to the length of the line through which the current has to pass. Thus, the battery which supplies the currents that are sent through the coils of the instrument at Edinburgh consists of 60 cells, but one-sixth of that number suffices for some of the short lines. The instrument almost exclusively used is the Morse recorder, and Wheatstone's automatic punching machine and transmitters are in constant employment. There are also some examples of other instruments to be seen in operation, such as the Hughes type printing telegraph, the American sounder, a few A, B, C, dial instruments, and a solitary specimen of a double-needle instrument. Upwards of 30,000 messages pass through this office each day.

But the most striking achievements in connection with telegraphy are the great submarine lines which unite the Old and New Worlds. Morse and Wheatstone about the same time (1843) independently experimented with sub-aqueous insulated wires, and their success gave rise to numerous projects for submarine lines. How far any of these might have been practical need not here be discussed, but it fortunately happened that some years after this, the electrical properties of gutta-percha were recognized, and this material, so admirably adapted for forming the insulating covering of

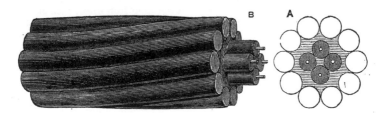

FIG. 297.—*Submarine Cable between Dover and Calais.*

wires, was taken advantage of by Brett and Co., who obtained the right of establishing an electric telegraph between France and England, and they succeeded in laying down the first submarine cable. This cable extended from Dover to Cape Grisnez near Calais, and the experiment proved successful ; but, unfortunately, the cable was severed within a week by the sharp rocks on which it rested near the French coast. It proved, however, the excellent insulating property of the new material, and demonstrated the possibility of submarine telegraphic communication. Another cable was manufactured, in which the gutta-percha core was protected by a covering of iron wires laid specially on the exterior, and thus combining greater security with a far larger amount of tenacity. A view and section of this —the first practically successful submarine cable—are given in Fig. 297 of the real size. It has four separate copper wires, each insulated with a covering of gutta-percha, and the whole was spun with tarred hemp into the form of a rope, and protected with an outer covering of ten of the thickest iron wires wound spirally upon it. The cable when complete was 27 miles in length, and each mile weighed 7 tons. This cable was laid in 1851, and from that time it has been in constant use, with the exception of a few interruptions from accidental ruptures. Its success immediately led to the construction of other cables connecting England with Ireland, Belgium, Holland, &c. In 1855 the practicability of an Atlantic cable was no longer doubted, and £350,000 were soon subscribed by the public for the project. A cable was manufactured weighing 10 tons to the mile, and in August, 1857, 338 miles of it had been successfully paid out by the ships when the cable parted. Better paying-out apparatus was now devised —self-releasing brakes were constructed, so that the cable should not be exposed to too great a strain; and in 1858 another cable, requiring a strain of 3 tons to break it, was manufactured, and the laying of it commenced in mid-ocean—the *Mægera* and *Agamemnon* going in opposite directions, and paying out as they proceeded. Twice the cable was severed, twice the ships met and repaired the injury ; but the third time, when they were 200 miles apart, the cable again broke. But again the attempt was repeated, and this time success crowned the effort ; for on the 5th of August the two continents were telegraphically connected. Unfortunately the electric continuity failed after the cable had been a month in use.

Seven years elapsed before another endeavour was made ; but the experience gained in the unsuccessful attempt was not lost ; and in 1865 another cable had been constructed, and the *Great Eastern* was employed in laying it. In this the conductor was composed of seven copper wires twisted into one strand, covered with several layers of insulating material, and covered externally with eleven stout iron wires, each of which was itself protected by a covering of hemp and tar. This cable was 2,600 miles long, and contained 25,000 miles of copper wire, 35,000 miles of iron wire, and 400,000 miles of hempen strands, or more than sufficient to go twenty-four times round the world. It was carefully made, mile by mile, formed into lengths of 800 miles, and shipped on board the *Great Eastern* in enormous iron tanks, which weighed, with their contents, more than 5,800 tons. This cable was manufactured by Messrs. Glass and Elliot, at Greenwich, to whom the iron wire for the outer covering was furnished by Messrs. Webster and Horsfall, of Birmingham. Fig. 298 represents the workshops with the iron wire in process of making. The great ship sailed from Valentia on the 23rd of July, 1865, and the paying out commenced. Constant communication was kept up with the shore, and signals exchanged with the instrument-room at Valentia, which is represented in Fig. 299, where, among various instruments invented by Sir W. Thompson, may be seen his mirror galvanometer. After several mishaps, which required the cable to be raised for repairs after it had been laid in deep water, the *Great Eastern* had paid out about 1,186 miles of cable, and was 1,062 miles from Valentia, when a loss of insulation in the cable was discovered by the electricians on board. This indicated some defect in the portion paid out, and the usual work of raising up again had to be once

FIG. 298.—*Making Wires for Atlantic Telegraph Cable.*

more resorted to. During this process the cable parted, and Fig. 300 shows the scene on board the *Great Eastern* produced by this occurrence, as represented by an artist of the " Illustrated London News" who accompanied the expedition. The broken cable was caught several times by grapnels, and raised a mile or more from the bottom, but the tackle proved

37

unable to resist the strain, and four times it broke; and after the spot had been marked by buoys, the *Great Eastern* steamed home to announce the failure of the great enterprise. For this 5,500 miles of cable had altogether been made, and 4,000 miles of it lay uselessly at the bottom of the ocean, after a million and a quarter sterling had been swallowed up in these attempts.

FIG. 299.—*The Instrument-Room at Valentia.*

But these disasters did not crush the hopes of the promoters of the great enterprise, and in the following year the *Great Eastern* again sailed with a new cable, the construction of which is shown of the actual size, in Fig. 301. In this there is a strand of seven twisted copper wires, as before, forming the electric conductor; round this are four coatings of gutta-percha; and surrounding these is a layer of jute, which is protected by ten iron wires (No. 10, B.W.G.) of Webster and Horsfall's homogeneous metal, twisted spirally about the cable; and each wire is enveloped in spiral strands of Manilla hemp. The *Great Eastern* sailed on the 13th of July, and on the 28th the American end of the cable was spliced to the shore section in Newfoundland, and the two continents were again electrically connected. They have since been even more so, for the cable of 1865 was eventually fished up, and its electrical condition was found to be improved rather than injured by its sojourn at the bottom of the Atlantic. It was spliced to a new length of cable, which was successfully laid by the *Great Eastern,* and was soon joined to a Newfoundland shore cable. There were now two cables connecting England and America, and one connecting America and France has since been laid. At the present time upwards of 20,000 miles of submerged wires are in constant use in various parts of the world.

Certain interesting phenomena have been observed in connection with submarine cables, and some of the notions which were formerly entertained

FIG. 300.— *The Breaking of the Cable.*

as to the speed of electricity have been abandoned, for it has been ascertained that electricity cannot properly be said to have a velocity, since the same quantity of electricity can be made to traverse the same distance with extremely different speeds. No effect can be perceived in the most delicate instruments in Newfoundland for one-fifth of a second after contact has been made at Valentia; after the lapse of another fifth of a second the received current has attained about seven per cent. of its greatest permanent strength, and in three seconds will have reached it. During the whole of this time the current is flowing into the cable at Valentia with its maximum intensity. Fig. 302 expresses these facts by a mode of representation which is extremely convenient. Along the line O X the regular intervals of time in tenths of seconds are marked, commencing from O, and the intensity of the current at each instant is expressed by the length of the upright line which can be drawn between O X and the curve. The curve therefore exhibits to the eye the state of the current throughout the whole time. If after nearly a second's contact with the battery the cable be connected with the earth at the distant end, the rising intensity of the current will be checked, and then immediately begin to decline somewhat more gradually than it rose, as indicated by the descending branch of the curve in Fig. 302. A little reflection will show the unsuitability for such currents of instruments which require a fixed strength to work them. We may remark that, supposing a receiving instrument were in connection with the Atlantic Cable which required the maximum strength of the received current to work it, the sending clerk would have to maintain contact for three seconds

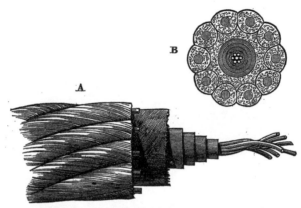

FIG. 301.—*Atlantic Telegraph Cable,* 1866.

before this intensity would be reached, and then, after putting the cable to earth, he would have to wait some seconds before the current had flowed out. Several seconds would, therefore, be taken up in the transmission of

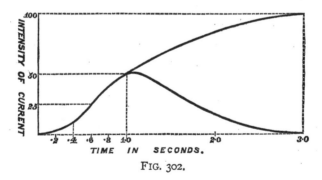

FIG. 302.

one signal, whereas by means of the mirror galvanometer about one-fourteenth of this time suffices, and the syphon recorder will write the messages twelve times as fast as the Morse instrument. The cause of the gradual rise of the current at the distant end of a submarine cable must be sought for in the fact that the coated wire plays the part of a Leyden jar, and the electricity which pours into it is partly held by an inductive action in the surrounding water. The importance of Sir W. Thomson's inventions as regards rapidity of signalling, upon which the commercial success of the Atlantic Cable greatly depends, will now be understood.

By furnishing the means of almost instantaneous communication between distant places, the electric telegraph has enabled feats to be performed which appear strangely paradoxical when expressed in ordinary language. When it is mentioned as a sober fact that intelligence of an event may actually reach a place before the time of its occurrence, a very extraordinary and startling statement appears to be made, on account of the ambiguous sense of the word *time.* Thus it appears very marvellous that details of events which may happen in England in 1876 can be known in America in 1875, but it is certainly true; for, on account of the difference of longitude between London and New York, the hour of the day at the latter place is about six hours behind the time at the former. It might, therefore, well happen that an event occurring in London on the morning of the 1st of January, 1876, might be discussed in New York on the night of the 31st of December, 1875. There are on record many wonderful instances of the celerity with which, thanks to electricity, important speeches delivered at a distant place are placed before the public by the newspapers. And there are stories in circulation concerning incidents of a more romantic character in connection with the telegraph. The American journals not long ago reported that a wealthy Boston merchant, having urged his daughter to marry an unwelcome suitor, the young lady resolved upon at once uniting herself to the man of her choice, who was then in New York, *en route* for England. The electric wires were put in requisition; she took her place in the telegraph office in Boston, and he in the office in New York, each accompanied by a magistrate; consent was exchanged by electric currents, and the pair were married by telegraph! It is said that the merchant threatened to dispute the validity of the marriage, but he did not carry this threat into execution. The following *jeu d'esprit* appeared a short time ago in " Nature," and, we strongly suspect, has been penned by the same hand as the lines quoted from " Blackwood," on page 508.

ELECTRIC VALENTINE,

(Telegraph Clerk ♂ to Telegraph Clerk ♀.)

" 'The tendrils of my soul are twined
  With thine, though many a mile apart;
And thine in close-coiled circuits wind
  Around the magnet of my heart.

" 'Constant as Daniell, strong as Grove;
  Seething through all its depths like Smee;
My heart pours forth its tide of love,
  And all its circuits close in thee.

" ' Oh tell me, when along the line
  From my full heart the message flows,
What currents are induced in thine?
  One click from thee will end my woes !'

" Through many an Ohm the Weber flew,
  And clicked this answer back to me—
' I am thy Farad, staunch and true,
  Charged to a Volt with love for thee.' "

[NOTE BY THE EDITOR.—*Ohm,* standard of electric resistance ; *Weber,* electric current; *Volt,* electro-motive force ; *Farad,* capacity (of a condenser).]

THE TELEPHONE.

OF more recent invention than any of the classes of instruments already mentioned for electrical communication at a distance is the telephone, which differs widely from the rest in many notable particulars. Though the telephone completely realized what had for years before been the dream of physicists, the first announcement of its capabilities was received, even by the scientific world, with some pause of incredulity; but when its powers were demonstrated, it created no small sensation. It has now, within a few years afterwards, become so familiar as an appliance of ordinary life and business, that people in general are less impressed by the wonder of it than were their fathers half a century ago by the electric telegraphs of Wheatstone and of Morse. Like all other inventions, it was led up to by preceding discoveries and tentative efforts. It will be unnecessary here to trace those successive steps with minuteness, or to attempt to adjust the claims of merit or priority that have been put forward for different inventors, but a notice of some of the stages in the evolution of this wonderful contrivance may be of interest. If the reader has no previous knowledge of the physical nature of sounds in relation to music, and especially to articulate speech, he should now refer to the brief explanation given in a subsequent chapter, at the commencement of the section on the Phonograph. He should, however, bear in mind that in that explanation are included some acoustical discoveries of a later date than some of the inventions we are here to speak of, or, at least, the real causes of which give other qualities than pitch to sound, had not been fully demonstrated when the notion of the electric telephone was conceived.

When the electric telegraph came into use and it was found possible to use it for communication of intelligence to great distances, it is not surprising that the further problem of transmitting by electricity, not signals merely, but audible speech, should be suggested. Perhaps the first scientific person who avowed a belief in the possibility of doing this, and even indicated the direction in which the solution of the problem was to be sought, was a Frenchman of science, M. Charles Bourseul. In 1854, he pointed out that sounds are caused by vibrations, and reach the ear by like vibrations of the intervening medium, and, although he could not say what took place in the modifications of the organs of speech by which syllables are produced, he inferred that these syllables could reach the ear only by vibrations of the medium, and that if these vibrations could be reproduced the syllables would be reproduced. He suggests that a man might speak near a flexible disc, which the vibrations of his voice would throw into oscillatory movements that could be caused to make and break a battery circuit, and that, at a distance, the currents might be arranged to produce the like vibrations in another disc. The weak point of this scheme was the want of any suggestion as to the mode in which this last effect was to be produced. Even when this part of the problem was solved in a few years afterwards, as we shall presently see, it was musical —and not articulate—sound that could be transmitted by an arrangement, using make and break contacts. The reader, who has understood what has been said of electrical currents, and also the account of the compounded vibrations in articulate sounds introduced into our section on the phonograph, should have little difficulty in seeing this must necessarily be the case, for the contacts could only give the succession of the vibrations by currents of equal intensity, and could not, like the yielding wax of the phonograph cylinder, correspond with their relative intensities. M. Bourseul pointed out advantages which would arise from the transmission of speech by electricity, such as simplicity of apparatus and facility in use —for, unlike the telegraph, no skilled operators would be needed—to signal messages, or time spent in spelling out the words letter by letter. He says that he had made some experiments, which promised a favourable result, but demanded time and patience, and that he is certain that, in a more or less distant future, speech will be transmitted by electricity, so that what is spoken in Vienna may be heard in Paris. One cannot help thinking that if M. Bourseul had but pursued his experiments a little longer, he would not improbably have achieved the invéntion of the speaking telephone, for which the world had to wait twenty years longer. As it is, we cannot but admire his scientific foresight and his confidence in the ultimate realization of his idea.

But before this came to pass, an intermediate stage was reached in the apparatus contrived by M. Reiss, a schoolmaster of Friedrichsdorf, who, in 1860, solved the problem of electrically transmitting musical tones. So far as concerned the reproduction of the sounds, this telephone was founded upon a discovery, made in 1837, by an American physicist, named Page, which was this : At the moment a bar of iron is magnetized, by sending a current through a coil surrounding it, as shown in Fig. 265, a slight but sharp click is heard. The transmitting apparatus was, in principle, Mr. Scott's phonoautograph (described in the section on the phonograph), which had been invented in 1855. The tracing style of this was replaced in Reiss' apparatus by a small disc of platinum, connected by a very light spring of the same metal with a binding-screw for the battery connection. Nearly in contact with the little disc was a platinum point, so arranged that the slightest oscillation of the membrane would bring them into actual contact and thus close the circuit. Worthy of remark is the very primitive nature of the materials with which Reiss made his first experimental apparatus. The receptacle for the voice was simply a large bung hollowed out into a conical cavity, and the membrane was supplied by the skin of a German sausage, while the clicking bar of the receiver was a stout knitting needle, surrounded by a coil of covered copper wire and stuck into the bridge of a violin, which, by acting as a sounding board, made the clicks produced in the needle distinctly audible. M. Reiss finally produced his telephone in the form shown in Fig. 302a, where I is the receiver ; B, the voltaic battery ; $I\,I$, the receiver ; $c\,c$ is a coil of insulated wire, surrounding a slender iron rod, mounted on the supports, $f\,f$, which rest on the sounding board, $g\,g$. The transmitter consists of the hollow box, A, provided with a trumpet-mouthed opening in one side and having at the top a circular piece cut out, across which is stretched a membrane with the little disc of platinum, n, fixed in its centre. When a person applying his mouth to A sings into the box, the membrane is thrown into vibrations corresponding with the notes, and at each vibration a contact is made and a click is emitted from the distant sounding box. The tones are concentrated by covering this box with the perforated lid. It was afterwards found that a trumpet mouth fitted into the receiver was still more effective. Reiss tried to use his arrangement for transmitting speech, but without success, although occasionally a syllable could be very indistinctly heard. An instrument, with springs so nicely adjusted that slight vibrations did not separate the platinum from actual contact, but merely caused change of pressure, has indeed been made to convey articulate sounds, although the arrangement was not essentially different from that of M. Reiss. This mode of action is, however, a different thing, and we shall presently see that very effective speech transmitters have been constructed by applying it in a more refined way. This musical telephone could give the pitch of the sounds in the song but not their quality (*timbre*), and the receiver added to the main system of vibration other sets that belongéd to itself, the result being a shrill and by no means pleasing tone, recalling that of a penny trumpet. Messrs. C. and L. Wray afterwards effected some considerable improvements in M. Reiss's telephone, with the object of intensifying the effects and producing better tones.

A further step towards the speaking telephone may be illustrated by

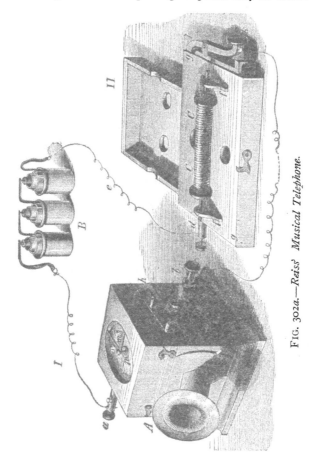

FIG. 302*a.—Reiss' Musical Telephone.*

an earlier invention of Mr. Graham Bell, a native of Scotland, who had settled in the United States. Mr. Bell's inventions, it may be mentioned, were by no means the results of fortunate accidents or of unsought and spontaneous flashes of conception, but rather the outcome of long, patient and systematic studies. His father, Mr. Alexander Melville Bell, of Edinburgh, had assiduously cultivated acoustic science, and had in conjunction with his son, undertaken special researches into the mechanism of the organs of speech, the elements of articulate speech in different languages, and the musical components of vocal sounds. When Graham afterwards pursued these studies in the light of the fuller investigation carried out by Helmholtz, he was naturally led to the application of electricity to acoustic transmission. After some experiments in the production of vowel sounds by combinations of electric tuning forks, he invented a telephone for reproducing musical sounds at a distance, which was a great improvement on that of Reiss, and involved another principle, which indeed is the same as that utilized in his more mature invention of the speaking telephone. As a like explanation of the action would apply

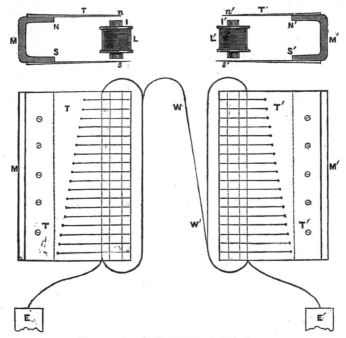

FIG. 302*b.*—*Bell's Musical Telephone.*

in both cases, the reader will find his advantage in following the observations we have to make on the earlier instrument. This consisted of what was virtually two sets of electric tuning forks, each set being acted upon by one electro-magnet. Fig. 302*b* will suffice to show the general form of the arrangement. A plate of steel is bent twice at right angles longitudinally, and is magnetized so that any transverse slice of it would constitute an ordinary horse shoe magnet. This is seen endways in Fig. 302*b* at M, and N. and S. will indicate the north and south poles respectively. To each limb of this broad magnet is attached a plate of steel, T, cut into teeth, just in the same way as the steel plate in a common musical box or mechanical piano, except that the teeth are not pointed. These are tuned to give severally in pairs the notes of the musical scale when thrown into vibration. Between the prongs of the series of tuning forks thus formed is an electro-magnet, L, made of a bar of soft iron, I, wound longitudinally by a coil, one end of which makes an earth connection at E and the other is connected by the wire, W W', to complete the circuit through the coil of the distant apparatus. It will be observed that the receiving and transmitting instruments are exactly alike. Now, suppose one of these teeth is struck or otherwise thrown into vibration, the result will be, since the free ends of the teeth are magnetic poles, that alternating electric currents will be generated in the coil of the electro-magnet (see page 509), and these will flow through the entire circuit, including the coil of the distant instrument, where the magnetism generated will alternately attract and repel the polar extremities of the teeth in the steel plate. It will be understood, of course, that the fellow prong of the fork will vibrate also, and will simultaneously approach to and recede from the soft iron core, so that being of opposite polarity, the effect on the electro-magnet will be doubled. The action on the distant electro-magnet will be a rapid series of reversals of the polarities of the core, and hundreds of times in every second the ends of the steel teeth will be alternately attracted and repelled. But not all of these will thereby be thrown into vibration—only the one pair which were tuned into unison with the former can and will respond to that particular series of impulses, and the consequence will be that the same note will be emitted by the receiving instrument. If two or more notes of the transmitter be simultaneously thrown into vibration, the same notes will be heard from the receiver, for each series of currents will flow along the wire independently, just as if the other did not exist, and each will produce its particular effect on the transmitter. In this way an air played on the one instrument is heard also from the other, with all its accents and combinations. But more than this, if a tune be played on a musical instrument near the sender, or if a song be sung, the air will be reproduced by the distant receiver. The reason of this is that the steel tongues take up, or are thrown into movement by, the vibrations that have the same periodicity. The manner in which a vibratory body responds to impulses of its own periodicity may be easily shown by exposing the wires of a piano and raising the dampers, when, if a note be sung near the instrument, it will be found that a number of the wires respond, namely, those that are capable of vibrating synchronously with the constituent vibrations of the voice, for neither a voice nor a sounding wire gives forth one simple system of vibrations, the audible effect being due to the superposition or composition of several diverse elementary systems. With the same arrangement another experiment may be made, as an illustration of a matter important for our subject. Let the different vowels be sung to the piano-wire on the same note or pitch, and in the responses to each a difference of the quality of the sound will be noticed, although the piano will not distinctly give back the vowel itself. It would, however, do so if a number of its wires were strung with certain definite relations in pitch to that of the fundamental note and in unison with the voice components of the vowel sound.

It has been said above that two systems of electrical currents of different periodicity would flow along one wire independently of each other, but it should be explained that this takes place by a composition of the currents, for it is evident that at any given instant the wire can only be in one of three conditions, viz. : (1) with no current flowing ; (2) with a current in the positive direction ; (3) with a current in the negative direction. Such must always be the case, and, therefore, it should be clearly understood how this is consistent with the superposition of currents of different periodicities, a matter which the diagram, Fig. 302*c*, is intended to illustrate. Suppose the *flow of time* to be represented by the dotted lines from *a* to *b*, the whole length of which we may call $\frac{1}{100}$th of a second, and that the current passing through the wire is represented in intensity and direction by the plain lines ; the intensity by distance above or below the dotted line ; the direction being positive where the plain line is above, and negative when it is below the dotted straight line, and of course no current at all occurs at the instant when the change of direction takes place. The line A will thus represent alternating currents, rising and sinking in intensity, and changing from one direction to the

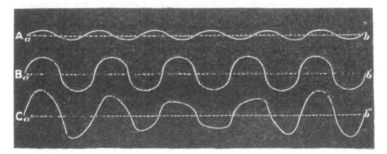

FIG. 302*c.*—*Superposition of Currents.*

other, going through 600 regularly recurring phases in one second of time. Similarly, B may represent another series of currents, having here a periodicity of 500 in one second of time. These are here supposed to have greater intensity than the former. If the two currents are sent through one wire their effects are superposed, so that the actual electrical state of the wire would be represented by the curve C, which is compounded from the two others, and where it will be observed the rise and fall of the current, its maxima and minima, no longer recur at regular intervals within the space of the $\frac{1}{100}$th of the second, the whole of that period being taken up by a less regular series of changes, the cycle being repeated only 100 times in the second. The same diagram might serve to illustrate the motions of, say, a particle of air or the drum of the ear in acoustic vibration, the distances above and below the straight line being taken to represent the displacements from the position of rest on one side and the other. If the sounds of an organ or piano consisted of only these primary vibrations, B would roughly* represent the movements of the

---

* The lines A and B in the diagram have not harmonic ordinates.

wires, the air and the drum of the ear, when the
note si_3 was sounded alone; A when the note
re_4 was more faintly sounded alone, and then C,
if these notes were sounded together, would correspond with the movements of the drum of the
ear. The movements it actually makes when we

si_3 re_4

hear speech, or even a single musical note, are,
however, a thousand-fold more complex, for no musical instrument gives
out a note with a single set of vibrations, the fundamental one being always
accompanied by other sets diversely related to it, according to the class of
instrument. In some cases, fifteen or sixteen sets of vibrations have been
distinguished along with the fundamental note, without exhausting the
possible number. Of a like order of complexity will be the currents
which the wire of a speaking telephone must convey, and the difference
between the undulatory nature of the currents in Bell's musical telephone
and any produced by mere make and break contacts, as in Reiss' arrangement, will be obvious, and recognized as an important step towards the
solution of the problem of transmitting speech. When Mr. Bell invented
his instrument, he was seeking for a method of simultaneously transmitting
by one wire several messages by audible *signs* merely; and by the
method used in his musical telephone this is practicable, for all that
would be required
would be pairs of
transmitters and
receivers, each ad-
justed to one single
particular note.
Another point that
should be noted is
that in the Bell
musical telephone
no battery is used,
for the currents are
those generated by
magneto - electric

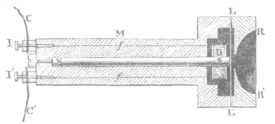

FIG. 302d.—*Bell's Speaking Telephone.*

induction, and the circuit through the wires and coils are completed by
earth connections.

In passing from the invention of the musical to that of the speaking
telephone, Mr. Bell passed from the more complex to the more simple
instrument, for of all apparatus by which communication can be carried
on at a distance, the Bell speaking telephone is one of the simplest. He
had only to make its vibrating disc of Scott's phono-autograph into a magnetized body, capable of producing currents in an electro-magnet coil in
the same way as did the vibrating plates in his musical telephone. The
Bell speaking telephone was publicly exhibited for the first time at Phila-
delphia, in 1876, and was shown the same year to the British Association
by Sir William Thomson, who pronounced it the wonder of wonders.
For the first time in England, the instrument in a still simpler form was
exhibited by Mr. Preece, at the Plymouth meeting of the British Association in 1877, and of nearly the same construction as is still often used,
although, as we shall presently see, for battery telephones the
transmitting apparatus is now made of larger dimensions, of a different
shape and on a different principle. We shall describe the simple form in

which transmitter and receiver are identical, each consisting externally of a small cylindrical wooden or ebonite box, and with a handle three or four inches in length of the same material. Fig. 302d is a section of the instrument where N S is a cylindrical steel magnet, on one end of which is wound the small coil B, made of fine silk covered copper wire, the extremities of which pass through the handle M at ff, and are connected by the binding screws $I\,I'$ with the line wire C C'. Close to the coil covered end of the magnet is a very thin diaphragm of iron, L L', and when this is thrown into vibration by the voice speaking into the trumpet-mouth opening, R R', its movements produce currents in the coil according to the principles that have already been explained, for it will be observed that the iron disc is magnetized by the inductive action of the permanent magnet N S. These currents passing through the coil of the receiving instrument raise or lower the intensity of the magnetic force in it, so that the distant disc reproduces the vibrations of the transmitter. Such is at least an obvious explanation of the action of this very simple arrangement ; but from a number of experiments and observations that have been made with modifications of the instruments, it would appear that other and much more complex phenomena concur in producing the effects. It has indeed been suggested—and the idea is supported by numerous experiments—that, in these telephonic transmissions of speech, vibrations are concerned which are not at all of the mechanical kind we have been dealing with in these explanations, but are *molecular*.

The Bell telephone is used by speaking distinctly before the mouthpiece of the transmitter, while the listener at the other end of the line applies the mouth-piece of his instrument to his ear, and one wire is sufficient with good earth connections, although sometimes a second wire is employed to complete the circuit. It is also found advantageous to have two instruments in the circuit at each end, so that one may be held to the ear while the operator is speaking through the other. In this way, a rapid conversation can be carried on with the greatest ease, or again, an instrument may be held at each ear, by which arrangement the words are more distinctly heard. It is not necessary to shout, as this has no effect, but to speak with a clear intonation, and some voices are found to suit better than others. The vowel sounds are best transmitted, except that of the English e, which, with the letters g, j, k, and q, are always somewhat imperfectly transmitted. A song is very distinctly heard, both in the words and the air, and the voice of the person singing is readily recognized. Several instruments may be included in one circuit at different stations, so that half a dozen persons may take part in a conversation, and questions and answers may be understood even when crossing each other. If two distinct telephone circuits have their wires laid for a certain distance (two miles) near each other, say a foot or more apart, and without any connection whatever, listeners at the end of the one line will hear the conversation exchanged through the other line. Other forms of the instruments have been arranged, by which a large audience may hear sounds produced at a distance, as, for instance, when a cornet-à-piston was played in London, it was heard by thousands of people assembled in the Corn Exchange at Basingstoke.

It would be impossible within our limits to even briefly describe the great number of improvements and modifications of Bell's system that were devised by various persons soon after the invention was brought out, and many additional complications were introduced into some of the arrangements. Advantage was also taken to a greater or less extent of another principle affecting the strength of electric currents, to which we have now to call the reader's attention, and to exemplify by one of the simplest instruments, leaving detailed accounts of the various forms in which it has been applied to be found in special treatises. The reader should first turn back to page 400, where he will see an expression of the strength of a battery current. It will be observed that the current may be increased or diminished by diminishing or increasing R, the external resistance, without changing the other terms. Now M. Du Moncel discovered, as far back as 1856, that an increase of pressure between two conductors in contact, and conveying a current, caused a diminution of the electrical resistance, and this discovery was utilized for telephonic purposes by Mr. Edison in his invention of the carbon transmitter (1876). In this there is no magnet, and a stretched membrane may take the place of the metallic plate, although a circle of photographers' ferro-type plate gives better results. A pad of india-rubber, cork, or other material is fixed on the plate, and rests upon a carbon disc, which again is in contact with a metallic conductor. Between the latter and the carbon the current from a constant battery passes. When the plate is thrown into vibration by speaking into the mouth-piece, the variations of pressure conveyed to the carbon cause variations in the resistance of its electrical contact, and thus a series of undulations are produced in the current, and these affect the electro-magnet of a Bell receiving instrument in the circuit as before, so that the sounds are reproduced. It is now time to say a word about the share in the invention of the speaking telephone which has been claimed by Mr. Elisha Gray, also of the United States, who, at the time Mr. Bell applied for the patent for his instruments, produced drawings and descriptions of a plan he had devised for transmitting speech by undulating electrical currents, and it has been admitted that the plan he had conceived was perfect in principle. He proposed to use a battery current, and his receiving instrument was nearly the same as Bell's. The undulations of the current were also determined, as in Edison's telephone, by changes in the external resistance, but this was effected in a different, though equally simple manner. To a membrane stretched across the lower end of a short wide tube that formed the mouth-piece of the transmitter, and was placed vertically, was attached a piece of platinum wire, conveying the current and dipping into a liquid of moderate conductivity, but not quite touching another platinum electrode fixed at the bottom of the vessel containing the liquid. The space of liquid traversed by the current being thus varied by the oscillations of the membrane, the resulting variations of the resistance produced the requisite undulations in the intensity of the current. Both Mr. Bell and Mr. Gray applied for patents on the 14th February, 1876, but the American Patent Office recognized the claim of the former as prior.

Du Moncel's observation was applied by Mr. Hughes in the construction of an instrument, which he named the *microphone*. This was in the same year that Edison had brought out his carbon telephone, and a certain similarity, resulting from the identity of the principle employed, led to an acrimonious controversy on what were supposed to be rival claims. But the microphone differs so much in arrangement and performance from the other instrument as to constitute a distinct invention. The instrument, if it may be so called, is simplicity itself, in the form represented in Fig. 302*e*, which is one of the most sensitive. There,

C and C are two small blocks of carbon, fixed on a small upright piece of wood. Two cup shaped cavities are hollowed out in the carbon blocks, and these serve to hold loosely, in a nearly vertical position, a small rod of gas retort carbon pointed at the ends. This rod is only about one inch in length, and the lower end merely rests on the bottom of the cup in C′, while the other is capable of moving about in the upper cavity, the vertical position being nearly maintained in a state of unstable equilibrium. The carbons are in the circuit of a voltaic cell or small battery, B, in the line through a Bell receiving instrument, which may be at a distance. When the microphone is to be used, it is placed on a table with a cushion or several folds of wadding beneath its base. If the receiver be applied to the ear of a listener, he will distinctly hear every word pronounced by one speaking near the microphone, even in a low tone ; but a loud voice may be heard when the speaker is 20 or 30 feet from the instrument. The minutest vibrations conveyed to the stand are perceived at the receiver as loud noises. The tread of a fly walking over the board, S, is heard like the tramp of a horse, and the ticks of a watch are audible in the receiver when the ear is several inches away from it. The slight

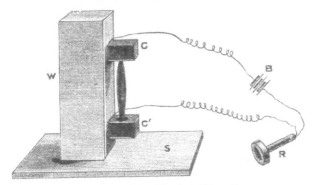

FIG. 302e.—*Mr. Hughes' Microphone.*
(B *and* R *are merely diagrammatic.*)

touch of a feather on the stand is distinctly audible, and a current of air impinging upon it is reproduced as the noise of a stream of water. The microphone is, in fact, the most sensitive detector of vibrations that is known, and its employment as a transmitter has brought the telephone to its present perfection. It has been constructed in an endless variety of forms, according to the purposes for which it is intended, and its simplicity is as wonderful as its extreme sensitiveness. We will further illustrate these qualities by an experiment of Mr. Willoughby Smith's on the same principle. Instead of the two carbon blocks, he laid on the table, in parallel positions, two small rat tail files, and completed the circuit by a third file, laid across the others at right angles. This arrangement constituted so sensitive a transmitter that the listener at the distant Bell receiver could hear even the faint sound of the speaker's breathing. Even three common pins, similarly crossed, make an effective transmitter. The feebleness of the variations in the current requisite to make the Bell receiver produce sounds is extraordinary, and a very weak battery current is sufficient, even under the circumstances of ordinary practical use. Still more remarkable is the fact that in favourable conditions the microphone is capable of transmitting sound without any battery at all, but merely with connections to earth, when the ticking of a watch placed upon the stand has been distinctly heard at the distance of nearly one-third of a mile, and speech, also, has been transmitted with unusual distinctness with the battery left out and merely a few drops of water placed at the carbon contacts; indeed, it is said that, even without the water, the voice may be heard. This effect has been attributed to the carbons and water forming a battery themselves, and in the latter to the moisture of the speaker's breath supplying the fluid element. But, again, the microphone will not only transmit speech, but, under certain arrangements, it will reproduce it (when one of the carbon electrodes is attached to a membrane), although the result is less distinct than with the Bell receiver. It is, however, not so easy to explain how mere variations of current intensity can produce the effect where there can be no magnetic attractions and repulsions. We must, no doubt, look for the cause in some other property of electric currents. The transmitters used in various lines of telephonic communication, erected by the Post Office or by companies in Great Britain, are generally applications of the principle of the microphone, and not of that of either Mr. Bell's or Mr. Edison's original instrument. But more recently, Mr. Edison has most ingeniously adapted variations of sliding friction, as modified by the action of the undulatory current on a liquid electrolyte between the sliding surfaces to the production of a loud speaking telephonic receiver—that is, one by which the sounds are made audible to a large assembly. From this instrument, the notes of a cornet-à-piston, played in Brighton, have been distinctly heard throughout a large hall in London.

Another curious transmitter is formed of a fine jet of water traversed by an electric current. Acoustic vibrations are easily set up in the jet, and these modify its conductivity so as to produce corresponding undulations of current intensity.

It would take long to point out the many scientific applications of so sensitive an instrument as the microphone with its Bell receiver. As a medium for conveying speech to a distance, whether for purposes of peace or war, its use is sufficiently obvious. Some curiosities of musical transmission have been noticed, and such experiments are repeated from time to time with increasing success. It has been applied to many purposes in surgery and medicine. In many cases of deafness it has made conversation easy. Even the passage of the molecules of gases, when diffusing through porous partitions, Mr. Chandler Roberts has by its means made audible. The distances to which speech can now be transmitted are considerable, as conversations have been carried on by persons nearly 300 miles apart.

LIGHTHOUSES.

W HO does not regard with interest the lighthouses which at night throw their friendly beams across the sea, to guide the mariner in his course, and warn him of perils from sunken rock or treacherous shoal? The modern lighthouse, with its beautiful appliances, is entirely the result of the applied science of our age; and it affords a fine example of the manner in which experiments, carried on to determine natural laws apparently of an abstract character and without any obvious direct utility, give rise to inventions of the highest importance and most extended usefulness. The lofty structures which were erected near certain ancient harbours, and of which the Pharos of Alexandria is the most memorable example, burned on their summits open fires of wood; and whatever beacons existed from that time down to the end of last century were merely blazing fires of wood or coal. The lighthouses of the South Foreland, which were established in 1634, displayed coal fires until 1790, and the lighthouses in the Isle of Man were first illuminated with oil only in 1816. Down to the beginning of the present century, therefore, the modern lighthouses showed no improvement on the ancient plan. Even the Tour de Cordouan, at the mouth of the Garonne river, which was completed in 1610, and is one of the most famous of modern lighthouses, from its great height (200 ft.), and the care which has always been given to render it efficient, showed down to 1780 merely a fire of billets of wood, the upward loss of the light being diminished by a rude reflector in the form of an inverted cone. In the improved means of obtaining artificial light, and in the admirable optical apparatus by which that light is utilized, we find the vast superiority of modern lighthouses. But these are sometimes erected on isolated, and almost submerged, rocks, exposed to the fury of the waves. The difficulties which have to be overcome in their construction cause some lighthouse towers to rank among the best specimens of engineering skill. We may, therefore, consider under the present head —the towers; the sources of light; the optical apparatus and its accessories.

One of the best-known lighthouses on the English coast is that on the Eddystone .Rock, about 14 miles S.S.W. from Plymouth. The structure which now* stands upon this rock was the work of Smeaton, and was completed in 1759. The stones forming the lower courses of this tower, which is represented in Fig. 303, half in section and half in elevation, are dovetailed into the rock itself and into each other. The masonry is carried up in a solid mass for about 12 ft., the stone used being granite, which also constitutes the whole of the exterior masonry. The four upper apartments are formed with arched roofs, the side-thrust of which is counteracted by iron chains surrounding the tower. These chains, which are bedded in lead,

FIG. 303.—*The Eddystone Lighthouse.*

were placed in their positions while hot, and by their contraction bound the structure together with great force. The masonry of the tower is 68 ft. high, and this is surmounted by the light-room, the total height from the lowest course of stonework to the gilt ball at the top being 94 ft., or nearly half that of the London Monument. The diameter at the base is 26 ft., and that at the top 15 ft. The light-room is of an octagonal shape, and is made of iron framework, glazed with thick plate glass. Below this are two store-rooms, a kitchen, and a bed-room. The Eddystone has now breasted the storms of more than a hundred years, and it remains as firm as the rock it is built on. Fig. 304 is a picture of this noble lighthouse, with the British fleet passing close to it, during a furious gale on the 22nd of October, 1859, or exactly a century after the completion of the structure. The incident of the man in the water, which occupies the foreground, is not an imaginary one, for it is recorded that the *Trafalgar* stopped in the midst of the storm to pick up a man who had fallen overboard. For eighteen hours the ships encountered the fury of the tempest, keeping out at sea in

---

*Smeaton's tower proving unsafe, has since been taken down and replaced, in 1882, by one from Mr. Douglass' design.

open order throughout the night. They wore in at dawn, came up the Channel in line of battle, steamed into Portland, and took up their anchorage without the loss of a sail, a spar, or a rope-yarn.

FIG. 304.—*The Eddystone in a Storm.*

The lighthouse tower on the Bell Rock is 100 ft. high, 42 ft. in diameter at the base, and 15 ft. at the top. The Inchcape Rock, on which it is placed, is the scene of Southey's ballad of "Ralph the Rover," and the lighthouse here is one of the most serviceable on the Scottish coast. for the dangerous spot on which it is placed lies in the direct track of all vessels entering the Firth of Tay from the German Ocean. The rock is submerged at spring tides to the depth of 12 ft. The tower bears a close resemblance in shape to that of the Eddystone : it is circular and faced with massive blocks of granite. The lower part, to the height of 30 ft., is solid, and the door is reached by a bronze ladder. The building contains five apartments, and a cistern for storing fresh water for the use of the keepers, who have sometimes to remain in their solitary situation for six or eight weeks together, the weather preventing the possibility of any communication with the shore.

Still loftier than the tower on the Bell Rock is that which rises in the midst of the Skerryvore Reef, 12 miles from Tyree, a small island off the coast of Argyleshire. This building may be taken as a typical specimen of a detached lighthouse, and the difficulties overcome in its construction attest the skill of the engineer, Mr. Alan Stevenson, who has written a highly interesting account of the work. The rocks here are of *gneiss*, an extremely hard formation, and their surfaces are worn as smooth as glass by the action of the water. On one of a numerous series of these small islets, where only a narrow strip of rock, a few feet wide, remains above the surface at high water, and this divided by rugged lumps into narrow gullies, through which the sea constantly rushes, the lighthouse is built. The work was commenced in 1838 by the erection of a temporary wooden barrack on piles at a little distance from the site chosen for the foundation. In a gale during the winter the whole of this structure was swept away in one night. Another, more strongly secured, was built the following summer, and in this Mr. Stevenson and his men remained sometimes for fourteen days together, the weather preventing any passage to or from the shore : here the men were sometimes awakened from their hard-earned repose by the water pouring over the roof, and by its rushing through the crevices, while the erection swayed and reeled on its supports. Mr. Stevenson relates that one night the men became so alarmed for the stability of their shelter that some descended, and sought in cold and darkness a firmer footing on the rocks. Two summers were occupied in cutting the foundations, and the blasting of the rock in so narrow a space was an operation attended with no little danger. A small harbour had to be formed at the rocks for the vessels bringing the ready-prepared stones of the building from the quarries, where also piers were built expressly for the shipment of the materials. In designing his tower, the engineer preferred to oppose the force of the waves by the weight of his structure, rather than to rely on dovetailed or joggled-jointed stones. Measurements were made of the force of the waves, which at Skerryvore was sometimes equivalent to a pressure of 4,335 lbs. on the square foot; and calculations based on these measurements showed that the mere weight of the superstructure would amply suffice to keep the stones immovable. Nearly 59,000 cubic feet of stone were used, or about five times the quantity contained in the Eddystone Lighthouse, and the total cost of the building was £87,000.

The use of iron, as a building material advantageously replacing stone, has extended to lighthouses, and many have been constructed entirely of cast and wrought iron, or partly of iron and partly of gun-metal, which is not readily acted on by the sea spray. Such lighthouses are cheap, easily and quickly erected, strong enough to bear shocks and vibrations, and proof against fire, lightning, and earthquakes. The lighthouse on Morant Point, Jamaica, is made of iron, cast in England ; and it was erected in a few months at a cost of one-third of that of a stone tower of the same altitude. Its height is 105 ft., and the shaft is formed of iron plates in segments of 10 ft. high, which are bolted together at their flanges. At Gibbs Hill, Bermuda, is a lighthouse 130 ft. high, constructed in the same manner.

So inefficient, inconvenient, and uncertain were the lamps or other means of artificial illumination known up to nearly the beginning of the present century, that nothing better could be found for the Eddystone Lighthouse for forty years after its erection than tallow candles stuck in a hoop—a means of illumination which would scarcely now be tolerated even in a booth at a village fair. To M. Argand, a Frenchman, we are indebted for the first great improvement in lamps. The admirable invention which bears his name is, as everybody knows, an oil lamp with a tubular wick, which occupies the annular space between two metallic tubes, in such a manner that a current of air rises through the inner tube, and thus reaches the interior of the flame. This current, and the current which supplies the exterior, are increased by surrounding the flame with a tall glass chimney; and a contraction of the chimney, just above the flame, aids greatly in distributing the air, so as to insure the complete combustion of the oil. In the original lamp the supply of oil to the flame depended on the capillary attraction in the meshes of the wick. M. Carcel applied clockwork to continuously pump up the oil into the burner, so that, by overflowing, it was maintained at an invariable level. This arrangement added greatly to the intensity and steadiness of the light; and, on account of the uniformity of its flame, the Carcel lamp has been selected as a standard to which, in France, photometric determinations are referred.

The power of the Argand lamp, as employed in lighthouses, is greatly increased by the plan of employing several concentric wicks instead of one. Between these wicks there are, of course, open spaces, through which the air obtains access to the flame, and the current of air is made more rapid by the use of a very tall chimney. The large amount of heat produced by the combustion of so much oil in a small space is partly carried off by the excess of oil which is made to overflow the burner—about four times the quantity consumed being constantly pumped up into the burner for this purpose. Lighthouse lamps are made with two, three, and four wicks; and the oil is forced up in the burners either by clockwork or by the pressure of a piston loaded with a weight. The following table gives the sizes of the burners and the illuminating powers of the lamps:

Order of Light.	Number of Wicks.	Diameter in inches.	Intensity of Light in Carcel Lamps.
1	4	$3\frac{1}{2}$	23
2	3	$2\frac{5}{16}$	15
3	2	$1\frac{3}{4}$	5

The quantity of oil consumed in these lamps is less than that proportional to the increase of the light:—for example, although the four-wick lamp gives twenty-three times the light of the simple Argand, it only consumes nineteen times the quantity of oil. The oil used in these lamps is colza; but experiments have been made with a view of introducing petroleum, which has the advantages, not only of being cheaper and uncongeal-able by cold, but of giving a whiter and more brilliant light. Hitherto, however, this substance has answered only with lamps of one wick.

Coal-gas has been applied to the illumination of lighthouses, and as it gives a light of great brilliancy and steadiness, when consumed in proper burners, it has certain advantages over oil lamps, which have caused it to be employed in situations where a supply can be readily obtained. The light produced by lime, ignited by the combustion of coal-gas or hydrogen mixed with oxygen, has also been suggested; but this plan is not without risk of interruptions and of dangerous accidents, and it has been considered inadvisable to entrust the apparatus to the persons who commonly take charge of lighthouses.

The electric light has been very successfully applied in certain light-houses, since the mode of producing steady currents by magneto-electric* machines has come into use. The lighthouses at the South Foreland have been thus illuminated by a machine constructed by Mr. Holmes in 1862. A very powerful electric light is exhibited from the lighthouse on Cape Grisnez; and the adoption of this source of light has been extending, as it is far more intense than any other artificial light, and can be sent in more concentrated beams across the sea, on account of its being emitted from a space which is practically a point. These circumstances cause the beams of the electric light to possess greater power of penetrating the atmosphere than those from any other source.

But it is perhaps in the optical apparatus of lighthouses that the greatest improvements and most admirable inventions are to be found. When only the blaze of an open fire furnished the guide to the mariner, the means resorted to in order to throw across the sea the light which issued from the flames upwards or landwards, appear to have been of the rudest kind, even where such attempts were made at all. The inverted cone on the Tour de Cordouan has been already mentioned, and we read of cases in which screens of sheet brass were placed on the landward side, to throw back the light seaward.

Here it may be proper to examine the conditions which determine how the light can be made most available for the guidance of the mariner. Everybody knows that the light from a luminous body spreads out from it in all directions equally. Thus, if we simply place an electric light on a tower such as that on the Bell Rock, but few of the luminous rays can benefit the mariner: namely, those which fall upon the sea or are directed to the horizon. A much larger portion of the light will stream upwards and be lost in space; another part will descend towards the base of the tower, and be equally wasted. Again, if the situation of our lighthouse were on the shore of the mainland, all the light which passes landwards, whether horizontally or not, would be entirely lost for our purpose. Even if, in the case of an isolated lighthouse, we can send out all the light in a nearly level zone over the sea to the horizon, the intensity of the illumination will diminish, on account of the widening space, as the distance increases. The question, therefore, arises whether it is possible to send the whole of the light in one unbroken beam, not liable to this kind of enfeeblement, so that the only loss it can experience may be absorption by the imperfectly transparent atmosphere.

There are two means of gathering up all the otherwise useless beams, and sending them in such a direction as to reach the eye of the distant mariner. The one is by reflection from mirrors, and the other by refrac-

---

* Now superseded by the dynamo-electric machines.

tion through lenses. The apparatus employed in the first process is termed *catoptric*, and in the latter *dioptric*.

When a luminous point is placed at the focus of a parabolic mirror, all the rays which fall upon the mirror are reflected by it in a direction parallel to its axis, so that they form a cylindrical beam. This is the method which was adopted in the first improvements effected in lighthouses. The parabolic reflector was first used at the Tour de Cordouan in 1780, and soon afterwards metallic reflectors became the ordinary appliances of lighthouses, and they are still largely used. Such reflectors are made of sheet copper, thickly plated with silver, about 6 oz. of this metal being applied to 16 oz. of copper. They are formed by carefully beating a circular sheet of the plated copper into a concave shape, which is finally brought to the exact curve by the aid of gauges, and is then turned and polished. The largest of these reflectors have a diameter of 2 ft. at the *mouth*, as it is termed, for the reflector comes forward in advance of the lamp, the chimney and burner passing through openings in the metal. The flame of the lamp occupies such a position that its brightest part is in the focus of the mirror; but since the focus is a point merely, whereas the flame has a certain magnitude, it follows that the want of coincidence of the other luminous points with the focus produces a certain divergence in the reflected rays, so that the beam is not accurately cylindrical. This, however, is far from being a disadvantage practically, for it has the effect of widening a little the strip of sea illuminated by the beams. But all that portion of the light which escapes from the mouth of the mirror without being reflected is radiated in the ordinary manner, and is practically lost. We shall presently see how even this light may be gathered up and brought into the main beam.

Let us suppose a number of such reflectors, each with its own lamp, placed in a horizontal circle, so as to throw their beams towards different points of the compass. If eight lamps were so placed, eight beams of light would stream out across the water, like eight spokes of a wheel; eight sectors would, however, be left unilluminated, and for ships in these spaces the lighthouse would be virtually non-existent: its rays could only reach vessels within the eight narrow strips traversed by the beams. If we double the number of reflectors in the circle, or if we arrange another series of eight in a circle above or below the others, so that a lamp in the second circle coincides vertically with an interval in the first, the effect will be that we shall have sixteen beams, and sixteen dark sectors, instead of eight; that is, only a very small part of the expanse of water will receive the benefit of the light. It must be remembered that the breadth of the cylindrical beam would not be greater than the diameter of the mirrors, and that the space illuminated by it has the same breadth at all distances; or rather, that this is nearly the case, for the light does not all issue precisely from the focus of the mirror. Thus, even if we use a very great number of mirrors, we shall succeed in illuminating but an extremely small proportion of the sea horizon. This evil is met by giving a horizontal rotatory motion to the reflectors, causing the beams to sweep over the whole expanse of the waters; and thus from every ship the light will be visible for an instant. The rotation is produced by clockwork, duly regulated, so that an uniform motion is obtained. The regular appearances and eclipses of the light prevent the mariner from mistaking for a lighthouse a bright star near the horizon or an accidental fire on the coast; and, further, it being necessary that the lighthouses along any particular coast should be readily distinguishable from each other, it becomes easy, by assigning to each light-

LIGHTHOUSES.

house a different period of revolution, to individualize them, so that the mariner shall be in no danger of confounding one with another.

But when the lighthouses on a certain extent of coast are numerous, this mode of distinguishing them becomes inconvenient, as mistakes might easily be made in small differences of time ; and it would be inexpedient to keep long intervals of darkness. Hence other methods have been re-- sorted to in addition—such as red lights, or lights alternately red and white. The following are the distinctions made use of among the Scottish light- houses, including the double lighthouses, which give a leading line to the navigator :

1. Fixed lights.
2. Revolving lights.
3. Revolving, with red and white beams alternately.
4. Revolving, with alternately two white beams and one red.
5. Revolving, with alternately two red beams and one white.
6. Flashing, in which the light increases and decreases at regular intervals.
7. Intermittent, in which, by means of a revolving screen, the light is abruptly cut off and exhibited.
8. Double fixed lights.
9. Double revolving lights, which appear and disappear at the same instant.

The efficiency of reflectors depends on the state of polish of the surface, and even with the most brilliant polish there is a very large loss of light : in the ordinary condition of lighthouse reflectors, it is found that one-half of the light is lost at the surface of the mirrors. An attempt was made in England, about the beginning of the present century, to substitute glass lenses for mirrors. But it was found that, in spite of the loss occurring in reflection, the mirrors produced a more intense beam. No doubt the person who made the attempt did not observe the true conditions of the problem. It was Fresnel, the illustrious Frenchman, whose name has already been mentioned in these pages, who successfully solved the pro- blem. He saw that it would be necessary to give the lenses a short focal length, and at the same time to have their diameters very great. The dimensions required by these conditions far exceeded any that could be given to lenses formed in the ordinary manner ; and even if they could be so formed, the great thickness of glass which would be necessary would diminish the transparency, and unduly increase the weight of the apparatus to the detriment of the revolving apparatus. An idea now occurred to Fresnel's mind, which, although similar to previous projects, he conceived independently, and was undoubtedly the first to carry out. This was the idea of the *lentille à échelons*, or " lens in steps." The construction of this will be understood from Fig. 305, where *a b* is a section of a lens in steps, and the dotted line, *c*, shows the thickness an ordinary lens of the diameter *a b* would have. Fresnel kept only the marginal part of such a lens ; and inside of the ring formed by this, he fitted the margin of a second large lens having the same focal distance ; inside of this another ring, and so on ; and in the centre a large lens of moderate thickness. He also placed above and below the lens the concentric prisms, *e e'* and *f f'*, which, by re- fraction and total reflections (see page 399), send the rays parallel to the axis of the lens. Fresnel also contrived methods of economically grinding such lenses and prisms with precision.

Fresnel saw that it would be useless to apply lenses in lighthouse illumi- nation unless the intensity of the light given out by the single-wick Argand lamps then in use could be considerably increased, without much enlarging

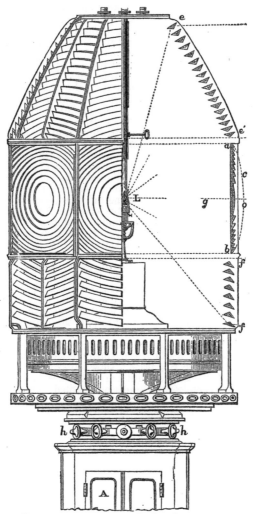

FIG. 305.—*Revolving Light Apparatus.*

the flame. Accordingly he devoted himself, in conjunction with his friend Arago, to this preliminary consideration. Their studies and experiments led them to the construction of the lamp with several concentric wicks—by which a brilliancy of light is obtainable twenty-five times greater than that of the single-wick Argand. The light which the improved lamp, when combined with Fresnel's lenses, could send to the horizon, was equivalent to that which would be given by the united beams of 4,000 Argand lamps without optical apparatus ; and it was eight times greater than any which could be produced by the reflectors then in use. The first apparatus constructed on Fresnel's plan was placed on the Tour de Cordouan in July, 1823.

France led the van in the erection of the most perfect lighthouses in the world, and it was not until 1835 that, by the strenuous advocacy of Mr. Alan Stevenson, a dioptric apparatus was employed in a British lighthouse; but at the present time Fresnel's principle has been adopted in the majority of British lighthouses. Fig. 305 is a part elevation, with the section, of a catadioptric apparatus of the first class. In plan it is a regular octagon, and it sends out eight beams, which are directed to the horizon, and made to sweep over the sea by its regular rotation, produced by clockwork contained in the case, A. The whole frame is very accurately balanced, and turns on its bearings, and the rollers, h, h, with great smoothness and steadiness. The moving power is given by the descent of a weight attached to a chain or cord, which is wound round a barrel. One train of wheels is connected with apparatus for regulating the speed, and to this an indicator is attached which registers the number of revolutions made in an hour. There is also a contrivance of some kind for maintaining the motion while the weight is being wound up. The reader will observe that all the light of the lamp, L, is utilized, except that which is directed towards the base and top of the apparatus—a quantity less than one-fifth of the whole. About 45 per cent. of the light emitted by the lamp falls on the refracting lenses ; $22\frac{1}{2}$ on the upper reflecting prisms ; and $13\frac{1}{2}$ on the lower reflecting prisms. The brightest part of the flame is placed so that the beams from it are directed towards the sea horizon, and the space between the horizon and the neighbourhood of the lighthouse receives ample light from the other parts of the flame. Thus a ship, or any part of the sea within the range of the lighthouse, will see the light appearing at regular intervals, as one after another of the eight beams passes across it, the intervals being one-eighth of the time in which the apparatus completes its revolution. The zones of totally reflecting prisms, shown at $e\,e'$, $f\,f'$, Fig. 305, were not adopted in British lighthouses until 1844, when the Skerryvore light was exhibited with the complete apparatus represented in the drawing.

The optical apparatus for lighthouses is constructed of certain sizes, adapted to the different situations in which it is to be used. The apparatus we have just described is made in six forms, according to the *order* of light required. The first three orders are for sea lights, the rest for harbour lights ; and the following are the dimensions of the apparatus for each order of revolving or fixed lights :

Order.	Height in Inches.	Internal Diameter in Inches.	Number of Reflecting Prisms.	
			In Upper Zone.	In Lower Zone.
1	$106\frac{1}{2}$	$72\frac{1}{2}$	18	8
2	$83\frac{1}{4}$	55	16	4
3	$61\frac{1}{2}$	$39\frac{1}{2}$	11	4
4	29	$19\frac{3}{4}$	5	4
5	$21\frac{3}{4}$	$14\frac{3}{4}$	5	4
6	$17\frac{1}{2}$	12	5	4

When a revolving apparatus of the above description is erected on shore, a reflector of suitable shape and dimensions is placed on the landward side of the lamp, so as to throw its rays back upon itself and towards the lenses which are directed seaward.

Fresnel also constructed glass apparatus for fixed lights. If we require to send the light equally towards the horizon in all directions at once, the problem is capable of solution, either by a proper form of glass apparatus or by a proper form of mirrors. Suppose the section, $e\,c\,f$, Fig. 305, to revolve about a vertical axis passing through the lamp, it would sweep out a form which, when executed in glass, would spread out all the light falling upon it into one horizontal sheet. Fresnel was obliged to content himself with an approximation to this shape, formed by a prismatic frame of many sides, containing straight horizontal bars of glass, having the section $e\,c\,f$. The light is not quite uniformly distributed by such apparatus, but the difficulty and expense attending the formation of prismatic rings were very great when Fresnel constructed this apparatus. Such rings can now be produced economically and accurately, and therefore the fixed-light apparatus is now constructed of circular glass rings, mounted in sections in such a manner that a vertical section through the axis of the apparatus would cut them in the form represented at $e\,c\,f$. Instead of forming the metal framework in which the glass is mounted with vertical ribs, it is made with the ribs placed somewhat diagonally, in order that the dark sectors which would be produced by the shadows of upright ribs may be avoided. It should be understood that the forms of the glass in each side of the octagonal apparatus represented in the figure are produced by the revolution of the same section, $e\,c\,f$, about the horizontal axis, $d\,g$.

An ingenious promoter of the catoptric system has contrived to solve the same problem by mirrors. The form of these may be understood by the aid of Fig. 306, which, however, relates to another contrivance. Suppose that the lines A B, A′ B′, are turned about C D as an axis, all three preserving their relative positions, A B and A′ B′ would sweep out two parabolic cones, which would have the property of reflecting in a horizontal direction all rays falling upon them from a lamp placed at L. But glass, as a material for lighthouse apparatus, has so many advantages over metal that it is probable that metallic reflectors will soon be entirely obsolete. The polish of the metal is very readily destroyed, and as it is constantly liable to be tarnished, the frequent cleaning required is apt to produce a scratched state of the surface, even when great care is used. Far greater accuracy of form can be imparted to glass than to metal reflectors. And then there is the great loss of light occurring at even the most highly polished surfaces of metal: a loss which is far greater than that occasioned by the refraction and reflections of the glass apparatus. There are cases, however, in which it is desirable to throw the whole of the light into one beam, and this cannot be done without reflecting the light from one side. Mr. Alan Stevenson contrived an excellent apparatus for this purpose, and the diagram, Fig. 306, will explain its nature. L is a point representing the source of light, A B, B′ A′, a parabolic metallic mirror. All the rays between L A and L B, and all between L A′ and L B′—that is, all those which fall upon the mirror—will be reflected parallel to L G; but those between L B and L B′ would escape from the mouth of the mirror, B B′, as a diverging cone. This is prevented by placing the lens, H I, the focus of which is at L, so as to convert the diverging cone, I L H, into the cylindrical beam, E H I F; and thus half the light emitted from the luminous point is sent in one direction. A hemispherical reflector, C K D, of which L is the centre, receives the other half, which is thus thrown back through L, and then follows the same course as the direct rays. For the metallic reflector, C K D, Mr. Stevenson afterwards substituted a system of glass zones, of which O P Q represents the sections. These had the same effect as the metallic reflectors, without the loss of light occasioned by the latter. The inner surface of the glass, C K D, is hemispherical, and the prismatic zones are such as would be produced by turning the section about L K (or C D) as an axis. The dotted lines show the course of a ray of light, L *m*, which, meeting the hemispherical surface perpendicularly, passes straight through it, and is totally reflected at *m* by the inclined surface, and again at *n*, so that it returns to L by the path *n* L. Reflecting glass prisms were also substituted for the metallic mirror, A B, B′ A′, and thus the use of metal has been entirely dispensed with in this apparatus. This light has been termed by Mr. Stevenson the *holophotal* (ὅλο, *entire*, φως, *light*). Such an apparatus will form the intensest beam that a given source

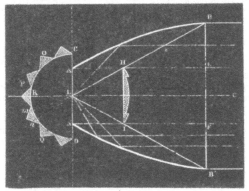

FIG. 306.—*Stevenson's Holophotal Light.*

of illumination can yield. On the other hand, when a fixed light is distributed to the whole horizon simultaneously, the illuminating power of the source is taxed to the utmost. These two cases may be considered the extreme modes of disposing of the light, while the parcelling of it into several beams, as effected by the apparatus represented in Fig. 305, is an intermediate mode.

It may be interesting to mention that the holophotal light at Baccalieu, in Newfoundland, is visible in clear weather from another point 40 miles distant. So long a range as this is seldom possible at sea, on account of the rounded form of the earth rendering it necessary to raise the light nearly 1,000 ft. above the water, if it is required to be visible at 40 miles' distance. A shorter distance generally suffices for the requirements of the navigator ; and therefore lighthouse towers rising from the water are seldom carried to a greater height than something between 100 ft. and 150 ft. A light elevated 100 ft. above the water would be seen from the deck of a vessel 14 miles distant, and from the masthead a much greater distance.

The optical apparatus of a lighthouse is protected by an outer metal framework glazed with thick plate glass. This framework is made of iron, or of gun-metal—the latter being preferred on account of the frequent painting which iron needs in order to preserve it from corrosion. The glass is carefully fitted into the framework, so as to avoid exposure to strains from the shocks and vibrations to which a lighthouse is exposed. The keepers are always provided with a store of panes of glass, ready for fitting into their places in case of accidents. Sometimes the glass is broken by large sea-birds dashing against it, and by pebbles which are thrown up by the waves, or driven by the wind against the panes. It is the interior of this lantern which forms the light-room already spoken of. Great pains have been bestowed on the proper ventilation of these light-rooms, as not only must the air have access to the lamp to supply the flame, but the carbonic acid which escapes from the chimney of the lamp must be promptly removed. Another serious inconvenience of an ill-ventilated light-room would be the condensation, in the inner surface of the plate glass, of the aqueous vapour, which is also a product of the combustion.

The lenses and circular prisms for lighthouses are usually made of crown glass, and are ground by fixing them on a large revolving iron table, on which they are bedded in plaster of Paris and cemented by pitch—great care being taken to place them in the exact position required, for only about one-eighth of an inch is allowed for grinding down to shape the glass as it comes from the moulds. Sand, emery, and finally rouge, are used with water for the grinding and polishing processes. The cost of the optical apparatus alone of a light of the first order, like that shown in Fig. 305, amounts to upwards of £1,500. The lenses and prisms are very carefully adjusted in their framework after this has been fixed, and no plan of testing the adjustment has been found more efficient than that of viewing the sea horizon through them from the position which the flame will occupy.

The men to whom the charge of a lighthouse is confided undertake a duty involving the gravest responsibilities, and demanding unremitting care. In those lighthouses where a number of reflectors are hung upon a revolving frame, the extinction of one lamp may not be a matter of much consequence ; but where only one lamp is used, life and death depend upon its burning. To isolated lighthouses—such as those of Skerryvore and the Bell Rock—four keepers are appointed, and one of these is always on shore on leave, so that the men may be relieved at intervals ; for it has been found that a residence in these lonely towers cannot be continued long together without bad effects. The duties of the lighthouse-keepers must be performed with the greatest regularity. The glasses of the light-room and the optical apparatus are carefully cleaned every morning ; the lamps are supplied with oil, the wicks trimmed or renewed, the machinery oiled and adjusted, and everything prepared in readiness for the evening. At sunset the lamps are lighted, and one keeper takes his watch until midnight, when he is relieved by another, who maintains the vigil till sunrise, when the lamps are extinguished.

The expediency of the regulation appointing three men to be always at the lighthouse may be illustrated by an incident which occurred about the beginning of the present century at the lighthouse on the " Smalls," a rock in the Bristol Channel. Two keepers held watch over the light on that rock, which for months together is sometimes cut off from all communication with the shore. At the time alluded to, after the weather had for two weeks prevented access to the lighthouse, it was rumoured among the seafaring men of the neighbouring ports that something was wrong at the " Smalls," for a signal of distress had been observed ; but the boats could not go within speaking distance, although many attempts were made to reach the rock. The relatives of the men became anxious, and night after night watched for the light. But the light never failed to appear at the proper hour. After four months came calmer weather, and then a boat brought to shore one lightkeeper alive, the other dead. What the former felt when he found his comrade to be dying in their dreadful isolation, or what his emotions were when he found himself there alone with the lifeless body, is not recorded. But the thought occurred to him that he must not commit the body to the waves, lest any suspicion of foul play might fall upon himself. He therefore contrived a sort of coffin for the dead man, and dragging it up to the gallery of the lighthouse, tied it there. Punctually and faithfully for four long months did he perform all the duties of his position, keeping watch from twilight till dawn in that lonely light-room, while his ghastly charge remained there within sight. But he came on shore strangely altered—a sad, silent, gloomy, worn man—so that even his intimate friends hardly knew him.

Here we close this brief account of the modern lighthouse, and of its beautiful appliances, by which Science "has given new securities to the mariner," in addition to those with which she furnished him when she showed him the use of the compass, supplied him with the chronometer, and placed the sextant in his hands. How anxiously must the seaman who has been prevented by unfavourable skies from ascertaining his exact position, and has been trusting to the log and the compass to work his reckoning, scan the horizon for the first glimpse of the hospitable light beacon, which seems to say that the country he is approaching has been watching for his coming, and welcomes him to its shores.

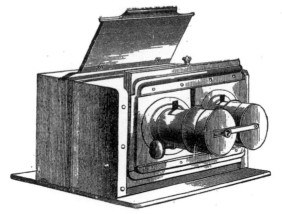

PHOTOGRAPHY.

N O other of our nineteenth century inventions is at once so beautiful, so precious, so popular, so appreciated as photography. It is exercising a beneficial influence over the social sentiments, the arts, the sciences of the whole world—an influence not the less real because it is wide-spread and unobtrusive. The new art cherishes domestic and friendly feelings by its ever-present transcripts of the familiar faces, keeping fresh the memory of the distant and the dead; it keeps alive our admiration of the great and the good by presenting us with the lineaments of the heroes, the saints, the sages of all lands. It gratifies, by faithful portrayals of scenes of grandeur and beauty, the eyes of him who has neither wealth nor leisure for travel. It has improved pictorial art by sending the painter to the truths of nature; it has reproduced his works with marvellous fidelity ; it has set before the multitude the finest works of the sculptor. It is lending invaluable aid to almost every science. The astronomer now derives his mathematical data from the photograph ; by its aid the architect superintends the erection of distant buildings, the engineer watches over the progress of his designs in remote lands, the medical man amasses records of morbid anatomy, the geologist studies the anatomy of the earth, the ethnologist obtains faithful transcripts of the features of every race. To the mind of an intelligent reader numberless instances will present themselves, not only of the utility of photography in the narrower sense of the term, but of its higher utility in ministering to our love of the beautiful in art and in nature.

Effects produced by chemical changes to which the rays of the sun give rise are matters of common observation. The fading of the colour in the portions of a fabric which are exposed to the light is a familiar instance ; and the bleaching of linen under the influence of sunshine in the presence of moisture is a well-known operation. Decompositions produced by light in certain compounds of silver soon attracted the attention of chemists, and the remarkable activity of the solar rays in causing the combination of hydrogen and chlorine gases has been even made the means of measuring the intensity of light. When equal volumes of these two gases are mixed together in the dark, they may be kept for an indefinite period without change, provided only that the mixture be preserved from access of light. But the instant it is exposed to the direct rays of the sun, or to an intense light, such as that of burning magnesium, the two gases suddenly unite with a loud explosion, in which the glass vessel containing them is shattered into atoms. The product is an intensely acid invisible gas, called hydrochloric acid ; and if the mixture is exposed to the diffused light of day, instead of the direct rays of the sun, then the production of hydrochloric acid will take place gradually, and with a rapidity depending on the intensity of the light.

Of vastly more importance than the small operations of the laboratory and the bleach-field are the changes which the sun's rays silently and unobtrusively effect in the vegetable world. The chemical effect of light here appears to reside in its power of separating oxygen from substances with which it is combined. The green parts of plants absorb from the atmosphere the carbonic acid gas, which is constantly produced by the respiration of men and animals, and by combustion, and other processes. Under the influence of sunshine, this carbonic acid is decomposed within the tissues of the plant; the oxygen is restored to the atmosphere; the carbon with which it was united is retained to build up the structure of the plant. In a similar manner light separates the oxygen from the hydrogen of water, and the former gas is given off by the leaves, while the hydrogen enters into the composition of the plant. The carbon, which forms so large an element in the food of plants, is chiefly obtained in this way; and the abundance of the supply of oxygen thus thrown into the atmosphere may be inferred from the fact that a single leaf of the water-lily will in the course of one summer give off nearly eleven cubic feet of oxygen. But for this continual restoration of oxygen to the atmosphere, animal life would soon disappear from the face of the earth. It is the office of the vegetable world not only to furnish a supply of organic matter as food for animals, but when the materials of that food have been converted into oxidized products in the animal system, and returned to the atmosphere as carbonic acid and aqueous vapour, the sunshine, acting on the vegetable structure (chiefly on the delicate tissue of the leaf), tears apart the oxygen and the other substance. These are, therefore, once more capable of combination, by which they may again supply the animal with heat and the other energies of life.

Those actions of light which have been last referred to are called by the chemist *reducing actions*, a term which he applies to the cases in which a compound is made to part with its oxygen or other similar element: when the remaining ingredient is a metal, the operation by which the other has been removed is always called *reduction*. On the other hand, the inverse operations by which oxygen, chlorine, &c., are fixed upon other bodies, are distinguished as processes of *oxidation*. Light is the means of determining each of these kinds of changes, according to the conditions and the nature of the substances exposed to its action. Thus moist chloride of silver will retain its white colour if preserved in the dark; but if exposed to sunlight, it quickly acquires a violet tint, which deepens in intensity until it has become black. The dark matter was formerly admitted to be silver ; for it was known that the finely divided metal has this appearance, that during the process the compound gives off chlorine, and that when nitric acid is poured upon the darkened matter, reddish fumes are given off, exactly as when the acid acts upon pure silver. The use of silver nitrate as a marking-ink for linen depends upon a similar alteration of the salt within the fibres; and the same reduction takes place when to a solution of the nitrate in water organic matter is added. If a piece of white silk be dipped into a solution of chloride of gold, and exposed to the sun's rays while still wet. the silk becomes first green, then purple, and finally a film of metallic gold will be found overspreading its surface. Many other chlorides and analogous compounds are similarly affected by sunlight. On the other hand, chlorides, as we have already seen, and oxygen, fix on hydrogen and on organic substances with greater energy under the influence of light. A large series of chemical compounds are obtained by means of the augmented affinity of chlorine for hydrogen induced by the rays of the sun.

It was in availing himself of an action of the latter class that, in 1813, Joseph Nicéphore Niepce* established photography; for he was the first to obtain a permanent sun-picture. Twelve years before this, Wedgwood and Davy had copied paintings made on glass, and the profiles of objects, the shadows of which were projected upon a piece of white paper, or white leather, saturated with a solution of nitrate of silver. The images so obtained could not be fixed, as no means was then known of removing the silver salts which had not been acted upon during the exposure; and the pictures soon blackened in every part when exposed to the light. The application of the *camera obscura*, and the fixing of the image so obtained, define the commencement of the art of photography. The process of Niepce, which was termed *heliography*, was conducted by smearing a highly polished metallic plate with a certain resinous substance known as " bitumen of Judæa," and this was exposed to the image formed in the camera for some hours. The action of the light was such, that the resin, which before exposure was soluble in oil of lavender, became insoluble in that substance. Hence, on treating the plate after exposure with that solvent, only the deep shadows dissolved away, the lights being represented by the undissolved resin. The brightly polished parts of the plate, which were uncovered by the removal of the resin, appeared dark when made to reflect dark objects, while the resin remaining unchanged on the plate appeared light in comparison.

In 1826 a French artist, named Daguerre,† who had already made some reputation as a painter of dioramas, entered into a sort of partnership with Niepce, into whose process he introduced some improvements ; but, dissatisfied with the slowness of this proceeding, he invented a process of his own, by which pictures of great beauty could be produced with all the shadows, lights, and half-tints faithfully rendered ; while the time of exposure in the camera was reduced to twenty minutes. In this process the burnished surface of silver formed the shadows. A plate of copper, coated with pure silver, had the silvered surface polished to the highest degree, and it was then exposed to the vapour of iodine until a thin yellow film had been produced uniformly over the silver. It was then placed in

---

the camera; and, although when withdrawn no image was perceptible, a latent image was nevertheless present; for when the plate was exposed to the vapour of mercury, that substance attached itself to the parts of the plate in proportion as they had been acted upon by the light. Means were adopted by Daguerre for fixing the picture; and after his processes had been made public in 1839, several important improvements were proposed by other persons. By using bromine as well as iodine the sensitiveness of the plates was so much increased that the time required for exposure was reduced to two minutes, so that about the year 1841 portraits began to be taken by this process.

The world at large, which profits most by great inventions, has little idea at what cost of intense application, concentrated thought, and heroic perseverance, such discoveries are made. What his discovery must have cost Daguerre may be inferred from an anecdote related by J. Baptiste Dumas, the distinguished French chemist and statesman. At the close of one of his popular lectures in 1825 —*fourteen years before Daguerre had perfected his process*—a lady came up to him and said, "Monsieur Dumas, I have to ask you a question of vital importance to myself. I am the wife of Daguerre, the painter. He has for some time let the idea possess his mind that he can fix the images of the camera. Do you, as a man of science, think it can ever be done, or is my husband mad?" "In the present state of our knowledge we are unable to do it," replied Dumas; "but I cannot say it will always remain impossible, or set down as mad the man who seeks to do it." The French Government, with an honourable recognition of the merits of Daguerre, and of Niepce who had passed away poor and almost unknown, awarded to the former a pension•of 6,000 francs (£240), and to Isidore Niepce, the son of the latter, a pension of 4,000 francs, one-half to be continued to their widows.

But Daguerre's process had no sooner been brought to perfection than it began to be supplanted by a rival method, devised by an Englishman, Mr. Fox Talbot, who had published his process six months before that of Daguerre was given to the world, and who, therefore, was unacquainted with the details of the latter. The first of Mr. Talbot's publications contained only an improved mode of preparing a sensitive paper for copying prints, by applying them to it and causing the light to pass through the paper of the print, so that the parts of the sensitive paper protected by the opaque black lines were not acted upon by the light. The paper was first dipped in a solution of chloride of sodium, and then in one of nitrate of silver, the result being the formation in the pores of the paper of chloride of silver, a substance much more quickly affected by light than the nitrate of silver used by Davy and Wedgwood. The impression so obtained was a *negative*, that is, the lights and shades of the original were reversed; but when this negative was again copied by the same process, it produced a perfect copy of the original print, for the lights and shades were of course reversed from those in the negative proof. Thus from one negative any number of positive or natural copies could be produced; and this point in Mr. Talbot's invention is one great feature of photography as now practised. In 1841, Mr. Talbot obtained a patent for a process he called the *Calotype*, but which, in his honour, has since been known as the Tabotype. A sheet of paper is soaked, first in a solution of nitrate of silver, and then in one of iodide of potassium, by which it becomes covered with iodide of silver; it may then be dried. It is prepared for the camera by brushing it over with a solution of gallic acid containing a little nitrate of silver. By this last process its sensitiveness is greatly increased, and an exposure in the camera for a few seconds, or minutes, according to the power of the light, suffices to impress the paper with a latent or invisible image, which reveals itself when the paper is treated with a fresh portion of the gallic acid mixture. The Tabotype is the foundation of the methods of photography now in general use ; but, before we describe these, it may be proper to mention some other substances which have been found sensitive to light, and to discuss the nature of the invisible images which are first produced in these processes.

The *art* of photography has outstripped the *science*—in other words, the nature and laws of the chemical actions by which its beautiful effects are produced are not yet clearly understood, and some quite recent discoveries seem to show that we have yet much to learn before a complete theory of the chemical action of light can be proposed. Some results which have been established may be mentioned, as they show those curious effects of light to be more general than would be supposed from a description of photographic processes dependent on silver salts only. It has been found that certain acids, certain salts, and certain compounds containing only two elements—of which one is a metal—have a tendency to split up, or resolve themselves into their several constituents, when exposed to the action of light. On the other hand, chlorine, bromine, and iodine exhibit, under the same conditions, an exalted affinity for the hydrogen of organic matters. These tendencies concur when the compounds above referred to are associated with organic materials, as in photography. Solution of nitrate of silver is blackened when it is exposed to light on a piece of paper which has been dipped into the solution ; but a piece of white unglazed porcelain similarly treated shows no change. A solution of nitrate of uranium in pure water is not changed by light ; but a solution of the same salt in alcohol becomes green, and deposits oxide of uranium. The reducing action of the light is insufficient of itself to accomplish the decomposition of the salt in the first case ; but the presence of the organic matter determines this decomposition in the second case. Bichromate of potassium is by itself not easily decomposed by light ; but when it is mixed with sugar, starch, gum, or gelatine, the sunbeams readily reduce it. It is remarkable that the gelatine, gum, or starch becomes insoluble by thus taking up oxygen, and the gelatine loses its property of swelling up in water. We shall presently see the advantages which have been drawn from these circumstances.

It is not necessary that the light should act upon both the organic substance and the oxidizing substance at the same time. If paper impregnated with iodide of silver and gallic acid be placed in the camera, the image soon appears ; but if, as in the Talbotype, the iodide of silver only be acted upon by the light, no image is perceptible on withdrawing the paper from the camera. The action of the light has nevertheless imparted to the silver salt a tendency to reduction ; for when the paper is afterwards dipped into a solution of gallic acid, the image immediately appears. In order to distinguish these two actions, the substance which receives and preserves the latent impression from the light is called the *sensitive* substance, and that which reveals the latent image is termed the *developing* substance. A considerable number of substances having this relation to each other have been observed, and the following table of instances—cited by Niepce de Saint-Victor, the nephew of the original inventor—will give some idea of their variety :

Sensitive Substances in the paper exposed to the action of the Light.	Developing Substance.	Results.
None, *i.e.*, plain paper.	A salt of silver	Black image.
Nitrate of silver, or iodide of silver.	Gallic acid, or sulphate of iron.	Black image.
Nitrate of uranium	Water	By prolonged action of light, a grey image of protoxide of uranium ; the image disappears when paper is kept in the dark, but shows itself again in the light.
	Red prussiate of potash ..	Intensely red positive image ; becomes blue by sulphate of iron.
Nitrate of uranium and tartaric acid.	Nitrate of silver or chloride of gold.	Unchangeable images— resembling those of ordinary photographs.
Chloride of gold.	Nitrate of uranium, sulphate of iron, sulphate of copper, bichloride of mercury, salt of tin.	. . . . . .
Gallic acid	Sulphate of iron	Blue-black image.
	Red prussiate of potash ..	Blue image.
Red prussiate of potash.	Water, bichloride of mercury, gallic acid, salt of silver, salt of cobalt.	Blue image, hastened by acids and by heat.
Bichloride of mercury.	Protochloride of tin, soda, potash, sulphide of sodium.	. . . . . .
Chromic acid, or bichromate of potash.	Salts of silver	Purple-red positive image.
Starch............	Blue litmus..................	Red image.
	Iodide of potassium	Reddish brown image.
	White indigo	Blue positive image.
	Campeachy wood	Red positive image.

These are only a few of the instances in which actions of this kind have been observed. It is remarkable that the order of the first two columns in this table may be inverted without changing the result. Thus, instead of exposing iodide of silver to the light and developing the image with gallic acid, one may expose a paper saturated with gallic acid solution, and develop with iodide of potassium and nitrate of silver. The first reaction noted in the table deserves some remark: it is not peculiar to paper, but is common to most organic materials, such as albumen, collodion starch, fabrics, and indeed to organic matters in general, provided they are not of a black colour. Tartaric acid, sulphate of quinine, and nitrate of uranium increase this sensibility. The paper which has been impressed preserves its undeveloped image for a prolonged period if kept in darkness ; and it has been found that one piece of paper can impart the image to another by simple contact in the dark. What is still more remarkable, the invisible impressions on a piece of paper may be transferred to another not in contact by merely placing it opposite the first, and separated by an interval of a quarter of an inch. No satisfactory explanation of these phenomena has been advanced, but many conjectures have been made. One of these supposes that some unknown intermediate products are formed, which are, in the case of the latent image on paper, very oxidizable ; but in the case of silver salts, &c., very reducible, so that the addition of a silver salt in the first case, and of organic matter in the second, only completes the phenomena by ordinary chemical action. Niepce de Saint-Victor, however, found that a surface of freshly broken porcelain alone will receive a latent impression from light, and will reduce in those places sensitive salts of silver. He believes that the light in these latent images is simply stored up, and that its energy remains fixed to the surfaces until the occasion of its producing a chemical action.

When a pure solar spectrum is made to fall upon paper rendered sensitive by silver salts, the effect is observed to be greatest near the Fraunhofer line H (No. 1, Plate XVII.), and it is prolonged with decreasing intensity beyond the violet end of the spectrum, while towards the other end it terminates about the line F. When other sensitive substances are used, the range of photographic power in the spectrum is modified. It has been found that when a daguerrotype plate which has been impressed by the light in the camera is afterwards exposed to the red or yellow rays of the spectrum, it loses its property of condensing the mercurial vapours. This destruction of photographic impression by red or yellow light has a practical application of great importance, for it permits the processes of preparing paper and plates to be carried on in a laboratory lighted by windows having yellow or red, instead of the ordinary colourless, glass. Thus we see that it is by no means the whole of the solar rays which are concerned in producing photographic images ; nay, there are some which even tend to destroy the impressions produced by others. The fact that it is not the light, but only certain rays in the sunbeam, may be proved very conclusively by an experiment with a glass bulb filled with a mixture of equal volumes of hydrogen and chlorine gases. When such a bulb is exposed to the light of the sun or of burning magnesium, which is made to reach it by passing through a piece of *red* glass, no explosion takes place ; but if the bulb be covered only with a piece of *blue* or *violet* glass, the explosion is produced just as quickly as if it were exposed to the unaltered rays.

The visible spectrum obtained in the experiment described on page 318 is far from constituting the only radiations which reach us from the sun. For invisible beams of heat, less refrangible than the red rays, are found beyond the red end of the spectrum ; and another invisible spectrum stretches far beyond the violet end, formed of rays recognized only by their chemical activity. It is these which effect photographic actions, and though they are in part more highly refrangible than any of the rays producing the visible spectrum, a large portion are refracted within its limits, so that the maximum of photographic action in a spectrum is usually near

the violet end. When we wish to examine the spectrum of the heat rays,
it is necessary to replace the glass prism by one made of rock salt, for
glass absorbs these heat rays. It also intercepts a great part of the most
refrangible rays ; for when a prism of *quartz* is substituted for the glass
one, the spectrum becomes greatly extended at the violet end. The dark
Fraunhofer lines which cross the visible spectrum are represented also in
great numbers in the invisible spectrum : in photographs of the *ultra-
violet* rays more than 700 dark lines have been counted. It has been
proposed to employ quartz lenses in the photographic camera; but there
is reason to believe that the increased transparency of such lenses for the
chemical rays would be counterbalanced by certain disadvantages attending the use of quartz.

The beauty of the images which are formed in the camera obscura long
ago gave rise to the desire of fixing them permanently. We know how
perfectly photography has already satisfied that desire, so far as the *forms*
are concerned. The very perfection of the results obtained in this direction increases our regret at our inability to fix also the colours, and secure
the picture, not in grey or brown tones of reduced silver, but with all the
glowing hues of nature. An observation made by Herschel, Davy, and
others, seemed at one time to hold out hopes of a possible realization of
chromatic photographs. It was noticed that the images developed upon
chloride of silver, of the different parts of the solar spectrum, partook
somewhat of the colours of the rays which produced them. Edmond
Becquerel made a plate of polished silver, placed in dilute hydrochloric
acid, form the positive pole of a battery. The plate thus became coated
with an extremely thin layer of chloride of silver, which, as its thickness
augmented, exhibited the series of colours due to the action of light on
thin films. The operation was stopped when the plate had become of a
violet colour for the second time ; it was then washed, dried, polished
with the finest tripoli, and heated to 212° F., the whole of these operations
having been carried on in the dark. When this plate was exposed for
about two hours to the solar spectrum, fixed by proper appliances which
counteracted the apparent motion of the sun, the luminous rays were found
to have impressed the plate with their respective colours. The yellow was
somewhat pale, but the red, green, and violet were exhibited in their true
tints. A theoretical explanation has been advanced, which supposes that
yellow light, for example, renders the surface of the plate on which it fall₁
peculiarly capable of receiving and transmitting vibrations corresponding
to those of yellow light. Just as a stretched cord responds to its own
musical note, the modified plate gives back, out of all the vibrations which
fall upon it in ordinary light, only those of which it has itself acquired the
periodicity. But since the plate has not lost its sensitiveness to take on
other rates of vibrations, it receives other impressions, which first weaken
and then overcome the former, and, therefore, the colour necessarily
vanishes. This kind of difficulty seems to be a necessary concomitant of
every attempt in this direction ; and all the hopes founded on results yet
obtained have been disappointed by the rapid fading of the images.

The comparative cheapness and convenience of Talbot's process, and
especially the facilities which it afforded for the multiplication of proofs,
gave an immense impulse to photographic art. But the irregular and
fibrous structure of paper prevented the attainment of the beautiful sharpness of outline and clear definition of detail which the plates of Daguerre
presented. Sir John Herschel suggested the use of glass plates coated

with sensitive photographic films, and Niepce de Saint-Victor succeeded in fixing upon glass layers of albumen (white of egg) containing the silver salts, a method which is still used to some extent. The art received, however, its greatest stimulus from the improvements which ensued on the application of *collodion* to this purpose. Collodion (κολλα, glue; in allusion to its adhesiveness) is the name which has been given to a solution in ether of gun-cotton, or of a substance nearly allied to it. Its employment was suggested by Le Grey of Paris, but the late Mr. Archer was the first to carry the idea into practice, and the process which he described in "The Chemist," in 1851, is virtually that which is now almost universally adopted. This process has now been tested, for nearly a quarter of a century, by the united experience of photographers all over the world, and it is agreed that it is surpassed by no other, for it secures every quality which a photograph can possess.* The minor details of the method can be, and are, infinitely varied; scarcely two experienced photographers will be found working the process in identically the same manner throughout. Before giving an outline of the collodion process, it may be well to say something respecting the chief instrument of photography—the camera.

The ordinary photographic camera is almost too well known to require

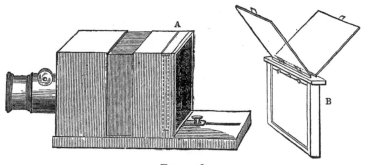

FIG. 308.

description. In its simplest form, Fig. 308, it is merely a rectangular box, in front of which is placed the lens, which slides in a tube, that its position may be adjusted so as to bring the rays to a focus on the surface of a piece of ground glass at the opposite end. This glass is fitted into a light frame, which slides in grooves, so that it can be raised vertically out of its position, and replaced by another frame, B, which contains a recess for the reception for the sensitive plate, and a sliding screen which protects it from light until the right moment. When this frame is placed in the camera, the sensitive surface occupies the same position as that of the ground glass, and the sliding screen is drawn up the moment before the operator removes from the front of the lens a cap which he places there after adjusting the focus. The sliding screen is usually made with a narrow strip at the lower part, joined to the rest by a hinge, so that when it has been drawn up it may be retained in its position, and placed out of the way, by being folded down horizontally. There is commonly provision for two plates in one frame, the slides, &c., being doubled, and the plates placed back to

---

* (1875) But see below, page 541.

back, as shown at B, Fig. 308. The camera is usually made in two parts, as shown in the figure, that at the back sliding within the other, so that a wider range for adjustment is obtained, and the same camera may even be used with lenses of different focal lengths. Many improvements have been made in the camera, by which it has been rendered more portable, and capable of more adjustments to suit varying circumstances. Fig. 309 represents a " bellows " or folding camera, which appears to supply every requirement for the studio. It is copied from Messrs. Negretti and Zambra's catalogue, as are also the other figures of photographic apparatus here given. Fig. 307 represents a camera for taking stereoscopic views, fitted

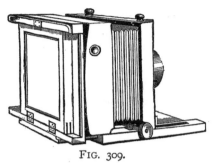

FIG. 309.

with two lenses, so that the two views are taken simultaneously on one plate.

No piece of apparatus used by the photographer is of so much importance as the lens; for good pictures cannot be obtained without well-defined, sharp images on the sensitive plate, and these images must have sufficient intensity to produce the required amount of chemical action in a short space of time. The formation of an image by means of a lens which is thickest at the centre is tolerably familiar to everybody; for most persons must have noticed that the lens of a pair of spectacles, or of an eye-glass, will produce an inverted image of the window-frame on a sheet of white paper, held a certain distance behind the lens. But the diagrams by which the paths of the rays are usually represented seem to convey a false impression to an ordinary reader, who usually goes away with the idea that somehow three rays are sent off by the object, and that one goes through the middle of the lens, and the other two meet it and produce an image. Let us suppose that, by means of a circular eye-glass, the image of a window is projected on a piece of white paper: a straight line passing through the centre of the glass perpendicular to its plane will meet the window and image each at a certain point. The point in which it meets the image is the *focus* of *innumerable* rays, which issue from the point in the window; that is, of the whole light sent out in every direction by the point a certain portion falls upon the lens, and by the refraction it undergoes in passing through it, the rays are again brought together at the point in the image. Thus the original point in the object is the apex of a solid cone of rays (if we may say so), of which the lens is the base, and the point in the image is the apex of another cone, having also the lens as its base. These cones would be termed *right* cones, because their bases are perpendicular to their *axes*, or central lines. But they represent the rays from only *one point* of the object. Let us now consider how the image of another point is formed, say one in the highest part of the object which forms an image on the screen. Those rays which are sent out by this point, and fall upon the lens, form now an *oblique* cone, of which the lens is the base, and the central ray will pass through the middle of the lens and continue its journey on the other side with little or no change of direction, forming also the axis of another oblique cone, constituted of the refracted rays, all of which will meet together at the lowest part of the image. Similar cones of incident and refracted rays, all having the lens as base, and all of them cones more or less oblique, will be formed by the light from each point of the object. Thus, the rays which issue from each point are brought together again in a series of points which have the same position with regard to each other, and collectively form an inverted image.

On carefully looking at the image, say of a window-frame, formed by a simple lens, the reader will observe two defects. The first is that the image cannot be made equally clear and well defined at the centre and at the edges : the adjustment which gives clear definition of one part leaves the other with blurred outlines. The second defect, which is best seen with large lenses, consists in coloured fringes surrounding the outlines of the objects. This depends upon the unequal refrangibility of the various rays, but it is obviated in *achromatic* lenses, which are formed of two or more different kinds of glass, so adapted that the refracting power of the compound lens is retained, and the most powerful rays of the spectrum are brought to a common focus. Such are the lenses always used in the photographic camera, and the skill of the optician is taxed to so combine them as to obtain, not only the union of the principal rays in one focus, but the greatest possible flatness of field in the image, the largest amount of light, the widest angle without distortion of the picture, and other qualities.

Photographers have even been so fastidious in the matter of lenses as to require all the perfection of finish which is given to the object-glasses of

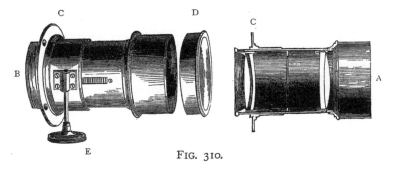

FIG. 310.

astronomical telescopes. Mr. Dallmeyer has made photographic lenses which cost upwards of £250 ; but it is doubtful whether the pictures formed by these would show any marked superiority over those produced by lenses costing only one-fifth of that amount. Fig. 310 shows the construction of the combination usually employed for taking photographic portraits. A is a section showing the forms and positions of the different lenses ; B is an external view of the brass mounting of the lens. It is provided with a flange, C, which is attached by screws to the woodwork of the camera ; and within the short tube, of which this is a part, slides the tube carrying the lenses, being furnished with a rack and pinion moved by the milled head, E. D is a cap for covering up the front of the sliding tube. A slit in the tube admits of plates of metal, perforated with circular openings, being inserted.

The openings are of various sizes; and these "stops" or diaphragms enable the operator to regulate the amount of light; and to cut off when required the rays passing through the marginal parts of the lens.

It now remains to describe in a few words a method of photography which was, and still is, much practised, namely, the *collodion process.* The collodion solution is prepared by dissolving one part of pyroxylin (gun-cotton) in ninety parts of ether and sixty of alcohol. The pyroxylin for this purpose may be obtained by steeping cotton-wool for a few minutes in a mixture of nitre and sulphuric acid, with certain precautions which need not here be mentioned. To the solution of collodion is added a certain quantity of iodide of potassium, or of iodide of ammonium; and sometimes other substances also are mixed with the solution with a view of increasing the sensitiveness of the plate when ready for exposure. Some of the collodion solution is poured on a well-cleaned plate of glass, which is placed horizontally; it spreads over the plate, and the excess having been poured back into the bottle, the evaporation of the liquids leaves the glass covered with a thin uniform transparent film, which firmly adheres. The next operation is to render the plate sensitive by means of the "silver bath." This is a neutral solution of nitrate of silver, one part to fifteen of pure water, which is placed in a trough of glass or porcelain, Fig. 311. By the aid of a proper support the plate is introduced quickly and steadily into the solution, immediately after the collodion film has been formed on its surface. In two or three minutes the layer of collodion becomes impregnated with iodide of silver, and when taken out of the bath, the plate exhibits a creamy-looking surface. The operation of sensitizing the plate by the silver bath must be performed in a room to which no light has access, except that which has passed through *red* or *yellow glass,* or a semi-transparent yellow screen.

The plate is now ready for immediate exposure in the camera. It is placed in the dark slide, in which it is conveyed to the camera; and there the image of the object is allowed to fall upon it for a time, which varies, according to the intensity of the light and the nature of the object, from 3 seconds to 45 seconds. The slide is withdrawn from the camera, and taken again to the "dark" room, *i.e.,* where only *yellow* or *red* light can reach it. If the plate be now examined, it will be found to present no trace of an image. A latent one, however, exists; and it is developed by pouring over the plate a solution of pyrogallic acid—one part to 480 of water, with commonly a little alcohol and acetic acid added. When it is desired to intensify the image still more, a few drops of the nitrate of silver solution is added to the *developing solution* immediately before pouring it on the plate. When the picture has become sufficiently distinct, it is washed with pure water, and then immersed in a strong solution of hyposulphite of soda. The last operation is termed by photographers "fixing" the picture, and the substance employed in it is invaluable to the art. It acts as a ready solvent of all the salts of silver which remain on the plate; and the discovery of this property of the hyposulphites by Sir J. Herschel, in 1839, marked an era in photography. The picture is then thoroughly washed in cold water, in order that the hyposulphite of soda may be entirely dissolved out. It is then dried, warmed before a fire, and finally the film is covered with a coat of transparent varnish, by which it is protected from mechanical injury. The image here is *negative*—that is, the strongest lights of the object appear as the darkest tints in the picture, and *vice versâ.* From it any number of *positive* pictures may be obtained by means of the sensitive paper prepared with chloride of silver as in Fox Talbot's plan.

As it is a tedious, and perhaps, in some cases, an impossible operation to completely remove all traces of silver salts and hyposulphites from photographs, they have frequently been found to fade ; but this is rarely the case with well-prepared specimens.   Processes have, however, been devised by which absolute permanence is secured for the photograph. One of the best of these is known as the Carbon Printing Process, and, as improved by Mr. Swan, it is thus practised :

A solution of gelatine is coloured by the addition of Indian ink, or any other pigment which will give the desired tone.   This solution is spread over sheets of paper which are then dried.   In this condition the paper may be preserved for any length of time without any special precautions. When it is required for use, it is floated, with the gelatine-covered side downwards, in a solution of bichromate of potash, and then dried ; but these operations must be carried on in the dark.   The paper is exposed under a negative photograph, with which its prepared side is in contact. The effect of the light is to render insoluble the gelatine on all those parts on which it has fallen, and this action extends to a depth in the layer proportionate to the intensity of the illumination.   The object is, therefore, to wash away all the *soluble* gelatine and the colour with which it is mixed; but this soluble gelatine is mainly on the side of the film which is in contact with the paper.   The gelatine surface is therefore made to adhere to another piece of paper by means of some substance insoluble in water ; and when this has been done, the whole is immersed in warm water. Then the soluble gelatine is soon dissolved ; the first paper floats off, and the insoluble gelatine, holding the Indian ink or other colouring matter in its substance, remains attached by the cement.   As the thickness of the layer rendered insoluble is in proportion to the intensity of the light passing through each part of the negative, the picture will be presented in all the proper gradations of light and shade.

The " wet collodion " process, that has been described on the preceding page, maintained an almost undisputed hold for more than twenty years in the practice of photography in all branches, and it was not until after the publication of the first edition of the present work that a new era in the art was commenced by the introduction of what is known as the *dry plate gelatino-bromide process*, to which the present enormous popularity of photography as a recreative art is due.   The difficulties of manipulation, the necessity for extensive experience, and for special and cumbersome appliances were obstacles it at once removed.   And not only so, but the whole scope of the art was extended ; for work that was before supposed impracticable, even to the most expert professional photographer, became the amusement of the amateur.   Here, we may remark in passing, that photography is greatly indebted for this, and many other improvements, to the enthusiasm of the amateur, which has accelerated the development of the art to a remarkable extent.   The collodion process itself admitted

FIG. 311.

of being modified as a dry plate method, by coating the film with a preservative solution of tannin, gum, albumen, or other substance, and then drying the plates, of course in a dark place. This plan made it possible to practise out-door photography with ease, and such plates were, at one time, much used for landscape photography, but they have now been almost superseded by the gelatine plates. It was Mr. Kennet, who, in 1874, first introduced the use of sensitive emulsions of gelatine, and the advantages offered by their use, caused them to be soon adopted by land-scape and amateur photographers. In 1878, Mr. Bennet showed, that these plates could be made wonderfully rapid in their action, so that portraits, etc., could be taken by them in an unprecedentedly short time. The preparation of the dry gelatine plates was then commenced on a large scale, and these were found so convenient, and reliable in use, that they were adopted by the professional photographers, who had hitherto adhered to the wet collodion and silver bath, from long habit and established associations. The collodion processes are, however, still much used, and are preferred by many to the gelatine plates ; indeed, it is admitted, that only by the former can certain desirable qualities of negatives be obtained, which are of great importance in some applications of the art.

There are, it need hardly be said, many modifications of the processes recommended for preparing gelatino-bromide dry plates, and each manu-facturer of the various kinds offered for sale has, no doubt, his own special plan and formula. In all, a very fine and carefully selected quality of gelatine is the medium in which the sensitive salts are embedded. An "emulsion" is prepared by adding to warm gelatine solution exactly determined quantities of solutions of certain compounds, of which a bromide (usually bromide of potassium) and silver nitrate are the essential ones, together with a small proportion of iodide of potassium. Minute quantities of iodine, hydrochloric acid, etc., are also often pre-scribed as additions. The mixture has to be heated, at the boiling temperature, for three quarters of an hour, then cooled, and mixed with more gelatine solution, or, instead of using acid and iodine and boiling, a little ammonia is added. When cold and set, the gelatine is washed with cold water, while squeezed through canvas, or after it has been cut into thin strips. It is then drained, dissolved at a gentle heat, and filtered warm. The clean glass plates are coated over with it, at the temperature of 120° F., and are set aside in a perfectly horizontal position until the gelatine has set, when they are placed for twenty-four hours in a drying cupboard, maintained at 80° F. It will be understood that these operations are conducted in a room where no light enters, except through a frame of ruby-coloured glass, and the plates, when dry, are carefully packed and stored in light-tight boxes. They are marvellously sensitive, and receive the photographic impression in about one-sixtieth ($\frac{1}{60}$th) of the time required for wet collodion plates. Half a second exposure in the camera may be sufficient to impress the image of a well lighted land-scape, even when a very small stop is used, and it is not unusual to employ for extra sensitive plates, a so-called "instantaneous shutter," when the exposure may be no more than $\frac{1}{80}$th to $\frac{1}{100}$th of a second, and yet obtain a perfectly strong image. Dry plates are manufactured in vast numbers in many large establishments, and the operations are carried on to a great extent by the aid of machinery, by which the plates are uniformly coated and automatically carried into drying chambers, etc.

If photography were popular before the introduction of the dry gelatino-bromide plates, it has since become a hundred-fold more so. Indeed, the camera is now seen everywhere, and few are the family circles in which at least one amateur practitioner of the art is not to be found ; indeed, the technical terms of the art have become "Familiar in their mouths as household words." The daguerrotype, notwithstanding its cost, had no sooner become a practicable process for taking likenesses, than it began to supersede miniature painting, and how rapidly it rose into general favour may be inferred from the fact that, in 1850, ten years after its introduction, it was estimated that in the United States of America, at least ten thousand persons had made it their profession, and, probably half as many more were occupied in making and selling chemicals, plates, cameras, lenses, mounting cases, and other apparatus connected with its practice. Such being the demand for photographic portraits, at the period when the sitter had, as we have already seen, to remain motionless for two whole minutes in sunlight, we can hardly be surprised at the increased popularity the art has acquired in the last decade, when a picture can be produced with one-hundredth the length of sitting, and at about the same reduction of cost. It may here be mentioned, that Daguerre's process is still occasionally used for special purposes ; it was, for instance, the method selected for obtaining the photographic records in the expedition sent out by the French Government, in 1874, to observe the transit of Venus.

The dry plate processes have given an immense impulse to landscape photography, and travellers have been able to bring back authentic representations of the scenery and inhabitants from every part of the globe. This advantage arises from the fact that having the camera, and its appurtenances, the tourist or traveller is not obliged to carry anything about with him except his plates, and when these have once been exposed in the camera, and stowed away in light-tight boxes, the latent images may be developed months, or even years, afterwards. But glass plates are heavy, and are liable to accidental breakage. Inventive ingenuity has been actively at work for the past few years, to find a means of obviating these remaining inconveniences. The first method adopted was to employ paper instead of glass, as a support for the sensitive gelatine film. The paper, having been cut to the proper size, is placed on a *film-carrier*, which is usually a thin plate of ebonite, by which the paper is kept flat. These carriers take the place of the glass plates in the ordinary dark slide, and after exposure in the usual way, the papers are removed in the dark room and made up into light-tight packages, where, of course, a large number will occupy but a small space, and the weight of them be wholly negligible. Many persons make use of this arrangement, which has the advantages of simplicity and of requiring no special apparatus. But an improvement was soon brought out, which consists in substituting for the carriers and pieces of sensitive paper a continuous roll of the material. For this purpose a special piece of apparatus, called the roll-holder, is made to take the place of the dark slide at the back of the camera. The arrangement will be readily understood from Fig. 311a. The figure shows the apparatus in section, but only the disposition of the principal parts, most of the mechanical details being omitted. R R' are two metallic or wooden rollers, which admit of being readily put in their places and taken out. Upon one of these, R, the full length of the material is previously wound, and the free end is passed over another roller, *r'*, and across the opening at E O, where the exposure is made. There is in front of this a dark slide (not here shown) to be drawn up when everything is ready for uncovering the lens. Immediately behind the paper is a flat plate of ebonite, E, or a smooth black board, the object of which is to keep the material quite flat as it passes over the opening to the roller, *r′*, which guides it to the roll, R′, on which it is wound as required. S S′ are two small rollers always pressed by springs against the rolls to prevent the turns working loose. There is a registering apparatus outside in connection with one of the rollers, *r*, or *r′*, to show when the proper length of material has been wound across the opening for a new exposure ; and at the same time a mark is automatically made on the paper to indicate where the negatives are to be separated for development by cutting the paper. Some forms of the apparatus also call the operator's attention to the sufficient winding of the roll by an audible signal, a stroke on a little bell tells that everything is ready for a new exposure. In some cases the number of exposures already made is registered by figures that appear on the outside. The paper in these processes is used only as a temporary support ; for after the negative has been developed in the ordinary way, the sensitive gelatine film is removed from it and made to adhere firmly on a plate of clear glass, from which

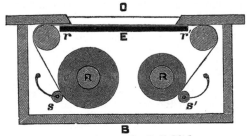

FIG. 311*a.—The Roll-Slide.*

prints are taken as usual. The operations required for the transferring require considerable dexterity of manipulation, and to both the paper and the glass special preparations have to be applied, before and after the transference of the film. This plan, therefore, of "stripping films" involves so great a number of delicate and somewhat troublesome operations that very many photographers have preferred to encounter the labour and risks of carrying about with them the more easily manageable glass plates. But what if some grainless, transparent substance could replace the paper in these rolls so that the negatives might be ready for printing from when merely developed and fixed? Many trials have been made to find this desideratum. A material sufficiently translucent, even, and of tenacity enough to bear the stretching strain between the rollers has, it is believed, been discovered in a very singular substance previously used for other purposes. The reader is no doubt familiar with it as the substitute for ivory in combs, knife handles, and other small articles. It is called *celluloid,* and is a composition the principal ingredients of which would never be guessed from its appearance—namely gun-cotton and camphor ! This material is prepared in a plastic condition that enables it to be shaped into any required form. It can be drawn into threads or rolled out into very thin films. Thin plates of it have been used in photography as a substitute for glass, for the sake of lightness, before its employment as a transparent film in the roll-holders. We have now at length the equipment of the travelling photographer reduced to the utmost conceivable limits of lightness and compactness. Thus the complete apparatus required for taking hundreds of pictures of a good size need not be more than a few pounds in weight, and can easily be carried in the hand. But even quite small negatives can now be very readily printed in a few seconds on paper, with an enlargement of many times the original dimensions. The resources of the photographic art appear indeed to be endless ; but a mere statement of even the more interesting of these would lead us beyond our limits, and descriptions of the details of manipulation are out of our province altogether. But a few of the more recent applications and developments of the art scarcely or not at all alluded to in the foregoing pages should receive some attention.

The extraordinary sensitiveness of the gelatine-bromide film which makes it possible to impress on it a photographic image in the merest fraction of a second of time, enables us to take pictures of objects in rapid motion. Express trains at their highest speed have been successfully photographed, and so has almost every moving object in nature. The photographs that have been taken of men, of birds, horses, and other animals in every phase of their most rapid actions, have solved many disputed and perplexing problems as to the nature of their movements, and sometimes the solutions have been of a very unexpected kind. Taking a photographic "shot" at a bird has become almost more than a figure of speech ; for there are contrivances by which a bird on the wing may be aimed at with the lens, and hit off on the sensitive plate with a certainty surpassing that of the fowling-piece. There are also photographic repeaters by which six or more successive photographs of the bird, etc., can be taken in a single second. Mr. Muybridge has published a number of such photographs of the horse, and by projection of the different images on a screen from a magic lantern, in rapid succession, he has been able to reproduce the visual appearance of horses trotting, leaping, galloping, etc., on the principle of the zoetrope (page 399). Photography has afforded wonderfully delicate observations in many departments of science, by recording phenomena too rapid for the eye to seize, or too recondite for direct perception. A few examples may be mentioned. First, the advantage of photographing the lines of spectra, such as those described in our article on the spectroscope, will at once suggest themselves, and accordingly this method of recording spectra has been largely used, and in the hands of Mr. Lockyer, Dr. Draper, and others has been successfully applied to the study of the solar and stellar spectra. But more than this, it is the sensitive photographic plate that has enabled us to explore the region of the solar spectrum lying far beyond its visible limits in the red and in the violet rays. The ultra-violet portion of the spectrum is shown photographically to be occupied by multitudes of the thin insensitive spaces—breaks in the continuity of the active rays—which are impressed on the photographic print as black lines, similar in every respect to the lines mapped out in the visible spectrum by Fraunhofer. It is known by these that the ultra-violet spectrum, produced by glass prisms, extends to a distance beyond the last visible rays of nearly double the space occupied by the colour spectrum. The principal lines, or rather the greater groups of lines in the invisible spectrum, are distinguished by the capital letters of the alphabet, in continuation of Fraunhofer's method, beginning from H and nearly exhausting the letters of the alphabet to designate them. These are photographed *in the dark ;* for all the solar beams that are allowed to enter the stereoscope are first passed through blue glass of such a depth that every kind of emanation capable of affecting the human eye is intercepted.

Another extremely interesting example of the application.of the art to scientific research is celestial photography. An image of the sun may be impressed on a sensitive plate in an ordinary camera, in an amazingly short space of time, but such image is much too small to show any of the markings on the disc of our luminary, even when the image is magnified, for its diameter is only about $\frac{1}{10}$th of an inch for each 12 inches of the focal length of the lens. In order to obtain an image of 4 inches diameter, a lens of 40 feet focal length must therefore be used. The first attempts in solar photography appear to have been made in France, in 1845, and the solar prominences were daguerrotyped in 1851 ; but it was not until 1860, that Mr. De La Rue succeeded in obtaining some beautiful negatives of the phenomena presented in an eclipse of the sun, and was thus enabled to determine a great astronomical problem, by showing that the red flames, or prominences, really belonged to the sun itself. 'At the present time, photographs of parts of the sun's disc are regularly taken at Kew, and other observatories, without the very long and heavy telescopes, which introduced many mechanical difficulties into the operation ; for, by means of Foucault's siderostat, the great lens and the photographic apparatus can be used in one fixed position. The siderostat is an instrument on which a flat mirror, made of glass worked to a perfect plane and silvered externally, is caused by clockwork to follow the motion of the sun, so that the reflected beams can be projected in any required direction unchangeably, and, therefore the image of the sun (or other heavenly bodies) viewed in the mirror, is absolutely stationary. The lens, carried in a short tube, has its axis directed to this image, just as it would be pointed at the luminary itself. In solar photography, the exposure is made through a very narrow slit in an opaque screen, which is caused to move rapidly in front of the image. Very fair photographic images of the sun, of several inches diameter, can, however, be obtained with an ordinary telescope of five feet or so focal length, by substituting a small photographic lens and camera in the eye-piece, and by enlarging the image in printing.

As early as 1840, Dr. Draper succeeded in daguerrotyping the moon, but it was not until 1851, that lunar photographs, obtained by Professor Bond, another American astronomer, were first exhibited in England. Many other distinguished experimenters have since successfully turned their attention to this subject, such as Dancer, of Manchester, Secchi, Crookes, Huggins, Phillips, and De La Rue. The latter, and also Mr. Fry, by photographing the moon, at different periods of her libration, have obtained very beautiful and interesting *stereoscopic* prints of our satellite, in which she presents to the eye the roundness and solidity of a cannon ball. Mr. Rutherford, in America, had an object glass of $11\frac{1}{4}$ inches diameter, made expressly with correction for the chemical rays, and with this instrument he has produced some of the finest photographs of the moon that have yet been taken. Reflecting telescopes, which have the advantage of uniting all the rays in one focus, have been used with excellent results, and it is said that some taken with the great reflector at Melbourne, where also the atmospheric conditions are very favourable, are almost perfect.

Excellent photographs of the planets have also been taken by Mr. Common and others ; but they are of course small, and have contributed so far, much less to our astronomical knowledge than those already mentioned. Very different are the results obtained in what, a short time ago, appeared a less promising field. The image of a so-called fixed star, in even the most powerful telescopes, presents itself as a mere luminous point, and this is the case whether the star is one of the brightest or one of the least conspicuous. The telescopic appearance is simply a more or less brilliant point. The various degrees of brightness which distinguish one star from another (*stella enim a stellâ differt in claritate*), and which the unassisted eye attributes to difference of size, led, long before the invention of telescopes, to a classification of them accordingly. The brightest stars are said to be of the 1st "magnitude," those of the next inferior degree of brilliancy, of the 2nd "magnitude," and so on, down to the 6th, which includes the faintest star discernible by an acute eye under favourable circumstances. But stars too faint to be thus seen came into view in the field of the telescope, and therefore those of the 7th magnitude, and beyond, are termed *telescopic* stars, and each additional power given to the instrument brings others in view that previously were invisible. The classification has been carried down to the 18th or 20th magnitude, which expresses the limit of visibility with the most powerful telescopes yet constructed. In the methods hitherto employed for this classification, there is necessarily much that is arbitrary and vague, and it is quite common to find a different magnitude assigned to the same star by different authorities. Now the photographic plate enables the astronomer to determine the relative brightness of stars quite definitely. Everyone knows that the time required to impress an image on the sensitive plate is longer, as that image is less luminous. Hence, by finding the time required for the images of different stars to be impressed, we have a measure of their relative luminosities. Suppose the image of a group of stars is allowed to act on a plate for, say, 5 seconds, we should find only the brightest stars represented. If a second plate have double the exposure given, it would be impressed by the images of not only the brightest stars of the group, but also by those of the next degree of brilliancy ; and a third plate exposed for 20 seconds would show more stars than the two former exposures. So that plate after plate might be exposed under the same group for successively longer and longer intervals indefinitely. Exposures extending over hours have been made, notably by Mr. Common in England, and by Mr. Gill at the Cape of Good Hope, showing not only how magnitude may be determined to any extent, and the heavens most accurately mapped out, but with this very remarkable result :—*thousands of stars, invisible even in the most powerful telescopes, are portrayed in the photographs.* Let us consider for a moment the significance of this fact with regard to the new space-exploring powers it has placed in the hands of science. The number of stars visible to the unassisted eye in the whole expanse of the heavens has been variously estimated, but the figures usually given lie between 3,000 and 4,000, and the highest estimate for the most acute eyesight, under the most favourable atmospheric conditions, places the limit at 5,000. The brightest star in the heavens is Sirius, and Sir. J. Herschel ascertained that its light is about 324, that of an average star of the 6th magnitude. Taking the average luminosities of stars of the first six magnitudes, Sir W. Herschel, from his own observations, represents their relative brightness by the following figures : 100; 25; 12; 6; 2; 1. The different degrees of brightness seen is, probably, due to the following three causes, combined in various proportions : (1) the different sizes of these luminaries themselves ; (2) differences in their intrinsic luminosity ; and, (3) differences in their distances from us. And it is also extremely probable that the last is generally by far the largest factor of the three. It has been found by photometrical experiments, that the light we receive from the sun is 20,000,000,000 (twenty thousand million) times more than that of Sirius. If we suppose Sirius to be in reality only as large and as bright as our sun, it follows that its distance from us must be no less than 13,433,000,000,000 miles. The distance of stars of the 16th magnitude has been estimated to be such that their light—travelling at the rate of 185,000 miles per second—takes between five and six thousand years to reach us. For a long time no sensible parallax could be discovered in any of the fixed stars ; that is, no change in their positions was discernible when viewed from points 183,000,000 miles apart, namely from the extremities of a diameter of the earth's orbit. In other words, if we suppose the line of the length just mentioned to form the base of a triangle, having a star at its vertex, the angle formed by the sides is so small that the most refined instruments failed to measure it. In recent times, however, the parallax of a few stars—about a dozen or so—has been detected and approximately measured. The greatest observed parallax belongs to in α the constellation of the Centaur, a star of the first magnitude, 30° from the south pole of the heavens, and of this the parallax amounts to but a little more than nine-tenths of a second of angular measurement, corresponding with a distance of nearly 20,000,000,000,000 miles, a space which takes light 3½ years to pass over. This star is, therefore, believed to be the nearest of any to our system. The smallest parallax that has been measured in any of these few stars is a fraction of a second of angle corresponding with a distance twenty times greater than the other, and requiring seventy years for light to traverse it. Now, as the photographic plate shows us stars of magnitudes indefinitely smaller even than the telescopic sixteenth, we cannot but marvel at the manner in which the light travelling from these suns in the immeasurable depths of space, and taking untold thousands (nay, millions, it may be) of years in its journey is yet able so to agitate the atoms of our silver compounds that images of things that will themselves, probably, never be seen by mortal eyes are presented to our view. A circumstance requiring explanation will occur to the reader's mind in connection with stellar photography ; and that is, how does it happen that, if the image of a star is a mere point, it nevertheless impresses the plate as a visible dot ? It is probably because the point is a centre whence the photographic influence radiates laterally on the plate to a small but yet sensible distance.

Among the cosmic objects presented to our observation there are none more fully charged with interest and instruction than the *Nebulæ.* These are faintly luminous patches, in some few cases visible to the naked eye, but for the most part telescopic. The milky way, which extends round the celestial sphere, is a very conspicuous phenomenon of the same kind. A few other hazy, cloudlike patches are seen in various parts of the heavens, visible on a clear moonless night when the eye is directed towards the proper quarter. The well known group of the Pleiades sometimes presents this appearance, but most persons are able by the unassisted vision to discern in it a group of six stars at least, and an opera-glass or ordinary hand telescope easily resolves the object into a cluster of 20 or 30 distinct stars. Telescopes of higher powers bring more stars into view, and as many as 118 have been counted in the group. There are several other groups of this kind perceptible to the naked eyes merely as diffused patches of light, but resolvable by the telescope into thickly clustered groups of minute stars ; but in many of the resolvable nebulæ the separate stars appear spread on a back-ground of diffused luminosity. Again, there are other nebulæ which telescopes of the highest powers we possess fail to resolve at all. Not only has the photographic method shown stellar components of some of these last, but it has depicted the form of nebulæ never seen at all, and whose existence was previously unknown and unsuspected. For example, the photograph has revealed the existence of a back-ground of nebulous patches to the stars of the Pleiades—a thing that had never before been suspected, although the group has been repeatedly observed by the most powerful telescopes. Those who are at all acquainted with astronomy, will understand the significance of this discovery for the science. The results already obtained afford a marvellous support to the famous speculation known as the nebular hypothesis. And as the forms of these objects are accurately shown for us by their own light, changes in their appearance may thus be detected as time goes on which may serve to lift the above named theory into the region of demonstrated truth. The nebulæ which neither telescope nor camera can resolve are such as the spectroscope proves to be masses of glowing gas or vapour.

It has been already mentioned that the light from these immeasurably distant stars and nebulæ is so faint that the most sensitive photographic plates have to be exposed for hours. This would be a matter of no difficulty if the clockwork mechanism by which the apparatus is made to follow the apparent motion of the heavens could be constructed with absolute perfection. But as this is not obtainable, even with the most careful workmanship, and the smallest jar or irregularity would distort and confuse the images, this source of disturbance is eliminated in the following manner : attached to the photographing apparatus and driven with it is a telescope, provided with cross wires, and through this an observer views some star during the whole period of the exposure, his business being to keep the image of the star accurately on the cross wire, which he is enabled to do by having the means of slightly modifying the movement of the clock-work. In the Paris Exhibition of 1889 were shown many very fine large photographic prints of nebulæ (notably of great nebula in Orion), which have recently been obtained in this manner, and those nebulæ that had been photographically resolved had the stellar components marked with wonderful distinctness. Comets and meteorites have been photographed, and even the *aurora borealis* and the lightning's path have been brought within the camera's ken.

Space would fail us to describe the many applications now found for photography in microscopy, in medicine and surgery, in anthropology, in commerce, and in the arts. It is obvious also from the improvements that are continually made, that many of these applications have not yet received their full developments. Photography has been enlisted into the service of the army and navy, and regular courses of instruction in the art are given in their training schools. A well equipped photographic waggon now accompanies every army corps, and in almost every ship of war, some proficient operator is to be found. By an ingenious combination of photography, aerostatics and electricity, it is possible to obtain with perfect safety accurate information of the disposition of an enemy's forces and fortifications. A small captive balloon is sent up, to which is attached a camera. At a height of a few hundred yards, the balloon is practically safe from any projectiles, and in its cable are interwoven two electric wires by which currents are conveyed to electro-magnets, which produce all the movements required for any number of exposures. Jurisprudence has found its account in recognizing the art, for the photograph is received in evidence for proving identity, etc. The administration of the criminal law takes advantage of the art to secure the likeness of prisoners for future identification, and the modern instan-taneous process renders unnecessary the subjects' concurrence with the operation. Again, if the "hue and cry" has to be raised for an individual "wanted" for any offence, and a photographic likeness of him is pro-curable, thousands of copies can be made of it in a few hours, by night as easily as by day, and distributed to every police station in the whole country.

Modern processes now enable us to obtain prints from negatives in as many seconds as a few years ago hours were required, and this by artificial light. A process of printing lately introduced and yielding artistic results which deserve to find more general favour, is that called the *platinotype.* Instead of the ordinary print produced on lightly glazed paper by the reduction of silver compounds, and of questionable permanency, the image is formed in the paper by metallic platinum, the most changeless of all possible substances under ordinary influences. The pictures are of a rich velvety black, with soft gradations, and the surface is without glaze or glare. The print has, in fact, the appearance and all the best qualities of the most highly finished mezzotint engraving, combined with the minute fidelity characteristic of the photograph. The problem of producing a photograph in colours, permanently showing nature's tints in all their gradations, has still a great fascination for some experimenters, and startling announcements are made from time to time of some discovery in this direction. It does not appear, however, that any success has really been arrived at, beyond the results long ago obtained by Becquerel as described on page 614; and, indeed, as our knowledge of the science of the subject increases, the less likely does the possibility of photograph-ing colours appear. It is, however, never safe to lay down the limits of discovery in science.* Note that precisely in the matter of rendering colour even in its due gradation of tone or luminous intensity, the photo-graph is quite untruthful. Everybody has noticed how unnaturally dark and heavy the foliage of trees appears in the prints; if we suppose a lady in a blue dress, with yellow trimmings, to sit for her portrait, the photograph will show her in a white dress with black trimmings; a sitter with light yellow or auburn hair will appear of quite a dark complexion; if you photograph a lemon and a plum together, the latter will probably come out lighter than the former; or if a daffodil be the subject, the flower will be drawn in tones much darker than the leaves. This in-correctness of tone relations can, however, be greatly lessened by the device of reducing the quantity of the blue rays, by interposing a piece of optically plane yellow-tinted glass, by using the sensitive plates tinted with certain coal-tar dyes, which are now prepared and sold under the name of "ortho-chromatic plates," or by both methods combined.

If any illustration were needed of the great popularity now attained by the practice of photography, reference might be made to the large number of periodicals devoted to the subject, and appearing weekly, fortnightly,

* See page 630.

quarterly or annually, in every civilised country, and also to the multitudes of societies that have been formed for the promotion of the art. In Great Britain alone there are now at least 150 such societies in active operation, and they are correspondingly numerous elsewhere. If, when we consider all that has been accomplished up to the present time, with the jubilee year of photography scarcely passed, and observe the increasing numbers of its cultivators guided by the explanations of its phenomena that science is beginning to furnish, we can expect a corresponding progress in the next fifty years, then the centenary may be reached with a roll of achievements that could we know them now we should think marvellous.

As already remarked elsewhere, the practical side of photography has outstripped the theoretical one, for so far its progress has been much less indebted for processes and technic to the direct guidance of science than almost any other of our Nineteenth Century acquisitions, such as telegraphy, electric lighting, etc. The materials employed, and the mode of manipulation, have certainly *not* been deduced from previous knowledge of the nature of light or from the laws of chemistry, although when, by repeated trials and happy guesses, the right direction had been found, the field into which it led could be more easily explored under the direction of chemistry and physics. But even yet the fundamental principle, or the precise nature of the action of light on certain compounds, has not been definitely made out, and although some theories on the subject have been proposed, no one has been generally accepted as an adequate explanation of the known facts, and still less have any quantitative relations been established for these actions. The photographer cannot compose a formula for the composition of his emulsions and developers from assured data like those that enable the chemist to weigh out with accuracy the constituents that go to produce a required compound.

The attainment of permanency in its products, which, by several processes, photography can now boast of, is one of its triumphs, and will tend greatly to enlarge the sphere of its utility. For example, we have a public institution, known as the National Portrait Gallery, in which it is sought to gather together and preserve the likenesses of the most eminent Englishmen, and presentments of such of far less fidelity than photographic portraits are eagerly sought after. It has been suggested that something like a National Gallery of *permanent photographic portraits* of the chief men of their time would be a fitting and acceptable legacy to the public of the future. This idea has much to recommend it, particularly as authentic likenesses would thus be secured for the nation beyond the chance of loss.

Photography has been applied in preparing blocks in relief for printing along with letterpress in the same way as wood-cut blocks. The process has the great advantage of producing in a wonderfully short time a perfect facsimile of the artist's drawing without the intervention of any engraver. A plate of zinc, brass, or copper, coated with a dried film of bichromated albumen, is exposed to light under the transparent negative of a drawing in pure line, that is, one having in it only lines of uniform colour throughout. The parts of the film reached by the light, which correspond with the lines of the original design, are rendered insoluble, while the rest can readily be removed by water. These unprotected parts have then to be removed by the action of acids, but these are used alternately with the application to the plate of certain compositions, the purpose of which is to prevent lateral erosion of the lines in relief before the requisite depth of the metal has been removed.   Fig. 147f is the reproduction of a pen-and-ink sketch by this or some similar process.   But nature and the ordinary photograph show us graduated tones which ordinary printers' ink cannot really reproduce, inasmuch as it is incapable of gradation, and can give the *effect* of gradation only by such devices as are mentioned on page 642 (last sentence).   Now, the photograph cannot yield a printing-block until its continuous tones are broken up into lines or dots.   Not a few methods of doing this have been contrived, but that which is by far the most commonly used, and is most successfully practised on the commercial scale, is simple in principle, although in actual working it calls for much experience and skill.   The negative is taken upon a wet collodion plate, in front of which, within the camera, and at a very short distance (say $\frac{1}{30}$th inch) from the film, is a transparent *screen*, bearing two sets of parallel opaque lines at right angles to each other.   These lines are mechanically ruled with the utmost regularity, and are separated by only very small intervals.   There may be from 80 to 200 of them in the space of one inch, according to the class of work required.   The effect of this is that the light reaches the photographic film through a series of minute transparent squares, the sides of which may be only from the $\frac{1}{140}$th to the $\frac{1}{400}$th of an inch in length.   Now it is found that the brighter lights from the original positive, after passing these small appertures, spread so as to more or less cover the opposite parts of the negative, while the feebler lights, from the shades of the original, impress the plate to a less degree, the developed image in these showing, perhaps, merely a small dot or, in the very darkest parts, a blank.   In this way, then, may the photographic negative be obtained with a granulated texture following in graduation the tones of the original.   After this, the rest is easy, for the process of exposing a metal plate, coated with a sensitive film under the negative, and of etching it with acids, etc., is essentially the same as in the foregoing.   Such is the *half-tone process*, which is now so largely superseding wood and other engraving.   It is unnecessary to describe technical details here, such as the employment of *bitumen of Judæa* as the coating for the metal plate, or how the image must be reflected into the lens from a mirror to avoid a reversal in the final print, etc.   There are endless modifications of the processes briefly mentioned above, and some of these are guarded as valuable trade secrets.   Several of the illustrations in this work are prepared by the half-tone process, of which plates I., IV., V., etc., are examples, and they should be examined with a strong lens, in order that the different rendering of the light and the dark parts may be compared.

PHOTOGRAPHY IN COLOURS.

I T is the statement as to the futility of assigning limits to scientific discovery that has been justified by facts.   The preceding edition of this work was not long in the hands of its readers before the solution of the problem of photography in colours was announced from Paris, where, at the close of 1890, the physicist M. Lippmann had succeeded in photographing the solar spectrum in its natural colours, and at the beginning of 1891, he was able to exhibit at the Academy of Science untouched photographs of a stained glass window in three colours, of a dish of oranges and red flowers, and of a gorgeously coloured parrot, all in their natural tints. The method employed had no apparent relation to that of Becquerel, but was of the simplest, and, moreover, one which any reader who has followed the first few pages of our section on the "Causes of Light and Colours" will have little difficulty in completely understanding, if he has devoted a little attention to Fresnel's interference experiment. M. Lippmann took a photographic plate, coated to a greater depth than usual with a gelatine film containing the sensitive salts of silver, and in the camera this plate was exposed with the glass towards the lens, while at the other side of the film was a metallic reflecting surface, namely, quicksilver. Supposing a ray of red light to enter the glass and traverse the film, it would be reflected from the metallic surface, and would meet the direct ray within the substance of the film, with a difference of length of path that would produce the interferences already described, and so give rise to alternate lines or bands of darkness and brightness. It would, of course, be in the lines of maximum brightness that the silver would be first deposited by the photographic action, and these microscopically fine lines or striæ of silver would give back, from ordinary light, a colour corresponding to the waves of red light that produced them. Similarly with the other colours. Anyone may observe the production of colour from ordinary white light in the iridescent tints of mother-of-pearl, where the effects are due to the varying distances of fine edges of the layers of the substance. If an impression is taken from a piece of mother-of-pearl by solid paraffin, or by white wax, or even by common red sealing-wax, the colours will seem to be adhering to the impression, but the operation may be repeated times without number. It is the distance apart of the lines or striæ that determinates the colour, and this is always some definite multiple of the wave lengths, given on p. 411, for the various colours. M. Lippmann's products are true colour photographs, and they form a new and elegant experimental demonstration of the doctrine of luminiferous undulations.

The colour effects of nature have also been reproduced by taking photographs of the same scene through coloured glass. Thus a screen of yellow glass will intercept the blue and the red rays, and the sensitive film will be impressed with images of objects containing yellow rays only, and that in proportion to the quantity of these rays that enter into any given tint. Similarly with images taken through red and blue glasses. The positives from these partial images being projected by three optical lanterns on the same space on a screen, and each being coloured by passing through tinted glasses like the original, the superposed images thus combined give a very lively impression of the natural colours in all their gradations.

Among the many processes for reproducing photographs by non-photographic processes, some have been more or less successfully combined with colour printing. Some of these productions are very effective, and are more attractive to many persons than the monochromatic tints of ordinary photographs.

FIG. 312.—*Portrait of Aloysius Senefelder.*

PRINTING PROCESSES.

AS it is beyond contradiction that printing is one of those inventions which have most influenced the progress of mankind, so it will be admitted that certain modern processes, by greatly facilitating the operations, and vastly extending the resources, of the art, possess an interest and importance surpassed by few of the subjects we have discussed. In a former article the reader has been made acquainted with the steam printing-press and other applications of machinery by which the impressions of a form of type, or of a pattern, can be rapidly multiplied. Here we have to describe some ingenious methods of preparing the forms or originals for letterpress and other printing, and certain beautiful processes for multiplying drawings, engravings, and pictures.

STEREOTYPING.

THIS term is applied to the process of obtaining the impression of a form of movable types, or of a woodcut, on a plate of metal which can be printed from. These plates, after the required number of copies

632

have been printed, can be stored away; and they are ready for use whenever another issue of the work is required. When the pages that are to be stereotyped have been set up in ordinary type, there are several methods by which the stereotype plates may be obtained from them; or rather, there are several materials used to form the matrix or mould in which the metal is cast. When plaster of Paris is used, the form is first slightly oiled, to prevent adhesion of the plaster; a thin mixture of plaster and water is then poured upon the form, which is surrounded by a raised rim, to retain the plaster. The thin plaster is carefully led into all the recesses of the type, and then some thicker material is poured on. The plaster soon sets, and is lifted off the type, and, after drying, is ready to receive the molten metal of which the stereotype plate is formed. This metal is an easily fusible alloy of lead, antimony, and other metals, which takes the form of the mould with great accuracy, and is, when solid, sufficiently hard to print from.

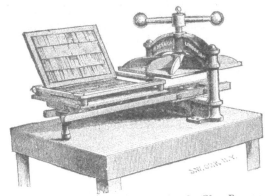

FIG. 313.—*Press for Stereotyping by Clay Process.*

Another plan is to make use of prepared clay, spread upon an iron plate, for the formation of the mould. The face of the type is brushed with benzine, the plate with the clay is laid upon it, and pressure is applied. The whole is then dried in a slow oven, and the clay, when detached from the type, is ready to form the mould. The advantages of the clay process are that the type does not require to be afterwards cleaned from oil, and that the material does not fill up the deeper spaces of the form, so that a thinner stratum of metal suffices to form the stereotype plate.

A third mode of obtaining the mould has been already mentioned in connection with the Walter Printing Press (page 313), in the working of which the *papier maché* process is ingeniously made to supply the curved stereotype plates for the cylinders. This process is also largely used for other newspaper presses, and sometimes for bookwork, as it forms an invaluable means of expeditiously obtaining a number of stereotype plates from the movable types. This production of a number of similar forms makes it possible to strike off a very large number of copies in a short time, for many presses can be employed simultaneously. For the paper process a number of sheets of tissue-paper are pasted together, and the moist paper is laid upon the form; then the operator, by light strokes of a brush, beats down the paper into the hollows of the type, beginning at the centre of the page, and going towards the margins. A sheet of stout unsized paper, called "plate paper," constitutes the upper layer; and when the whole has been well beaten down upon the type, pressure is applied by means of a screw acting upon a plate of iron covering the whole. In this condition a gentle heat, produced by steam, is made to completely dry and harden the paper matrix, which is very soon fit to be used for casting the metal. The apparatus for this purpose consists of a hollow iron table, within which steam is made to circulate. On this the form is placed, and the platen is pressed down upon it by means of a screw. In many cases the platen also is heated by steam, to accelerate still further the drying of the matrix, which is effected in about four or five minutes. One paper matrix, by careful use, will serve for the production of a series of casts without receiving any damage from the molten metal, as this is fusible at a low temperature.

The mould for casting flat stereotype plates from the paper matrix is made of iron, and has parallel surfaces, which admit of being so adjusted that the thickness required in the plates may be obtained very nearly. The paper matrix is laid on the horizontal iron bed of the mould; gauge-bars are adjusted, which retain it in its position; and then the second plate is folded down—the distance between that and the paper being determined by the gauge-bars. The cover is secured by clamping-screws, and then the mould is turned upright to receive the metal, which is removed, when solid, after the mould has been turned back into its horizontal position.

However the stereotype plates have been produced, it is necessary accurately to adjust their thickness by planing off some of the material from the back. The edges have also to be cut and trimmed to the exact dimensions required by the press. Various machines have been devised for effecting all these operations with accuracy and dispatch. The plates are afterwards mounted on wooden or metal blocks to bring them to the height of ordinary type.

A fourth method of producing plates for the same purpose as the stereotype plates already described is by *electrotyping*. This method appears to have been introduced as early as 1840, but the first results were not without imperfections. Now, however, this plan is almost universally applied to bookwork and woodcut illustrations. Many of our popular illustrated periodicals have so large a circulation that the wooden blocks would necessarily be spoiled by being used in steam presses long before they had yielded the required number of impressions; and the method has also the great advantage of securing the original engraving from the chance of accidental damage, by which a block is sometimes irretrievably injured. Hence woodcut illustrations are now always printed from electrotype copies of the engraved blocks, whether the work itself be printed from movable type or not. But the electrotype or stereotype process is always resorted to in the case of a work, whether illustrated or not, when it is foreseen that a re-issue will be demanded. These processes are also of great advantage to the practical printer, because when the pages set up in type have received their final corrections, he can take the casts, and then the type may be distributed—that is, returned to the cases ready for the compositors to use for other work.

The electrotype process is almost as simple as those for producing stereotype plates by casting, and its productions excel these by their great durability and extreme exactness of reproduction. We may take it for granted that the reader is familiar with the fact that ordinary letterpress characters and woodcuts are printed from forms, in which the black portions are in *relief.* For woodcuts the artist makes the drawing, in reversed position, on a block of finely-grained boxwood, in which the fibres of the wood are perpendicular to the surface. The engraver hollows out all the parts which in the impression remain white, while all the parts which are to receive the ink and produce the black parts of the impression must be left at the original level. The wooden blocks thus engraved would serve to produce a certain number of impressions, which could be taken off by careful hand-printing without perceptible damage to the block. But the pressure necessary for printing inevitably crushes the projecting parts of the block ; and the impressions, after a certain number, lose their sharpness. This is especially the case in machine printing ; but not only does the electrotype cast present a surface capable of bearing hard usage much better than those of the hardest wood, but even if the number of impressions required should wear out the metal plate, it can easily be replaced by another cast from the original block.

The mould which serves to give the electrotype cast may be made either of gutta-percha softened by a gentle heat and applied to the wood, or of wax. In either case a powerful pressure is applied, in order to force the yielding substance to take the forms of the engraved block or of the metal type. Wax is now generally preferred ; the yellow wax used for this purpose is melted, and poured into a shallow pan ; when it has become solid, it is sprinkled over with finely-powdered pure blacklead, which is brushed over the surface, and then the excess is removed by blowing with bellows made for the purpose. Thus prepared, the wax is placed over the type-form or wooden block in a powerful press, sometimes worked by hydraulic power ; but more frequently a *toggle* press is employed, in which the pressure is given by a screw and crank-wheel acting on two *elbow joints*, or *toggles*. For the information of non-mechanical readers it may be stated that a "toggle" consists of two bars jointed together, and placed *nearly* in a straight line : when a pressure is applied to the joint, tending to bring the rods still more nearly into a straight line, their extremities are thrust apart with a great force, which increases indefinitely as the rectilinear position is approached. In the electrotyper's press there are two toggles constructed of very broad bars, or rather thick plates, for they have nearly the width of the bed of the press. With this machine a very powerful and regular pressure is applied ; and the wax in a few minutes takes a sharp impression, embracing all the most delicate details of the work, and becomes at the same time very hard. The impression, of course, has hollows corresponding to the projections of the wooden block or type-form, and *vice versâ*. The face of the wax mould is now very carefully and completely blackleaded, a soft brush being used in the process. It is then placed in the solution of sulphate of copper, and the blacklead receives a deposit of copper, in the manner explained in a former page (498). In about forty or fifty hours a firm, compact deposit, about as thick as the finger-nail, covers the blackleaded surface, forming a perfect reproduction of even the most minute details of the engraved block or letterpress form.

The next operation has for its object the removal of the thin shell of copper from the wax. This is effected by exposing the mould to a gentle heat by immersing it in hot water, or by placing it on a hollow iron table which is heated by steam. The wax is run off into a proper receptacle for future use, and any portion adhering to the copper is removed by the action of naphtha or of a solution of potash. The thin copper shell is then tinned on the back, and an alloy of lead with some tin and antimony, forming the *backing metal*, is poured on it, to the depth of about one-eighth of an inch. When this has become solid the backing is planed so that the compound plate may have a certain regular thickness, and that the back surface may be parallel to the face. The edges are cut by a circular saw and trimmed by machine-tools, and the plate is rendered perfectly even, and adjusted with the greatest possible exactness to the required thickness. It is prepared for the press by being screwed down upon a block of wood of a certain thickness, so that the face of the plate may have the same height as common type, the screws passing through the margin or other hollow parts of the face of the cast. No more enduring surface than the copper of these electrotype casts, backed up by the hard alloy, has yet been discovered.

LITHOGRAPHY.

TO Aloysius Senefelder, a musician attached to one of the theatres in Munich, whose portrait appears at the head of this article, is due the invention of the art of lithography. It is said he used to arrange his musical compositions on a kind of slates, formed of flakes of the limestone which is found in the neighbourhood of Munich. One day a memorandum which he had made in this manner happened to fall into a slop-bucket full of greasy water; on withdrawing the piece of stone, he noticed with surprise that the grease had attached itself to the characters, while the rest of the stone remained quite clean. Such an incident might have happened to each one of a thousand men, and its significance might not be perceived; but it suggested great possibilities to Senefelder, who, applying himself for some years with ingenuity and perseverance to experiments with the Munich limestone, became, in the year 1800, the inventor of a new art. Though he was no chemist, and was unskilled in mechanics and in drawing, yet within four years from his first observation he had succeeded in finding the proper materials for his crayons and the appropriate acids for acting on the stone, in contriving a suitable press for taking the impressions, and in producing samples of lithographic work in various styles of art. He endeavoureed to keep his processes secret, and having obtained the exclusive right of exercising his invention in his own country, he attempted to carry on all the operations himself. Little by little, however, the general nature of the process became known, and although the details were jealously concealed, ingenious persons in France and elsewhere, by force of experiment, succeeded in re-inventing the art for themselves, and Senefelder never profited by his invention as he should have done.

The first lithographic press in London was established by Mr. Hullmandel in 1810. The value of lithography as a means of multiplying works of art was soon afterwards proved by the publication of a magnificent series of picturesque delineations of the quaint architecture of the old towns of Flanders and Germany, drawn on the stone by Samuel Prout. The late Mr. J. D. Harding largely contributed to the popularity of lithography by the landscapes which he drew on the stone, and thus placed in the hands of every one, prints in which all the freedom and force of the artist's work were secured. The French designers excel in fine-art lithography, and many beautiful productions of their crayons have been published in every department of pictorial illustration.

The best lithographic stones come from Germany; but for some kinds of work stones from other localities are used, on account of their less cost. Thus, in England, a stone yielded by the white lias formation near Bath has been found to possess the requisite qualities. The stones for lithography are prepared in much the same way as slabs of marble are polished; that is to say, by rubbing one slab against another with sand and water. When the stones have thus been brought to a plane surface, they are finished according to the purpose for which they are intended. If they are intended to receive written characters, they are polished to a very smooth surface by means of pumice-stone. But if they are to take drawings, then a certain uniform grain is given by means of finely-sifted sand, the operation being performed in a similar manner to that in which the stones are dressed, only pressure is not applied to the upper stone. The stones, after being washed and dried, are carefully covered on their prepared surfaces with thin paper, and are sent out for use.

When the stone is employed to reproduce written characters, or drawings imitating those done with a pen, *lithographic ink* is made use of with an ordinary pen, a ruling-pen, a fine brush, or a pen which the lithographer makes for the occasion out of thin metallic plates. The composition of the ink varies much : the usual ingredients are wax, gum-mastic, gum-lac, soap, and lampblack. This composition forms a solid, which is rubbed down with water to a thick liquid when required for use. The characters have, of course, to be written on the stone in a reversed position, and the lithographer acquires the habit of doing this with neatness and dexterity. He is provided with a looking-glass for viewing his work, in order to see the effect which will be given by the impression, for the looking-glass shows the characters in their usual position, just as the image of ordinary writing seen in it is reversed, showing, in fact, the very appearance the characters present on the stone. For a drawing, a *lithographic crayon* is used, made of wax, soap, grease, lampblack, and other ingredients. With this the drawing is made on the stone exactly as on paper, save the necessary reversals.

When the design has been placed on the stone, a liquid containing nitric acid and gum is poured over it. This liquid acts on all the parts of the stone not protected by the ink or crayon : they are thus rendered incapable of receiving printing-ink, while the protected parts have the impression more strongly fixed; for when the stone has been well washed with water, and turpentine has afterwards been applied, so that all the matter used in marking the design is dissolved away, the seemingly obliterated characters re-appear when—after the stone has been lightly wiped with a damp sponge—the roller charged with printer's ink is applied. The ink is taken up by the stone only at those places which have not been acted on by the acid. The impression is obtained by laying a sheet of damp paper on the inked stone and applying pressure by means of a roller, under which the stone passes. The stone is moistened with water after each impression before the inking-roller is again applied.

The lithographic stone, like other originals used in printing, is liable to deteriorate when large numbers of impressions are taken from it. This would be a serious drawback in lithography, but for a method of renewing the impression, which renders it unnecessary for the artist to retouch his work. This is the process of *transferring*, which is practised by the aid of a certain kind of paper specially prepared by a coating of paste. On this a proof is taken from the original drawing on the stone, and the still moist sheet is then applied to another stone, with the face downwards, and passed under the press. The effect of the pressure is to cause the adherence of the layer of paste to the stone; and when the paper has been thoroughly wetted at the back, it may be removed, leaving the paste still adhering to the stone, with the impression beneath it. When water is applied, the paste is washed off, while the ink of the impression remains attached to the stone, there reproducing the design drawn on the first stone. The transferred design is treated in exactly the same manner as the original drawing, acid being poured over the stone, &c., and the impressions obtained by the same method of successively sponging, inking, and pressing. The transferred drawing may be made to yield another transfer, and so on indefinitely; but when a large number of impressions from one design are required, it is usual to make at once from the original as many transfers to separate stones as will yield the required number of impressions without deterioration. In this way as many as 70,000 copies have been taken from a single drawing without their showing any marked difference in the character of the impressions.

The transfer process is also applied to place on the stone characters which have been written with a pen in the ordinary manner on prepared paper. In this way a person's handwriting is so accurately reproduced in the impressions that it is often very difficult to detect the interposition of the lithographic stone, and the impression often passes as the immediate production of the writer's pen. It is obvious that drawings etched with the pen on transfer-paper can be printed from in the same manner. And line engravings, which have been originally produced by cutting hollow lines on polished plates of copper, can be printed lithographically by transferring an impression to the stone. By transfer also the impressions of raised types or of woodcuts can be printed from the stone when desirable.

A beautiful and important application of lithography to the reproduction of pictures in colours has been so successfully carried out that a new branch of the art, termed *chromo-lithography*, now gives facsimiles of water-colour drawings and of paintings in oil. The copies of water-colour drawings especially are remarkable for their artistic qualities, and it is undeniable that these cheap reproductions of good paintings have done much to extend the knowledge of art. It is not contended that a chromo-lithograph, for example, after one of old William Hunt's rustic figures, or birds' nests with banks of primroses, can possess the wonderful refinement of the original; but it will nevertheless convey much of the artist's sentiment. Such transcripts of the works of our best artists adorn the homes of thousands who have never perhaps had the opportunity of even seeing the painter's original handiwork. In many a remote settlement in distant colonies, as in many an English home, the chromo-lithograph is the brightest of the household art treasures.

The principle of chromo-lithography consists in printing on the same paper with inks of various colours from different stones successively, so as to produce, by the juxtaposition and superposition of the various tints, the effect of a coloured drawing or painting. The artistic effects of the best chromo-lithographs require a great number of printings for their production, in some cases as many as twenty different stones being employed.

he stones and colours for such productions require true artists to prepare them, persons who can thoroughly understand and enter into the spirit of the original work. The first operation consists in the preparation of a faithful but spirited outline of the original, etched on transfer-paper, from which the outline is placed on a lithographic stone. This sketch we have called an outline, but it is in reality something more ; for it should suggest all the markings and limits of tints which belong to the original. This first sketch has some points marked on the margin by dots or crosses, which serve to secure true register in the subsequent processes ; that is, the impressions of the successive tints are so placed on the press that these points coincide in each impression.

From the first stone as many impressions of the sketch are transferred in light ink to other stones as there are colours required in the reproduction. To each colour a special stone is assigned, on which the lithographer, guided by the slight impression of the sketch, draws with the ordinary black crayon the form which that colour is to produce on the paper. Much artistic skill and judgment are required to do this in such a manner as to obtain a clear and harmonious final result. The gradations of the colours, and their blendings by superposition, must be carefully regarded. When the form and limits of each colour have been skilfully laid down upon its own stone, the surface is acted on by the acid, it is washed, the ink is dissolved off by turpentine, the stone is sponged, and the roller charged with ink of the appropriate tint is passed over it. The ink, as before, adheres only to the parts over which the crayon has passed, and an impression may be drawn off. Each of the other stones is similarly treated, and when the whole are ready, a proof is taken by giving the same sheet of paper the whole series of impressions in their proper order and colours, with the greatest possible accuracy of register. If any alterations appear desirable, they are made accordingly, by aid of certain devices which need not be here described, and when a satisfactory result has been obtained, the printing of the whole series of impressions is proceeded with. When the number of these is very large, transfers of each stone are taken as in ordinary lithography, only with certain extra precautions for obtaining precision in the register.

The brilliant effects produced by using gold and silver in lithography are obtained by using a kind of varnish, instead of coloured ink, for printing those parts where the metal is to appear. When this varnish has acquired a certain stickiness by partial drying, powdered gold or silver is applied, and this attaches itself only to the varnish ; when the sheet is dry it is passed under a burnished steel roller, the pressure of which imparts a brilliant lustre to the metal.

A method of colour-printing, in some respects resembling that of chromo-lithography, is practised by printing in variously coloured inks from a series of wooden blocks. This admits of far greater expedition in working off the impressions than the process with stones. The gradations of the coloured inks and powdered tints are produced in the same manner as those of ordinary woodcuts in black and white ; and when the colours are well chosen, and care is taken to secure the accurate superposition of the impressions, very pleasing effects can be produced by this means. The coloured prints which are from time to time issued as supplements to the "Illustrated London News" are produced by this process, and are no doubt well known to the reader. Our plate of spectra, No. XVII., is an example of another method of printing in colours.

OTHER PROCESSES.

I N recent times a great number of printing processes have been devised, but only a few have found their way into practical use, and some of these have scarcely been so extensively applied as their merits appear to deserve : either because the public demand has been insufficient to bring these inventions into common use, or the cost of working them has been too great. There is no doubt of their scientific success, whatever may be their commercial value as competing with cheaper and readier methods. We shall first describe the plan which has been termed *Nature Printing*.

This process is applicable only to certain objects which possess, or may be made to assume, a flat form. It has been most successfully applied to botanical specimens, the impressions of the leaves, flowers, and other parts of plants being given with an accuracy and minuteness of detail which the finest work of an engraver could never attain. In fact, the prints may be examined with a microscope, and they then reveal the minute structure of the object with wonderful clearness and delicacy. The notion of nature printing originated with M. Auer, the Superintendent of the Imperial Printing Office at Vienna; but the process was introduced into England, with certain improvements, by Mr. H. Bradbury. Supposing the object to be printed is a plant or the frond of a fern, it is first thoroughly dried by being pressed between folds of blotting-paper by means of a screw-press. The paper is changed several times, and, when necessary, the drying is accelerated by a gentle heat. When the specimen is perfectly dry, it requires very careful handling, for it is then generally extremely brittle. It is laid upon a sheet of pure soft lead, the face of which has been formed into a perfectly even surface, smooth and bright as a mirror. Mr. Bradbury encountered some difficulties in attempting to produce a surface of this kind, for small irregularities of the lead surface showed themselves; but Mr. James Wood succeeded in preparing for him a machine by which the lead is planed and polished in one operation. The object having been carefully laid upon the bright and smooth surface of the lead, a powerful pressure is applied by passing the plate between a pair of polished steel rollers. The effect of this is to embed the plant in the soft metal, which thus receives even the most delicate markings of the object. The next operation is the careful and patient removal of the object from the plate ; and as this is very brittle, it will be easily understood that it does not in general come away entirely, but portions will be left embedded in the metal. The skill of the operator is shown by destroying these by means of a blowpipe-flame, without in the least fusing the lead, which would of course ruin the impression.

When the whole has been removed, the leaden plate will have been engraved, as it were, by the object itself; and in this state the plate will yield impressions with ink in the same manner as an engraved copper plate. But in the soft metal the image would soon be obliterated, and therefore a facsimile of its impression is obtained in copper by the electrotype process. For this end the lead is covered with a varnish, except on the face, and tnus the deposit of copper takes place only where it is required, and the current of electricity is continued until a proper thickness of deposit has been obtained. This electrotype has all the hollow forms of the lead plate in relief, and it is used only for the preparation of another electrotype. For this purpose its face is brushed over with fine, pure blacklead, in order to prevent the deposit from becoming incorporated with it, while the rest of the plate is varnished. When it is placed in the electrotyping solution the copper is deposited on the blackleaded face, and the action is continued until the layer of metal has acquired the thickness of one-eighth of an inch. It is then removed from the matrix, and is ready for the printer, who deals with it in the ordinary manner of copperplate printing, except that he uses a softer paper, and this is forced by the pressure into the depressions in the plate, so that the impression is really embossed on the paper. Coloured inks are also used instead of black; for instance, to the leaves green-coloured ink is applied, and to the stems, &c., brown ink.

Several works on certain branches of natural history have been very appropriately illustrated in this way; among these, perhaps, no more beautiful example is to be found than in "The Ferns of Great Britain and Ireland," with text by Lindley and Moore. The merits of the nature-printing process appear to be the accuracy of outline in the flat form, and the delicacy of detail in parts projecting from the surface. The impressions cannot present artistic or natural shading in the objects; for the depth of colour will be in proportion to the projection of the part, whereas in nature the darkest shades are seen in the deepest recesses.

A copper plate, cut in the ordinary manner—as a line engraving, for example—soon deteriorates, as the pressure applied for each impression taken from it tends to close up the lines. It has therefore been necessary, where a plate has to yield a large number of impressions, to make use of steel instead of copper. But the electrotype has given the means of multiplying indefinitely facsimiles of engraved copper plates, so that in many cases a number of these are prepared, and used so long as they continue to yield clear impressions, the original plates engraved by the artist only furnishing the matrix. The mode of reproducing the plates by electrotyping from the original engraved plates is identical with that just described for obtaining the plates for nature printing from the leaden plates.

Another process of wider interest, and producing very beautiful results, is known as the Woodbury printing process, from the name of its inventor. It is a mode of photographically forming a picture in relief, from which printing blocks are obtained in much the same manner as in the nature-printing process. But the subject which is thus printed is a photograph; and it is only because in the actual production of the impression on paper the agency of light is not called into play that it is not described under the head of photography, for it is an ingenious mode of causing the photograph to engrave its own image on a metal plate. It is founded on a fact which has already been noticed, namely, the insolubility of gelatine which has been mixed with a bichromate and exposed to the action of light. Mr. Woodbury has obtained the best results with a solution of Nelson's opaque gelatine, 1 oz. of which is dissolved in 5 oz. of water, and to each ounce of the solution 15 grains of ammonium bichromate are added. When a layer of this mixture, which is of course prepared in the dark, is exposed to the action of light under a negative photograph, the gelatine is rendered insoluble under those parts of the negative through which the light passes, that is, in the parts corresponding with the dark shades in the original object, and the depth of the layer thus rendered insoluble in each part will depend on the relative thickness of the silver deposit in the negative photograph. Thus, in the half-tints the insoluble layer will not be so deep as under the parts of the negative through which the light passes without interruption. But the differences of depth will appear when the soluble

41

gelatine has been dissolved away on the side of the layer which is farther from the negative. Hence, Mr. Woodbury spreads his layer of bichromated gelatine on a sheet of plate-glass, previously coated with collodion, and when the gelatine has become dry, the double film is detached from the glass and exposed under a negative, the collodion side being uppermost and in contact with the photograph. After exposure the film is temporarily attached to another piece of glass, by means of a solution of India-rubber, and is then immersed in warm water, which quickly dissolves the soluble parts of the gelatine. Thus a counterpart in relief of the photograph is obtained. This is allowed to dry, and the next operation consists in obtaining an impression from it in metal : this Mr. Woodbury at first obtained by electric deposition, but he has discovered a much more expeditious process, which one would hardly have supposed possible before actual trial. The dry hard gelatine is placed upon a flat, truly-surfaced steel plate, with the collodion surface downward, a plate of soft metal is placed upon the gelatine, and the whole is subjected to a pressure of about four tons per square inch in a hydraulic press. In one minute a perfect impression of the gelatine relief, down to the smallest detail, is formed in the soft metal; and, strangely enough, the delicate sculpture which the light has executed on the gelatine is not in the least injured, but will stamp its image on an indefinite number of metal plates in the same manner.

The reader will understand that the impressed plate of metal now bears a hollow sculpture representing the image of the original object from which the negative photograph was taken, the darkest shades of the object being represented by the deepest depressions in the plate, while the highest lights are represented by portions of the metal at the level, or nearly so, of the surface of the plate. From this plate the prints on paper are obtained as follows : The plate is placed horizontally, with its impressed face upwards, and a quantity of a certain kind of ink is placed upon it. The composition of this ink, if ink it may be termed, is one of the ingenious parts of this elegant process. It is made of gelatine, coloured with some suitable transparent or semi-transparent pigments, and it is poured on the plate in a warm and fluid state, and in quantity more than sufficient to fill all the hollows. A sheet of paper is placed over the plate, and a moderate pressure is applied, when the excess of ink is squeezed out and escapes. That which remains in the hollows of the plate, becoming set by cooling, adheres to and is removed with the paper, giving in each part a force of tint proportional to its quantity, that is, according to the depth of the hollow in the plate. The paper is laid aside to dry, and although the picture has at first a certain relief, yet the gelatine ink dries down, the picture becoming so flat that no difference of the surface is perceptible. It will be observed that this mode of printing rests upon a distinctly new principle—namely, the production of shades and gradations of tints by the varying quantity of the ink laid upon the different parts of the paper. The method is in this respect identical with that by which the water-colour painter produces his gradations; for the colour is applied in transparent layers, and the depth of the tint produced depends upon the mass of the pigment laid on, and is greater or less according as the white of the paper is more or less visible through the film of colouring matter. The gradations of tint in wood and steel engraving and in lithographs are dependent upon quite another principle—namely, the varying distribution of spots, patches, or lines in black ink of uniform intensity. The Woodbury print has all the detail and clearness of the photograph, together with a certain softness, produced by the transparency of the colouring matter, not found in the ordinary photographic print. The method admits of any desired tint being given to the prints, and these are perfectly unchangeable by light. Thus the result is a print which secures every good quality of a photograph without any of the unpleasant ones, such as hardness, harsh tints, opacity, fugaciousness. The prints may be taken on plates of glass, and they then form beautiful transparencies. Such prints constitute most admirable slides for the magic lantern, since the semi-transparent colouring matter, and the soft gradations, produce charming effects.

Another ingenious invention of Mr. Woodbury's provides a means of making the sunbeam engrave a mezzotint copper plate from a photograph. The action of light on bichromated gelatine is here again taken advantage of. A film is prepared similar to that used in the above-described Woodbury process proper, but the gelatine is mixed with some powdered or granular material, so that it may give rise to a granulated texture in the resulting plate. This film is treated exactly in the same way as before with regard to exposing, washing with warm water, drying, &c. The product is a very thin sheet, having a mezzotint-like surface, with more or less grain according to the action of the light. The white parts are perfectly freed from the granular matter by the solution of the gelatine, while in the darkest parts there is the greatest accumulation. The dry film in this condition is pressed into soft metal, and by a double process of electrotyping and subsequent facing with steel, a plate is obtained fit for printing at the copper-plate press. The firm of Messrs. Goupil and Co., of Paris, extensively employ this process for the preparation of the illustrations in that elegant publication, " The Portfolio." Another method of photographic engraving lately projected by Mr. Woodbury is the following: a plate of steel is covered with a layer of gelatine, mixed with a certain proportion of gum and glucose, and dried in a dark room. This is exposed to the action of light under a transparent photograph on glass. When afterwards this gelatine layer is breathed upon, the moisture attaches itself to the portions which have not been acted on by the light, and these become more or less sticky. Sand or emery sifted to three different degrees of fineness is then sprinkled over the plate, beginning with the coarsest, which attaches itself to the most sticky parts. The less sticky parts are incapable of retaining these larger particles; while the finest sand, which is sprinkled on last, is held by parts of the plate that are even very slightly sticky; but the places where the light has been intense are dry, and none of the sand adheres. The gelatine layer is then completely dried, and the plate, being covered with another of soft metal, is placed in a press, by which a granular impression is produced on the soft metal, and this may then be copied in copper by the electrotype process. The larger particles of sand produce deeper depressions in the plate, and thus a gradation of tint is obtained.

Amongst other applications of the gelatine relief devised by Mr. Woodbury is that of producing a watermark in paper. A very delicate relief is firmly attached to a plate of steel or zinc, and when paper is rolled in contact with these plates, it receives an impression of the design, all the delicate half-tints being represented in the slight opacity of the paper. Mr. Woodbury is at present engaged in perfecting a method for wedding his own process to that of chromo-lithography, by first printing the different tints on the paper, and then transferring the Woodbury prints to the top of the colours. The transparency of the gelatine and ink is such that the most brilliant effects are attainable in this way.

Bichromated gelatine is also the agent employed in *photolithography*, the image of a negative photograph being thus rendered insoluble in a layer of gelatine spread on the stone, which is acted on by acids, &c., in the usual way, after the soluble portions have been removed by water. As there are also methods of using the lithographic process with plates of zinc instead of stones, so there are processes of impressing the image photographically upon the zinc. Of the general nature of the processes of *zincography*, *photolithography*, and *photozincography* the reader will now probably be able to form some idea, but the details need not here be described. The last two, and some other processes for printing photographic effects mechanically, all labour under the defect of imperfectly rendering the *half-tints* of a picture. This remark does not apply to the Woodbury process. The photo-lithographic process gives marvellous results in cases where no gradations are required. Thus a whole page of the *Times* newspaper may be lithographed in a space not exceeding half of this page, and although the characters may be indistinguishable to the naked eye, a lens will show them perfectly. Similarly, we may obtain within the compass of an octavo page a photo-lithograph of one of Hogarth's large engravings, which will show every touch of the original artist's *burin*.

There is reason to hope that the time is not far distant when all our tedious mechanical methods of reproducing drawings by wood or steel engravings will be superseded by processes which will give us absolute facsimiles of every touch of the artist's pencil; and when some process, giving all the delicacy and truthfulness of Mr. Woodbury's prints, will supply us with faithful transcripts of nature for book illustration at a cost not exceeding that of the ordinary methods. So far as relates to one style of drawing, these requirements appear to be nearly realized in the process termed the *graphotype*, which reproduces mechanically, in the form of a metal plate with all the lines in relief, a design which the artist has etched on a flat surface. This is effected in the following manner : Chalk is powdered very finely, and sifted through wire gauze having very narrow meshes. A quantity of this is spread upon a smooth plate of metal, and subjected to an intense pressure by means of an hydraulic press. The particles of the chalk cohere into a mass, having sufficient firmness to admit of its surface being drawn upon in the same manner as a block of boxwood. The drawing is effected with an ink composed of lampblack and glue, a finely-pointed camel's-hair brush being employed ; but the shades must be produced by lines and strokes as in wood engraving. When the ink is quite dry, the surface is rubbed with a fitch brush or with velvet ; and by this brushing the particles of chalk not protected by the inked strokes are loosened and carried off. In a short time the chalk between the strokes becomes quite hollowed out ; and when a depth of about one-eighth of an inch has been attained, every line remains standing in relief exactly as in an engraved wood block. A strong solution of silicate of potash is then poured upon the chalk, which its chemical action converts into a kind of stone without in any way altering the forms. Although this artificial stone is quite hard, so that impressions may at once be taken from it, yet it is incapable of enduring the wear and tear of the printing-press. Accordingly a mould is taken from it, and this is made, by some of the processes of casting or electrotyping already described, to furnish a metal stereotype plate.

THE LINOTYPE MACHINE.

AMONG recent inventions in connection with printing, the *linotype machine* calls for special mention. In this machine a great number of actions are combined and co-ordinated with the utmost ingenuity, but such mechanism does not lend itself to popular description, and we must confine ourselves to a statement of what it effects, recommending

FIG. 313*a.—The Linotype Machine.*

the reader to avail himself of some opportunity of seeing the apparatus at work. It will not then be needful to give details of every one of the very numerous parts, which present in the *ensemble* a great ap-. pearance of complication, the more so that much ingenuity has been exerted to make the machine compact, which is a practical point of great importance.   The disposition of parts is not, therefore, that which is cal-culated to show each movement clearly to the spectator, but that by which the least space is occupied.   The machine is driven by belting from a main shaft, turned by a steam-engine, gas-engine, electro-motor, or other regular source of power, and rotated at such a rate that the main pulley of the machine itself (14½ inches in diameter) shall make about 60 re-volutions per minute.   Fig. 313*a* shows the general aspect of the machine and seat for the one operator required, but as we are not undertaking a detailed and complete description of the whole mechanism, no letters of reference are given ; but the reader will be able, from the following diagrams, to identify the more important parts, and form a general idea of their action and purpose.   In this machine great use is made of the contrivances called *cams*, several of which may be observed in the sketch towards the side of the machine on the left, being fixed on and turning with its main shaft.   They consist of plates, or open rims of various forms, which move levers, etc., in any required way, and at any required period of the revolution.

The linotype is not a type-setting and type-distributing machine, but one in which the form is stereotyped line by line ; hence its name of *linotype*.   The mould, or matrix, is made up of a number of brass matrices, each of which consists of a flat plate having on its edge a letter incised.

FIG. 313*b*.—*A Matrix.*

One of these is represented on Fig. 313*b* wherein *a* is the hollow letter. At the upper end the plate is cut into a number of notches like the teeth of a saw, only that some of the teeth have their points cut off, leav-ing steps, as it were, with faces parallel to the longer edges of the matrix. There may be seen one of these at *b*, and on the opposite side of the V-shape, three may be observed.   The number and arrangement of the cut-away notches is different for the matrix of each letter (or sign), and special to it.   The meaning of this will be seen presently.   The diagram Fig. 313*c* will help us to see how these matrices are assembled by touches of the finger on the required letters as marked on the keyboard at D. The matrices are assorted and stored in separate channels in the "matrix" magazine, A, a portion of its cover being here represented as broken off in order to show the channels.   It will readily be understood that, by a system of levers connected with each key, the corresponding matrix is released by means of an escapement (B′), and falls down one of the channels E on to the travelling belt F, which conveys it to compos-

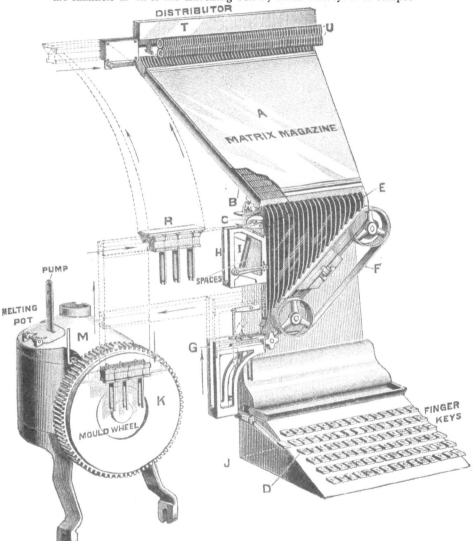

FIG. 313*c.*—*Diagram of Movements.*

ing stick G, in which the matrices successively assemble in the order to constitute a line (Fig. 313*c*), in which observe that the several words are separated by spaces formed by long wedges of steel, the thick ends of which hang down considerably below the line of matrices.   These are dropped one by one from a store at I (Fig. 313*c*), when required, by a touch on the key-bar J ; two of them are shown in position in the assembling stick G.   In Fig. 313*a* a bell is seen in front of the keyboard, and this is automatically rung by a mechanical device when the line of matrices is approaching in length to that allotted to the work.   At this point the

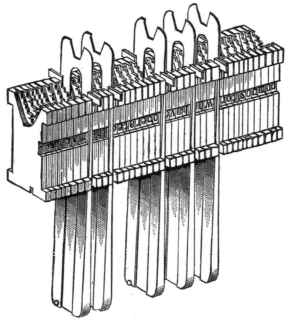

FIG. 313*d.—A Line of Matrices.*

operator has to consider whether he can complete the line with another, or with how many syllables of a word, and he touches the keys of the re-quired letters.   The assembling stick then contains all the matrices com-paratively loosely packed side by side, for the words are as yet separated by only the thin edges of the space wedges.   A touch of the operator on a lever brings into play another part of the mechanism by which the com-posed line is bodily lifted a short way, then moved horizontally, and conveyed to the " mould wheel," in which there is a slot, adjustable in length and width, and the line is here firmly pressed against the face of the wheel in such a way that the slot coincides with the line of hollow letters on the edges of the matrices, as shown in Fig. 313*b*.   This moulding arrangement is not the least ingenious device in this machine, and well deserves attention. Before the moulding takes place, but while the line is in its place, the wedge spaces are pushed up through the matrices by another portion of the mechanism, and thus the line is immediately "justified," as the printers term it ; that is, the wedges rise up, separating the words, more or less, until the line has exactly its assigned length, and the words are, at the same time, separated by equal spaces. A melting-pot behind the mould-wheel contains a quantity of fusible metal, resem-

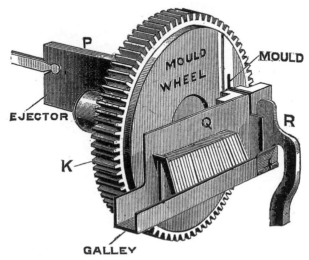

FIG. 313*e.*—*A Finished Line entering Galley*

bling stereotype metal, which is maintained at just the temperature of fluidity by a regulated gas burner. At the right moment a plunger is forced into the fluid mass, causing it to rise through a kind of spout to the level of the slot in the wheel, and be forced through that into the line of letters. The metal instantly solidifies in the mould, the line of matrices is removed on a bar to a new position at R, Fig. 313*c*, and the wheel then makes a quarter of a turn, bringing the mould from the horizontal into a vertical position (Fig. 313*e*). The linotype is subjected to the operation of certain knives (not shown), by which it is pared smoothly to the exact thickness and height required, and finally ejected, as shown in Fig. 313*e*, dropping in its proper order into a receiving galley. The line, as completed, has the shape represented in Fig. 313*g*, and a number of these lines assembled constitute a "form," answering all the purposes of the ordinary forms consisting of separate type. These last, after having served their purpose, must be "distributed," that is, each single letter must be returned to the case from which it was taken by the compositor ; but the linotype form, after use, is simply returned to the melting-pot for its metal to be recast into new forms. The forms can, of course, remain standing for any length of time at the mere expense of keeping the metal unemployed. One advantage of the linotype is that the printing is all done from new clean-faced forms, instead of the old and dull-faced characters of ordinary type that have been much used, but have to be resorted to under ordinary circumstances.

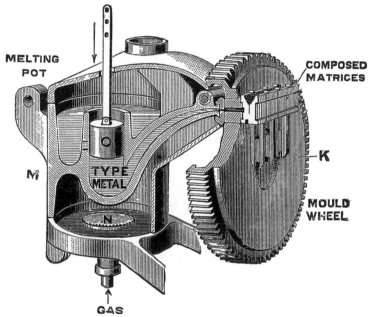

FIG. 313*f.*—*The Melting Pot and Mould Wheel.*

It may occur to the reader that errors in linotype would be much more difficult of correction than those occurring with the ordinary type composed by hand. If by chance a wrong matrix appears in the line, this can be changed by hand at once ; but supposing that the operator overlooks some error in reading the assembled line, which, observe, he reads with the characters arranged as they will appear in the impression, or that he has misread his manuscript, and the line is cast, assembled into a " form " with the rest, and then in the printed proof the error is discovered, how is it to be rectified ? Simply by removing the faulty linotype from the form, and casting a new one. This is so quickly and easily done that it has been found by actual test between linotype and ordinary type matter containing the same defects, that the former could be corrected in less than one-third of the time required for the latter.

We left the line of matrices at R (Fig. 313*c*), and we must now indicate the method by which each is automatically returned to its own magazine, an operation for which much ingenious mechanism has been contrived, of which the details cannot be well described in this place. The line having reached R, the space wedges are disengaged from it and removed to their receptacle at I, while the matrices become engaged by their teeth in the grooves of a horizontal bar, and then the bar is grasped by a lever which lifts it up to the distributing arrangement at the top of the machine, where the teeth of the matrices come to the exact level of the grooves of the distributor bar T. The line is then pushed laterally, the sides of the

FIG. 313*g.*—*The Finished Line.*

matrices become engaged in the hollows of two parallel screws U, by which, while suspended only by such of their inclined teeth as the corresponding groove of the distributor can support, they are made to slowly travel along from left to right until each reaches a certain point, namely, that at which its sustaining V grooves on the bar are interrupted by cuts which permit it to drop into its own special magazine. A little consideration will show how, by various combinations of the notches on the matrix, and corresponding cuts at the right places in the grooves of the bar, each

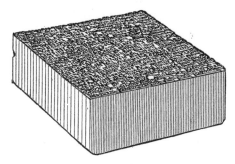

FIG. 313*h.*—*Lines assembled into a " Form."*

matrix may be made to move along until it reaches a determinate place, and there dropped. Compare Fig. 313*b* and Fig. 313*i*. Each matrix thus again deposited in its proper magazine has completed the circuit of the machine, or, at least, has passed from the bottom of its magazine to the assembling stick, hence to the mould, and, by the distributor, finds its way back to the top of its magazine, whence, in its turn, it will descend to perform again the same duty.

It must be understood that, beyond the operator's touches on the key-

board, and that required to send off the assembled line to the moulding
apparatus, all the actions are done automatically without the interference
of the operator, who, while one line is getting moulded, raised up, and
distributed, calmly proceeds with the composition of the following one.

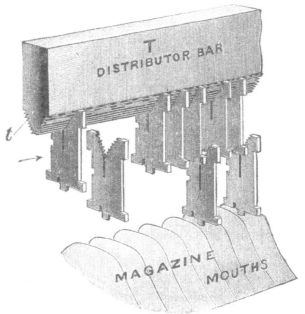

FIG. 313*i*.—*Matrices dropping into Magazine.*

The rate at which the work is produced is very great. One good
operator with one machine can, it is said, turn out, hour by hour, matter
that would be equivalent to two and a half pages of this book, arranged
solid or without break.   There are, of course, record performances of exceptional operators who have completed more than twice as much as this
in a single hour.

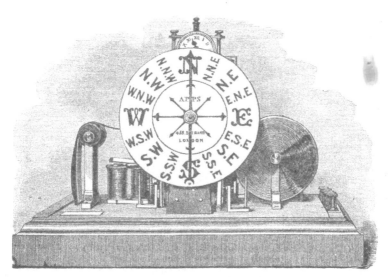

FIG. 314.—*Recording Anemometer.*

RECORDING INSTRUMENTS.

S IR JOHN HERSCHEL, in enumerating at the close of his inesti-
mable "Discourse on the Study of Natural Philosophy" the causes
of the rapid development of the physical sciences in modern times, assigns
a prominent place to the improvement of scientific apparatus, especially
of those instruments by which exact measurements or observations are
made. The accurate and elaborate instruments which serve for the delicate and precise determinations and observations of modern science require
for their production a very advanced state of mechanical art, such as is
indicated by the perfection of the tools we described in a former article;
and these tools are themselves, on the other hand, the outcome of accurate knowledge, and another proof of the interaction between science and
practical art. Since precise observations and accurate measurements form
the essential bases of every science, its progress will be accelerated by
every improvement in its instruments which increases their delicacy and
exactness. Indeed, hardly any branch of knowledge becomes entitled to
be called a science until it rests upon quantitative data of some kind.
Chemistry was nothing but a confused collection of vague notions until the
exact determinations of the balance were employed, and the proportions
of the substances combining or separating in chemical actions were found
to be related by certain simple and very definite laws. In all branches of
inquiry there is the same necessity of quantitative comparisons: lengths,

angles, surfaces, volumes, masses, durations must be compared with standards of their own kind; motions, forces, pressures, temperatures, lights must be measured. The case of chemistry shows the line along which other sciences are advancing. Physiology has made great strides since instruments of precision have been used in its investigations, and as some of these are of the kind we here propose to treat of, they will be described in the sequel. To recording instruments, meteorology is also largely indebted for the remarkable progress which it is making, and which will soon place this branch of knowledge in a condition to supply the most striking illustration of the difference between a science founded on accurate measurements and a mass of vague observations.

The obvious advantage of a recording instrument (say, for example, such a one as that represented in Fig. 314, which registers the force and direction of the wind) is that the results are obtained without the immediate attention of an observer, and they can be continuously recorded at every instant, day and night; but there is another and yet greater advantage in certain kinds of instruments which write their own records, in the fact that they can be made to register results which would altogether escape direct observation. It is said that a practised astronomical observer will correctly record the time of a phenomenon to nearly the tenth of a second; but there are cases in which we may desire to estimate time to the thousandth part of a second or less. An investigation of M. Foucault has already been named in which a far less interval of time was concerned (page 387); but the recording instruments we have to mention here are of use for enabling us to make certain instantaneous actions mark the time of their occurrence with the greatest precision, and also for enabling us to note the variations in actions which are too rapid to be directly observed in their various phases.

Fig. 315 is a diagram which will serve to explain the method in which the height of the barometer and the thermometer are registered in the ingenious *metereograph*, invented by Professor Hough, of Dudley Observatory. The contrivance has the advantage of performing the operation for both instruments, with a single piece of mechanism and on the same sheet of paper. The diagram is not intended to indicate the actual arrangement of the parts of the apparatus, but merely to explain the principle of its action. Let A represent a cylinder about 6 in. in diameter and 7 in. high, covered with a sheet of paper, ruled with certain lines, some parallel to the axis, and others perpendicular to those. This drum revolves by clockwork, controlled by a pendulum, at a certain regular rate of, say, one turn in seven days. B is a metallic bar or lever, about 2 ft. in length, mounted on an axis or fulcrum at C. At D is a pencil or style projecting from the extremity of the bar opposite the centre of the drum, but not in actual contact with the paper. E and F are platinum wires attached to the lever at about 3 in. distance from the fulcrum, C; E passes into the open tube of a mercurial thermometer, G, and F into the shorter branch of a syphon barometer, H. The clockwork has other offices to perform besides turning the drum, A, on its axis; and one of these is to alternately elevate and depress the lever, B, every half-hour. If the end, F, be depressed, it is plain that the wire will come into contact with the metallic float, which is supported by the mercury and follows its movements. If, therefore, wires from a battery, K, including an electro-magnet, I, in their circuit, be connected with the bar at C, and with the mercury at H, when the wire at F touches the float, the current will pass and the armature of the electro-magnet will be attracted. The movement of the armature is so arranged that it causes a blow to be given to the end of the bar D, so that the pen there marks a dot on the drum, thus indicating its height at the time, and therefore that

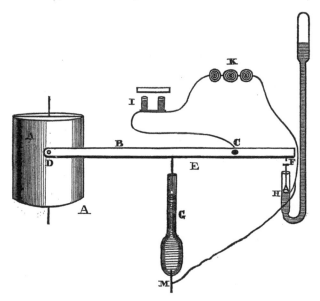

FIG. 315.—*Registration of Height of Barometer and Thermometer.*

of the mercury in H. When the lever is depressed at the other end, the wire, E, similarly completes the circuit through the mercury in the thermometer, and the height of the latter can be known from the dot which is similarly impressed on the lower part of the paper. These movements may be made with almost any degree of precision required. The clockwork is also made to raise the hammers which strike the pen against the drum, at the instant the electric current passes.

The instrument, as actually constructed, registers also the height of a wet-bulb thermometer, by another wire requiring a lower depression of the lever to bring it into contact with the mercury in a wet-bulb thermometer. A complete double motion of the lever requires one hour, and in that interval the heights of the barometer and both thermometers are each recorded once. The wet and dry-bulb thermometers are registered within a minute of each other, and half an hour elapses between the barometer and thermometer records.

Another invention of Professor Hough's is a barometer which marks a continuous pencil-line on a revolving cylinder, by which the variations of the mercury are shown for every instant of the day. Another part of the arrangement is a machine for automatically printing on paper in ordinary characters the height of the mercury to the thousandth part of an inch.

A very simple and trustworthy record of thermometric and barometric heights is obtained by photography at Kew and elsewhere. A sheet of sensitive paper passes horizontally at a uniform speed behind the tube of the instrument, so that the only light it can receive must pass through the glass. A lamp is placed in front, and a portion of the paper is protected from its rays by the mercury, while those which pass through the tube above the mercury make their impression on the paper, and thus record the indications of the instrument.

Fig. 314 represents part of another ingenious meteorological instrument invented by Mr. J. E. H. Gordon, and made by Mr. Apps. It is an electrical anemometer, for indicating and registering the direction and force of the wind. The apparatus consists of an external portion, which is of course fixed on some high and exposed part of the building; and the indicating and registering instrument, which communicates with the former only by insulated wires connected with a galvanic battery, and which may be placed on any convenient table within the house. The registering apparatus in this instrument is very neat and compact, and the reader will no doubt be able to form a sufficiently good idea of its nature from the portion which is visible in the cut, and from the knowledge of similar apparatus he may have derived from the descriptions already given in the article on the electric telegraph.

Modes of making phenomena record the time and duration of their own occurrence are now much used in all scientific investigations; and in connection with the electric *chronograph* or *chronoscope* which we are about to describe, few more efficient or elegant methods of "interrogating nature" —to use Bacon's phrase—have yet been devised. The reader who has never seen an instrument of this kind will be the better able to understand its principle by a simple illustration, which may very easily be made a practical one by himself if he has a tuning-fork at hand. Let him fix the tuning-fork firmly into a board in an upright position, by inserting the part usually held in the hand into a hole in the board; and then attach to the fork, by means of a little bees'-wax, a short bristle, which is to project from the extremity of one prong in a direction perpendicular to the plane in which the prongs vibrate. He has now only to provide himself with a piece of glass a few inches square in order to obtain a record of the vibrations of the fork when sounding. By the help of another piece of board it will be easy to arrange a guide by which the piece of glass can be made to fall down by its own weight in a plane parallel to the prongs, and in such a manner that the free end of the bristle shall just touch its surface during the whole time of its descent. Now let the surface of the glass be blackened in the flame of a candle. If the glass be allowed to slide down when the fork is not vibrating, the end of the bristle, by removing the lampblack from the surface as the glass falls, would trace out a vertical line. If, on the other hand, the blackened surface were itself not moved, but simply brought into contact with the end of the bristle, while the fork was sounding, there would be marked only a very short horizontal line, corresponding with the extent of the vibratory movements of the prong. When the glass is allowed to fall while the prong is in motion, the combination of the horizontal movements of the bristle, and the vertical one of the glass, will produce a waved line, which will exhibit perfectly regular curves if the glass has been moved with uniform velocity. It is plain that if the time taken by the glass to pass in front of the bristle were accurately known, the number of movements per second executed by the prong of the fork could be found. On the other hand, if the rate of vibration of the fork be known, the time occupied in the passage of the glass may accurately be read off. If this simple experiment be understood, the principle of the electric chronograph will be clear. Substitute for the sliding glass a cylinder covered with white glazed paper which has been coated with lampblack in the same manner; suppose the cylinder to revolve at a uniform rate, while a tuning-fork is similarly writing its vibra-

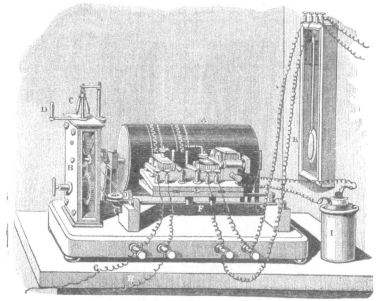

FIG. 316.—*The Electric Chronograph.*

tions on the surface of the paper, and let the same mechanism which turns the cylinder, slowly draw the sounding-fork along a straight slide parallel to the axis of the cylinder. The waved line will not form a complete circle on the surface of the cylinder, but will be traced out in a spiral, owing to the combined motions of the fork and the cylinder. As the number of movements per second of a vibrating body emitting a given note are accurately determined and perfectly regular, the waved line on the cylinder thus furnishes an exact measure of small intervals of time, the utility of which will presently be seen.

Fig. 316 represents the apparatus as actually constructed. A is the cylinder covered with the blackened paper, and driven by clockwork contained in the case, B, the rate of movement being regulated by the conical pendulum at C, so as to be approximately uniform. D is a lever for starting and stopping the movement. The clockwork also causes the carriage, E, to slide along the bars, F. This carriage bears three electro-magnets,

42

and to the armature of each a fine pointed strip of metal is attached so as just to touch the surface of the cylinder. The movement of the armatures takes place in a direction parallel to the axis of the cylinder, and they are acted on by independent currents, each electro-magnet having its own binding-screws for the attachment of wires. The armature of one of these is a steel bar, which vibrates in the manner of the prong of a tuning-fork, at a rate known by the note it gives out. Sometimes, for delicate experiments, the movements of this armature are checked and controlled by introducing into its electric circuit a contact-breaker, formed of a real tuning-fork, the vibrations of which are maintained by electro-magnets. Another electro-magnet on the carriage, E, can be connected with the pendulum of a standard clock, so that seconds may be also marked on the cylinder, as larger units in the division of the time. In an electric chronograph, which was exhibited at the International Exhibition of 1862, an ingenious and excellent mode of making and breaking the electric contacts from the standard pendulum was adopted. Two short vertical glass tubes, placed side by side, had each near its lower end a small horizontal branch; these branches were placed in a line with each other, with their ends in very close proximity. The larger tubes were filled with mercury which flowed into the horizontal branches, and the two streams joined in the narrow space between the ends of the tubes. Although the mercury was here unsupported by the glass, and surrounded by air only, it did not run down, for the space was so small that the cohesion of the mercury itself sufficed to keep the drop hanging between the two open ends of the tubes. A very thin sheet of mica was carried by an arm of the pendulum, so placed that at each complete oscillation the mica entered the space between the two tubes and divided the mercury, and thus all electrical communication between the two reservoirs was cut off. The mica remained during one beat of the pendulum in this position; and at the commencement of the return beat was withdrawn, allowing the divided columns of mercury to flow together again, and complete the electric circuit. This admirable make-and-break arrangement acted with the greatest regularity: the mercury was not spilt, as might have been expected, there was no friction, and any oxide formed by the spark which passed when the current was interrupted was removed by the mica. The pendulum provided with such an arrangement allows the current to pass, and interrupts it, during alternate seconds, and the result is that the cylinder is marked with a regularly divided broken line, thus, ‾‿_‿‾‿_‿‾‿_, the establishment and interruption of the current at the end of each second being marked with great sharpness and precision.

The third electro-magnet of the apparatus represented in Fig. 316 is acted on by currents through the wires, G, H. The point attached to its armature traces a plain spiral line on the revolving cylinder, except at the instant when the current is established or interrupted. And the phenomenon to be timed is in some way made to accomplish the making and breaking of this circuit. This may be perhaps better understood by an example. It has sometimes happened that the boxes employed in the pneumatic dispatch stick fast in the tubes, and resist all efforts to dislodge them by manœuvres with the compressed or rarefied air, or other means. In such a case it becomes necessary to ascertain with tolerable accuracy the position of the obstruction, so that the tube may be cut at the right place and the obstacle removed. The known velocity of sound has been ingeniously used for this purpose; the electric chronograph being made the means of ascertaining, to a small fraction of a second, the time required for the report of a pistol to be propagated through the air of the tube and reflected back from the obstruction. An elastic membrane is spread over the open extremity of the pneumatic tube ; this membrane has in the centre a little disc of platinum, in electric connection with one of the wires, G, H. The other wire passes to a galvanic cell or battery, and the return wire from the battery is connected with a platinum point, the distance of which from the disc can be adjusted with nicety by means of a screw, so that the circuit may be complete only when the platinum point and the platinum disc are in contact. The screw is adjusted so as to bring the point as near as possible to the disc without actual contact. The chronograph cylinder having been set in motion, a small pistol is fired into the pneumatic tube through a side opening. Sound-waves of alternate compression and rarefaction pass along the tube, and are reflected backwards and forwards many times in succession between the obstacle and the membrane. By these the membrane is alternately forced out and drawn in, making and breaking the electric contact accordingly, and thus causing the point of the electro-magnet to describe in the cylinder an indented line, the intervals of which indicate the time the sound requires to traverse the tube to the obstruction ; and thus the position of the latter may be known with sufficient accuracy.

A very interesting application of the electric chronograph is to the measurement of the velocities of projectiles. The science of gunnery has acquired an exactness unknown before electricity was made to carry messages from the cannon-ball in its swiftest flight, and to write the record of its own course. Instruments for thus measuring the velocities of projectiles have been contrived by several electricians, among whom Wheatstone appears to have been the first. The principle in most of these chronographs is precisely the same as that on which the apparatus represented in Fig. 316 is constructed. The action of the projectile which is electrically indicated is the severing of a slender wire, which is stretched from side to side of a wooden frame, so that it passes continuously backwards and forwards in parallel lines. Thus a kind of screen is formed, through which the missile must pass, and in its passage must rupture the wire. If the wire conveys a current of electricity, this current is therefore interrupted at the moment the ball passes. Sometimes the immediate effect of the breaking of the wire is mechanical, as in Wheatstone's arrangement, where the wire is stretched by a weight over a series of pulleys, and attached to a contact-maker, which completes the circuit when it is set free by the rupture of the wire. A similar arrangement in the screens has been proposed by Mr. Siemens for the establishment of circuits in connection with charged Leyden jars, the sparks of the discharges being made to take place at the surface of a revolving cylinder of polished steel, where the place is shown by the spot they leave on the metal. M. Pouillet's chronoscope dispenses with the revolving cylinder, and measures the *duration* of the current established by the projectile at one part of its course and cut off at another, by the arc through which the needle of a galvanometer is impelled.

The instrument which has been most employed in this country by artillerists is that invented by Professor Bashforth. Its indications are extremely accurate, for readings may be taken to the two-thousandth part of a second. From ten to fifteen screens are placed in the path of the projectile at distances asunder which may vary from 15 ft. to 150 ft., but which, of course, are carefully measured. Each screen is formed of a wire carrying an independent current, which also circulates in the coils of an electro-magnet. The ten electro-magnets have styles attached to their armatures in such a manner that when all the currents are passing, ten parallel spiral lines are traced on the surface of a vertical revolving cylinder, about 12 in. long and 4 in. diameter. The cylinder is covered with a sheet of highly-glazed paper coated with lampblack ; and when the sheet is removed, the lampblack may be fixed on the paper if desired, and the record made permanent. The equal intervals of time are marked on the same cylinder by another electro-magnet connected with a pendulum beating half-seconds. The axis of the revolving cylinder, &c., carries a heavy flywheel, which is set in motion by the hand at the rate of about three revolutions in two seconds, and the rest of the mechanism is thrown into gear just before the signal to fire is given. The arrangement of the electric connections at the screens is such that at the moment of the rupture of the wire, the circuit is broken for an instant only. At this moment, the iron of the corresponding electro-magnet ceasing to attract its armature, the latter is drawn away by a spring ; but the re-establishment of the current immediately brings it back, and the style continues to trace the same spiral line as before. The passage of the ball through the screen is therefore marked by a little notch in the spiral line, thus, ———ᴧ———, and the point where the deviation begins indicates the time at which the ball passed. In order to read off this time, a straight-edge is applied to the cylinder parallel to its axis ; and by means of a scale sliding upon the straight-edge the distances between the notches on the several spirals are compared with those between the pendulum marks.

To record the instant at which a projectile passes determined points in a cannon's bore, lateral plugs are screwed in, each having, just projecting into the bore, a small steel ball, which, pressed outwards by the passing projectile, causes a cutter to divide the primary wire of a Ruhmkorff coil, whereupon the spark that passes in the secondary circuit leaves its record on a uniformly moving disc. Each plug has its own battery, coil, and disc.

A special feature of recording instruments may be exemplified by certain applications of the principle to the investigation of physiological actions. A skilled physician is often able to detect in the pulse of his patient certain characteristics besides the mere rate, which are highly significant as regards the condition of the circulatory system. The range of these indications has been greatly extended by an instrument invented by MM. Chauveau and Marey, by which the pulse is made to write down a graphic representation of its action. The patient's arm having been placed on a suitable support, a little stud covered with soft leather is lightly pressed against the artery by a spring. The stud is in contact with the shorter end of a very light lever, the other extremity of which is furnished with a point, which registers its movements on a cylinder of blackened metal, made to rotate and advance longitudinally by clockwork; or the record is taken on strips of flat smoked glass. As the motion is much magnified by the lever, every variation in the pressure of the blood in the artery during the beat of the pulse is distinctly and faithfully indicated. From the line so traced the physician may obtain infallible data for judging of the condition of the heart, the action of its valves, &c. It is marvellous to observe the manner in which the curves of the *sphygmograph*, as the instrument is termed, change their form when certain drugs are administered : the change in some cases occurs immediately, so that the eye can detect by the inspection of the sphygmographic curve almost the instant at which the

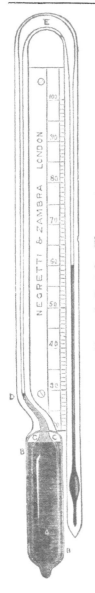

FIG. 317.—*Negretti's Deep-Sea Thermometer.*

drug was introduced into the system, and the nature of its action on the heart.

Another instrument which is doing good service in the hands of medical investigators is the *spirograph*, in which the rise and fall of the chest in breathing are similarly traced by the motions of a lever. In this instrument a small pad, which presses on the chest, communicates its movements to an elastic membrane, which, like the skin of a drum-head, covers one end of a cylindrical box maintained in a fixed position relatively to the person of the patient. The air in this box is in communication, by means of a flexible tube, with the interior of another similarly closed box; the elastic membrane of the latter acts against the short end of a lever, which is made to register its movements as in the sphygmo-graph, for the compression of the air caused by the rise of the chest is conveyed to the second box through the flexible tube. The curves furnished by this instrument also give valuable indications, and exhibit marked changes under any influence in the least degree affecting the respiratory system.

The value of a self-registering instrument for solving problems, the intricacy of which is increased by the multiplicity and rapidity of the actions to be observed, cannot be better illustrated than by the success with which Professor Marey has thus studied some complicated actions of locomotion, as related in his extremely interesting work entitled, "La Machine Animale," a translation of which has appeared in "The International Series." The action of the horse in the various paces, walking, trotting, galloping, &c., has been an endless subject of discussion, with no other data than the shoe-marks left in soft ground, and the general appearance of the animal's movements to an observer. But M. Marey —by means of elastic bags containing air, communicating the pressure through flexible tubes, so as to move little levers, which write their traces on a revolving cylinder impelled by clockwork, and carried by the rider,—has completely and finally settled all the points in dispute. It is now definitely known how the horse's feet are placed on the ground in each of his paces, and the actual and relative time that each foot remains down. The instruments are also made to register the vertical movements of the animal, so that a complete record of its motion can be obtained.

It was long a difficulty to obtain data as to the temperature of the sea at great depths below the surface. It is obvious that the ordinary maximum and minimum registering thermometers would not give the temperature at any particular depth to which they might be submerged, but merely the temperature of the warmest or coldest stratum of water through which they passed in their descent or ascent. No plan which has been devised to obviate these difficulties appears to have been attended with success, until a quite recent invention of Messrs. Negretti and Zambra supplied the desideratum, and furnished a convenient instrument for trustworthy determination of the temperature of the ocean at any required depth. The same firm long ago constructed thermometers for deep-sea soundings, with bulbs protected from the pressure of the water by an outer covering of thick glass surrounding the delicate bulb of the thermometer, between which and the outer casing a space was left, partially filled with mercury, so that heat might readily pass to or from the inner bulb without the latter being exposed to the superincumbent pressure. The new recording deep-sea thermometer differs, however, from all other registering thermometers by containing mercury only, without alcohol, or springs, or other removable indices, and, consequently, it is free from liability to derangement. The following is the description of the instrument :

In the first place it must be observed that the bulb of the thermometer is protected so as to resist the pressure of the ocean, which varies according to depth, that of 3,000 fathoms being something like 3 tons pressure on the square inch. The new instrument is in snape like a syphon with parallel legs, all in one piece, and having a continuous communication, as shown in Fig. 317. The scale of the thermometer is pivoted on a centre, and being attached in a vertical position to a simple apparatus (which will be presently described), is lowered to any depth that may be desired. In its descent the thermometer acts as an ordinary instrument, the mercury rising or falling according to the temperature of the stratum through which it passes; but so soon as the descent ceases, and a reverse motion is given to the line, so as to pull the thermometer towards the surface, the instrument turns once on its centre, first bulb uppermost, and afterwards bulb downwards. This causes the mercury, which was in the left-hand column, first to pass into the dilated syphon-bend at the top, and thence into the right-hand tube, where it remains, indicating on a graduated scale the exact temperature at the time the thermometer was turned over. The cut shows the position of the mercury *after* the instrument has been turned on

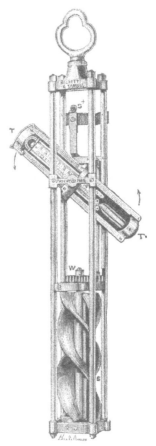

FIG. 318.— *Negretti's Deep-Sea Thermometer, general arrangement.*

its centre. A is the bulb; B the outer coating or protecting cylinder; C is the space of rarefied air, which is reduced if the outer casing be compressed; D is a small glass plug, as in Negretti and Zambra's maximum thermometer, which at the moment of turning cuts off the mercury in the tube from that in the bulb, thereby ensuring that none but the former can be transferred into the indicating column; E is an enlargement made in the bend, so as to enable the mercury to pass quickly from one tube to another in revolving; and F is the indicating tube, or thermometer proper. When the thermometer is put in motion, as soon as the tube has acquired a slightly oblique position, the mercury breaks off at the point D, runs into the curved and

FIG. 319.—*The Atmospheric Recording Instrument.*

enlarged portion, E, and eventually falls into the tube, F, as the instrument resumes its original vertical position.

The contrivance for turning the thermometer over at the bottom of the sea may be described as a vertical propeller, to which the instrument is pivoted. So long as the instrument is descending the propeller is lifted out of gear and revolves free; but as soon as the ascent commences, the action is reversed, the propeller falls into gear with a pinion connected with the thermometer, and by these means the thermometer is turned over, and after one turn it remains locked and immovable. The engraving, Fig. 318, shows the general arrangement, T being the thermometer, S a metal screw connected with the frame of the thermometer by a wheel-and-pinion movement at W; S × is the stop for arresting the movement of the thermometer when it has made one complete turn.

The atmospheric recording thermometer (Fig. 319) differs from the deep-sea thermometer by not having the double or protected bulb, as it is not required to resist pressures. In this form of the instrument, the thermometer is turned over by a simple clock movement, which can be set to any hour that may be desired. It is fixed on the clock, and when the hand arrives at the hour determined upon, and to which the clock has been set as an alarum clock is set, a spring is released, and the thermometer turns over as before described. A wet and dry-bulb hygrometer is also arranged on the same plan. For observatories, or where it is important to obtain hourly or half-hourly records of the temperature, twelve or more thermometers are placed on a frame, and these are turned over by clockwork one after the other at every hour or half-hour as required.

The reader can hardly fail to perceive that powerful aid to the investigation of the laws of nature must be afforded by such instruments as we have described. And we have but taken an example here and there of the scientific uses of the recording principle, selecting those that are most readily understood, or that are connected with matters coming home to the business and bosom of every one. The science of meteorology does not deal with subjects which furnish merely amusing speculation for the hour. Forecasts of storms and cyclones would often save many lives and much valuable property; and our dependence upon meteorological conditions cannot be more forcibly illustrated than by reference to the disastrous floods which this year (1875) desolated some districts of France. Meteorology has received a great impulse from the introduction of recording instruments; and the vast number of results which are now hourly recorded must lead to the certain development of the science, and its reduction to exact laws. For even the winds obey laws—laws as definite as those which control the motions of the planets; and could we but take into account *the whole* of the circumstances upon which the movements and other conditions of the atmosphere depend, we should be able to forecast the weather with the same certainty as—thanks to the great and simple law of gravitation—we predict eclipses or other astronomical phenomena. Already, by aid of the telegraph, it is often possible to send a day's warning of approaching storms to localities lying in their probable track. The Signal Service, which is a Department of the United States War Office, has a corps of meteorological observers spread over the length and breadth of the States, who send every eight hours, to a Central Office in Washington, a report of the force and direction of the wind, height of the barometer, &c. The officer at Washington sends back by telegraph to the public press a synopsis of each day's weather, and points out what weather will probably follow; but if any city or port be threatened with a storm, special telegrams are sent. Thus, a warning of the approach of a great storm, which entered the American continent at San Francisco on the 22nd Feb., 1871, was sent to Cheyenne, Omaha, and Chicago, twenty-four hours before the storm reached these cities, which it was foreseen lay in its track. Although the hurricane did much damage at some of these places, it would probably have been far more destructive had not the inhabitants been prepared for its approach.

An elegant form of *barograph* or recording barometer has been brought out, which is small, but sufficiently accurate for all ordinary purposes. It is founded on the *aneroid*, which, as everybody knows, is an instrument for indicating atmospheric pressure by the changes of form it produces in a thin circular metallic box, partially exhausted of air. The ordinary form of the aneroid is very sensitive and portable (sometimes it is made only the size of a small watch), and bears an index needle moving over a graduated disc; which arrangement is in the barograph aneroid, replaced by a long lever, carrying an ink tracing point in contact with the face of a cylinder that is caused by clockwork to make one revolution per week. On the cylinder is spread a printed paper diagram, divided by lines for each day of the week and each hour of the day, and on this the tracing point marks a continuous curve, showing all the fluctuations of the barometric pressure. The diagram is removed at the end of the week, and a fresh form adjusted to the cylinder. The impressed papers thus form a permanent and continuous record, from which the height of the barometer, at any given moment, may be read off.

THE PHONOGRAPH.

EVERYTHING yet contrived in the way of recording instruments is eclipsed in wonder and interest by one which is among the latest marvels of the age. It is a recording instrument, and more than a recording instrument, for it can reproduce to the senses the very phenomena it records; and these same phenomena are the most familiar in their effects, and, at the same time, so subtle and delicate, that the impressions they convey are not generally thought of otherwise than in connection with our finest intellectual and emotional perceptions. We are alluding to the *phonograph,* which can register for us music and song, and articulate human speech in all their tones and modulations, and, like an aerial spirit, address them to the ear again, as often as we wish, and thus

> Inform the cell of hearing, dark and blind;
> Intricate labyrinth, more dread for thought
> To enter than oracular cave,
> Strict passage through which sighs are brought,
> And whispers for the heart, their slave;
> And shrieks, that revel in abuse
> Of shivering flesh; and warbled air,
> Whose piercing sweetness can unloose
> The chains of frenzy, or entice a smile
> Into the ambush of despair;
> Hosannas pealing down the long-drawn aisle,
> And requiems answered by the pulse that beats
> Devoutly, in life's last retreats!

In order that the reader may understand the action of the phonograph, it is necessary that he should know something of the science of sound. Then we must remember that this word is commonly used to express sometimes those sensations of which the ear is the organ, and at other times the external cause of those sensations. It is with the former meaning that we use such expressions as "a sweet sound"; with the latter, such phrases as "sound travels." It will not be necessary to speak of the physiology of the organ of hearing; but attention should be directed to the different kinds of audible perceptions we can distinguish, let us suppose, when listening to a song: First, there is the pitch, or the notes in the musical scale, which, by their particular sequence, constitute the air or melody. Second, there is the degree of loudness or lowness of the notes. Third, the enunciation, or those differences by which we distinguish, for example, the vowels *a, e, i, o, u,* one from another. Fourth, the quality of the voice by which we can distinguish between two vocalists singing the same vowel on the same note with equal loudness. Observe that these four kinds of sound perceptions are independent one of another. The last kind of difference may also be well illustrated by the instance of musical instruments of different kinds sounding the same note, in which case the difference of the quality or *timbre* is readily recognized. We have now to show the nature of the mechanical movements outside of us which act on the ear and reach our perception as sounds, giving the distinguishable impressions that have been enumerated ; and for the present we shall consider the case of such musical sounds as those just referred to.

That the source of a sustained sound is an elastic body in a state of vibration is a fact of which, in most cases, we are easily made aware by the evidence of sight and touch; as a bell, a violoncello string, a pianoforte wire, or a tuning-fork. On p. 656 is described a simple method by which a tuning-fork may be made to write down its own vibrations, and the more exact plan of recording them on the surface of blackened paper on a revolving and advancing cylinder has also been referred to. By the intervention of appropriate apparatus, a similar record may be obtained from all sounding bodies. From observations of this kind, and others in which totally different methods are used for counting the number of vibrations per vibrations made in a given time, it is known that the *pitch* of the sound or note depends on the rapidity of the vibrations — the pitch rises with the number of them per second, and the relationship between the notes of a musical scale depends entirely on these numbers. Thus, when the vibrations for the eight notes of an octave are counted, the numbers always have this *proportion,* beginning from the lowest note —24, 27, 30, 32, 36, 40, 45, 48. Thus of the two notes—

as produced on musical instruments tuned to the concert pitch of the present day, the lower corresponds with 264 complete vibrations per second, the higher with 528. It will be observed, too, that all the harmonies are determined by some simple ratio in the rates of vibration : the interval of the *fourth* is 3, 4 ; that of the *fifth* is 2, 3, etc. Another easily discoverable fact is that the loudness of the sound depends upon the amplitude of the vibrations. This is sufficiently obvious by a few experiments with a tuning fork ; and by close examination of such tracings as have been mentioned, we shall soon become aware of another circumstance — namely, that the vibrations not simple, but that the larger or general movement has one or more sets of small vibrations within it. In Fig. 319*a*, A is the curve that would be traced by the tuning-fork in a state of simple vibration ; B and C are tracing such as are given by a fork in two of its modes of vibration. The fork gives out its proper or fundamental note in both cases ; but the ear recognizes a difference in the quality of the sound due to the smaller and more numerous vibrations. Differences of the same kind are recognized in the notes produced by different musical instruments ; but these are usually more complicated, and their forms are characteristic of the particular quality of the tone, which is thus shown to be due to the superposing of several related systems of vibration upon the fundamental one. Thus three out of the four qualities of sound recognized by the ear have had their physical causes assigned. As for the fourth—namely, the distinction we perceive among the different vowels sung on the same note—it has a physical origin identical with the last. For since parts of the vocal organs assume

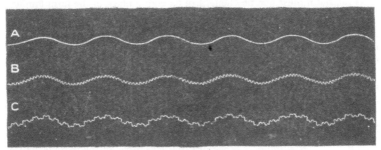

FIG. 319*a*.—*Traces of Vibrations of a Tuning-Fork.*

different positions in enunciating the different vowels, they constitute for the time being so many varied musical instruments, and the graphical traces of the sounds (for they can be obtained) show a corresponding modification. Here, in Fig. 319*b*, for example, are represented the tracings of the vibrations given to the air in various vowel sounds. It is also through the vibrations conveyed by the air to the little membrane called the drum of the ear that the sensations of sound are received, and of the nature of these vibrations a few words must be said presently. In considering the different qualities of sound, we have so far confined ourselves to sustained musical notes, as, for instance, the vowel sounds in singing. This has been done to show the relations of rapidity of vibration to pitch and for simplicity of illustration of the superposition of vibrations, etc. Two other remarks must be added—viz., that the vibrations of musical tones are *isochronous* —that is, whether the note be loud or soft, the same time is taken up in each vibration corresponding with the same fundamental note. Other vocal sounds than sustained vowel notes are found to be due to

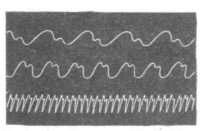

FIG. 319*b*.—*Phonautographic Tracings of Different Vowel Sounds.*

still more complicated combinations of vibrations of shorter duration, and *noises*, as distinguished from musical sound, are formed also by the superposition of a greater or less number of systems of vibrations, the rapidities of which are wanting in harmonic relations such as we have pointed out belong to the musical scale. Even in noises, however, there is often one or more predominant systems of vibrations which a musical ear can detect. If the reader will hold a pencil or pen-holder at one end and tap with it on the edge of the table, passing in quick succession to parts progressively nearer where the pencil is held, he will hardly fail to recognize a rising pitch in the little noises.

A word or two must be said as to the way in which sound is transmitted through the air. This progression is commonly spoken of as a wave motion, but it must not be thought of as taking place in the form familiar to us as waves on water ; still less must the reader confound it with the sinuous lines shown in the graphical representations of vibrations given in

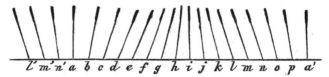

FIG. 319*c.—Diagram.*

the figures. It is rather a series of rapid pulsations of the particles of the air taking place in the direction in which the sound is propagated, and resembling waves on water only by presenting periodical phases in uniform succession. The difference may be illustrated from what may be seen in a field of wheat when the wind is blowing over it. The stalks bend down, and rise again when the breeze has passed, and thus the general appearance of the waves of the sea is produced. If we confine our attention, however, to the motion of the several ears of the wheat in a file of stalks, we shall obtain a clearer notion of what takes place in the so-called waves of sound. The positions of the stalks at some one instant of time may be represented by the diagram, Fig. 319C. Each stalk is swinging backwards and forwards like an inverted pendulum, and the successive phases of these vibrations bring the adjacent ears nearest to each other about *i*,

FIG. 319*d.—Phases of Sound Waves.*

and farthest apart at *a* and *a'*. The places of these, and of all the intermediate degrees of approximation and retreat, pass along the file. Instead of the ears of wheat swinging on their elastic stalks, suppose particles of air approaching and receding by virtue of the elasticity by which they resist compression and recover from it, and you will obtain an elementary idea of what takes place in the transmission of sound. Fig. 319*d* is a picture of a column of air acted upon by a tuning-fork. The swiftly advancing prong is compressing the air in front of it, and in swinging back it will tend to leave a vaccum behind it by which the air is partially rarefied ; and these alternate condensations and rarefactions will travel along through the air by virtue of its elasticity, and the mechanical action' by which they are able to agitate a stretched membrane (or other elastic body), so that its vibrations will correspond with them in period and magnitude, may be easily understood. The vibration produced is a simple one, but any number of other systems may pass at the same time, and each one will be propagated as if the rest did not exist, just as we may see different systems of undulations moving on the surface of water. It should be observed that the velocity of propagation is the same whatever may be the period or the magnitude of the vibrations. The high and the low, the loud and the soft notes of a piece of music played at a distance, all take the same time in reaching the ear. Light as are the particles of air, the mechanical actions which a number of them carrying strong vibratory impulses will produce, may be illustrated by the rattling of window panes by a loud peal of thunder, and may be bodily felt by a person standing close to a very large bell while the hour is striking.

We have referred to instruments for registering sound, and even vocal sounds, before anything has been said of the construction of the phonograph, and it is, in fact, many years since the problem was solved of recording the vibrations produced by speech. Mr. Leo Scott, in 1856, invented an instrument, called the *Phonautograph*, which did this. It consisted of a cone of sheet zinc like a large ear-trumpet, across the smaller end of which was stretched a membrane, having attached to it a very light style, which left a record of the vibrations of the membrane on a blackened cylinder properly disposed to receive the tracing. When any sound was produced near the open end of the cone, the impulses reflected from its internal surface were concentrated on the membrane, throwing it into corresponding vibrations. Now, this process could be reversed if the tracing could be made to give back again to the style its original movements, these transferred to the membrane would throw the air within the cone into corresponding vibrations, and the sounds that gave rise to the tracing would be reproduced. Yet Mr. Scott seems to have suggested no such possibilities for his instrument ; but a few years after the invention of the phonautograph, M. Cros deposited at the Academy of Sciences in Paris a sealed paper, which was opened after Mr. Edison had patented the phonograph (1877), and found to contain suggestions of how this might be done, but describing no experiments in which any approach had been made towards realizing the conditions laid down. To Mr. Edison belongs the honour of solving the problem by the invention of the phonograph, which was patented by him in 1877. The device which he happily hit upon for converting the phonautograph into a phonograph was very simple in principle, and consisting merely in substituting a sheet of tinfoil for the blackened paper in Scott's apparatus, the mechanism required for reproducing articulate human speech was thus found, contrary to all expectations that had previously been entertained, to be essentially of a remarkably simple character, for the arrangement of the parts was even more direct than in the phonautograph itself. This is not derogatory to the merit of the inventor, for every invention depends upon something previously attained, and the discovery of suitable materials for the various parts of the machine, and the many delicate adjustments of their forms and disposition to secure the required object, demanded the application of very remarkable experimental skill. The phonograph differs from the phonautograph by giving up what it has registered in the original form and material, and thus it is a speaking machine. It is a speaking machine which *reproduces* articulate speech, not produces it. Much ingenuity has been devoted to the construction of speaking machines which should be capable of *producing* the sounds of the human voice. By throwing into vibration the air contained in cavities of certain shapes, it was long ago found possible to produce sounds closely resembling those of a voice singing particular vowels, and a real speaking machine that could articulate words was exhibited in America before the phonograph had been brought out. It was constructed by Mr. Faber, and formed a very complicated arrangement, in which all the organs of human speech were imitated. There were bellows acting like the lungs ; a larynx with various diaphragms, a mouth with movable tongue and lips, and a tube to resemble the cavity of the nose. The positions and connections of these parts were determined by levers acted from a key-board, like that of a piano, and by certain pedals. By moving these in proper

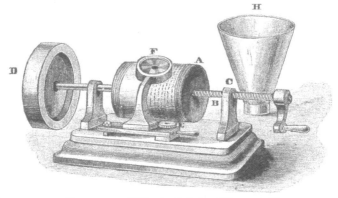

FIG. 319*e.*—*Edison's Original Phonograph.*

order, the machine pronounced words distinctly enough, but in a strange drawling tone. So like, however, were the sounds to those of the human voice, that some accused the exhibitors of imposition, and unjustly credited ventriloquism instead of mechanism with the results. It will be observed that it is the function of the phonograph to reproduce, not produce, human speech, and the mechanical arrangements of the instrument are simplicity itself compared with Faber's speaking machine.

Fig. 319*e* shows the form of the phonograph as designed by Mr. Edison in 1877. It had a brass cylinder (A) upon which a narrow helical groove was cut, and was mounted upon an axle (B), having a narrow screwthread corresponding with the groove on the cylinder, and working in the upright (C), so that when the handle was turned the cylinder revolved, and at the same time advanced in the direction of its axis. A heavy flywheel (D) was attached, in order that the rate of motion might be nearly uniform. A sheet of tinfoil, or of very thin copper, was wrapped round the brass cylinder, and on this metallic foil rested the steel point attached to the vibrating diaphragm, which was mounted in the ring (F). This point was always adjusted so as to be over the helical groove in the cylinder, and made to touch the tinfoil with a regulated pressure. **E**

shows the manner of firmly supporting the diaphragm in such a manner
that it could be readily removed from the cylinder when the latter had to
be covered afresh with tinfoil, or the cylinder adjusted for reproducing
the sounds.   The relation of the diaphragm and point to the tinfoil is
shown in Fig. 319*f*, which represents the apparatus in section.   The tracing point (*t*) is not attached directly to the vibrating diaphragm, but to an
adjustable spring (*s*), and interposed between the spring and the thin
metallic diaphragm is a little pad, formed of a ring of small india-rubber
tubing.   When the mouthpiece (M) is spoken into, the sound vibrations
reach the diaphragm (*g g*) through the opening (*o*), and the movements
thus communicated to the point (*t*) which indents the tinfoil to various

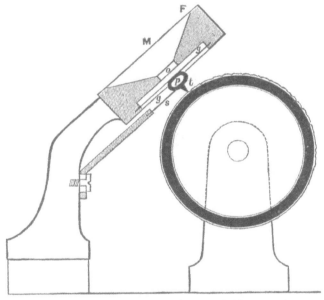

FIG. 319*f.—Diagrammatic Section of Phonograph.*

depths, and with varying frequency, as the handle is turned, bringing
the whole length of the groove in succession to be operated upon.   When
the instrument is required to reproduce the speech so easily recorded, all
that is necessary is to allow the indentations to re-act on the point that
made them.   The cylinder is re-adjusted to the tracing point at the end
at which it began, the cylinder is set in motion, and the traces made on
the tinfoil move the point up and down, the vibrating disc (*g*) following
its movements, and thus communicating to the air a system of impulses
which are the counterparts in period, force and succession of those that
originally entered at *o*.   It was usual to attach a conical mouthpiece to the
ring (F) in order to concentrate the reproduced sounds, which might then
be heard in all parts of a large room.   When, in reproducing the sounds,

the cylinder was turned with the same velocity as when the words were spoken, the pitch of the voice issuing from the instrument was the same. If it were turned quicker, the pitch was raised ; if slower, lowered.   The words registered on the tinfoil could be reproduced two or three times, but with decreasing distinctness, as the tracings gradually become obliterated.   However, the sheet could be removed from the cylinder, and the speech reproduced at any place at any time afterwards by means of a similar instrument, and a method of stereotyping was proposed for preserving the records.   The original phonograph was greatly improved

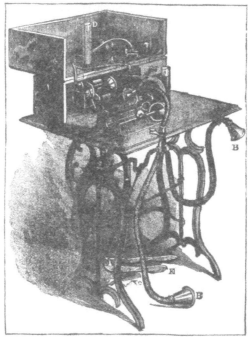

FIG. 319g.—*The Graphophone.*

when well regulated clockwork was used for imparting motion instead of the winch.   Mr. Edison contrived a modification of the machine, which made it much easier of manipulation, by substituting for the cylinder a flat plate on which a spiral groove was cut.   The plate was turned by clockwork, while the vibrating point was made to follow the groove from the centre to the circumference.   The phonograph in its original form reproduced speech with peculiarities of its own.   The quality was metallic, and reminded one of the intonation of the street Punch.   It will easily be understood that the disc itself must necessarily have its own systems of vibration, and these will be further modified by the action of all the other parts.   Mr. Edison's expectations of the capabilities of the instrument not being realized, he turned his attention, after several unsuccessful attempts at its improvement, to the electric light and other subjects, at the same time declaring his conviction that the perfection of the instrument would be but a matter of time ; in fact, within a very few years afterwards such improvements were made on Mr. Edison's instrument as went far to justify this prophecy. These were the work of Dr. Chichester Bell and Mr. Tainter, who, after long continued experiments, found in paraffin wax, with a small admixture of some other substances, a better material for receiving the impressions. A cutting style made to act upon this cuts out a fine groove, the bottom of which is not a series of indentations, but a continuous wavy curve, representing every degree of inflection the vibrating diaphragm. In the new form of the instrument, FIG. 319*g*, which was called the *graphophone*, to distinguish it from Edison's, the cylinder does not move forward : it is the diaphragm that advances parallel to the revolving axle. The cylinder is driven by a treadle, like a sewing-machine, and there is an ingenious arrangement by which the speed is controlled so that it can be maintained quite uniform. The movement of cylinder and style can be instantly arrested by touching a button, and as readily re-started. Quite recently Mr. Edison has returned to improving the phonograph by using rather thick, solid cylinders of wax, which are previously prepared for use by the instrument itself paring them down to a truly cylindrical and perfectly uniform surface, the result being a great increase of clearness in the speech and tones of music. Mr. Edison's new instruments are driven by small electro-motors, and the speed is regulated by a centrifugal governor. It is said that these wax cylinders are capable of giving out the same record for a thousand times without perceptible sign of deterioration ; and when the cylinders are required to receive a fresh impression, a former one can easily be pared off. The machines can be arranged so as to sound loud enough to be heard by a large assembly ; but the quality of the tones of speech or music is most perfect when conveyed from the receiving chamber in front of the diaphragm to one or both of the auditor's ears by means of a short elastic tube. Half a dozen persons can thus hear the record on the cylinder with such marvellous distinctness as to be able to recognize the tones of a known voice. The very latest form of the instrument, as it has just left Mr. Edison's hands, is represented in Fig. 319*h*. In this one single very small diaphragm serves both for recording and reproducing the sounds. This is made of extremely thin glass, to which is attached a small projection made of celluloid, which acts on a bar that carries the recording point. The configuration of this point is most ingenious and peculiar, for it is, in fact, double, one part being shaped like a gouge, which cuts into the walls of the minute depression traced on the wax cylinder, while a style-shaped part impresses the wax with punctured indentations. The shaping of its forms is a difficult and delicate operation, for they are very small and are cut in sapphire. The reproduced speech given out by this instrument is said to possess the properties of sharpness and clearness in a remarkable degree. The machine is provided also with a sapphire cutting edge, by which an old record may be pared off by the very motion of the cylinder in receiving a new one. This phonograph is put in movement by ingenious mechanical devices, for giving uniform rotation from such motive power as may be supplied by the foot or by water or by clockwork. This improved instrument lays claim to practical utility, and its manufacture will, it is stated, be shortly commenced on a large scale.

43

Quite recently there has been established in America a big manufactory of phonographs in the form of a toy, which is sure to become very popular everywhere. Here they turn out daily several hundred *real speaking dolls*, which contain clockwork actuating a phonographic cylinder impressed with the words of some childish story or simple rhyme, such as " Jack and Jill," " Mary Had a Little Lamb," etc. Each doll of course repeats its little tale as often as the clockwork is set going. These toys are adapted to all nationalities ; for, besides many English, there are a number of French, German, Italian, etc., girls employed doing nothing all day long but addressing appropriate words to each little automaton's waxen cylinder.

The capabilities of the phonograph suggest some curious applications

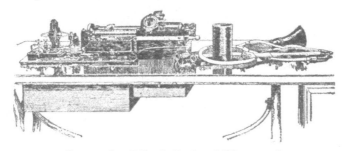

FIG. 319*h.—Edison's Perfected Phonograph.*

that may be made of it. For example, the songs of a fine singer may thus, in all their modulations, reach people in distant lands, or be made audible to future generations. Thousands of people in England have heard with their ears, through Mr. Edison's instruments lately brought over by Col. Gournaud, songs and speeches, and pieces of concerted music, sung, said, or played in America months before. Music can be bottled up, so to speak, without the consent of the originators ; and, indeed, it is said that an eminent *prima donna* has applied for an injunction to restrain certain *phonographers* from reproducing her vocal triumphs with their instruments. A speech of Mr. Gladstone's, delivered in England, has been phonographically heard in New York with great applause. There is no reason but what, with a loud speaking phonograph uttering an orator's very words and tones, while instantaneous photographs of his successive gestures and attitudes are projected on a screen, a true and lively impression of his eloquence might be conveyed centuries after his decease. One is almost led to speculate as to the consequences if these nineteenth century inventions had been antedated by a few thousand years : what stores of knowledge we might now possess ! and how pleasant it would be thus

> To hear each voice we feared to hear no more !
> Behold each mighty shade revealed to sight,
> The Bactrian, Samian sage, and all who taught the right !

FIG. 320.—*The Domestic Aquarium.*

AQUARIA.

———◆———

UNDER the date of May 28th, 1665, the curious gossiping diary of Samuel Pepys contains this entry : " Thence to see my Lady Pen, where my wife and I were shown a fine rarity ; of fishes kept in a glass of water, that will live so for ever—and finely marked they are, being foreign." This doubtless refers to the now well-known gold fishes, which about the time alluded to were introduced into Europe from China, where they had probably been for ages reared and kept in captivity, chiefly for the sake of ornament. Perhaps the reader may be disposed to think that, therefore, the aquarium cannot be distinctively a nineteenth century invention, nor at all a modern invention, in principle at least; but merely the " glass of water," or the globe of gold fish on a larger scale. Such a notion would be quite incorrect, for the principles which are embodied in the modern aquarium were not recognized and applied until quite recently. Aquatic animals kept for a period in vessels in which the water is changed from time to time cannot be considered as properly forming an aquarium. The beauty and value of a well-regulated aquarium depend not merely on the opportunities it affords of studying the habits of the animals ; the spectacle it presents has a far wider interest, as illustrating and confirming the conclusions of science regarding certain great principles which govern the whole animal and vegetable life of this terraqueous globe. Perhaps in the whole range of nature nothing is more wonderful than the direct interdependence of animal and vegetable life, and the exact balance between them, which preserves the composition of the atmosphere unchanged. The constituents of the atmosphere have an immediate relation to both forms of life. No animal can live without a supply of oxygen gas, which it absorbs and replaces by carbonic acid gas. The latter, on the other hand, is absorbed by plants, for these, under the influence of light, decompose the carbonic acid, returning the oxygen to the atmosphere, thus purifying the air by again fitting it for the respiration of animals.

It might be supposed that animals which live entirely beneath the surface of water are removed from the influence of atmospheric oxygen, and that they form exceptions to this law. But such is not the case, for water absorbs and holds in solution a certain quantity of air, the oxygen of which is taken up by aquatic animals. In the lower forms of animals inhabiting water, the absorption of this vital element takes place at the general surface of the body ; but in the more highly organized creatures there are special organs appropriated to this purpose, of which the gills of a fish may be cited as a typical example. The giving out of carbonic acid is an action as universal in the animal world as the absorption of oxygen, and all aquatic animals tend to charge the water in which they live with this gas. Fish, or any other water animals, will soon die if they are placed in water from which all the air has previously been expelled by boiling, or by placing under the receiver of an air-pump. In this case the creature dies from want of oxygen ; but it would also die, even if supplied with oxygen, were the poisonous carbonic acid emitted by itself allowed to accumulate in the liquid. In nature, this carbonic acid forms the food of aquatic plants and sea-weeds, and these restore oxygen to the water. If a bunch of watercresses be placed in a bottle filled with water, and exposed to strong sunshine, the leaves may soon be seen covered with small bubbles of gas. This gas may be collected and examined by a suitable arrangement of the bottle, and it will be found to be pure oxygen.

The merit of having first imitated the plan of nature for the preservation of aquatic animals appears to belong to Mr. Ward, the inventor of the "Wardian cases" for ferns and other plants. He, in 1841, formed in London a fresh-water aquarium, in which, for the first time, the animals were kept in a healthy condition by the compensating action of plants. Mr. Gosse, Dr. Price, and others, made experiments with marine animals and plants, about 1850. Mr. Mitchell, who was then secretary to the Zoological Society of London, saw about this time a small aquarium on the balancing principle at Dr. Bowerbank's, and this suggested the erection of the fish-house in the Zoological Gardens, Regent's Park. This was opened in 1853, being the first public aquarium ever constructed. The tanks remain at the present time in nearly their original condition, and this aquarium has been remarkable, not only as predecessor of the many public aquaria which have since been erected, but for having given rise to a movement in favour of aquaria as domestic establishments. The setting-up of household aquaria became almost the rage of the day, and so many books and magazine articles devoted to the subject appeared during the ten years following the establishment of the Regent's Park aquarium, that the literature of the subject is quite considerable. Mr. Gosse showed how water for marine aquaria could be produced by adding to fresh water the solid constituents of sea-water ; and, in the marine aquaria of some inland towns far distant from the sea, this artificial sea-water is the only kind used. After the establishment of the Regent's Park aquarium, public aquaria were opened successively in Dublin, Galway, Edinburgh, Scarborough, Weymouth, the Crystal Palace, Brighton, Manchester, and Southport ; and on the continent at Paris, Hamburg, Hanover, Boulogne, Havre,

Brussels, Cologne, Vienna, and Naples; also in North America at San Francisco, and in other places. The general interest in public aquaria, and especially marine aquaria on the large scale, seemed to increase as the comparative failure of the domestic tanks lessened the taste for them. The causes of the failure so often attending the attempt to maintain aquaria on the small scale arise partly from the amateur naturalist's want of exact knowledge, and the great amount of attention and care required, and partly from the inherent difficulties of the subject. An aquarium, even on the largest scale, and with every appliance that science can suggest, only represents, after all, *a few* of the conditions which actually exist in nature; but in small vessels, with a limited quantity of water, without the continual motion of the liquid, which belongs naturally to seas and streams, and with circumstances of light and temperature widely different from those which are obtained in nature, it is not surprising that the success of domestic aquaria should be but very partial, and that the taste for them should have declined accordingly.

Many public aquaria proved commercial failures; but we select for special description two which have been thoroughly efficient, and are remarkable for size, reputation, and successful management. The arrangements at these two institutions as regards the aëration and renewal of the water are, however, quite different. Some plan by which the same sea-water might be supplied with oxygen, and kept in a clear and pure condition, was necessary for the very existence of the inland marine aquarium at the Crystal Palace, whereas the position of Brighton made the natural sea-water more available. The success of the former method at the Crystal Palace Aquarium, under the judicious system adopted by Mr. W. A. Lloyd, the superintendent, perhaps renders this aquarium one of the most interesting, in a scientific point of view, of any yet in operation. The water here is never changed by the addition of sea-water; but fresh water is added as required, simply to supply the loss by evaporation; and any solid constituents which the animals may abstract from the water as material for their shells is replaced, so that the ordinary composition of sea-water is maintained. This is merely imitating Nature, for the evaporation from the surface of the sea is compensated by the fall of rain and the influx of rivers, the latter constantly bringing in the various salts held in solution. The following particulars regarding the Crystal Palace Aquarium are derived from Mr. Lloyd's excellent handbook, which contains not only clear descriptions of the inhabitants of the tanks, but interesting historical notices and a well-written disquisition on the principles which should regulate the construction and management of aquaria.

THE CRYSTAL PALACE AQUARIUM.

THE building was commenced in July, 1870, and was opened in August, 1871. It was designed by Mr. Driver, of Victoria Street, and presents an admirable simplicity, which entirely accords with the purpose for which it was erected. The whole available space has been occupied, and nothing has been wasted on unmeaning or fantastic embellishments. Even the decorative shams, in which ordinary painters delight, have been excluded. No part of the walls or of the woodwork is painted to look like marble, or even to imitate oak. The building, which is about 400 ft. long and 70 ft. broad, is situated at the north end of the Palace, partially occupying the site of the portion which was so unfortunately burnt down in 1866. It is but one storey high, and besides a large reservoir beneath the floor, holding 100,000 gallons of sea-water, there is a series of sixty tanks, with thick plate-glass fronts, which collectively contain 20,000 gallons of water. This water, weighing over 1,000,000 lbs., was brought from the coast and conveyed to the Palace by the Brighton Railway Company at a very moderate rate. For many weeks after the water was placed in the reservoir and tanks it was very turbid, from taking up the lime used in their construction and in that of the rockwork. In this condition it was very alkaline; but the lime was slowly precipitated by the carbonic acid of the air, the water became clear, and vegetation appeared in the tanks. The great capacity of the reservoir facilitates the cleansing of the water; for, supposing that the water in one of the tanks, holding, say, 6,000 gallons, became turbid from any cause, the water from this tank could be run off into the reservoir, where its mixture with the much larger quantity would not sensibly affect the purity of the mass, from which within half an hour the tank could again be filled.

All the tanks are constantly receiving water from the clear and cool reservoir below, in which there are no animals, so that the motion of the water in the tanks, like that of the ocean, is incessant. The water issues from the pump at a rate (indicated by a counter) of from 5,000 to 7,000 gallons per hour. The pump is worked by a steam engine of three horse-power, and the machinery requires the unremitting attention of three engineers, who succeed each other by turns, each working for eight hours. Two sets of the machinery—pumps, steam engines, and boilers—are provided, one being always kept in reserve, ready for use in case of any accident. Even in winter, when, from the lower temperature, the water contains the largest amount of oxygen, it is found that the stopping of the circulation of the water for only a few hours occasions manifest discomfort to some of the animals. The water is poured into the two centre tanks in an equally divided stream, and by a simple fall of a few inches from tank to tank it flows by two routes to the lowest tank, from which it passes into the reservoir below. This incessant circulation of the water constantly exposes fresh surfaces to the action of the air, by which oxygen is absorbed. But besides this, other small streams of water are made to forcibly enter the tanks from jets, by which a large quantity of air is carried down in very small bubbles. The removal of carbonic acid is accomplished by the vegetation which spontaneously makes it appearance in sea-water under suitable circumstances. It has been found quite unnecessary to introduce purposely any kind of sea-weeds, for the spores of low forms of vegetation are always present in the water, and they develop rapidly under the stimulus of light. Indeed, one of the difficulties of aquarium management is to avoid this excessive vegetation by limiting the light as much as possible, and yet leave sufficient illumination for the observation of the animals. The amount of light falling upon each tank is very carefully attended to at the Crystal Palace, and where it cannot be diminished sufficiently to check the overgrowth of vegetation, without at the same time interfering with a proper view of the animals, certain molluscs and fishes which live upon *algæ* are put into the tanks to consume them. This spontaneous vegetation is so vigorous that a comparatively small quantity suffices to remove from the water all the carbonic acid which it may derive from the animals and decomposing matters.

It should be mentioned that at this aquarium the water is never filtered, but its clearness is obtained merely by the perfect system of circulation. The unused food and excrementitious matters are oxygenated by the air which the water abundantly holds in solution—thanks to the surface exposed in its constant circulation, the injection of the jets of water carrying minute bubbles of air into the mass of water, and the gas given off by the vegetation. The whole process of purification is therefore chemical, and the success and excellent adaptation of the system may be judged from the fact that the water seen in masses 9 ft. deep appears perfectly clear and bright. The building is very cool in summer : even in extremely hot weather the temperature of the air within it is never higher than 68° F., and that of the water in the tanks never exceeds 63°. In winter the temperature of

FIG. 321.—*The Opelet (Anthea cereus).*

the air is maintained by hot-water pipes at from 60° to 65°, and the temperature of the water at about 55°. On winter evenings the aquarium is illuminated with gas, and the habits of many nocturnal animals can then be conveniently studied.

"All the animals in this aquarium," says Mr. Lloyd, "have to be fed constantly ; and as for the sea-anemones—of which there are in the aquarium over 5,000 individuals—every one of them has a morsel of food proportioned to its size, and according to the condition of the water, given it at frequent intervals with a pair of wooden forceps by an attendant who makes this his sole occupation—as these flower-like creatures, being so non-locomotive as to be almost absolutely fixed, cannot pursue their food, or in an aquarium obtain it in any other manner. They are here deprived of the action of the waves, which in the actual ocean brings them nutriment, which is arrested by their outspread and waving tentacles. The food consumed by a few of the animals now present in the aquarium is vegetable, consisting of green weeds (*Ulva, Porphyra, Enteromorpha, &c.*), but by far the greater number have animal food given them. This consists of shrimps, alive or dead, crabs, mussels, oysters, and fish, but they are never fed on butcher's meat."

The creatures known as "sea-anemones" are well represented in the Crystal Palace Aquarium. The observer cannot fail to be struck by their resemblance to flowers, from the radiated arrangement of their tentacles, and the beautiful colours they often exhibit. The opelet (*Anthea cereus*), Fig. 321, is perhaps the most beautiful among British species, and is a conspicuous denizen of the Aquarium, where its long green tentacles, tipped with lilac, are commonly seen expanded or twisting about like so many snakes. These tentacles are stretched out in search of food, and when by chance an unlucky shrimp or other suitable prey merely touches a tentacle, it is seized and held with remarkable pertinacity, the rest of the tentacles closing round it. The mouth of the creature, placed in the centre of the disc, then expands to an extraordinary size, and the prey is quickly lodged in the capacious digestive sac of the *actinia*, where the soft parts are soon dissolved, and the hard indigestible residue is ejected by the mouth. The tentacles of *Anthea*, and of other species belonging to the same sub-division of the animal kingdom, are furnished with an immense multitude of curious organs, which consist of cells or minute bags, containing coiled up within them a slender highly elastic filament. When these cells are compressed, the filament shoots out of its capsule to a surprising length ; and it has been supposed that the adhesive power of the tentacles depends upon these filiferous capsules ; while it is not improbable that some virulent fluid is also emitted from the cells, for the victims appear as if paralysed almost as soon as they are seized. Our knowledge of these animals has been largely extended by the opportunities of observing their habits which are afforded by marine aquaria.

FIG. 322.—*The Viviparous Blenny (Zoarces viviparus).*

To obtain the variety of animals requisite for stocking a public aquarium is by no means an easy matter ; for the animals must be good specimens, in a healthy condition, uninjured by their capture or transport from the sea. The Crystal Palace Aquarium Company have at Plymouth a large pond, which communicates with the sea at every tide ; and this, under the superintendance of the company's resident agent, serves as a store for animals. Similar arrangements exist at Southend, Weymouth, Tenby, and other places. The specimens are brought to Sydenham by fast trains— special facilities being afforded by the railway companies for this purpose. The mode of carrying the animals depends upon their nature, and is sometimes a matter of no little difficulty. All fishes, except perhaps eels and blennies, must be carried in a sufficient bulk of water ; and then the due oxygenation of the water and the removal of the carbonic acid can be but very imperfectly accomplished. A considerable mass of water is absolutely necessary in such cases, and the difficulties and cost of the transit are much increased by its weight. In warm weather the quantity of oxygen retained in the water is materially diminished, and under such circumstances the creatures would soon perish. On the other hand, in very cold weather the temperature may be so far reduced below that suited to their habits that death may also result from this cause. Crabs, lobsters, sea-anemones, sea-urchins, and similar animals can in general be carried without being immersed in a mass of water. These animals are placed in

FIG. 323.—*The Lancelet (Amphioxus lanceolatus).*

layers of wet sea-weed contained in baskets, so that the air has access to the moisture which covers the bodies of the animals, which is prevented from drying up by the humidity. As in this case the small quantity of water exposes a very large surface to the air, oxygen is plentifully supplied. Mr. Lloyd points out that it is owing to the readiness with which mere films of water are aërated, that it has been found possible to convey to Australia the eggs of salmon and trout, and hatch them there. They could not have been carried in water, but they were successfully conveyed when surrounded by a cool and very moist atmosphere. This mode of transmission is much more economical and convenient than the plan of carrying the creatures in water, and it is therefore resorted to whenever the organization of the animal permits.

Specimens of a very remarkable creature are, or lately were, exhibited at this aquarium in the lancelet (*Amphioxus lanceolatus*), Fig. 323, which animal itself is a comparatively recent discovery. It is about 2½ in. long, and although it is fish-like in form, it presents so many points of structure common to lower animals, that it is looked upon by naturalists as a link between the molluscs and the fishes—being the lowest of the latter in organization. The creature can hardly be said to possess a skeleton, the tissues representing that structure are so soft. It has no definite brain, but it possesses olfactory and optic organs of a rudimentary kind.

THE BRIGHTON AQUARIUM.

THE Brighton Aquarium, already so well known as a place of popular resort, is a structure of considerable architectural pretensions, and is the largest establishment of the kind in existence. The idea of this undertaking appears to have originated with Mr. E. Birch, the engineer of the actual structure, who, having, in 1866, visited the aquarium at Boulogne, perceived that the construction at Brighton of a marine aquarium on a very extended scale offered every promise of commercial success. The promoters, in 1868, obtained from Parliament an act authorizing them to acquire a certain site for the aquarium, but imposing such limits as to the height of the structure that it was necessary to place the greater part of the building below the level of the ground, a matter involving considerable engineering difficulties. The aquarium is situated close to the Chain Pier and immediately below the cliff, the building being protected from the waves by a strong sea-wall of concrete and Portland stone. The building was definitely opened in August, 1873, while the meeting of the British Association was being held in the town. Its length is no less than 715 ft., and its average width 100 ft. The predominant element in the architectural style of the building is Italian. The following particulars as to the arrangement and dimensions of the various parts of the building are derived from the official guide-book:

Entering the gates at the western end, the visitor finds himself at the top of a flight of granite steps leading to the entrance court, 60 ft. by 40 ft. The front elevation of the building is 18 ft. high, and consists of five arches, with terra-cotta columns and enrichments. On the frieze running round the sides are the appropriate words, "And God said, Let the waters bring forth abundantly the moving creature that hath life." On the northern side of the entrance court is the restaurant; and on the southern side a series of niches ornamented with vases. From this outer court, the entrance hall, which is 80 ft. by 45 ft., is approached through three doors. This is furnished with reading-tables and supplied regularly with periodicals, journals, and telegrams; while between the pillars supporting the roof are handsome pedestals, surmounted with large glass vases containing the smaller interesting marine and fresh-water animals, which would be lost to view in the larger tanks. In one of the recesses facing the entrance are four microscopes, in which specimens illustrative of subjects in natural history connected with the aquarium are constantly exhibited. To the north of the hall lie the general manager's offices, the retiring-rooms, kitchen, &c.; and eastwards, in a direct line with the restaurant, is the entrance to the western corridor of the aquarium proper. This corridor, which contains a great many tanks, is the longest of any: it extends 220 ft., and is broken by a central vestibule, 55 ft. by 45 ft. The roof, which is groined, is constructed of variegated bricks, and rests upon columns of Bath stone, polished serpentine marble, and Aberdeen granite, the carved capitals of the columns having appropriate marine subjects.   On each side are placed the first two series of tanks, twenty-one in number.   These increase in dimensions from 11 ft. by 10 ft. upwards, the largest measuring over 100 ft. in length by 40 ft. in width, and holding 110,000 gallons of sea-water.   This colossal tank is the largest in the building, and is devoted to the exhibition of porpoises, turtles, and other animals of large size.   The next largest tank, 50 ft. by 30 ft., is immediately opposite.

The eastern end of the western corridor opens upon the conservatory, which serves as an approach to the rockwork, fernery, and picturesque

FIG. 324.—*Sea-Horses* (*Hippocampus brevirostris*).

cascade, and also to the eastern corridor. Some artificial rockwork, skirting the north side of the conservatory, is traversed by a stream of water, broken up at intervals so as to form numerous little bays and ponds, and utilized for the reception of seals and the larger reptilia. In the side-space between the conservatory and the second or eastern corridor are six octagonal table-tanks, of elegant design, for the exhibition of some of the smaller and more rare marine animals, and, at the eastern extremity, apparatus which serves to illustrate the hatching and development of trout and salmon. The entire length of this second corridor is about 160 ft., one side of the eastern portion, which is 90 ft. by 23 ft., being devoted to the exhibition of freshwater animals. At the end of the corridor are situated the curator's offices and the naturalists' room, fitted with open tanks and all necessary appliances; and the engines, pumps, &c., for supplying the water, and keeping it constantly aërated.

The system adopted for aërating the water at the Brighton Aquarium is quite different from that used at the Crystal Palace. In the former the water is pumped directly from the sea into reservoirs formed under the floors, and capable of holding 500,000 gallons, which can be filled in ten hours. From these the water is pumped up into the tanks as required; but there is no general circulation through the system of tanks and reservoirs.

Each tank is treated independently, and its water is aërated and kept moving by the injection of air at the lower part, effected by steam power.

The popularity of the Brighton Aquarium may be judged of from the fact that the average daily number of visitors is about 9,000, and that on some occasions nearly twice that number pass the turnstiles. Among the specialities of the establishment are herring and mackerel, which it has hitherto been considered impossible to preserve in confinement for any length of time. They are now thriving well in the Aquarium, although these fishes are both extremely impatient of confinement. The herring feed readily upon small shrimps, in catching which they display a wonderful activity. Fig. 324 shows the curious fish called the "sea-horse" (*Hippocampus*), from the singular resemblance of the front part of the body to a horse's head, or, at least, to that form which conventionally represents the "knight" among a set of chessmen. The tail of the creature is prehensile, and enables it to cling to seaweeds and other bodies. The sea-horses have thriven well in the Brighton Aquarium, and also in that at the Crystal Palace. The latest novelties are the *Proteus* from the dark caves of Adelsburg, axolotls from Mexico, the mud-fish (*Lepidosiren annectans*) from the Gambia, and the telescope-fish from Shanghai. Some of these creatures are of great interest from the circumstance of their forming the connecting-links between fishes and reptiles.

FIG. 325.—*Proteus anguinus.*

There are, therefore, now on view at the Brighton Aquarium specimens of three species of animals possessing a high interest for naturalists and others—not so much because their existence has been discovered in recent times, as because they are illustrations of the great law of gradation which exists in nature. Their position in the scale of organization is so intermediate between reptiles and fishes, that naturalists have not entirely agreed as to the kingdom to which these ought to be assigned. Fig. 326 represents *Lepidosiren annectans*, which has gills covered by flaps, and not exposed as they are in ordinary amphibious animals; and is provided with four fins, or rudimentary legs, according as the reader may choose to call them. The creature's nostrils do not communicate with the mouth, but are merely two blind sacs, as in fishes. The *Proteus anguinus*, shown in Fig. 325, is an eel-like creature, only met with in the subterranean waters of the Grotto of the Maddalena at Adelsburg. It has four imperfectly developed legs, and gills reduced to mere fringe, while there are lungs extending nearly the whole length of the abdomen. The optical organs are entirely undeveloped, being represented merely by two specks. The axolotl, Fig. 327, inhabits certain Mexican lakes, and is remarkable for preserving, through the whole period of its life, the gills for aquatic respiration, which other *amphibia* possess in the tadpole stage only.

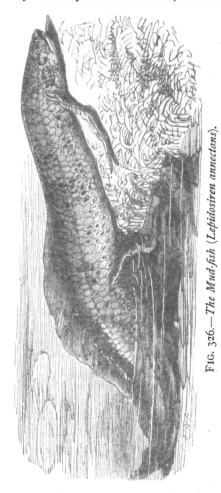

FIG. 326.—*The Mud-fish (Lepidosiren annectans).*

The mania for domestic aquaria which was at its height some years ago, and the great popularity of public marine aquaria wherever they are properly managed, express the real interest which is felt in the varied forms of animal life, of which the aquarium affords the opportunity of observing new and unknown phases. The progress of the science which treats of the organization of the animal kingdom has made rapid strides during the present century. Among the remarkable truths which have been acquired is the fact of the unity of the plan which pervades the animal kingdom. Each kind of animal has much in common with the kind above it, and with the kind below it: a certain community of organization pervades the whole, which is knit into one by the gradational forms which may be ob-

FIG. 327.— *The Axolotl.*

served connecting, like links of a chain which cannot be broken, the more defined modifications from each other.   It is their position in the scale of organization which, in the eyes of the philosophic naturalist, gives so much interest to some of the forms of life which have been figured above.

FIG. 328.—*Sorting, Washing, and Digging at the South African Alluvial Diamond-Fields.*

GOLD AND DIAMONDS.

———◆———

IT need hardly be said that gold and diamonds are named under nineteenth century discoveries in relation to the newly-found fields which have yielded these highly-prized substances in remarkable abundance.

GOLD.

THIS precious metal is met with in nearly all parts of the world, and its splendid colour, high lustre, the ease with which it may be wrought, and its property of ever remaining untarnished, have caused it to be greatly esteemed for ornamental purposes from the earliest historical ages. No doubt the store set upon gold is derived from its suitability for decorative uses ; and its comparative scarcity enhances the regard in which it is held. Its use, as a standard of value, is justified by the general estimation in which it is held, and by the fact that the amount of labour required to obtain the metal is on the whole tolerably uniform. It is one of the few metals which are found in nature in the uncombined state, but its separation from the materials with which it is associated requires the performance of a certain amount of work, in whatever form the metal may occur. Its general distribution is another advantage attending its selection as the standard of value. It occurs in England and Wales ; in France, in Spain, in France, in Hungary, in Piedmont, and in other parts of Europe ; in various localities in Asia ; in both divisions of the New World ; in the remaining quarter of the globe, where it was obtained even in very ancient times, for South-East Africa was probably the locality to which a naval expedition was despatched by King Solomon—"they came to Ophir, and fetched from thence gold." Australia also has, in the last half of our century, yielded much gold.

Gold is never met with in regular veins, but in primitive or igneous rocks, or in deposits formed by the disintegration of these. In Australia the metal is associated with quartz, in slate rocks geologically equivalent to the Cambrian formations of England and Wales ; and in California it is also chiefly found in material which has been formed by the wearing down of quartz and granite rocks. Before the discoveries in California and Australia most of the gold in circulation was obtained from auriferous iron pyrites. The first finding of gold in California occurred in September, 1847, when a Mr. Marshall, the proprietor of a saw-mill on the Sacramento River, observed some glistening grains among the sand in his mill-race. The news soon spread, and the inhabitants of the town of San Francisco, then numbering about two hundred persons, were greatly excited thereby. When it became known that gold was really to be found, multitudes flocked to California, the population of San Francisco rapidly increased, and at the present day the city contains nearly a thousand times as many inhabitants as it did at the time gold was first discovered. The annual value of the metal found in California averaged about £23,000,000 for ten years after 1851 ; but this subsequently declined to less than half in 1872.

Sir Roderick Murchison, the distinguished geologist, pointed out the great probability of the existence of gold in Australia many years before the precious metal was actually found. It has, however, been stated that gold was met with in Australia so long ago as 1788. Considering the mode in which the metal occurs, it seems strange that the emigrants who occupied the auriferous districts as agriculturists did not long ago discover the riches which Nature had scattered over the surface of the soil. No doubt, their attention was too much devoted to their sheep and cattle to notice the glittering particles which might be seen in the water-courses, and it would probably never enter their minds that the eagerly desired metal could lie exposed to view on the surface of the land. But the announcement of the discoveries in California induced men to look at the soil more attentively, and in April, 1851, Mr. Hargreaves appears to have found at Bathurst the first gold met with in Australia. Four months afterwards the metal was also picked up at Ballarat, Victoria, and the gold-fields so discovered proved even richer than those of Sydney.

The effect of this discovery on the colony of Victoria proved marvellous. The population, which in 1851 was 77,000, had in 1867 become 660,000 ; in the same period the land under cultivation expanded from 57,000 acres to 631,000, and the value of property rose enormously when the grazier's estimate of its worth was replaced by that of the miner. The authorities of the colony from the first regulated the mining operations by enactments defining the rights of the miners to the "claims," as the allotments of land for working upon are termed; and thus disorder and lawlessness were almost unknown. Fig. 329 will give the reader a notion of the appearance of a miners' settlement in the Australian gold-fields in the earlier period.

FIG 329.– *Gold Miners' Camp.*

The fundamental rocks in the colony of Victoria belong to the oldest series of strata. They answer to the Silurian formation which exists in Cumberland, Wales, and Scotland. Although the strata of the rocks are much bent, and they have been worn down by the action of water, they are as a whole but little altered, consisting chiefly of sandstones and shales. These strata are interpenetrated by innumerable veins of quartz, which vary in thickness from $\frac{1}{16}$ in. to 150 ft. It is in these quartz veins that the gold is seen in its original matrix. The metal is sometimes in the form of grains or flakes, or in moss-like threads, embedded in the quartz; sometimes in the form of well-defined crystals, sometimes in rough lumps or *nuggets.* Fig. 330 shows three of the various modes in which the gold is found disseminated through quartz. Overlying the more ancient rocks with their auriferous quartz veins are various rocks of different ages; and as these have been in part formed by the wearing down of the older rocks, they also are in general auriferous, and contain the gold in detached pieces, varying in size from particles of fine dust to the huge nugget, containing 2,280 oz., or nearly £10,000 worth of pure gold, which was found at Dunolly.

The soil, which has been formed by the disintegration of masses of auriferous quartz, is full of gold, so that a patch of such soil 12 ft. square has been known to yield 30 oz. of gold by a very rough kind of washing to the depth of 1 ft. Soil of this kind has been carried down by rivers and streams ages ago; and the lighter particles having been carried off by the water, while the gold, from its greater specific gravity, remained at the bottom of

44

the stream, the sands and gravel of these river-beds are very rich in gold. In many instances the ancient water-courses have been entirely covered by igneous rocks, such as basalt, which have flowed over the land in a

FIG. 330.—*Gold in Rocks.*

molten state. The gold-miner often finds his reward in burrowing beneath these basalts and lavas, following the bed of the ancient river, and recovering its long-buried treasures.

The methods of carrying on the gold-seeking operations vary according to the nature of the deposit which is worked and the resources of the miner. The simplest, which was that most practised in the early days of the gold-fields, consists in throwing into a tub several shovelsful of the surface soil, and in pouring in water while the contents of the tub are stirred about with a spade. The lighter matters are washed away, but the gold by its great specific gravity remains behind. An improvement in this, but still a very

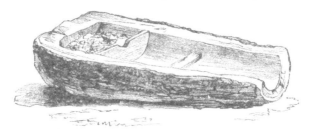

FIG. 331.—*" Cradle" for Gold-washing.*

rude process, is practised by aid of the *cradle*, Fig. 331, which is merely the trunk of a tree, hollowed out, and provided with transverse partitions and ribs. The auriferous earth is thrown into the upper compartment, which is then filled with water. The cradle is rocked, so that the water may wash away all but the gold and the heavy stones. Any particles of the former which may be carried out of the head of the cradle will be stopped by the ribs which cross the lower part. Machines for puddling by horse-power are now in use, and other contrivances have superseded the tub and the cradle in *surface-washing*. The auriferous earth obtained by excavating the soil from pits is washed in a similar manner, as is also the material reached by penetrating the deeper tertiary deposits, and by driving adits or tunnels along the ancient river-beds beneath the layers of basalt.

A mode of washing accumulations of auriferous earths by streams of water is employed where circumstances are favourable. A long inclined channel is constructed, and lined with boards; or, when the natural inclination of the soil requires it, a long trough is constructed and supported on trestles. The trough is made of sawn boards, 1½ in. thick, in sections 12 ft. long, and it has a width of about 1 ft., the sides being from 8 in. to 2 ft. high. The inclination of the troughs is from 8 in. to 24 in. in 12 ft., and depends upon the abundance of the water : the more water, the steeper is the slope. The bottoms of the troughs are crossed by a number of transverse bars, which arrest the auriferous particles in their descent. The *sluice*, or series of troughs, may be from 50 ft. to several hundreds or even thousands of feet in length, and the cost from £100 to £8,000. The earth is thrown in at the upper part of the trough, and it is gradually washed down, the water being allowed to flow in some cases by night as well as by day, but commonly in the daytime only, as the troughs must be watched, to see that they do not become choked up, and the soil washed out by the overflowing water. The *run* goes on for six or ten days, and then the current is stopped for a *cleaning-up*, which occupies from half a day to a day. For this operation the stream of water is stopped, and quicksilver is used to dissolve the grains of gold from the sand, &c., collected by the *riffle-bars*. The quicksilver is afterwards expelled from the amalgam by heat, and the gold remains as a porous mass.

Sometimes, instead of shovelling the earth into the troughs, it can be washed out of its position into suitable channels by means of a powerful jet of water. This mode of working, which is termed *hydraulic jet sluicing*, offers great advantages where the natural conditions admit of its adoption. In this plan, instead of bringing the auriferous earth to the water, the water is brought to the earth by a flexible pipe, like the hose of a fire-engine, from a reservoir about 200 ft. higher, and the stream is directed upon the material by a nozzle. This powerful jet of water is used to separate and carry away the earth to the head of a system of channels and troughs, like those already described. The hose has a diameter of 8 in., but the orifice of the nozzle from which the water issues is contracted, in order to increase the force of the jet. The hydraulic jet sluicing requires from three to six men to work it, and the material of a hill can be carried into the sluices in less time than a hundred persons could do it by spades. Immense quantities of earth are removed in this way, and fatal accidents are not infrequent from the falling masses burying the men who carry the pipe. The force of the jet of water itself is another source of danger, for broken limbs and even fatal injuries have often been caused by it. The number of accidental deaths occurring in hydraulic jet sluicing operations in the colony of Victoria is reported to average about 60 in a year. Material which has been worked before often yields a considerable amount of gold when the operations are repeated; and in localities favourable to the hydraulic jet system, the work can be carried on with little labour. In this way three men have been known to extract in one week from dirt washed for the third time, gold of the value of £330.

The gold which is embedded in quartz and other minerals, as shown in Fig. 330, is obtained by crushing the material in stamping machines, which are usually constructed with logs of wood shod with iron. In another form of crushing-mill two large cast-iron rollers are used instead of stampers.

From the crushed material the particles of gold are extracted by amalga-
mation with mercury, which is afterwards removed by distillation.

The richness of the Victoria gold-fields may be inferred from the fact
that, up to the year 1868, 36,835,692 oz. had been obtained, the value of
which is no less than £147,342,767. The total value of the gold then
annually obtained throughout the whole world is estimated at about 20
millions of pounds sterling. When gold was found so plentifully in California and Australia, it was supposed by some that its value as a monetary
standard would be affected. This has not happened, altnough the prices
obtained for the metal by its producers were considerably lower in the
last decade of the century than about 1867. The total annual output of
gold throughout the world is of course variable, and no doubt there are
also variations in the demand; but, so far, the fluctuations have been
relatively small, and there has been no such depreciation by excess of
production as in the case of silver. Yet the increased production of gold
after the discoveries made about the middle of the century was beyond
precedent. It has been estimated that between 1850 and 1875 the total
value amounted to £600,000,000, showing an annual average twelve times
greater than that for the period between 1700 and 1850. Between 1875
and 1890 there was a falling off in the supply, the annual average becoming only £20,000,000 in value. But since the last-named date there has
been a rise year by year, and at the close of the century the value of the
gold produced throughout the world in one year may not be less than
£40,000,000.

As already remarked, the distribution of gold is world-wide; and it
has happened in recent times, that just as one source of supply has
shown signs of failure, other fields have been discovered and have
attracted thousands of eager seekers to new regions. So, when the
Californian supply was falling off, there came the rush to Australia,
where easily worked alluvial deposits or rich veins continued for years
to reward the toil of the gold-finder, though in an ever-lessening degree,
until in 1886 or 1887 the centre of attraction was shifted. But at a later
period fresh discoveries in Australia again raised the productiveness of
that quarter; and still more recently, the announcement of the existence
of much auriferous deposit in the valley of the Yukon River (Klondyke),
and in various localities of British Columbia, drew thousands to desolate
and undeveloped districts, in spite of the extremities of hardship and
destitution that might be endured.

The discovery of 1886 takes us to South Africa, a region with which
also our next section is mainly concerned, and the scene of an activity
unprecedented in the annals of gold-mining. From circumstances immediately connected with our present subject, the close of our century
finds public attention intensely occupied with affairs at the austral extremity of the "dark continent." The history of South Africa, from the
time when, in 1486, the tempest-driven Portuguese mariner, Bartholomew
Diaz, first struck its shores at the promontory he named the "Cape of
Storms" (*Cabo Tormentoso*), and when, eleven years afterwards, the
celebrated Vasco de Gama sailed round it on his memorable voyage to
India, is one which, in many respects, presents features of peculiar
interest. It is not our province to enter into details of these, but it may
be stated that the "Cape of Good Hope"—the more auspicious designation which the King of Portugal substituted for that of Diaz—was,
towards the end of the seventeenth century, colonized by Dutch and

some French settlers, and afterwards Table Bay became a regular port of call for Dutch, English, and other ships trading to India. The Cape was taken possession of by the British in 1795, but restored to the Dutch in 1803, only to be three years after (1806) resumed by England, under whose rule "Cape Colony" has since remained. The abolition of slavery in all British dominions, enacted in 1833, was the occasion of great dissatisfaction to the descendants of the Dutch settlers, who inhabited isolated farmsteads, their possessions consisting chiefly of great herds of cattle, tended by slaves. These people, or at least the majority of them, resolved to quit the confines of British territory, and seek fresh fields and pastures new in unoccupied regions north of the Orange River, so that from 1835 to 1838 there was a continued "exodus of the emigrant farmers." The story of the following years, with its exciting events and the vacillating policy of successive British Governments, must be perused elsewhere : suffice it to say here, that the settlements of the "emigrant farmers" had by 1854 established themselves into two separate States, nominally recognising Great Britain as the "paramount power," but practically independent of it ; for, at the last-named date, the autonomy of "The Orange River Free State" was acknowledged, and two years before that another section of the Boers, *i.e.* of the "emigrant farmers," who had settled beyond the Vaal River, was absolved from British allegiance, and, restricted only by a claim of certain suzerain powers, was constituted into "The South African Republic," of which the precise boundaries were at length determined by the "Convention of London" in 1884. This territory is usually called for shortness the *Transvaal,* and here in 1854 the existence of gold was first announced; but the Boer authorities at once prohibited further prospecting, fearing, and perhaps with reason, that the winning of the precious metal within their bounds might disturb their pastoral quietude. The Boer character has been the unique product of a race withdrawn for two centuries from contact with European and civilized culture, living in widely separated dwellings with scarcely other associations than cattle and enslaved blacks. The Boer is described as of a type which draws away from the enterprising man of modern times towards the primitive patriarch centred in his flocks and herds : he hates innovations, and greatly distrusts strangers ; he would rather keep, in a box under his bed, any money he may possess than employ it, or his own energies, in developing the immense mineral resources of his territory, in which are included not only gold, but copper, silver, lead, iron, and abundance of coal.

After some years the prohibition against the exploitation of gold in the Transvaal was withdrawn, and several localities in the Republic subsequently became small capitals of gold-mining industry. The most notable were Leydenburg and Barberton, at which latter place as many as 10,000 gold-seekers were congregated when the discovery of 1886 drew most of them away. These communities were formed almost entirely by the influx of people from beyond the Boer boundaries, mainly, of course, English-speaking people from the Cape Colony, Australia, America, etc. Their operations were hampered by the Transvaal legislation, and impeded by the absence of adequate means of communication, which was a characteristic feature of the Boers' unprogressiveness ; nevertheless, gold-mining has been pursued in some of these localities ever since, though with varying fortune. What drew nearly all the gold-seekers of the Transvaal and of adjoining regions at once to the north side of the Vaal River was the discovery there of *real gold mines*. This was at a district within the Transvaal territory, named Witwatersrandt (= White-waters-ridge), the designation which has been reduced by abbreviation or affection to " The Randt"—or, *anglice*, the Rand.

It is singular that although the Randt district had been explored by expert prospectors between 1877 and 1891, the outcrops of the auriferous reefs entirely escaped their notice. But when the first hint of the existence of these deposits was bruited abroad, it was the Kimberley men who were foremost in surveying the spot. By the Kimberley men we mean those who so soon had by lucky chance lighted in 1870 upon the rich and apparently inexhaustible diamond mines, as related in our next section. By the time of the announcement of gold-finds on the north side of the Vaal River many of these men had become rich—very rich indeed. If they had been so disposed they might then have returned to their native countries with enviable fortunes, but it was just as the affairs of their diamond companies had been settled by consolidation on a satisfactory basis, and the spirit of discovery and adventure was still strong upon them, that they resolved personally to explore the new El Dorado. To a wild desolate region they proceeded, enduring there and on the track thither the like discomforts they had experienced in their earlier quest. But they took with them experts provided with all appliances for ascertaining the prospective value of the alleged discovery, and, when convinced of its reality, they purchased from the Boer possessors their land at the price demanded, and it was not long before they had chemists and engineers at work, having, at the cost of making their own roads, had brought to the spot the necessary machinery and appliances. The usual influx of workers, builders, speculators, etc., followed, and in a wonderfully short space of time a town sprang up where in 1886 there had been only a single poor farm. The town grew rapidly to the dimensions of a city inhabited by 150,000 people. Its name is Johannesburg.

In his book on South Africa, the late Lord Randolph Churchill, describing Johannesburg in 1891, says that it has much of the appearance of an English manufacturing town, but without noise, smoke, or dirt. " The streets are crowded with a busy, bustling, active, keen, intelligent-looking throng. There are gathered together human beings from every quarter of the globe, the English possessing an immense predominance. The buildings and general architecture of the town attain an excellent standard, style having been consulted and sought after, stone and bricks the materials, corrugated iron being confined to the roofs, solidity, permanence, and progress being the general characteristics."

The Randt mines having drawn into the South African Republic great numbers of enterprising workers, who have acquired wealth and built cities, it would have been expected that they would have been permitted to acquire the ordinary rights of citizenship. The Boers' character, however, has been manifested by their refusal of such rights, and by their exacting grievous imposts from these *Uitlanders* (Out-landers, or strangers), who, being for the most part of English race, are finding the injustice too hard to be borne, and greatly strained relations between the Transvaal and Great Britain have again supervened. Indeed, the situation has become so serious that it is feared actual war may result, and that is why, in almost the last year of our century, people are looking anxiously at the position of affairs in South Africa.

The geological conditions of the Randt are these : the upper series of beds in the *Karoo formation*, which extends over the greater part of South Africa, consist of quartzose strata, and in the district in question these are much broken, faulted, and variously inclined. They are inter-stratified with beds of sandstone and with the layers of gold-bearing conglomerate, of which last there are several parallel to one another and not far separated, ranging in their several thicknesses from 6 inches to 6 feet, the thickest being known as the main reef. These reefs form an oval basin,—that is, they dip with varying angle towards a centre, and crop out at their up-turned edges. Johannesburg is situated nearly 6,000 feet above the sea-level, on an elevated ridge, along which for 30 miles east, and nearly the same westwards, the northern outcrop extends, curving towards the south, while the southern edge of the basin appears in the Orange Free State, where it has been traced for a distance of 130 miles. There a shaft, sunk to the great depth of 2,400 feet, found the main reef with undiminished richness. The outcrop of the reefs stretches east and west for 130 miles, and the distance between north and south is 30 miles. From such data it has been inferred that the reefs contain altogether not less than 450 million pounds worth of gold. The con-glomerate of these reefs consists of rounded quartz pebbles (which contain no gold), and pieces of sandstone and of argillaceous material, the whole cemented together into a very hard mass by iron pyrites. This last is the matrix in which the gold exists, in the form, for the most part, of minute scarcely visible crystals. To a depth of from 50 to 150 feet, air and moisture have acted on the pyritic matter, and the material of the reef becoming in consequence easily disintegrated, has yielded by mere mechanical treatment most of its gold, whereas by the same operations on the underlying hard, tough conglomerate, only about half its gold could be obtained. Hence, after breaking up the ore, the pyritic matter is sorted out and transported to the stamp battery, reduced to powder, from which about five-eighths of the contained gold is removed by quick-silver. The residue is concentrated by washing in a special machine called the "Frue vanner," and the *concentrates*, after roasting in order to oxidize base metals, are subjected to the action of chlorine gas, by which the gold is converted into a soluble chloride, from the solution of which it is precipitated by ferrous sulphate. The *tailings*, *slimes*, and other residues are further acted on by a solution of potassium cyanide, which dissolves the minute remaining particles of gold, and from the solution the metal is obtained by electrolysis. By these supple-mentary chemical processes the total of the gold recovered from the ore is raised to 90 per cent. or more of all that chemical analysis shows to exist.

When it is said that the reefs are arranged in a basin-like form, it must be understood that this applies to their general disposition, for the regularity of geometrical shape does not belong to geological basins. There are considerable variations in the inclinations of the reefs : at some places they are nearly vertical, but generally they dip towards the centre at various angles, a slope between 25° and 45° being quite usual ; and the inclination becomes less and less the deeper they go, so that it is presumed that the beds are level towards the centre of the basin. In the Randt the vertical shaft is rather the exception, the entrance to the mine usually following the inclination of the reefs, and the trucks of ore are drawn up sloping rails. From the inclined adits horizontal galleries are excavated right and left at various depths by which the main reef is worked, and there are cross cuts by which the reefs to the north and south may be reached. The most active district of the Randt is that which extends eastward of Johannesburg, where a long succession of tall chimneys and winding headgears together with the other appurtenances are visible. But there is nothing of a picturesque character about a gold mine, more than is presented by the aspect of an ordinary colliery.

The importance of the Randt gold-fields does not consist in the actual richness of the crude material, which indeed in places here and there cannot be profitably worked,—in mining parlance, it is not "pay ore." It is rather the great ascertained extent of these gold-bearing beds and the general persistence of their character throughout that give to the Randt its unique character amongst metalliferous workings. This contrasts with the comparative uncertainty attending the exploitation of auriferous quartz veins, which occur in detached unconnected patches, that often end suddenly where least expected. There are in the Randt nearly one hundred companies working mines, and of these there are many that pay very handsome dividends on their original capital. A few pay 100 per cent., while a considerable number distribute 25 per cent. and upwards ; so that some of these Gold Companies are amongst the richest and most influential financial houses in the world. The Randt is second only to the United States in the quantity it adds annually to the world's production.

DIAMONDS.

IN ancient times, and down to a comparatively late period, the only region from which were derived all the diamonds that found their way to Europe, was India, where Golconda was long celebrated for the productive mines in its neighbourhood, and for the high estimation in which fine specimens of their yield were held. In the seventeenth century these mines employed 60,000 persons, it is said ; and in other districts of India diamond-seeking has also been carried on from time immemorial. A gradual decrease in the finds of Indian diamonds has long been observed, and the supremacy the East had so long enjoyed as the purveyor of gems was in the earlier part of the eighteenth century transferred to another hemisphere. In 1727 the diamond was first discovered in Brazil ; or rather, we might say, was then first discerned there. For the gold-seekers in washing the sands of certain Brazilian rivers had found numberless specimens which they either threw aside as worthless, or, seeing them prettier pebbles than the rest, used them as counters in their card games ; their true nature was not recognized, because the rough diamond has by no means the attractive appearance of the cut and polished brilliant flashing with refractive radiance. It must have been these last, and not diamonds in their natural state, that presented themselves to the imagination of the poet when he penned the line—

Or deep with diamonds in the *flaming* mine.

The announcement of some diamonds having been found in America had no effect on the prices in the Indian market, but the exports that soon after came from Brazil in great abundance quite changed the conditions of the trade, for in the first fifty years their value was estimated at no less than £12,000,000 sterling. As already stated, the presence of diamonds in Brazil was not recognized until 1727, and then by the accident of one Lobo, an inhabitant of the gold district of Minas Geräes, who had been in India and had seen rough diamonds there, observing the resemblance ; he took some of the Brazilian stones to Lisbon, where their identity with the products of the Indian mines was established. But the European dealers, alarmed lest this discovery should depreciate the value of their stocks of Indian gems, spread a report that the so-called diamonds from Brazil were but the refuse of the Indian mines that had been sent to Brazil. This had the effect of stopping for a time the sale of the Brazilian diamonds ; but the traders in these were not above taking a hint from their rivals—*fas est et ab hoste doceri*—for they carried their diamonds to Bengal, and there sold them as Indian stones at Indian prices. For nearly one hundred and forty years after this Brazil was by far the most productive diamond region in the whole world, and especially after 1754, when diamond-seekers congregated by thousands in the very rich fields of Bahia, a district of Brazil. Nor have the places above mentioned been by any means the only localities in Brazil where diamond-finders have been at work ; but the production has decreased and has lost its relative importance by the South African discoveries that about 1870 caused an entire change in the diamond industry, and the high prices of the Brazilian gems no longer capable of being maintained, the fall in value has rendered the workings less remunerative than formerly. We may now pass over with mere mention, discoveries of diamondiferous districts in North America, Australia, and elsewhere.

While rejecting as entirely inapplicable and inexcusable by any stretch of poetic licence the epithet *flaming* for the diamond mine, we must question whether the word *mine*, that as the customary word we have continued to use, does not convey an equally false notion of the nature of the workings to which hitherto reference has been made. For these in most cases are nothing more than holes, very much like gravel pits in the side of a hill. The diamonds which have so far been in question are usually found among alluvial sands or gravels, the water-worn fragments of disintegrated rocks. These are in many cases carried down by rivers, and the diamonds under such circumstances are very frequently accompanied by gold ; indeed, it is the search for gold that has in many cases led to their discovery. In the dry season of the year, which extends from April to October, the lessened currents of certain of the Brazilian streams are diverted from their course into canals, so as to leave dry the bed of the stream, and here the mud is dug out to the depth of six or eight feet or more, and transported near the washing huts, these operations being continued throughout the dry season. When this is over the digging is necessarily interrupted by great volumes of water that fill the rivers and streams, and the diamond-seekers devote their attention to washing the mud that has been collected. About one cwt. of this is placed in a long trough, and water is made to flow in, while the negro labourer stirs up the mass with his hands, until the water runs off clear, all the particles of mud having been washed away. The residual gravel is then very carefully examined, stone by stone, and any diamonds found are handed to the overseer, who watches all proceedings from an elevated seat. These Brazilian diamonds are mostly of a small size: occasionally, but very rarely, stones of quite exceptional value are found, but perhaps not one in 10,000. Formerly when in the Brazilian fields a negro slave found one of 18 carats, or more (18 carats=72 grains), he not only obtained his freedom, but was rewarded with gifts, and for the finding of smaller stones commensurate rewards were given. The value of a diamond of the larger sizes depends upon so many adventitious circumstances that it would not be easy for any one to state the money's worth of an 18-carat stone, but, considering too that the price increases in a more rapid ratio than the weight, we may to some extent draw an inference from the published values in 1867 of smaller Brazilian brilliants, perfectly white, pure, and flawless, when one of 5 carats (20 grains) in weight was priced at £350. As the rough diamond gives a brilliant of only half its weight, we may from the above assume an 18-carat stone to be worth in its finished state at least £1,000.

It may well be asked what are the qualities possessed by the diamond which have caused it to be so highly valued as an adornment all the world over ; and here it will be proper to invite the reader's attention to the chemical as well as to the physical character of the diamond. The most obvious and attractive quality of the cut brilliant is its unsurpass-able lustre, which is due to its high refractive power. In a section of our article on light the subject of refraction has been dealt with, and an explanation given of the *index of refraction*. That of the diamond is the highest known, being 2·50 to 2·75 ; other precious stones have indices ranging from 1·58 to 1·78 ; those of glass and of quartz are between 1·50 and 1·57. It follows from the known laws of refraction that the *limiting* or critical angle is less for diamond than for other substances, as, for example, glass, for the posterior surface of a diamond will totally reflect all the light that falls upon it at any angle with the normal greater than 24° ; glass will totally reflect only when the incidence is greater than about 42° : hence the diamond reflects from its farther surface about 64 per cent. of rays that glass similarly situated would allow to pass outwards without reflection.

Another property in which the diamond excels all other substances is hardness. It is the hardest substance in nature ; for a diamond will *scratch* every other, but by none can it be *scratched*, except by another diamond. Not but that by the application of a file the edges of a diamond or brilliant may be notched and broken ; but this would be through sheer mechanical force tearing the substance, and would be a test of *brittleness*, not of *hardness*. These two properties have not unfrequently been confounded, as when it was foolishly prescribed as a test for the genuineness of a diamond, that it should be placed on an anvil and struck with a hammer. No doubt many good and valuable stones have been sacrificed by this ignorant treatment. The hardness of the diamond does not prevent its being reducible to powder when so required. Again, diamonds are sometimes in such a condition of internal strain that very slight shocks are sufficient to cause them to separate into fragments. We read of diamonds that are suspected to be in this condition being packed for transmission within raw potatoes. The extreme hardness of the diamond secures it from all those accidental abrasions and injuries to which softer materials are liable, so that it does not deteriorate by age or use. It is unaffected also by any chemical substances.

In chemical composition the diamond is pure carbon, one of the most commonly diffused of the elementary bodies, as it enters into the constitution of the atmosphere, of all organic bodies, and of a vast number of mineral substances. Carbon in a less pure form also occurs naturally as *graphite, plumbago* or *black lead,* and in other conditions comes into ordinary use as already explained in our article on Iron. It was only towards the end of the eighteenth century that the composition of the diamond was demonstrated by the celebrated French chemist, Lavoisier, who actually burnt a diamond in oxygen gas, and found the resulting product to be carbonic acid gas, identical with that obtained by similarly burning a piece of charcoal. Soon afterwards another French chemist, Clouet, confirmed Lavoisier's conclusion by producing *steel* from pure iron and diamond heated together, an experiment of much significance when considered in the light of the remarkable relation between these substances, which is one of the latest discoveries of our century. It should be observed that Clouet's result implies a fusion of the diamond as well as of the iron in the act of entering into chemical combination.

Like nearly every solid substance of definite chemical composition, this pure carbon takes the crystalline form. The phenomena of crystallization are of the highest interest and beauty, for in them we see shapeless matter fashioning itself into definite and often perfect geometrical solids, as if it had been wrought by the hand of some mathematical artist. Every substance forms crystals of some one shape when the conditions are identical, and one essential condition for any crystallization is that the particles should be capable of free movement in arranging themselves, and this condition can occur only when the substance is in the state of liquid or of gas. Crystals are commonly deposited from solutions when the solvent evaporates or is cooled down ; or they are formed when a fused substance solidifies. In either case the crystals are the larger and more perfect as they are allowed the greater time to form. Now, carbon in any of its conditions has been found to be absolutely infusible and insoluble, and therefore the origin of the diamond has long been a puzzle to scientific men, very diverse surmises having been propounded on this subject. Some have thought it was separated from carbonic acid by the action of heat, or of electricity ; others, that the carbon had been gasified by subterranean heat ; others, among whom were Newton and the German chemist, Liebig, believed that heat had nothing to do with it, but that the crystals slowly separated from vegetable matters (hydro-carbons) in the process of decomposition under some unknown conditions ; others, that the diamond crystallized out from liquid carbonic acid, holding under pressure some unknown form of carbon in solution ; others, that carbon was ejected by volcanic action in a fused state ; and so on. We hope to show that the problem has at length been solved, and how.

The shapes of the natural crystals of the diamond must not be confounded with those of the cut brilliants. The most frequently met with of the former is the octahedron, or eight-sided figure, such as would result from two square pyramids joined base to base, the triangles forming the sides of the pyramids being of such a height that the three pairs of opposite points are equidistant one from another, so that the octahedron enclosed in a cube would have an apex in the middle of each surface of the cube. There are other shapes of diamond crystals, but they are all related to the cube, that is, they are all obtainable from the cube by successively slicing off edges and angles. The natural diamond sometimes has as many as 48 faces formed by such a process. This will easily be understood by the reader if he will take a cube of common soap and perform on it these operations *gradually* with a sharp knife, taking care always to make the new faces he produces equally inclined to the adjoining ones.   He may begin by cutting off a tiny piece from one corner of the cube, forming a small equilateral triangle ; then let him do the same at two opposite corners, and again at all the eight corners. Then he should make the cuts larger and larger, always producing equal sized equilateral triangles so long as these can be formed.   In every case he will have shaped out such forms as belong to diamond crystals. Instead of this, he may pare off one or more edges of the cube, or he may in various ways combine the two operations, and he will probably be surprised at the variety of forms producible in this manner, all derived from the original cube and all representing possible forms of natural diamonds, and indeed those of any substance that crystallizes in the *cubical system*.   A model of the diamond octahedron can be readily made from the description already given, and the whole series of operations will constitute an elementary but very instructive lesson in the science of crystallography.

Diamonds are liable to occur with every imaginable distortion, so as to be scarcely recognizable by their external form.   A very pure smooth uncut diamond, belonging to the Rajah of Mattam in Borneo, is shaped exactly like a pear, two inches in length.   By the way, battles have been fought for the possession of this gem, and it is said that £200,000 was vainly offered for it.   The diamond, notwithstanding its hardness, splits with comparative ease in certain planes, and by such cleavage (a property common to all crystals) the octahedral form commonly emerges.   It was not until the middle of the fifteenth century that the art of cutting the diamond into regular facets was practised, and this can be done only by the aid of diamond powder, prepared by crushing fragments and faulty stones in a hard steel mortar.   The first operation is to split the stone by its natural cleavage, and the rough facets so produced of two diamonds are ground together until they are quite smooth.   The grinding of other facets and the polishing are effected on horizontal discs of steel making 2,000 revolutions per minute, and overspread with diamond powder mixed with olive oil.

The external surface of the diamond in its natural state is often very rough, the stone being always coated with a more or less opaque crust, so that its translucent interior is concealed or veiled ; but when the reflection from its inner surfaces pierces this veil it glows as if lighted from within, giving that peculiar appearance which is called its "fire." The surfaces of the diamond crystals are very often curved instead of being flat, and the dodecahedral shape, when this is the case, takes on an almost globular appearance.   Diamonds of all colours are found, as well as the highly esteemed colourless stones.   Yellow ones of various tints are frequent,—orange, brown, and pink are not very rare ; but red, green, blue, and black are almost unique, at least in a condition to form large and perfect gems, and are accordingly much prized.   The black diamonds found in Borneo are so hard that ordinary diamond powder has no effect whatever upon them ; they have to be manipulated with their own dust. The nature of the substances that impart these colours to the diamond has never been made out ; they must be excessively small in quantity. When a diamond is burnt in air or in oxygen gas by aid of a large burning glass or otherwise, an extremely minute quantity of ash remains, and this often retains the shape of the stone, in the form of a most delicate network ; and of the composition of the ash, this much has been made out : it contains silica and *iron.* We shall find that the presence of the last named element, although but in the merest trace, is not without significance.

The purely utilitarian uses of the diamond are few, but of importance. The most familiar is in the glazier's tool for cutting glass, and in connection with this we may mention a fact not generally known, namely, that though any point of a diamond will *scratch* glass, it is only by a natural point of the crystal, and that point of a certain shape, that glass can be *cut.* Another kind of diamond, valueless as a gem, has been turned to good account in Major Beaumont's invention, described in our section on Rock Drilling Machines, to which the reader is referred. Minute diamonds are employed for writing on glass, for very fine engraving, etc.

Having now said sufficient about diamonds in general to give the reader an interest in the subject, and yet but little more than was needed to impart the information necessary for following the further development of the theme, we approach the discoveries in this connection which have specially distinguished our century. We must transfer the reader's attention to South Africa, and if he can refer to any recent map of that region, particularly to one showing its physical features, it will be of advantage.

In 1867 some children, playing near the banks of the Orange River, found what they thought to be merely a pebble prettier than the rest. A neighbour seeing the stone in the children's possession, obtained it from their mother for a trifle. It passed through several hands, and was bought at last by the Governor of the colony for £500. The discovery shortly afterwards of other diamonds in the same locality attracted numbers of persons to the district, and especially to the banks of the Vaal River, which speedily became the scene of a great search for diamonds. Though this search was confined to merely the surface of the soil, it was attended with considerable success, and many fine diamonds rewarded the diligence of the eager seekers. One of the most remarkable stones for its great size, which equalled that of a walnut, was discovered by a Kaffir. When this gem had finally reached the hands of Messrs. Hunt and Roskell, of London, its value was estimated at no less than £25,000. News of these discoveries having spread, a rush set in for the diamond-fields of the Vaal River, and the banks of this stream soon presented an animated spectacle. Europeans flocked to the spot, London jewellers sent agents, and the inevitable Jews appeared on the scene to purchase the precious gems from the lucky finders. It turned out that many of the larger stones had a slightly yellow tinge, varying in different specimens from the palest straw to a decided amber colour, and, as this detracted greatly from their value, no little disappointment and loss were sometimes experienced when the gems came to be sold in London and Paris.

One of the first settlements which sprang up on the banks of the Vaal River was a place called Pniel, of which the reader may form some idea from Fig 332, which is copied from a sketch actually taken from the windows of Jardine's Hotel. It was then only a little straggling village, chiefly of wooden sheds or corrugated iron erections, with but two or three more substantial structures. The diamonds which were found in this neighbourhood were obtained from gravel which lay on the slopes of the hills rising from the river. The mode of conducting the search for diamonds in these gravels was simple enough. The first operation was the washing of the material, in order to remove sand and dirt, and this process was usually performed at the margin of the river, where the gravel was brought down in carts and deposited in a suitable place, at which a cradle was erected. The cradle was simply a strong wooden framing sustaining sieves of wirework or perforated metal, placed one above the other, those at the top having the largest meshes, so that the lowest would only permit sand or very small pebbles to

Fig. 332.—*Pniel, from Jardine's Hotel (c. 1870).*

pass through. The cradle was capable of receiving a rocking movement, and while the gravel was thus sorted, water was freely poured on the uppermost layer, so that the stuff was in a short time thoroughly cleansed and sorted. When this had been accomplished, the gravel was thrown in successive lots on a table, at which the digger sat and rapidly examined it for diamonds by help of a flat piece of wood or iron (see Fig. 328). The larger gems were readily detected, and indeed could be picked out from among the pebbles on the sieve before the stuff was thrown on the sorting-table. Crystals of quartz, which sometimes glisten among the mass, often excited groundless delight in the bosom of the inexperienced worker.

On the payment of certain fees, the digger obtained a "claim,"—that is, he acquired the right of working an assigned portion of the soil. But if the claim had been left unworked for a week, it might be, in mining parlance, "jumped"—that is, any person might take possession of it, or jump into it, on procuring a proper licence.

Since the first rush of diamond-seekers to the river-banks, the stones were abundantly found elsewhere, namely, at the "dry diggings," where the soil, dug out with a pick or shovel, was sifted first through rough sieves, afterwards through sieves having fine wire meshes. The sieve, in such cases, was often suspended by thongs of hide between two upright poles, in the manner represented in Fig. 333. The miner was thus enabled to swing the sieve rapidly about, until the sand and dirt were separated, when the remaining gravel was emptied on the sorting-table in the manner before described. As the idea was formerly entertained that diamonds lie only on or near the surface of the soil, the early miners seldom penetrated more than a foot or two beneath the surface. But it was discovered that, so far from it being true that diamonds are present in superficial deposits only, the finest stones are met with at considerable depths to which no defined limit can be assigned; thus in sinking a well large diamonds have been found at 100 ft. below the surface. When these facts became known, many of the abandoned claims were worked over again down to a depth of 30 ft. or 40 ft. .

FIG. 333.—*Sifting at the "Dry Diggings"* (*c.* 1870).

The rapid rise of localities under such conditions may be illustrated by the case of Du Toit's Pan, which is the centre of a dry-digging district, and grew in a wonderfully short space of time from nothing to be a town having several large hotels, two churches, several public billiard-rooms, a hospital, and a theatre. In 1871 the *claims* at this place, each 30 ft. square, sold at prices varying from £1 to £50—the person who worked a claim paying also a small monthly sum for the licence. But those who were lucky enough to have obtained the first possession of the claims at another famous dry-digging locality, named Colesberg Kopje, at the cost of only the licence at 10s. per month, must have been still more fortunate, and have realized an enormous percentage on their investments ; for, four months afterwards the ruling prices at the last-mentioned place were £2,000 and £4,000 per claim. This great increase in value cannot be wondered at, if the accounts related of the value of the diamonds found here are true. For instance, it is stated that one individual, who just before the great rush had bought a claim for £50, found in it diamonds worth £20,000. Colesberg has become a populous town, with good buildings and regularly laid-out streets, while a great camp of tents and other temporary structures still surround it on all sides.

At all the towns above-mentioned newspapers were published, relating chiefly to matters interesting to the miners—giving, for example, lists of "finds," with the names of the lucky finders. It is curious that the term "diamondiferous" has, in these localities, come to be used as a general term denoting excellence of any kind. Thus, when it is desired to apply an epithet of superlative praise to a pickaxe or to a piece of furniture, this significant adjective is made use of ; and a salesman in the diamond-fields will not hesitate to speak of *diamondiferous* coats and trousers !

FIG. 334.—*The Vaal River, from Spence Kopje* (*c.* 1870).

It will be seen that the early diamond-seekers at the Cape followed very primitive methods, by simply washing in sieves the gravel and sand shovelled out of the river banks ; and indeed, it was only when, about 1871, they began to dig deeper that their working seems entitled to be called mining. The " dry-digging" operations began at the since famous Du Toit's Pan, by the circumstance of a Boer farmer finding to his great surprise diamonds sticking in the walls of his house, which had been built of mud. When the locality of this mud was examined by digging, more diamonds were found ; and when the excavation was continued downwards, still more. At this place and at four others, all within a circuit of less than four miles diameter, have been developed the richest diamond *mines* in the world, throwing into the shade the produce of all the river gravel washings ; and what is still more remarkable, showing no signs of exhaustion after nearly thirty years of working, but rather the contrary. The locality soon presented a scene of the most active industry, and it was not long before the town sprang up which has since become celebrated all the world over—Kimberley, the diamond capital. Kimberley is situated at the northern part of the British territory known as Cape Colony, not far within its boundary, and about 14 miles from the Vaal River, in Lat. 28° 43′ S., Long. 24° 46′ E. It lies in a north-easterly direction from Cape Town, at a distance of about 550 miles. When the existence of diamonds at the Cape became known, a great influx of strangers seeking fortune set in to a land that had failed to offer the attractions to colonists that America and Australia did. Before the establishment of the overland route opened a more direct way to India, China, and Australia, Cape Colony owed whatever importance it had to its position as a provisioning and coaling station for ships and steamers. As a British settlement it was little regarded, and its somewhat somnolent condition would have been deepened by the opening of the Suez Canal in 1869, had not the diamond discovery in that very same year brought about a great change. But the early diamond-seekers found their land of promise a wilderness without roads and without habitations, for the development of civilization did not then extend far from the coast. It is true, that here and there, at great distances apart, a few primitive missionary stations might be found, like that of Pniel shown in one of our cuts, which also represent the inhospitable aspect of the country. One cannot but admire the pluck of the adventurers, who, though unversed in their quest, encountered in its prosecution prolonged toils and many hardships. But they were young men, and their perseverance gained its reward. They came from all parts : from Britain, from America, from Australia, from Germany, even from Russia.

The finding at Du Toit's Pan, and at contiguous places, of diamonds at some depth below the surface of the soil, led to geological examinations of the district, which ultimately resulted in discoveries of the highest interest and importance, as will now be explained, with first a few words about the external features of the country.

A traveller directing his steps northward from the sea-shore at almost any part of the southern coast of Cape Colony will be faced by several successive ranges of mountains, or what will appear to be such, running more or less parallel to the coast, and of no great elevation. When he has reached the summits of these heights he will not find corresponding declivities on the northern side, but nearly level plains, bounded northwards by other similar ranges. Supposing him to set out at a point, say, 150 miles east of Cape Agulhas (the most southern point of Africa), he will, about 50 miles from the shore, have reached the top of the third of the great escarpments which rise up like the stages of a gigantic terrace, and having thus gained the ridge of the Black Mountains, he will see one of these almost level plains stretching before him a breadth of 80 miles, for the most part arid and inhospitable, with a much greater length east and west, and bounded on the north by a portion of the range of elevations that in an almost unbroken line runs through Cape Colony to Delagoa Bay nearly parallel with the coast, at a distance from it between 100 and 150 miles. This extensive plain is known as the *Groot Karoo* (Great Karoo),—*karoo* being the generic name for such plains in South Africa. After crossing the Great Karoo, our traveller, on mounting the last far-reaching step of the Brobdignagian staircase, may find himself on the summit of the Nieuveldt Mountains, at an altitude of nearly 10,000 feet above the sea-level, attained in several widely separated stages within a distance of 140 miles. From the summit of these elevations there is no descent by terraces northwards, but the high table-land or plateau stretches away for hundreds of miles, descending by only a gentle slope towards the Orange River, but maintaining an average altitude of nearly 6,000 feet, and extending far beyond the Orange River towards the Equator. Kimberley is situated about 50 miles north of the Orange River, and 4,042 feet above the sea-level.

It was soon observed that the Karoos had common geological characters, consisting in a certain series of shales, coal, limestones, etc., and this series naturally came to be called the "Karoo formation," just as we in England speak of the Wealden formation, etc.; and it was found that it extended over a greater part of Central South Africa, covering an area of at least 200,000 square miles, with an estimated thickness of 5,000 feet. The reader need not imagine that a boring nearly a mile in depth had to be made for the ascertainment of this last dimension, if he will remember what has been said in the last paragraph about escarpments of the rocks looking everywhere towards the coast. There is reason to believe that these beds were originally the sedimentary deposits of a vast fresh-water lake, or inland sea, far back in geological times. But here we need only concern ourselves with the development of the Karoo beds about Kimberley. There the ground is covered by a sandy soil of a red colour, for it contains much iron. Below this there is a layer of decomposed basalt, also containing much iron, its thickness varying from 20 to 90 feet. This lies upon a bed of very combustible shale, with carbon and iron pyrites, 250 feet thick, which from its great development here is known as Kimberley shale; then, after a conglomerate stratum 10 feet thick, is found a very hard compact rock, resembling hornblende, extending 400 feet downwards, and resting on another hard rock of quartz, also 400 feet in depth. These beds are *nearly* horizontal, but dipping a little towards the north. In speaking of them collectively we may use the local term of the miners and call them "the reef."

Now, there are a few certain spots near Kimberley, and two or three elsewhere, in which the strata forming "the reef" are not found, but something quite different. These may be compared to large dry wells, extending vertically downwards to unknown depths, which have been filled up with matters from below. They are called *pipes*, but they are uncommonly large ones; for though of a somewhat irregular circular or oval shape, their diameters range from 200 to 500 feet. Nor must it be supposed that the enclosing reef presents itself as a smooth wall, as the name "pipe" might suggest. These *pipes* are true diamond mines. They are believed to have been formed by an eruptive action originating from below at a great depth, and this was not by the escape of red-hot lava or other molten rocks, but by that of steam or other gases. It is known that the eruptive forces acted from below, for the edges of some of the strata are seen in places in the walls of the reef that surround the pipes to be turned a little *upwards*. It is known that the erupted matter was not molten lava or rock, for the shale and other strata show none of those changes of character near the pipes which would have resulted from igneous action, and for the same reason the gas or steam that escaped by these pipes could not have been highly heated. It must therefore have forced its way through the strata by enormous tension or pressure, and this either at one terrific outburst or possibly by the gradual enlargement of smaller volcanic chimneys. These blow-holes are filled with a mixture of subterranean débris, as if mud had been forced up from below, carrying with it an extraordinary variety of rock fragments and crystallized minerals. These are embedded in a mass of a bluish-green colour much resembling indurated clay (but nearly as hard as ordinary sandstone), and this on long exposure to the weather crumbles down to a yellow friable substance. More than eighty different kinds of minerals of the volcanic class have been found in this *breccia*, as it is termed by geologists, and it is remarked that these fragments could not have been exposed to any great heat, for their edges show no signs of fusion. There are also embedded in the agglutinating substance large masses of the surrounding strata, sometimes having an area of several thousand square feet, and these are called in miners' parlance "floating reef." The cementing material is named "blue ground," and the same when crumbled down by exposure is known as "yellow ground." These colours are due to oxides of iron, which in the unaltered ground give the blue-green tint, being lower oxides ; but are converted by absorption of oxygen into yellow and higher oxides. The upper part of the pipes is filled to a depth of about 70 feet with "yellow ground," produced by the penetration of atmospheric influences. Blue ground and yellow ground alike contain diamonds, and the yield of these is pretty regular at all depths in the same mine (some have been explored down to nearly 2,000 feet), although it varies considerably from one mine to another, and in some the east side is often richer than the west. Thus in one load (1,600 lbs.) of *ground* from Du Toit's Pan, in 1890, the quantity of diamonds found averaged less than 2 grains (0·5 carat), while Kimberley yielded 1·25 to 1·5 carats (5 to 6 grains). It is singular that the stones from mines quite close together are so distinctly different in character, that the Kimberley merchants can tell at once the source of any particular parcel. This would indicate that the blue mud was not forced up the several pipes at one and the same time, carrying with it diamonds from one birthplace.

The existence of the diamondiferous pipes is pointed out by no indication on the surface, which is covered nearly uniformly with the red sandy soil already spoken of ; although indeed the site of the Kimberley mine was marked by a slight elevation, and that of Du Toit's Pan by one of the depressions there called *pans*, which, at least in the wet season, are receptacles for surface water. The Wesselton mine, which was found only in the last decade of the century, about a mile from Du Toit's Pan, also showed a surface depression, and that had been utilised as a depositing place for dry rubbish. At a later period the "Leicester mine" was accidentally discovered 40 miles away. At Jagersfontein, in the Orange River Free State, 60 miles from the Kimberley mines, is another pipe which yields the finest diamonds of any, commanding prices nearly the double of those paid for the De Beers and Kimberley gems, being in fact their nearest commercial rival. The proprietorship of the Kimberley group having in 1889 become united in the hands of one company, known as the "De Beers Consolidated Mines," this company is able practically to control the diamond market, as it has sometimes turned out in a year as much as 3 million carats of diamonds, which sell for about £3,500,000. Up to the end of 1892, 10 tons of diamonds had been derived from these mines, representing a value of £60,000,000 sterling. In 1895, the De Beers Company sold diamonds to the amount of £3,105,958, the total expense of working for that year being £1,704,813, —the net profit was £1,401,145. The effect of consolidating all the Kimberley diamond interest into the De Beers Company has been to give an almost complete monopoly to this last, which has however found it advantageous to restrict its production to an annual output of about £3,000,000 in value, as the putting of a larger quantity of diamonds on the market would cause lowering of their price, and a diminution of the profits all round. The reason is, that though the world at large annually spends between 4 and 4½ million pounds sterling in the purchase of diamonds, yet it would not by a reduction in their price be induced to spend proportionately more. The company are sufficiently supplied by only two of their mines, the Kimberley and the De Beers, the expenses of working these being also relatively smaller than is the case with the others. It may be of interest to compare the quantities of diamonds that have so far been produced from the world's greatest fields, leaving out Borneo, the Ural Mountains, Australia, etc., as comparatively insignificant. Estimated produce of India, from the remotest period, 10 million carats; of Brazil (since 1728), 12 million carats; of South Africa, in only 19 years, 57 million carats.

At the time of the discovery of the Kimberley mine (July 1871) it was divided into about 500 claims, each 31 feet square, and between these were roadways across; but when the claims were excavated to a depth of 100 feet or more the roadways became unsafe, and, the "blue ground" underneath them being too tempting always to be left for their support, they began to fall in, and the mine was often threatened with ruin from this cause. The state of things became still worse when the unsupported walls of the "pipe" itself began to collapse, so that by 1878 a quarter of the claims were buried in the ruins of the reef. These falls continued, and although very large sums were year after year expended in removing the fallen reef, the cost amounting in 1882 to 2 million pounds sterling, it was found at last that very few of the claims could be regularly worked, and when in 1883 a tremendous fall of 250,000 cubic yards of reef took place, covering half the area of the mine, it became necessary to adopt another mode of working, namely, a regular system of underground mining. Vertical shafts were sunk at a considerable distance from the pipe itself, and tunnels from these carried through at different levels, with a system of galleries so arranged that all the "blue ground" is removable without danger to the miners. The whole mine is illuminated by electric lights, and the different kinds of labour are carried on by distinct sets of workmen, some of whom drill holes for the reception of dynamite cartridges, others shovel the material into trucks, others again wheel the trucks along tram lines, which converge to a space where their contents are discharged into skips holding four truck loads, in which they are hoisted to the surface at about the rate of 400 loads per hour. This goes on day and night, the miners working in three shifts of eight hours each. About 8,000 persons are employed, 6,500 of whom are blacks.

Fig. 334a is a sketch section of the Kimberley diamond mine, approximately to scale, and a glance at this will elucidate the foregoing description. The thick vertical and horizontal lines show the positions of the shafts and galleries that have at various times been excavated, the lowest gallery being connected with a shaft a considerable distance from the pipe, towards the right, but out of the range of the sketch. The fringed lines at the top, with dates, give some idea of the forms of the excavations until the final fall of reef that determined the resort to subterranean working.

When the "blue ground" has come to the surface, how are the diamonds to be extracted from the hard mass? how can a stone of a few grains weight be found amongst 1,600 lbs. of miscellaneous matter—a thing perhaps not larger than a peppercorn in four cubic feet of compact material? The "blue ground" is spread out on levelled and carefully prepared areas called "depositing floors," and there, after a few months' exposure, all but the very hardest pieces crumble down, the atmospheric action being accelerated by turning the material over with harrows, and by occasional waterings. The "blue ground" from the De Beers mine requires at least six months of this treatment, and it contains a certain

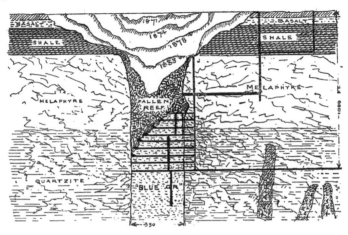

FIG. 334*a.*—*Sketch Section of the Kimberley Diamond Mine.*

proportion of refractory lumps that would not disintegrate in perhaps less than two years. These lumps are coarsely crushed between rollers, and the fragments are spread over slowly moving tables, from which any larger diamonds are picked off; the fragments left go through smaller crushers, and are subjected to still greater concentration. The depositing floors of the De Beers mine are laid out as rectangles, 600 yards long by 200 yards wide, each holding about 50,000 loads. They occupy several square miles, and as the "blue ground" spread upon them is always one of the most valuable assets of the company, the quantity of it forms an important item in the balance-sheets, and the amount that can be realized from it can be estimated with sufficient closeness, on account of the nearly uniform distribution of the diamonds. Thus in June 1895, the 3,360,256 loads then on the floors were put down as equivalent to nearly 1 million pounds sterling. When the "ground," thoroughly weathered, has become yellow and friable, it is transferred to the washing machinery, by which about 99 per cent. of the original non-diamondiferous material is removed, and, thus concentrated, the gravel is together with the mechanically crushed material submitted to the action of a machine called the *pulsator*, where the gravel is first assorted into sizes by being turned about within an inclined iron cylinder perforated with several stages of round holes of diameters successively of 2, 3, 4 and 6 sixteenths of an inch. The pieces that are too coarse to pass through the largest holes are taken to the *sorting house* direct; but the stones that have passed through the cylinder drop according to their sizes into four separate sieves called at Kimberley *jigs*, from the well-known mining term *jigger*, applied to a man who washes ores in a sieve. The several jigs into which passes the now assorted gravel have screens with meshes corresponding to the holes in the cylinder ; and by a very ingenious arrangement the concentration is carried to the point at which the diamonds can be individually picked out. The "jigs" themselves do not move, but all over the meshes of the screen is spread a layer of leaden bullets, which prevent a too rapid passage through the screens, while the material is kept moving in water, by that liquid *pulsating* or emerging in quickly succeeding gushes from below the meshes, and thus carrying off the lighter matters, while those of greater specific gravity, including the diamonds, work their way downwards between the bullets and through the meshes, and are received in boxes which are periodically carried to the sorting house.

When the now much concentrated diamondiferous gravel reaches the sorting house, the remaining operation consists merely in picking the diamonds out. But simple as this operation is, it has to be conducted systematically. In the sorting house are long tables covered with plates of iron, and placed in a good light. Upon these is thrown the wet gravel, but not promiscuously ; the different sizes being set apart, the sorter spreads out the heap before him with a flat piece of zinc, picks out the diamonds and drops them into a small box. Only white men in whom confidence can be placed are allowed to deal with largest sized material, for this offers the strongest temptation to purloiners, as in this of course the most valuable stones are met with. This material, after the first search, is submitted to the scrutiny of another person, to see that no diamond has been overlooked ; but the smaller assortments are examined by blacks, who are closely supervised by white men. The value of the diamonds occasionally sorted out in a single day may reach £10,000.

At the diamond mines little trust is reposed in the honesty of the blacks. Below ground and above ground they work under the constant surveillance of white men, and they live in "compounds" which are spacious areas—perhaps of 20 acres in extent—enclosed by lofty iron fences, and containing long rows of corrugated iron erections divided into rooms, each appropriated to a score of natives. Food, etc., is supplied from a store at less than ordinary prices, and the company find fuel and water gratis, and provide a well equipped hospital and medical attendance. There are swimming baths, and ample recreation grounds for dancing, etc. The natives of each of the many tribes keep by themselves apart, and follow their own fancies. They receive good wages, and some of them save money. They are not allowed out of the "compound" or the mine, except to work on the depositing floor, which they do under guard. They accept their restrictions voluntarily, making agreements for a certain term, three months being the least. Those who leave, as many do to spend their earnings, often "not wisely but too well," usually return. The depositing floors are surrounded by fences 7 feet high, unscalably and impenetrably armed with barbed wire ; and as here robbery would have the readiest chance, where the largest stones might be met with, extraordinary precautions are taken, watch and ward being maintained by day and by night. Not more vigilantly did Cerberus keep the entrance of Pluto's domain, nor the wakeful dragon guard the golden apples of the Hesperides, than the patrols observe the depositing floors. At night powerful electric searchlights are made to play across the enclosures, so that unauthorized movements can scarcely escape detection. Besides these provisions against theft, the laws of the

Colony prohibit any attempt at illicit dealing in diamonds, under a penalty of two years' penal servitude.

The maximum penalties for contravention of the Diamond Laws are, however, much more severe, and that to an extraordinary degree. Thus any unlicensed dealer is liable to a fine of £1,000, or fifteen years' imprisonment, or both. And the authorized dealers are required to keep a most minute record of all their transactions, to send a copy of it every month to the head of the police, and to produce it when required. It is needless to say that extraordinary precautions are taken to prevent the native workmen from secreting diamonds. And any person even finding a diamond, and neglecting to report the circumstance to the proper quarter at once, is liable to the pains and penalties above mentioned.

The "blue ground" was at first supposed to be the original home of the diamond, within which it had somehow taken its shape. But no satisfactory explanation was forthcoming as to the state of the carbon before its solidification into the crystalline form. The more general opinion has been in favour of a volcanic origin due to very high temperature ; and although the "blue ground" itself is clearly not the ordinary erupted matter of volcanoes due to igneous fusion, the geology points to the district having been the scene of very active and extensive volcanic energies at more than one remote period, for the bed of the Karoo inland sea has been several times covered by level sheets of molten matter extruded somewhere from below ; but not through the "pipes," which were blown out ages afterwards. The strata of basalt and of hornblendic mineral, which extend horizontally over great areas in the Karoo formation, are of igneous origin, as are also some nearly vertical dykes of trap rock, about 7 feet wide, that are found traversing the "blue ground" in certain directions. These intrusive dykes are of course more recent than the formation of the blue ground, and that is itself later than the production of the pipes. The fact of many fragments of crystals being found in the "blue ground" does not comport with the theory that supposes it to be the matrix ; and besides this, many of the diamonds show scratches, and as these are producible only by other diamonds, it would appear that they must all have travelled in company, some part of their journey at least.

Carbon in any form is quite infusible at the highest temperature we have hitherto been able to produce, although an incipient softening under the influence of the electric arc has been suspected. Professor Dewar, an English chemist, basing his data on analogies with other substances, and on purely theoretical grounds, has calculated that the melting temperature of carbon is near 3,600° C. (6,512° F.), and that it cannot remain in a liquid state at a temperature exceeding 5,527° C., when its vapour would have a tension equivalent to a pressure of 15 tons on the square inch. So far as these deductions are correct, both the melting point of carbon and the boiling point of its liquid must lie within the range of temperature expressed by 3,600° C. and 5,527° C. The most intense heat we can produce is that developed in the electric arc discharge, and an eminent French chemist and metallurgist, M. Moissan, by employing special arrangements and very powerful currents, has thus been able to obtain in his "electric furnace" a temperature estimated at 3,500° C., which nearly approaches the lower of the above-mentioned limits, and he has thereby produced many new and unexpected chemical combinations of refractory elements. Among the most striking of his results is the formation artificially of real crystalline diamonds. He found that carbon is freely dissolved by several of the metals in fusion at the temperature of the electric furnace. When the carbon separated from the metals, as they cooled and became solid, it was always in the condition of graphite. The carbons of the electric poles were readily attacked by molten iron, and it was from the solution of carbon in iron that Moissan prepared his diamonds. The fact of carbon thus combining with iron was of course no discovery, as the reader already knows ; and the resulting combination was found, on allowing the metal to cool, to be simply cast iron, the greater part of the carbon separating out in the graphitic form. But M. Moissan, having studied the conditions of the Kimberley mines, and recognizing the probability of the diamonds having taken their origin at very great depths, where the pressure due to the weight of superincumbent strata would be immense, was struck with the idea of *pressure* being in some way a factor in their formation ; and it occurred to him that the carbon might separate from its liquid condition in the iron in the crystalline, and not in the graphitic form, if the solidification could be effected under great pressure. The apparently insurmountable difficulty of applying an enormous pressure to a small quantity of molten iron (half a pound) yielded to the experimenter's ingenuity. He took advantage of the circumstance that cast iron at the moment of solidification expands, a property upon which depends its use for many purposes. If then the fused mass were suddenly cooled on the outside, we should have a shell of solid iron enclosing a nucleus of still fluid metal, which, on cooling in its turn, would tend to expand, and by so doing would exert a great pressure within the shell by which it was confined. At first Moissan plunged his glowing crucible into cold water, but a method of more rapidly cooling it was to immerse it in melted lead. It seems a strange proceeding to cool the crucible by surrounding it with hot metal, yet the *difference* of the temperatures was sufficient to produce the desired effect, the cooling contact of water not really operating on the intensely heated body, which becomes separated from the liquid by a coating of *steam.* When the mass of iron was dissolved off, diamonds of all kinds were found in the residue, and, though extremely small, some crystals were perfect in shape and colour ; every variety that occurred in the mines being found reproduced in tiny size. There was also some graphite in the residue. Many more crystals of "pure water" were obtained by the lead-cooling than by the water-cooling, as the former process gave some flawless cubes and octahedra. The largest of the set was only $\frac{1}{50}$ inch across, and although of perfect form when first extracted, within the course of three months it had spontaneously split up into fragments.

There was evidently no danger of M. Moissan's manufacture of diamonds from coke causing consternation at Kimberley ; though it would not be without interest to speculate upon the consequences had the French *savant* achieved the greater triumph of turning out carbon crystals in every respect equal to the productions of nature's own laboratories. What a drop there would have been in the shares of the De Beers Mines Consolidated ! What heaviness of heart would have fallen upon those great ladies who exult in the exclusive possession of priceless tiaras and precious necklaces flashing with the resplendent gems ! From a scientific point of view, M. Moissan's fabrication of even those minute crystals, which so soon spontaneously crumbled into fragments, is a distinct and valuable success ; for, notwithstanding their diminutive size and instability, they show us that art has so far succeeded in imitating the processes of nature, that some of her secrets have been revealed. Though we know the exact chemical composition of all kinds of crystallized minerals, very very few of these have we been able to imitate artificially. Nor is this to be wondered at ; for nature's resources are immense compared with ours : she can command temperatures unlimited by which to form her solutions or liquefactions ; prodigious pressures to keep them close ; and time immeasurable—geological time—in which to let them cool, and their particles freely coalesce into geometric forms. Human agency, being obviously unable to reproduce, even on the smallest scale, such conditions as attended the deposition and slow cooling of the earth's crust, may not hope to rival the products of the planet's prime. So the fair owners of the earth-born gems may possess their souls in peace, free from any fear of the chemists' crucibles ; and the Kimberley Diamond Companies are not likely to suffer panics from the results of scientific researches, and probably will continue to pay their handsome dividends for time indefinite.

But curiously enough, a discovery of the latest years of our century has revealed the existence of diamonds in a region not mapped by the most advanced of geographers—a region which indeed cannot be defined by degrees of latitude and longitude. In the recesses of an unquestionable meteorite—one of those celestial lumps of iron of which mention has been made in the earlier pages of this volume—real diamonds have been found. These quite resembled the products of M. Moissan's experiments, being extremely small, but including clear and perfectly shaped crystals, associated with black ones, and also with much graphite in more or less definite forms. So very limited, however, could be the quantity of diamonds obtainable from this hitherto unsuspected source, that even if they rivalled in quality the finest stones from the South African mines, it might be difficult to form a " Company" for their *exploitation*. Still, there is the possibility of some one falling in with a little meteorite containing some mature full-sized carbon crystals, and such a one might be considered equally fortunate with the finder of the famous Australian nugget " Welcome " (£25,000). The association of diamonds with the ferruginous matter of the "blue ground" in the Kimberley pipes, their crystallization out of iron in M. Moissan's experiments, and their presence in iron meteorites, would seem to point to special relations between the two elements, iron and carbon. Some of these relations are exemplified in another way by the profound modification effected in the physical properties of iron, by its combination with a very small quantity of carbon, as in some kinds of steel ; or again, by the differences between white cast iron and grey cast iron, as determined by the condition of the carbon in each.

FIG. 335.—*Portrait of Sir Humphrey Davy.*

NEW METALS.

THE chemistry of the nineteenth century can boast of a series of discoveries more brilliant and more numerous than ever belonged to any other science within a like period. And the advantage to the world must have been great, for chemistry more directly than any other branch of knowledge ministers to the useful arts and promotes the comfort and well-being of society. The science itself, as it now exists, is almost the creation of the present age. But its recent developments cannot be here discussed; nor, of the immense number of new products with which it has enriched the world, can more than a very few be brought under the reader's notice in the remaining pages of the present work. Among the most striking of the remarkable series of discoveries by which Sir Humphrey Davy penetrated the mysteries of matter was the isolation of the alkali metals—a circumstance which marks an important era in the history of chemistry. That the alkalies were oxides of unknown metals had indeed been previously surmised by chemists, from the fact of their behaving like metallic oxides in neutralizing and combining with acids to form the class of compounds called *salts.* All attempts to decompose these alkalies had proved

714

fruitless until Davy separated the metal *potassium* from potash, in 1807. When, however, this alkali had once been proved a compound, more correct ideas were introduced into chemical science; the nature of other alkalies and earths was explained in like manner, and new and powerful re-agents were placed in the hands of the chemist.

Davy first obtained potassium by exposing to the action of the voltaic current a fragment of potash which had become moist on the surface by exposure to the air. The battery was formed of the then unprecedented combination of two hundred pairs of 6-inch plates on Wollaston's plan, which was constructed for the Royal Institution of London. The heat produced by the passage of the current fused the potash, and globules of metallic potassium were separated at the negative wire. This method yielded the metal in very small quantities only, and at a great cost. Gay Lussac and Thenard soon afterwards found that potassium could be obtained more cheaply and in greater abundance when fused potash was made to flow over iron-turnings heated to whiteness in a gun-barrel, and the hydrogen and potassium vapour were passed into a cooled receiver, in which the latter body was condensed. The metal is now obtained by heating potassium carbonate with charcoal. For this purpose it suffices to heat crude tartar in a covered vessel from which air is excluded. The tartar is first calcined in a crucible until all combustible vapour has been driven off. The charred mass, which now consists of potassium carbonate mixed with finely-divided carbon, is then broken into lumps and quickly introduced into a wrought-iron retort, which is heated in a furnace to nearly a white heat. A receiver in the form of a flat iron box, 12 in. long, 5 in. wide, and $\frac{1}{4}$ in. deep, is adapted to the neck of the retort, and is kept cooled by the application of a wet cloth on the outside. The potassium thus obtained is not pure, and it must be distilled in an iron retort, as otherwise a powerfully detonating compound is apt to be formed by a portion of the metal combining with carbonic oxide.

Immediately after his discovery of potassium Davy obtained sodium in the same manner, and Gay Lussac and Thenard also procured it by the same process they used for the sister metal. Sodium is now extracted on the manufacturing scale for use as an agent in the reduction of two other metals, of which we shall have to speak. A mixture of dried sodium carbonate, powdered charcoal, and chalk is heated in wrought-iron cylinders, about 4 ft. long, 5 in. internal diameter, and $\frac{1}{2}$ in. thick. The chalk takes no part in the chemical action, but is added in order to give the sodium carbonate when it fuses a pasty consistence, and thus prevent the separation of the charcoal. A number of these iron cylinders are set in a reverberatory furnace; but they are coated with fire-clay and enclosed in earthenware tubes, to prevent their destruction by the intense heat. To one end of each cylinder a receiver is adapted, of the form and dimensions already described for potassium. The other extremity is closed by an iron plug, luted with fire-clay. When the charge in a cylinder is exhausted, a fresh one is introduced by removing the plug, taking out the residue, and inserting a new supply of the mixture made up in a canvas bag. The operation is therefore continuous, and the metal obtained is nearly pure, as sodium does not exhibit the same tendency as potassium to form compounds with carbonic oxide.

Potassium and sodium are extremely soft metals ; they are lighter than water, upon which they float, at the same time rapidly decomposing that compound by displacing half the hydrogen, which is set on fire by the heat.

The instant a piece of potassium touches the surface of water, a violet flame bursts forth ; but with sodium no flame appears unless the metal is dropped on warm water, or prevented from swimming about. Since these metals are thus capable of displacing hydrogen from its combination with oxygen at ordinary temperatures, it follows that they must have a powerful affinity for oxygen ; and, indeed, they can only be preserved in rock oil, for they rapidly combine with the oxygen of the air. The great attraction of these metals for oxygen, and for chlorine and other similar bodies, induces the chemist to employ them for separating such bodies from their combination with other metals. Sodium is generally employed for this purpose, as being far cheaper than potassium.

Among the sixty-nine elementary or undecomposable substances which, variously combined, constitute the whole material of our planet, so far as we are acquainted with it, no fewer than fifty-six are metals. Of these fifty-six metals very few are found in a free or uncombined state, like the gold described in the last article. On the contrary, the whole of the metallic elements of the globe, with insignificant exceptions, exist in nature in a state of combination with one or more of the other thirteen non-u etallic substances. In this condition they form the stony masses which are termed the ores of the more common metals, and they constitute also the earths, the metallic bases of which were, until recent times, unsuspected and unknown. Davy followed up his discovery of the metals of potash and soda by experimental demonstrations that the earths *alumina, magnesia,* and others, were really oxides of metals ; and when the nature of these substances had once been established, chemists soon devised means for readily obtaining their metallic bases in an isolated form. The new metals which have been thus isolated all deserve the attention of the chemist ; and the general reader will probably also regard with interest the processes by which two of these new metals, for which practical applications have been found, are extracted, and the properties which have caused them to be produced on the commercial scale. These are *aluminium,* the metallic base of common clay ; and *magnesium,* the metallic base of common magnesia, and Epsom salts, and a constituent of dolomite, or magnesian limestone.

Aluminium was first isolated by Œrsted, in 1827, by decomposing its chloride by means of potassium. The chlorine leaves the aluminium to combine with the potassium, and thus the former is set free. Wöhler effected some improvements in Œrsted's process, and he first obtained the metal in malleable globules. It is, however, to Deville that we are indebted for the invention, in 1854, of a process which admitted of application on a manufacturing scale. He obtains chloride of aluminium by mixing alumina (the oxide of the metal) with powdered charcoal made into a paste with oil, and heating the mixture in a tubular earthenware retort, like those sometimes used in the manufacture of coal-gas, while a current of dry chlorine is made to pass through the vessel. The charcoal combines with the oxygen, forming carbonic oxide, a permanent invisible gas ; and the aluminium unites with the chlorine, giving rise to aluminium chloride, which, being volatile, sublimes into a chamber lined with glazed tiles, in which it condenses as a yellow translucent mass. The metal is reduced from the chloride in the following manner : A tube of hard glass, about an inch and a half in diameter, is placed over a furnace, or chaffing-dish, as shown in Fig. 336, where D C is the tube, and G G an iron pan for containing the red-hot charcoal. Into the part of this tube marked E, about half a pound of dry aluminium chloride has previously been introduced, and is kept in its place by plugs of asbestos. A current of dry hydrogen gas, perfectly free from air, is passed through the tube ; the gas being generated in the vessel, A, and in B passed over some substance which removes from it all moisture. The aluminium chloride is then gently heated by placing red-hot charcoal beneath it, so that any hydrochloric acid it may contain may be expelled. A long narrow porcelain tray, or "boat," containing pieces of sodium, F, is then introduced into the tube ; and, the current of hydrogen being still maintained, heat is applied to the part of the tube containing the sodium, and the aluminium chloride is made to distil over by a regulated heat. As it passes over the sodium, it

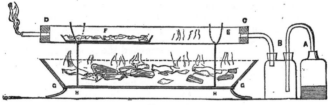

FIG. 336.

is reduced with a vivid glow. The aluminium is set free, and collects in the tray with the double chloride of sodium and aluminium which is produced by the reaction. The tray is removed and more strongly heated in a porcelain tube through which a current of hydrogen is passing, and the metal is thus obtained in globules.

Messrs. Bell, of Newcastle, undertook the manufacture of aluminium by a system founded on this process. The first step is the preparation of pure alumina, which may be obtained by igniting ammonia alum, or by precipitating from a solution of alum free from iron, or from sodium aluminate made from the mineral called *bauxite*. The precipitate of hydrated alumina, mixed with charcoal and common salt, is made into balls and dried. These balls, which are about as large as an orange, are placed in upright earthenware retorts, which are heated to redness, while a current of dry chlorine is passed through them. The volatile double chloride of aluminium and sodium distils over, and is condensed in chambers lined with earthenware. This substance is mixed with powdered fluor-spar, or with *cryolite* (itself a compound of aluminium), which serves as a flux ; and small pieces of sodium are interspersed throughout the mixture. The proportions are ten parts of the double chloride, five of fluor-spar, and two of sodium. This mixture is thrown upon the hearth of a reverberatory furnace, and the doors are shut to exclude air. A very intense action occurs: the chlorine, quitting the aluminium, seizes on the sodium, and their combination is attended by an enormous increase of temperature. The fused aluminium is run off from the furnace together with the slags which are produced by the operation. In this way, with a furnace having a hearth 16 ft. square, about 16 lbs. of aluminium can be obtained in one operation.

Rose, the eminent German metallurgist, prefers to obtain aluminium from cryolite, which is a compound of sodium, aluminium, and fluorine, found in large quantities in Greenland. It is powdered and mixed with common salt, and with the mixture a certain quantity of sodium cut into small pieces is uniformly mingled. The whole is thrown into a heated crucible, previously lined with a fused mixture of cryolite and salt, and more of the same mixture is poured upon the contents of the crucible, which is then covered and exposed to a red heat for two hours. The aluminium generally collects into buttons, which may be easily melted together by heating them in a crucible with common salt.

It will be obvious, from the preceding account of the processes of extracting aluminium, that the cost of the metal must depend upon that of sodium ; and the same remark will apply to the case of magnesium. It is interesting to observe how the price of the alkaline metals has decreased as improved processes have been devised, and as the scale of production has increased with the commercial demand for the article. Prepared by Gay Lussac and Thenard's process, these metals were produced in but small quantities, and were sold at £5 per oz. When the mode of reducing them by charcoal came into operation, the price fell to 30*s.* per oz.; and the researches of Deville so far improved the processes, that in 1854 sodium could be procured for 5*s.* per oz. Mr. Gerhard, of Battersea, subsequently manufactured sodium, so that it can now be retailed at less than 1*s.* per oz. The price of aluminium before Deville's investigations was about 24*s.* per oz., but now the metal can be purchased at about one-eighth of that cost. [1875.]

Aluminium is a white malleable metal, in colour and hardness not unlike zinc. Its colour is not so white as that of silver, as it has a marked bluish tint. It can be rolled into very thin sheets, and by rolling it becomes harder and more elastic. It can also be drawn into fine wire. It is remarkably sonorous, and a suspended bar gives out a clear musical note when struck. Perhaps no property of aluminium more strikes a person, who examines the metal for the first time, than its lightness. It is, in fact, only two and a half times as heavy as water, while zinc is seven times, silver ten and a half times, and gold more than nineteen times as heavy as water. It retains its lustre in dry or in moist air for any length of time, and at all ordinary temperatures. It is not acted upon by nitric or sulphuric acids, but is attacked by hydrochloric acids and by alkaline solutions with great energy. It has great rigidity and tenacity, and can be turned, chased, and filed with the greatest ease, and without clogging the tools. In the Paris Exhibition,[*] M.Christofle showed spoons and forks and a cup made from it ; and it may be mentioned, as showing the hardness and strength of the metal, that the cup could be allowed to fall on a stone pavement without being indented. The metal gives a good impression by casting ; and by striking under a die, some admirable medals have been produced in it. Aluminium has hitherto been chiefly used for ornamental articles, and for purposes where lightness and rigidity are desirable, such as in the tubes of telescopes, opera glasses, beams of balances, &c. Its unalterability and admirable working qualities have also caused it to be used for cheap trinkets and ornaments—such as watch-cases, bracelets. combs, seals, penholders, candlesticks, &c. It is, however, incapable of receiving the lustrous polish of silver, as it has a decidedly blue tint, so that it will probably never replace silver for ornamental plate ; but it would be a good material for egg and mustard-spoons, as it is quite unaffected by the sulphur compounds which so readily tarnish silver. It has been suggested that if aluminium could be procured cheaply enough, " its hardness, lightness, and incapability of rusting would render it admirably adapted for the helmets and cuirasses of the cavalry; it would make splendid field-guns, as strong as the present ones, and not one-third of their weight; and, in sheets. it might serve as an incorrodible roofing, far lighter and more durable than zinc. It would admirably replace copper, if not silver, for the purpose of coinage. A crown-piece in aluminium would hardly weigh more than a shilling in silver, or a piece the size of a penny about as much as a copper farthing. The same qualities of lightness, hardness, and incorrodibility also excellently fit it for the beams of delicate balances, and for the small weights used by the analytical chemist. It would make admirable utensils for the more delicate operations of cooking—replacing the copper ones, which render pickles and soups so poisonous. It is extremely sonorous, and would make capital bells."

Some difficulty in working the metal has occurred from the want of any suitable solder. This difficulty has been overcome by electrolytically coating the metal with copper at the place where it has to be united with others, and then soldering the copper in the ordinary manner. Aluminium readily forms alloys with copper, silver, and iron. The alloys with copper vary in colour from white to golden yellow, according to the proportion of the metals. Some of these alloys are very hard and possess excellent working qualities. The alloy of copper with 10 per cent. of aluminium, which is called *aluminium bronze*, has been manufactured by Messrs. Bell in considerable quantities. It is made by melting a quantity of very pure copper in a plumbago crucible, and when the crucible has been removed from the furnace, the solid aluminium is dropped in. An extraordinary increase of temperature then occurs: the whole mass becomes white hot, and unless the crucible be made of a highly refractory material, it is fused by the heat developed in the combination of the two metals, although it may have stood the heat necessary for the fusion of copper.

The qualities of aluminium bronze have been investigated by Lieut.-Col. Strange, who finds that the alloy possesses a very high degree of tensile strength, and also great power of resisting compression. Its rigidity, or power of resisting cross strains, is also very great; in other words, a bar of the alloy, fixed at one end and acted on at the other by a transverse force tending to bend it, offers great resistance,—namely, three times as much as gun-metal. An advantage attending the use of the alloy for many delicate purposes is found in its small expansibility by heat; it is therefore well adapted for all finely-graduated instruments. It is very malleable, has excellent sounding properties, and resists the action of the atmosphere. It works admirably with cutting tools, turns well in the lathe, and does not clog the files or other tools. It is readily made into tubes, or wires, or other desired forms. The elasticity it possesses is very remarkable; for wires made of it are found to answer better for Foucault's pendulum experiment than even those of steel. These admirable qualities would seem to recommend the alloy for many applications in which it might be expected to excel other metals. It appears, however, that the demand for it has not met the expectations of the manufacturers, and the production has been somewhat diminished of late, although it is used to some extent for chains, pencil-cases, toothpicks, and other trinkets. When more than 10 per cent. of aluminium is added to the copper, the alloy produced is weaker; and if the proportion is increased beyond a certain extent, the bronze becomes so brittle that it may be pulverized in a mortar.

The metal *magnesium* was first prepared, in 1830, by the French chemist Bussy, by a process similar to that by which Deville obtained aluminium.

Bussy heated anhydrous magnesium chloride with potassium in a porcelain crucible; and when the vessel had cooled, and the soluble residue had been dissolved out by water, the metal was found as a grey powder, which could be melted into globules. The recognition of the metal as the base of magnesia is, however, due to Davy. About a quarter of a century after Bussy's discovery Deville having shown that sodium could be substituted for potassium in such reductions, the metal became more cheaply producible, and soon afterwards Bunsen and Roscoe pointed out its value as a source of light. Mr. Sonstadt devoted himself to the elaboration of a method of working Deville's process on the large scale, and he succeeded in establishing a company in Manchester for the manufacture. The process as carried on at the company's works in Salford is thus described in the " Mechanics' Magazine," 30th August, 1867:

"Lumps of rock magnesia (magnesium carbonate) are placed in large jars, into which hydrochloric acid in aqueous solution is poured. Chemical action at once ensues : the chlorine and the magnesium embrace, and the oxygen and carbon pass off in the form of carbonic acid. The result is magnesium in combination with chlorine, and the problem now is how to dissolve this new alliance—to get rid of the chlorine and so obtain the magnesium. First, the water must be evaporated, which would be easy enough if not attended with a peculiar danger. To get the magnesium chloride perfectly dry it is necessary to bring it to a red heat; but this would result in the metal dropping its novel acquaintance with chlorine and resuming its ancient union with oxygen. To avert this re-combination, the magnesium chloride whilst yet in solution is mixed with sodium chloride (*i.e.*, common salt), and thus fortified, the aggressions of oxygen whilst drying are kept off. The mixture is exposed in broad open pans over stoves, and when sufficiently dry, the double salt is scraped together and placed in an iron crucible, in which it is heated until melted, whereby the last traces of water are driven off. It is then stowed away until required in air-tight vessels, to prevent deliquescence. Here comes in that curious metal, sodium, also discovered by Davy. Five parts of the mixed magnesium and sodium chlorides, mingled with one part of sodium, are placed in a strong iron crucible with a closely-fitting lid, which is then screwed down. The crucible is heated to redness in a furnace, and its contents being fused, the sodium takes the chlorine from the magnesium. When the crucible has been lifted from the fire and allowed to cool, the lid is removed and a solid mass is discovered, which, when tumbled out and broken up, reveals magnesium in nuggets of various sizes and shapes, bright as silver."

The crude metal also presents itself in the crucible as small grains, and even as a black powder. The whole is carefully separated from the refuse; it is purified by distillation in a current of hydrogen gas; and it is afterwards melted and cast into ingots. Magnesium is a very light metal, its specific gravity being only 1·743; that is, it is only one and three-quarter times heavier than water. When heated in the air it takes fire, and is rapidly converted into the oxide, magnesia. In the form of wire or of narrow ribbon, it burns easily in the air, producing a light of dazzling brilliancy, which among artificial modes of illumination is rivalled only by the electric light. This is the chief use at present made of the metal. Lamps have been contrived for burning the wire in such a manner as to obtain a steady light, the wire being pushed forward at a regulated rate by clockwork. The magnesium light is rich in the rays which act upon sensitive photographic plates, and it has been successfully employed in obtaining photographs of dark interiors, such as vaults or caverns, and for the exploration of mines and other dark places. The brilliancy of the firework displays which can be produced by magnesium far surpasses that obtainable by any other material used by the pyrotechnist. In such exhibitions balloons are sent up having burning magnesium attached to them; and the metal in the state of filings is also mixed with other materials. But magnesium is still a very costly metal, and while the firework-makers find it too expensive for common use, they complain that its brilliancy in occasional displays dulls by contrast the effect of the ordinary fireworks, with which the spectators are no longer satisfied.

Magnesium wire is not produced by drawing, as the metal is not ductile. The wire is formed by a method identical with that used in the fabrication of the leaden rope for making bullets (p. 330); that is to say, the metal is forced in a heated and softened state through a small opening in an iron cylinder. The intensity of the magnesium light has been measured by Bunsen and Roscoe. They say that 72 grains of magnesium, when properly burnt, evolve as much light as 74 stearine candles burning for ten hours, and consuming 20 lbs. of stearine. Lamps in which magnesium may be steadily burnt are made by Mr. F. W. Hart, of London. In the more elaborate forms of these lamps, there are springs and wheels for pushing forward the magnesium ribbon, or a strand of magnesium wire, into the flame of a spirit-lamp; while at the same time the magnesium wire is made to revolve on its axis, in order to overcome its tendency to bend down, which would be a great disadvantage when the light is used for optical apparatus. But for ordinary purposes a much simpler arrangement suffices: the magnesium ribbon or wire is coiled on a drum, from which it is drawn off by passing between two little rollers, which are turned by hand. The wire or ribbon is drawn off the drum by the rollers, and pushed forward through a guiding tube, which brings it into the apex of the flame of a spirit-lamp. In this simpler form of lamp the rate is, of course, directly dependent on the person who turns the winch of the feeding-rollers; but in the automatic lamp there are appliances for adjusting the rate; the suitable speed must be first found by trial, and then the apparatus is to be regulated accordingly. By means of these lamps photographs can be taken as quickly as with sun-light, on account of the abundance of chemically-active rays given out by the burning magnesium. It has been found that an equivalent of magnesium, in combining with oxygen, liberates a larger amount of heat than the equivalent quantity of any other metal, not excluding even potassium. Magnesium forms alloys with several other metals, such as lead, tin, mercury, gold, silver, platinum. All these alloys are brittle, and have a granular or crystalline fracture. They are too readily acted on by air and moisture to be of any service in the arts. The alloy of 85 parts of tin with 15 of magnesium is hard and brittle; its colour is lavender, although both constituents are white, or nearly so.; and it decomposes water at ordinary temperatures. Both metallic magnesium and aluminium furnish useful re-agents to the scientific chemist. The latter metal, when fused, dissolves boron, silicon, and titanium, and on cooling deposits these elements in the crystalline form, this being the only known process for artificially preparing them in the crystalline state.

Since the above paragraphs were written, the price of sodium has been further greatly reduced, and it can now (1890) be purchased in bulk at about 4s. per lb. This cheapness has brought the substance into use for the reduction of other metals and one consequence has been a great fall in the price of aluminium. At Salindres, in France, the process of obtaining this metal that has been described on page 587, has been in use for many years, during which considerable quantities of aluminium have been produced, the output for 1882 being stated as 5,280 lbs. Aluminium has lately been prepared by a company at Wallsend-on-Tyne from *cryolite*, a mineral which is found only in Greenland, but occurs there in great abundance. Cryolite is a double fluoride of aluminium and sodium, and the processes for its reduction consist in fusing it with common salt in a reverberatory furnace, drawing off the mixture into an iron vessel, and stirring into the fused mass a certain quantity of sodium. This produces a violent action, on the cessation of which the slag is poured off, and the metallic aluminium is found as a "button" at the bottom of the converter. For obtaining a purer metal, the fusion is made in crucibles, and the sodium is added in two operations without removing the crucible. The yield of aluminium is about 8 per cent. of the weight of cryolite, and three parts of sodium are required to furnish one part of aluminium. Another large manufactory of aluminium is in operation at Oldbury, near Birmingham. There is a special difficulty in the metallurgy of aluminium, arising from the fact of the qualities of the metal being much deteriorated by the presence of a very small amount of foreign matters such as iron, silicon, &c., at the same time that no process has been found for purifying the product from these substances. If the aluminium is to be pure it must be so prepared at the first. Electrolysis has been proposed as a means of reducing the compounds, and obtaining the metal free from admixtures. Experiments seem to show that the dynamo-electric machine may be applied to this purpose, as well as to the reduction of sodium compounds, when certain practical difficulties arising from the chemical energies of the liberated substances have been overcome. What is called the "electric" furnace has been successfully used in the production of aluminium bronze. It is a rectangular iron box, 5 feet long, 1 foot deep, and 15 inches wide, with electrodes formed of rods of carbon 30 inches long and 3 inches in diameter. It is charged with a mixture of 25 parts of corundum (native crystallized oxide of aluminium), 12 parts of carbon, and 50 parts of granulated copper. This is covered at the top by lumps of charcoal, and a lid is fastened over the whole. The current from a powerful dynamo is sent through the carbons, and in about ten minutes the copper is melted, when the electrodes, at first only a few inches apart, are moved to an increased distance, and the strength of current increased. The corundum is reduced, the aluminium alloying itself with copper, and the oxygen combining with the carbon to form carbon monoxide, which is driven off. The resulting alloy is cast into ingots, its percentage of aluminium ascertained, and then it is melted with enough copper to produce aluminium bronze (page 719). The price of aluminium, which was as already stated about 3*s.* per ounce in 1875, has been so much reduced that the metal may now (1890) be purchased for 11*s.* or 12*s.* per lb. We may therefore expect to see wider applications of its excellent qualities. Though the price per lb. is still much higher than that of copper—22 or more times as much—the metal is so much lighter that a lb. of aluminium occupies nearly 3⅓ times the space of a lb. of copper, so that, taking bulk for bulk, aluminium is only about seven times as dear as copper. [1890.]

When first introduced by Deville, in 1854, aluminium cost £20 per lb.; but its prospective value for application in the arts was recognised, and in two or three years afterwards it was put on the market at 40s. per lb.

It was then, as already remarked, applied to many purposes where lightness is desirable, such as for the tubes of telescopes, opera-glasses, the mounting of photographic lenses, &c. And in 1888, when the production of sodium had been cheapened and applied to the separation of aluminium, the price of the latter metal fell to 18s. per lb. In the meantime, the cheap electricity of the dynamo caused attention to be again directed to the original electrolytic method; but many difficulties in detail had to be overcome in applying this process on the commercial scale. At length the sodium process was superseded; and by the beginning of 1890, a Swiss company was producing aluminium at 11s. per lb. In the course of the following year they succeeded in bringing the price down to 2s. per lb.; and again three years later, namely at the beginning of 1894, they could offer the metal at 1s. 7d. per lb. The conditions required for effecting this great reduction were found in driving the dynamo machinery by water-power, and in an abundant supply of cryolite at moderate cost. This cheapness of production at once placed the Swiss company in the position of being the largest and most successful aluminium manufacturers in the world. so that in 1892 they had realised a net profit of £21,563, paying their shareholders 8 per cent., and, further, in 1893, the net profit was half as much again, and the dividend was increased to 10 per cent. A British aluminium company has recently been formed in London for acquiring the rights of working all the processes of the successful Swiss company, purchasing outstanding English patents, amalgamating with certain existing companies, and for working the *bauxite* deposits in Ireland, &c., &c. There is every reason to believe that an important result of this enterprise will be a still further reduction in the price of this metal, and consequently a great extension of its applications. And now (September, 1895) we have already heard of a further reduction in the price of this metal, which, at the present time, can be purchased in bulk for about 1s. 6d. per lb.

FIG. 337.—*Portrait of Mr. Thomas Hancock.*

INDIAN-RUBBER and GUTTA-PERCHA.

INDIAN-RUBBER.

RESEARCHES into the history of the human race in remote ages have revealed the fact, that before man knew how to extract metals from their ores, his only implements were formed of stone ; and before he became acquainted with iron, there was an intermediate period in which the more easily obtained metal, copper, had to serve as the material for all tools and weapons. Hence archæologists speak of the stone age, the bronze age, and the age of iron. If we were obliged to name the nineteenth century after the material which distinctively serves in it for the most extensive and varied uses, surely we should call it the Age of Indian-rubber !

The industrial application of Indian-rubber is entirely modern. The substance itself appears, however, to have been known to the natives of Peru from time immemorial, and to have been used for the preparation of some kind of garments. Although the first specimens were sent to Europe so long ago as 1736, and the substance was from that time submitted to many investigations, no other use was found for it up to the year 1820 than to efface from paper the marks made by pencils. From this it derives the name by which it is commonly known. It has also been called "gum elastic," and *caoutchouc* from the Indian name. Crude caoutchouc is the product obtained by the spontaneous solidifying of the milky juice of certain tropical plants—such as the *Hævea elastica, Jatropha elastica,* and the *Siphonia cautshu.* The first grows chiefly in South America, and in the basin of the Amazon forms immense forests. At a certain season each year bands of persons, called "*seringarios,*" armed with hatchets, visit these forests for the purpose of extracting the caoutchouc. They make incisions into the trunk, and the milky juice immediately runs out, and drops into a vessel placed to receive it, and attached to the tree by means of a lump of clay. In about three hours the juice ceases to flow, and the *seringario* collects the products of the incisions in one large vessel. By dipping a board into this vessel, it becomes covered with the juice; and when this is allowed to dry, the caoutchouc remains as a thin brownish yellow layer. The caoutchouc is not dissolved in the juice, but is merely suspended in it ; and to hasten the drying and coagulation of the liquid, the board is warmed over a smoky fire made with green wood. When alternate immersions and drying have covered the board with a sufficient thickness of caoutchouc, the layer is slit open with a knife, and the board is withdrawn. This is the best kind of crude caoutchouc, because it is free from all admixture of foreign bodies except the carbon derived from the smoky flame. The *bottle* Indian-rubber is moulded on pear-shaped lumps of clay, which are covered with successive layers of the milky juice ; when a sufficient thickness has been attained, the clay is removed by soaking in water.

Up to 1820, as already mentioned, Indian-rubber was used only for effacing pencil-marks, and about that time a piece half an inch square sold for two shillings and sixpence. But the extreme elasticity and extensibility of this singular substance was attracting the attention of practical men in England, Scotland, and France. One of the earliest patents obtained in this country for applications of caoutchouc was taken out by Mr. Thomas Hancock, of Newington, in 1820. This gentleman has written an account of the Indian-rubber manufacture from the commencement, and the book is extremely interesting from the clear and simple manner in which the inventor describes how he effected one improvement after another in his processes and machinery. Mr. Hancock had, previous to his turning his attention to Indian-rubber, no acquaintance with chemistry ; but he was skilled in mechanical engineering and the use of tools, and this knowledge proved to be precisely the kind most valuable for dealing with the first stages of caoutchouc manufacture. His first patent was for the use of Indian-rubber for the wrists of gloves, for braces, for garters, for boots and shoes instead of laces, and for other similar purposes. The rings for the wrists of gloves, &c., were simply cut from the bottle Indian-rubber by machinery the patentee himself contrived for that purpose. Mr. Hancock next arranged an apparatus for flattening the raw Indian-rubber by warmth and pressure, so as to make it available for the soles of boots, &c. He relates the practical difficulties he had to encounter in his operations, and the manner in which he overcame them. He soon noticed and utilized the fact that two clean freshly-cut surfaces of caoutchouc, when pressed together, cohere and unite perfectly. This further led him to devising a machine by which all the waste cuttings and parings might be worked up. This machine consisted of a cylinder revolving within a cover, both being

provided with steel teeth, by which the pieces of caoutchouc placed between them were torn into shreds, and then kneaded into a solid coherent
mass of homogeneous texture. The first machine of this kind made by
Mr. Hancock would work up about 1 lb. of Indian-rubber ; but now machines on the same principle are in use operating on more than 200 lbs. of
material at once, and turning it out on a roll 6 ft. long, and 10 in. or 12 in. in
diameter.

While Hancock was thus successful in mechanically working Indian-
rubber, Macintosh, of Glasgow, found means of effecting its solution by
coal-naphtha, and he obtained, in 1823, a patent for the application of his
discovery to the fabrication of waterproof garments. Waterproof cloth, or
" Macintosh," is prepared by varnishing one side of a suitable fabric with
a solution of caoutchouc, or by covering one side of a cloth with a thin
film, and then bringing it into contact with a second piece similarly prepared—the two caoutchouc layers becoming incorporated when the double
cloth is passed between rollers. Other solvents for Indian-rubber have
been discovered in ether, chloroform, sulphide of carbon, and rectified
turpentine. By treatment with these liquids it swells up, and eventually
dissolves, producing a viscid ropy mass, which, by evaporation of the
solvent, leaves the caoutchouc with all its original elasticity. By the use
of these last-named solvents the persistent and diagreeable odour occasioned by coal-oil is avoided. Mr. Hancock relates that when the manufacturers had overcome all obstacles, and had succeeded in producing
thin, light, pliable, and perfectly waterproof fabrics, they had to encounter
another quite unexpected difficulty—the tailors set their faces against the
new material, and could not be induced to make it up ! The manufacturers were, therefore, obliged themselves to fashion waterproof garments,
and retail them to the public. This, however, turned out to be a benefit,
for the seams were made waterproof, so as to exclude even the little water
which would otherwise pass in by capillary attraction at the stitches.

It will now be observed that there are two distinct modes of working
caoutchouc : by dealing, viz., with the solid material, or with the solutions.
Thus, from a solid disc of caoutchouc long ribbons of the material may be
cut, and these ribbons, by being passed between a set of circular knives,
may be divided into a number of square threads. These threads may then
be drawn out to six or ten times their length ; and, if wound and maintained in this state of tension for forty-eight hours in a warm place, they
will lose their condition of tension, and their elongated form will become
their natural or unstrained one. In this manner are the Indian-rubber
threads prepared, which, covered with silk or other material, form elastic
fabrics such as those used in the sides of boots. The circumstance of
caoutchouc, when heated for some hours at a temperature a little above
the boiling-point of water, retaining whatever form it has during the heating, is the basis of methods of obtaining thin sheets and other forms of the
material. Tubes are made by forcing the heated caoutchouc through an
annular opening by application of great pressure ; it sets in cooling, preserving a section corresponding with the orifice through which it issues.
In another mode of forming tubes, a paste composed of caoutchouc, oxide
of zinc, and lime, is formed into sheets, which are cut into strips. The strips
are folded longitudinally, and the edges are cut together at an angle of
45° with the surface, so that the cut surfaces may meet each other when
the strip is rolled on a mandril to give it a cylindrical form. A slight
pressure suffices to solder together the cut surfaces, and the tube is then
" vulcanized " by a process to be presently described.

The dissolved caoutchouc serves to prepare waterproof garments, round threads, sheets of Indian-rubber, &c. Fabrics are coated with the solution by pouring it on the material as it is passing horizontally from a roller. A straight-edge, under which the charged cloth passes, distributes the caoutchouc in a uniform layer, the thickness of which is regulated by the space between the knife-edge and the fabric. When sulphide of carbon is the solvent used, its evaporation is complete in about ten minutes, but with other solvents two or three hours are required. The caoutchouc is usually mixed with lampblack before being spread on the cloth, and the article is finished by giving the Indian-rubber layer a coat of gum-lac varnish. Sheets of Indian-rubber are obtained by spreading fifteen or twenty layers over a cloth, which is afterwards detached by wetting it with a solvent.

Threads of circular section are manufactured from a paste of caoutchouc, made by dissolving that substance in sulphide of carbon mixed with 8 per cent. of alcohol. This paste is placed in a cylinder, out of which it is forced by a piston through a number of circular holes, whence it issues in the form of filaments. These are received upon a stretched cloth, which moves along, carrying the parallel threads, until the sulphide of carbon has evaporated.

A modification of caoutchouc, possessing very valuable qualities for many purposes, was discovered by Mr. Charles Goodyear, and largely applied by him in the United States to the fabrication of waterproof boots. In 1842 these boots were imported into Europe, and it was seen that this form of the material had the advantages of not sticking to other bodies at any ordinary temperatures, and of preserving its elasticity even in the coldest weather, whereas ordinary Indian-rubber becomes rigid by cold. The cut edges of this variety of caoutchouc do not cohere by pressure. Mr. Goodyear attempted to keep his process a secret; but Mr. Hancock, having soon detected the presence of sulphur in the American preparations, set to work to discover how that substance was made to combine with the caoutchouc. He succeeded, and he obtained a patent for sulphurizing Indian-rubber before the original inventor had applied for one. Mr. Hancock found that a sheet of caoutchouc immersed in melted sulphur at 250° F., gradually absorbed from 12 to 15 per cent. of its weight of sulphur; and, further, that this does not in any way alter its properties. When, however, the sulphurated substance was for a few minutes exposed to a temperature of 300°, it acquired new qualities, which were precisely those of the modification employed by Mr. Goodyear for his impervious boots. This transformation effected by sulphur Mr. Hancock called *vulcanization;* and *vulcanized Indian-rubber* is now employed in nearly all the innumerable applications of caoutchouc, provided the presence of sulphur is not absolutely objectionable. Goodyear's process consists in mixing the sulphur with the caoutchouc, the suitable proportion (7 to 10 per cent.) having been determined beforehand, and the sulphur ground up with the Indian-rubber in the masticating machine, or disseminated through the viscid liquid if a solution is used, or dissolved in the solvent employed. This gives better results than Hancock's process, because the sulphurization is more uniform, and this method is therefore more largely employed. When the various articles have been fabricated in the ordinary manner from the mixture of caoutchouc and sulphur, they are enclosed in vessels, where they are submitted for two or three hours to the action of steam under a pressure of nearly 4 atmospheres, so that the steam may have a temperature of about 280° F. A still easier method, due to Mr. Parkes, consists in steeping the articles (which in this case should be thin) in a solution of one part of chloride of sulphur in sixty of bisulphide of carbon. The object becomes vulcanized by simple exposure to the air, without the aid of heat. But this process is said to be liable to cause the article afterwards to become brittle. The addition of oxide of zinc, carbonate of lead, and other substances, is found to yield a product better adapted for certain purposes than one in which only sulphur is used.

The list of applications of vulcanized Indian-rubber would be a very long one ; but as a great number of these applications must be known to every-body, it will be unnecessary to specify them. It has lately been used for carriage-springs, for the tires of wheels, and for the rollers of mangles. Its employment in the construction of portable boats, pontoons, life-buoys, dresses for divers and for the preservation of life at sea, air-tight bags and cushions, air and water beds, cushions of billiard-tables, are a few of the thousand instances of its utility which might be quoted.

When the proportion of sulphur mixed with the caoutchouc is increased to 25 or 35 per cent., another product having qualities entirely different from those of vulcanized Indian-rubber is obtained when the mixture is heated. This is the jet-black substance termed *ebonite* or *vulcanite*, which is made into such articles as combs, paper-knives, buttons, canes, portions of ornamental furniture, and plates of electrical machines. It is in many cases an excellent substitute for horn and for whalebone, while for insulating supports, &c., in electric apparatus, it is unrivalled. It has a full black colour and takes a bright polish ; and it may be cut, or filed, or moulded. It is very tough, hard, and durable. In the transformation of Indian-rubber into vulcanite, the temperature must be somewhat higher than that required for the production of the vulcanized Indian-rubber. The caoutchouc used is very carefully purified before it is incorporated with the sulphur ; and the yellow paste formed by the mixture is subjected to the contact of steam at a temperature of about 310°.

GUTTA-PERCHA.

GUTTA-PERCHA is a substance very like Indian-rubber in its chemical properties, having the same composition, although in outward appearance very different. It was first sent to Europe in 1822, but did not become an article of commerce until 1844. It is the solidified juice of a tree (*Isonandra percha*) which abounds in Borneo and Malacca. The trunk of the tree grows to a diameter of 6 ft., but as timber it is valueless. When an incision is made through the bark and into the wood, a milky juice flows out, which speedily solidifies. Gutta-percha is a very tough substance, but is without the elasticity of Indian-rubber. It differs from the latter, too, in becoming softened by a gentle heat, and it will then readily take and retain any impressions with great sharpness and fidelity. Thus beautiful mouldings and other ornamental objects are easily made. It also has the valuable quality of welding when softened by heat. It is a non-conductor of electricity, and it is largely used for covering telegraph-wires, and especially for forming an insulating coating in submarine cables. It seems to have become known precisely at the time it was required for this purpose, and the success of ocean telegraphy is largely owing to its valuable properties. It is employed as a substitute for leather in soling shoes and boots, and in forming straps and bands for driving machinery; also in the preparation of tubes used for conveying liquids, and for speaking-tubes. Dilute mineral acids have no action upon it, and hence it is especially valuable for making bottles to contain hydrofluoric acid, which attacks glass. A drawback to the use of gutta-percha is its tendency to become oxidized when exposed to light and air, by which it entirely loses its power of becoming plastic by heat, and is converted into a brittle substance. But in the dark, or under water, it may, however, be preserved for an indefinite period without change.

Mr. Charles Hancock, in 1847, patented a machine for cutting the gutta-percha into slices. In this machine there is a circular iron plate, with three radial slots, in which knives are fixed somewhat in the manner of the cutting tool of a spokeshave. The lumps of gutta-percha drop against these knives as the plate is driven round, and the material is cut into slices, which have a thickness determined by the projection which has been given to the blades. Sometimes an upright chopper is used, with straight or curved blades. These slices are immersed in hot water, until they are softened, and they are then subjected to the action of rollers armed with toothed blades, called "breakers," and also to the action of the mincing cylinder, which is furnished with radiating blades, and revolves partly immersed in the water. The material is carried out of the hot water to these machines by endless webs mounted on rollers. The breakers and mincing cylinders make about 800 revolutions per minute. The gutta-percha, thus reduced to fragments, is carried forward again by endless webs into cold water, where it is thoroughly washed and separated from the impurities, which fall to the bottom, while the lighter gutta-percha floats on the surface of the water.

Gutta-percha, like caoutchouc, can be combined with sulphur. The best product is obtained when a small proportion of sulphur is used along with some metallic sulphide. Mr. C. Hancock uses 48 parts of gutta-percha, 1 of sulphur, and 6 of antimony sulphide. These ingredients are thoroughly mixed and put into a boiler, where they are heated under pressure for an hour or two. Another method of treating gutta-percha was also devised by Mr. C. Hancock, who found that when this strange substance was exposed to nitric oxide gas (which is given off when nitric acid acts on copper) it became quite smooth, and acquired an almost metallic lustre, losing also all its stickiness. Another modification is formed by treating gutta-percha with chloride of zinc; and yet another by the action of a solvent, such as turpentine, a sulphide, sulphur, and carbonate of ammonia, employed simultaneously. Mr. Hancock mixes all these materials together in a "masticator," and then applies heat to them while confined in a vessel under pressure. The product of these operations is a very singular modification of gutta-percha, in which the material assumes a spongy, elastic condition, and in this form it is used to form the stuffing of sofas, easy chairs, &c. Among the purposes to which gutta-percha has been applied besides the general one of waterproof tissues and fabrics, may be named the formation of straps, belts, bandages, cups, and other vessels, rollers for cotton-spinning machinery, hammers of pianofortes, cards for wool-carding, hammercloths, life-preservers, and trusses.

Gutta-percha is made into strips, bands, cords, or threads of any required section, by passing sheets of suitable thickness between rollers provided with grooves and cutting-edges. For strips and bands the sheets are passed through the machine cold, and divided by the cutting-edges. But for round cords or threads sheets are supplied to the rollers from a receptacle in which they acquire a temperature of about 200° F. The material is forced to take the form of the grooves, the operation in this case being analogous to that of rolling iron bars. The gutta-percha cords are received as they issue from the rollers in a tank of cold water, from which they pass on, to be wound on reels or drums. It is obvious that cords of any section may be formed by making use of grooves of suitable shape. Tubes of gutta-percha are made by forcing the softened substance out of an annular orifice : it is received into vessels filled with cold water. Telegraph-wires are covered by a similar process—the copper wire being made to pass through the centre of a circular opening with the gutta-percha surrounding it. Picture-frames, &c., are made by forcibly pressing warm gutta-percha into the warmed moulds. Gutta-percha tubing is largely used everywhere for the speaking-tubes by which persons in remote apartments of even the largest building can converse. This is one of the labour-saving inventions of our day. It must have struck every one who has seen these speaking-tubes in operation in a large establishment, what a vast amount of running to and fro they save, and how much they expedite business by the convenient means they afford of giving orders and directions to persons in distant apartments. This tubing is also used for the conveyance of liquids, and it has been proposed to employ it instead of the ordinary leaden piping used for carrying water. It may seem to the reader that gutta-percha is too fragile a material to resist the pressure to which water-pipes are exposed. But, judging from some experiments made by the engineer of the Birmingham Waterworks, the power of gutta-percha tubing to resist pressure is quite extraordinary, and far beyond what would be supposed. The tubes experimented on had diameters of $\frac{3}{4}$ and $\frac{1}{4}$th of an inch respectively. The water from the main, where the pressure was that caused by a head of 200 ft., was in communication with these pipes for several weeks, and they were found unaltered in any way. In order to test the strength of the tubes, and find the greatest pressure they would bear, the engineer then had them connected with a hydraulic proving-pump ; and here, when exposed to the highest pressure at which the ordinary water-pipes were tested, namely, to 250 lbs. on the square inch, they also remained intact. The pressure was afterwards increased to 337 lbs., but without any damage to the tubes.

The increasing importance of gutta-percha may be inferred from the continually augmented importation of the crude substance into this country. In 1850 only 11,000 cwt. were imported, but the quantity has increased year by year ; and in 1872 we received nearly 46,000 cwt. The demand is still increasing ; but there is reason to apprehend that under the stimulus of a rising market, the producers have collected the gutta-percha wastefully and with great destruction of the trees, so that it is not improbable that if the demand still increases, there may be a gutta-percha famine. The concreted juices of certain other trees have been proposed as substitutes for gutta-percha. None of these have as yet come into practical use. The increase in a few years of the quantity of Indian-rubber imported into the United Kingdom is perhaps more extraordinary. From the tables given in Mr. Hancock's book, it appears that our imports of caoutchouc were 853,000 lbs. in 1850, but by 1855 they had amounted to 5,000,000 lbs.

FIG. 338.—*Portrait of Sir James Young Simpson, M.D.*

ANÆSTHETICS.

THE discovery which is indicated by the somewhat unfamiliar word *
which heads this article is perhaps the greatest which has ever been
made in connection with the science of medicine. At least, there is no
other discovery of modern times which has so largely and directly contributed to the assuagement of human suffering. Nay, in this respect there
is perhaps in the whole annals of the healing art no other which can rival
it, if we except that famous one of Jenner's which has arrested the ravages
of small-pox. During the last thirty years, all the more formidable operations of the surgeon have been, in almost every case, performed with a happy
unconsciousness on the part of the patient. In unconsciousness, induced
by the same means, has relief also been found for severe suffering arising
from other causes. The substances which are denoted by the word "anæsthetics" differ from the drugs which the older surgeons sometimes administered before an operation, in order to lull the patient's sense of pain.
They differ in their nature and in the mode of their administration; by
the certainty and completeness of their action; by the entirely transient
effects they produce, which pass off without leaving a trace.

To the great chemist whose name has already been mentioned as the

* From α (αν), privative, and αισθητικος, capable of perceiving or feeling.

discoverer of the metals of the alkalies and alkaline earths we are indebted for the first of the remarkable class of bodies we are about to discuss. The first work that Davy published had for its title "Researches, Chemical and Philosophical, chiefly concerning Nitrous Oxide and its Respiration." This was in the year 1800, when the philosopher had hardly completed his twenty-first year. The work caused no little sensation in the scientific world, and it was in consequence of the reputation he acquired by these researches that Davy was appointed to the chemical professorship at the Royal Institution. Davy was not the original discoverer of nitrous oxide, but he first entered upon a full investigation of its properties, and announced the singular effect produced by its inhalation. The kind of transient intoxication and propensity to laughter which it exites have obtained for this compound the familiar name of *laughing gas*. Davy had by experiment on his own person proved the anæsthetic properties of this gas, for he had a tooth painlessly extracted when under its influence, and he says in the work above named that "as nitrous acid gas seems capable of destroying pain, it could probably be used with advantage in surgical operations where there is no effusion of blood." Davy's observations and suggestions were destined to lie barren for nearly half a century, but they nevertheless formed the basis of the great results which have since been attained.

Before proceeding farther, it will perhaps be well to make the unscientific reader acquainted with the chemistry of nitrous oxide. We may presume that he knows that atmospheric air is a *mixture* of the two invisible gases, nitrogen and oxygen (the small quantity of carbonic acid also present need not now be considered). When a known quantity of air is passed over red-hot copper turnings, contained in a tube, the whole of the oxygen is seized upon by the copper, and only the nitrogen issues from the tube, and may be collected. Some of the copper is thus converted into oxide, and the increase of the weight of the tube's contents shows the weight of oxygen contained in the air, while the weight of nitrogen may be known from the volume collected. In this way the chemist analyses atmospheric air, and determines that 100 parts by weight of dry air contain about 79 of nitrogen and 21 of oxygen; or, by measure, about four times as much of the former as of the latter. Now, chemists are acquainted with no fewer than five *different* substances which contain nothing but nitrogen and oxygen. These substances are either gases, or can be changed into the gaseous form by heat, and they can all be analysed in the same manner as air. The results of such analyses show in 100 parts by weight of each substance the following proportions of its constituents:

	No. 1.	No. 2.	No. 3.	No. 4.	No. 5.
Nitrogen......	63·64	46·67	36·84	30·44	25·93
Oxygen	36·36	53·33	63·16	69·56	74·07

In casting the eye over this table, no relation will probably be detected between the five cases. But if we write down, not the quantities of nitrogen and oxygen contained in 100 parts of each compound, but the quantity of oxygen which in each compound is united to some fixed quantity of nitrogen, we shall at once detect a remarkable law : thus, taking 28 as the fixed weight of nitrogen, for reasons which need not be here explained :

	No. 1.	No. 2.	No. 3.	No. 4.	No. 5.
Nitrogen......	28	28	28	28	28
Oxygen	16	32	48	64	80
or	16×1	16×2	16×3	16×4	16×5

Chemists have a sort of shorthand method of expressing the composition of substances, which may be conveniently illustrated by the case before us. Let it be agreed that the letter N shall not only represent nitrogen, but always *fourteen* parts by weight—grains, ounces, &c., &c.,—of nitrogen; and that, similarly, O shall stand for *sixteen* parts by weight of oxygen. It is plain that the composition of the compound No. 2 may be represented by simply writing down " NO ;" and that of No. 4, in which there is just double the proportion of oxygen, by " NOO." But to avoid an unnecessary repetition of the same symbol, when it has to be taken more than once, a small figure is written after and a little below it. Thus, for OO, " O_2 " is written. The proportional composition of each of the five compounds will now be obvious from the following symbols :

No. 1.	No. 2.	No. 3.	No. 4.	No. 5.
N_2O	NO	N_2O_3	NO_2	N_2O_5

These symbols may be regarded as merely a compendious expression of the composition of each substance—as a shorthand statement of the *facts* of analysis. But to the majority of chemists the symbols have a deeper significance; for they are taken as representing the *atoms* of each element which enter into each smallest possible particle of a compound ; they express a certain theory of the ultimate constitution of matter. Thus, if we suppose that there exist indivisible particles of nitrogen and of oxygen, and that each smallest particle, or *molecule,* of the compounds under consideration is constituted of a certain definite and invariable number of each kind of atoms ; and, further, if we suppose that an *atom* of oxygen is heavier than one atom of nitrogen in the proportion of 16 to 14, or 8 to 7, we shall have a simple theoretical *explanation* of the relations in the proportions already pointed out. In fact, these would result from the simplest combinations of the two kinds of *atoms;* and we can picture each one of the smallest particles of the several bodies as thus constituted :

No. 1.	No. 2.	No. 3.	No. 4.	No. 5.
N_2O	NO	N_2O_3	NO_2	N_2O_5

The black circles represent nitrogen atoms, and the open ones oxygen atoms; the symbols are placed below in order that their relation to the supposed atomic constitution may be obvious at a glance. While the symbol of a compound must always accord with its percentage composition, the latter does of itself determine the symbol or formula. A number of other circumstances, which cannot here be discussed, are taken into account as evidence of the constitution of the molecule.

This digression on chemical formulæ will, it is hoped, enable the general reader, who may not previously have been acquainted with them, to perceive their significance, instead of passing them over as unintelligible cabalistic letters when they occur in the following pages. With this object, it may be added that the elements, hydrogen, carbon, and chlorine, are respectively represented by H, C, and Cl; and that the proportional quantities, which are also implied in the symbols, and are those by which H, C, and Cl combine with other bodies, are 1, 12, and 35·5 respectively. Another point which should be understood is that the properties and behaviour of a chemical compound are different, and usually extremely different, from those of any of its constituents. This is well illustrated in the subject we are considering. Atmospheric air is a *mixture* (not a compound) of nitrogen and oxygen gases, and all its properties are intermediate between those of its ingredients taken separately. Nitrous oxide, N_2O, has properties not possessed by either constituent separately. For example, it is very soluble in water, whereas oxygen is very slightly so, and nitrogen still less. The other compounds we have referred to differ widely from nitrous oxide and from each other in their properties.

Nitrous oxide is an invisible gas, having a slightly sweetish taste and smell. It is dissolved by water, which, at ordinary temperatures, takes up about three-fourths of its volume of the gas. By cold and great pressure the gas may be condensed into a colourless liquid. The gas is obtained in a pure state by gently heating the salt called ammonium nitrate, which is formed by neutralizing pure nitric acid with carbonate of ammonia. The action which occurs may be explained thus: the hydrogen of the ammonium unites with a portion of the oxygen of the nitric acid, forming water, whilst the remainder of the oxygen combines with the nitrogen. As chemical actions are regarded as either separations or unions of atoms, they can be expressed by what is called a *chemical equation*, the left-hand side of which shows the arrangement of the atoms before the action, and the right-hand side the arrangement after it, the sign of equality being read as " produce" or " produces." But the validity of the equations, like that of the symbolic formulæ, is quite independent of the existence of atoms; for the equation always rests on certain facts, namely, the relations between the quantities of the substances which enter into, and those which are produced by, a chemical action. Thus, in the present case the action may be symbolically expressed as follows:

$$H_4N\ NO_3 \quad = \quad 2H_2O \quad + \quad N_2O$$

Ammonium nitrate.   Water   Nitrous oxide.

The equation expresses the fact that every 80 parts by weight of ammonium nitrate, which are used in this reaction, split up into 36 of water and 44 of nitrous oxide.

No attempt seems to have been made to turn Davy's suggestion to practical account; but in courses of chemical lectures at the hospitals and elsewhere the peculiar physiological properties of nitrous oxide have, since Davy's announcement, always been demonstrated by some person inhaling the gas. In the medical schools the students often operated on a comrade who was under the influence of nitrous oxide to the extent of bestowing sundry pinches and cuffs, which fully proved the anæthestic qualities of the nitrous oxide. In 1818 Faraday pointed out the similarity between the effects of *ether* and of nitrous oxide, and from that time Professor Turner regularly included among the experiments of his course of chemistry the inhalation of the vapour of ether by one of the students. This was done by simply pouring a little ether into a bladder of air, and by means of a tube drawing the mixed air and vapour into the mouth. Until 1844 the effects of nitrous oxide and of ether vapour remained without application, although thus continually demonstrated in lectures. At the close of that year, Mr. Horace Wells, a dentist, of Hartford, Connecticut, U.S.A., witnessed the usual experiments with nitrous oxide at a public lecture. At his request the lecturer attended at Mr. Wells's residence the following day, to administer to him the nitrous oxide, in order that he might try its efficacy in annulling pain, for he was himself to have a tooth extracted by a brother dentist. His exclamation on finding the operation painlessly over was, " A new era in tooth-pulling !" Mr. Wells continued his experiments on the use of nitrous oxide in dental operations, but he did not apparently obtain uniform results, for he pronounced its effects uncertain, and he gave it up. On the occasion when Mr. Wells's tooth was extracted, Dr. W. T. G. Morton was present, and he soon afterwards found that under the influence of ether vapour, teeth might be painlessly extracted and surgical operations performed. Dr. Morton attempted to conceal the substance he used under the name of " letheon," for which he obtained a patent. But the well-known and characteristic odour of ether declared the nature of the "letheon ;" and Dr. Bigelow having in consequence tried ether, found it to produce all the effects of "letheon." So the matter was no longer a secret. Dr. Morton was, therefore, the person who first applied ether vapour, and the extraction of a tooth was the occasion of its first application. This was in 1846. It was used for the first time in England on the 19th of December, 1846, also for the extraction of a tooth ; and two days afterwards Mr. Liston, the eminent surgeon, performed the operation of amputating the thigh while his patient was under the influence of ether. The employment of ether in surgical operations quickly spread, and its administration in hospitals became general throughout Europe and America.

The chemical constitution of ether, and its relation to alcohol, may be indicated by the following formulæ :

$$\underset{\text{Water.}}{\text{HOH}} \qquad \underset{\text{Alcohol.}}{\text{HO}(C_2H_5)} \qquad \underset{\text{Ether.}}{(C_2H_5)O(C_2H_5)}$$

If we suppose one of the hydrogen atoms in the molecule of water to be removed and replaced by the group (C_2H_5), the result is alcohol. If, now, (C_2H_5) be substituted in the alcohol for the remaining atom of hydrogen, we get a particle of *ether*. Ether was discovered in 1540, and described as sweet oil of vitriol, but its real nature was first pointed out by Liebig. It is prepared by distilling a mixture of sulphuric acid and alcohol. It is a colourless transparent liquid, extremely volatile, and possessing a peculiar and powerful odour. It evaporates so rapidly that a drop allowed to fall from a bottle on a warm day may be converted into vapour before it reaches the ground. When its vapour is inhaled in sufficient proportion mixed with air, it soon produces a complete insensibility to pain. In the case of a full-grown man who inhales air containing 45 per cent. of the vapour, about 2 drams per minute of the liquid are consumed. The air is allowed to stream over the surface of the liquid in a proper apparatus, where it takes up the vapour, and the two pass through a flexible tube to a piece fitting over the mouth and nostrils of the patient. The effects produced are progressive, and may be thus described :

For about two minutes after the beginning of the inhalation, the patient retains his mental faculties, and has some power of controlling his movements, but in a confused and disordered manner. At the end of the third minute he is unconscious ; there are no voluntary movements, but muscular contractions may agitate the frame. At the end of the fourth minute, the only perceptible movements are the motions of the chest in respiration. If the inhalation be discontinued at the end of the fourth minute, when 1 oz. of ether will have evaporated, similar stages are passed through in reverse order during recovery. The condition reached at the end of the fourth minute continues about two minutes ; the intermediate state lasts three or four minutes ; the condition of confused intellect and will, about five minutes. This is succeeded by a feeling of intoxication and exhilaration, which continues for ten or fifteen minutes. It was probably this excitement of the system produced by ether which has caused it to be superseded—in Britain, at least—in about twelve months after its adoption, by *chloroform.*

Chloroform appears to have been independently discovered in 1831, by Soubeiran, and by an American chemist, Guthrie. It is usually procured by distilling a mixture of bleaching powder, spirits of wine, and water. Chloroform is a colourless volatile liquid, of an odour much more agreeable than that of ether. Its composition is represented by $CHCl_3$. The merit of having first applied the singular properties of this substance to the alleviation of human suffering belongs to the late Sir J. Y. Simpson, of Edinburgh. Its use as an anæsthetic was apparently suggested to this eminent professor by Mr. Waldie, of Liverpool. It was first applied at Edinburgh on the 15th November, 1847 ; and when its efficacy had been proved, it was soon extensively used, and in Europe, at least, almost entirely superseded ether, as being more rapid and certain in its action, not producing injurious excitement, and being pleasanter to inhale. A notion prevailed that chloroform was not only more powerful in its operation than ether, but also more safe. In January, 1848, its administration, however, proved fatal to a patient ; and since then a certain number of casualties of this kind have occurred with chloroform, ether, and other anæsthetics.

The patient is often made to inhale the vapour of chloroform by merely holding before his mouth and nostrils a sponge or handkerchief, on which a small quantity of the liquid has been poured. Dr. Snow contrived an apparatus for administering the vapour with more regularity. A metal box adapted to the shape of the face is made to cover the mouth and nostrils. This piece has two valves, one of which admits the air and vapour from an elastic tube connected with the apparatus containing the chloroform, and prevents its return ; the other valve is a flap opening outwards, which allows the expired air to escape. There is also an adjustment for admitting directly into the mouthpiece more or less atmospheric air.

The sensations first experienced when chloroform is inhaled are said to be agreeable. Many persons have described the feeling as resembling rapid travelling in a railway carriage ; there is a singing in the ears, and when the power of vision ceases, and the person is no longer conscious of light, the sensation is that of entering a tunnel. After this there is a lessened sensibility to pain ; and in the next stage the unconsciousness to outward impressions is deeper, but the mental faculties, though impaired, are not wholly suspended, for the patient may speak, and usually dreams something which he afterwards remembers. When the person is still more under the influence of the chloroform, no voluntary motions take place, although there may be some inarticulate muttering. Dr. Snow describes several conditions which may be observed in patients undergoing operations under the influence of chloroform. First, the patient may preserve the most perfect quietude without a sign of consciousness or sensation ; this is the most usual condition. Second, he may moan, or cry, or flinch under the operation, without, however, having the least memory of any pain when he recovers. Third, the patient may talk, laugh, or sing during the operation ; but what he says is altogether devoid of reference to what is done. Fourth, he may be conscious of what is taking place, and may look on while some minor operation is proceeding, without feeling it, or without feeling it painfully. This is often the condition of the patient as the effect is passing off, while some smaller operation is still proceeding. Fifth, the patient may complain he is being hurt ; but afterwards, when the effect of the chloroform has passed off, he will assert that he felt no pain whatever. When the chloroform has been inhaled for but a short time, the patient becomes conscious in about five minutes after its discontinuance ; but with a longer inhalation the period of unconsciousness may last for perhaps ten minutes. The return of consciousness takes place with tranquillity : not unfrequently the patient's first speech, even after a serious operation, often being an assertion that the chloroform has not taken effect.

In the strongest degree of ether and chloroform effects, all the muscles of the body are relaxed ; the limbs hang down, or rest in any position in which they are placed ; the eyelids droop over the eyes, or remain as they are placed by the finger ; the breathing is deep, regular, and automatic ; there is often snoring, and this is, indeed, characteristic of the deepest degree of unconsciousness ; the relaxation of the muscles renders the face devoid of expression, and with a placid appearance, as if the person were in a sound natural sleep. He is perfectly passive under every kind of operation. The breathing and the action of the heart proceed all the while with unimpaired regularity. It is, however, known by experiments on animals that if the inhalation be prolonged beyond the period necessary to produce these effects, the respiratory functions are interfered with by the insensibility extending to the nerves on which they depend. The breathing of an animal thus treated becomes irregular, feeble, or laborious, and death ensues. However nearly dead from inhalation of ether vapour the animal may be, provided respiration has not actually ceased, it always recovers when allowed to breathe fresh air. Of course, the etherization is never carried to this stage with human beings.

Air containing 2 grs. of chloroform in 100 cubic inches suffices to induce insensibility ; but 5 grs. in 100 cubic inches is found a more suitable proportion. Dr. Snow, who strongly disapproved of the uncertain and irregular mode of administering chloroform on a handkerchief or sponge, contrived the inhaling apparatus already described. The air before reaching the mouth and nostrils of the patient passes through a vessel containing bibulous paper moistened with chloroform. This vessel he surrounds with water at the ordinary temperature of the air, in order to supply the heat absorbed by the conversion of the liquid into vapour, so that the formation

47

of the latter may go on regularly. The same thoughtful arrangement formed part of the ether-inhaler he had previously contrived.

The extraordinary effects of ether and chloroform have introduced new and important facts into psychological science, and have illustrated and extended some of the most interesting results of physiological research. Let us trace the action of these substances, and explain it as far as may be. Nitrous oxide, ether vapour, and chloroform vapour are all soluble in watery fluids. The lungs present a vast surface bathed by watery fluids, and therefore these gases are largely absorbed ; and by a well-known process, they pass directly into the blood, through the delicate walls of the capillary vessels. The odour of ether can be detected in any blood drawn from persons under its influence. Ether, or chloroform, thus brought into the general current of the circulation, is quickly carried to all parts of the body, and thus reaches the nerve-centres. On these it produces characteristic effects by suspending or paralysing nervous action : why or how this effect takes place is unknown. The nervous centres are not all acted upon in an equal degree—some require a larger quantity of the drug to affect them at all. The parts of the nervous system first affected are the cerebral lobes, which are known to be the seat of the intellectual powers. The *cerebellum* —the function of which there is reason to believe is the regulation and co-ordination of movements—is the next to yield to the influence. Then follow the spinal nerves, which are the seat of sensibility and motive power. This is as far as the action can safely be carried : the nervous centre called the *medulla oblongata*, which is placed at the junction of the brain and the spinal cord, still performs its functions—one of the most important of which is to produce the muscular contractions that keep the respiratory organs in action. We have seen, by the effects of further etherization in animals, that when this part of the system is affected, the animal dies from a stoppage of the respiration.

But, unfortunately, there have been instances in which death has been caused by the administration of ether and chloroform even under the most skilful management. But these occurrences were not the result of the inhalation having been carried so far as to stop respiration : in some cases the patient has died before the first stage of insensibility. These fatal cases have all been marked by a sudden paralysis of the heart—that organ has abruptly ceased to act. Why in these, certainly a very small percentage of patients, the action of the drug should at once take effect on the heart has not yet been explained. The rhythmic action of the heart depends upon nervous centres enclosed within its own substance, so that this organ is to a certain extent independent; but it is connected with the other nervous centres by the branches of a remarkable nerve which proceeds from the *medulla oblongata*, and also by another set of nerves which come from the chain of ganglia called the *sympathetic nerve*. The nerve connecting the heart with the *medulla* is a branch of that called the *pneumo-gastric*, and it is a well-established fact that the action of the heart may be arrested by irritation of this nerve. The comparatively few fatalities which have attended the use of anæsthetics may, therefore, be due either to an immediate action on the nerve-centres of the heart, or possibly to a mediate action through the *medulla* and the pneumo-gastric nerve.

Soon after the introduction of ether the use of nitrous oxide was discontinued by the dentists, on account of the apparent uncertainty of its action. Within the last few years, however, its employment in the extraction of teeth has been revived by Dr. Evans, of Paris, who found that to insure certainty in its action, the great point is the inhalation of the gas in a pure state and without admixture of air. Nitrous oxide seems now to be extensively used by dentists, and thus Davy's experiment of 1800 is repeated and verified daily in thousands of cases, and to the great relief of hundreds who probably never heard his name.

Other bodies, such as amylene (C_5H_{10}), carbon tetrachloride (CCl_4), &c., have been tried as substitutes for ether and chloroform; but having been found less efficacious or more dangerous, their use has been abandoned. It might be instructive to reflect how much unnecessary pain would have been spared to mankind had ether and chloroform been known and applied at an earlier age. We know not what other beneficent gifts chemistry may yet have in store for the alleviation of suffering, but it is unlikely that even ether and chloroform are her *derniers mots.* It should be remembered that the chemists who discovered and examined these bodies were attracted to the work by nothing but the love of their science. They had no idea how invaluable these substances would afterwards prove. The chemist of the present day, whose labour is often its own reward, may be cheered and stimulated in his toil by the thought that while no discovery is ever lost, but goes to fill its appropriate place in the great edifice of science, even the most apparently insignificant truth may directly lead to invaluable results for humanity at large.

What strange things the ancient thaumaturgists might have done had they been possessed of the secret of chloroform or of nitrous oxide ! What miracles they would have wrought—what dogmas they would have sanctioned by its aid ! But the remarkable effects produced by the inhalation of certain gases or vapours were not altogether unknown to the ancients —although these effects were then attributed to anything but their real cause. It is related that a number of goats feeding on Mount Parnassus came near a place where there was a deep fissure in the earth, and thereupon began to caper and frisk about in the most extraordinary manner. The goatherd observing this, was tempted to look down into the hole, to see what could have caused so extraordinary an effect. He was himself immediately seized with a fit of delirium, and uttered wild and extravagant words, which were supposed to be prophecies. The knowledge of the presumed divine inspiration spread abroad, and at length a temple in honour of Apollo was erected on the spot. Such was the origin of the famous Oracle of Delphi, where the Pythoness, the priestess of Apollo, seated on a tripod placed over the mysterious opening, delivered the response of the god to such as came to consult the oracle. It is stated by the ancient writers, that when she had inhaled the vapour, her eyes sparkled, convulsive shudders ran through her frame, and then she uttered with loud cries the words of the oracle, while the priests who attended took down her incoherent expressions, and *set them in order.* These possessions by the spirit of divination were sometimes violent. Plutarch mentions a priestess whose frenzy was so furious, that the priests and the inquirers alike fled terrified from the temple ; and the fit was so protracted that the unfortunate priestess herself died a few days afterwards.

FIG. 339.—*A Railway Cutting.*

EXPLOSIVES.

THE illustration above will serve to remind the reader of the great importance of explosive agents in the operations of civil industry. By reason of the more impressive and exciting spectacles which attend the use of such agents in warfare, we are rather apt to lose sight of their far more extensive utility as the giant forces whose aid man invokes when he wishes to rend the rock in order to make a road for his steam horse, or in order to penetrate into the bowels of the earth in search of the precious ore. A little reflection will show that if such work had to be done with only the pickaxe, the chisel, and the crowbar, the progress would be painfully slow; and railway cuttings through masses of compact limestone, like that represented in Fig. 339, for example, would be well-nigh impossible. The formation of cuttings and tunnels, and the removal of rocks in mining operations, are not the only service which explosive agents render to the industrial arts; there is, besides other uses which might be enumerated, the preparation of foundations for buildings, bridges, harbours, and lighthouses. The use of gunpowder in all such operations as those which have been referred to is too well known to require description. But of late years gunpowder has been to a great extent superseded for such purposes by two remarkable products of modern chemistry, called *gun-cotton* and *nitro-glycerine.* Military art has also benefited by at least one of these

740

products; and the use of charges of gun-cotton for torpedoes has already been described and illustrated in these pages.

It is not a little curious that the two most terribly powerful explosives known to science should be prepared from two most harmless and familiar substances. The nice, soft, clean, gentle cotton-wool, in which ladies wrap their most delicate trinkets, becomes, by a simple chemical transformation, a tremendously powerful explosive; and the clear, sweet, bland liquid, glycerine, which they value as a cosmetic for its emollient properties, becomes, by a like transformation, a still more terrifically powerful explosive than the former. It is, perhaps, even more curious that having undergone the transformation which confers upon it these formidable qualities, neither cotton-wool nor glycerine is changed in appearance. The former remains white and fleecy; the latter is still a colourless syrupy-looking liquid.

The fibres which form cotton, linen, paper, and wood, are composed almost entirely of a substance which is known to the chemist as *cellulose* or *cellulin*. That this substance, as it exists in the fibres of linen and in sawdust, could be converted into an explosive body by the action of nitric acid, appears to have been first observed by the French chemist, Pelouze, in 1838. The action with cellulose in the form of cotton-wool was more fully examined by Professor Schönbein, of Basle, who, in 1846, first described the method of preparing *gun-cotton*, and suggested some uses for it. He directs that one part of finely-carded cotton-wool should be immersed in fifteen parts of a mixture of equal measures of strong sulphuric and nitric acids; that after the cotton has remained in the mixture for a few minutes, it should be removed, plunged in cold water, and washed until every trace of acid has been removed, and then carefully dried at a temperature not exceeding the boiling-point of water.

After Professor Schönbein had demonstrated the power of the new agent in blasting, and its projectile force in fire-arms, its manufacture on a large scale was undertaken at several places. Messrs. Hall commenced to make it at their gunpowder works at Faversham, and a manufactory was also established near Paris. In July, 1847, a fearful explosion of gun-cotton occurred at the Faversham works, which was believed to have been caused by the spontaneous detonation of that substance. This induced Messrs. Hall to discontinue the manufacture as too dangerous; and they even destroyed a large quantity of the product which they had in hand by burying it in the ground. The making of gun-cotton was soon afterwards discontinued also by the French, who did not find the substance to possess all the qualities fitting it for military use. The Prussian Government also began to make gun-cotton; but the experiments were put a stop to by the explosion of their factory. An eminent artillery officer in the Austrian service, General von Lenk, undertook a thorough examination of the manufacture and properties of gun-cotton for military purposes. He introduced several improvements into the processes of the manufacture; and the Austrian Government established works at Hïrtenberg, with a view to the adoption of gun-cotton as a substitute for gunpowder in fire-arms. It has some undoubted advantages over powder, for it neither heats the gun nor fouls it, and it produces no smoke. Notwithstanding this the Austrians have not abandoned the use of gunpowder in favour of gun-cotton.

Gun-cotton, as a military agent, has a strenuous advocate in Professor Abel, who presides over the Chemical Department of the British War Office. To this gentleman we are indebted for great improvements in the manufacture of gun-cotton, and for a more complete investigation of its properties. Professor Abel's processes were put in practice at a manufactory which the Government established at Waltham Abbey ; and Messrs. Prentice also set up works at Stowmarket.

Some details of the mode in which the manufacture of gun-cotton was carried on at Stowmarket may be of interest. The cotton was first thoroughly cleansed and carefully dried; and these operations are of great importance, for unless they are well performed, the product is liable to explode spontaneously. The cotton was then weighed out in charges of 1 lb., and each charge was completely immersed in a separate vessel, containing a cold mixture of sulphuric and nitric acids. After a short immersion the cotton was removed from the liquid, and with about ten times its own weight of acids adhering to it, each charge was placed in a separate jar, where it was allowed to remain for forty-eight hours. The vessels were kept cool during the whole period by being placed in a trough through which cold water was flowing. On removal from the jars, the cotton was freed from adhering acid by being placed in a centrifugal drying machine. It was then drenched with a large quantity of cold water, and dried, washed again in a stream of cold water for forty-eight hours, and the operations of alternately washing for forty-eight hours and drying were repeated eight times. The drying was effected by placing the material in cylinders of wire-gauze, which were whirled round by a steam engine at the rate of 800 revolutions per minute, so that the water was expelled by centrifugal force. The cotton was next reduced to a pulp by a process similar to that which is employed in paper-making, and the moist pulp was rammed into metallic cylinders by hydraulic pressure, in order that it might be brought into forms suitable for use in blasting, &c. The pulp was put into these moulds while wet, but the water was nearly all expelled by the compression. The cylinders of gun-cotton thus obtained were then covered with paper-parchment, and finally dried at a steam temperature, with many precautions. The compression of the cotton pulp, by bringing a large quantity of the material into a smaller bulk, causes a greater concentration of the explosive energy, and this is a matter of great importance in blasting.

We may now consider what chemistry has to teach concerning the nature of the action by which cotton-wool is converted into gun-cotton. Cotton itself is nearly pure cellulose. The chemical composition of cellulose may be represented most simply by the formula $C_6H_{10}O_5$. Nitric acid is a powerful oxidizing agent, and is constantly used in chemistry to fix oxygen in various substances; but another kind of action exerted by nitric acid in certain cases consists in the substitution of a portion of its atoms for hydrogen, by which the residue of the particle of nitric acid is converted into water. The formula for nitric acid may be written $HO\,NO_2$, and it will be seen that by changing NO_2 for H, water, HOH, would be produced. This is precisely the kind of action which occurs when cellulose is converted into *nitro-cellulose*. Two or three, or more, atoms of hydrogen may be taken out of cellulose, and replaced by two or three, or more, groups NO_2, and the result will be a different kind of *nitro-cellulose*, according to the number of atoms in the molecule replaced by NO_2. Several varieties of gun-cotton are known, these being doubtless the result of the differences here alluded to. The action producing di-nitro-cellulose is represented by this equation :

$$\underset{\text{Cellulose.}}{C_6H_{10}O_5} \;+\; \underset{\text{Nitric acid.}}{2HNO_3} \;=\; \underset{\text{Di-nitro-cellulose.}}{C_6H_8(NO_2)_2O_5} \;+\; \underset{\text{Water.}}{2H_2O.}$$

The equation shows that water is produced by the reaction, and the sulphuric acid which is used in the preparation performs no further part than to take up this water, which would otherwise go to dilute the rest of the nitric acid. The union of sulphuric acid and water is attended with great heat, hence the necessity of cooling the vessels in making the gun-cotton. Quite other products would be formed if the mixture became heated.

The action of nitric acid on glycerine is of the same kind as that on cellulose. When glycerine is allowed to drop into a cooled mixture of nitric acid and sulphuric acid, the eye can detect little or no difference between the appearance of the liquid which collects in the bottom of the vessel and the glycerine dropped in. The product of the action is, however, the terrible *nitro-glycerine*, a heavy, oily-looking liquid, which explodes with fearful violence. Even a single drop placed on a piece of paper, and struck on an anvil, detonates violently and with a deafening report. The chemical change which is effected in the glycerine $(C_3H_8O_3)$, is the substitution of three NO_2 groups for three of hydrogen, producing $C_3H_5(NO_2)_3O_3$, or tri-nitro-glycerine. The general reader may perhaps marvel that the chemist should be able not only to count the number of atoms which go to make up the particles of a compound body, but to say that they are arranged so and so : that the atoms do not form an indiscriminate heap, but that they are connected in an assignable manner. The reader is no doubt aware that these compound particles are extremely small, and he may reasonably wonder how science can pronounce upon the structure of things so small. He may be more perplexed to learn that a calculation made by Sir W. Thompson shows that the particles of water, for instance, cannot possibly be more than the $\frac{1}{250000000}$th of an inch in diameter, and may be only 1-20th of that size. The truth is that the very existence of atoms and molecules is an assumption. Like the undulatory ether, it is an hypothesis which is adopted to simplify and connect our ideas, and not a demonstrated reality. But the atomic hypothesis has so wide a scope that some philosophers hold the existence of atoms and molecules as almost a known fact. Be that as it may, the chemist in assigning to a body a certain *molecular formula* really does nothing but express the results of certain experiments he has made upon it. With one re-agent it is decomposed in this manner, with another in that. By certain treatment it yields an acid, a salt; so much carbonic acid, such a weight of water, is acted on or remains unaltered; gives a precipitate or refuses to do so. Such are the *facts* which the chemist conceives are co-ordinated and expressed by the formula he gives to a substance. The best formula is that which accords with the greatest number of the properties of the body—which includes as many of the facts as possible. It follows, therefore, that a formula which aims at expressing more than the mere percentage composition of the body—which, in the language of the atom hypothesis, seeks to represent the mode in which the atoms are grouped in the molecule, but which in reality represents only reactions, is written according as the chemist considers this or that group of reactions more important. These remarks might be illustrated by filling this page with the different formulæ (a score or more) which have been proposed as representing the constitution (*reactions ?*) of one of the best-known of organic compounds, namely, acetic acid.

Whether atoms really exist, and their arrangement in the particles of bodies can be deduced from the phenomena, or not, the fact is undeniable that these ideas have given chemists a wonderful grasp of the facts of their science. The consistency and completeness of the explanation afforded by these theories are ever being extended by modifications which enable them to embrace more and more facts. Some of the properties of the substance we are now considering confirm in a remarkable manner the theoretical views which are expressed in its constitutional formula. We may first consider the nature of gunpowder, and by comparing it with nitro-glycerine, endeavour to explain the greater power of the latter substance. Gunpowder is a mixture of charcoal, sulphur, and nitre, the latter constituting three-fourths of its weight. Nitre supplies oxygen for the combustion of the charcoal, which is thus converted into carbonic acid, and the sulphur, which is added to increase the rapidity of the combustion, is also oxidized. The products of the action are, however, numerous and complicated, but the important result is the sudden generation of a quantity of carbonic acid, nitrogen, carbonic oxide, hydrogen, and other gases, which at the oxidizing temperatuie and pressure of the air would occupy a space 300 times greater than the powder from which they are set free; but the intense heat attending the chemical action dilates the gases, so that at the moment of explosion they would occupy a space at least 1,500 times greater than the gunpowder. The materials of which gunpowder is composed are finely powdered, in order that each portion shall be in immediate contact with others, which shall act upon it. Plainly, the more thorough the incorporation of the materials—that is, the more finely ground and intimately mixed they are—the more rapid will be the inflammation of the powder.

Looking now at the crude formula of nitro-glycerine, $C_3H_5N_3O_9$, the reader will remark that the molecule contains more than sufficient oxygen to form carbonic acid with all the carbon atoms, and water with all the hydrogen atoms; for the C_6 in *two* molecules of nitro-glycerine would take only O_{12} to form $6CO_2$; and the H_{10}, to be converted into $5H_2O$, would only need O_5; thus there would be an excess of oxygen. Now, it may occur to the reflective reader that in every molecule of nitro-glycerine the carbon and hydrogen are already associated with as much oxygen as they can take up: that they are, in fact, already burnt, and that no further union is possible. But from chemical considerations it has been deduced that in the nitro-glycerine molecule the oxygen atoms, except only three, which are partially and *imperfectly* joined to carbon, are united to nitrogen atoms only. The constitution of the molecule may be represented by arranging, as below, the letters which stand for the atoms, and by joining them with lines, which shall stand for the bonds by which the atoms are united.

$$
\begin{array}{ccccccc}
O & & H & H & H & & O \\
\| & & | & | & | & & \| \\
N-O-C-C-C-O-N \\
\| & & | & | & | & & \| \\
O & & H & O & H & & O \\
& & & | & & & \\
& & O=N=O & & &
\end{array}
$$

We see here that the hydrogen atoms are completely, and the carbon atoms partially, detached from the oxygen atoms; and therefore these atoms are in the condition of the separated carbon and oxygen atoms in gunpowder. Only the pieces of matter which lie side by side in gunpowder are in size to the molecules of nitro-glycerine as mountains to grains of sand. The mixture of the materials is then so much more intimate in nitro-glycerine, since atoms which can rush together are actually within the limits of the molecules; and these molecules have such a degree of minuteness, that 25 millions, at least, could be placed in a row within the length of an inch. We know that the finer the grains and the more intimate the mixture, the quicker will gunpowder inflame; but here we have a mixture far surpassing in minute subdivision anything we can imagine as existing in gunpowder. Hence the combination in the case of nitro-glycerine must be instantaneous, whereas that in gunpowder, quick though it be, must still require a certain interval. If it take a thousandth of a second for the gases to be completely liberated from a mass of gunpowder, and only one-millionth of a second for a vast quantity of carbonic acid, nitrogen, and steam to be set free from nitro-glycerine, the destructive effect will be much greater in the latter case. Again, the volume of the gases liberated from nitro-glycerine in its detonation have at least 5,000 times the bulk of the substance. We have entered into these chemical considerations, at some risk of wearying the reader, with the desire of affording him a clue to the singular properties of nitro-glycerine and gun-cotton, which we are about to describe.

The nature of the chemical changes which may be set up in an explosive substance, and the rapidity with which these changes proceed throughout a mass of the material, are greatly modified by the conditions under which the action takes place. If a red-hot wire be applied to a small loose tuft of gun-cotton, it goes off with a bright flash without leaving any smoke or any other residue. Thus, when the substance is quite unconfined, no explosion occurs. If the cotton-wool be made into a thread, and laid along the ground, it will burn at the rate of about 6 in. per second; if it be twisted into a yarn, the combustion will run along at the rate of 6 ft. per second; but if the yarn be enclosed in an Indian-rubber tube, the ignition proceeds at the rate of 30 ft. in a second. If to a limited surface of gun-cotton, such as one end of a length of gun-cotton yarn, a source of heat is applied—the temperature of which is high enough to set up a chemical change, but not high enough to inflame the resulting gases (carbonic oxide, hydrogen, &c.) —the cotton burns comparatively slowly, rather smouldering than inflaming. If, however, a *flame* be applied, the gun-cotton flashes off with great rapidity, because the heat applied sets fire to the gaseous products of the chemical action. But if the gun-cotton be confined so that the gases cannot escape, the combustion becomes rapid however set up. The reason is that if the gases escape into the air, they carry off so much of the heat produced by the smouldering gun-cotton, that the temperature does not rise to the extent required to produce the flaming ignition, in which the products are completely oxidized. If a mass of gun-cotton be enclosed in a capacious vessel from which the air has been removed, and the gun-cotton be ignited by means of a wire made hot by electricity, the cotton will at first only burn in the slow way without flame; but as the gases accumulate and exert a pressure which retards the abstraction of heat accompanying their formation, the temperature will rise and attain the degree necessary for the complete and rapid chemical changes involved in the flaming combustion. Thus, the more resistance is offered to the escape of the gases, the more rapid and perfect is the combustion and explosive force produced by the ignition. Now, the explosion of gun-cotton has been found to be too rapid when it is packed into the powder-chamber of a gun, for its tendency is to burst the gun before the ball has been fairly started. Hence a material like gunpowder, in which the combustion is more gradual, is better suited for artillery. The ignition of gunpowder, though rapid, is not instantaneous, and therefore we can speak of it as more or less gradual. Indeed, in even the most violent explosives, some time is doubtless required for the propagation of the action from particle to particle. This extreme rapidity of combustion, and consequent rending power, which is so objectionable in a gun-chamber, makes gun-cotton a most powerful bursting charge for shells, and, when it is enclosed in strong receptacles, for torpedoes also.

But by the researches of Nobel, Professor Abel, and others, it has been discovered—and this is, perhaps, the most remarkable discovery in connection with explosives—that gun-cotton, nitro-glycerine, and other explosive bodies, are capable of producing explosions in a manner quite different from that which attends their ignition by heat. The violence of this kind of explosion is far greater than that due to ordinary ignition, for the action takes place with far greater rapidity throughout the mass, and is, indeed, practically instantaneous. It appears to be produced by the mere mechanical agitation or vibrations which are communicated to the particles of the substance. Turning back to the representation of the molecule of nitro-glycerine on page 744, it will not be difficult to imagine that this may be an unstable kind of structure ; that the atoms of oxygen are prevented from rushing into union with those of hydrogen and carbon only by the interposition of the nitrogen ; and that an agitation of the structure might shake all the atoms loose, and leave them free to re-combine according to their strongest affinities. Nitro-glycerine is by no means so ready to *inflame* as is gun-cotton : it is said that the flame of a match may be safely extinguished by plunging it into the liquid ; and when a sufficient heat is applied to a quantity of the liquid in the open air, it will burn quietly and without explosion. Even when nitro-glycerine is confined, the application of heat cannot always be made to produce its explosion ; or, at least, the circumstances under which it can do so are not accurately known, and the operation is difficult and uncertain. On the other hand, nitro-glycerine explodes violently even when freely exposed to the air if there be exploded in contact with it a *confined* charge of gunpowder, or a detonating compound such as fulminating powder. Gun-cotton possesses the same property of exploding by concussion, which appears indeed to be a general one belonging to all explosive bodies. According to recent researches, even gunpowder is capable of a detonative explosion. A mass of gunpowder confined with a certain proportion of gun-cotton, which is itself set off by fulminate of mercury, is said to exert an explosive force four times greater than that developed by the ignition of the gunpowder in the ordinary manner. It has also been found that *wet* gun-cotton can be exploded by concussion, and the force of the explosion is unimpaired even when the material is saturated with water. This makes it possible to use gun-cotton with greater safety, as it may be transported and handled in the wet condition without risk, and it preserves its properties for an indefinite period without being deteriorated by the water. Some experiments illustrating the extraordinary force of the detonative explosions of gun-cotton and nitro-glycerine will give the reader an idea of their power.

A palisade, constructed by sinking 4 ft. into the ground trunks of trees 18 in. in diameter, was completely destroyed in some experiments at Stowmarket by the explosion of only 15 lbs. of gun-cotton. Huge logs were sent bounding across the field to great distances, and some of the trees were literally reduced to match-wood. The gun-cotton, be it observed, was simply laid on the ground exposed to the air. The destructive powers of nitro-glycerine are even greater. A tin canister, containing only a few ounces of nitro-glycerine, is placed, without being in any way confined, on the top of a smooth boulder stone of several tons weight ; it is exploded by a fuse containing fulminating powder, which is fired from a distance by electricity. There is a report, and the stone is found in a thousand fragments. The last experiment shows one of the advantages of nitro-glycerine over gunpowder as a blasting material, beyond its far greater power, which is about ten times that of gunpowder. A charge of gunpowder inserted in a vertical hole tends to force out a conical mass, the apex of which is at the space occupied by the charge. With nitro-glycerine, and also with gun-cotton, which last has almost six times the force of gunpowder, a powerful rending action is exerted *below* as well as above the charge. Again, in blasting with gunpowder the charge must be confined, and the hole is filled in above the charge with tightly rammed materials, forming what is termed the *tamping.* But nitro-glycerine requires no tamping : a small, thin metallic core containing the charge is simply placed in the drill-hole, or the liquid itself is poured in, and a little water placed above it. The effect of the explosion of nitro-glycerine in " striking down," when apparently no resistance is offered, will seem very strange to the reader who is oblivious of certain fundamental principles of mechanics. The force of the explosion is due entirely to the sudden production of an enormous volume of gas, which at the ordinary pressure would occupy several thousand times the bulk of the material from which it is produced. This gas, by the law of the equality of action and reaction, presses down upon the stone with the same force that it exerts to raise the superincumbent atmosphere. The pressure of the gas at the moment of its liberation is enormous ; but the atmosphere cannot instantaneously yield to this, for time is required to set the mass of air in motion, and the wave of compression advances slowly compared with the rapidity of the explosion. Hence the air acts, practically, like a mass of solid matter, against which the gases press, and which yields less readily than the rock, so that the blow which is struck takes visible effect on the latter. Now, with gunpowder, the evolution of gas is less rapid, the atmosphere has time to yield, and the reaction has not the same violence. The rapidity of the evolution of gas from the exploding nitro-glycerine is so great, that the gases, though apparently unconfined, are not so in reality ; for the atmosphere acts as a real and very efficient tamping.

When nitro-glycerine first came into use for blasting purposes, it was used in the liquid form under the name of " blasting oil ;" but the dangers attending the handling of the substance in this state are so great, that it is now usual to mix the liquid with some powdered substance which is itself without action, and merely serves as a vehicle for containing the nitro-glycerine. To mixtures of this kind the names " *dynamite,*" " *dualine,*" " *lithofracteur,*" &c., have been given.

It is now hardly necessary to point out that the discovery of these new explosives has largely extended our power over the rocks, enabling works to be executed which would have been considered impracticable with less powerful agents. It is true that the most fearful disasters have been accidentally produced by the new explosives ; but such occasional devastation is the price exacted for the possession of powerful agents. And just as in other cases—steam, for example—where great forces are dealt with, so these new powers must be managed with unceasing care, and placed in the hands of only skilful and intelligent men.

The products of the combustion of gunpowder are not all gaseous, but include solid compounds, such as carbonate and sulphate of potassium. It is these that give rise to the smoke seen when a gun is discharged, and which, in rapid firing, soon obscures the sight of the objects aimed at. They are also the causes of the fouling of the bore. Gun-cotton is quite unexceptionable in these respects, and that prompted the attempts made soon after its introduction to use it instead of gunpowder in fire-arms. But the explosion of gun-cotton was found too sudden and violent for guns and rifles, so that many serious accidents in consequence occurred. The next thing done was to lessen the rapidity of the explosion by using gun-cotton mixed with ordinary cotton, or twisted in threads round some inert substance—in fact, to mitigate the violence of the shock by some mechanical disposition of the material. The introduction of rapid firing guns and repeating rifles forced on the problem of a smokeless powder; and as the plan of replacing nitrate of potassium, in ordinary gunpowder, by nitrate of ammonium was found to be attended with loss of the keeping quality of the powder, other materials, such as picric acid, which forms also the basis of the explosive called *mélinite*, have been proposed. The composition of *mélinite* was long a mystery, and that of the smokeless powder adopted by the French was so carefully concealed that many experiments had to be made by other nations to discover some similar preparation, which was found possible by combining certain substances with gun-cotton so as to modify the violence of its explosion, and produce a manageable material having the required properties. The British Government, after many experiments and much careful testing, decided to adopt *cordite*, made of nitro-glycerine, in which, by the aid of volatile solvent, di-nitro-cellulose is dissolved, together with a little mineral oil. The semi-fluid composition, forced through a round hole or die, comes out like a thread or cord, which the evaporation of the volatile solvent leaves with very much the appearance of common brown window-cord. This material has the several advantages of keeping well, of being *uniform* in its propulsive powers, of being capable of imparting as high a velocity as a much larger charge of the ordinary black gunpowder, while at the same time exercising a less pressure on the chase of the gun.

It will have become obvious from the preceding paragraphs that, according to the conditions under which an explosive is to be used, selection must be made of the most suitable. For example, the substances employed for propelling projectiles from guns must not have the violent rending power of certain others, which, by this very property, are most useful for blasting operations; and, again, although explosives of this last kind are inadmissible as projectile agents, they are of the kind best adapted for use in shells where it is the disruptive action that is required. Also, in blasting operations, the explosive has to be adapted to the nature of the work, and it has been found that a substance which has worked well in driving a heading for a tunnel through one kind of rock may prove both slow in progress, and more costly in expenditure, when some different kind of rock is reached. Besides this, regard must be had in blasting operations to the nature of the effect required, which is in some instances a shattering of the rock into fragments, in others a detachment of it in masses. Thus, in the working of a slate quarry, the explosive used must not be of a nature to shiver the rock into useless splinters, but must operate in such a manner that compact masses may be separated from the mountain side in a condition suitable for cleaving, by appropriate tools, into numberless broad laminæ, which, trimmed into rectangular shape, constitute our well-known roofing slates. The blasting used on a coal seam must be so conducted as to yield the material as much as possible in big lumps or cobbles rather than in slack. When granite is blasted for the purpose of obtaining building stones, the explosive must be one that, by its comparatively slow action, divides the compact rock into the largest possible blocks. On the other hand, when granite is blasted merely with the object of removing it, as when a tunnel has to be driven through a mass of it, the most disintegrating agent is then the best. The common popular expressions by which the two classes of explosives just referred to are distinguished are "high explosives" and "low explosives." Dynamite may be taken as a type of the former, and gunpowder a type of the latter. As will be gathered from what is to follow, no definite separation between these classes can be fixed, but in a general way it may be said that, where a destructive, rather than a propelling or pressure effect is required, the explosive used is one brought into operation by a concussive or detonating priming, and acting mostly by detonation within itself, such as dynamite, &c.

Whereas, up to nearly the middle of the nineteenth century, gunpowder was practically the only explosive in use for either civil or military purposes, the close of the century can show a list of several hundred preparations that have been proposed or actually used in its stead. The names by which these are put forward are expressive sometimes of an ingredient in their composition, such as "ammonia dynamite," "cellulosa," "mica powder," "dynamite au carbon," "dynamite de boghead," &c.; and sometimes the inventor's name, as "So-and-so's powder or explosive"; sometimes of the strength of the mixture under various fanciful names, such as "dynamite," "heraklin," "vigorite," &c., &c.; sometimes the names relate to the appearance of the compound, as "white gunpowder," "blasting gelatine," &c., &c.; and sometimes to other circumstances, such as "pudrolithe," "saxifragine," "safety powder," &c., &c. A very long list might be given of the substances severally used in these various compositions. It will be sufficient to indicate the general nature of the several classes into which the new explosives may be divided. By turning back to p. 746, the reader will be reminded of the composition of gunpowder, and of the part played therein by the nitre (nitrate of potassium). Now a considerable number of the recently patented explosives are simply modified gunpowders, which all contain some nitrate, replacing wholly, or in part, the nitrate of potassium, while sulphur is an ingredient of nearly all, and in many, the charcoal of gunpowder is partly or wholly replaced by other carbonaceous materials, such as saw-dust, coal-dust, tan, starch, paraffin, lycopodium, graphite, peat, flour, bran, &c. Certain mineral salts enter into the composition of some, such as sulphate of iron, carbonate of copper, sulphide of antimony, &c., &c.

In another class of the newer explosives chlorate of potassium takes the place of the nitrate as the oxygen supplying material, with similar variations as to the carbonaceous matter as are referred to above. Yellow prussiate of potash and sugar sometimes replace both the charcoal and sulphur of gunpowder in this class. Explosives of this chlorate class are usually dangerous to manufacture, and are often very sensitive, and also liable to changes by keeping, which render them still more dangerous.

The next class of preparations brings us to the ' high explosives," properly so called, and it is among these that most notable preparations are met with. Of all the explosive nitro-compounds, gun-cotton was the first practically employed (*vide* p. 741); but very soon after nitro glycerine was discovered by Sobrero when working in Pelouze's labora tory. This afterwards became known as " blasting oil," but it was many years before nitro-glycerine came into use as an explosive, namely, when, about 1860, Nobel, a Swedish engineer, had established factories for its production as an agent for blasting. At first there were difficulties and dangers attending its use, and it was only when Nobel had discovered the detonation method of setting free its tremendous energy that the new era of "high explosives" really commenced. Between 1860 and 1870 such a number of appalling catastrophes occurred in the handling of the new "blasting oil" that in several European countries its use was entirely prohibited. And, in England at least, this prohibition remains, for "in a liquid state this explosive cannot be sold in, or imported into this country. It is manufactured under the strict provision that it is forthwith made up into dynamite or some kindred licensed explosive." * * * "The only source, practically speaking, of nitro-glycerine on a commercial scale in this country is the factory of Nobel's Explosive Co. (Ltd.) at Ardeer, in the county of Ayr." * Nitro-glycerine being so ex-tremely dangerous to handle in the liquid form led Nobel to propose its use in an altered condition, by causing it to be absorbed by some inert porous material, the most suitable being a siliceous earth found in Germany, and there known as *kieselguhr*, of which one part will absorb three times its weight of liquid nitro-glycerine. Here we have the original dynamite, but now other substances are used for absorbing the liquid, and there are, indeed, dynamites of two different classes :

1. Dynamites with inert absorbents.
2. Dynamites with absorbents which are themselves combustible, or explosive.

Of the latter class there are endless varieties. One that has latterly been much used is called "blasting gelatine," and is practically a com-bination of nitro-glycerine and nitro-cotton, this last ingredient being a less nitrated cellulose than gun-cotton. Blasting gelatine contains a very large percentage of nitro-glycerine (93-95 per cent.), and has the appear-ance of stiff jelly of a pale yellow colour. It may be of interest to remark that this second class of dynamites admits of well-defined sub-divisions according to the nature of the absorbent, as :

(*a*) Charcoal, or other simple carbonaceous material.
(*b*) Gunpowder, or other nitrate or chlorate mixtures.
(*c*) Gun-cotton, or other nitro-compounds.

* Major Cundill, H.M.'s Inspector of Explosives.

FIG. 340.—*View on the Tyne.*

MINERAL COMBUSTIBLES.

———◆———

CERTAIN mineral combustibles may fairly claim attention in a work treating of the discoveries of the nineteenth century, not because these bodies have been known and used only in recent times, but for other reasons. The true nature of coal—that most important of all combustibles—its relation to the past history of the earth, and to the present and future interests of mankind ; the work it will do ; the extent of the supply still existing in the bowels of the earth ; the innumerable chemical products which it yields—are subjects on which the knowledge gained during the present century forms a body of discovery of the most interesting and important kind. Another substance we have to mention, though not a modern discovery, has lately been found in far greater abundance, and is now so largely used for various purposes, that it has become an important article of commerce.

COAL.

MOST persons know, or at least have been told, that coal is fossil vegetable matter,—the long-buried remains of ancient forests. But probably many receive the statement, not perhaps with incredulity, but with a certain vague notion that it is, after all, merely a daring surmise. And, indeed, nothing is at first sight more unlike stems, or leaves, or roots of plants than a lump of coal. Then everybody knows that coal is found thousands of feet beneath the surface of the earth, whereas plants can grow only in the light of the sun. One begins to understand the matter only when the teachings of geology have shown him that, so far from the crust of the earth being, as he is apt to suppose, fixed and unchangeable, it is in a state of constant fluctuation. Changes in the levels of the ground are always going on : in one place it is rising, in another sinking ; here a

FIG. 341.—*Fossil Trees in a Railway Cutting.*

tract of land is emerging from the ocean, there a continent is subsiding beneath the water. The extreme slowness with which these changes proceed causes them to escape all ordinary observation. The case may be compared to the hour-hand on the dial, which a casual spectator might pronounce quite stationary, since the observation of a few seconds fails to detect its movement. As the whole period comprehended in human annals counts but as a second of geological time, it cannot be wondered at that it required a vast accumulation of facts, and much careful and patient deduction from them, before a conclusion was reached so apparently contradictory of experience. It is, indeed, startling to learn that "the sure and firm-set earth" is in a state of flow and change. Even the "everlasting hills" give evidence that their materials were collected at the bottom of the sea, and we know that the water which runs down their sides is slowly but surely carrying them back particle by particle. Of the magnitude of the changes which the surface of the earth has undergone in times past, and which are still imperceptibly but constantly proceeding, the ordinary experience of mankind can of itself give no example. But such changes have sufficed to entomb a vast quantity of relics of the innumerable forms of vegetation which flourished and waved their branches in the sun, ages upon ages before the advent of man.

It may be thought impossible that vegetable matter should have so changed as to become a dense, black, glistening, brittle mass, showing no obvious forms of leaves or texture of wood. But no one who has seen how a quantity of damp hay closely pressed together will, after a time, become heated and change in colour to black, can have any difficulty in comprehending how chemical and mechanical actions may completely alter the aspect of vegetable matter. We have, however, the most direct evidence of the vegetable origin of coal in the numberless unquestionable forms of trees and plants met with in all coal strata. Sometimes the trunks of the trees fossilized into stony matter are found upright in the very situation in which they grew. Thus in Fig. 341 is represented the appearance exhibited by the trunks and roots of some fossil trees, which were exposed to view in the formation of a railway cutting between Manchester and Bolton. In every coal-field also beautiful impressions of the stems and leaves of plants are met with—one common form of which is shown in Fig. 342. Most of the plants so found belong to the flowerless division of the vegetable kingdom. Some are closely allied to the ferns of the present day—to the common "mare's-tail" (*Equisetum*), to the club-moss, and to other well-known plants. The firs and pines of the coal age are scarcely distinguishable from existing species. If a fragment of ordinary coal be ground to a very thin slice—so thin as to be transparent—and placed under the microscope, it will show a number of minute rounded bodies, which are, there is good reason to believe, nothing else than the spores or seeds of plants, closely resembling the existing club-mosses. The spores of the club-moss (*Lycopodium*) are so full of resinous matter, that they are used for making fireworks and the flashes of lightning at theatres. It is, therefore, extremely probable that the bitumen of coal is due to the resin of similar spores, altered by the effects of subterranean heat. The

FIG. 342.—*Impression of Leaf found in Coal Measures (Pecopteris).*

48

FIG. 343.—*Possible Aspect of the Forests of the Coal Age.*

immense abundance of these little spores in the coal is a proof that they accumulated in the ancient forests as the mosses grew, and therefore the matter of coal was not accumulated under water or washed down into the sea ; for these little spores are extremely light, and they cannot be wetted by water, and therefore they would have floated on the surface, and would not have been found so diffused throughout the coal. Fig. 343 is a picture of the possible aspect of the ancient forests of the coal age. In the humid atmosphere which probably prevailed at that period, the large tree-ferns and gigantic club-mosses, which are conspicuous in the picture, must have flourished luxuriantly.

The immense importance of coal for domestic purposes will be obvious from the fact that it is estimated that in the United Kingdom alone no less than 30,000,000 tons are annually consumed in house fires. Another great use of coal is in the smelting, puddling, and working of iron, and this probably consumes as much as our domestic fireplaces. Then there is the vast consumption by steam engines, by locomotives, and by steamboats. Another purpose for which coal is largely used is the making of illuminating gas ; and to the foregoing must also be added the quantity which goes to feed the furnaces necessary in so many of the arts—such as in the manufacturing of glass, porcelain, salt, chemicals, &c. The quantity of coal raised in Great Britain was not accurately known until 1854, when it was ordered that a register should be kept, and an annual return made. The following figures, in round numbers, are the returns published up to 1873. The table is continued in Note A.

Year.	Coal raised, in Tons.	Year.	Coals raised, in Tons.
1854	64,661,000	1864	92,787,000
1855	64,453,000	1865	98,150,000
1856	66,645,000	1866	101,630,000
1857	65,395,000	1867	104,500,000
1858	65,008,000	1868	103,141,000
1859	71,979,000	1869	107,427,000
1860	83,208,000	1870	110,289,000
1861	85,635,000	1871	117,352,000
1862	83,638,000	1872	123,497,000
1863	88,292,000	1873	127,017,000

The first return showed our annual produce to be 64,661,000 tons. The amount did not greatly vary until 1859, when there was an increased production of nearly seven millions of tons ; in 1860 a further increase of eleven millions of tons more. Since then the quantity annually raised has been increasing. Comparing the quantity which has been raised in any year after 1863 with that raised ten years before, we see that the increase in ten years is nearly half as much again ; or, that at the present rate of increase the amount annually raised doubles itself at least every twenty years. Now, the question arises, How long can this go on? However great may be the store of coal, it must sooner or later come to an end. Is it possible to calculate how long our coals will last? and what are the results of such calculations? These calculations have been made ; but there are great discrepancies in the results, for the estimates of the amount of available coal still remaining vary greatly, and different views are held regarding the rate of consumption in the future. A very liberal estimate, by an excellent authority, of the quanuty of coal remaining under British soil, makes it 147,000 millions of tons. With a consumption stationary at the present rate, this will last 1,200 years ; with an increase of consumption of 3,000,000 tons a year, 276 years ; but if the consumption continues to increase in the same geometrical ratio it has hitherto followed, the supply will scarcely last 100 years. It cannot be conceived, however, that this last will be the real case, for the increasing depth to which it will be necessary to go will soon cause a great increase in the cost, and thus effectually check the consumption. Great Britain will, however, be compelled to retire from the coal trade altogether, by the cheaper supplies which other countries will yield, long before the absolute exhaustion of her own coalfields. It is calculated that the coal-fields of North America contain thirteen times as much as those of all Europe put together. Coal is also found abundantly in India, China, Borneo, Eastern Australia, and South Africa; and it is believed that these stores will supply the world for many thousand years.

Seeing, then, that our supply of coal has a limit, and that at the present increasing rate of consumption, the chief source of the wealth of Great Britain must necessarily be exhausted in a few more centuries, it behoves us to turn our mineral treasures to the best account, and to adopt every possible means of obtaining from our coal its whole available heat and force. The amount of avoidable waste of which we are guilty in the con-

FIG. 344.—*The Fireside.*

sumption of coal is enormous. This is especially the case in its domestic use, where probably nineteen-twentieths of the heat produced is absolutely thrown away—sent off from the earth to warm the stars. In England people look upon the wide open fireplace as the image and type of home comfort. No doubt there are, from long use and habit, many pleasing associations which cluster round the domestic hearth ; but we, to whom it is given to "look before and after," must think what it takes to feed that wide-throated chimney. All but a very small fraction of the heat thus escapes, and is lost to man and the world for ever ; and surely we shall deserve the curses of our descendants if we continue recklessly to throw away a treasure which, unlike the oil in the widow's cruse, is never renewed—for there is no contemporaneous formation of coal. Thanks to the enhanced price of coal during the last few years, some attention has been directed to contrivances for the economical consumption of coal in its domestic, as well as in its manufacturing, applications.

A time, however, will sooner or later come, when the whole available coal shall have been consumed. What will then be the fuel of the engines, and steamboats, and locomotives of the future? The reader may think that then it will only be necessary to burn wood. But wood is already being consumed from the face of the earth much more rapidly than it is produced. How, then, can it be available when coal fails? The truth is, we are now consuming not merely the wood which the sun-rays are building up in our own time, but in hewing down the forests we are using the sun-work of a century, while in coal we have the forests of untold ages at our disposal—the accumulated combustible capital stored up during an immense period of the earth's existence. Upon this stored-up capital we are now living, our current receipts of sun-force being wholly inadequate to meet our expenditure. The coal is the sun-force of former ages; and it is from this we are now deriving the energy which performs most of our work. George Stephenson long ago declared that his locomotives were driven by sunshine—by the sunshine of former ages bottled up in the coal. And he was right. The mechanical energy of our steam engines, and the chemical energy of our blast furnaces, are derived from the combustion of vegetable matter, in which the heat and light of the sun—our present sun or that of the coal ages—are in some way stored up. The burning of wood or coal is, chemically, the reverse action to that performed by the sunlight : by the former carbon and oxygen are united, by the latter they are separated.

We foresee, then, a future period—however distant may be that future—in which the world's capital shall have been exhausted, and the energies which are now employed in doing the world's work will no longer be available. But the reader will perhaps think that by improvements in the steam engine, and in other ways, means will be found of getting more and more work out of coal. It is true that we obtain from coal only a *fraction* of its available energy ; but the whole work which could, by any possible process, be done by the combustion of coal is *definite and limited*, although its amount is large. A pound of coal burnt in one minute sets free an amount of energy which would, if it could all be made available, do as much as 300 horses working in the same time. But, again, the reader may think, even if at some distant future the supplies of fuel for the steam engines of our remote posterity should fail, that before that time some other form of force than steam or heat engines will have been discovered—some application of electricity, for example. Now, it will appear, from principles which will be discussed in a subsequent article, that not only is there no probability of such a discovery, but that now, when the relations of the whole available energies of the globe have been traced and defined, Science can find no ground for admitting such a possibility under the present condition of the universe.

PETROLEUM.

WHEN coal is heated in closed vessels, there are given off, as we shall presently see, a number of gaseous and volatile products—many being compounds of carbon and hydrogen—which condense to liquids or solids at ordinary temperatures. Carbon is by far the largest constituent of coal, which commonly contains only about 10 per cent. of other substances, although the proportions vary very widely. Another important constituent of coal is its hydrogen, and the value of coal as a source of heat depends almost entirely upon the carbon and hydrogen it contains. Carbon is one of the most remarkable of all the elements of the globe for its power of entering into an enormous number of compounds. Thus, for example, the compounds of carbon with only hydrogen are innumerable; but they are all definite, and their composition is expressible by the admirable system of chemical symbols, of which the reader has now, it is hoped, some definite notion. Perhaps these hydro-carbons are among the best evidences which could be adduced that modern science has obtained a grasp of certain conceptions which have a real correspondence with the actual facts of nature, even as regards the intimate constitution of matter. This is not the place to enter into a complete exposition of this subject. We may, however, invite the reader's attention to a few simple facts. A very large number of compounds of carbon and hydrogen are known. If the percentage compositions of these be compared together, it is only the eye of a most expert arithmetician which can detect any relation between the proportions of the constituents in the various compounds. The chemist, however, by associating such of these compounds as resemble each other in their general properties, finds that they can be arranged in series, in which the composition is accurately expressed by multiples of the proportions: $C = 12$, $H = 1$. And not only so, the different series themselves form a series of series, having a simple relation to each other. Thus, confining ourselves to some of the known hydro-carbons, we have the following:

A	B	C	D	E	F
$C\ H_4$	$C\ H_2$				
C_2H_6	C_2H_4	C_2H_2			
C_3H_8	C_3H_6	C_3H_4	C_3H_2		
C_4H_{10}	C_4H_8	C_4H_6	C_4H_4	C_4H_2	
C_5H_{12}	C_5H_{10}	C_5H_8	C_5H_6	C_5H_4	C_5H_2
C_6H_{14}	C_6H_{12}	C_6H_{10}	C_6H_8	C_6H_6	C_6H_4
&c.	&c.	&c.	&c.	&c.	&c.
C_nH_{2n+2}	C_nH_{2n}	C_nH_{2n-2}	C_nH_{2n-4}	C_nH_{2n-6}	C_nH_{2n-8}

This table might be indefinitely extended, but enough is given to enable the intelligent reader to discover the laws connecting these formulæ. The series headed B, it will be observed, have all the same percentage composition. Why, then, one formula rather than another? The answer to this question is the statement of a theoretical law upon which the whole science of modern chemistry is based; for it has the same relation to that science as gravitation has to astronomy. It is a matter of fact that all gases, whatever their chemical nature, expand alike with the same application of heat, and all obey the same law, which connects volumes and pressures. These are very remarkable uniformities, for gases in this respect exhibit the most decided contrast to liquids and solids. The volume of each solid and of each liquid has its own special relations to temperature and pressure; here there is endless diversity. The volumes of all gases have one and the same relation to temperature and pressure: here there is absolute uniformity. As an explanation of these and other facts relating to gases, Amedeo Avogadro, in 1811, put forward this hypothesis—*Equal volumes of all gases, under like circumstances of temperature and pressure, contain the same number of molecules.* This hypothesis was revived by Ampère a few years later, and sometimes is called his. A necessary consequence of this law is that the weights of the molecules of gases are proportional to their densities or specific gravities. Hence when the percentage composition of a hydro-carbon has been determined, by burning or oxidizing it in such a manner as to obtain and weigh the products, carbonic acid and water, the next thing the chemist does is to obtain the weight of a volume of the gas. The number of times this exceeds the weight of hydrogen gas, under the same conditions, expresses how many times the molecule is heavier than the hydrogen molecule. Now, the chemist's unit of weight in these inquiries is the weight of a single *atom* of hydrogen; and, as there are grounds for believing that the *molecule* of hydrogen consists of two atoms of that substance, its weight $= 2$. Now, if the molecule of marsh gas, the first hydro-carbon in the above list, has the composition assigned, it will be $12+4=16$ times heavier than the *atom* of hydrogen, and $\frac{16}{2}=8$ times heavier than the *molecule* of hydrogen. Hence, if Avogadro's law be correct, marsh gas should be just eight times heavier than hydrogen gas; which is really the fact. The formula expressing the composition of the molecule of a hydro-carbon, or of any chemical compound whatever, is always so fixed that the same relations may hold; and almost the first thing a chemist does in examining a new body is to endeavour to obtain it in the state of gas.

The first four members of the series headed A are gases at ordinary temperatures, the fifth is a gas at temperatures above the freezing-point, and a liquid at lower temperatures; the next following members are liquids which boil (that is, are converted into gases) at temperatures rising with each additional carbon atom about 20° F. The specific gravities and boiling-points of these liquids augment as we pass from one hydro-carbon to the next, and the lower members of the series are solids, fusing at temperatures higher and higher as the number of carbon atoms is greater. Similar gradations of properties are exhibited by the other series of hydro-carbons. Petroleum or rock-oil is the name given to liquid hydro-carbons found in nature, and consisting chiefly of compounds belonging to the series marked A in the above list. Some varieties of petroleum hold in solution other hydro-carbons, and in some cases paraffin is extracted from the oils by exposing the liquid to cold, when the solid crystallizes out. Paraffin is a solid belonging to the B series, and it is for the most part obtained by heating certain minerals.

Deposits of liquid hydro-carbons, perhaps formed by a kind of natural subterranean distillation from coal or other fossil organic matter, exist in various localities. These deposits have long been known and utilized at Rangoon, in Burmah, and on the shores of the Caspian Sea. At Rangoon the mineral oil is obtained by sinking wells about 60 ft. deep in a kind of clay soil, which is saturated with it. The oily clay rests upon a bed of slate also containing oil, and underneath this is coal. It may be supposed that subterranean heat, acting upon the coal, has distilled off the petroleum, which has condensed in the upper strata. This petroleum, when distilled in a current of steam, leaves about 4 per cent of residue, and the volatile portion contains about one-tenth of its weight of a substance (paraffin) which is solid at ordinary temperatures. After an agitation with oil of vitriol, and another distillation, *rock oil* or *naphtha* is obtained, which, however, is still a mixture of several distinct chemical compounds. Mineral oils have also been found in China, Japan, Hindostan, Persia, the West India Islands, France, Italy, Bavaria, and England. In one of the Ionian Islands there are oil-springs which have flowed, it is said, over 2,000 years.

But it is the recently discovered and extremely copious springs and wells in Pennsylvania and Canada which have given a vastly extended importance to the trade in mineral oil. Rock oil is now used in enormous quantities as the cheapest illuminating oil, and that which furnishes the most intense light. Its consumption as a lubricating oil for machines has also been very large. Mineral oil was occasionally found at various places in the United States, and sometimes used by the inhabitants of the locality before the recent discoveries; but it was not until August, 1859, that it was met with in large quantities. About this time a boring which was made at Oil Creek, Pennsylvania, reached an abundant source, for 1,000 gallons a day were drawn from it for many weeks. The news of the discovery of this copious oil-spring spread rapidly : thousands of persons flocked to the neighbourhood in hopes of easily making a fortune by "striking oil." Before the end of 1860 more than a thousand wells had been bored, and some of these had yielded largely. The regions of North America in which petroleum has been found cover a large part of the States, and comprise Pennsylvania, New York, Ohio, Michigan, Kentucky, Tennessee, Kansas, Illinois, Texas, and California. In the vicinity of Oil Creek the bore-holes are usually about 3 in. or 4 in. in diameter, and are often 500 ft. deep, and even 800 ft. is not uncommon. To make a bore-hole 900 ft. deep, and procure all the requisites—steam engines, barrels, &c., for pumping the oil— costs about $5,000. In 1869 many of these wells still yielded regularly 300 barrels a day, but the supply has not continued with the same abundance. One of the luckiest wells flowed at its first opening at the rate of about 25,000 barrels a day. The apparatus used for working the oil-wells is very simple—a rude derrick, a small steam engine, a pump, and some barrels and tubs being all that is necessary. Fig. 345 will give the reader an idea of the scene presented by a cluster of oil-wells in the Oil Creek region. Oil Creek received its name before the petroleum trade was established, from the oil found floating on the surface of the water. It is on the Alleghany River, about 150 miles above Pittsburg, and here at its mouth is situated Oil City. There is a wharf in Pittsburg for the oil traffic, and the barrels are brought down the river in flats, or the oil is poured into very large flat boxes, divided into compartments, which are then closed, and the boxes floated down in groups of twenty or more. The refining process consists in placing the crude oil into a large iron retort, connected with a condenser formed of a coil of iron pipes, surrounded by cold water. Heat is applied, and the lighter hydro-carbons (naphtha) come over first. After the naphtha, the oils which are used for illuminating purposes distil off. A current of steam is then forced into the retort, and this brings over the heavy oils which are used for greasing machinery. A black tarry oil yet remains ; and, finally, after the separation of this, a quantity of coke. The products are subjected to certain processes of purification, which need not here be described. The magnitude of the American oil trade may be inferred from the fact that in the second year of its existence, from January to June, 1862, more than 4,500,000 gallons were exported from four seaports. This can hardly be wondered at, considering the extremely low

FIG. 345.—*View on Hyde and Egbert's Farm, Oil Creek.*

price at which this excellent illuminating and lubricating agent can be produced. Refined petroleum can be bought at Pittsburg for 16 cents. per
gallon. It is believed by some that the supplies of petroleum which exist
in various localities are so abundant that they will furnish illuminating oils
to the whole world for centuries.

PARAFFIN.

IN the course of some researches on the substances contained in the tar,
which is obtained by heating wood in close vessels, Reichenbach found
a white translucent substance, to which he gave the above name, because
it was not acted upon by any of the ordinary chemical reagents, such as
sulphuric acid, nitric acid, &c. This substance, which is composed of
carbon and hydrogen only, is not unlike spermaceti; it is colourless, translucent, and without smell or taste. But when slightly warmed, it becomes
very plastic, and may then be moulded with the greatest ease—and in this
respect it differs from spermaceti. Paraffin melts at from 88° to 150° F.,
to a colourless liquid, which is so fluid that it may be filtered through paper
like water, and at a higher temperature it can be distilled unchanged.
Paraffin does not dissolve in water, and is but slightly soluble in alcohol.
In ether, naphtha, turpentine, benzol, and sulphide of carbon, it dissolves
very readily. When heated with sulphur, it is decomposed: the sulphur
seizes upon its hydrogen, sulphuretted hydrogen is given off, and the
carbon is separated; and this action has been proposed as a ready means
of obtaining pure sulphuretted hydrogen for laboratory use. It is probable

that paraffin is a mixture of various hydro-carbons, having a composition expressible by the formula, $C_n H_{2n}$; for different specimens fuse at different temperatures, according as the paraffin has been obtained from one or the other source.

In the year 1847, Dr. Lyon Playfair directed the attention of Mr. James Young, then of Manchester, to a dense petroleum which issued from the crevices of the coal in a Derbyshire mine. It was soon found that this substance yielded a distillation—a pale yellow oil—which, on cooling, deposited solid paraffin. Mr. Young, recognizing the importance of this discovery, had an establishment at once erected on the spot, and the work of extracting paraffin was carried on until the supply of the petroleum had become nearly exhausted. Forced to seek for other sources of paraffin, Mr. Young was fortunate enough, after many trials, to discover that a species of bituminous coal, which occurs at Boghead, near Bathgate, in the county of Linlithgow, yielded by distillation annually large quantities of paraffin. In 1850 he procured a patent for "treating bituminous coals to obtain paraffin, and oil containing paraffin, therefrom." This method consisted in distilling the coal in an iron retort, gradually heated up to low redness, and kept at that temperature until the volatile products ceased to come off. Under this patent, Mr. Young developed the manufacture of paraffin into a new and important branch of industry. The oil which first comes over in the distillation of the Boghead mineral is largely used for illuminating purposes under a variety of names besides that of *paraffin oil*, which term is, we believe, chiefly applied to a less volatile portion, extensively used for lubricating machinery, and consisting of liquid hydro-carbons of the same percentage composition as solid paraffin, which substance it also holds in solution. Mr. Young's process consisted in placing the mineral in a retort encased in brickwork—an arrangement which caused the temperature of the retort to be more uniform than if the heat of the furnace had been applied to it directly. The retorts were placed vertically, and they were fed with the mineral by a hopper at the top. The products of the distillation passed through a worm tube surrounded by cold water into a cooled receiver. The result of the first distillation was a crude oily matter, differing from tar in being lighter than water, and in not drying-up when exposed to the air. This crude oil was then several times alternately treated with sulphuric acid and caustic potash, and distilled; and when about two-thirds of the oil had been separated from the rest, as an oil for burning and lubricating purposes, the residue yielded paraffin, or "paraffin wax," as it is sometimes called. It is estimated that in Scotland no less than 800,000 tons of shale are annually distilled for mineral hydro-carbons, with a consumption of 500,000 tons of fuel. It is believed that about 25,000,000 gallons of crude oil are thus obtained, and from this 350,000 gallons of illuminating oil, 10,000 tons of lubricating oil, and 5,800 tons of solid paraffin are produced. Among the products exhibited in the International Exhibition of 1862, was a block of beautifully translucent paraffin, of nearly half a ton weight.

Paraffin is also obtained on the continent by distilling a variety of coal termed *lignite*. The tar which comes over is distilled, until nothing but coke remains. The condensed products are then treated with caustic soda, in order to remove carbolic acid and other substances. After washing with water, the oils are treated with sulphuric acid, in order to remove basic substances. The oil is again washed, and is then rectified by another distillation. The products which successively come over are, if necessary, separated by being collected in different vessels ; but sometimes they are mixed together, and sent into the market as illuminating oils under various names, such as " photogen," " solar oil," &c. Oils having a specific gravity about 0·9 are collected apart, and are placed in tanks in a very cool place. In the course of a few weeks the solid paraffin, which is dissolved in the other hydro-carbons, crystallizes out. The liquid oils are drawn off, and the crude paraffin, which is of a dark colour, is freed from adhering oil by a centrifugal machine, and afterwards by pressure applied by hydraulic power. It then undergoes several other processes of purification before it is obtained as a colourless translucent solid.

Several thousand tons of paraffin are annually consumed for making candles, which is the most important application of the material. The variation in the fusing-points of different specimens is doubtless due to admixtures in greater or less proportion of other more easily fusible hydro-carbons. It was on account of the imperfect separation of these that the candles first made from paraffin were so liable to soften and bend, and felt greasy to the touch. Paraffin for candle-making is sometimes mixed with a certain proportion of other substances, such as palmitic acid, &c. Among the patented applications of paraffin are the lining of beer-barrels, and the preserving of fruits, jams, and meat. Some kinds of paraffin are also used in the manufacture of matches.

Liebig once expressed a wish that coal-gas might be obtained in a solid form : " It would certainly be esteemed one of the greatest discoveries of the age if any one could succeed in condensing coal-gas into a white, dry, odourless substance, portable and capable of being placed in a candlestick or burned in a lamp." Now, it is curious that paraffin has nearly the same composition as good coal-gas : it burns with a bright and smokeless flame, and beautiful candles are formed of it, which burn like those made of the finest wax. When the fused paraffin first assumes the solid form, it is transparent like glass ; and if it could be retained in that condition, we might have the pleasing novelty of transparent candles. But the particles seek to arrange themselves in crystalline forms, and the substance soon takes on its white semi-opaque appearance.

The great richness of the Boghead mineral in paraffin, which appears to exist in it ready formed, prevented for many years any successful competition by the working of other sources of supply. But paraffin is an abundant constituent of Rangoon petroleum, and considerable quantities may be obtained by distilling peat, and other fossil substances. All petroleums and paraffins are, in fact, mixtures of a number of hydro-carbons, which in many cases cannot be entirely separated from each other. The accidents which have from time to time occurred with some of these combustibles, and have caused legislative enactments with regard to them, are due to the imperfect removal by distillation of the more volatile bodies, which rise in vapour at ordinary temperatures. Explosions of the hydro-carbons can occur only under the same circumstances as with coal-gas ; that is to say, the application of a flame to a mixture of the vapour with atmospheric air.

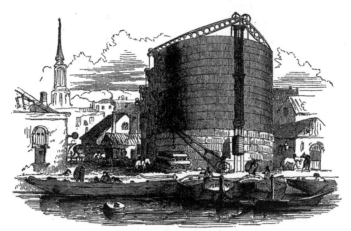

FIG. 346.—*View of the City of London Gas-works.*

COAL-GAS.

WHEN coal is burning in a common fire, we may see jets of smoky gas issuing from the pieces of coal before they become red hot. This vapour, coming in contact with flame in another part of the fire, may often be observed to ignite, thus supplying an instance of gas-lighting in its most elementary form. In the ordinary fire the air has free access, and the inflammable gases and vapours continue to burn with flames more or less bright, and when these have ceased the cabonaceous portion continues afterward to glow until nearly the whole has been consumed, except the solid residue which we call the ashes. These ashes in general contain a portion of unconsumed carbon, mixed with what is chemically *the ash,* namely, certain incombustible salts, constituting the white part of the ashes. If, however, we heat the coal in a vessel which prevents access of air, and allows the gases to escape, the coal is decomposed much in the same way as when it is burnt in the open fire; but the products formed are no longer burnt, the supply of oxygen being cut off. Every one knows the familiar experiment of filling the bowl of a common clay tobacco-pipe with powdered coal, then covering it with a dab of clay, and placing it in a fire. The gas which soon comes from the stem of the pipe does not take fire unless a light be applied, when it may be seen to burn with a bright flame, and after the flow of gas has ceased, nearly the whole of the carbon of the coal will be found unconsumed in the bowl of the pipe. This simple experiment illustrates perfectly the first step in the manufacture of coal-gas, namely, the process of heating coal to redness in closed vessels, by which operation the substances originally contained in coal are destroyed, and their elements enter into new combinations.

These elements are few in number; for, except the very small portion which remains as incombustible white ash, coal is constituted of carbon, hydrogen, oxygen, nitrogen, and a little sulphur. All the varied and interesting products obtained by the destructive distillation of coal are combinations of two or more of these four or five elements. Illuminating gas is far from being the only product when coal is heated without access of air; for of the numerous substances volatized at the red heat of the gas-retort a great number are not only incapable of affording light, but liable to generate noxious compounds when burnt. Besides this there are numerous bodies which, though leaving the retort in the gaseous form, immediately assume the liquid or solid state at ordinary temperatures. All such substances must be separated before permanent gases are obtained fit for illuminating purposes and capable of being carried through pipes to distant places. Thus an important part of the apparatus for gas manufacture consists in arrangements for separating the condensible bodies, and for removing useless or injurious gases from the remainder.

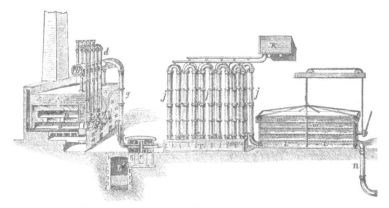

FIG. 347.—*Section of Gas-making Apparatus.*

The products resulting from the destructive distillation of the coal may, therefore, be classified as—*a*, solids left behind in the retort; *b*, solids and liquids condensed by cooling the vapours which issue from the retort; *c*, coal-gas—a mixture of gases from which certain useless and noxious constituents must be removed. Fig. 347 is intended to give a diagrammatic view of the apparatus employed in the generation, purification, and storage of gas, the various parts being shown in section. A is the furnace containing several retorts, of which B is one. From each retort a tube, *d*, rises vertically, and curving downward like an inverted U, it enters a long horizontal cylinder, *f*, half filled with water, beneath the surface of which the open end of the recurved tube dips. The cylinder containing water passes horizontally along the whole range of furnaces in the gas-works, and is known as the *hydraulic main*. It is here, then, the tar and the moisture first condense, and the pipe is always kept half full of these liquids, so that the ends of the pipes, *d*, from the retorts, dipping beneath its surface, form traps or water-valves, which allow any retort to be opened without per-mitting the gas to escape. As the tar accumulates in the hydraulic main, it flows over through a pipe, *g*, leading downwards into the *tar-well*, H. The gases take the same course ; but while the tar flows down the vertical tube, R, the gases pass on through *j* into the condensers or refrigerators. Gas cannot escape from the open end of the tube, for it is always closed by the liquids—tar and ammoniacal liquor—which accumulate and flow over the top of the open inner vessel into the cistern, S, from which they are drawn off from time to time by the stop-cock, I. Although when the gas has arrived at this cistern much of its tar and ammoniacal vapours have been condensed, a portion is still retained by reason of the high tempera-ture of the gas ; and this has to be removed before it is permitted to enter the purifier. This is the object of passing the gas through the series of pipes, *j j*, forming the *condenser*. These are kept cool by the large surface they expose to the air, and, when necessary, cold water from the cistern, K, may be made to flow over them. The tar and other liquids condense in the iron chest, T, which is so divided by partitions as to compel the gas to pass through the whole series of tubes ; and as the liquid accumulates, it also overflows into the tar-well. The cooled gas then enters the purifier, L L, in which are layers of slaked lime placed on a number of shelves. By contact with the extensive surface of slaked lime the gas has its sulphur-etted hydrogen, carbonic acid, and some other impurities, removed ; and it then, through the tube *n*, enters the gasholder, in which it is stored up for use.

Hydrated oxide of iron is now much used for purifying coal-gas. The oxide is mixed with sawdust, and placed in layers 10 in. thick. Sulphide of iron and water is formed ; and when the mixture has ceased to absorb any more, it is removed and exposed to a current of air ; the hydrated oxide is thus reproduced and sulphur set free. The process may be re-peated many times in succession, until the absorbent power is impaired by the accumulation of sulphur.

The gasholder—or " gasometer," as it is often improperly named—is an immense cylindrical bell, made of wrought iron plates, and inverted in a tank of water, in which it rises or falls. It is counterpoised by weights attached to chains passing over pulleys, so as to press the gas with a small force in order to drive it along the main, which communicates with the pipes sup-plying it to the various consumers. The pressure impelling the gas through the mains does not in general exceed that of a column of water two or three inches high.

It will be necessary, after this slight outline describing the essential parts of the apparatus, to enter more fully into the details of the several parts.

The retorts are constructed of wrought iron, cast iron, or earthenware, and in shape are cylindrical, with a diameter of 12 in. to 18 in., or more, and a length of 6 ft. to 10 ft. Though sometimes circular in section, other forms are commonly used—such as the elliptical, and especially the ◠-shaped. The retorts are closed except at the mouth-end, Fig. 348, from the top of which rises the stand-pipe, A, which has usually a diameter of about 5 in. When the charge has been introduced, the mouth is closed by a plate of iron, B, closely and securely applied by means of a screw, C, as shown in the figure—a perfectly tight joint being obtained by a luting of lime mortar spread on the part of the lid which comes into contact with the mouth of the retort. The retorts are always set horizontally in the furnace—each furnace usually including a set of five retorts. The charge of coals is introduced on a tray of sheet iron adapted to the size of the retort, which, when properly pushed in, is inverted so as turn out the contents, and then withdrawn.

The time required to completely expel the volatile constituents from the charge in a gas retort varies very much, because there are great diversities in the composition of the different kinds of coal employed. Some varieties of coal, such as cannel, are easily decomposed, and the operation may be complete in about three hours ; while other kinds may require double that time. The quantity of gas procurable from a given weight of coal also varies according to the kind of coal made use of. Thus, while a hundredweight of cannel may give 430 cubic feet of gas, the same weight of Newcastle coals will yield but 370 cubic feet. The nature of the gases given off from a retort will be different at the different stages of the operation.

FIG. 348.— *The Retort.*

The scene presented by the retort-house of a large gas manufactory, when viewed at night, is a singular spectacle. The strange lurid gleams which shoot out amid the general darkness as the retorts are opened to withdraw the coke, and the black forms of the workmen partially illuminated by the glare, or flitting like dark shadows across it, form a picture which might engage the pencil of a Rembrant. In Fig. 348*a* is depicted the retort-house at the Imperial Gas Works, King's Cross. Here the retorts are arranged in several tiers—the coal being brought, and the coke withdrawn, by the aid of an iron carriage running on rails parallel to the line of furnaces.

In the process of heating, a proper regulation of the temperature is of the highest importance. It is found that when the retorts are heated to bright cherry-red, the best results are obtained. At a lower temperature a larger quantity of condensable vapours are given off, which collect in the gasholders and distributing pipes as solid or liquid, and occasion much inconvenience, while the quantity of gas obtained is decreased. On the other hand, if the temperature be too high, some of the gases are decomposed, and the quantity of carbon contained in the product is so much diminished as seriously to impair the illuminating power. Again, every second the gases after their production remain in the red-hot retort diminishes their light-giving value ; for those hydro-carbons on which the luminiferous power of the gas depends, are then liable to partial decomposition ; a portion of their carbon is deposited on the walls of the retort in a dense layer, gradually choking it up, while the liberated hydrogen does

FIG. 348a.—*Retort House of the Imperial Gas-Works, King's Cross, London.*

not add to the illuminating but to the heating constituents of the gas. A plan has been patented by Mr. White, of Manchester, for rapidly removing the illuminating gases from the retort by sweeping them out by means of a current of what has been termed "water gas." This water gas is produced by causing steam to pass over heated coke, and is a mixture of carbonic acid, carbonic oxide, and hydrogen. Though only two of these are combustible gases—and even they do not yield light by their combustion, and, by adding to the bulk of the gas, serve rather to dilute it—yet it has been found that in some cases twice the amount of light is obtainable by White's process than the same weight of coal supplies when treated in the ordinary manner.

The hydraulic main, as already mentioned, being kept half full of tar into which the lower ends of the dip-pipes descend, prevents the gas from escaping through the stand-pipes when the lid of a retort is removed for the introduction of a fresh charge. The hydraulic main is from 12 to 18 in. diameter, and the dip-pipes pass into it by gas-tight joints. Various forms of purifiers are in use besides the simple one already mentioned. Some of these have arrangements for agitating the gas with a purifying liquid by mechanical means, the motion being supplied by a steam engine.

The gasholder, as it sinks in the water of the cistern, presses with less force on the contained gas, and unless this inequality of pressure were counteracted there would be very unequal velocities in the flow of gas from the burner. The equality of pressure is obtained by making the weight of the chains by which the gasholder is suspended equal to half the weight the gasholder loses in the same length of its motion. Gasholders are also constructed without chains or counterpoises, as these are found to be unnecessary where the height of the gasholder does not exceed half its width. In such cases, especially when the vessel is very large, the difference of pressure at the highest and lowest position is quite inconsiderable, and nothing more is necessary than that upright guides or pillars be placed to preserve the vertical motion of the vessel. Another improvement, which enables a lofty gasholder to be used without increasing the depth of the tank, consists in forming the gasholder of several cylinders, which slide in and out of one another like the draw-tubes of a telescope. Each cylinder has a groove formed by turning up the iron inside the rim, and at the top of the next cylinder the edge is turned outwards so as to drop in the groove or channel, which thus forms a gas-tight joint, for it is of course filled with water as it rises. The pressure is, however, more accurately regulated by an apparatus called the *governor*, through which the gas passes in before it enters the mains. The construction and action of the regulator will be understood from Fig. 349, where A represents a kind of miniature gasholder, inverted in the cistern, B. From the centre of the interior of the bell hangs a cone, C, within the contracted orifice of the inlet-pipe. If this cone be drawn up, the size of the orifice, D, is reduced, and, on the other hand, by its descent it enlarges the opening through which the gas passes outward. By properly adjusting the weights of the counterpoise, E, such a position of the cone may be found that the gas passes into the mains at an assigned pressure. Suppose, now, that from any cause the pressure of gas in F increases, that pressure acting upon the inverted bell, A, causes it to rise and carry with it the cone, which, by narrowing the orifice of the outlet, checks the flow of gas. Similarly, a decrease of pressure in the mains would be followed by the descent of the cone, and consequently freer egress of gas. In hilly towns it is necessary

49

to fix regulators of this kind at certain heights in order to equalize the pressure. It is found that a difference of 30 ft. in level affects the pressure of gas in the same main to about the same amount as would a column

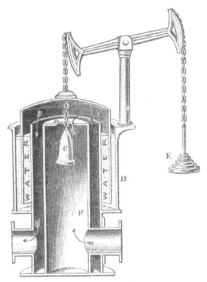

Fig. 349.—*The Gas Governor.*

of water one-fifth of an inch high, the pressure being least at the lowest point.

Coal-gas is a mixture of several gases, and these may be classified as, first, the light-giving gases, or those which burn with a luminous flame; secondly, gases which burn with a non-luminous flame, and which therefore contribute to the *heat*, and not to the light, of a gas-flame, and have the effect of diluting the gas; third, gases and vapours which are properly termed impurities, as they are either incombustible or by their combustion give rise to injurious products. Of the first kind the principal is olefiant gas, a gas which burns with a brilliant white flame without smoke. It is a compound of hydrogen and carbon, six parts by weight of carbon being combined with one part by weight of hydrogen. Besides olefiant gas other gaseous hydrocarbons are found in smaller quantities. These contain a larger proportion of carbon than olefiant gas. The second class contains hydrogen, light carburetted hydrogen, and carbonic oxide. Hydrogen is one element of water, of which it forms one-ninth of the weight. It burns with a flame giving singularly little light, but having intensely heating power; in fact, one of the brightest lights we can produce is obtained by allowing the flame of burning hydrogen to heat a piece of lime. Light carburetted hydrogen, like olefiant gas, is a compound of hydrogen and carbon, but the proportion of carbon to hydrogen is only half what it is in olefiant gas, namely, three parts to one. This gas enters largely into the composition of coal-gas, and occurs naturally in the coal seams, being, in fact, the dreaded *fire-damp* of the miner. It is much lighter than olefiant gas, for while that gas is of nearly the same specific gravity as atmospheric air, light carburetted hydrogen is only a little more than half that specific gravity. It is this ingredient of coal-gas which renders it so light as to be available for inflating balloons. It burns with either a bluish or a slightly yellow flame, yielding hardly any light. Olefiant gas and the other luminiferous hydro-carbons, when exposed to a bright red heat, split up for the most part into this gas and carbon. This explains the importance of rapidly removing the gas from the retort in which it is generated, a point which has been referred to above. Carbonic oxide is a gas which one may often see burning with a pale blue flame above the glowing embers of a common fire, the flame giving, however, little light. It is a compound of carbon and oxygen, containing only one-half the quantity of oxygen which its carbon is capable of uniting with, and therefore ready to unite with another proportion, which it does in burning, carbonic acid being the product.

The third class of constituents of coal-gas—the impurities—are those which the manufacturer strives to remove by passing the gas over lime, milk of lime, oxide of iron, &c. Sulphuretted hydrogen, a compound of sulphur and hydrogen, has an extremely nauseous odour resembling that of rotten eggs. It is always formed in the distillation of coal, and if not removed from the gas in the process of purification, it has a very objectionable effect ; for one product of its combustion is sulphurous acid, and in a room where such gas is burnt much damage may be done by the acid vapours; for example, the bindings of books, &c., soon become deteriorated from this cause. The detection of sulphuretted hydrogen in coal-gas is quite easy, for it is only necessary to hold in a current of the gas a piece of paper dipped in a solution of the acetate of lead. If in a few minutes the paper becomes discoloured the presence of sulphuretted hydrogen is indicated.

But the *bête noire* of the gas-maker is a substance called "sulphide of carbon," which is formed whenever sulphur and carbonaceous matters are brought together at an elevated temperature. Sulphide of carbon is, in the pure state, a colourless liquid, of an intensely offensive odour, resembling the disagreeable effluvia of putrefying cabbages. The liquid is extremely volatile, and coal-gas usually contains some of its vapour. When too high a temperature is used in the generation of the gas, it contains a large quantity of this deleterious ingredient, especially if the amount of sulphur contained in the coal is at all considerable. This sulphide of carbon vapour is very inflammable, and one product of its combustion is a large quantity of sulphurous acid. This substance cannot be removed from coal-gas by any process sufficiently cheap to admit of its application on the large scale. It is said, however, that by passing the gas over a solution of potash in methylated spirit, the sulphide of carbon vapour can be completely got rid of. The price of these materials renders the process available in special cases only, where the damage done by the sulphurous acid would be serious, as in libraries, &c. Besides the impurities we have already enumerated, many others are present in greater or less quantity. Carbonic acid—the gas resulting from the complete combustion of carbon—should be entirely removed by the lime purifiers, as the presence of even a small percentage detracts materially from the illuminating power. This gas is not inflammable and cannot support combustion. It has decided acid properties, and readily unites with alkaline bases forming carbonates :

it is upon this behaviour that its removal by lime depends. The illuminating power of coal-gas containing only 1 per cent. of carbonic acid is reduced thereby by about one-fifteenth of its whole amount.

The proper mode of burning the gas so as to obtain the maximum amount of light it is capable of yielding requires a compliance with certain physical and chemical conditions. The artificial production of light depends upon the fact that by sufficiently heating any substance, it becomes luminous, and the higher the temperature the greater the luminosity. The light emitted by solid bodies moderately heated is at first red in colour; as the temperature rises it becomes yellow, which gradually changes to white when the heat becomes very intense. The widest difference exists, however, in the temperature required to render solids or liquids luminous, and that needed to cause gases to give off light. In all luminous flames the light is emitted by *solid* particles highly heated. Every luminous gas-flame contains solid particles of carbon, as may be easily shown by the soot deposited on any cold body—such as a piece of metal—introduced into the flame. On the other hand, the flame of burning hydrogen, which produces only aqueous vapour, furnishes no light, but a heat so intense, that a piece of lime introduced into the jet becomes luminous to a degree hardly supportable by the eye. The conditions requisite, therefore, for burning illuminating gas are, first, just such a supply of air as will prevent particles of carbon from escaping unconsumed in the form of smoke, and yet not enough to burn up the carbon before it has separated from the hydrogen, and passed through the flame in the *solid* state; second, the attainment of the highest possible temperature in the flame, compatible with the former condition. When the supply of oxygen is not in excess, the hydrogen of the gaseous hydro-carbon appears to burn first; the carbon is set free, and its solid particles immersed in the flame of the burning hydrogen are there intensely heated; but ultimately reaching the outer part of the flame, they enter into combination with the oxygen of the air, producing carbonic acid; or if present in excessive quantity, they are thrown off as smoke. If the purpose of burning the gas is to obtain heating effects only, this is accomplished by supplying air in such quantities, that the carbon enters into combination with oxygen in the body of the flame, without a previous separation from the hydrogen with which it is combined. In this case a higher temperature is attained, and the flame is wholly free from smoke; so that vessels of any kind placed over it remain perfectly clean and free from the least deposit of soot. The last result is of great advantage in chemical processes, especially where glass vessels require to be heated, for the chemist retains an uninterrupted view of the actions taking place in his flasks and retorts.

FIG. 350.—*Bunsen's Burner.*

No better illustration of the nature of the combustion in a gas-flame can be found than is furnished by Bunsen's burner, Fig. 350, now universally employed as a source of heat in chemical laboratories. In this burner the gas issues from a small orifice at the level of *a*, near the bottom of the tube, *b*, which is open at the top, and is in free communication at the bottom with openings through which air enters and mixes with the gas, as they rise together in the tube and are ignited at the top. If the pressure of the gas be properly regulated, the flame does not descend in the tube, but the mixture burns at the top of the tube, producing a pale blue flame incapable of emitting light, but much hotter than an ordinary flame, for the combustion is much quicker. If the openings at *a* be stopped, the supply of air to the interior of the tube is cut off, and then the gas burns at the top of the tube, *b*, in the ordinary manner, giving a luminous flame. Ordinary gas-jets burning in the streets, at open stalls or shops, may be seen on a windy night to have their light almost extinguished by the increased supply of oxygen, carried mechanically into the body of the flame, the white light instantly changing to pale blue. The disappearance of the light in such cases is due, as in Bunsen's burner, to the supply of oxygen in sufficient quantity to combine at once with the carbon as well as the hydrogen of the hydro-carbons.

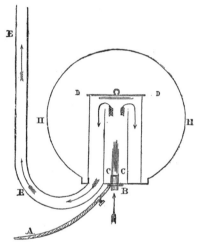

FIG. 351.—*Faraday's Ventilating Gas-burner*

The burners now chiefly used for the consumption of coal-gas for illuminating purposes are the bat's-wing, the fish-tail, and various forms of Argand. The bat's-wing burner is simply a fine slit cut in an iron nipple, and it produces a flat fan-like flame. The fish-tail is formed by boring two holes so that two jets of gas inclined at an angle of about 60° infringe on each other and produce a flat sheet of flame. The Argand, in its simplest form, consists of a tubular ring perforated with a number of small holes from which the gas issues. Many modifications of this kind of burner have been devised, in all of which a glass chimney is requisite to obtain a current of air sufficient to consume the gas without smoke, and it is important that the height of the chimney should be adapted to the amount of light required if the gas is to be used economically. Argand burners are specially advantageous where a concentrated light is required. Fig. 351 represents a ventilating gas-burner, contrived by Faraday, the object being to remove from the apartment the whole of the products of the combustion of the gas. A is the pipe conveying the gas to the Argand burner, B, the flame of which is enclosed in the usual cylindrical glass chimney, C C, open at the top. This is enclosed in a wider glass cylinder closed at the top by a double disc of talc, D D, and opening at its base into the ventilating tube, E E. The direction of the currents produced by the heat of the flame is shown by the arrows. The whole is entirely enclosed by a globe of ground glass. Means are provided for regulating the draught in the pipe, E E, which, when heated, creates of itself a strong current of air through the apparatus.

The illuminating power of coal-gas may be measured directly by comparing the intensity of the light emitted by a gas-flame consuming a known quantity of gas per hour with the light yielded by some standard source. The standard usually employed is a spermaceti candle burning at the rate of 120 grains of sperm per hour. It is not necessary that the candle actually used should consume exactly this amount, but the consumption of sperm by the candle during the course of each experiment is ascertained by the loss of weight, and the results obtained are easily reduced to the standard of 120 grains per hour. An instrument is used for determining the relative intensities of the illumination, called Bunsen's photometer. It consists of a graduated rule, or bar of wood or metal, about 10 ft. long. At one end of this bar is placed the standard candle, at the other is the gas-flame. A stand slides along the rule supporting a circular paper screen at the level of the two flames, and at right angles to the line joining them. This paper screen is made of thin writing-paper, which has been brushed over with a solution of spermaceti, except a spot in the centre, or, more simply, a grease-spot is made in the middle of a piece of paper. In consequence the paper surrounding the spot is much more transparent; yet when it is placed so that both sides are equally illuminated, a spectator will not perceive the spot in the centre when viewing the screen on either side. When the screen has been placed by trial in such a position between the two sources of light, it is only necessary to measure its distance from each flame in order to compute the number of times the illuminating power of the gas-flame exceeds that of the candle. This computation is based on the fact that the intensity of the light from any source diminishes as the square of the distance from the source. Thus, if a sheet of paper be illuminated by a candle at 2 ft. distance, it will receive only one-fourth of the light that would fall upon it were its distance but 1 ft., and if removed to 3 ft. distance it has only one-ninth of the light. In the instrument used for measuring the illuminating power of gas the rule is graduated in accordance with this law, so that the relative intensities may be read off at once. The gas passes through a meter for measuring accurately the quantity per minute which is consumed by the burner, and there is also a gauge for ascertaining the pressure. Another mode of estimating the illuminating power of coal-gas is by determining the quantity of carbon contained in a given volume. For, in general, the richness of the gas in carbon is a fair index of the quantity of its luminiferous constituents. This may be readily effected by exploding the gas with oxygen, and measuring the amount of carbonic acid produced. Still more accurate determinations of the illuminating value of gas may be obtained by a detailed chemical analysis.

The illuminating power of any gas is so calculated that it represents the number of times that the light emitted by a jet of the gas, burning at the rate of 5 cubic feet per hour, exceeds the light given off by the standard sperm candle burning 120 grains of sperm per hour. For example, when it is said that the illuminating power of London gas is 13, it is meant that when the gas is burnt in an ordinary burner at the rate of 5 cubic feet per hour, the light is equal to that given by thirteen sperm candles burning together 13 × 120 grains per hour. The quality of gas varies very much, as it depends upon the kind of coal employed, and upon the mode in which the manufacture is conducted. The following are the results of experiments made to determine the illuminating power of the gas supplied to several large towns :

	Candles.		Candles.
London	12·1	Carlisle	16·0
Paris	12·3	Liverpool	22·0
Birmingham	15·0	Manchester	22·0
Berlin	15·5	Glasgow	28·0

The relative quantities of tar, ammonia water, and coke yielded in various gas manufactories also vary very considerably for the same reasons.

In the early days of gas illumination the consumers were charged according to the number of burners; but this arrangement proved so unsatisfactory that the *gas-meter* became a necessity, and already in 1817 meters had been devised, which were not essentially different from those now in use. Although gas is used in so many houses, there are few persons who have any notion of the mechanism of the gas-meter. Our space will not allow full details of the construction, but the following particulars may be mentioned. In the ordinary " wet " meter there is a drum divided into four compartments by radiating partitions; this drum revolves on a horizontal axis, and the lower half of the drum, or rather more, is beneath the surface of water contained in the case, the water being at the same level inside and outside the drum. The gas enters one of the closed chambers formed between the surface of the water and a partition of the drum. Its pressure tends to increase the size of the chamber, hence the drum revolves. The preceding division of the drum being filled with gas, this is driven into the exit pipe by the motion of the drum, as it is included in a space comprised between the water and a partition. Each division in turn comes into communication with the gas-main, and as it is filled passes on towards the position in which a passage is opened for it to the exit-pipe. Each turn of the drum, therefore, carries forward a definite quantity of gas, and the only thing necessary is a train of wheels, to register the number of revolutions made by the drum. The " wet " meter is much inferior in almost every respect to the " dry " meter, in which no water is used. The principle of the " dry " meter is very simple. The gas pours into an expanding chamber, partly constructed of a flexible material, and which may be compared to the bellows of a circular accordion. The expansion is made to compress another similar chamber, already filled with gas, which is thus forced through the exit-pipe. When the first chamber has expanded to a definite volume, it moves a lever, and this reverses the communications. The expanded chamber is now opened to the exit-pipe, and the other to the entrance-pipe, and so on alternately. A train of wheels registers the number of movements on a set of dials.

Recent years have brought no essential changes in the methods of gas making, although of course improvements in many minor details of the processes and of the apparatus have been effected. These demand no description at our hands, as they are of interest only to those concerned with the actual technology of gas-making, nor need some of the later forms of burners for using the gas be noticed, as these are sufficiently familiar. They really do effect a considerable economy in the consumption of gas, especially in cases where a more powerful light is required. But the reader will have already learnt from a foregoing section on Electric Lighting that the importance of gas as an illuminant is already on the wane. Indeed, it will not be too much to say that, before the close of the present century, every town will have its streets, and still more certainly, all its places of public assembly, such as theatres, concert halls, churches, libraries, &c., fitted with installations for electric illumination, and even in shops and private houses, it is probable that before long, gas will be superseded by the electric light. Some of the disadvantages of burning gas have already been referred to, and the danger attending its accidental escape into apartments is illustrated by the yearly tale of victims to suffocation and violent explosions. The inherent disadvantage of gas used as an illuminant, is the enormous quantity of heat produced by its combustion, compared with the amount of light evolved. The absolute quantity of heat required to render a body highly luminous is really very small, for masses of matter almost inappreciable become very luminous, provided only that their *temperature* be sufficiently raised. Thus, for example, the few residual particles of gas in a Geissler's tube (p. 431) become incandescent by electrical discharges, while the number of them is too small to sensibly heat the glass vessel, and the very attenuated carbon filament in an electric glow lamp suffices by the mere contraction and concentration of the current within it raising its temperature high enough, to diffuse a brighter light than a large gas-flame. This explains the fact alluded to elsewhere, that if instead of burning the gas we use it in a gas engine, driving a dynamo connected with an electric light installation, we shall obtain a much greater luminous effect. As there is no combustion, the surrounding air is neither heated nor deteriorated with gaseous products and smoke.

Without any rivalry from the electric-light, gas, as a domestic luminant, has now met with a competitor on the ground of cheapness in the mineral oils mentioned in the preceding article. If these could be deprived of their unpleasant odour, and a perfectly safe lamp contrived for burning them, it would be only under very favourable conditions that gas could compete with them on the score of economy. But of late years two applications of gas to other purposes than to illumination will have been observed. First to heating, for warming, cooking, and other domestic purposes, and also in various processes in the arts. In all the appliances so used, the principle of Bunsen's burner (p. 722) is adopted, and stoves, fireplaces, and kitchen-ranges, heated by gas have obvious advantages in their greater cleanliness and readiness. The other new application of gas is as a motive power in the gas engine, by which a very convenient supply of mechanical energy is afforded. There can be little doubt that in the future, gas will be greatly used for these purposes, and perhaps be for them consumed as largely as at present. A singular thing in the history of gas-manufacture is the great value that the bye-products have attained, that is to say, the ammoniacal liquor, the coke, and especially the tar. So many valuable substances are now derived from this last, that even if coal should cease to be destructively distilled for gas, the operation would still be largely carried on if only for the tar.

A jet of hydrogen gas burning in a dark room is all but invisible, yet no gas can give so intense a heat. The lime-light, which no doubt is perfectly familiar to everyone as an illuminant in magic lantern projections, is simply a jet of mixed hydrogen and oxygen gases directed on a piece of lime, which is rendered incandescent by the heat. The flame of the Bunsen burner (p. 772) is distinguishable only by a very pale blue colour, and it is impossible to discern objects, or to read by its light in an otherwise dark room. But if a piece of thin platinum wire formed into a coil, as by twisting round a pencil, be introduced into the flame, the wire will glow with great brilliancy, and its thickness will seem much increased. It will, in fact, emit so much light that reading by its glow becomes easy. This shows that, as already stated, a solid will give off light at a temperature which scarcely suffices to make a gas visible. Thus a Bunsen burner flame can be made to give light simply by putting into it some incombustible solid, which itself incapable of suffering any chemical change under the conditions, nevertheless becomes luminous by merely acquiring the temperature of the almost invisible heated gas. The cause of the luminosity of the ordinary gas burner, as compared with the almost invisible Bunsen burner flame, has, indeed, been already explained on a previous page, but the phenomenon is again, by the experiment just referred to, brought clearly before our attention ; and it becomes obvious that substances other than the carbon of the hydrocarbon constituents of the coal gas will emit rays of light. Chemical analysis shows that by far the larger proportion of the constituents of ordinary coal-gas consist of gases which do not themselves produce luminous flames, and that, taking 16 candle-gas, about 10 candles of the illuminating power is due to compounds of which the gas does not contain more than 4 per cent. Nearly half the bulk of purified coal-gas is hydrogen, which itself gives no light whatever when burnt ; marsh-gas, which burns with only a slight luminosity, forms 35 per cent. of ordinary coal-gas ; and there is usually present about 7 per cent. of carbonic oxide, which in burning gives only a pale blue flame. This shows that by far the greatest product of the combustion of coal-gas is not light but heat. The flame of hydrogen is much the hottest known, and as that gas enters so largely into the composition of coal-gas, and the complete combustion of all the other constituents takes place when the gas is previously mixed with air, as in the Bunsen burner, we are provided with an economical means of obtaining high temperatures. But coal-gas was in the first instance intended to provide us with a cheap illuminant, and although for some time the gas itself was very impure, and it was long before the crude appliances for burning it were superseded by contrivances giving steadier and more brilliant lights, such as the Argand and the regenerative burners. It is only quite recently that the full illuminating possibilities of coal-gas have been developed by the happy notion of converting the heating power of its flame into light-giving power, by the simple plan of suspending a suitable solid over the hot but non-luminous Bunsen burner.

The manner in which an effective method of doing this was discovered is not a little curious. The construction of the ordinary incandescent electric lamp, Fig. 280*h*, involves the necessity of enclosing the carbon filament in an exhausted glass bulb; and it was when Auer von Welsbach was engaged in attempting to find some substance that could be brought into incandescence by the electric current, and yet be incombustible even in the open air, that his investigations led to the invention we have now to describe—an invention apparently destined to give a new lease of. life to coal-gas illumination.

It is singular also that Welsbach, in seeking for the most suitable materials for heating to incandescence in the Bunsen burner flame, should find them in certain very rare minerals, containing a group of

FIG. 351*a.*

FIG. 351*b.*

elements formerly of interest only to the scientific chemist, and up to that time devoid of any practical applications. The names of these elements, the oxides of which are called "earths," will, of course, be strange to non-chemical readers, but we give their names, with the remark that the nearest familiar substance they at all resemble is aluminium, of which the oxide, or "earth," is alumina. These rare metals, the oxides of which are the materials of the Welsbach "mantle," are all discoveries of the present century, or nearly so, and they are called lanthanum, zirconium, thorium, cerium, didymium, yttrium, erbium, &c. They occur as silicates or phosphates very sparingly, and in a few localities in Norway; but some of them have now been found more abundantly in America. The minerals, from which for the most part the oxides are obtained, are called *monazite, orthite,* and *thorite.* It was found áfter many trials that a blend of these earths in certain proportions gives a mantle that yields a pure

FIG. 351*c.*

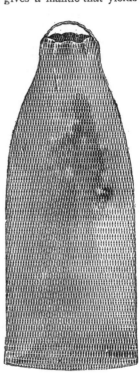

FIG. 351*d.*

white light, while any preponderancé of one or anóther would impart some tint to the light. This proper blending of the constituents forms a great improvement on the first mantles, which generally shed a greenish light.

The *mantles* are made by an ingenious process, in which a network of cotton thread is knitted into the form of a tube; this is cut up into suitable lengths, and a piece attached to form the top. The network is then saturated with a solution of the nitrates of the rare earths abovementioned, and dried on glass rods. After this a loop of asbestos thread is passed through the top, by which the *mantle* may be attached to its support. The mantle is now shaped to the required form, and the cotton thread burnt off, when a thin skeleton of the oxides is left reproducing the form of the original network. The mantle is again strongly heated, and after cooling is dipped into a solution of collodion, dried, and carefully laid in a box. The collodion serves to strengthen the mantle sufficiently for transit, for it is very frail, and would otherwise be liable to fall in pieces by slight shocks. Fig. 351*d* is a full-sized representation of the completed mantle, and Fig. 351*e* shows it mounted on the burner, where a rather small flame is allowed for the first time to play upon it, by which the collodion is quickly burnt off, and then the chimney-glass is placed over it, as in Fig. 351*a*. In the earlier forms of lamp the lighting of the gas was a matter requiring some delicacy of manipulation, for a rude shock, or an awkward touch might cause the mantle to crumble into ruin, but now the makers fit their lamps with a by-pass by which a very small flame is maintained within the lamp ready to light up the gas when that is fully turned on. (Fig. 351*c*.) The makers have also now made the lamps available for street lighting, to which the fragility of the mantle was formerly an obstacle, as it was liable to collapse by the tremor of the traffic. This risk has been obviated by providing a spring to support the mantle at the base. (Fig. 351*b*.)

The qualities of the Welsbach lamp have been examined by competent persons, and from the statements they supply, we extract the following particulars. The light is, for the same gas consumption, seven times that of an ordinary gas burner; more than four times that of an Argand burner; more than twice that of a "regenerative" lamp. It follows, of course, that, light for light, the products of combustion, such as carbonic acid, heat, &c., amount to only something like ⅛th of those produced by ordinary burners, and the consumption of the gas is perfect, there being absolutely no smoke. Though the mantles have to be renewed about three times a year, when the burners are in constant use, the total cost, light for light, is only ⅓th of that of ordinary burners. The light of the Welsbach burner is whiter than ordinary gaslight. It is rich in the blue rays, and, therefore, more like daylight, permitting well the comparison of shades of colour, and it is excellently suited for workers

FIG. 351*e*.

with the microscope, &c. This new gas-lighting must also be a great boon to photographers using artificial illumination, for the actinic power is, with the same visual illumination, nearly twice that of the ordinary gas flame.

FIG. 352.—*Apparatus for making Magenta.*

COAL-TAR COLOURS.

———◆———

COAL-TAR is an exceedingly complex material, being a mixture of a great number of different substances. The following table shows the chemical name of many of the substances obtainable from the coal-tar. It must not be supposed that these substances exist ready formed in the coal, and that they are merely expelled by the heat. We can understand better how heat, acting upon an apparently simple substance like coal, and one containing so few elements, is able to produce so large a variety of different bodies, if we remember that heat is the agent most often employed to effect chemical changes, and that from even two elements, variously combined, bodies differing entirely from each other are producible.

SUBSTANCES FOUND IN COAL-TAR.

a. COMPOUNDS OF CARBON AND HYDROGEN.

Hydrides of amyl, hexyl, heptyl, nonyl, and decyl.
Amylene, hexylene, heptylene, octylene, nonylene, decylene; (paraffin).
Benzol, toluol, xylol, cumol, cymol.
Naphthalene.
Anthracene.
Pyrene.
Chrysene.

b. COMPOUNDS OF CARBON, HYDROGEN, AND OXYGEN.

Phenol, cresol, phlorol.
Rosolic acid, brunolic acid.

c. COMPOUNDS OF CARBON, HYDROGEN, AND NITROGEN.

Aniline, toluidine:
Pyridine, picoline, lutidine, collidine, parvoline, coridine, rubidine, viridine.
Leucoline, lepidine, cryptidine.
Cespitine, pyrrol.

This list contains only the names of substances which have actually been found in the coal-tar, and it is certain that a number of products must have escaped notice. It is obvious, too, that by using coal of different kinds, and by varying the temperature and pressure at which the operation of distilling the coal is effected, we shall probably be able to increase the number of possible constituents of coal-tar almost indefinitely. The list above presents to the non-chemical reader a string of quite unfamiliar names; but, though the system of nomenclature in chemistry is far from perfect, yet each of these names has a meaning for the chemist beyond the mere designation of a substance. The chemical name aims at showing, or at least suggesting, the composition of a body and the general class to which it belongs. This may be illustrated by the names of hydro-carbons in the above list. The five compounds headed by benzol have many properties in common, and each one is entirely different in its chemical behaviour to those which follow amylene. The Greek numerals enter into the names of the latter, in order to express, in this case, the number of atoms of carbon which are supposed to be contained in each ultimate particle of the body. We write down in parallel columns the names of these two classes of bodies, together with the *symbols* which represent their composition, reminding the reader that the letter C represents carbon; the letter alone indicating *one* atom of that element, but, when followed by a small figure, it implies that number of carbon atoms; in like manner H, N, and O represent atoms of hydrogen, nitrogen, and oxygen respectively.

Hexylene	C_6H_{12}	Benzol	C_6H_6
Heptylene	C_7H_{14}	Toluol	C_7H_8
Octylene	C_8H_{16}	Xylol	C_8H_{10}
Nonylene	C_9H_{18}	Cumol	C_9H_{12}
Decylene	$C_{10}H_{20}$	Cymol	$C_{10}H_{14}$

If these lists be carefully examined, it will be observed that there is a regular progression in the constituent atoms, so that each set of substances forms a series, the differences being always the same. The various bodies contained in the coal-tar are separated from each other by taking advantage of the fact that each substance has its own boiling-point; that is, there is a certain temperature, different for each body, at which it will rise into vapour quickly and continuously. Benzol, for example, boils at 82° C., toluol at 114° C., and phenol at 188° C.; so that, if we apply heat to a mixture of these three substances, the benzol will boil when the temperature reaches 82°, and will pass away in vapour, carrying off heat, so that the temperature will not rise until all the benzol has been driven off; then, when the temperature reaches 114°, the toluol will begin to come off, but not until that has all passed over into the receiver will the temperature rise above 114°; and the phenol remaining will distil only at 188°.

Another mode of separating bodies when mixed together is by treating them with a liquid which acts on, or dissolves out, some of the constituents, but not the rest. The coal-tar, as it is received from the gas-works, is placed in large stills, capable, perhaps, of holding several thousand gallons, and usually made of wrought iron. Stills sufficiently good for the purpose are commonly constructed from the worn-out boilers of steam engines. The application of heat, of course, causes the more volatile substances to come over first. These are condensed and collected apart until products begin to come off which are heavier than water. The first portion of the distillate, containing the lighter liquids, is termed " coal naphtha." The process is continued, and heavier liquids come over, forming what is called in the trade the " dead oil." Pitch remains behind in the retort, from which it is usually run out while hot, but sometimes the distillation is carried a step further.

The chief colour-producing substances contained in coal-tar are benzol, toluol, phenol, naphthalene, and anthracene. The aniline which is present in the tar is very small in amount, and if this ready-formed aniline were our only supply, it would be impossible to make colours from it on an industrial scale. The first of the above-named substances, benzol, was discovered by Faraday, in 1825, in liquid produced by strongly compressing gas obtained from oil. He called it bicarburet of hydrogen; but afterwards another chemist, having procured the same body by distilling benzoic acid with lime, termed it *benzine*. It readily dissolves fats and oils; and is used domestically for removing grease-spots, cleaning gloves, &c., and in the arts as a solvent of india-rubber and gutta-percha. It is a very limpid, colourless liquid, very volatile, and, when pure, is of a peculiar but not disagreeable odour. It boils at 82° C., and, cooled to the freezing-point of water, it solidifies into beautiful transparent crystals, a property which is sometimes taken advantage of to separate it in a state of purity from other liquids which do not so solidify.

Benzol is very inflammable, and its vapour produces an explosive mixture with air. The vapour, which is invisible, will run out of any leak in the apparatus, like water, and flow along the ground. Accidents have occurred from this cause, and a case is on record in which the vapour having crept along the floor of the works, was set on fire by a furnace forty feet away from the apparatus, the flame, of course, running back to the spot from which the vapour was issuing. Benzol is a dreadful substance for spreading fire should it become ignited, for, being lighter than water, it floats upon its surface, and therefore the flames cannot be extinguished in the ordinary way. The discovery of the presence of benzol in coal-tar was made by Hofman in 1845. It is obtained from the light oil of coal-tar by first purifying this liquid by alternately distilling it with steam and treating with sulphuric acid several times. The product so obtained is a colourless liquid, sold as " rectified coal naphtha, which, however, has again to be several times re-distilled with a careful regulation of the temperature, so that the benzol may be distilled off from other substances, boiling at a somewhat higher temperature, with which it is mixed. Even then the resulting liquid (commercial benzol) contains notable quantities of toluol. If benzol be added in small quantities at a time to very strong and warm nitric acid, a brisk action takes place, and when after some time water is added, a yellow oily-looking liquid falls to the bottom of the vessel. The

FIG. 353.—*Iron Pots for making Nitro-Benzol.*

benzol will have disappeared, for nitric acid under such circumstances acts upon it by taking out of each particle an atom of hydrogen, which it replaces by a *group* of atoms of nitrogen and oxygen, and, instead of benzol, we have the yellow oil, *nitro-benzol.* Chemists are accustomed to represent actions of this kind by what is called a *chemical equation*, the left-hand side showing the symbols representing the constitution of the bodies which are placed together, and the right hand the symbols of the bodies which result from the chemical action. Here is the equation representing the action we have described:

$$\underset{\text{Benzol.}}{C_6H_6} \; + \; \underset{\text{Nitric acid.}}{NO_2OH} \; = \; \underset{\text{Nitro-benzol.}}{C_6H_5(NO_2)} \; + \; \underset{\text{Water.}}{HOH}$$

Nitro-benzol has a sweet taste and a fragrant odour. It is known in commerce under the names of artificial oil of bitter almonds and essence of mirbane, and it has been used for perfuming soap. The chemical action between benzol and concentrated nitric acid is so violent that, when nitro-benzol first had to be manufactured on the large scale, great difficulty was experienced on account of the serious explosions which occurred. The apparatus now used in making nitro-benzol on the large scale is represented in Fig. 353, which shows some of the cast-iron pots, of which there is usually a long row. These pots are about 4½ ft. in diameter, and the same in depth. Each is provided with a stirrer, which is made to revolve by a bevil-wheel, *c*, on its spindle, working with a pinion on a shaft, *b*, driven by a steam engine. A layer of water is kept on the tops of the lids, the water being constantly passed in and drawn off through the pipes, *d*, in order to keep it cool. For the chemical action is, as usual, attended with heat, which vaporizes some of the benzol, but the cold lid re-condenses the

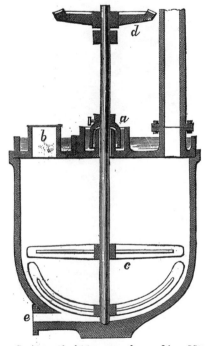

FIG. 354.—*Section of Apparatus for making Nitro-Benzol.*

vapour, which would otherwise escape with the nitrous fumes that pass off by the pipe, *a*. There is at *e* an opening, through which the material may be introduced, and in the bottom of the vessel is an aperture through which the products may be drawn off. Fig. 354 shows a section of one of the cast-iron vessels, and exhibits the mode in which the spindle of the stirrer passes through the lid. In the cup, *a*, filled with a liquid, a kind of inverted cup, which is attached to the spindle, turns round freely. It would not do to choose water for the liquid in this cup, for water would, by absorbing the nitrous fumes, form an acid capable of attacking and destroying the spindle. Nothing has been found to answer better for this purpose than nitro-benzol itself. The charge introduced into these vessels is a mixture

50

of nitric and sulphuric acids together with the benzol. During the action, which may last twelve or fourteen days, no heat is applied, for the mixture becomes hot spontaneously, and in fact care must be taken that it does

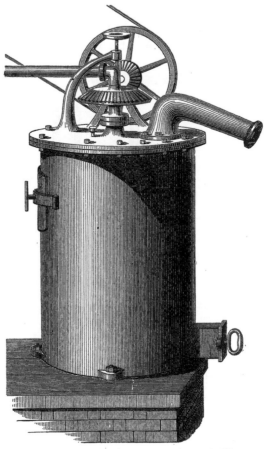

FIG. 355.—*Apparatus for making Aniline.*

not become too hot. The nitro-benzol thus obtained is purified by washing with water and solution of soda.

If nitro-benzol were brought into contact with ordinary hydrogen gas, no action whatever would take place. But it is well known to chemists that gases which are just being liberated from a compound have at the *instant of their generation* much more powerful chemical properties than they possess afterwards. Gases in this condition are said to be in the *nascent state.*

If we submit nitro-benzol to the action of nascent hydrogen we find a remarkable change is produced. This change consists, first, in the hydrogen robbing the nitro-benzol of all its oxygen atoms; second, in the addition of hydrogen to the remainder; third, in some re-arrangement of the atoms, by which a new body is formed. Not that these changes are successive, or that we actually know the movement of atoms, but we are thus able to form ideas which correspond with the final result. The new substance is named *aniline*. It is regarded by chemists as a base; that is, a substance capable of neutralizing and combining with an acid to form a *salt*. Its composition is represented by the symbols $C_6 H_5 H_2 N$. Aniline was found in coal-tar in 1834, and even its colour-producing power was noticed, for its discoverer named it *kyanol*, in allusion to the blue colour it produced with chloride of lime. Later it was obtained by distilling indigo with potash, and hence received its present name from *anil*, the Portuguese for indigo. The quantity of aniline contained in the tar is quite insignificant.

Aniline is prepared from nitro-benzol on the large scale by heating it with acetic acid and iron filings or iron borings, a process which rapidly changes the nitro-benzol into aniline. The equation representing the change is—

$$C_6 H_5 NO_2 \quad + \quad H_6 \quad = \quad C_6 H_5 H_2 N \quad + \quad 2H_2 O.$$

Nitro-benzol.   Hydrogen.   Aniline.   Water.

The operation is effected in the apparatus represented in Fig. 355. It consists of a large iron cylinder, within which works a paddle on a vertical revolving spindle, which, being hollow, is also a pipe to convey high pressure steam within the apparatus. Fig. 356 is a section of the hollow spindle, in which f is the pivot at the bottom of the cylinder on which it turns; d is the stirring paddle; e is an aperture admitting the steam from the pipe, c, forming the shaft of the paddle, which is made to revolve by the bevil-wheel. The steams enters by the elbow-pipe, which has a nozzle ground to fit the head of the vertical revolving pipe, upon which it is pressed down by the screw. When the materials have been introduced into the cylinder, the stirrer is set in motion, and superheated steam is sent down the pipe; the aniline is volatilized and passes with the steam through the

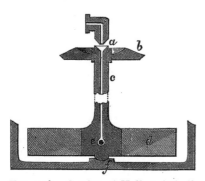

FIG. 356.—*Section of Hollow Spindle, —Aniline Apparatus.*

pipe, which is connected with a worm surrounded by cold water. The aniline is purified by another distillation over lime or soda. When pure, aniline is a colourless, somewhat oily-looking liquid, of a feeble aromatic odour. Under the influence of light and air it becomes of a brownish tint, in which condition it usually presents itself in commerce. It scarcely dissolves in water, but is readily soluble in alcohol, ether, &c.

It was Mr. Perkin who, in 1856, first obtained from aniline a substance practically available for dyeing. Let it be noticed that when Mr. Perkin discovered *aniline purple,* he was not engaged in searching for dye-stuffs, but was carrying on a purely scientific investigation as to the possibility of artificially preparing quinine. With this view, having selected a substance into the composition of which nitrogen, hydrogen, and carbon enter in exactly the same proportions as they occur in quinine, but differing from it by containing no oxygen, he thought it not improbable that by oxidizing this body he might obtain quinine. In this he was disappointed, for the result was a dirty reddish-brown powder. Being desirous, however, of understanding more fully the nature of this reddish powder, he proceeded to try the effects of oxidation on other similarly constituted but more simple bodies. For this purpose he fortunately selected *aniline,* which, when treated with sulphuric acid and bichromate of potash, he found to yield a perfectly black product. Persevering in his experiments by examining this black substance, he obtained, by digesting it with spirits of wine, the now well-known "aniline purple." Mr. Perkin, having determined to make the aniline purple on the large scale, patented his process, and succeeded in overcoming the many obstacles incident to the establishment of a new manufacture requiring as its raw material products not at that time met with as commercial articles. The process is now carried on on the large scale by mixing sulphuric acid and aniline in the proportions in which they combine to form the sulphate of aniline, and dissolving by boiling with water in a large vat. Bichromate of potash is dissolved in water in another large vat. When both solutions are cold, they are mixed together in a still larger vessel and allowed to stand a day or two. A fine black powder settles on the bottom of the vessel in large quantities; this is collected in filters, washed with water, and dried. This powder is not aniline purple alone, but a mixture of this with other products, presenting a very unpromising appearance; but when it has been digested for some time with diluted methylated spirit, all the colouring matter is dissolved out, and is obtained from the solution by placing the latter in a still, where the spirit is distilled off and collected for future use, while all the colouring matter remains behind, held in solution by the water. From this aqueous solution the mauve is thrown down by adding caustic soda. It is collected, washed, and drained until of a pasty consistence, in which condition it is sent into the market. It can be obtained in crystals, but the commercial article is seldom required in this form, as the additional expense is not compensated by any superiority in the practical applications of the colour. Mauve is readily soluble in spirits of wine, but not very soluble in water. Its tinctorial power is so great that *one-tenth of a grain* suffices to impart quite a deep colour to a gallon of water. Silk and woollen fabrics have an extraordinary attraction for this colouring matter, which attaches itself very firmly to their fibres. If some white wool is dipped into even a very dilute solution, the colour is quickly absorbed. Mauve is more permanent than any other coal-tar colour, being little affected by the prolonged action of light.

Mauve is chemically a salt of a base which has been termed "mauveine." Mauveine itself is a nearly black crystalline powder, which forms solutions of a dull blue-violet tint, but when an acid is added to such a solution the tint is at once changed to purple. Mauveine is a powerful base, displacing ammonia from its compounds. The commercial crystallized mauve is the acetate of mauveine.

The process by which Mr. Perkin orginally obtained mauve from aniline evidently depends upon the well-known oxidizing property of bichromate of potash, and experiments were accordingly made with other oxidizing bodies and aniline ; in fact, patents were taken out for the use of nearly every known oxidizing chemical. Three years after Mr. Perkin's discovery of mauve, M. Verguin, of Lyons, obtained, by treating *crude* aniline with chloride of tin, the bright red colouring matter now known as magenta. It was found also that crude aniline, when treated with other metallic chlorides, nitrates, or other salts, which are oxidizing agents less powerful than bichromate of potash, yields this bright red colouring matter. A process patented by Medlock, in 1860, in which arsenic acid is the oxidizing agent, has almost entirely superseded, in England at least, all the others yet proposed for the manufacture of magenta. It is not a little remarkable that magenta would not have been discovered had M. Verguin and others operated on *pure* aniline instead of on the ordinary commercial article. For it was found subsequently by Dr. Hofman that pure aniline cannot be made to yield magenta : the presence of another body is necessary. A reference to the table of coal-tar constituents will show that there is a hydro-carbon named "toluol." This substance is of a similar nature to benzol, and has a boiling-point so little above that of benzol, that in the rough methods of separation usually employed, a notable quantity of toluol is carried over with the benzol, and is always present in the commercial article. In the processes which benzol undergoes for conversion into aniline, the toluol accompanies it in a series of parallel transformations, resulting in the production of a base termed " toluidine "—similar to aniline —being, however, in its pure state a solid at ordinary temperatures. We write down the symbols representing the composition of the bodies formed in the two cases in order to clearly show this :

Benzol	C_6H_6	Toluol	C_7H_8
Nitro-benzol	$C_6H_5(NO_2)$	Nitro-toluol	$C_7H_7(NO_2)$
Aniline	$C_6H_5NH_2$	Toluidine	$C_7H_7NH_2$

This aniline prepared from commercial benzol always contains some toluidine ; and it is essential for the production of magenta that this substance should be operated on along with the aniline. Whether the presence of some toluidine is also necessary for the production of mauve and other colours is not yet known, but they are always prepared from commercial benzol. It is certain that pure aniline yields no magenta, neither does pure toluidine ; but a mixture supplies it in abundance. For the preparation of magenta the best proportions for this mixture would be about three parts of aniline to one of toluidine ; but, in practice, it is not necessary to obtain the two substances separately, as benzol, mixed with a sufficient quantity of toluol, may be obtained by regulating the distillation. The apparatus used in the production of magenta is shown in Fig. 352. It consists of a large iron pot set over a furnace in brickwork, and having a lid with a stuffing-box, through which passes a spindle carrying a stirrer. A bent tube rises from the lid, and is connected with a worm surrounded by cold water, for the purpose of condensing the aniline which is vapourized in the process. The aniline, containing a due amount of toluidine, is mixed in this apparatus with about one and a half times its weight of a saturated solution of arsenic acid (H_3AsO_4). The fire is lighted and kept up for several hours : water first, and lastly aniline, distil over. When the operation is ended, steam is blown through the apparatus, thus carrying off an additional portion of aniline. The crude product is then boiled with water, the solution filtered, and common salt added, which precipitates an impure magenta. This is afterwards dissolved and recrystallized several times. The crystals of this magenta—like those of many of the coal-colour products—have a peculiar greenish metallic lustre ; they dissolve in warm water, forming a deep purplish-red solution. The chemical composition of magenta has been investigated by Dr. Hofman, who found it to be a salt of an organic base, to which he gave the name of " rosaniline." This rosaniline is easily obtained from magenta by addition to its solution of an alkali. While all its salts are intensely coloured, rosaniline itself is a perfectly colourless substance, becoming reddened by exposure to the air, as it absorbs carbonic acid, thus passing to the condition of a salt. Rosaniline, then, displays its chromatic powers only when it is combined with an acid. This property is sometimes shown at lectures in a striking manner by dipping a piece of paper into a colourless solution of rosaniline, and exposing it to the air, when, as the rosaniline absorbs carbonic acid, the paper changes from white to red. A more elegant form of the same experiment is to dip a white rose into a solution of rosaniline containing a little ammonia. As the ammonia escapes, or is expelled by a current of warm air, the same kind of action occurs, and the white rose changes to red— as if by magic, the emblem of the House of York is transformed into the badge of Lancaster ! The chemical nature of rosaniline is regarded as analogous to that of ammonia—it is, in fact, looked upon by chemists as a sort of ammonia, in each particle of which some atoms of hydrogen have been replaced by certain *groups* of carbon and hydrogen atoms—some of these groups being derived from the aniline and others from the toluidine. The particular salt of rosaniline which constitutes the crude product of the action on the aniline and toluidine, depends on the substance employed to effect the oxidation. If a chloride, the resulting product is chloride of rosaniline ; if a nitrate, it is the nitrate ; and so on. The magenta which is formed in the first instance by the process we have described is an arseniate of rosaniline ; but in the subsequent processes, it is converted into the chloride— the salt usually sold as magenta. Other salts of rosaniline are made on the large scale— especially the acetate, the beautiful crystals of which have the advantage of being very soluble.

Magenta attaches itself strongly to animal fibres, but the colour is somewhat fugacious under the action of sunlight. It is used not only as a dye, but more largely as the raw material from which a number of other beautiful colours are obtained. For this reason it is manufactured on an enormous scale, thousands of tons being produced annually, and the money value of the colour produced from it must be reckoned by thousands of pounds. Yet aniline was a few years ago merely a curiosity never met with out of the laboratory of the scientific chemist. It is stated that a single firm now makes more than twelve tons of aniline weekly, and on its premises may be seen tanks, in each of which 30,000 gallons of magenta solution is depositing its crystals. If a salt of rosaniline be heated with aniline, the colour changes gradually through purple to blue. while ammonia is at the same time given off. This is the colour known as aniline blue, " bleu du Lyons," &c. In its preparation it has been found that the best results are obtained by employing the salt of some weak acid—acetate of rosaniline, for example—and pure aniline, that is, aniline free from toluidine. The operation is conducted in iron pots very similar to those used in making magenta, but smaller. These pots are not set over a fire, but a number of them are placed in a large vessel containing oil, by which they can be maintained at a regulated temperature when the oil is heated. The crude product undergoes several purifications, and the aniline blue is supplied in commerce in powder, or dissolved in spirits of wine. It is insoluble in water, and this has been an obstacle to its employment; but recently a similar substance has been obtained in a soluble form, and is extensively used for dyeing wool, under the name of " Nicholson's blue." Other blues have been similarly prepared, and from the same two substances, magenta and aniline, a colour known as " violet imperial" was formerly made in very large quantities, but it has been superseded by the colours about to be described. It may be well to mention that these blues and violets have been found to contain bases formed of rosaniline, in which one, two, or three atoms of hydrogen are replaced by the group C_6H_5. This group of atoms will be noticed to belong to aniline, and chemists have named it phenyl, and, therefore, bases of these coloured salts are respectively named phenyl-rosaniline, di-phenyl-rosaniline, tri-phenyl-rosaniline. But Dr. Hofman found that other groups of atoms besides C_6H_5 may be made to take the place of H in rosaniline. By acting on rosaniline or its salts with iodides of ethyl, C_2H_5I, or iodide of methyl, CH_3I, he obtained a beautiful series of violets, of which many shades could be produced,.varying from red-purple to blue. These are the colours so well known as Hofman's violets, and are prepared on the large scale by heating a solution of magenta (chloride of rosaniline) in alcohol or wood spirit, with the iodide of ethyl or the iodide of methyl. The nature and proportions of the ingredients are regulated according to the tint required. The vessels are hermetically closed during the heating, which is accomplished by steam admitted into a steam-jacket surrounding the vessel. The crude product has to be separated from the substances with which it is mixed, and the colouring matter is finally obtained, presenting in the solid state the peculiar semi-metallic lustre so characteristic of these products. Like the other colours, Hofman's violets are salts of *colourless* bases, which, as indicated above, are substitution products. of rosaniline. The tints they produce incline to red, violet, or blue, according as one, two, or three hydrogen atoms are replaced by the ethyl or methyl groups. Colours have also been obtained from mauve and iodide of ethyl—for example, the dye known in commerce as " dahlia." Other colours are procured from magenta by treating it with various compounds: one such is the " Britannia violet," discovered also by Mr. Perkin, who procures it from magenta and a hydrocarbon-bromide derived from the action of bromine or common turpentine. This is a very useful colour, and is largely used in dyeing and printing violets, of which any shades may be obtained.

Another derivative of rosaniline is the aniline green. It is obtained by dissolving the rosaniline salt in dilute sulphuric acid, adding crude *aldehyde* (a substance obtained by acting with oxidizing agents on alcohol). The mixture is heated until a sample dissolves in acidulated water with a blue tint; it is poured out into boiling water containing in solution hyposulphite of sodium, boiled, the liquid filtered; and the green dye, if required in the solid state, is precipitated by carbonate of sodium. Aniline green dyes wool and silk, the latter especially, of a magnificent green; perhaps as beautiful a colour as any of the coal-tar series, and one which has the singular advantage among greens of looking as beautiful in artificial light as in daylight. The manner in which this dye was discovered is somewhat curious. It is related by Mr. Perkin of a dyer, named Chirpin, that he was trying to render permanent a *blue* colouring matter, which had been found could be produced from rosaniline by the action of aldehyde and sulphuric acid. After a number of fruitless attempts at fixing it, he confided his perplexities to a photographic friend, who evidently thought that if it was possible to fix a photograph, anything else might be fixed in like manner, for he recommended his confidant to try hyposulphite of sodium. On making the experiment, however, the dyer did not succeed in fixing his blue, but converted it into the splendid aldehyde green. Like other colouring matters we have described, this is a salt of a colourless base containing sulphur. Like rosaniline, the colourless base takes on the characteristic colour of its salts by merely absorbing carbonic acid from the air.

Again, by a modification of the process for producing the Hofman violets, another green of an entirely different constitution may be obtained. It is bluer in tint than the former, and is much used for cotton and silks, under the name of "iodine green."

In the manufacture of magenta there is formed a residuum or bye-product, consisting of a resinous, feebly basic substance, from which Nicholson obtained a dye, imparting to silk and wool a gorgeous golden yellow colour. This dye cannot be obtained directly, but is always produced in greater or less quantity when magenta is made on the large scale, and is separated during the purification. By first dyeing the silk or wool with magenta, and then with this dye, which is commercially known as "phosphine," brilliant scarlet tints are obtained. The yellow colours have been found to be salts of a base termed chrysaniline, a sort of chemical relative of rosaniline, as may be seen in comparing the formulæ which represent their constitution; with which we place also the symbol for another substance obtained by submitting rosaniline to the influence of nascent hydrogen. This body, *leucaniline*, again yields rosaniline very readily when the hydrogen is removed by oxidizing agents. It will be noticed that the three bodies form a series the members of which differ only by H_2, thus indicating their close relationship.

$$C_{20}H_{17}N_3 \quad \ldots\ldots\ldots\ldots \quad \text{Chrysaniline.}$$
$$C_{20}H_{19}N_3 \quad \ldots\ldots\ldots\ldots \quad \text{Rosaniline.}$$
$$C_{20}H_{21}N_3 \quad \ldots\ldots\ldots\ldots \quad \text{Leucaniline.}$$

Some idea will have been obtained from the foregoing particulars of the great colour-supplying capabilities of aniline; but we have not yet exhausted the utility of this interesting substance. It is probable that the letters on the page now under the reader's eye owe their blackness to an aniline product. For after all the salts furnishing the lovely tints we have mentioned have been extracted, there is in their manufacture a final residuum, and from this an intense black is obtained, which is largely used in the manufacture of printing-ink.

We have mentioned *phenol* as a substance yielding colours. Phenol is the body now so well known as a disinfectant under the name of "carbolic acid," a name given to it by its discoverer, Runge, who prepared it from coal-tar, in 1834. Phenol forms colourless crystals, which dissolve to some extent in water, and very readily in alcohol. It is a powerful antiseptic, that is, it arrests the process of putrefaction in animal or vegetable bodies, and it is also highly poisonous. The constitution of phenol is given by the formula $C_6H_5\,OH$, in which the reader will recognize the same group of atoms already indicated as entering into the aniline derivatives. From some of these phenol may in fact be obtained, and although it cannot be formed *directly* from benzol, phenol can be made to furnish benzol. When crude phenol is treated with a sulphuric acid and oxalic acid, a substance is obtained which presents itself as a brittle resinous mass of a brown colour, with greenish metallic lustre. This substance is called *rosolic acid* by chemists, but in commerce it is known as *aurine*, and is used for dyeing silk of an orange colour, which, however, is not very permanent. But by heating rosolic acid with liquid ammonia, a permanent red dye is procured which has been termed *peonine*, and has been much used for woollen goods. But it lately had the reputation of exerting a poisonous action, producing blistering and sores when stockings or other articles dyed with it were worn in contact with the skin. It is now, therefore, less extensively employed. *Coralline*, another body identical with or very similar to the former, is similarly prepared from rosolic acid by heating it with ammonia under pressure.

Again, by heating coralline with aniline, a blue dye, known as " azurine," or " azuline," was formerly made in large quantities ; but it has been supplanted by the aniline blues already described.

When phenol is acted upon by nitric acid new compounds are produced, standing in the same relation to phenol as nitro-benzol does to benzol. The final result of the action of nitric acid on phenol is *picric acid*, called also " carbazotic acid," and, more systematically, " tri-nitro-phenol ;" for it is regarded as phenol in which three of the hydrogen atoms have been replaced by the group NO_2 thus, $C_6H_2(NO_2)_3$ OH. It forms bright yellow-coloured crystals, and its solution readily imparts a bright pure yellow colour to wool, silk, &c. It received the name of picric acid ($\pi\iota\kappa\rho\sigma\varsigma$, *bitter*) from the exceedingly bitter taste of even an extremely diluted solution. It is said that picric acid is employed as an adulterant in bitter ale instead of hops. Now, the colouring power of picric acid is so great, that even the minute quantity which could be used to impart bitterness to beer is recognizable by dipping a piece of white wool into the beer, when, if picric acid be present, the wool acquires a clear yellow tint. Besides its employment as a yellow, it is useful for procuring green tints by combination with the blues. Picric acid again furnishes, by treatment with *cyanide of potassium*, a deep red colour, consisting of an acid which, when combined with ammonia, furnishes a magnificent colouring material—which is, in fact, *murexide*, a dye identical with the famous Tyrian purple of the ancients, and formerly obtainable only from certain kinds of shell-fish.

Naphthaline—another of the colour-yielding substances of coal-tar—is, like benzol, a hydro-carbon, but one belonging to quite another chemical series. Its formula is $C_{10}H_8$, and it has an interest to chemists altogether apart from its industrial uses, from having been the subject of the classic researches of the French chemist, Laurent—researches which resulted in the introduction of new and fertile ideas into chemical science, contributing largely to its rapid progress. Naphthaline forms colourless crystals, which, like camphor, slowly volatilize at ordinary temperatures, and are readily distilled in a current of steam. It is thus sufficiently volatile to escape complete deposition in the condensers of the gas-works, and to be partly carried over into the mains, where its collection occasions some trouble. Nitric acid acts upon naphthaline in a manner analogous to that in which it acts on benzol, forming nitro-naphthaline, which, in its turn, submitted to the action of iron filings and acetic acid, is transformed into a base called " naphthylamine." The salts of naphthylamine are coloured products which, in some cases, have been found available as dyes. There is a crimson colour, and a yellow largely used under the name of

"Manchester yellow," for imparting to silk and wool a gorgeous golden yellow colour. Another coloured derivative of naphthaline, called "carminaphtha," was discovered by Laurent in the course of his researches.

It would be easy to fill this volume with descriptions of the properties, and modes of preparing the numerous colouring matters that have been obtained from coal-tar products. In order to give the reader an idea of the extent to which the tar products have been made to minister to our sense of the beautiful, a list is here given of the principal colouring matters from these sources that have been employed in the arts. The various names under which a product has been commercially known are in most cases given. It must be understood that the same name is frequently applied to products chemically distinct, and some of the names which appear as synonyms may also in reality indicate different substances.

LIST OF COAL-TAR COLOURS.

I. Colours Derived from Aniline and Toluidine.

Blues and Violets.

Mauve, aniline purple, Perkin's violet, violine, mauve, rosaniline, indisine, &c.

Aniline blue, rosaniline blue, Hofman's blue, bleu de Paris, bleu de Lyons, bleu de Mulhouse, bleu de Mexique, bleu de nuit, bleu lumière, night blue.

Hofman's blue.

Nicholson's blue, soluble blue.

Hofman's violet, rosaniline violet.

A long series of red and blue violets, bearing Hofman's name and distinguished in commerce by adding R or B, according to the redness or the blueness of the tint, ranging from RRRR to BBBB.

Dahlia.

Toluidine blue.

Violet de Paris.

Mauvaniline.

Violaniline.

Regina blue, opal blue, bleu de Fayolle, violet de Mulhouse.

Britannia violet.

Violet imperial.

And many others.

Reds.

Aniline red, new red, magenta, solferino, anileine, rougé, roseine, azaline, Rubine, rubine imperial.

Chrysaniline red.

(The above are all salts of rosaniline.)

Xylidine, tar red, soluble red.

Yellows.

Chrysaniline, phosphine, aniline yellow, yellow fuschine.

Chrysotoluidine.

Dinaline.

Field's orange.

Greens.

Aldehyde green, aniline green, viridine, emeraldine.
Iodine green, iodide of methyl green, iodide of ethyl green.
Perkin's green.

Browns.

Havanna brown.
Bismarck brown, aniline brown, Napoleon brown, aniline maroon.

Greys and Blacks.

Aniline grey, argentine.
Argentine black.

II.—COLOURS DERIVED FROM PHENOL.

Blues and Violets.

Isopurpuric acid, Grénat.
Azuline, azurine.

Reds.

Picramic acid.
Coralline, peonine.
Red coralline.

Yellows.

Picric acid, carbazotic acid.
Aurine, rosolic acid.

Green.

Chloropicrine.

Browns.

Picrate of ammonia.
Isopurpurate of potash.
Phenyl brown, phenicienne.

III.—COLOURS DERIVED FROM NAPHTHALENE.

Reds.

Pseudoalizarine, naphthalic red.
Roseonaphthaline, carminaphtha.

Yellows.

Binitronaphthaline, naphthaline yellow, golden yellow, Manchester yellow.
And others.

The introduction of aniline colours into dyeing and calico-printing has caused quite a revolution in these arts, the processes having become much more simple, and the facilities for obtaining every variety of tint largely increased. The arts of lithography, type-printing, paper-staining, &c., have also profited by the coal-tar colours. For such purposes the colour is prepared by fixing it on alumina, a process in which much difficulty was at first experienced, for the colours are themselves almost all of a basic nature. The desired result is now attained by fixing them on the alumina with tannic or benzoic acid. These lakes produce brilliant printing-inks, which are extensively used. The aniline colours are also employed for coloured writing-inks, tinted soaps, imitations of bronzed surfaces, and for a variety of other purposes.

Not many years ago coal-tar was a valueless substance : it was actually given away by gas-makers to any one who chose to fetch it from the works. It was then "matter in the wrong place;" but Mr. Perkin's discovery led to its being put in the right place, and it has become the raw material of a manufacture creating an absolutely new industry, which has developed with amazing rapidity. This industry dates from only 1856, and in 1862 the annual value of its products was more than £400,000. Dr. Hofman, in reporting on the coal-tar colours shown at the Paris Exhibition of 1867, computed the value at that time at about £1,250,000, although the products were much cheaper than before. Large manufactories have been established in Great Britain, in France, Germany, Switzerland, America, and other countries. The possibility of such an industry is an interesting illustration of the manner in which the progress made in any one branch of practical science may lead to unexpected developments in other quarters. The quantity of aniline obtained from coal-tar is very small compared to the amount of coal used, as may be seen from the following table, in which the respective weights of the various products required in the manufacture of *mauve* are arranged as given by Mr. Perkin for the produce of 100 lbs. of coal.

	lbs.	oz.
Coal	100	0
Coal-tar	10	12
Coal-tar naphtha	0	$8\frac{1}{2}$
Benzol	0	$2\frac{3}{4}$
Nitro-benzol	0	$4\frac{1}{4}$
Aniline	0	$2\frac{1}{4}$
Mauve	0	$0\frac{1}{4}$

From this we may perceive that had not the manufacture of gas been greatly extended, so as to yield a large aggregate produce of tar, the requisite supply for the manufacture of aniline would not have been attainable ; and the industrial application of the previously worthless bye-product reacts upon gas manufacture by cheapening the price of that commodity, thus tending still more to extend its use.

Although anthracene has already been named as one of the colour-producing substances found in coal-tar, we have not in the list of coal-tar colours included the colouring matter which anthracene is capable of yielding. The reason is that this case stands apart in some respects from the rest. The colours derived from aniline and the other substances already enumerated are instances of the production of bodies not found in nature —mauve, magenta, &c., do not, so far as we know, exist in nature. Their artificial formation was a production of substances absolutely new. The colour of which we have now to treat is, on the other hand, found in nature, and from its occurrence in the *rubia tinctoria,* the roots of that plant have for ages been employed as a source of colour, and are well known in this country as "madder." The plant is grown largely in Holland, in France, in the Levant, and in the south of Russia.* Madder is used in enormous quantities for dyeing reds and purples : the well-known " Turkey red " is due to the colouring matter of this root. The total annual value of the madder grown is calculated to reach nearly 2½ million pounds sterling. More than forty years ago it was discovered that the madder-root yielded a colouring substance, to which the name of " alizarine " was bestowed, from *alizari*, the commercial designation of madder in the Levant. The aliza-rine does not exist in the fresh root, but is produced in the ordinary pro-cesses of preparing the root and dyeing with it, in consequence of a peculiar decomposition or fermentation. Alizarine may be procured from dried madder by simply submitting it to sublimation, when beautiful orange needle-shaped crystals of alizarine may be obtained. It is nearly insoluble in water, but readily dissolves in hot spirits of wine. Acids do not dissolve it, but potash dissolves it freely, striking a beautiful colour ; with lime, barytes, and oxide of iron, it forms purple lake, and with alumina a beauti-ful red lake. According to Dr. Schunck, of Manchester, to whose investi-gations we are indebted for much of our knowledge of madder, the root contains a bitter uncrystallizable substance called " rubian," which, under the action of certain ferments, and of acids and alkalies, is decomposed into a kind of sugar, and into alizarine and other colouring matters. The ferment, which in the process of extracting the colouring matter from the roots causes the formation of alizarine, is contained in the root itself.

We have already seen how an investigation relating to a question of pure chemical science accidentally led Mr. Perkin to the discovery of mauve—the precursor of the long range of beautiful colours already de-scribed. The mode of artificially preparing alizarine, so far from being an accidental discovery, was sought for and found in 1869 by two German chemists, Graebe and Liebermann. The researches of these chemists were conducted in a highly scientific spirit. Instead of making attempts to pro-duce alizarine by trying various processes on first one body, then another, to see if they could hit upon some tar product, or other substance, which would yield the desired product, they began by operating analytically on alizarine itself. Just as a mechanic ignorant of horology, required to make a watch, would be more likely quickly to succeed in his task by taking a watch to pieces to see how it is put together, than if he had tried all man-ner of arranging springs and wheels until he hit upon the right way ; so these chemists set themselves to take alizarine to pieces, in order to see from what materials they might be able to put it together. They decom-posed alizarine, and among the products found a hydro-carbon identical in all its properties with *anthracene.*

Anthracene was discovered in coal-tar by Laurent in 1832, and its properties were investigated by Anderson in 1862. It may be remarked that such investigations were not conducted with a view to any industrial uses of anthracene, but merely for the sake of chemistry as a science. Certainly no one could have supposed at that time that the slightest rela-tion existed between anthracene and madder. Anthracene is a white solid hydro-carbon, which comes over only in the last stages of the distil-lation of coal-tar, accompanied by naphthaline, from which it is easily separated by means of spirits of wine, by which the naphthaline is readily

---

* The natural Order to which the madder plant belongs is interesting from the number of its members which supply us with useful products. That valuable medicine, quinine, is obtained from plants belonging to this family, as is also ipecacuanha, and other articles of the *materia medica.* *Coffea arabica*, which furnishes the coffee-berry, is another member.

dissolved, but the anthracene scarcely. Anderson, in 1861, discovered, among other results, that anthracene, $C_{14}H_{10}$, by treatment with nitric acid became changed into oxy-anthracene, $C_{14}H_8O_2$; and this reaction we shall see is a step in the process of procuring alizarine from anthracene. Phenol, as already mentioned, can be made to yield benzol, by a process of deoxidization. With a view to similarly obtaining a hydro-carbon from alizarine, Graebe and Liebermann passed its vapours over heated zinc filings, and thus produced anthracene from alizarine. It now remained to find a means of reversing this process, that is, so to act on anthracene as to produce alizarine, and this was effected by treating anthracene with bromine, forming a substance which, on fusing with caustic potash, yielded *alizarate of potash*, from which pure alizarine resulted by treatment with hydrochloric acid. A much cheaper method was, however, necessary for manufacturing purposes, and it was found in a process by which oxy-anthracene, $C_{14}O_8H_2$, is treated at a high temperature with strong sulphuric acid, and the product so formed heated with a strong solution of potash, yielding alizarate of potassium as before. Many other interesting substances appear to be formed in the reactions, but the nature of these bodies has as yet been imperfectly investigated. No doubt whatever can be entertained of the identity of natural with artificial alizarine ; and the production of this substance, the first instance of a natural colouring matter made artificially, may be regarded as a great triumph of chemical science. It was not long ago supposed that the chemical bodies found in plants or animals, or produced by vital actions, could not possibly be formed by any artificial process from their elements. The laws which presided at their formation were, it was conceived, wholly different from those which governed the chemicals of the laboratory, for they were held to act exclusively under the influence of a mysterious agent, namely, "vital force." It was supposed, for example, that from pure carbon, oxygen, and hydrogen, no chemist would ever be able to produce such a compound as acetic acid. Accordingly the domain of chemical science, previous to the end of the first quarter of the present century, was divided by an impassable barrier into the two regions of organic and inorganic chemistry. Now, however, the chemist is able to build up in his laboratory from their very elements a great number of the so-called *organic* bodies. And it is quite possible to do this in the case of alizarine ; that is, a chemist having in his laboratory the elements, hydrogen, carbon, oxygen, &c., could actually build up the substance which gives its value to madder.

The quantity of anthracene procurable from coal-tar is, unfortunately, comparatively small, for it is found that from the distillation of 2,000 tons of coal only one ton of anthracene can be obtained. The use of artificial alizarine would doubtless entirely supplant the employment of madderroot if anthracene could be obtained in larger quantities; and the change would be highly advantageous to this country, for as no madder is grown in Great Britain, and we consume nearly half the whole annual growth, it follows that every year a million pounds sterling go out of the country for this commodity. When anthracene is produced from coal in sufficient abundance, this sum will be available for the support of our own population. In the meantime, the manufacture of artificial alizarine is restricted only by the supply of its raw material.

The foregoing paragraphs of the present article, which were written for the first edition of this work, not long after the introduction of artificial alizarine, require some supplementary reference to the subsequent progress of discovery and to the increased importance of the manufacture of the coal-tar colours on the large scale. Since the first introduction of alizarine as a commercial product, the substance has received much attention from chemists. The constitution of the body called above *oxy-anthracene* is now better understood, and its chemical relationship is more clearly indicated by the systematic name of *anthraquinone*, which it now bears. The process of the manufacture of alizarine has received some advantageous modifications, and the artificial product may now be said to have entirely displaced the madder-root in dyeing. But, more than this, chemists have found means of preparing a number of "derivatives" of alizarine, many of which are either colouring matters or are easily converted into such. Nearly thirty of these substances have been described, and several of them have found extensive industrial applications. We may mention *alizarine blue*, C_{17} H_9 NO_4 and another substance, produced by combining that with *sodium bi-sulphite*, and having the formula C_{17} H_9 NO_4. $2N_a$ H SO_3. This last, manufactured largely, and sold under the name of "*alizarine blue S.*," is remarkable for being one of the most permanent of all colouring matters. It is said to be a faster colour than even indigo blue, which, indeed, it is rapidly replacing in dyeing, where it is applicable to cotton with a chromium mordant and to silk with one of alumina. Two other colouring matters have also been derived from anthracene, and are much used in dyeing ; one is commercially named *anthracene purple*, the other is *anthracene green*, which supplies the calico printer with very fast shades of olive-green.

Several of the substances enumerated in the list of coal-tar colours, in pages 689 and 690, are now but little used, or altogether abandoned in dyeing and calico printing, because either their beautiful hues prove too fugitive, or other bodies of the same class can be produced at a much cheaper rate. The range of choice is now of the amplest, for chemical discovery has been wonderfully active, but in many cases the real nature and relationship of the artificial colouring matters enumerated above have only quite recently been made out. Mauve (now called *rosolane*), for example, the oldest of all the colour-tar colours, and one which, as we have seen, was manufactured on an extensive scale many years ago, is now scarcely made at all, because much cheaper violets have taken its place. The science of the tinctorial substances has lately taken a much more distinct form, and this knowledge has borne fruit for industrial purposes. It would be out of place here to review what has been done in this way, but a few facts will show the richness of the field. It was only in 1886 that the true chemical constitution of a class of coal-tar derivatives, called *azines*, was first made out. They present themselves as pale yellow or orange coloured crystallized solids, which melt at a comparatively high temperature and may be distilled without decomposition. Although highly coloured substances themselves, before they are converted into fast dyes they require further treatment, which introduces into their molecules another group of atoms. An almost indefinite number of such compounds are theoretically possible, but from only a very few of them many useful dye stuffs are now prepared on the large scale. Amongst the most important of these are "neutral red," "neutral violet," and two other violet colouring matters, "red dyestuff," "fuchsia," "giroflé," "Magdala red," "indazine" and "Basle blue."

Among the colouring matters before enumerated are "aniline yellow" and "Bismarck brown." Their real nature was not understood until a few years ago ; and though the use of the aniline yellow itself has been abandoned on account of its fugacity, the substance has been found a most prolific parent, which has supplied dye stuffs of the most diverse and brilliant hues. These form what chemists term the *azo colours*, and they have been manufactured in great variety and on a very large scale. In 1876, the class of them called *chrysoidines* was introduced, and again, in 1878, *tropæolines*. Great numbers of different azo colours have been sent into commerce under various names, such as "butter yellow," "*crocein scarlet*," "*Biebrich scarlet*," "*Congo red*," "*Bordeaux G*.," "fast red," &c., &c. About 140 of these azo dyes have been described, and the commercial importance of this one class of compounds alone may be inferred from the fact of no fewer than 200 patents having been taken out for processes relating to their manufacture in the eleven years from 1878 to 1888.

It would not be difficult to fill this book with instances of the way in which the resources of modern life have been increased by chemistry alone, a science almost entirely the creation of the present century. Many of the processes of manufacture in which chemistry is applied to the production of articles of everyday use have been so often described, that they may be assumed to be already so well known as to offer few elements of novelty to the general reader, whose interest would also be likely to flag if he were carried over a long range of even the brilliant discoveries that are so delightful and instructive for the special students of this science. There is no parallel to the rapidity of the progress made by the younger branch of the science which concerns itself with the chemistry of one element—namely, carbon and its various combinations, and it is from these carbon compounds that our examples have been drawn. In the explosives, we have some of these compounds supplying resistless forces for rending rocks, and furnishing in warfare the most dreadful powers of destruction. In anæsthetics, we see beneficent applications of others in alleviating suffering and annulling pain ; and again we have just shown how richly another set of them can minister to our sense of beauty. The discussion of these topics has afforded an opportunity for bringing before the reader some of the laws or summarized statements of experimental facts, and also some of those symbolical conceptions of the constitution of compounds, which together furnish the clues that guide the chemist through the vast labyrinth of the endless transformations of matter. The results attained show that the notions expressed by such words as *atom, molecule, compound radical, structural formula*, etc., have a true representative correspondence with something in the actual constitution of bodies.

THE GREATEST DISCOVERY OF
THE AGE.

THE indulgent reader who may have followed the course of the foregoing pages, will perhaps peruse the title of this article with some
little bewilderment. His attention has been drawn to one after another
of a series of remarkable and important discoveries, and he will naturally
wonder what can be the discovery which is greater than any of these.
Now, a discovery is great in proportion to the extent and importance of
the results that flow from it. These results may be immediate and practical,
as in the case of vaccination ; or they may be scientific and intellectual, as
in Newton's discovery of the identity of the force which draws a stone to
the ground with that which holds the planets in their orbits. Such discoveries as most enlarge our knowledge of the world in which we live, by
embracing in simple laws a vast field of phenomena, are precisely those
which are most prolific in useful applications. If we admit, as we must,
the truth of Bacon's aphorism, which declares that " Man, as the minister
and interpreter of nature, is limited in act and understanding by his obser-

vation of the order of nature ; neither his understanding nor his power extends farther,"* then it would be easy to show that the discovery of which we have to treat, more than any other, must be of immense practical service to mankind in every one of the ways in which a knowledge of the order of nature can be of use, viz. :—" First, In showing in how to avoid attempting impossibilities. Second, In securing us from important mistakes in attempting what is, in itself, possible, by means either inadequate or actually opposed to the end in view. Third, In enabling us to accomplish our ends in the easiest, shortest, most economical, and most effectual manner. Fourth, In inducing us to attempt, and enabling us to accomplish, objects which, but for such knowledge, we should never have thought of undertaking." †

A great principle, like that which we are about to explain to the reader, is too vast in its bearings for its discovery and elaboration to have been the work of an individual. This truth, and indeed the whole of our knowledge, is but the result of the development and growth of pre-existing knowledge. In fact, every discovery, however brilliant—every invention, however ingenious, is but the expansion or improvement of an antecedent discovery or invention. In strictness, therefore, it is impossible to say where the first germ of even our newest notions may be found. Our latest philosophy can be shown to be the result of progressive modifications of ideas of remote ages. Hence every great truth, every grand invention, has in reality been the offspring of many minds ; but we record as *the* discoverers and inventors those men who have made the longest strides in the path of progress, and whose genius and labours have overcome obstacles defying ordinary efforts.

The extent of the field which is covered by the principle we have in view is so vast—embracing, as it does, the whole phenomena of the universe— that it will not be possible to do more within our limits than give the reader a general notion of the principle itself. It may be useful to instance a truth which has a similar generality and significance, and which has also acquired the force of an axiom, because it is verified every hour. It is that greatest generalization of chemistry, affirming that in all its transformations *matter is indestructible,* and can no more be destroyed than it can be called into being at will. This truth is so well established, that some philosophers have asserted that an opposite state of things is *inconceivable.* But it was not always known ; and there are at the present day untutored minds which not only believe that a substance destroyed by fire is utterly annihilated, but what they find *inconceivable* is the continued existence of the substance in an invisible form. The candle burns away, its matter vanishes from our view ; but if we collect the invisible products of the combustion, we find in them the whole substance of the candle in union with the atmospheric oxygen. We may, in imagination, follow the indestructible atoms of carbon in their migrations, from the atmosphere to the plant, which is eaten by the animal and goes to form its fat, and from the tallow, by combustion, back into the atmosphere again. The notion of the real identity of matter under changing forms has been expressed by our great dramatist in a well-known passage, which is remarkable for its philosophic insight, when we consider the age in which it was written :

---

* " Homo naturæ minister et interpres, tantum facit et intelligit quantum de naturæ ordine re vel mente observaverit : nec amplius scit aut potest."—*Novum Organum, Aphor.* x.
†Sir J. Herschel.

HAMLET. To what base uses we may return, Horatio! Why may not imagination trace the noble dust of Alexander, till he find it stopping a bung-hole?

HORATIO. 'Twere to consider too curiously to consider so.

HAMLET. No, faith, not a jot; but to follow him thither with modesty enough, and likelihood to lead it. As thus: Alexander died, Alexander was buried, Alexander returneth to dust; the dust is earth; of earth we make loam; and why of that loam, whereto he was converted, might they not stop a beer-barrel?

> Imperial Cæsar, dead, and turned to clay,
> Might stop a hole to keep the wind away;
> O, that the earth, which kept the world. in awe,
> Should patch a wall to expel the winter's flaw!

Now the greatest discovery of our age is that force, like matter, is inde-structible, and that it can no more be created than can matter. The reader may perhaps think the statement that we cannot create force is in contradiction to experience. He will be disposed to ask, What is the steam engine for but to create force? Do we not gain force by the pulley, the lever, the hydraulic press? And are not tremendous forces produced when we explode gunpowder or nitro-glycerine? When the principle with which we are here concerned has been developed and stated in accurate terms, it is hoped the reader will see the real nature of these contrivances. We are, however, aware that it is quite impossible within the limits of a short article to do much more than indicate a region of discovery abounding with results which may be yet unfamiliar to some. Into this, if so minded, they should seek for further guidance, which they will pleasantly find in the pages of Dr. Tyndall's " Heat considered as a Mode of Motion," and in a little work by Professor Balfour Stewart, entitled " The Conservation of Energy," and quite fascinating from the clearness and simplicity of its style. We may continue our humble task of merely illustrating the general nature of this, in reality the most important, subject which we have had occasion to bring under the reader's notice.

Perhaps the first step should be to point out the fact of the various forces of nature — mechanical action, heat, light, electricity, magnetism, chemical action — being so related that any one can be made to produce all the rest directly or indirectly. Some examples of the conversion of one form of force into another occur in the foregoing pages. Thus, on page 485 an experiment is described in which electricity produces a mecha-nical action; electricity is also shown, on page 496, to produce heat; on page 491 chemical action; on page 501 magnetism. Then, as instances of the inverse actions, there is on page 488, in the first paragraph on "Electric Induction," an account of the mode in which mechanical movements may give rise to electricity; and in the experiments in pages 508, 509, and particularly in the account of the Gramme machine, page 511, it is shown how mechanical movements can, through magnetism, produce electricity. The voltaic element, page 491, and the galvanic batteries, are instances of chemical action supplying electricity. On page 518 a striking instance is mentioned of changes in the forms of force. Every lighted candle is a case of chemical action giving rise to light; and interesting examples of the inverse relation are referred to on page 608. On page 168 is represented the conversion of arrested motion into heat and light. We have, indeed, sufficient examples to arrange a series of these conversions of forces in a circle. Thus, chemical action (oxidation in the animal system) supplies muscular power, this sets in motion a Gramme machine, the motion is con-verted into electricity, the electricity produces the electric light, and light causes chemical action, and with this the cycle is complete. In the steam engine heat is converted into mechanical force, and many cases will present themselves to the reader's mind in which mechanical actions give rise to heat. The doctrine of a mutual dependence and convertibility among all the forms of force was first definitively taught in England by Mr. (now Justice) Grove, in 1842; and almost simultaneously Dr. Meyer promulgated similar views in Germany. Mr. Grove subsequently embodied his doctrine in a treatise, called "The Correlation of the Physical Forces," which has seen several editions.

But this teaching included much more than a mere connection between the various forces, for it extended to quantitative relations. It declared that a given amount of one force always produced a definite amount of another that a certain quantity of heat, for example, would give rise to a certain amount of mechanical action, and that this amount of mechanical action was the *equivalent of the heat* which produced it, and would in its turn reproduce all that heat. These last doctrines, however, rested on a speculative basis, until Mr. James Prescott Joule, of Manchester, carried out a most patient, laborious, and elaborate experimental investigation of the subject. His labours placed the truth of the numerical equivalence of forces on a foundation which cannot be shaken; and he accomplished for the principle of the indestructibility of force what Lavoisier did for that of the indestructibility of matter—he established it on the incontrovertible basis of accurate and conclusive experiment. His determination of the value of the *mechanical equivalent of heat* especially is a model of experimental research; and subsequent investigators have, by diversified methods, confirmed the accuracy of his results. A great part of his work consisted in finding what quantity of heat would be produced by a given quantity of *work.*

Before we proceed to give an indication of one of Dr. Joule's methods of making this determination, we may point out that if a weight be raised a certain height, the work which is done in raising it will be given out by the weight in its descent. If you carry a 1 lb. weight to the top of the London Monument, which is 200 ft. high, you perform 200 units of work. When the weight is at the top, the work is not lost; for let the weight be attached to a cord passing over a pulley, and it will, as it descends, draw up to the top another 1 lb. weight.* If you drop the weight so that it falls freely, it descends with a continually increasing velocity, strikes the pavement, and comes to rest. Still your work is not lost. The collision of the weight and the pavement develops *heat,* just as in the case of the experiment depicted on page 168, but to a less degree—the increase of temperature might not be sensible to the touch, but could be recognized by delicate instruments. Your work, then, has now changed into the form of heat—the weight and the pavement are hotter than before. This heat is carried off by contiguous substances. But still your work is not lost, for it has made the earth warmer. The heat, however, soon flows away by radiation from the earth, and is diffused into space. The final result of your work is, then, that a certain *measurable* quantity of heat has been sent off into space. Is your work now finally lost? Not so: in reality, it is only diffused throughout the universe in the form of radiant heat of low intensity. Yet it is lost for ever for useful purposes; for from this final form of diffused heat there is no known or conceivable process by which heat can be gathered up again.

Dr. Joule arranged paddles of brass or iron, so that they could turn freely in a circular box containing water or quicksilver. From the sides of the box partitions projected inwards, which contained openings that permitted the divided arms of the paddle to pass, and preventing the liquid from

---

* See Note B, at the end.

moving *en masse,* thus caused a churning action when the paddle was turned. Now, every one who has worked a rotatory churn knows that a considerable resistance is offered to this action; but every one does not know that under these circumstances the liquid becomes warmer. It was Dr. Joule's object to discover how much the temperature of his liquid was raised by a measured quantity of work. He used very delicate thermometers, and had to take a number of precautions which need not here be described; and he obtained the definite quantity of work by the descent of a known weight through a known distance, a cord attached to the weight being wound on a drum, which communicated motion to the paddle. The experiments were conducted with varying circumstances, to avoid chances of error, and were repeated very many times until uniform and consistent indications were always obtained. The result of the experiments showed that 772 units of work (foot-pounds) furnished heat which would raise the temperature of 1 lb. of water from 32° to 33° F., which is the unit of heat. This number, 772, is a constant of the greatest importance in scientific and practical calculations, and is called "*the mechanical equivalent of heat.*" The amount of *work* it represents is sometimes called a "Joule," and is always represented in algebraical formulæ by "J." Mr. Joule's first paper appeared in 1843, and soon afterwards various branches of the subject of "The Equivalence and Persistence of Forces" were taken up by a number of able men, who have advanced its principles along various lines of inquiry. Among the most noted contributors to this question we find the names of Sir William Thomson, Helmholtz, James Thomson, Rankin, Clausius, Tait, Andrews, and Maxwell.

In the steam engine the case is the inverse of that presented by the above-named experiment of Dr. Joule's. Here we have heat producing work. Now, the quantity of steam which enters the cylinder of a steam engine may be found, and the temperature of the steam can be determined, and from these the amount of *heat* which passes into the cylinder per minute, say, can be calculated. A large portion of this heat is, in an ordinary engine, yielded up to the condensing water, and another part is lost by conduction and radiation from the cylinder, condenser, pipes, &c. But both these quantities can be estimated. When the amount is compared with that entering the cylinder in the steam, a difference is always found, which leaves a quantity of heat unaccounted for. When this quantity is compared with the *work* done by the engine in the same interval (which work can be measured as described on page 10), it is always found that for every 772 units of work a unit of heat has disappeared from the cylinder. The numerical relation between work and heat which is established in these two cases has been tested in many quite different ways; and, within the limits of experimental errors, always with the same numerical result. But equally definite quantitative relations are known to exist among all the other forms of force; and the manner in which these are convertible into each other has already been indicated, although want of space prevents full illustration of this part of the subject. It may, however, be seen that each form of force can be mediately or immediately converted into mechanical effect, hence each is expressible in terms of work. That is to say, we can assign to a unit of electricity, for example, a number expressing the work which it would do if entirely converted into work; and the same number also expresses the work which would be required to produce the unit of electricity. An ounce of hydrogen in combining with 8 oz. of oxygen produces a certain measurable quantity of heat. If that heat, say $= H$, were all converted into work, we now know that the work would = HJ. Hence we can express a definite chemical action in terms of *work*. The same is generally true of all physical forces, though in some cases, such as light, vital action, &c., the quantitative relations have not yet been definitely determined.

Since, then, all the forces with which we are acquainted are expressible (though the exact relations of some have yet to be discovered) in terms of work, it is found of great advantage to consider the power of doing work as the common measure of doing all these. Thus, if we define *energy* as that which does, or that which is capable of doing, work, we have a term extremely convenient in the description of some aspects of our subject. Thus we can now speak of the *energies of nature*, instead of the *forces*. And all forces, active or passive, may be summed up in one word—*energy*. And, further, the great discovery of the conservation of forces under definite equivalents, may be summed up very briefly in this statement— THE AMOUNT OF ENERGY IN THE UNIVERSE IS CONSTANT. To make this statement clear requires that a distinction between two forms of every kind of energy be pointed out. To recur to the example before imagined : if you carry the 1 lb. weight up the Monument, and deposit it on the ledge at the top, it might lie there for a thousand years before it was made to give back the work you had performed upon it. That work has been, in a·manner, *stored up* by the *position* you have given to your weight. Now, in taking up the weight, you expended energy—you really performed work : that is an instance of energy in operation, and may be termed "actual energy." In what form does the energy exist during the thousand years we may suppose your weight to lie at the top of the Monument ? It is ready to yield up your work again at any moment it is permitted to descend, and it possesses therefore during the whole period a *potential energy* equal in amount to the *actual energy* you bestowed upon it. A similar distinction between actual and potential energy exists with regard to every form of force. If by any means you separate an atom of carbon from an atom of oxygen, you exert actual energy. The process is analogous to carrying up the weight. The atoms when separated possess *potential* energy,—they can rush together again, like the weight to the earth, and in doing so will give out the work which was expended on their separation. A parallel illustration might be drawn from electrical force.

A typical example of the storing up of energy is furnished by a cross-bow. The moment a man begins to bend the bow he is doing work, because he pulls the string in opposition to the bow's resistance to a change in its form ; and it is plain that the amount of energy thus expended is measurable. Suppose, now, the bow has been bent and the string caught in the notch, from which it is released by drawing the trigger when the discharge of the bow is desired. The bow may be retained for an indefinite period in the bent condition, and in this state it possesses, in the form of *potential energy*, all the work which has been expended in bending it, and which it will, in fact, give out, in some way or other, whenever the trigger is drawn. To fix our ideas, let us suppose that to draw the string over the notch required a pull of 50 lbs. over a space of 6 in. ; that is equivalent to $50 \times \frac{1}{2} = 25$ units of work. Now let the bow be used to shoot an arrow weighing $\frac{1}{4}$ lb. vertically upwards. The height in feet to which the arrow will rise multiplied into its weight in pounds will be the work done upon it by the bow. Now, we say that experiment proves that in the case supposed the arrow would rise just 100 ft., so that the work done *by* the bow ($\frac{1}{4} \times 100 = 25$) would be precisely that done *upon it*. For the sake of simplicity, we keep this illustration free from the mention of interfering causes, which have to be considered and allowed for when the matter is put to the real test of quantitative experiment. The instance of the cross-bow brings into notice a highly instructive circumstance, which is this: the bow, which it may have taken the strength of a Hercules to bend, will shoot its bolt by the mere touch of a child on the trigger. In the same way, when a man fires a gun, he merely permits the *potential* energy contained in the charge to convert itself into *actual*, or *kinetic*, energy. The real source of the energy, in the case of the child discharging the cross-bow, is the muscular power of the man who drew it; the real source of the energy in exploding gunpowder is the separation of carbon atoms from oxygen atoms, and that has been done by the sun's rays, as truly as the string was pulled away from the bow by muscular power. If we turn our attention to nitro-glycerine or to nitro-cellulose, we can, by following the chemical actions giving rise to these substances, in like manner trace their energies to our great luminary. The unstable union by which oxygen and nitrogen atoms are locked up in the solid and liquid forms of nitro-cellulose and nitro-glycerine is also the work of the sun; for nitrogen acids, or rather nitrates, are produced naturally under certain electrical and other conditions of the atmosphere, which are due, directly or indirectly, to the sun's action; and they cannot be formed artificially, except by imitating the natural conditions, as by passing electric sparks through air, &c.

It will now be understood, as regards the wonderful relations between animal and vegetable life, which have already been alluded to more than once, how the sun, by expending actual energy, separates atoms of carbon from atoms of oxygen in the leaves of plants, and confers upon these a position of advantage, *i.e.*, potential energy; and how animals, absorbing the separated carbon in the form of food, and inhaling the separated oxygen in the air they breathe, cause the conversion of the potential into actual energy, which appears in the heat, movements, and vital functions of the animal body. In coal we have the energy which plants absorbed from the sun ages ago, stored up in a potential form. The carbon atoms are ready to rush into union with oxygen atoms, and convert their energy of position into the energies developed by chemical action, viz., heat, light, &c. Energy is thus constantly shifting its form from actual to potential, and *vice versâ*, and exhibiting itself under the various transformations of force, as when sun-force changes to chemical action, chemical action to heat, heat to electricity, &c. Energy is, indeed, the real modern PROTEUS—constantly assuming different shapes, difficult to grasp if not held in fetters; now taking on the form of a lion, now of a flame of fire, a whirlwind, a rushing stream. As sober, literal matter of fact we catch glimpses of energy under these very forms.

The greatest discovery of the age has, as already indicated, immediate and important practical bearings. The amount of thought which, even in the present day, is devoted by unscientific mechanics to the old problem of perpetual motion is far greater than is generally supposed. The principle of the conservation of energy shows that this is an impossibility; that the inventor who seeks to create force might just as well try to create matter; that the production of a perpetually moving self-sustaining machine is as far removed from human power as the bringing into existence of a new planet. In force, as in matter, the law is inexorable—*ex nihilo nihil fit*. Again, knowing the definite amount of energy obtainable from the combustion of a pound of coal, we can compare the amount we actually procure from it in our steam engines with this theoretical quantity as the limit towards which our improvements should bring us continually nearer, but which we can never exceed, or, indeed, even reach. The schemers of perpetual motion are not the only class of speculators who pursue objects which are incompatible with our principle. There are many who seek to accomplish desirable ends by inadequate means: who, for example, are aiming perhaps to accomplish the reduction of ores by a quantity of fuel less than that mechanically equivalent to the work, or who conceive that by adding to coal some substance which itself is unchanged, an indefinitely greater amount of heat may be liberated by the combustion.

Enough has been said to show that the energies of animal life can be traced to the sun as their source. The sun builds up the plant, separating oxygen from carbon. The animal—directly or mediately by devouring other animals—takes the carbonaceous matter of the plant, and reunites it with oxygen. In the plant the sun winds up the spring which gives life to the animal mechanism ; for the winding-up of a spring and the separation of the atoms having chemical affinities are alike instances of supplying potential energy. In the animal there is a running-down of the potential into actual energy. It is plain also that of the total energy radiated from the sun in every direction, the earth receives but a very small part $(\frac{1}{2300000000})$. By far the larger part is diffused into space, where, for all such purposes as those with which we are concerned, it is lost. The heat which the sun sends out in a year is calculated to be equal to that which would be produced by the combustion of a layer of coal 17 miles thick over the whole surface of the luminary. Is the sun, then, a flaming fire ? By no means. Combustion is not possible at its temperature ; and as we know the substances which enter into its composition are the same as those we find in the earth, we know that the chemical energies of such substances could not supply the sun's expenditure. Passing over as unsatisfactory an explanation which might occur to some minds—namely, that the sun was created hot at the beginning, and has so continued—there are two theories which attempt to account for the sun's heat. One is that of Meyer, who supposes the heat is due to the continual impact of meteorites drawn to the sun by its gravity ; and the other is that of Helmholtz, who attributes the heat to the continual condensation of the substance of the sun. Helmholtz calculates that a shrinking of the sun's diameter by only $\frac{1}{10000}$th of its present amount, would supply heat to last for two thousand years ; while the condensation of the substance of the sun to the density of the earth would cover the sun's expenditure for 17,000,000 of years. There is great probability that both theories may be correct, and that the cause of the sun's heat may be considered as due in general terms to aggregation of matter, by which the original potential energy of position is converted into the actual energy of heat and light. Now, however immense may be our planetary system, the sun being continually throwing off this energy into space, there must come a time when the supplies of meteorites will fail, and when the great globe of the sun will have shrunk to its smallest dimensions. We see, then, that heat and light are produced by the aggregation of matter ; the heat and light are radiated into space ; the small fraction intercepted by our globe is the source of almost every movement—the original stuff, so to speak, out of which all terrestrial forces are made. The sun produces the winds, the thunderstorms, the electric currents of the Aurora, the phenomena of terrestrial magnetism, and is the source of vegetable and animal life. The waves, the rains, the mountain torrents, the flowing rivers, are the work of the sun's emanations.

In the illustration of the energy expended on raising a weight afterwards dropped, we traced that energy into the final form of heat of a low temperature radiated into space. It would be easy to show that all energy ultimately takes the same form. Now, although it is easy to convert work into heat, there is no conceivable process by which uniformly-diffused heat can again be made to do any kind of work. The case may be compared to water, which in moving down from a higher to a lower level may be made to perform any variety of work. But when all the water has passed down from the higher level to the lower, it can no longer do any work. Whenever work is done by the agency of heat, there is always a passing from a higher temperature to a lower—a transference of heat from a hotter body to a colder. If the condenser of the steam engine had the same temperature as the steam, the machine would not work. Not only do all the energies in operation on the face of the earth continually run down into the form of radiant heat sent off by the earth into space; but our sun's energy, and that of the suns of other systems, are also continually passing off into space; and the final effect must be a uniform diffusion of heat in a universe in which none of the varied forms of energy we now behold in operation will be possible, because all will have run down to the same dead level of uniformly-diffused heat. This startling corollary from the principle of the conservation of energy has been worked out by Sir W. Thomson under the title of "The Dissipation of Energy." It leads us to contemplate a state of things in which all light and life will have passed away from the universe—a condition which the poet's terrible dream of darkness, "which was not all a dream," seems to shadow forth—

> "The bright sun was extinguished, and the stars
> Did wander darkling in the eternal space,
> Rayless and pathless; and the icy earth
> Swung blind and blackening in the moonless air.
> *   *   *   *   *   *
>   The world was void,
> The populous and the powerful was a lump,
> Seasonless, herbless, treeless, manless, lifeless—
> A lump of death—a chaos of hard clay.
> The rivers, lakes, and ocean all stood still,
> And nothing stirred within their silent depths.
> *   *   *   *   *   *
> The waves were dead; the tides were in their grave,
> The Moon, their mistress, had expired before;
> The winds were withered in the stagnant air,
> And the clouds perished; Darkness had no need
> Of aid from them—She was the Universe."

The doctrine of this persistence and dissipation of energy completely harmonizes with the grand speculation termed the "nebular hypothesis," which regards the universe as having originally consisted of uniformly diffused matter, which, being endowed with the power of gravitation, aggregated round certain centres. This process is still going on; and, according to modern speculations, light and life and motion are but manifestations of this primæval potential energy being converted into actual energy, and degrading ultimately into the form of universally-diffused heat. To quote the closing sentences of the eloquent passage in which Professor Tyndall concludes the work mentioned above, "To nature nothing can be added, from nature nothing can be taken away; the sum of her energies is constant, and the utmost man can do in the pursuit of physical truth, or in the applications of physical knowledge, is to shift the constituents of the never-varying total. The law of conservation rigidly excludes both creation and annihilation. Waves may change to ripples, and ripples to waves; magnitude may be substituted for number, and number for magnitude; asteroids may aggregate to suns, suns may resolve themselves into floræ and faunæ, and floræ and faunæ melt in air : the flux of power is eternally the same. It rolls in music through the ages, and all terrestrial energy—the manifestations of life as well as the display of phenomena—are but the modulations of its rhythm."

The discoveries to which we have here endeavoured to attract the reader's attention thus give rise to conceptions of the utmost grandeur and interest. We see that the sum of Nature's energies is constant ; that all the manifestations of force are but the transference of power from one position to another. And we have recognized the material source of all our terrestrial energies in the sun. Two theories have already been mentioned by which it is sought to account for the sun's heat—the meteoric theory of Meyer and Thomson, and the shrinkage theory of Helmholtz. These both assume gravitation as the primal force from which the supply of heat and other energies must be drawn, and they assume also that the laws of radiation and of the degradation of temperature in the transformation of heat into other forces, as we find them operating at the earth's surface, are equally in action in every region of space. Hence is deduced that conception of the final state of the universe as one of merely equally diffused temperature admitting of no further transformation. This speculation presents the *universe* in the aspect of a clock, now indeed going, but when once run down, incapable of ever being again wound up. There seems in this view a want of symmetry, so to speak ; we miss the feeling of harmonious *rhythm* to which Tyndall refers. There is, however, another cosmic theory, well supported by accumulating facts, which assigns to gravitation a less important part in the production of solar heat and in the evolution of worlds, and it is one which supplies also a basis for the explanation of such phenomena as aerolites, comets, variable stars, the inclination of planets' axes to their orbits, the proper motion of our sun, and that of the so called fixed stars, of all of which the nebular hypothesis fails to give any account ; while, on the other hand, the *impact theory*, as it has been named, includes the other, and goes beyond it. The reader who desires to pursue this subject may be referred to Croll's book on Stellar Evolution.

In the last few paragraphs we have been dealing with speculations as much as with discoveries. But indeed the former are the offspring of the latter, as certainly as one invention becomes the parent of others. The human mind never rests contented with the knowledge and mastery of nature actually gained, but ever seeks to pass beyond and attain still greater power. The volume we are now bringing to a close has given but brief and imperfect indications of specimens, taken here and there, of what has been done during the short period of one century. We may draw an augury for the future of man's dominion from the powers his Promethean spirit has already grasped :

> " The lightning is his slave ; heaven's utmost deep
> Gives up her stars, and like a flock of sheep
> They pass before his eye, are numbered, and roll on !
> The tempest is his steed, he strides the air ;
> And the abyss shouts from her depth laid bare,
> " Heaven, hast thou secrets? Man unveils me ; I have none.''

NOTES A AND B.

Note A—Continuation of Table on page 755, showing the quantity of Coals raised annually in Great Britain.

Year.	Coal raised in Tons.	Year.	Coal raised in Tons.
1874	126,590,108	1885	159,351,418
1875	133,306,458	1886	157,518,482
1876	134,125,166	1887	162,121,576
1877	134,179,968	1888	169,935,219
1878	132,612,063	1889	176,916,724
1879	133,720,393	1890	181,614,280
1880	146,969,409	1891	185,479,126
1881	154,184,300	1892	181,786,871
1882	156,499,977	1893	164,325,795
1883	163,737,327	1894	188,277,525
1884	160,757,779		

Note B—CONSERVATION OF ENERGY.—Page 804.

THE statement here should have been more explicit, as it has reference to a state of things not to be realised in practice. Like the well-known "first law of motion," it can neither be demonstrated *à priori*, nor proved by any direct and simple experiment. The first law of motion asserts that a body in motion, not acted on by any external force, will continue to move in a straight line, and with a uniform velocity. Now we cannot place a body in such a position that it will not be acted upon by some external forces ; but the more we lessen the effect of external forces, the more nearly is the motion straight and uniform. Similarly in the case supposed, the intention is to show that the weight carried up is in a position to do just as much work as was done upon it. We must suppose several impracticable but conceivable conditions in order to eliminate considerations which do not concern the theoretical question ; we must suppose the cord to be weightless and absolutely devoid of rigidity ; the pulley to have no mass or inertia, that is to require no force to set it in motion, and to move without any friction ; the air to offer no resistance ; and the force of gravity to be uniform throughout the space. Some approximation to these conditions is practicable, as, for example, the pulley might be the lightest possible, and turn on friction wheels, the cord might be the finest silk thread, and so on. But it is not the influence of these external forces we are considering, but only the energy due to the position of the raised weight. Assuming, therefore, the disturbing conditions absolutely eliminated, it is not difficult to see that no downward force or pressure, however small, could be applied for ever so short a time, to the upper weight without setting the system in motion. The motion would be an accelerated one so long as the force was applied, it would become uniform when the force ceased to act ; it would have a velocity proportionate to the force. In any case, after a time the descending weight would reach the ground, and for our point of view it is quite immaterial whether the time occupied by the movement were 5 minutes or 5,000 years, for be it observed, time does not enter into the definition of *work* as it does into that of "horse-power." Then by pushing the conceived conditions to their limits, we may see that without considering any question of conversion of motion into heat, the raised weight can, in theory at least, give back again the energy spent upon it.

INDEX.

813

EDINBURGH : PRINTED BY MORRISON AND GIBB LIMITED